FUNDAMENTALS OF COLLEGE
PHYSICS

SECOND EDITION

VOLUME ONE

PETER J. NOLAN

State University of New York
at Farmingdale

WCB **Wm. C. Brown Publishers**

Dubuque, IA Bogota Boston Buenos Aires Caracas Chicago
Guilford, CT London Madrid Mexico City Sydney Toronto

S0-CBU-065

Book Team

Editor *Megan Johnson*
Developmental Editor *Tom Riley*
Production Editor *Carol M. Besler*
Photo Editor *Lori Hancock*
Permissions Coordinator *Karen L. Storlie*
Art Processor *Joyce Watters*
Publishing Services Coordinator *Barbara J. Hodgson*

Wm. C. Brown Publishers
A Division of Wm. C. Brown Communications, Inc.

Vice President and General Manager *Beverly Kolz*
Vice President, Publisher *Jeffrey L. Hahn*
Vice President, Director of Sales and Marketing *Virginia S. Moffat*
Vice President, Director of Production *Colleen A. Yonda*
National Sales Manager *Douglas J. DiNardo*
Marketing Manager *Amy Halloran*
Advertising Manager *Janelle Keeffer*
Production Editorial Manager *Renée Menne*
Publishing Services Manager *Karen J. Slaght*
Royalty/Permissions Manager *Connie Allendorf*

Wm. C. Brown Communications, Inc.

President and Chief Executive Officer *G. Franklin Lewis*
Senior Vice President, Operations *James H. Higby*
Corporate Senior Vice President, President of WCB Manufacturing *Roger Meyer*
Corporate Senior Vice President and Chief Financial Officer *Robert Chesterman*

Copyedited by Patricia Steele

Cover/interior design by Rokusek Design

Cover/part opener images © Fundamental Photographs

The credits section for this book begins on page C–1 and is considered
an extension of the copyright page.

Copyright © 1993, 1995 by Wm. C. Brown Communications, Inc. All
rights reserved

A Times Mirror Company

Library of Congress Catalog Card Number: 94–70898

ISBN Fundamentals of College Physics   Casebound: 0–697–23138–0
ISBN Fundamentals of College Physics, Volume 1   Paperbound: 0–697–24392–3
ISBN Fundamentals of College Physics, Volume 2   Paperbound: 0–697–24393–1

No part of this publication may be reproduced, stored in a retrieval
system, or transmitted, in any form or by any means, electronic,
mechanical, photocopying, recording, or otherwise, without the
prior written permission of the publisher.

Printed in the United States of America by Wm. C. Brown Communications, Inc.,
2460 Kerper Boulevard, Dubuque, IA 52001

10 9 8 7 6 5 4 3 2 1

This book is dedicated to my sons—Thomas, James, John Michael, and Kevin—the joy of my life.

Contents In Brief

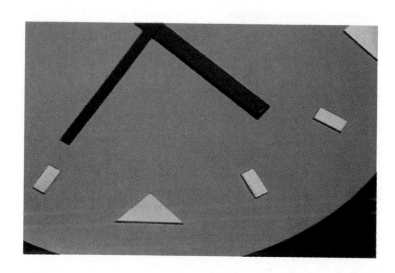

Contents

v

PART TWO

Vibratory Motion, Wave Motion, and Fluids, 280

10 Elasticity, 283

11 Simple Harmonic Motion, 297

12 Wave Motion, 321

13 Fluids, 367

Preface

This book is an outgrowth of more than twenty-five years of teaching college physics. The original manuscript has been used in the classroom for several of them, and the first edition is used in over one hundred colleges and universities throughout the country. Because the book is written for students, it contains a great many of the intermediate steps that are often left out of the derivations and illustrative problem solutions in many traditional college physics textbooks. Students new to physics often find it difficult to follow derivations when the intermediate steps are left out. In addition, the units of measurement are carried along, step by step, in the equations to make it easier for students to understand. This book does not require calculus; the only prerequisites are high school algebra and trigonometry. In fact, a short review of trigonometry is given in chapter 2, before the discussion of the components of a vector.

This text gives a good, fairly rigorous, traditional college physics coverage. Instructors are expected to choose those topics they deem most important for the particular course. Students can read on their own the detailed descriptions found in those chapters, or parts of chapters, omitted from the course. Unfortunately, many interesting and important topics in modern physics are never covered in college physics courses because of lack of time. These chapters, especially, are written in even more detail to enable students to read them on their own. Even years after taking the course, students can read these sections for their own edification and enjoyment. This is one of the reasons that students should never sell any of their college textbooks. They are an investment for a lifetime of reference, illumination, and relaxation.

The organization of the text follows the traditional sequence of mechanics, wave motion, heat, electricity and magnetism, optics, and modern physics. The emphasis throughout the book is on clarity. The book starts out at a very simple level, and advances as students' understanding grows.

Color has been used extensively throughout, especially in the diagrams, to help students visualize the material. To standardize the colors used, each main part of the text uses a specific color scheme. This color scheme is spelled out in the Color as a Study Aid sections at the beginning of each part. On a few occasions it was necessary to depart from the standards to avoid confusion. In addition, some standard colors, like red and blue, may mean one thing in one Part but something quite different in another Part. This is unavoidable because there are not enough different colors to account for every physical quantity. However, the colors are consistent within any one Part of the book.

There are a large number of diagrams and illustrative problems in the text to help students visualize physical ideas. Important equations are highlighted to help students find and recognize them. A summary of these important equations is given at the end of each chapter.

Students sometimes have difficulty remembering the meanings of all the vocabulary associated with new physical ideas. Therefore, a section called The

Language of Physics, found at the end of each chapter, contains the most important ideas and definitions discussed in that chapter.

To comprehend the physical ideas expressed in the theory class, students need to be able to solve physics problems for themselves. Problem sets at the end of each chapter are grouped according to the section where the topic is covered. Problems that are a mix of different sections are found in the Additional Problems section. If you have difficulty with a problem, refer to that section of the chapter for help. The problems begin with simple, plug-in problems to develop students' confidence and to give them a feel for the numerical magnitudes of some physical quantities. The problems then become progressively more difficult and end with some that are very challenging. The more difficult problems are indicated by a dagger (†). The starred problems are either conceptually more difficult or very long. However, just because a problem is starred is no reason to avoid attempting its solution. Many problems at the end of the chapter are very similar to the illustrative problems worked out in the text. When solving these problems, students can use the illustrative problems as a guide. However, students should be warned that physics cannot be learned by memorizing the exhaustive set of illustrative problems. These problems are only a guide to foster greater understanding. To facilitate setting up a problem, the Hints for Problem Solving section, which is found before the problem set in chapter 3, should be studied carefully.

New to the second edition of this book is a section called Interactive Tutorials, which uses computer spreadsheets in a unique and exciting way to solve physics problems. These Interactive Tutorials can be found at the end of the problems section in each chapter. They are also available on a computer disk for both the instructor and the student. More details on these Interactive Tutorials can be found in the section "A Special Note to the Student" at the end of the Preface.

A series of questions relating to the topics discussed in the chapter is also included at the end of each chapter. Students should try to answer these questions to see if they fully understand the ramifications of the theory discussed in the chapter. Just as with the problem sets, some of these questions are either conceptually more difficult or will entail some outside reading. These more difficult questions are also indicated by a dagger (†).

A word about units. Many recent college physics texts use only the International System of Units (SI units). Although working exclusively in SI units is a desirable goal, we do live in a world where other systems of units, particularly the British engineering system, are continuing to be used. Failure to show students how to work with these other units, or how to solve problems regardless of the units employed, does them a disservice. In addition, from the point of view of pedagogy, it is beneficial in the learning process to build on knowledge already possessed by students. For example, when we say a car is moving at 55 miles per hour, students immediately have a feel for this motion. With time, they will have that same feel for the car moving at 88.6 km/hr. Hence, both systems of units are found at the beginning of this book. As we progress through the book, however, the emphasis switches to the SI units, until, from about chapter 16 on, the SI units are used almost exclusively. This approach weans students from the British engineering system to the International System (SI) of units, and at the same time prepares students to convert to any unit or system of units when necessary.

However, for those instructors who prefer to work only in the International System of Units, for this second edition I have taken approximately 60 problems that are presently in the British Engineering System of Units and duplicated them so that they are now also available to be used in the International System of Units. This will give a greater choice of problems for those who wish to work only in SI units. These duplicated problems are not an exact duplicate in that the numbers used are not just a conversion from one system to another, but instead the numbers are totally different. Hence, an instructor wishing to use both systems can assign both problems if he or she desires.

Also new to this second edition is an additional 165 new physics problems, all in SI units, when added to the 60 new duplicated problems gives a total of 225 new physics problems, all in SI units. There are also 129 new computer Interactive Tutorials, giving a grand total of 354 new problems in this second edition.

The levels of college physics courses vary greatly, depending on the backgrounds of the students and the institution where the course is taught. Some instructors like to introduce the concepts of vector multiplication in the chapter on vectors and utilize these concepts throughout the course. Others feel that this introduces unnecessary mathematical difficulties. To satisfy both instruction preferences, vector multiplication is found in appendix F and can be used throughout the book if desired.

Scattered throughout the text, at the ends of chapters, are sections entitled "Have you ever wondered . . . ?" These are a series of essays on the application of physics to areas such as meteorology, astronomy, aviation, space travel, the health sciences, the environment, philosophy, traffic congestion, sports, and the like. Many students are unaware that physics has such far-reaching applications. These sections are intended to engage students' varied interests but can be omitted, if desired, without loss of continuity in the physics course.

The relation between theory and experiment is carried throughout the book, emphasizing that our models of nature are good only if they can be verified by experiment.

Concepts presented in the lecture and text can be well demonstrated in a laboratory setting. To this end, *Experiments In Physics, second edition* by Peter J. Nolan and Raymond E. Bigliani is also available from Wm. C. Brown Publishers. *Experiments in Physics* contains some 55 experiments covering all the traditional topics in a physics laboratory. New to the second edition of *Experiments in Physics* is a complete computer analysis of the data for every experiment. Your WCB Sales Representative can arrange to have any of your own department's lab exercises combined with those experiments you desire from the Nolan and Bigliani second edition and bound into your own personal laboratory manual.

A Bibliography, given at the end of the book, lists some of the large number of books that are accessible to students taking college physics. These books cover such topics in modern physics as relativity, quantum mechanics, and elementary particles. Although many of these books are of a popular nature, they do require some physics background. After finishing this book, students should be able to read any of them for pleasure without difficulty.

Finally, we should note that we are living in a rapidly changing world. Many of the changes in our world are sparked by advances in physics, engineering, and the high-technology industries. Since engineering and technology are the application of physics to the solution of practical problems, it behooves every individual to get as much background in physics as possible. *You can depend on the fact that there will be change in our society. You can be either the architect of that change or its victim, but there will be change.*

The Teaching Package

A very extensive teaching package supports the use of *Fundamentals of College Physics,* second edition. The author and the publisher have worked closely with your colleagues to develop this material.

For the Instructor

Instructor's Solutions Manual You can find worked-out solutions to every problem in the textbook in the *Instructor's Solutions Manual* offered with *Fundamentals of College Physics.*

Transparency Set A set of over 100 full-color acetate transparencies is available to adopters of this textbook.

Instructor's Resource Manual This package item includes selected transparency masters, documentation for the more than 125 Interactive Tutorials and spreadsheet diskettes that accompany the text. The tutorials are available on both the IBM Windows and Macintosh platforms.

Test Item File More than 4,000 classroom-tested test items are available to adopters of this textbook. The test items are available on the computer test generating program MicroTest for both the IBM and Macintosh platforms.

Physics Videodisc Over 600 lab demonstrations are available from the Wm. C. Brown Publishers videodisc library. These professionally produced experiments are considered the best available lab demonstrations on the market. Please call your local Wm. C. Brown Sales Representative for more information.

Experiments in Physics, second edition, by Nolan and Bigliani This laboratory text for the algebra-based introductory physics course is the first and only laboratory textbook that incorporates computer spreadsheets to calculate and analyze data obtained in the lab. This manual is completely customizable. Contact your Wm. C. Brown Sales Representative for more information.

For the Student

Student Solutions Manual Approximately twenty percent of the problems in the textbook can be found in the *Student Solutions Manual* to accompany *Fundamentals of College Physics.* Every solution is completely worked out for the student.

Physics Student Study Art Notebook A notebook of key physics art work is available packaged with the new edition of *Fundamentals of College Physics.* This notebook provides full-color line art from the text, with space for note taking by the student. This unique notebook helps students study for exams and take clearer, more accurate classroom notes.

Student Study Guide An excellent *Student Study Guide,* written by Lewis O'Kelly of Memphis State University, is available to accompany *Fundamentals of College Physics.* This study guide reinforces the concepts present in the main text and provides at-home quizzes and problem solving techniques.

Acknowledgments

I would like at this time to thank Eileen and my children for their love, understanding, patience, and encouragement throughout the long stages of writing and rewriting this book. I would also like to thank the following individuals for their encouragement: Dr. Lloyd Makarowitz, Chairman of the Physics Department at SUNY Farmingdale; Linda Rennie, our secretary; and all the members of the Physics Department at SUNY Farmingdale.

I would also like to thank all the very friendly and helpful people at Wm. C. Brown Publishers who helped in the production of this book. In particular, I would like to thank Executive Editor Jeffrey Hahn, Physics Editors Jane Ducham and Megan Johnson, Developmental Editor Tom Riley, and Production Editors Carol Besler and Julie Wilde. I would like to thank Jerry Hopper and all his people at Diphrent Strokes, Inc., who entered the realm of high technology and did all the beautiful artwork on computers. This will certainly be the way of the future. I would like to give a very special thank you to Pat Steele for the magnificent job of copy editing this book. Her help was indispensable. I would also like to thank the following individuals for their contribution to the overall text package: *Student Study Guide* by Lewis O'Kelly, Memphis State University; *Student's Solutions Manual and Instructor's Solutions Manual* by Ronald Cosby, Ball State University, Joel Levine, Orange Coast College, and Paul Morris, Abilene Christian University; *Test Item File* by Lloyd Makarowitz, SUNY-Farmingdale. Finally, I wish to acknowledge and thank the following reviewers for their helpful criticisms and suggestions: Joel M. Levine, Orange Coast College; Russell A. Roy, Santa Fe Community College; Paul Morris, Abilene Christian University; C. Sherman Frye, Northern Virginia Community College; Franklin Curtis Mason, Middle Tennessee State University; Franklin D. Trumpy, Des Moines Area Community College; Lewis B. O'Kelly, Memphis State University; Ronald M. Cosby, Ball State University; Paul Feldker, Florissant Valley Community College; J. W. Northrip, Southwest Missouri State University; Michael J. Matkovich, Oakton Community College; Michael K. Garrity, St. Cloud State University; and Phyllis A. Salmons, Embry-Riddle Aeronautical University.

A Special Note to the Student

"One thing I have learned in a long life: that all our science measured against reality, is primitive and childlike—and yet it is the most precious thing we have."

Albert Einstein
as quoted by Banesh Hoffmann in
Albert Einstein, Creator and Rebel

The language of physics is mathematics, so it is necessary to use mathematics in our study of nature. However, just as sometimes "you cannot see the forest for the trees," you must be careful or "you will not see the physics for the mathematics." Remember, mathematics is only a tool used to help describe the physical world. You must be careful to avoid getting lost in the mathematics and thereby losing sight of the physics. When solving problems, a sketch or diagram that represents the physics of the problem should be drawn first, then the mathematics should be added.

Physics is such a logical subject that when a student sees an illustrative problem worked out, either in the textbook or on the blackboard, it usually seems very simple. Unfortunately, for most students, it is simple only until they sit down and try to do a problem on their own. Then they often find themselves confused and frustrated because they do not know how to get started.

If this happens to you, do not feel discouraged. It is a normal phenomenon that happens to many students. The usual approach to overcoming this difficulty is going back to the illustrative problem in the text. When you do so, however, do not look at the solution of the problem first. Read the problem carefully, and then try to solve the problem on your own. At any point in the solution, when you cannot proceed to the next step on your own, peek at that step and only that step in the illustrative problem. The illustrative problem shows you what to do at that step. Then continue to solve the problem on your own. Every time you get stuck, look again at the appropriate solution step in the illustrative problem until you can finish the entire problem. The reason you had difficulty at a particular place in the problem is usually that you did not understand the physics at that point as well as you thought you did. It will help to reread the appropriate theory section. Getting stuck on a problem is not a bad thing, because each time you do, you have the opportunity to learn something. Getting stuck is the first step on the road to knowledge. I hope you will feel comforted to know that most of the students who have gone before you also had these difficulties. You are not alone. Just keep trying. Eventually, you will find that solving physics problems is not as difficult as you first thought; in fact, with time, you will find that they can even be fun to solve. The more problems that you solve, the easier they become, and the greater will be your enjoyment of the course.

Interactive Tutorials

For the last ten or more years, there has been an ongoing discussion in the physics community for the need to introduce the computer into the freshman physics class. However, to this day no one has been able to agree on how the computer should best be used. In those cases where a computer has been used, the traditional use has been to state the problem, write down a general equation, and give the result of the calculation, with all the in-between steps of the calculation missing. The student gets the result of the calculation, but does not see how it is attained. It is almost like magic. In this second edition of *The Fundamental of College Physics,* the computer is used in a highly effective and unique way. At the end of the problems section in each chapter, I have introduced a new section called **Interactive Tutorials.** The Interactive Tutorials are a series of physics problems, very much like the examples in the chapter and the problems at the end of the chapter. A computer disk is available to both the instructor and the student with each of these problems on the disk. These computer files are available for the IBM PC or compatible personal computer using the Lotus 1-2-3 computer spreadsheet or on a Macintosh computer using Microsoft's Excel computer spreadsheet.

What makes these Interactive Tutorials unique is the way the computer is used as an instructional tool. Figure 1 shows a typical tutorial for a problem in chapter 3 on Kinematics. When the student opens this particular spreadsheet, he or she sees the problem stated in the usual manner. That is, this problem is an example of a projectile fired at an initial velocity $v_0 = 53.0$ m/s at an angle $\theta = 50.0°$, and it is desired to find the maximum height of the projectile, the total time the projectile is in the air, and the range of the projectile. Directly below the stated problem is a group of yellow-colored cells labeled **Initial Conditions.** Into these yellow cells are placed the numerical values associated with the particular problem. For this problem the initial conditions consist of the initial velocity v_0, the initial angle θ, and the acceleration due to gravity g as shown in figure 1. The problem is now solved in the traditional way of a worked-out example in the book. Words are used to describe the physical principles and then the equations are written down. Then the in-between steps of the calculation are shown in light green-colored cells, and the final result of the calculation is shown in a light blue-green-colored cell. The entire problem is solved in this manner, as shown in figure 1. If the student wishes to change the problem by using a different initial velocity or a different launch angle, he or she then changes these values in the yellow-colored cells of the initial conditions. When the initial conditions are changed the computer spreadsheet recalculates all the new in-between steps in the problem and all the new final answers to the problem. In this way the problem is completely interactive. It changes for every new set of initial conditions. The Interactive Tutorials make the book a living book. The tutorials can be changed many times over to solve for all kinds of special cases.

One of the most troubling aspects of using a spreadsheet, however, is its inability to use subscripts or superscripts. Therefore, information contained on the spreadsheet can sometimes be confusing. As an example, when the textbook refers to an initial velocity, the notation v_0 is utilized. Since the spreadsheet cannot use subscripts, this will be written as vo. Other examples are v_{ox} = vox; y_{max} = ymax; M_1 = M1; M_2 = M2; and so on. The notation used for the subscript in the spreadsheet will, however, be the same notation used in the textbook. When in doubt about notation, refer to the stated problem in the textbook. For superscripts, the notation $\hat{}$ is used. Hence, the quantity x^2 in your textbook will be written on the spreadsheet as x^2. A term containing both subscripts and superscripts, such as v_y^2, will be written on the spreadsheet as vy^2. Powers of ten that are used in scientific notation are written with the capital letter E. Hence, the number 5280, written in scientific notation as 5.280×10^3, will be written on the spreadsheet as 5.280E+3. Because the spreadsheet cannot mix fonts in a particular cell, the notation Δx will be written on the spreadsheet as (delta) x, and angles such as θ will be written as (theta).

Chapter 3 Kinematics

Computer Assisted Instruction

Interactive Tutorial

79. A golf ball is hit with an initial velocity vo = 53.0 m/s at an angle (theta) = 50.0 degrees above the horizontal. (a) How high will the ball go? (b) What is the total time the ball is in the air? (c) How far will the ball travel horizontally before it hits the ground?

Initial Conditions

The magnitude of the initial velocity vo =	53 m/s
The angle (theta) =	50 degrees
The acceleration of gravity g =	9.8 m/s^2

The x-component of the initial velocity is found as

vox = vo cos(theta)

vox = (53 m/s) × cos (50)

vox = 34.07 m/s

The y-component of the initial velocity is found as

voy = vo sin(theta)

voy = (53 m/s) × sin(50)

voy = 40.6 m/s

(a) The maximum height ymax of the golf ball above the launch point is found from the kinematic equation

vy^2 = voy^2 − 2 g y

When y = ymax, vy = 0. Therefore

0 = voy^2 − 2 g ymax

Solving for the maximum height ymax above the launch point, we get

ymax = (voy^2)/2g

ymax = (40.6 m/s)^2 / 2 (9.8 m/s^2)

ymax = 84.1 m

(b) The total time tt the ball is in the air is found from the kinematic equation

y = voy t − 0.5gt^2

When t is equal to the total time tt, the projectile is on the ground and y = 0, therefore

0 = voy tt −.5 g tt^2

or dividing each term by tt we get

0 = voy −.5 g tt

upon solving for the total time tt we get

tt = 2 voy/g

tt = 2 × (40.6 m/s) / (9.8 m/s^2)

tt = 8.29 s

(c) The maximum distance the ball travels in the x-direction before it hits the ground, is found from the kinematic equation for the displacement of the ball in the x-direction

x = vox t

The maximum distance xmax occurs for t = tt. Therefore

xmax = vox tt

xmax = (34.07 m/s) × (8.29 s)

xmax = 282.28 m

Figure 1.

A typical Interactive Tutorial.

Finally, we should note that the spreadsheets are "protected" by allowing you to enter data only in the designated light yellow-colored cells of the Initial Conditions area. Therefore, the student cannot damage the spreadsheets in any way, and they can be used over and over again.

Another example of an Interactive Tutorial, shown in figure 2, is an Interactive Tutorial for a problem on an *RLC* series *AC* circuit from chapter 24. The initial conditions in the yellow-colored cells set the values of the resistance *R*, the inductance *L*, the capacitance *C*, the frequency *f*, and the applied voltage *V*. The spreadsheet then completely solves the problem for these initial values. The student can follow through all the in-between steps of the problem in the light green-colored cells of the spreadsheet and all the answers in the light blue-green-colored cells. The spreadsheet then draws a graph showing the phase relations for the voltage across the resistor, V_R, the voltage across the inductance V_L, the voltage across the capacitance, V_C, and the resultant voltage V in the circuit depending upon the initial values in the yellow cells. As special cases, the student can then let $L = 0$ in the yellow cell area, and observe the characteristics of an *RC* series circuit; then replace a value of *L* in the initial conditions and then let $C = 0$ and observe the characteristics of an *RL* series circuit. Finally, inserting values for *L* and *C* and letting $R = 0$, the student can observe the characteristics of an *LC* series circuit. In all the cases the spreadsheet calculates all the quantities with all the in-between steps, and shows the phase relations of the voltages.

Chapter 24 Alternating Current Circuits

Computer Assisted Instruction

Interactive Tutorial

Figure 2.

An Interactive Tutorial for Chapter 24 Alternating Current Circuits.

59. RLC series circuit. An RLC series circuit has a resistance R = 800 ohms, Inductance L = 2.00 H, capacitance C = 5.55 × 10^−6 F, and they are connected to an applied voltage V = 110 V, operating at a frequency f = 60 Hz. Find (a) the inductive reactance XL, (b) the capacitive reactance, XC, (c) the impedance Z of the circuit, (d) the current i in the circuit, (e) the voltage drop VR across the resistor, (f) the voltage drop VL across the inductor, (g) the voltage drop VC across the capacitor, (h) the total voltage drop across RLC, (i) the phase angle (phi), and (j) the resonant frequency fo. (k) Plot the curves of the voltage across R, L, and C individually and the voltage across the combination. Show the phase relations for all these voltages.

Initial Conditions

R =	800 ohms	f =	60 Hz
L =	2.00E+00 H	V =	110 V
C =	5.55E−06 F		

(a) The inductive reactance XL is found from equation 24.15 as

$$XL = 2 \text{ (pi) } f \, L$$

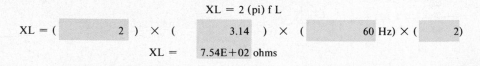

XL = 7.54E+02 ohms

(b) The capacitive reactance XC is found from equation 24.17 as

$$XC = 1/[2 \text{ (pi) f C}]$$

$$XC = (\boxed{1}) \quad / \quad (\boxed{6.28}) \times (\boxed{60} \text{ Hz}) \times (\boxed{5.55\text{E}-06})$$

$$XC = \boxed{4.78\text{E}+02} \text{ ohms}$$

(c) The impedance Z of the circuit is found from equation 24.18 as

$$Z = \text{Sqrt}[R^2 + (XL - XC)^2]$$

$$Z = \text{Sqrt}[(\boxed{800} \text{ ohms})^2 + ((\boxed{753.98} \text{ ohms}) - (\boxed{477.94} \text{ ohms}))^2$$

$$Z = \boxed{8.46\text{E}+02} \text{ ohms}$$

(d) The effective current in the circuit is found from Ohm's law, equation 24.19, as

$$i = V/Z$$

$$i = (\boxed{110} \text{ V}) \quad / \quad (\boxed{8.46\text{E}+02} \text{ ohms})$$

$$i = \boxed{1.30\text{E}-01} \text{ A}$$

(e) The voltage drop VR across the resistor is found from equation 24.13, as

$$VR = i\,R$$

$$VR = (\boxed{1.30\text{E}-01} \text{ A}) \times (\boxed{8.00\text{E}+02} \text{ ohms})$$

$$VR = \boxed{1.04\text{E}+02} \text{ V}$$

(f) The voltage drop VL across the inductor is found from equation 24.14, as

$$VL = i\,XL$$

$$VL = (\boxed{1.30\text{E}-01} \text{ A}) \times (\boxed{7.54\text{E}+02} \text{ ohms})$$

$$VL = \boxed{9.80\text{E}+01} \text{ V}$$

(g) The voltage drop VC across the capacitor is found from equation 24.16, as

$$VC = i\,XC$$

$$VC = (\boxed{1.30\text{E}-01} \text{ A}) \times (\boxed{4.78\text{E}+02} \text{ ohms})$$

$$VC = \boxed{6.21\text{E}+01} \text{ V}$$

(h) The total voltage across R, L, C in series is found from equation 24.11 as

$$V = \text{Sqrt}[VR^2 + (VL - VC)^2]$$

$$V = \text{Sqrt}[(\boxed{103.98} \text{ ohms})^2 + ((\boxed{98} \text{ ohms}) - (\boxed{62.12} \text{ ohms}))^2$$

$$V = \boxed{1.10\text{E}+02} \text{ V}$$

which is equal to the applied voltage as expected.

(i) The phase angle (phi) is found from equation 24.11 as

$$\text{(phi)} = \arctan[(VL - VC)/VR]$$

$$\text{(phi)} = \arctan[((\boxed{98} \text{ V}) - (\boxed{62.12} \text{ V})) \quad / \quad (\boxed{103.98} \text{ V})]$$

$$\text{(phi)} = \boxed{19.04} \text{ degrees}$$

(continued)

(j) The resonant frequency fo is found from equation 24.21 as

$$fo = 1/[2 \text{ (pi) sqrt(L C)}]$$

fo = (1   )   /   (6.28   ) × sqrt(2 H) × (5.55E−06)

fo =   4.78E+01 Hz

The angular frequency (omega) of the alternating current is related to the frequency of the current by

$$(\text{omega}) = 2 \text{ (pi) } f$$

(omega) = (2   )   ×   (3.14)   ×   (60 Hz)

(omega) =   3.77E+02 rad/s

The maximum value of the current, imax, is found from rearranging equation 24.5 as

$$\text{imax} = \text{ieff} / 0.707$$

imax = (0.13 A)   /   (0.71   )

imax =   0.18 A

The equation for the alternating current is given by

$$i = \text{imax} \sin[(\text{omega}) t]$$

i = (0.18   A) × sin[(3.77E+02 rad/s) t]

The data to plot the current as a function of time is calculated from this equation and the results are shown in the graph below.

The data to plot the voltage drop across the resistor as a function of time is calculated from the equation

$$VR = i R$$

$$VR = \text{imax } R \sin[(\text{omega}) t]$$

and the results are shown in the graph below.

The data to plot the voltage drop across the inductor as a function of time is calculated from the equation

$$VL = i XL$$

$$VL = \text{imax } XL \sin[(\text{omega}) t + (\text{pi})/2]$$

The term + (pi)/2 in the sine term is to show that VL leads VR by 90 degrees. The results are shown in the graph below.

The data to plot the voltage drop across the capacitor as a function of time is calculated from the equation

$$VC = i XC$$

$$VC = \text{imax } XC \sin[(\text{omega}) t - (\text{pi})/2]$$

The term − (pi)/2 in the sine term is to show that VC lags VR by 90 degrees. The results are shown in the graph below.

V is the sum of the voltages across R, L, and C in series and is given by

$$V = VR + VL + VC$$

and is shown in the graph below

Phase Relations in an RLC Series Circuit

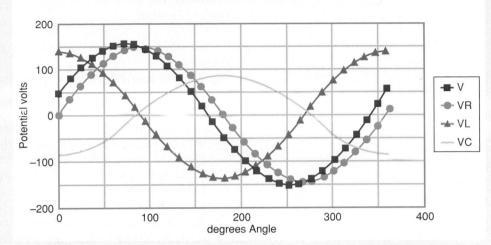

Since VR is in phase with the current i in the circuit, we can compare the phase relationship between the voltage V and the current i, by comparing the voltage V and the voltage VR. The phase angle can best be determined from this diagram by observing where the V curve intersects the horizontal axis, and where the VR curve intersects the horizontal axis. The difference between these two values represents the phase angle (phi), which was already calculated in part (i).

Special Cases
1) An RC series circuit. Let L = 0.
2) An RL series circuit. Let C = 0.
3) An LC series circuit. Let R = 0.

These Interactive Tutorials are a very helpful tool to aid in the learning of physics if they are used properly. The student should try to solve the particular problem in the traditional way using paper and an electronic calculator. Then she or he should open the spreadsheet, insert the appropriate data into the Initial Conditions cells, and see how the computer solves the problem. Go through each step on the computer and compare it to the steps you made on paper. Does your answer agree? If not, check through all the in-between steps on the computer and your paper and find where you made a mistake. Repeat the problem using different Initial Conditions on the computer and your paper. Again check your answers and all the in-between steps. Once you are sure that you know how to solve the problem, try some special cases. What would happen if you changed an angle? a weight? a force? and so forth. In this way you can get a great deal of insight into the physics of the problem and also learn a great deal of physics in the process.

You must be very careful not to just plug numbers into the Initial Conditions and look at the answers without understanding the in-between steps and the actual physics of the problem. You will only be deceiving yourself. Be careful. These spreadsheets can be extremely helpful if they are used properly.

Getting Started on the Computer

The **Interactive Tutorials** was written on Lotus 1-2-3 Release 4 for windows. The Windows environment is used because it is the easiest way to use a computer. It relies on the basic concepts of point and click. An electronic mouse is used to control the location of the pointer on the computer screen. By moving the mouse, you can place the pointer at any location on the screen. If you place the pointer at a menu item and click the left mouse button you activate that particular function of the menu.

Procedure for using the Interactive Tutorials disk that comes with this textbook.

1. Turn on your PC. After a moment or two, the *Program Manager* window appears on the screen.

2. Move the pointer with the mouse until the pointer is directly over the *Lotus Applications* icon on the screen. Double click the left mouse button. A new window will open showing you another series of icons.

3. Place the mouse pointer directly over the icon labeled *Lotus 1-2-3 Release 4* and again double click the left mouse button. After another moment the Lotus 1-2-3 Spreadsheet program will appear on your computer monitor screen in the form of a blank spreadsheet.

4. At this point, insert your Interactive Tutorials disk that comes with your textbook into drive B of your computer.

5. Using the mouse, point at the upper left-hand corner of your screen at the *File* menu of the Lotus spreadsheet. Click on *File* with the left mouse button and a menu will open, falling down on the left side of the screen. Move the pointer down to the file option *Open...* and then click on *Open...* with the left mouse button.

6. A new *Open File* dialogue box will open. On the lower right-hand side is a small box labeled *Drives:*. Click on the *Drives:* box and a submenu will drop down showing the drives *a:*, *b:*, *c:*, and *d:*. Click on the *b:* drive, and the computer will show you, in a box on the left-hand side of the *Open File* dialogue box, all the files on the b: drive of your computer. Because there are more files on the disk than can be displayed in the box, you cannot see all the files in this box. However, notice that to the immediate right of the box is a narrow box with an arrow pointing upward at the top of the box and another arrow pointing downward at the bottom of the narrow box. If you place and hold the pointer on the down arrow, the screen will scroll downward, displaying the additional files on the disk. If you place and hold the pointer on the up arrow the screen will scroll upward. In this way you can observe all the files on the disk. The files are listed by chapter and problem number. As an example, the file Ch3P79.wk4 is problem 79 in chapter 3. The suffix .wk4 is the notation that is used to identify this file as a Lotus computer spreadsheet, release 4.

7. Double click on the file of your choice and the spreadsheet will open for you. The spreadsheet will have all the necessary data for the problem already in it. If you wish to change any data in the yellow-colored cells of the Initial Conditions, place the pointer at the particular cell in the spreadsheet and click the left mouse button. This will activate that particular cell of the spreadsheet and you can enter any numbers you wish into that cell.

8. If you wish to open a new spreadsheet problem repeat steps 5 through 7. Good luck to you, and have fun with these spreadsheets.

THE LEARNING SYSTEM

More than 200 **new** problems have been extensively checked for accuracy.

†63. A 1.50-kg block moves along a smooth horizontal surface at 2.00 m/s. It then encounters a smooth inclined plane that makes an angle of 53.0° with the horizontal. How far up the incline will the block move before coming to rest?

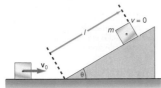

†64. Repeat problem 63, but in this case the inclined plane is rough and the coefficient of kinetic friction between the block and the plane is 0.400.

†65. In the diagram mass m_1 is located at the top of a rough inclined plane that has a length $l_1 = 0.500$ m. $m_1 = 0.500$ kg, $m_2 = 0.200$ kg, $\mu_{k_1} = 0.500$, $\mu_{k_2} = 0.300$, $\theta = 50.0°$, and $\phi = 50.0°$. (a) Find the total energy of the system in the position shown. (b) The system is released from rest. Find the work done for block 1 to overcome friction as it slides down the plane. (c) Find the work done for block 2 to overcome friction as it slides up the plane. (d) Find the potential energy of block 2 when it arrives at the top of the plane. (e) Find the velocity of block 1 as it reaches the bottom of the plane. (f) Find the kinetic energy of each block at the end of their travel.

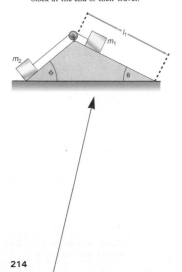

†66. If a constant force acting on a body is plotted against the displacement of the body from x_1 to x_2, as shown in the diagram, then the work done is given by

$$W = F(x_2 - x_1)$$
$$= \text{Area under the curve}$$

Show that this concept can be extended to cover the case of a variable force, and hence find the work done for the variable force, $F = kx$, where $k = 2.00$ N/m as the body is displaced from x_1 to x_2. Draw a graph showing your results.

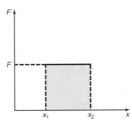

Interactive Tutorials

🖥 67. A projectile of mass $m = 100$ kg is fired vertically upward at a velocity $v_0 = 50.0$ m/s. Calculate its potential energy PE (relative to the ground), its kinetic energy KE, and its total energy E_{tot} for the first 10.0 s of flight.

🖥 68. Consider the general motion in an Atwood's machine such as the one shown in the diagram of problem 31; $m_A = 0.650$ kg and is at a height $h_A = 2.55$ m above the reference plane and mass $m_B = 0.420$ kg is at a height $h_B = 0.400$ m. If the system starts from rest, find (a) the initial potential energy of mass A, (b) the initial potential energy of mass B, and (c) the total energy of the system. When m_A has fallen a distance $y_A = 0.75$ m, find (d) the potential energy of mass A, (e) the potential energy of mass B, (f) the speed of each mass at that point, (g) the kinetic energy of mass A, and (h) the kinetic energy of mass B. (i) When mass A hits the ground, find the speed of each mass.

🖥 69. Consider the general motion in the combined system shown in the diagram of problem 42; $m_1 = 0.750$ kg and is at a height $h_1 = 1.85$ m above the reference plane and mass $m_2 = 0.285$ kg is at a height $h_2 = 2.25$ m, $\mu_k = 0.450$. If the system starts from rest, find (a) the initial potential energy of mass 1, (b) the initial potential energy of mass 2, and (c) the total energy of the system. When m_1 has fallen a distance $y_1 = 0.35$ m, find (d) the potential energy of mass 1, (e) the potential energy of mass 2, (f) the energy lost due to friction as mass 2 slides on the rough surface, (g) the speed of each mass at that point, (h) the kinetic energy of mass 1, and (i) the kinetic energy of mass 2. (j) When mass 1 hits the ground, find the speed of each mass.

🖥 70. Consider the general case of motion shown in the diagram with mass m_A initially located at the top of a rough inclined plane of length l_A, and mass m_B is at the bottom of the second plane; x_A is the distance from the mass A to the bottom of the plane. Let $m_A = 0.750$ kg, $m_B = 0.250$ kg, $l_A = 0.550$ m, $\theta = 40.0°$, $\phi = 30.0°$, $\mu_{k_A} = 0.400$, $\mu_{k_B} = 0.300$, and $x_A = 0.200$ m. When $x_A = 0.200$ m, find (a) the initial total energy of the system, (b) the distance block B has moved, (c) the potential energy of mass A, (d) the potential energy of mass B, (e) the energy lost due to friction for block A, (f) the energy lost due to friction for block B, (g) the velocity of each block, (h) the kinetic energy of mass A, and (i) the kinetic energy of mass B.

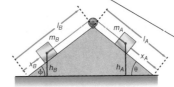

Daggers represent challenge problems.

Spreadsheet application exercises at the end of each chapter have been enhanced in number and quality.

Mechanics

Use of *large, full-color visuals in end-of-chapter problems* helps students with problem solving.

Diskette icons indicate a problem may be solved using a computer spreadsheet.

THE LEARNING SYSTEM

End-of-chapter "Language of Physics" features key concepts and terms that are defined and page-referenced.

The Language of Physics

Kinematics
The branch of mechanics that describes the motion of a body without regard to the cause of that motion (p. 39).

Average velocity
The average rate at which the displacement vector changes with time. Since a displacement is a vector, the velocity is also a vector (p. 39).

Average speed
The distance that a body moves per unit time. Speed is a scalar quantity (p. 41).

Constant velocity
A body moving in one direction in such a way that it always travels equal distances in equal times (p. 42).

Acceleration
The rate at which the velocity of a moving body changes with time (p. 43).

Instantaneous velocity
The velocity at a particular instant of time. It is defined as the limit of the ratio of the change in the displacement of the body to the change in time, as the time interval approaches zero. The magnitude of the instantaneous velocity is the instantaneous speed of the moving body (p. 45).

Kinematic equations of linear motion
A set of equations that gives the displacement and velocity of the moving body at any instant of time, and the velocity of the moving body at any displacement, if the acceleration of the body is a constant (p. 46).

Freely falling body
Any body that is moving under the influence of gravity only. Hence, any body that is dropped or thrown on the surface of the earth is a freely falling body (p. 51).

Acceleration due to gravity
If air friction is ignored, all objects that are dropped near the surface of the earth, are accelerated toward the center of the earth with an acceleration of 9.80 m/s² or 32 ft/s² (p. 52).

Projectile motion
The motion of a body thrown or fired with an initial velocity v_0 in a gravitational field (p. 55).

Trajectory
The path through space followed by a projectile (p. 56).

Range of a projectile
The horizontal distance from the point where the projectile is launched to the point where it returns to its launch height (p. 64).

Summary of Important Equations

"Summary of Important Equations" provides a quick reference of equations presented in each chapter.

Average velocity
$$\mathbf{v}_{avg} = \frac{\Delta \mathbf{r}}{\Delta t} = \frac{\mathbf{r}_2 - \mathbf{r}_1}{t_2 - t_1} \tag{3.32}$$

Acceleration
$$\mathbf{a} = \frac{\Delta \mathbf{v}}{\Delta t} = \frac{\mathbf{v} - \mathbf{v}_0}{t} \tag{3.33}$$

Instantaneous velocity in two or more directions, which is a generalization of the instantaneous velocity in one dimension
$$\mathbf{v} = \lim_{\Delta t \to 0} \frac{\Delta \mathbf{r}}{\Delta t}$$
$$v = \lim_{\Delta t \to 0} \frac{\Delta x}{\Delta t} \tag{3.8}$$

Velocity at any time
$$\mathbf{v} = \mathbf{v}_0 + \mathbf{a}t \tag{3.35}$$

Displacement at any time
$$\mathbf{r} = \mathbf{v}_0 t + \frac{1}{2}\mathbf{a}t^2 \tag{3.34}$$

Velocity at any displacement in the x-direction
$$v^2 = v_0^2 + 2ax \tag{3.16}$$

Velocity at any displacement in the y-direction
$$v^2 = v_0^2 + 2ay \tag{3.16}$$

For Projectile Motion

x-displacement
$$x = v_{0x}t \tag{3.38}$$

y-displacement
$$y = v_{0y}t - \frac{1}{2}gt^2 \tag{3.39}$$

x-component of velocity
$$v_x = v_{0x} \tag{3.40}$$

y-component of velocity
$$v_y = v_{0y} - gt \tag{3.41}$$

y-component of velocity at any height y
$$v_y^2 = v_{0y}^2 - 2gy \tag{3.48}$$

Range
$$R = \frac{v_0^2 \sin 2\theta}{g} \tag{3.47}$$

Questions for Chapter 3

"Questions" help students review key chapter concepts and prepare for "Problems."

1. Discuss the difference between distance and displacement.
2. Discuss the difference between speed and velocity.
3. Discuss the difference between average speed and instantaneous speed.
† 4. Although speed is the magnitude of the instantaneous velocity, is the average speed equal to the magnitude of the average velocity?
5. Why can the kinematic equations be used only for motion at constant acceleration?
6. In dealing with average velocities discuss the statement, "Straight line motion at 60 mph for 1 hr followed by motion in the same direction at 30 mph for 2 hr does not give an average of 45 mph but rather 40 mph."

7. What effect would air resistance have on the velocity of a body that is dropped near the surface of the earth?
8. What is the acceleration of a projectile when its instantaneous vertical velocity is zero at the top of its trajectory?
9. Can an object have zero velocity at the same time that it has an acceleration? Explain and give some examples.
10. Can the velocity of an object be in a different direction than the acceleration? Give some examples.
11. Can you devise a means of using two clocks to measure your reaction time?
†12. A person on a moving train throws a ball straight upward. Describe the

motion as seen by a person on the train and by a person on the station platform.
13. You are in free fall, and you let go of your watch. What is the relative velocity of the watch with respect to you?
†14. What kind of motion is indicated by a graph of displacement versus time, if the slope of the curve is (a) horizontal, (b) sloping upward to the right, and (c) sloping downward?
†15. What kind of motion is indicated by a graph of velocity versus time, if the slope of the curve is (a) horizontal, (b) sloping upward at a constant value, (c) sloping upward at a changing rate, (d) sloping downward at a constant value, and (e) sloping downward at a changing rate?

THE LEARNING SYSTEM

c. The image distance for the second lens, found from equation 27.26, is

$$\frac{1}{q_2} = \frac{1}{f_2} - \frac{1}{p_2} = \frac{1}{(-4.50\ \text{cm})} - \frac{1}{(-5.4\ \text{cm})}$$

$$q_2 = -27\ \text{cm}$$

Because q_2 is negative the final image is virtual.

"Have you ever wondered   ?"

An Essay on the Application of Physics

Nature's Camera—the Human Eye

Have you ever wondered how the human eye works? It is one of the greatest marvels of nature, but one that is usually taken for granted. The human eye is very much like a camera. It contains a lens, an iris diaphragm to let the light in, and its film is the retina at the back of the eye, figure 1. The lens focuses incident light onto the retina so a sharp image is observed. The iris diaphragm opens wide in subdued lighting and closes down in very bright light.

In the eye, the distance from the lens to the retina q, does not change as in a camera, instead the shape of the lens is changed by the muscles and ligaments of the eye. Hence, *the eye changes its focal length in order to bring objects at different object distances to a focus at the same image distance q, the distance from the lens to the retina. This changing of the focal length of the eye is called accommodation.* A schematic of the normal eye is shown in figure 2(a). The greatest amount of bending of light in the eye occurs at the cornea, because it is here where the greatest difference in the index of refraction occurs. The index of refraction of air is of course equal to one, whereas the index of refraction of the cornea is about 1.35. The index of refraction of the lens is about 1.44; that of the aqueous and vitreous humor is about 1.34.

Some eyes do not have the ability to change the shape of the lens as necessary and this is referred to as a defect in vision. One such defect is called **hyperopia, or farsightedness**, and is shown in figure 2(b). *A farsighted person can see distant objects*

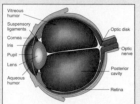

Figure 1
The human eye.

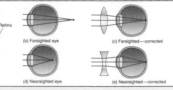

Figure 2
Defects of the human eye.

     Light and Optics

---

clearly, but objects a short distance away are blurred. The lens is not capable of bending the incoming light of near objects to a focus on the retina. Instead the light is focused behind the retina, leaving blurred vision. The defect is easily remedied by placing a converging lens in front of the eye, as shown in figure 2(c). The converging lens brings these incoming rays to a focus at the retina. Farsightedness is a defect that occurs to most people as they age. The lens hardens with age and the muscles weaken and are no longer capable of shaping the lens as needed.

Another defect is **myopia, or nearsightedness**, as illustrated in figure 2(d). *A nearsighted person can see nearby objects clearly, but distant objects are blurred.* The distant objects come to a focus in front of the retina leaving a blurred image on the retina. This defect can be eliminated by placing a diverging lens in front of the eye. The diverging lens causes the rays to converge more slowly and hence the image is formed farther back at the retina, as shown in figure 2(e).

The minimum distance from the eye at which an object can be seen distinctly is called the near point of the eye. For the average person, the near point is about 25.0 cm. That is, an object at a distance of 25.0 cm is the minimum distance that an object can be placed to give a distinct image on the retina. If the distance is decreased below this value, the image will be blurred for most people. Remember, this is only an average value, as we get older, the minimum distance of distinct vision increases. For example, the near point can be as close as 6.00 cm for a child and can increase to over a 100 cm for an elderly person.

Another common defect of the eye is called *astigmatism*. Astigmatism occurs in some people, because their eye is not completely spherical. That is, the radius of curvature of the eye in the vertical direction is not the same as the radius of curvature in the horizontal direction. Hence, vertical rays do not converge to the same position as horizontal rays. This defect is usually corrected with a lens of cylindrical curvature.

Although the magnification of a lens was defined in equation 27.22 as $M = h_i/h_o$, when viewing an object with the eye it is sometimes more desirable to talk about the angular magnification of the lens. Figure 3(a) shows an object at three distances from the unaided eye. The size of the object, as determined by the unaided eye, depends on the angle subtended by the object at the eye. Thus, when the object is at position 1, it subtends an angle θ_1; when it is at position 2, it subtends the angle θ_2; and when it is at position 3, it subtends the angle θ_3. As seen from the diagram, $\theta_1 < \theta_2 < \theta_3$. Thus, the closer the object is to the eye, the greater the angle subtended and the easier it is for the eye to see the object. However, the object cannot be moved closer to the eye than the near point, because that is the closest point that the eye can still see the object distinctly. In figure 3(b), the object is placed at the object distance $p_1 = 25.0$ cm in front of a normal eye. The height of the object h_o subtends an angle α at the lens of the eye. The image of the object is focused on the retina of the eye. This is the largest image that the unaided eye can make of the object. If the object is placed within the focal point of a converging lens, the object can be brought even closer

---

Example 27H.1

Range of focal length for the eye. The distance from the cornea to the retina in the human eye is about 20.0 mm. What should the focal length of the human eye be in order to see clearly an object (a) at infinity and (b) at the near point of most distinct vision, 25 cm?

Solution

a. For the object at infinity, the image is located at the principal focus. Hence

$$f = q = 20.0\ \text{mm} = 2.00\ \text{cm}$$

is the focal length of the eye when it views an object at infinity.

b. For the object at $p = 25.0$ cm, the focal length would have to be

$$\frac{1}{f} = \frac{1}{p} + \frac{1}{q} = \frac{1}{25.0\ \text{cm}} + \frac{1}{2.00\ \text{cm}} = 0.540\ \text{cm}^{-1}$$

$$f = 1.85\ \text{cm}$$

Thus, the eye only has to change its focal length by (2.00 cm − 1.85 cm = 0.15 cm) 15 mm to see its entire range of vision.

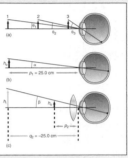

Figure 3
Angular magnification and the simple microscope.
A doctor examines a patient's eye.

Chapter 27   The Law of Refraction

$$M_A = \frac{\beta}{\alpha}$$

From figure 3(c), we see that

$$\tan \beta = \frac{h_o}{p_2}$$

whereas from figure 3(b), we see that

$$\tan \alpha = \frac{h_o}{p_1} = \frac{h_o}{25.0\ \text{cm}}$$

Let us take the ratio

$$\frac{\tan \beta}{\tan \alpha} = \frac{h_o/p_2}{h_o/25.0\ \text{cm}}$$

$$= \frac{25.0\ \text{cm}}{p_2}$$

Because the angles considered are small, we can use the small-angle approximation, setting the tangent of an angle equal to the angle itself. Hence,

$$\frac{\beta}{\alpha} = \frac{25.0\ \text{cm}}{p_2}$$

But this ratio is the definition of the angular magnification, equation 27H.1, hence

$$M_A = \frac{25.0\ \text{cm}}{p_2} \qquad \text{(27H.2)}$$

The value of p_2, found from the lens equation, is

$$\frac{1}{p_2} = \frac{1}{f} - \frac{1}{q_2}$$

$$\frac{1}{p_2} = \frac{1}{f} - \frac{1}{-25.0\ \text{cm}} = \frac{25.0\ \text{cm} + f}{(25.0\ \text{cm})f} \qquad \text{(27H.3)}$$

And the object distance for the image to be at the near point is

$$p_2 = \frac{(25.0\ \text{cm})f}{25.0\ \text{cm} + f} \qquad \text{(27H.4)}$$

Substituting equation 27H.4 into equation 27H.2, gives

$$M_A = \frac{\beta}{\alpha} = \frac{25.0\ \text{cm}}{p_2} = \frac{25.0\ \text{cm}}{(25.0\ \text{cm})f/(25.0\ \text{cm} + f)}$$

$$= \frac{25.0\ \text{cm}}{f} + 1 \qquad \text{(27H.5)}$$

Equation 27H.5 gives the angular magnification of a converging lens when the image is located at the near point of the eye, which is assumed to be 25.0 cm for the normal eye.

$$M_A = \frac{25.0\ \text{cm}}{f} + 1$$

$$= \frac{25.0\ \text{cm}}{10.0\ \text{cm}} + 1$$

$$= 3.5$$

The eye is more relaxed when viewing an object at infinity than at the near point. It is, therefore, sometimes more desirable to view an image at infinity than at the near point. For the image to be at infinity, the object must be placed at the principal focus. The lens equation, equation 27H.3, shows that if p is at infinity, then the object distance p_2 is equal to the focal length f. Hence substituting f for p_2 in equation 27H.2, gives

$$M_A = \frac{25.0\ \text{cm}}{f} \qquad \text{(27H.6)}$$

Equation 27H.6 gives the angular magnification of a simple microscope when the image is set for viewing at infinity.

Example 27H.3

The angular magnification of a lens when the image is at infinity. A magnifying glass of 10.0-cm focal length is used to view an object, such that the image is at infinity. Find the angular magnification of the lens.

Solution

The angular magnification, found from equation 27H.6, is

$$M_A = \frac{25.0\ \text{cm}}{f}$$

$$= \frac{25.0\ \text{cm}}{10.0\ \text{cm}} = 2.50$$

Notice that when the image is viewed at the near point the angular magnification is 3.5, whereas when viewed at infinity, it is 2.5. If viewing is for a long time, it might be desirable to view the image at infinity because it is easier on the eyes. If the magnification is the most important consideration, then the image should be viewed at the near point of the eye.

     Light and Optics

"Have You Ever Wondered?" boxes focus on the applied nature of physics by showing students how physics affects their everyday lives.

THE LEARNING SYSTEM

Highlighted equations reinforce textual material.

Example 7.5

Power expended. A person pulls a block with a force of 15.0 N at an angle of 25.0° with the horizontal. If the block is moved 5.00 m in the horizontal direction in 5.00 s, how much power is expended?

Solution

The power expended, found from equations 7.3 and 7.2, is

$$P = \frac{W}{t} = \frac{Fx \cos \theta}{t}$$

$$= \frac{(15.0 \text{ N})(5.00 \text{ m})\cos 25.0°}{5.00 \text{ s}} = 13.6 \ \frac{\text{N m}}{\text{s}} = 13.6 \text{ W}$$

When a constant force acts on a body in the direction of the body's motion, we can also express the power as

$$P = W = \frac{Fx}{t} = F\frac{x}{t}$$

but

$$\frac{x}{t} = v$$

the velocity of the moving body. Therefore,

$$P = Fv \qquad (7.4)$$

is the power expended by a force F, acting on a body that is moving at the velocity v.

Example 7.6

Power to move your car. An applied force of 5500 N keeps a car moving at 95 km/hr. How much power is expended by the car?

Solution

The power expended by the car, found from equation 7.4, is

$$P = Fv = (5500 \text{ N})\left(95 \ \frac{\text{km}}{\text{hr}}\right)\left(\frac{1 \text{ hr}}{3600 \text{ s}}\right)\left(\frac{1000 \text{ m}}{1 \text{ km}}\right)$$

$$= 1.45 \times 10^5 \text{ N m/s} = 1.45 \times 10^5 \text{ J/s}$$

$$= 1.45 \times 10^5 \text{ W}$$

7.4 Gravitational Potential Energy

Gravitational potential energy is defined as the energy that a body possesses by virtue of its position. If the block shown in figure 7.7, were lifted to a height h above the table, then that block would have potential energy in that raised position. That is, in the raised position, the block has the ability to do work whenever it is allowed to fall. The most obvious example of gravitational potential energy is a waterfall (figure 7.8). Water at the top of the falls has potential energy. When the water falls to the bottom, it can be used to turn turbines and thus do work. A similar example is a pile driver. A pile driver is basically a large weight that is raised above a pile

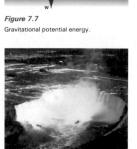

Figure 7.7
Gravitational potential energy.

Figure 7.8
Water at the top of the falls has potential energy.

Mechanics

In-text Examples are followed by Solutions and include intermediate steps.

---

that is to be driven into the ground. In the raised position, the driver has potential energy. When the weight is released, it falls and hits the pile and does work by driving the pile into the ground.

Therefore, whenever an object in the gravitational field of the earth is placed in a position above some reference plane, then that object will have potential energy because it has the ability to do work.

As in all the concepts studied in physics, we want to make this concept of potential energy quantitative. That is, how much potential energy does a body have in the raised position? How should potential energy be measured?

Because work must be done on a body to put the body into the position where it has potential energy, the work done is used as the measure of this potential energy. That is, the potential energy of a body is equal to the work done to put the body into the particular position. Thus, the potential energy (PE) is

$$\text{PE} = \text{Work done to put body into position} \qquad (7.5)$$

We can now compute the potential energy of the block in figure 7.7 as

$$\text{PE} = \text{Work done}$$

$$\text{PE} = W = Fh = wh \qquad (7.6)$$

The applied force F necessary to lift the weight is set equal to the weight w of the block. And since $w = mg$, the potential energy of the block becomes

$$\text{PE} = mgh \qquad (7.7)$$

We should emphasize here that the potential energy of a body is referenced to a particular plane, as in figure 7.9. If we raise the block a height h_1 above the table, then with respect to the table it has a potential energy

$$\text{PE}_1 = mgh_1$$

While at the same position, it has the potential energy

$$\text{PE}_2 = mgh_2$$

with respect to the floor, and

$$\text{PE}_3 = mgh_3$$

with respect to the ground outside the room. All three potential energies are different because the block can do three different amounts of work depending on whether it falls to the table, the floor, or the ground. Therefore, it is very important that when the potential energy of a body is stated, it is stated with respect to a particular reference plane. We should also note that it is possible for the potential energy to be negative with respect to a reference plane. That is, if the body is not located above the plane but instead is found below it, it will have negative potential energy with respect to that plane. In such a position the body can not fall to the reference plane and do work, but instead work must be done on the body to move the body up to the reference plane.

Figure 7.10
Changing potential energy.

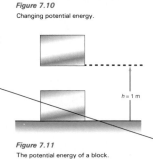

$h = 1$ m

Figure 7.11
The potential energy of a block.

Example 7.7

The potential energy. A mass of 1.00 kg is raised to a height of 1.00 m above the floor (figure 7.11). What is its potential energy with respect to the floor?

Solution

The potential energy, found from equation 7.7, is

$$\text{PE} = mgh = (1.00 \text{ kg})(9.80 \text{ m/s}^2)(1.00 \text{ m})$$

$$= 9.80 \text{ J}$$

Chapter 7 Energy and Its Conservation

FUNDAMENTALS OF COLLEGE
PHYSICS

SECOND EDITION

Mechanics

Color as a Study Aid

As a pedagogical aid, color has been extensively used throughout the book. The purpose of the color is not just to make the book look good, but rather to help the student visualize the material. The colors have been standardized according to the following code.

Chapter 2

For multiple vectors in a diagram each vector has a different color:

- First vector **a**
- Second vector **b**
- Third vector **c**
- Fourth vector **d**
- Resultant vector **R**
- Negative of a vector

The components of a vector are always a lighter shade of the same color of the original vector:

- Vector **a**
- *x*- and *y*-components of **a**
- Vector **R**
- *x*- and *y*-components of **R**

The *x*, *y*, and *z* coordinates are always black.

Chapter 3

The color code for vectors:

- Displacement vectors
- Velocity vectors
- Acceleration vectors
- Trajectories

The components of a vector are always a lighter shade of the same color of the original vector:

- Displacement vectors
- Components of displacement vector
- Velocity vectors
- Components of the velocity vector
- Acceleration vectors
- Components of the acceleration vector

Coordinates are always black.

Chapter 4

Color code for force vectors:

- Applied force vectors
- Components of applied force vector
- Tension force vectors
- Components of tension force vector
- Friction force vectors
- Normal force vectors
- Weight force vectors
- Components of weight force vector

- Reaction force vectors
- Centripetal force vector

Note that a different shade of green is used for the centripetal force vector so that it is not the same dark green that is used for the applied force vector.

Also note that a different shade of dark blue is used for the weight force vector so that it is not the same blue that is used for the velocity vector.

Chapter 5

The same color code is used as in the previous chapters. In addition:

- Lever arms

Chapter 6

The same color code is used as in the previous chapters. In addition:

- Arc *s* of a circle
- Circular orbits
- Elliptical orbits
- Gravitational force vector

Note that this is the same color as the dark blue of the weight vector.

Chapters 7 through 9

In all these chapters the same color code is used as in the previous chapters.

1 Introduction and Measurements

Earth rise from the moon.

In exploring nature, therefore, we must begin by trying to determine its first principles.

Aristotle

The method with which we shall follow in this treatise will be always to make what is said depend on what was said before.

Galileo Galilei

1.1 Historical Background

Physics has its birth in mankind's quest for knowledge and truth. In ancient times, people were hunters following the wild herds for their food supply. Since they had to move with the herds for their survival, they could not be tied down to one site with permanent houses for themselves and their families. Instead these early people lived in whatever caves they could find during their nomadic trips. Eventually these cavemen found that it was possible to domesticate such animals as sheep and cattle. They no longer needed to follow the wild herds. Once they stayed long enough in one place to take care of their herds, they found that seeds collected from various edible plants in one year could be planted the following year for a new crop. Thus, many of these ancient people became farmers, growing their own food supply. They, of course, found that they could grow a better crop in a warm climate near a readily available source of water. It is not surprising then that the earliest known civilizations sprang up on the banks of the great rivers: the Nile in Egypt and the Tigris and Euphrates in Mesopotamia. Once permanently located on their farms, these early people were able to build houses for themselves. Trades eventually developed and what would later be called civilization began.

To be successful farmers, these ancient people had to know when to plant the seeds and when to harvest the crop. If they planted the seeds too early, a frost could destroy the crop, causing starvation for their families. If they planted the seeds too late, there might not be sufficient growing time or adequate rain.

In those very dark nights, people could not help but notice the sky. It must have been a beautiful sight without the background street lights that are everywhere today. People began to study that sky and observed a regularity in the movements of the sun, moon, and stars. In ancient Egypt, for example, the Nile river would overflow when Sirius, the Dog Star, rose above the horizon just before dawn. People then developed a calendar based on the position of the stars. By their observation of the sky, they found that when certain known stars were in a particular position in the sky it was time to plant a new crop. With an abundant harvest it was now possible to store enough grain to feed the people for the entire year.

For the first time in the history of humanity, obtaining food for survival was not an all time-consuming job. These ancient people became affluent enough to afford the time to think and question. What is the cause of the regularity in the motion of the heavenly bodies? What makes the sun rise, move across the sky, and then set? What makes the stars and moon move in the night sky? What is the earth made of? What is man? And through this questioning of the world about them, **philosophy** was born—the search for knowledge or wisdom (*philos* in Greek means

Figure 1.1

The caveman steps out of his cave.

"love of" and *sophos* means "wisdom"). Philosophy, therefore, originated when these early people began to seek a rational explanation of the world about them, an explanation of the nature of the world without recourse to magic, myths, or revelation. Ancient philosophers studied ethics, morality, and the essence of beings as determined by the mind, but they also studied the natural world itself. This latter activity was called **natural philosophy**—the study of the phenomena of nature. Among early Greek natural philosophers were Thales of Miletus (ca. 624–547 B.C.), Democritus (ca. 460–370 B.C.), Aristarchus (ca. 320–250 B.C.), and Archimedes (ca. 287–212 B.C.), perhaps the greatest scientist and mathematician of ancient times.

For many centuries afterward, the study of nature continued to be called natural philosophy. In fact, one of the greatest scientific works ever written was by Sir Isaac Newton. When it was published in 1687, he entitled it *Philosophiae Naturalis Principia Mathematica* (*The Mathematical Principles of Natural Philosophy*).

*Natural philosophy, therefore, studied all of nature. The Greek word for "natural" is physikos. Therefore, the name **physics** came to mean the study of all of nature.* Physics became a separate entity from philosophy because it employed a different method to search for truth. Physics developed and employed an approach called the scientific method in its quest for knowledge.

The **scientific method** is the application of a logical process of reasoning to arrive at a model of nature that is consistent with experimental results. The scientific method consists of five steps:

1. Observation
2. Hypothesis
3. Experiment
4. Theory or law
5. Prediction

This process of scientific reasoning can be followed with the help of the flow diagram shown in figure 1.2.

1. *Observation.* The first step in the scientific method is to make an observation of nature, that is, to collect data about the world. The data may be drawn from a simple observation, or they may be the results of numerous experiments.

Figure 1.2

The scientific method.

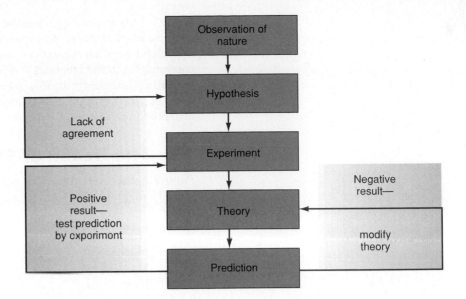

2. *Hypothesis.* From an analysis of these observations and experimental data, a model of nature is hypothesized. The dictionary defines a hypothesis as an assumption that is made in order to draw out and test its logical or empirical consequences; that is, an assumption is made that in a given situation nature will always work in a certain way. If this hypothesis is correct, we should be able to confirm it by testing. This testing of the hypothesis is called the experiment.

3. *Experiment.* An experiment is a controlled procedure carried out to discover, test, or demonstrate something. An experiment is performed to confirm that the hypothesis is valid. If the results of the experiment do not support the hypothesis, the experimental technique must be checked to make sure that the experiment was really measuring that aspect of nature that it was supposed to measure. If nothing wrong is found with the experimental technique, and the results still contradict the hypothesis, then the original hypothesis must be modified. Another experiment is then made to test the modified hypothesis. The hypothesis can be modified and experiments redesigned as often as necessary until the hypothesis is validated.

4. *Theory.* Finally, success: the experimental results confirm that the hypothesis is correct. The hypothesis now becomes a new theory about some specific aspect of nature, a scientifically acceptable general principle based on observed facts. After a careful analysis of the new theory, a prediction about some presently unknown aspect of nature can be made.

5. *Prediction.* Is the prediction correct? To answer that question, the prediction must be tested by performing a new experiment. If the new experiment does not agree with the prediction, then the theory is not as general as originally thought. Perhaps it is only a special case of some other more general model of nature. The theory must now be modified to conform to the negative results of the experiment. The modified theory is then analyzed to obtain a new prediction, which is then tested by a new experiment. If the new experiment confirms the prediction, then there is reasonable confidence that this theory of nature is correct. This process of prediction and experiment continues many times. As more and more predictions are confirmed by experiment, mounting evidence indicates that a good model of the way nature works has been developed. At this point, the theory can be called a *law of physics.*

This method of scientific reasoning demonstrates that the establishment of any theory is based on experiment. In fact, the success of physics lies in this agreement between theoretical models of the natural world and their experimental confirmation in the laboratory. A particular model of nature may be a great intellectual achievement but, if it does not agree with physical reality, then, from the point of view of physics, that hypothesis is useless. Only hypotheses that can be tested by experiment are relevant in the study of physics.

1.2 The Realm of Physics

Physics can be defined as the study of the entire natural or physical world. To simplify this task, the study of physics is usually divided into the following categories:

I. Classical Physics
   1. Mechanics
   2. Wave Motion
   3. Heat
   4. Electricity and Magnetism
   5. Light
II. Modern Physics
   1. Relativity
   2. Quantum Mechanics
   3. Atomic and Nuclear Physics
   4. Condensed Matter Physics
   5. Elementary Particle and High-Energy Physics

Although there are other sciences of nature besides physics, physics is the foundation of these other sciences. For example, astronomy is the application of physics to the study of all matter beyond the earth, including everything from within the solar system out to the remotest galaxies. Chemistry is the study of the properties of matter and the transformation of that matter. Geology is the application of physics to the study of the earth. Meteorology is the application of physics to the study of the atmosphere. Engineering is the application of physics to the solution of practical problems. The science of biology, which traditionally had been considered independent of physics, now uses many of the principles of physics in its study of molecular biology. The health sciences use so many new techniques and equipment based on physical principles that even there it has become necessary to have an understanding of physics.

This distinction between one science and another is usually not clear. In fact, there is often a great deal of overlap among them.

1.3 Physics Is a Science of Measurement

In order to study the entire physical world, we must first observe it. To be precise in the observation of nature, all the physical quantities that are observed should be measured and described by numbers. The importance of numerical measurements was stated by the Scottish physicist, William Thomson (1824–1907), who was made Baron Kelvin in 1892 and has since been referred to as Lord Kelvin:

> *I often say that when you can measure what you are speaking about, and express it in numbers, you know something about it; but when you cannot express it in numbers, your knowledge is of a meager and unsctisfactory kind.*

Figure 1.3

A thought experiment on temperature.

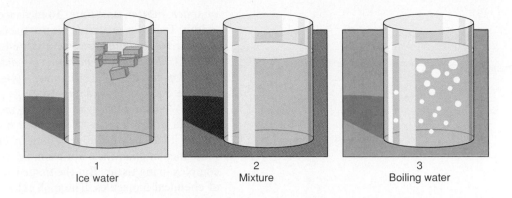

1	2	3
Ice water	Mixture	Boiling water

We can see the necessity for quantitative measurements from the following example. First, let us consider the following *thought experiment. (A thought ex periment is an experiment that we can think through, rather than actually per forming the experiment.)* Three beakers are placed on the table as shown in figure 1.3. In the first beaker, we place several ice cubes in water. We place boiling water in the third beaker. In the second beaker, we place a mixture of the ice water from beaker 1 and the boiling water from beaker 3. If you put your left hand into beaker 1, you will conclude that the ice water is *cold*. Now place your left hand into beaker 2, which contains the mixture. After coming from the ice water, your hand finds the second beaker to be hot by comparison. So you naturally conclude that the mixture is *hot*.

Now take your right hand and plunge it into the boiling water of beaker 3. (This is the reason that this is only a *thought experiment*. You can certainly appreciate what would happen in the experiment without actually risking bodily harm.) You would then conclude that the water in beaker 3 is certainly *hot*. Now place your right hand into beaker 2. After the boiling water, your hand finds the mixture cold by comparison, so you conclude that the mixture is *cold*. After this relatively "scientific" experiment, you find that you have contradictory conclusions. That is, you have found the middle mixture to be either hot or cold depending on the sequence of the measurements.

We can therefore conclude that in this particular observation of nature, describing something as hot or cold is not very accurate. Unless we can say numerically how hot or cold something is, our observation of nature is incomplete. In practice, of course, we would use a thermometer to measure the temperature of the contents of each beaker and read the hotness or coldness of each beaker as a number on the thermometer. For example, the thermometer might read, 0°C or 50°C or 100°C. We would now have assigned a number to our observation of nature and would thus have made a precise statement about that observation. This example points out the necessity of assigning a number to any observation of nature. The next logical question is, "What should we observe in nature?"

1.4 **The Fundamental Quantities**

If physics is the study of the entire natural world, where do we begin in our observations and measurements of it? It is desirable to describe the natural world in terms of the fewest possible number of quantities. This idea is not new; some of the ancient Greek philosophers thought that the entire world was composed of only four elements—earth, air, fire, and water. Although today we certainly would not accept these elements as building blocks of the world, *we do accept the basic principle that the world is describable in terms of a few fundamental quantities.*

When we look out at the world, we observe that the world occupies space, that within that space we find matter, and that space and matter exists within something we call time. So we will look for our observations of the world in terms of space, matter, and time. To measure space, we use the fundamental quantity of length. To measure matter, we use the fundamental quantities of mass and electrical charge. To measure time, we use the fundamental quantity of time itself.

*Therefore, to measure the entire physical world, we use the four **fundamental quantities** of length, mass, time, and charge.* We call all the other quantities that we observe derived quantities.

We have assigned ourselves an enormous task by trying to study the entire physical world in terms of only four quantities. The most remarkable part of it all is that it can be done. Everything in the world can be described in terms of these fundamental quantities. For example, consider a biological system, composed of very complex living tissue. But the tissue itself is made up of cells, and the cells are made of chemical molecules. The molecules are made of atoms, while the atoms consist of electrons, protons, and neutrons, which can be described in terms of the four fundamental quantities.

We might also ask of what electrons, protons, and neutrons are made. These particles are usually considered to be fundamental particles, however, the latest hypothesis in elementary particle physics is that protons and neutrons are made of even smaller particles called quarks. And although no one has yet actually found an isolated quark, and indeed some theories suggest that they are confined within the particles and will never be seen, the quark hypothesis has successfully predicted the existence of other particles, which *have* been found. The finding of these predicted particles gives a certain amount of credence to the existence of quarks. Of course if the quark is ever found then the next logical question will be, "Of what is the quark made?"

This progression from one logical question to the next in our effort to study the entire natural world is part of the adventure of physics. But to succeed on this adventure, we need to be precise in our observations, which brings us back to the subject at hand. If we intend to measure the world in terms of the four fundamental quantities of length, mass, time, and charge, we need to agree on some standard of measurement for each of these quantities.

1.5 The Standard of Length

The fundamental quantity of length is used to measure the location and the dimensions of any object in space. An object is located in space with reference to some coordinate system, as shown in figure 1.4. If the object is at the position P, then it can be located by moving a distance l_x in the x direction, then a distance l_y in the y direction, and finally a distance l_z in the z direction.

But before we can measure the distances l_x, l_y, and l_z, or for that matter, any distance, we need a standard of length that all observers can agree on. For example, suppose we wanted to measure the length of the room. We could use this text book as the standard of length. We would then place the text book on the floor and lay off the entire distance by placing the book end-over-end on the floor as often as necessary until the entire distance is covered. We might then say that the room is 25 books long. But this is not a very good standard of length because there are different sized books, and if you performed the measurement with another book, you would say that the floor has a different length.

We could even use the tile on the floor as a standard of length. To measure the length of the room all we would have to do is count the number of tiles. Indeed, if you worked at laying floor tiles, this would be a very good standard of length. The choice of a standard of length does seem somewhat arbitrary. In fact, just think of some of the units of measurement that you are familiar with:

The *foot*—historically the foot was used as a standard of length and it was literally the length of the king's foot. Every time you changed the king, you changed the measurement of the foot.

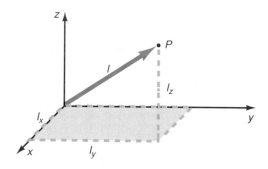

Figure 1.4

The location of an object in space.

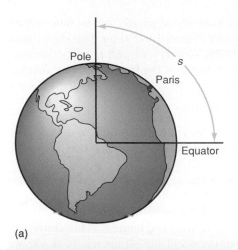

(a)

(b)

Figure 1.5

(a) The original definition of the meter.
(b) View of the earth from space.

The *yard*—the yard was the distance from the outstretched hand of the king to the back of his neck. Obviously, this standard of length also changed with each king.

The *inch*—the inch was the distance from the tip of the king's thumb to the thumb knuckle.

With these very arbitrary and constantly changing standards of length, it was obviously very difficult to make a measurement of length that all could agree on.

During the French Revolution, the French National Assembly initiated a proposal to the French Academy of Sciences to reform the systems of weights and measures. Under the guidance of such great physicists as Joseph L. Lagrange and Pierre S. de Laplace, the committee agreed on a measuring system based on the number 10 and its multiples. In this system, the unit of length chosen was one ten-millionth of the distance *s* from the North Pole to the equator along a meridian passing through Paris, France (figure 1.5). The entire distance from the pole to the equator was not actually measured. Instead a geodetic survey was undertaken for 10 degrees of latitude extending from Dunkirk, in northern France, to Barcelona, in Spain. From these data, the distance from the pole to the equator was found. The meter, the standard of length, was defined as one ten-millionth of this distance. A metal rod equal to this distance was made, and it was stored in Sèvres, just outside Paris. Copies of this rod were distributed to other nations to be used as their standard.

In time, with greater sophistication in measuring techniques, it turned out that the distance from the North Pole to the equator was in error, so the length of the meter could no longer represent one ten-millionth of that distance. But that really did not matter, as long as everyone agreed that this length of rod would be the standard of length. For years these rods were the accepted standard. However, they also had drawbacks. They were not readily accessible to all the nations of the world, and they could be destroyed by fire or war. A new standard had to be found. The standard remained the meter, but it was now defined in terms of something else. In 1960, the Eleventh General Conference of Weights and Measures defined the meter as a certain number of wavelengths of light from the krypton 86 atom.

Using a standard meter bar or a prescribed number of wavelengths indeed gives us a standard length. Such measurements are called *direct measurements*. But in addition to direct measuring procedures, an even more accurate determination of a quantity sometimes can be made by measuring something other than the desired quantity, and then obtaining the desired quantity by a calculation with the measured quantity. Such procedures, called *indirect measuring techniques,* have been used to obtain an even more precise definition of the meter.

We can measure the speed of light *c,* a derived quantity, to a very great accuracy. The speed of light has been measured at 299,792,458 meters/second, with an uncertainty of only four parts in 10^9, a very accurate value to be sure. Using this value of the speed of light, the standard meter can now be defined. On October 20, 1983, the Seventeenth General Conference on Weights and Measures redefined the meter as: *"The **meter** is the length of the path traveled by light in a vacuum during a time interval of 1/299,792,458 of a second."*

We will see in chapter 3 on kinematics that the distance an object moves at a constant speed is equal to the product of its speed and the time that it is in motion. Using this relation the meter is defined as

$$\text{distance} = (\text{speed of light})(\text{time})$$

$$= \left(\frac{299{,}792{,}458 \text{ meters}}{\text{second}}\right)\left(\frac{\text{second}}{299{,}792{,}458}\right) = 1 \text{ meter}$$

Hence the meter, the fundamental quantity of length, is now determined in terms of the speed of light and the fundamental quantity of time. The meter, thus defined, is a fixed standard of length accessible to everyone and is nonperishable. For everyone brought up to think of lengths in terms of the familiar inches, feet, or yards, the meter, abbreviated m, is equivalent to

$$1.000 \text{ m} = 39.37 \text{ in.} = 3.281 \text{ ft} = 1.094 \text{ yd}$$

For very precise work, the standard of length must be used in terms of its definition. For most work in a college physics course, however, the standard of length will be the simple meter stick.

The system of measurements based on the meter was originally called the metric system of measurements. Today it is called the **International System (SI) of units.** The letters are written SI rather than IS because the official international name follows French usage, "Le Système International d'Unités." This system of measurements is used by scientists throughout the world and commercially by almost all the countries of the world except the United States and one or two other countries. The United States is supposed to be changing over to this system also.

One of the advantages of using the meter as the standard of length is that the meter is divided into 100 parts called centimeters (abbreviated cm). The centimeter, in turn, is divided into ten smaller divisions called millimeters (mm). The kilometer (abbreviated km) is equal to a thousand meters, and is used to measure very large distances. Thus the units of length measurement become a decimal system, that is,

$$1 \text{ m} = 100 \text{ cm}$$
$$1 \text{ cm} = 10 \text{ mm}$$
$$1 \text{ km} = 1000 \text{ m}$$

A further breakdown of the units of length into powers of ten is facilitated by using the following prefixes:

$$\text{tera (T)} = 10^{12}$$
$$\text{giga (G)} = 10^{9}$$
$$\text{mega (M)} = 10^{6}$$
$$\text{milli (m)} = 10^{-3}$$
$$\text{micro } (\mu) = 10^{-6}$$
$$\text{nano (n)} = 10^{-9}$$
$$\text{pico (p)} = 10^{-12}$$
$$\text{femto (f)} = 10^{-15}$$

Students unfamiliar with the powers of ten notation, and scientific notation in general, should consult appendix B. Using these prefixes, the lengths of all observables can be measured as multiples or submultiples of the meter.

The decimal nature of the SI system makes it easier to use than the *British engineering system,* the system of units that is used in the United States. For example, compare the simplicity of the decimal metric system to the arbitrary units of the British engineering system:

$$12 \text{ inches} = 1 \text{ foot}$$
$$3 \text{ feet} = 1 \text{ yard}$$
$$5280 \text{ feet} = 1 \text{ mile}$$

In fact, these units are now officially defined in terms of the meter as

$$1 \text{ foot} = 0.3048 \text{ meters} = 30.48 \text{ centimeters}$$
$$1 \text{ yard} = 0.9144 \text{ meters} = 91.44 \text{ centimeters}$$
$$1 \text{ inch} = 0.0254 \text{ meters} = 2.54 \text{ centimeters}$$
$$1 \text{ mile} = 1.609 \text{ kilometers}$$

A complete list of equivalent measurements can be found in appendix A.

1.6 The Standard of Mass

The simplest definition of **mass** is that *mass is a measure of the quantity of matter in a body*. This is not a particularly good definition, but it is one for which we have an intuitive grasp. Mass will be redefined more accurately in terms of its inertial and gravitational characteristics later. For now, let us think of the mass of a body as being the matter that is contained in the sum of all the atoms and molecules that make up that body. For example, the mass of this book is the matter of the billions upon billions of atoms that make up the pages and the print of the book itself.

The standard we use to measure mass can, like the standard of length, also be quite arbitrary. In 1795, the French Academy of Science initially defined the standard as the amount of matter in 1000 cm³ of water at 0°C and called this amount of mass, one kilogram. This definition was changed in 1799 to make the kilogram the amount of matter in 1000 cm³ of water at 4°C, the temperature of the maximum density of water. However, in 1889, *the new and current definition of the **kilogram** became the amount of matter in a specific platinum iridium cylinder 39 mm high and 39 mm in diameter.* The metal alloy of platinum and iridium was chosen because it was considered to be the most resistant to wear and tarnish. Copies of the cylinder (figure 1.6) are kept in the standards laboratories of most countries of the world.

Figure 1.6

The standard kilogram mass.

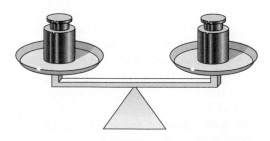

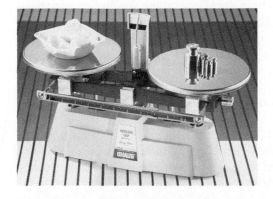

Figure 1.7

A simple balance.

The disadvantage of using this cylinder as the standard of mass is that it could be easily destroyed and it is not readily accessible to every country on earth. It seems likely that sometime in the future, when the necessary experimental techniques are developed, the kilogram will be redefined in terms of the mass of some specified number of atoms or molecules, thereby giving the standard of mass an atomic definition.

With the standard of mass, the kilogram, defined, any number of identical masses, multiple masses, or submultiple masses can be found by using a simple balance, as shown in figure 1.7. We place a standard kilogram on the left pan of the balance, and then place another piece of matter on the right pan. If the new piece of matter is exactly 1 kg, then the scale will balance and we have made another kilogram mass. If there is too much matter in the tested sample the scales will not balance. We then shave off a little matter from the sample until the scales do balance. On the other hand, if there is not enough matter in the sample, we add a little matter to the sample until the scales do balance. In this way, we can make as many one kilogram masses as we want.

Any multiple of the kilogram mass can now be made with the aid of the original one kilogram masses. That is, if we want to make a 5-kg mass, we place five 1-kg masses on the left pan of the balance and add mass to the right pan until the scale balances. When this is done, we will have made a 5-kg mass. Proceeding in this way, we can obtain any multiple of the kilogram.

To make submultiples of the kilogram mass, we cut a 1-kg mass in half, and place one half of the mass on each of the two pans of the balance. If we have cut the kilogram mass exactly in half, the scales will balance. If they do not, we shave off a little matter from one of the samples and add it to the other sample until the scales do balance. Two 1/2-kg masses thus result. Since the prefix kilo means a thousand, these half-kilogram masses each contain 500 grams (abbreviated g). If we now cut a 500-g mass in half and place each piece on one of the pans of the balance, making of course whatever corrections that are necessary, we have two 250-g masses. Continuing this process by taking various combinations of cuttings and placing them on the balance, eventually we can make any submultiple of the kilogram. The assembly of these multiples and submultiples of the kilogram is called a set of masses. (Quite often, this is erroneously referred to as a set of weights.)

We can now measure the unknown mass of any body by placing it on the left pan of the balance and adding any multiple, and/or submultiple, of the kilogram to the right pan until the scales balance. The sum of the combination of the masses placed on the right pan is the mass of the unknown body. So we can determine the mass of any body in terms of the standard kilogram.

The principle underlying the use of the balance is the gravitational force between masses. (The gravitational force will be discussed in detail in chapter 6.) The mass on the left pan is attracted toward the center of the earth and therefore pushes down on the left pan. The mass on the right pan is also attracted toward the earth and pushes down on the right pan. When the force down on the right pan is equal to the force down on the left pan, the scales are balanced and the mass on the right pan is equal to the mass on the left pan. Mass measured by a balance depends on the force of gravity acting on the mass. Hence, mass measured by a balance can be called *gravitational mass*. The balance will work on the moon or on any planet where there are gravitational forces. The equality of masses on the earth found by a balance will show the same equality on the moon or on any planet. But a balance at rest in outer space extremely far away from gravitational forces will not work at all.

1.7 The Standard of Time

What is time and how do we measure it? Time is such a fundamental concept that it is very difficult to define. We will try by *defining time as a duration between the passing of events.* (Do not ask me to define duration, because I would have to define it as the time during which something happens, and I would end up seemingly caught in circular reasoning. This is the way it is with fundamental quantities, they are so fundamental that we cannot define them in terms of something else. If we could, that something else would become the fundamental quantity.) As with all fundamental quantities, we must choose a standard and measure all durations in terms of that standard. To measure time we need something that will repeat itself at regular intervals. The number of intervals counted gives a quantitative measure of the duration. The simplest method of measuring a time interval is to use the rhythmic beating of your own heart as a time standard. Then, just as you measured a length by the number of times the standard length was used to mark off the unknown length, you can measure a time duration by the number of pulses from your heart that covers the particular unknown duration. Note that Galileo timed the swinging chandeliers in a church one morning by the use of his pulse, finding the time for one complete oscillation of the pendulum to be independent of the magnitude of that oscillation.

In this way, we can measure time durations by the number of heartbeats counted. However, if you start running or jumping up and down your heart will beat faster and the time interval recorded will be different than when you were at rest. Therefore, for any good timing device we need something that repeats itself over and over again, always with the same constant time interval. Obviously, the technique used to measure time intervals should be invariant, and the results obtained should be the same for different individuals. One such invariant, which occurs day after day, is the rotation of the earth.

It is not surprising, then, that the early technique used for measuring time was the rotation of the earth. One complete rotation of the earth was called a day, and the day was divided into 24 hours; each hour was divided into 60 minutes; and finally each minute was divided into 60 seconds. The standard of time became the second. It may seem strange that the day was divided into 24 hours, the hour into 60 minutes, and the minute into 60 seconds. But remember that the very earliest recorded studies of astronomy and mathematics began in ancient Mesopotamia and Babylonia, where the number system was based on the number 60, rather than on the number 10, which we base our number system on. Hence, a count of 60 of their base units was equal to 1 of their next larger units. When they got to a count of 120 base units, they set this equal to 2 of the larger units. Thus, a count of 60 seconds, their base unit, was equal to 1 unit of their next larger unit, the minute. When they got to 60 minutes, this was equal to their next larger unit, the hour.

Their time was also related to their angular measurements of the sky. Hence the year became 360 days, the approximate time for the earth to go once around the sun. They related the time for the earth to move once around the heavens, 360 days, to the angle moved through when moving once around a circle by also dividing the circle into 360 units, units that today are called *angular degrees*. They then divided their degree by their base number 60 to get their next smaller unit of angle, 1/60 of a degree, which they called a *minute of arc*. They then divided their minute by their base number 60 again to get an angle of 1 second, which is equal to 1/60 of a minute. The movement of the heavenly bodies across the sky became their calendar. Of course their minutes and seconds of arc are not the same as our minutes and seconds of time, but because of their base number 60 our measurements of arc and time are still based on the number 60.

What is even more interesting is that the same committee that originally introduced the meter and the kilogram proposed a clock that divided the day into 10 equal units, each called a *deciday*. They also divided a quadrant of a circle (90°) into a hundred parts each called the *grade*. They thus tried to place time and angle measurements into a decimal system also, but these units were never accepted by the people.

So the second, which is 1/86,400 part of a day, was kept as the measure of time. However, it was eventually found that the earth does not spin at a constant rate. It is very close to being a constant value, but it does vary ever so slightly. In 1967, the Thirteenth General Conference of Weights and Measures decided that the primary standard of time should be based on an atomic clock, figure 1.8. *The second is now defined as "the duration of 9,192,631,770 periods (or cycles) of the radiation corresponding to the transition between two hyperfine levels of the ground state of the cesium-133 atom."* The atomic clock is located at the National Bureau of Standards in Boulder, Colorado. The atomic clock is accurate to 1 second in a thousand years and can measure a time interval of one millionth of a second.

The atomic clock provides the reference time, from which certain specified radio stations (such as WWV in Fort Collins, Colorado) broadcast the correct time. This time is then transmitted to local radio and TV stations and telephone services, from which we usually obtain the time to set our watches.

For the accuracy required in a freshman college physics course, the unit of time, the second, is the time it takes for the second hand on a nondigital watch to move one interval.

Figure 1.8
The atomic clock.

1.8 The Standard of Electrical Charge

One of the fundamental characteristics of matter is that it has not only mass but also electrical charge. We now know that all matter is composed of atoms. These atoms in turn are composed of electrons, protons, and neutrons. Forces have been found that exist between these electrons and protons, forces caused by the electrical charge that these particles carry. The smallest charge ever found is the charge on the electron. By convention we call it a negative charge. The proton contains the same amount of charge, but it is a positive charge. Most matter contains equal numbers of electrons and protons, and hence is electrically neutral.

Although electrical charge is a fundamental property of matter, it is a quantity that is relatively difficult to measure directly, whereas the effects of electric current—the flow of charge per unit time—is much easier to measure. *Therefore, the fundamental unit of electricity is defined as the ampere, where "the ampere is that constant current that, if maintained in two straight parallel conductors of*

Table 1.1
Systems of Units

Physical Quantity	International System (SI)	British Engineering System
Length	meter (m)	foot (ft)
Mass	kilogram (kg)	slug
Time	second (s)	second (s)
Electric current	ampere (A)	
Electric charge	coulomb (C)	
Weight or force	newton (N)	pound (lb)

infinite length, of negligible circular cross section, and placed one meter apart in a vacuum, would produce between these conductors a force equal to 2×10^{-7} newtons per meter of length." This definition will be explained in more detail when electricity is studied in section 22.7. The ampere, the unit of current, is also defined as the passage of 1 **coulomb** of charge per second in a circuit. This represents a passage of 6.25×10^{18} electrons per second. Therefore, the charge on one electron is 1.60×10^{-19} coulombs.

1.9 Systems of Units

When the standards of the fundamental quantities are all assembled, they are called a *system of units*. The standards for the fundamental quantities, discussed in the previous sections, are part of a new system of units called the International System of units, abbreviated SI units. They were adopted by the Eleventh General Conference of Weights and Measures in 1960. This system of units refines and replaces the older metric system of units, and is very similar to it. Table 1.1 shows the two systems of units that will be considered.

Let us add another quantity to table 1.1, namely the quantity of weight or force. In SI units this is not a fundamental quantity, but rather a derived quantity. (A complete definition of the concepts of force and weight will be given in chapter 4.) For the present, let us add it to the table and say the weight and mass of an object are related but not identical quantities. As already indicated, mass is a measure of the quantity of matter in a body. The weight of a body here on earth is a measure of the gravitational force of attraction of the earth on that mass, pulling the mass of that body down toward the center of the earth. In the international system, the unit of weight or force is called the *newton,* named of course after Sir Isaac Newton.

An important distinction between mass and weight can easily be shown here. If you were to go to the moon, figure 1.9, you would find that the gravitational force on the moon is only 1/6 of the gravitational force found here on earth. Hence, on the moon you would only weigh 1/6 what you do on earth. That is, if you weigh 180 lb on earth, you would only weigh 30 lb on the surface of the moon. Yet your mass has not changed at all. The thing that you call you, all the complexity of atoms, molecules, cells, tissue, blood, bones, and the like, is still the same. Your weight would have changed, but not your mass. The difference between mass and weight will be explained in much more detail in a later chapter. The unit of weight or force, the newton, is only placed in the table now in order to compare it to the next system of units.

Figure 1.9
Your weight on the moon is very different from your weight on earth.

What has been said about systems of units so far is fine, but the system that most of you are familiar with has been only briefly mentioned. The system of units that you are accustomed to using is called the British engineering system of units (see table 1.1). In that system, the unit of length is the foot. (Recall that the unit of a foot is now defined in terms of the standard of length, the meter.) The unit of time is again the second.

In the British engineering system (BES), mass is not defined as a fundamental quantity; instead the weight of a body is described as fundamental, and its mass is derived from its weight. The fundamental unit of weight in the BES is defined as the pound with which we are all familiar. The unit of mass is derived from the unit of weight, and is called a *slug*. Whenever you hear or see the word pound it means a weight or a force, never a mass.

The history of some of the names of these units is quite interesting. The name slug came from the idea that when we try to move a large massive object and put it into motion, the more massive the object the more difficult it is to put that object into motion. The object is thus said to be sluggish. Therefore, the more massive the body, the more sluggish it is, and hence, it contains more slugs. This unit is often used in engineering, but not in our daily lives.

As you are probably aware, the United States is now in the process of changing from the British engineering system to the international system of units. The British engineering system will probably be obsolete within 10 to 20 years. (Even the British no longer use the British engineering system.) There may be a little confusion during the transition, but in the long run the international system is a better system because it is a much easier system to use.

However, problems have already arisen even before the SI units have been officially adopted. For example, if you go to the local supermarket and buy an average-sized can of beans, you will see printed on it "Net wt. 21 oz. (1 lb. 5 oz.) or 595 g." The business sector has already erroneously equated mass and weight by calling them the same name, grams or kilograms. What the businessman really means is that the can of beans has a mass of 595 grams. The weight of an object in SI units should be expressed in newtons. We will show how to deal with this new confusion later. In this book, however, whenever you see the word kilogram or gram it will refer to the mass of an object.

To simplify the use of units in equations, abbreviations will be used. All unit abbreviations in SI units are one or two letters long and the abbreviations do not require a period following them. The name of a unit based on a proper name is written in lower-case letters, while its abbreviation is capitalized. All other abbreviations are written in lower-case letters. The abbreviations are shown in table 1.1.

Most of the measurements used in this book will be in SI units, although many will be used in the British engineering system since this is the system you probably have the best feel for, because you grew up with it. Sometimes it will be necessary to convert from one of these systems of units to another. In order to do this, it is necessary to make use of a conversion factor.

1.10 Conversion Factors

A **conversion factor** is a factor by which a quantity expressed in one set of units must be multiplied in order to express that quantity in different units. The numbers for a conversion factor are usually expressed as an equation, relating the quantity in one system of units to the same quantity in different units. Appendix A, at the back of this book, contains a large number of conversion factors. An example of an equation leading to a conversion factor is

$$1 \text{ m} = 3.281 \text{ ft}$$

A conversion factor is also used to change a quantity expressed in one system of units to a value in different units of the same system or to a unit in another system of units. For example, if both sides of the above equality are divided by 3.281 ft we get

$$\frac{1 \text{ m}}{3.281 \text{ ft}} = \frac{3.281 \text{ ft}}{3.281 \text{ ft}} = 1$$

Thus,

$$\frac{1 \text{ m}}{3.281 \text{ ft}} = 1$$

is a conversion factor that is equal to unity. If a height is multiplied by a conversion factor, we do not physically change the height, because all we are doing is multiplying it by the number one. The effect, however, expresses the same height as a different number with a different unit.

Example 1.1

Converting feet to meters. The height of a building is 100.0 ft. Find the height in meters.

Solution

To express the height h in meters, multiply the height in feet by the conversion factor that converts feet to meters, that is,

$$h = 100.0 \text{ ft} \left(\frac{1 \text{ m}}{3.281 \text{ ft}} \right) = 30.48 \text{ m}$$

Notice that the units act like algebraic quantities. That is, the unit foot, which is in both the numerator and the denominator of the equation, divides out, leaving us with the single unit, meters.

___

The technique to remember in using a conversion factor is that the unit in the numerator that is to be eliminated, must be in the denominator of the conversion factor. Then, because units act like algebraic quantities, identical units can be divided out of the equation immediately.

Conversion factors should also be set up in a chain operation. This will make it easy to see which units cancel. For example, suppose we want to express the time T of one day in terms of seconds. This number can be found as follows:

$$T = 1 \text{ day} \left(\frac{24 \text{ hr}}{1 \text{ day}} \right) \left(\frac{60 \text{ min}}{1 \text{ hr}} \right) \left(\frac{60 \text{ s}}{1 \text{ min}} \right)$$

$$= 86{,}400 \text{ s}$$

By placing the conversion factors in this sequential fashion, the units that are not wanted divide out directly and the only unit left is the one we wanted, seconds. This technique is handy because if we make a mistake and use the wrong conversion factor, the error is immediately apparent. For example, let us go back to the first example and use the reciprocal of the conversion factor by mistake:

$$h = 100.0 \text{ ft} \left(\frac{3.281 \text{ ft}}{1 \text{ m}} \right) = 328.1 \frac{\text{ft}^2}{\text{m}}$$

It is immediately obvious that we have made a mistake because the final units do not represent the height in meters, the unit for which we were looking. Whenever something like this happens, we simply go back and use the correct conversion factor.

Mechanics

These examples are, of course, trivial, but the important thing to learn is the technique. Later when these ideas are applied to problems that are not trivial, if the technique is followed as shown, there should be no difficulty in obtaining the correct solutions.

1.11 Derived Quantities

Most of the quantities that are observed in the study of physics are derived in terms of the fundamental quantities. For example, the speed of a body is the ratio of the distance that an object moves to the time it takes to move that distance. This is expressed as

$$\text{speed} = \frac{\text{distance traveled}}{\text{time}}$$

$$v = \frac{\text{length}}{\text{time}}$$

That is, the speed v is the ratio of the fundamental quantity of length to the fundamental quantity of time. Thus, speed is derived from length and time. For example, the unit for speed in SI units is a meter per second (m/s).

Another example of a derived quantity is the volume V of a body. For a box, the volume is equal to the length times the width times the height. Thus,

$$V = (\text{length})(\text{width})(\text{height})$$

But because the length, width, and height of the box are measured by a distance, the volume is equal to the cube of the fundamental unit of length L. That is,

$$V = L^3$$

Hence, the SI unit for volume is m³.

As a final example of a derived quantity, the density of a body is defined as its mass per unit volume, that is,

$$\rho = \frac{m}{V}$$

$$= \frac{\text{mass}}{(\text{length})^3}$$

Hence, the density is defined as the ratio of the fundamental quantity of mass to the cube of the fundamental quantity of length. The SI unit for density is thus kg/m³. All the remaining quantities of physics are derived in this way, in terms of the four fundamental quantities of length, mass, charge, and time.

Note that the international system of units also recognizes temperature, luminous intensity, and "quantity of matter" (the mole) as fundamental. However, they are not fundamental in the same sense as mass, length, time, and charge. Later in the book we will see that temperature can be described as a measure of the mean kinetic energy of molecules, which is described in terms of length, mass, and time. Similarly, intensity can be derived in terms of energy, area, and time, which again are all describable in terms of length, mass, and time.

Learning physics at an early age.
PEANUTS reprinted by permission of UFS, Inc.

The Language of Physics

Philosophy
The search for knowledge or wisdom (p. 5).

Natural philosophy
The study of the natural or physical world (p. 6).

Physics
The Greek word for "natural" is *physikos*. Therefore, the word physics came to mean the study of the entire natural or physical world (p. 6).

Scientific method
The application of a logical process of reasoning to arrive at a model of nature that is consistent with experimental results. The scientific method consists of five steps: (1) observation, (2) hypothesis, (3) experiment, (4) theory or law, and (5) prediction (p. 6).

Fundamental quantities
The most basic quantities that can be used to describe the physical world. When we look out at the world, we observe that the world occupies space, and within that space we find matter, and that space and matter exists within something we call time. So the observation of the world can be made in terms of space, matter, and time. The fundamental quantity of length is used to describe space, the fundamental quantities of mass and electrical charge are used to describe matter, and the fundamental quantity of time is used to describe time. All other quantities, called derived quantities, can be described in terms of some combination of the fundamental quantities (p. 10).

International System (SI) of units
The internationally adopted system of units used by all the scientists and almost all the countries of the world (p. 12).

Meter
The standard of length. It is defined as the length of the path traveled by light in a vacuum during an interval of $1/299,792,458$ of a second (p. 11).

Mass
The measure of the quantity of matter in a body (p. 13).

Kilogram
The unit of mass. It is defined as the amount of matter in a specific platinum iridium cylinder 39 mm high and 39 mm in diameter (p. 13).

Second
The unit of time. It is defined as the duration of 9,192,631,770 periods of the radiation corresponding to the transition between two hyperfine levels of the ground state of the cesium-133 atom of the atomic clock (p. 15).

Coulomb
The unit of electrical charge. It is defined in terms of the unit of current, the ampere. The ampere is a flow of 1 coulomb of charge per second. The ampere is defined as that constant current that, if maintained in two straight parallel conductors of infinite length, of negligible circular cross section, and placed one meter apart in vacuum, would produce between these conductors a force equal to 2×10^{-7} newtons per meter of length (p. 16).

Conversion factor
A factor by which a quantity expressed in one set of units must be multiplied in order to express that quantity in different units (p. 17).

Questions for Chapter 1

1. Why should physics have separated from philosophy at all?
† 2. What were Aristotle's ideas on physics, and what was their effect on science in general, and on physics in particular?
3. Is the scientific method an oversimplification?
4. How does a law of physics compare with a civil law?
† 5. Is there a difference between saying that an experiment validates a law of nature and that an experiment verifies a law of nature? Where does the concept of truth fit in the study of physics?
6. How does physics relate to your field of study?
7. In the discussion of hot and cold in section 1.3, what would happen if you placed your right hand in the hot water and your left hand in the cold water, and then placed both of them in the mixture simultaneously?
8. Can you think of any more examples that show the need for quantitative measurements?
9. Compare the description of the world in terms of earth, air, fire, and water with the description in terms of length, mass, electrical charge, and time.
10. Discuss the pros and cons of dividing the day into decidays. Do you think this idea should be reintroduced into society? Using yes and no answers, have your classmates vote on a change to a deciday. Is the result surprising?
11. Discuss the difference between mass and weight.

Problems for Chapter 1

In all the examples and problems in this book we assume that whole numbers, such as 2 or 3, have as many significant figures as are necessary in the solution of the problem.

1. How many cubic centimeters are there in a cubic inch?
2. A liter contains 1000 cm³. How many liters are there in a cubic meter?
3. A speed of 60.0 miles per hour (mph) is equal to how many ft/s?
4. The density of 1 g/cm³ is equal to how many kg/liter?
5. How many seconds are there in a day? a month? a year?
6. Calculate your height in meters.
7. What is 90 km/hr expressed in mph?
8. How many square meters are there in 1 acre, if 1 acre is equal to 43,560 ft²?
9. How many feet are there in 1 km?

10. Express the age of the earth (approximately 4.6×10^9 years) in seconds.

11. The speed of sound in air is 331 m/s at 0°C. Express this speed in ft/s and mph.

12. The speedometer of a new car is calibrated in km/hr. If the speed limit is 55 mph, how fast can the car go in km/hr and still stay below the speed limit?

13. A floor has an area of 144 ft². What is this area expressed in m²?

14. A tank contains a volume of 50 ft³. Express this volume in cubic meters.

15. Assuming that an average person lives for 75 yrs, how many (a) seconds and (b) minutes are there in this lifetime? If the heart beats at an average of 70 pulses/min, how many beats does the average heart have?

16. A cube is 50 cm on each side. Find its surface area in m² and ft² and its volume in m³ and ft³.

17. The speed of light in a vacuum is approximately 186,000 miles/s. Express this speed in mph and m/s.

18. The distance from home plate to first base on a baseball field is 90 ft. What is this distance in meters?

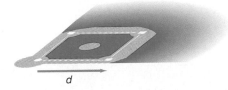

d

19. In the game of football, a first down is 10 yd long. What is this distance in meters? If the field is 100 yd long, what is the length of the field in meters?

20. The diameter of a sphere is measured as 6.28 cm. What is its volume in cm³, m³, in.³, and ft³?

21. The Empire State Building is 1245 ft tall. Express this height in meters, miles, inches, and millimeters.

22. A drill is 1/4 in. in diameter. Express this in centimeters, and then millimeters.

23. The average diameter of the earth is 7927 miles. Express this in km.

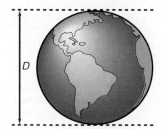

D

24. A 31-story building is 132 m tall. What is the average height of each story in feet?

25. Light of a certain color has a wavelength of 589 nm. Express this wavelength in (a) pm, (b) mm, (c) cm, (d) m. How many of these 589 nm waves are there in an inch?

26. Calculate the average distance to the moon in meters if the distance is 239,000 miles.

27. A basketball player is 7 ft tall. What is this height in meters?

28. The mass of a hydrogen atom is 1.67×10^{-24} g. Calculate the number of atoms in 1 g of hydrogen.

29. The Washington National Monument is 555 ft high. Express this height in meters.

30. The Statue of Liberty is 305 ft high. Express this height in meters.

31. Cells found in the human body have a volume generally in the range of 10^4 to 10^6 cubic microns. A micron is an older name of the unit that is now called a micrometer and is equal to 10^{-6} m. Express this volume in cubic meters and cubic inches.

32. The diameter of a deoxyribonucleic acid (DNA) molecule is about 20 angstroms. Express this diameter in picometers, nanometers, micrometers, millimeters, centimeters, meters, and inches. Note that the old unit angstrom is equal to 10^{-10} m.

33. A glucose molecule has a diameter of about 8.6 angstroms. Express this diameter in millimeters and inches.

34. Muscle fibers range in diameter from 10 microns to 100 microns. Express this range of diameters in centimeters and inches.

35. The axon of the neuron, the nerve cell of the human body, has a diameter of approximately 0.2 microns. Express this diameter in terms of (a) pm, (b) nm, (c) μm, (d) mm, and (e) cm.

36. The Sears Tower in Chicago, the world's tallest building, is 1454 ft high. Express this height in meters.

37. A baseball has a mass of 145 g. Express this mass in slugs.

38. One shipping ton is equal to 40 ft³. Express this volume in cubic meters.

39. A barrel of oil contains 42 U.S. gallons, each of 231 in.³. What is its volume in cubic meters?

40. The main span of the Verrazano Narrows Bridge in New York is 1298.4 m long. Express this distance in feet and miles.

41. The depth of the Mariana Trench in the Pacific Ocean is 10,911 m. Express this depth in feet.

42. Mount McKinley is 6194 m high. Express this height in feet.

43. The average radius of the earth is 6371 km. Find the area of the surface of the earth in m² and in ft². Find the volume of the earth in m³ and ft³. If the mass of the earth is 5.97×10^{24} kg, find the average density of the earth in kg/m³.

44. Cobalt-60 has a half-life of 5.27 yr. Express this time in (a) months, (b) days, (c) hours, (d) seconds, and (e) milliseconds.

45. On a certain European road in a quite residential area, the speed limit is posted as 40 km/hr. Express this speed limit in miles per hour.

46. In a recent storm, it rained 6.00 in. of rain in a period of 2.00 hr. If the size of your property is 100 ft by 100 ft, find the total volume of water that fell on your property. Express your answer in (a) cubic feet, (b) cubic meters, (c) liters, and (d) gallons.

47. A cheap wrist watch loses time at the rate of 8.5 seconds a day. How much time will the watch be off at the end of a month? A year?

48. A ream of paper contains 500 sheets of 8½ in. by 11 in. paper. If the package is 1 and ⅞ in. high, find (a) the thickness of each sheet of paper in inches and millimeters, (b) the dimensions of the page in millimeters, and (c) the area of a page in square meters and square millimeters.

Interactive Tutorials

⌨ 49. Conversion Calculator. The Conversion Calculator will allow you to convert from a quantity in one system of units to that same quantity in another system of units and/or to convert to different units within the same system of units.

2

Vectors

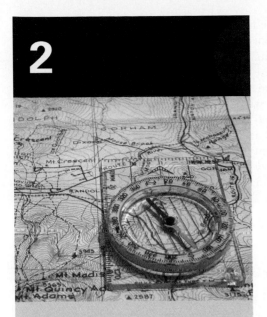

You need vectors to navigate.

Corollary I. A body, acted on by two forces simultaneously, will describe the diagonal of a parallelogram in the same time as it would describe the sides by those forces separately.

Isaac Newton

2.1 Introduction

Of all the varied quantities that are observed in nature, some have the characteristics of scalar quantities while others have the characteristics of vector quantities. *A **scalar** quantity is a quantity that can be completely described by a magnitude, that is, by a number and a unit.* Some examples of scalar quantities are mass, length, time, density, and temperature. The characteristic of scalar quantities is that they add up like ordinary numbers. That is, if we have a mass $M_1 = 3$ kg and another mass $M_2 = 4$ kg then the sum of the two masses is

$$M = M_1 + M_2 = 3 \text{ kg} + 4 \text{ kg} = 7 \text{ kg} \qquad (2.1)$$

*A **vector** quantity, on the other hand, is a quantity that needs both a magnitude and a direction to completely describe it.* Some examples of vector quantities are force, displacement, velocity, and acceleration. The velocity of a car moving at 50 miles per hour (mph) due east can be represented by a vector. Velocity is a vector because it has a magnitude, 50 mph, and a direction, due east.

A vector is represented in this text book by boldface script, that is, **A**. Because we cannot write in boldface script on note paper or a blackboard, a vector is written there as the letter with an arrow over it. A vector can be represented on a diagram by an arrow. A picture of this vector can be obtained by drawing an arrow from the origin of a cartesian coordinate system. The length of the arrow represents the magnitude of the vector, while the direction of the arrow represents the direction of the vector. The direction is specified by the angle θ that the vector makes with an axis, usually the *x*-axis, and is shown in figure 2.1. The magnitude of vector **A** is written as the absolute value of **A** namely $|\mathbf{A}|$, or simply by the letter A without boldfacing. One of the defining characteristics of vector quantities is that they must be added in a way that takes their direction into account.

2.2 The Displacement

Probably the simplest vector that can be discussed is the displacement vector. Whenever a body moves from one position to another it undergoes a displacement. *The displacement can be represented as a vector that describes how far and in what direction the body has been displaced from its original position.* The tail of the displacement vector is located at the position where the displacement started, and its tip is located at the position at which the displacement ended. For example, if you walk 3 mi due east, this walk can be represented as a vector that is 3 units long

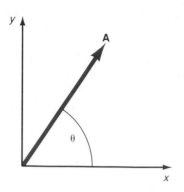

Figure 2.1

Representation of a vector.

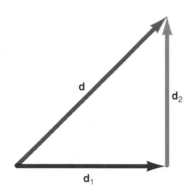

Figure 2.2

The displacement vector.

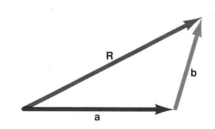

Figure 2.3

The addition of vectors.

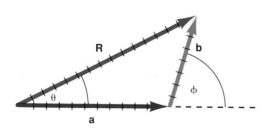

Figure 2.4

The graphical addition of vectors.

and points due east. It is shown as d_1 in figure 2.2. This is an example of a displacement vector. Suppose you now walk 4 mi due north. This distance of 4 mi in a northerly direction can be represented as another displacement vector d_2, which is also shown in figure 2.2. The result of these two displacements is a final displacement vector d that shows the total displacement from the original position.

We now ask how far did you walk? Well, you walked 3 mi east and 4 mi north and hence you have walked a total distance of 7 mi. But how far are you from where you started? Certainly not 7 mi, as we can easily see using a little high school geometry. In fact the final displacement d is a vector of magnitude d and that distance d can be immediately determined by simple geometry. Applying the Pythagorean theorem to the right triangle of figure 2.2 we get

$$d = \sqrt{d_1^2 + d_2^2}$$
$$= \sqrt{(3 \text{ mi})^2 + (4 \text{ mi})^2} = \sqrt{25 \text{ mi}^2} \qquad (2.2)$$

and thus,

$$d = 5 \text{ mi}$$

Even though you have walked a total distance of 7 mi, you are only 5 mi away from where you started. Hence, when these vector displacements are added

$$d = d_1 + d_2 \qquad (2.3)$$

we do not get 7 mi for the magnitude of the final displacement, but 5 mi instead. *The displacement is thus a change in the position of a body from its initial position to its final position. Its magnitude is the distance between the initial position and the final position of the body.*

It should now be obvious that vectors do not add like ordinary scalar numbers. In fact, all the rules of algebra and arithmetic that you were taught in school are the rules of scalar algebra and scalar arithmetic, although the word scalar was probably never used at that time. To solve physical problems associated with vectors it is necessary to deal with vector algebra.

2.3 Vector Algebra—The Addition of Vectors

Let us now add any two arbitrary vectors **a** and **b.** The result of adding the two vectors **a** and **b** forms a new **resultant** vector **R,** which is the sum of **a** and **b.** This can be shown graphically by laying off the first vector **a** in the horizontal direction and then placing the tail of the second vector **b** at the tip of vector **a,** as shown in figure 2.3.

The resultant vector **R** is drawn from the origin of the first vector to the tip of the last vector. The resultant vector is written mathematically as

$$R = a + b \qquad (2.4)$$

Remember that in this sum we do not mean scalar addition. The resultant vector is the vector sum of the individual vectors **a** and **b.**

We can add these vectors graphically, with the aid of a ruler and a protractor. First, we need to choose a scale such that a unit distance on the graph paper corresponds to a unit of magnitude of the vector. Using this scale, we lay off the length that corresponds to the magnitude of vector **a** in the x-direction with a ruler. Then, at the tip of vector **a,** place the center of the protractor and measure the angle ϕ that vector **b** makes with the x-axis. Mark that direction on the paper. Using the ruler, measure off the length of vector **b** in the marked direction, as shown in figure 2.4. Now draw a line from the tail of vector **a** to the tip of vector **b.** This is the resultant vector **R.** Take the ruler and measure the length of vector **R** from the diagram. This length R is the magnitude of vector **R.** Using the protractor, measure

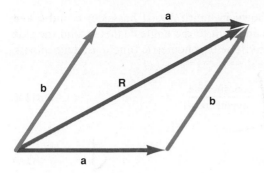

Figure 2.5

The addition of vectors by the parallelogram method.

Figure 2.6

The negative of a vector.

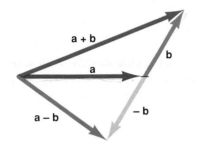

Figure 2.7

The subtraction of vectors.

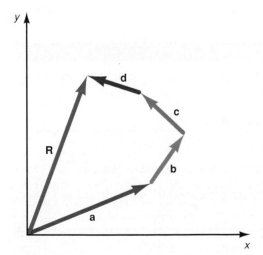

Figure 2.8

Addition of vectors by the polygon method.

the angle θ between **R** and the x-axis—this angle θ is the direction of the resultant vector **R.**

Although a vector is a quantity that has both magnitude and direction, it does not have a position. Consequently a vector may be moved parallel to itself without changing the characteristics of the vector. Because the magnitude of the moved vector is still the same, and its direction is still the same, the vector is the same.

Hence, when adding vectors **a** and **b,** we can move vector **a** parallel to itself until the tip of **a** touches the tip of **b.** Similarly, we can move vector **b** parallel to itself until the tip of **b** touches the tail end of the top vector **a.** In moving the vectors parallel to themselves we have formed a parallelogram, as shown in figure 2.5.

From what was said before about the resultant of **a** and **b,** we can see that the resultant of the two vectors is the main diagonal of the parallelogram formed by the vectors **a** and **b,** hence we call this process the **parallelogram method of vector addition.** It is sometimes stated as part of the definition of a vector, that vectors obey the parallelogram law of addition. Note from the diagram that

$$\mathbf{R} = \mathbf{a} + \mathbf{b} = \mathbf{b} + \mathbf{a} \qquad (2.5)$$

that is, vectors can be added in any order. Mathematicians would say vector addition is commutative.

2.4 Vector Subtraction—The Negative of a Vector

If we are given a vector **a,** as shown in figure 2.6, then the vector minus **a** ($-\mathbf{a}$) is a vector of the same magnitude as **a** but in the opposite direction. That is, if vector **a** points to the right, then the vector $-\mathbf{a}$ points to the left. Vector $-\mathbf{a}$ is called the negative of the vector **a.** By defining the negative of a vector in this way, we can now determine the process of vector subtraction. The subtraction of vector **b** from vector **a** is defined as

$$\mathbf{a} - \mathbf{b} = \mathbf{a} + (-\mathbf{b}) \qquad (2.6)$$

In other words, the subtraction of **b** from **a** is equivalent to adding vector **a** and the negative vector ($-\mathbf{b}$). This is shown graphically in figure 2.7.

2.5 Addition of Vectors by the Polygon Method

To find the sum of any number of vectors graphically, we use the polygon method. In the polygon method, we add each vector to the preceding vector by placing the tail of one vector to the head of the previous vector, as shown in figure 2.8. The resultant vector **R** is the sum of all these vectors. That is,

$$\mathbf{R} = \mathbf{a} + \mathbf{b} + \mathbf{c} + \mathbf{d} \qquad (2.7)$$

We find **R** by drawing the vector from the origin of the coordinate system to the tip of the final vector, as shown in figure 2.8.

Vectors are usually added analytically or mathematically. In order to do that, we need to define the components of a vector. However, to discuss the components of a vector, we first need a brief review of trigonometry.

2.6 Review of Trigonometry

Although we assume that everybody reading this book has been exposed to the fundamentals of trigonometry, the essential ideas and definitions of trigonometry will now be reviewed.

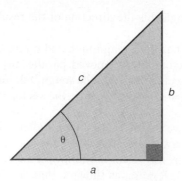

Figure 2.9

A simple right triangle.

Consider the right triangle shown in figure 2.9. It has sides a and b and hypotenuse c. Side a is called the side adjacent to the angle θ (theta), and the side b is called the side opposite to the angle θ. The trigonometric functions, defined with respect to this triangle, are

$$\text{sine } \theta = \frac{\text{opposite side}}{\text{hypotenuse}} \tag{2.8}$$

for the **sine function,** or

$$\sin \theta = \frac{b}{c} \tag{2.9}$$

and

$$\text{cosine } \theta = \frac{\text{adjacent side}}{\text{hypotenuse}} \tag{2.10}$$

for the **cosine function,** or

$$\cos \theta = \frac{a}{c} \tag{2.11}$$

and

$$\text{tangent } \theta = \frac{\text{opposite side}}{\text{adjacent side}} \tag{2.12}$$

for the **tangent function,** or

$$\tan \theta = \frac{b}{a} \tag{2.13}$$

Let us now review how these simple trigonometric functions are used. Assuming that the hypotenuse c of the right triangle and the angle θ that the hypotenuse makes with the x-axis are known, we want to determine the lengths of sides a and b of the triangle. From the definition,

$$\cos \theta = \frac{a}{c} \tag{2.11}$$

we can find the length of side a by multiplying both sides of equation 2.11 by c, that is,

$$a = c \cos \theta \tag{2.14}$$

Example 2.1

Using the cosine function to determine a side of the triangle. If the hypotenuse c is equal to 10.0 cm and the angle θ is equal to 60.0°, find the length of side a.

Solution

The length of side a is found from equation 2.14 as

$$a = c \cos \theta = 10.0 \text{ cm} \cos 60.0° = 10.0 \text{ cm} (0.500) = 5.00 \text{ cm}$$

(We assume here that anyone can compute the cos 60.0° with the aid of a hand-held calculator.)

To find side b of the triangle we use the definition of the sine function:

$$\sin \theta = \frac{b}{c} \tag{2.9}$$

Multiplying both sides of equation 2.9 by c we obtain

$$b = c \sin \theta \tag{2.15}$$

Example 2.2

Using the sine function to determine a side of the triangle. The hypotenuse c of a right triangle is 10.0 cm long, and the angle θ is equal to 60.0°. Find the length of side b.

Solution

The length of side b is found from equation 2.15 as

$$b = c \sin \theta = 10.0 \text{ cm} \sin 60.0° = 10.0 \text{ cm} (0.866) = 8.66 \text{ cm}$$

Therefore, if the hypotenuse and angle θ of a right triangle are given, the lengths of the sides a and b of that triangle can be determined by simple trigonometry.

Suppose that the lengths of sides a and b of a right triangle are given and we want to find the hypotenuse c and the angle θ of that triangle, as shown in figure 2.9. The hypotenuse is found by the **Pythagorean theorem** from elementary geometry as

$$c^2 = a^2 + b^2 \tag{2.16}$$

Hence,

$$c = \sqrt{a^2 + b^2} \tag{2.17}$$

The angle θ is found from the definition of the tangent function,

$$\tan \theta = \frac{b}{a} \tag{2.13}$$

Using the inverse of the tangent function, sometimes called the arctangent, the angle θ becomes

$$\theta = \tan^{-1} \frac{b}{a} \tag{2.18}$$

Example 2.3

Using the Pythagorean theorem and the inverse tangent. The lengths of two sides of a right triangle are $a = 5.00$ cm and $b = 8.66$ cm. Find the hypotenuse of the triangle and the angle θ.

Solution

The hypotenuse of the triangle is found from equation 2.17 as

$$c = \sqrt{a^2 + b^2} = \sqrt{(5.00 \text{ cm})^2 + (8.66 \text{ cm})^2} = 10.0 \text{ cm}$$

and the angle θ is found from equation 2.18 as

$$\theta = \tan^{-1} \frac{b}{a} = \tan^{-1} \frac{8.66 \text{ cm}}{5.00 \text{ cm}} = \tan^{-1} 1.73 = 60.0°$$

Therefore, if the lengths of the sides a and b of a right triangle are known we can easily calculate the hypotenuse and angle θ. We will repeatedly use these elementary concepts of trigonometry in the discussion of the components of a vector.

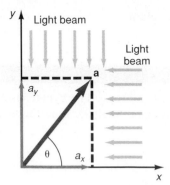

Figure 2.10

Defining the components of a vector.

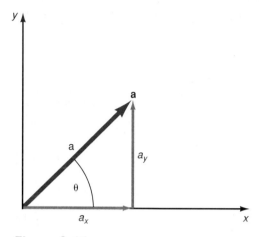

Figure 2.11

Finding the components of a vector mathematically.

2.7 Resolution of a Vector into Its Components

An arbitrary vector **a** is drawn onto an x,y-coordinate system, as in figure 2.10. Vector **a** makes an angle θ with the x-axis. To find the x-component a_x of vector **a**, we project vector **a** down onto the x-axis, that is, we drop a perpendicular from the tip of **a** to the x-axis. One way of visualizing this concept of a **component of a vector** is to place a light beam above vector **a** and parallel to the y-axis. The light hitting vector **a** will not make it to the x-axis, and will therefore leave a shadow on the x-axis. We call this shadow on the x-axis the x-component of vector **a** and denote it by a_x. The component is shown as the light red line on the x-axis in figure 2.10.

In the same way, we can determine the y-component of vector **a**, a_y, by projecting **a** onto the y-axis in figure 2.10. That is, we drop a perpendicular from the tip of **a** onto the y-axis. Again, we can visualize this by projecting light, which is parallel to the x-axis, onto vector **a**. The shadow of vector **a** on the y-axis is the y-component a_y, shown in figure 2.10 as the light red line on the y-axis.

The components of the vector are found mathematically by noting that the vector and its components constitute a triangle, as seen in figure 2.11. From trigonometry, we find the x-component of **a** from

$$\cos \theta = \frac{a_x}{a} \tag{2.19}$$

Solving for a_x, the x-component of vector **a** obtained is

$$a_x = a \cos \theta \tag{2.20}$$

We find the y-component of vector **a** from

$$\sin \theta = \frac{a_y}{a} \tag{2.21}$$

Hence, the y-component of vector **a** is

$$a_y = a \sin \theta \tag{2.22}$$

Example 2.4

Finding the components of a vector. The magnitude of vector **a** is 10.0 units and the vector makes an angle of 30.0° with the x-axis. Find the components of a.

Solution

The x-component of vector **a,** found from equation 2.20, is

$$a_x = a \cos \theta = 10.0 \text{ units} \cos 30.0° = 8.66 \text{ units}$$

The y-component of **a,** found from equation 2.22, is

$$a_y = a \sin \theta = 10.0 \text{ units} \sin 30.0° = 5.00 \text{ units}$$

What do these components of a vector represent physically? If vector **a** is a displacement, then a_x would be the distance that the object is east of its starting point and a_y would be the distance north of it. That is, if you walked a distance of 10.0 miles in a direction that is 30.0° north of east, you would be 8.66 miles east of where you started from and 5.00 miles north of where you started from. If, on the other hand, vector **a** were a force of 10.0 lb applied at an angle of 30.0° to the x-axis, then the x-component a_x is equivalent to a force of 8.66 lb in the x-direction, while the y-component a_y is equivalent to a force of 5.00 lb in the y-direction.

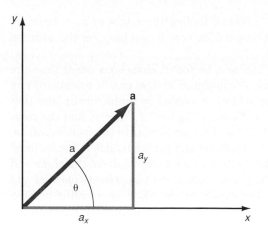

Figure 2.12

Determining a vector from its components.

2.8 Determination of a Vector from Its Components

If the components a_x and a_y of a vector are given, and we want to find the vector **a** itself, that is, its magnitude a and its direction θ, then the process is the inverse of the technique used in section 2.7. The components a_x and a_y of vector **a** are seen in figure 2.12. If we form the triangle with sides a_x and a_y, then the hypotenuse of that triangle is the magnitude a of the vector, and is determined by the Pythagorean theorem as

$$a^2 = a_x^2 + a_y^2 \tag{2.23}$$

Hence, the magnitude of vector **a** is

$$a = \sqrt{a_x^2 + a_y^2} \tag{2.24}$$

It is thus very simple to find the magnitude of a vector once its components are known.

To find the angle θ that vector **a** makes with the x-axis we use the definition of the tangent, namely

$$\text{tangent } \theta = \frac{\text{opposite side}}{\text{adjacent side}} \tag{2.12}$$

For the simple triangle of figure 2.12, the opposite side is a_y and the adjacent side is a_x. Therefore,

$$\tan \theta = \frac{a_y}{a_x} \tag{2.25}$$

We find the angle θ by using the inverse tangent, as

$$\theta = \tan^{-1} \frac{a_y}{a_x} \tag{2.26}$$

Example 2.5

Finding a vector from its components. The components of a certain vector are given as $a_x = 8.66$ and $a_y = 5.00$. Find the magnitude of the vector and the angle θ that it makes with the x-axis.

Solution

The magnitude of vector **a,** found from equation 2.24, is

$$a = \sqrt{a_x^2 + a_y^2} = \sqrt{(8.66)^2 + (5.00)^2}$$
$$= \sqrt{75.0 + 25.0}$$
$$= 10.0$$

The angle θ, found from equation 2.26, is

$$\theta = \tan^{-1} \frac{a_y}{a_x} = \tan^{-1} \frac{5.00}{8.66} = \tan^{-1} 0.577$$
$$= 30.0°$$

Therefore, the magnitude of vector **a** is 10.0 and the angle θ is 30.0°.

Note that the same values were used in this example as in example 2.4 to show that one process is indeed the inverse of the other. That is, we started with a vector with magnitude $a = 10.0$ and angle $\theta = 30.0°$, and found the components $a_x = 8.66$ and $a_y = 5.00$. Then we performed the inverse problem by giving

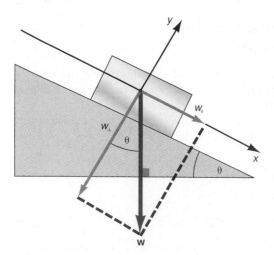

Figure 2.13

Components of the weight parallel and perpendicular to the inclined plane.

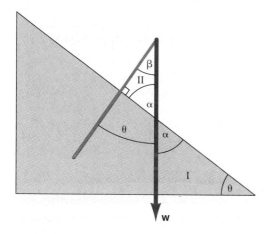

Figure 2.14

Comparison of two triangles.

the components $a_x = 8.66$ and $a_y = 5.00$ and finding the values of $a = 10.0$ and $\theta = 30.0°$. The techniques developed here will be very useful later for the addition of any number of vectors.

The components of a vector can also be found along axes other than the traditional horizontal and vertical ones. A coordinate system can be orientated any way we choose. For example, suppose a block is placed on an inclined plane that makes an angle θ with the horizontal, as shown in figure 2.13. Let us find the components of the weight of the block parallel and perpendicular to the inclined plane.

We draw in a set of axes that are parallel and perpendicular to the inclined plane, as shown in figure 2.13, with the positive x-axis pointing down the plane and the positive y-axis perpendicular to the plane. To find the components parallel and perpendicular to the plane, we draw the weight of the block as a vector pointed toward the center of the earth. The weight vector is therefore perpendicular to the base of the inclined plane. To find the component of $\mathbf{w}$ perpendicular to the plane, we drop a perpendicular line from the tip of vector $\mathbf{w}$ onto the negative y-axis. This length $w_\perp$ is the perpendicular component of vector $\mathbf{w}$. Similarly, to find the parallel component of $\mathbf{w}$, we drop a perpendicular line from the tip of $\mathbf{w}$ onto the positive x-axis. This length $w_\parallel$ is the parallel component of the vector $\mathbf{w}$.

The angle between vector $\mathbf{w}$ and the perpendicular axis is also the inclined plane angle θ, as shown in the comparison of the two triangles in figure 2.14. (Figure 2.14 is an enlarged view of the two triangles of figure 2.13). In triangle I, the angles must add up to 180°. Thus,

$$\theta + \alpha + 90° = 180° \tag{2.27}$$

while for triangle II

$$\beta + \alpha + 90° = 180° \tag{2.28}$$

From equations 2.27 and 2.28 we see that

$$\beta = \theta \tag{2.29}$$

This is an important relation that we will use every time we use an inclined plane.

Example 2.6

Components of the weight perpendicular and parallel to the inclined plane. A 100-lb block is placed on an inclined plane with an angle $\theta = 50.0°$, as shown in figure 2.13. Find the components of the weight of the block parallel and perpendicular to the inclined plane.

Solution

We find the perpendicular component of $\mathbf{w}$ from figure 2.13 as

$$\begin{aligned} w_\perp &= w \cos \theta \\ &= 100 \text{ lb} \cos 50.0° = 64.2 \text{ lb} \end{aligned} \tag{2.30}$$

The parallel component is

$$\begin{aligned} w_\parallel &= w \sin \theta \\ &= 100 \text{ lb} \sin 50.0° = 76.6 \text{ lb} \end{aligned} \tag{2.31}$$

One of the interesting things about this inclined plane is that the component of the weight parallel to the inclined plane supplies the force responsible for making the block slide down the plane. Similarly, if you park your car on a hill with the gear in neutral and the emergency brake off, the car will roll down the hill. Why? You can now see that it is the component of the weight of the car that is parallel to the hill that essentially pulls the car down the hill. That force is just as real as if a person were pushing the car down the hill. That force, as can be seen from equation 2.31, is a function of the angle θ. If the angle of the plane is reduced to zero, then

$$w_{\parallel} = w \sin 0° = 0$$

Thus, we can reasonably conclude that when a car is not on a hill (i.e., when $\theta = 0°$) there is no force, due to the weight of the car, to cause the car to move. Also note that the steeper the hill, the greater the angle θ, and hence the greater the component of the force acting to move the car down the hill.

2.9 The Addition of Vectors by the Component Method

A very important technique for the addition of vectors is **the addition of vectors by the component method.** Let us assume that we are given two vectors, **a** and **b**, and we want to find their vector sum. The sum of the vectors is the resultant vector **R** given by

$$\mathbf{R} = \mathbf{a} + \mathbf{b} \tag{2.32}$$

and is shown in figure 2.15. We determine **R** as follows. First, we find the components a_x and a_y of vector **a** by making the projections onto the x- and y-axes, respectively. To find the components of the vector **b**, we again make a projection onto the x- and y-axes, but note that the tail of vector **b** is not at the origin of coordinates, but rather at the tip of **a.** So both the tip and the tail of **b** are projected onto the x-axis, as shown, to get b_x, the x-component of **b.** In the same way, we project **b** onto the y-axis to get b_y, the y-component of **b.** All these components are shown in figure 2.15(a).

The resultant vector **R** is given by equation 2.32, and because **R** is a vector it has components R_x and R_y, which are the projections of **R** onto the x- and y-axes, respectively. They are shown in figure 2.15(b). Now let us go back to the original diagram, figure 2.15(a), and project **R** onto the x-axis. Here R_x is shown a little distance below the x-axis, so as not to confuse R_x with the other components that are already there. Similarly, **R** is projected onto the y-axis to get R_y. Again R_y is slightly displaced from the y-axis, so as not to confuse R_y with the other components already there.

Look very carefully at figure 2.15(a). Note that the length of R_x is equal to the length of a_x plus the length of b_x. Because components are numbers and hence add like ordinary numbers, this addition can be written simply as

$$R_x = a_x + b_x \tag{2.33}$$

That is, *the x-component of the resultant vector is equal to the sum of the x-components of the individual vectors.*

In the same manner, look at the geometry on the y-axis of figure 2.15(a). The length R_y is equal to the sum of the lengths of a_y and b_y, and therefore

$$R_y = a_y + b_y \tag{2.34}$$

Thus, *the y-component of the resultant vector is equal to the sum of the y-components of the individual vectors.* We demonstrated the addition of vectors for only two vectors because it is easier to see the results in figure 2.15 for two vectors than it would be for many vectors. However, the technique is the same for the addition of any number of vectors. For the general case, where there are many vectors, equations 2.33 and 2.34 for R_x and R_y can be generalized to

$$R_x = a_x + b_x + c_x + d_x + \cdots \tag{2.35}$$

and

$$R_y = a_y + b_y + c_y + d_y + \cdots \tag{2.36}$$

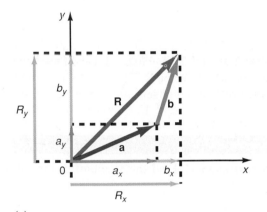

(a)

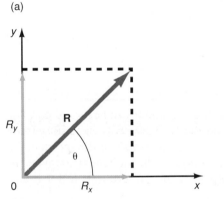

(b)

Figure 2.15

The addition of vectors by the component method.

The plus sign and the dots that appear at the far right in equations 2.35 and 2.36 indicate that additional components can be added for any additional vectors.

We now have R_x and R_y, the components of the resulting vector **R.** But if we know the components of **R,** we can find the magnitude of **R** by using the Pythagorean theorem, that is,

$$R = \sqrt{R_x^2 + R_y^2} \tag{2.37}$$

The angle θ in figure 2.15(b), found from the geometry, is

$$\tan \theta = \frac{R_y}{R_x} \tag{2.38}$$

Thus,

$$\theta = \tan^{-1}\frac{R_y}{R_x} \tag{2.39}$$

where R_x and R_y are given by equations 2.35 and 2.36. Thus, we have found the magnitude R and the direction θ of the resultant vector **R.** Therefore, the sum of any number of vectors can be determined by the component method of vector addition.

Example 2.7

The addition of vectors by the component method. Find the resultant of the following four vectors:

$$A = 100, \theta_1 = 30.0°$$
$$B = 200, \theta_2 = 60.0°$$
$$C = 75.0, \theta_3 = 140°$$
$$D = 150, \theta_4 = 250°$$

Solution

The four vectors are drawn in figure 2.16. Because any vector can be moved parallel to itself, all the vectors have been moved so that they are drawn as emanating from the origin. Before actually solving the problem, let us first outline the solution. To find the resultant of these four vectors, we must first find the individual components of each vector, then we find the x- and y-components of the resulting vector from

$$R_x = A_x + B_x + C_x + D_x \tag{2.35}$$
$$R_y = A_y + B_y + C_y + D_y \tag{2.36}$$

We then find the resulting vector from

$$R = \sqrt{R_x^2 + R_y^2} \tag{2.37}$$

and

$$\theta = \tan^{-1}\frac{R_y}{R_x} \tag{2.39}$$

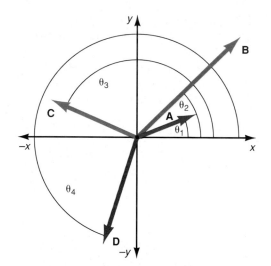

Figure 2.16

Addition of four vectors.

The actual solution of the problem is found as follows: we find the individual x-components as

$$
\begin{aligned}
A_x &= A \cos \theta_1 = 100 \cos 30.0° = 100(0.866) & &= 86.6 \\
B_x &= B \cos \theta_2 = 200 \cos 60.0° = 200(0.500) & &= 100.0 \\
C_x &= C \cos \theta_3 = 75 \cos 140° = 75(-0.766) & &= -57.5 \\
D_x &= D \cos \theta_4 = 150 \cos 250° = 150(-0.342) & &= -51.3 \\
\hline
& R_x = A_x + B_x + C_x + D_x & &= 77.8
\end{aligned}
$$

whereas the y-components are

$$
\begin{aligned}
A_y &= A \sin \theta_1 = 100 \sin 30.0° = 100(0.500) & &= 50.0 \\
B_y &= B \sin \theta_2 = 200 \sin 60.0° = 200(0.866) & &= 173.0 \\
C_y &= C \sin \theta_3 = 75 \sin 140° = 75(0.643) & &= 48.2 \\
D_y &= D \sin \theta_4 = 150 \sin 250° = 150(-0.940) & &= -141.0 \\
& & R_y = A_y + B_y + C_y + D_y &= 130.2
\end{aligned}
$$

The x- and y-components of vector $\mathbf{R}$ are shown in figure 2.17. Because R_x and R_y are both positive, we find vector $\mathbf{R}$ in the first quadrant. If R_x were negative, $\mathbf{R}$ would have been in the second quadrant. It is a good idea to plot the components R_x and R_y for any addition so that the direction of $\mathbf{R}$ is immediately apparent.

We find the magnitude of the resultant vector from equation 2.37 as

$$
\begin{aligned}
R &= \sqrt{R_x^2 + R_y^2} = \sqrt{(77.8)^2 + (130.2)^2} \\
&= \sqrt{23,004.8} \\
&= 152
\end{aligned}
$$

The angle θ that vector $\mathbf{R}$ makes with the x-axis is found as

$$
\begin{aligned}
\theta &= \tan^{-1}\frac{R_y}{R_x} = \tan^{-1}\frac{130.2}{77.8} = \tan^{-1} 1.674 \\
&= 59.1°
\end{aligned}
$$

as is seen in figure 2.17.

It is important to note here that the components C_x, D_x, and D_y are negative numbers. This is because C_x and D_x lie along the negative x-axis and D_y lies along the negative y-axis. We should note that in the solution of the components of the vector $\mathbf{C}$ in this problem, the angle of 140° was entered directly into the calculator to give the solution for the cosine and sine of that angle. The calculator automatically gives the correct sign for the components if we always measure the angle from the positive x-axis.[1]

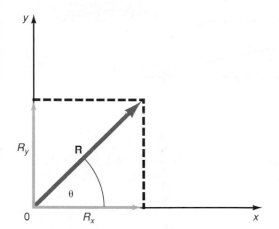

Figure 2.17

The resultant vector.

Example 2.8

The necessity of taking the wind velocity into account when flying an airplane. An airplane is flying due east from city A to city B with an airspeed of 150 mph. A wind is blowing from the northwest at 35 mph. Find the velocity of the airplane with respect to the ground.

Solution

The velocity of the plane with respect to the air is shown as the vector $\mathbf{v_{PA}}$ in figure 2.18. If there were no wind present, the plane would fly in a straight line from city A to city B. However, there is a wind blowing and it is shown as the vector $\mathbf{v_{AG}}$, the

---

1. We can also measure the angle that the vector makes with any axis other than the positive x-axis. For example, instead of using the angle of 140° with respect to the positive x-axis, an angle of 40° with respect to the negative x-axis can be used to describe the direction of vector $\mathbf{C}$. The x-component of vector $\mathbf{C}$ would then be given by $C_x = C \cos 40° = 75.0 \cos 40° = 57.5$. Note that this is the same numerical value we obtained before, however the answer given by the calculator is now positive. But as we can see in figure 2.16, C_x is a negative quantity because it lies along the negative x-axis. Hence, if you do not use the angle with respect to the positive x-axis, you must add the positive or negative sign that is associated with that component. In most of the problems that will be covered in this text, we will measure the angle from the positive x-axis because of the simplicity of the calculation. However, whenever it is more convenient to measure the angle from any other axis, we will do so.

(a)

Figure 2.18

When flying an airplane, the velocity of the wind must be taken into account.

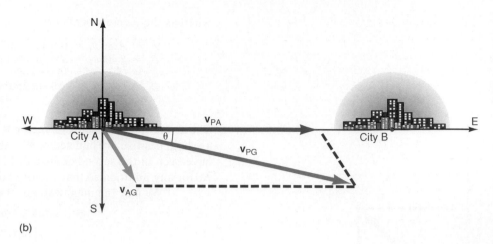

(b)

velocity of air with respect to the ground. This wind blows the plane away from the straight line motion from A to B. The total velocity of the plane with respect to the ground is the vector sum of $\mathbf{v_{PA}}$ and $\mathbf{v_{AG}}$. That is,

$$\mathbf{v_{PG}} = \mathbf{v_{PA}} + \mathbf{v_{AG}}$$

A wind from the northwest makes an angle of $-45°$ or $+315°$ with the positive x-axis. We find the x-component of the resulting velocity as

$$(v_{PA})_x = v_{PA} \cos \theta_2 = 150 \text{ mph} \cos 0° = 150 \text{ mph}$$
$$(v_{AG})_x = v_{AG} \cos \theta_1 = 35.0 \text{ mph} \cos 315° = \underline{24.7 \text{ mph}}$$
$$(v_{PG})_x = (v_{PA})_x + (v_{AG})_x = 174.7 \text{ mph}$$

While the y-component of the resulting velocity is

$$(v_{PA})_y = v_{PA} \sin \theta_2 = 150 \text{ mph} \sin 0° = 00.0 \text{ mph}$$
$$(v_{AG})_y = v_{AG} \sin \theta_1 = 35.0 \text{ mph} \sin 315° = \underline{-24.7 \text{ mph}}$$
$$(v_{PG})_y = (v_{PA})_y + (v_{AG})_y = -24.7 \text{ mph}$$

The magnitude of the resulting velocity of the plane with respect to the ground is

$$v_{PG} = \sqrt{[(v_{PG})_x]^2 + [(v_{PG})_y]^2}$$
$$= \sqrt{(174.7 \text{ mph})^2 + (-24.7 \text{ mph})^2}$$
$$= 176 \text{ mph}$$

The angle that the velocity vector $\mathbf{v_{PG}}$ makes with the positive x-axis is

$$\theta = \tan^{-1}\frac{(v_{PG})_y}{(v_{PG})_x}$$
$$= \tan^{-1}\frac{-24.7 \text{ mph}}{174.7 \text{ mph}}$$
$$= -8.05°$$

Thus the direction of the aircraft as it moves over the ground is 8.05° south of east. If the pilot does not make a correction, he or she will not arrive at city B as expected.

The concepts of the vector addition and vector subtraction have been discussed in this chapter. We will use these concepts throughout the book. The multiplication of vectors will be shown in appendix F.

The Language of Physics

Scalar
A scalar quantity is a quantity that can be completely described by a magnitude, that is, by a number and a unit (p. 23).

Vector
A vector quantity is a quantity that needs both a magnitude and direction to completely describe it (p. 23).

Resultant
The vector sum of any number of vectors is called the resultant vector (p. 24).

Parallelogram method of vector addition
The main diagonal of a parallelogram is equal to the magnitude of the sum of the vectors that make up the sides of the parallelogram (p. 25).

Sine function
The ratio of the length of the opposite side to the length of the hypotenuse in a right triangle (p. 26).

Cosine function
The ratio of the length of the adjacent side to the length of the hypotenuse in a right triangle (p. 26).

Tangent function
The ratio of the length of the opposite side of a right triangle to the length of the adjacent side (p. 26).

Pythagorean theorem
The sum of the squares of the lengths of two sides of a right triangle is equal to the square of the length of the hypotenuse (p. 27).

Component of a vector
The projection of a vector onto a specified axis. The length of the projection of the vector onto the x-axis is called the x-component of the vector. The length of the projection of the vector onto the y-axis is called the y-component of the vector (p. 28).

The addition of vectors by the component method
The x-component of the resultant vector R_x is equal to the sum of the x-components of the individual vectors, while the y-component of the resultant vector R_y is equal to the sum of the y-components of the individual vectors. The magnitude of the resultant vector is then found by the Pythagorean theorem applied to the right triangle with sides R_x and R_y. The direction of the resultant vector is found by trigonometry (p. 31).

Summary of Important Equations

Vector addition is commutative
$$\mathbf{R} = \mathbf{a} + \mathbf{b} = \mathbf{b} + \mathbf{a} \tag{2.5}$$

Subtraction of vectors
$$\mathbf{a} - \mathbf{b} = \mathbf{a} + (-\mathbf{b}) \tag{2.6}$$

Addition of vectors
$$\mathbf{R} = \mathbf{a} + \mathbf{b} + \mathbf{c} + \mathbf{d} \tag{2.7}$$

Definition of the sine
$$\text{sine } \theta = \frac{\text{opposite side}}{\text{hypotenuse}} \tag{2.8}$$

Definition of the cosine
$$\text{cosine } \theta = \frac{\text{adjacent side}}{\text{hypotenuse}} \tag{2.10}$$

Definition of the tangent
$$\text{tangent } \theta = \frac{\text{opposite side}}{\text{adjacent side}} \tag{2.12}$$

Pythagorean theorem
$$c = \sqrt{a^2 + b^2} \tag{2.17}$$

x-component of a vector
$$a_x = a \cos \theta \tag{2.20}$$

y-component of a vector
$$a_y = a \sin \theta \tag{2.22}$$

Magnitude of a vector
$$a = \sqrt{a_x^2 + a_y^2} \tag{2.24}$$

Direction of a vector
$$\theta = \tan^{-1}\frac{a_y}{a_x} \tag{2.26}$$

x-component of resultant vector
$$R_x = a_x + b_x + c_x + d_x \tag{2.35}$$

y-component of resultant vector
$$R_y = a_y + b_y + c_y + d_y \tag{2.36}$$

Magnitude of resultant vector
$$R = \sqrt{R_x^2 + R_y^2} \tag{2.37}$$

Direction of resultant vector
$$\theta = \tan^{-1}\frac{R_y}{R_x} \tag{2.39}$$

Questions for Chapter 2

1. Give an example of some quantities that are scalars and vectors other than those listed in section 2.1.
2. Can a vector ever be zero? What does a zero vector mean?
† 3. Since time seems to pass from the past to the present and then to the future, can you say that time has a direction and therefore could be represented as a vector quantity?
4. Does the subtraction of two vectors obey the commutative law?
5. What happens if you multiply a vector by a scalar?
6. What happens if you divide a vector by a scalar?
7. If a person walks around a block that is 800 ft on each side and ends up at the starting point, what is the person's displacement?
8. How can you add three vectors of equal magnitude in a plane such that their resultant is zero?
9. When are two vectors **a** and **b** equal?
†10. If a coordinate system is rotated, what does this do to the vector? to the components?
†11. Why are all the fundamental quantities scalars?
12. A vector equation is equivalent to how many component equations?
13. If the components of a vector **a** are a_x and a_y, what are the components of the vector $\mathbf{b} = -5\mathbf{a}$?
14. If $|\mathbf{a} + \mathbf{b}| = |\mathbf{a} - \mathbf{b}|$, what is the angle between **a** and **b**?

Problems for Chapter 2

2.7 Resolution of a Vector into Its Components and
2.8 Determination of a Vector from Its Components

1. A strong child pulls a sled with a force of 100 lb at an angle of 35° above the horizontal. Find the vertical and horizontal components of this pull.

2. A 50-N force is directed at an angle of 50° above the horizontal. Resolve this force into vertical and horizontal components.

3. A girl wants to hold a 25-lb sled at rest on a snow covered hill. The hill makes an angle of 15° with the horizontal. What force must she exert parallel to the slope? What is the force perpendicular to the surface of the hill that presses the sled against the hill?

4. A boy wants to hold a 68.0-N sled at rest on a snow-covered hill. The hill makes an angle of 27.5° with the horizontal. (a) What force must he exert parallel to the slope? (b) What is the force perpendicular to the surface of the hill that presses the sled against the hill?

5. A displacement vector, at an angle of 35° with respect to a specified direction, has a y-component equal to 150 ft. What is the magnitude of the displacement vector?

6. A plane is traveling northeast at 200 km/hr. What is (a) the northward component of its velocity, and (b) the eastward component of its velocity?

7. While taking off, an airplane climbs at an 8° angle with respect to the ground. If the aircraft's speed is 200 km/hr, what are the vertical and horizontal components of its velocity?

8. A car that weighs 3200 lb is parked on a hill that makes an angle of 23° with the horizontal. Find the component of the car's weight parallel to the hill and perpendicular to the hill.

9. A car that weighs 8900 N is parked on a hill that makes an angle of 43° with the horizontal. Find the component of the car's weight parallel to the hill and perpendicular to the hill.

10. A girl pushes a lawn mower with a force of 90 N. The handle of the mower makes an angle of 40° with the ground. What are the vertical and horizontal components of this force and what are their physical significances? What effect does raising the handle to 50° have?

11. A missile is launched with a speed of 1000 m/s at an angle of 73° above the horizontal. What are the horizontal and vertical components of the missile's velocity?

12. When a ladder leans against a smooth wall, the wall exerts a horizontal force **F** on the ladder, as

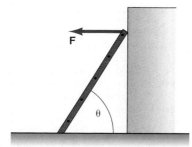

shown in the diagram. If F is equal to 50 N and θ is equal to 63°, find the component of the force perpendicular to the ladder and the component parallel to the ladder.

2.9 The Addition of Vectors by the Component Method

13. Find the resultant of the following three displacements; 3 mi due east, 6 mi east-northeast, and 7 mi northwest.

14. A girl drives 3 km north, then 12 km to the northwest, and finally 5 km south-southwest. How far has she traveled? What is her displacement?

15. An airplane flies due east at 200 mph straight from city A to city B. A northeast wind of 40 mph is blowing. (Note that all winds are defined in terms of the direction from which the wind blows. Hence, a northeast wind blows out of the northeast and blows toward the southwest.) What is the resultant velocity of the plane with respect to the ground?

16. An airplane flies due north at 380 km/hr straight from city A to city B. A southeast wind of 75 km/hr is blowing. (Note that all winds are defined in terms of the direction from which the wind blows. Hence, a southeast wind blows out of the southeast and blows toward the northwest.) What is the resultant velocity of the plane with respect to the ground?

17. Find the resultant of the following forces: (a) 30 N at an angle of 40° with respect to the x-axis, (b) 120 N at an angle of 135°, and (c) 60 N at an angle of 260°.

18. Find the resultant of the following set of forces. (a) **F₁** of 200 N at an angle of 53° with respect to the x-axis. (b) **F₂** of 300 N at an angle of 150° with respect to the x-axis. (c) **F₃** of 200 N at an angle of 270° with respect to the x-axis. (d) **F₄** of 350 N at an angle of 310° with respect to the x-axis.

Additional Problems

19. A heavy trunk weighing 150 lb is pulled along a smooth station platform by a 50 lb force making an angle of 37° above the horizontal. Find (a) the horizontal component of the force, (b) the vertical component of the force, and (c) the resultant downward force on the floor.

20. A heavy trunk weighing 800 N is pulled along a smooth station platform by a 210-N force making an angle of 53° above the horizontal. Find (a) the horizontal component of

the force, (b) the vertical component of the force, and (c) the resultant downward force on the floor.

21. Vector **A** has a magnitude of 50 ft and points in a direction of 50° north of east. What are the magnitudes and directions of the vectors, (a) 2**A**, (b) 0.5**A**, (c) −**A**, (d) −5**A**, (e) **A** + 4**A**, (f) **A** − 4**A**?

22. Given the two force vectors **F₁** = 20.0 N at an angle of 30.0° with the positive x-axis and **F₂** = 40.0 N at an angle of 150.0° with the positive x-axis, find the magnitude and direction of a third force that when added to **F₁** and **F₂** gives a zero resultant.

23. When vector **A**, of magnitude 5.00 m/s at an angle of 120° with respect to the positive x-axis, is added to a second vector **B**, the resultant vector has a magnitude R = 8.00 m/s and is at an angle of 85.0° with the positive x-axis. Find the vector **B**.

24. A car travels 100 mi due west and then 45 mi due north. How far is the car from its starting point? Solve graphically and analytically.

25. Find the resultant of the following forces graphically and analytically: 5 lb at an angle of 33° above the horizontal and 20 lb at an angle of 97° counterclockwise from the horizontal.

26. Find the resultant of the following forces graphically and analytically: 25 N at an angle of 53° above the horizontal and 100 N at an angle of 117° counterclockwise from the horizontal.

†27. The velocity of an aircraft is 200 km/hr due west. A northwest wind of 50 km/hr is blowing. (a) What is the velocity of the aircraft relative to the ground? (b) If the pilot's destination is due west, at what angle should he point his plane to get there? (c) If his destination is 400 km due west, how long will it take him to get there?

28. A plane flies east for 50.0 km, then at an angle of 30.0° north of east for 75.0 km. In what direction should it now fly and how far, such that it will be 200 km northwest of its original position?

†29. The current in a river flows north at 5 mph. A boat starts straight across the river at 8 mph relative to the water. (a) What is the speed of the boat relative to the land? (b) If the river is 2 mi wide, how long does it take the boat to cross the river? (c) If the boat sets out straight for the opposite side, how far north will it reach the opposite shore? (d) If we want to have the boat go straight across the river, at what angle should the boat be headed?

30. The current in a river flows south at 7 km/hr. A boat starts straight across the river at 19 km/hr relative to the water. (a) What is the speed of the boat relative to the land? (b) If the river is 1.5 km wide, how long does it take the boat to cross the river? (c) If the boat sets out straight for the opposite side, how far south will it reach the opposite shore? (d) If we want to have the boat go straight across the river, at what angle should the boat be headed?

†31. Show that if the angle between vectors **a** and **b** is an acute angle, then the sum **a** + **b** becomes the main diagonal of the parallelogram and the difference **a** − **b** becomes the minor diagonal of the parallelogram. Also show that if the angle is obtuse the results are reversed.

32. Find the resultant of the following three vectors. The magnitudes of the vectors are $a = 5.00$ km, $b = 10.0$ km, and $c = 20.0$ km.

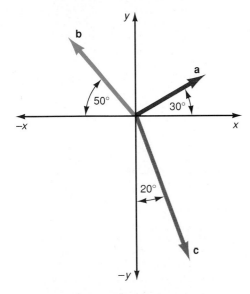

33. Find the resultant of the following three forces. The magnitudes of the forces are $F_1 = 2.00$ N, $F_2 = 8.00$ N, and $F_3 = 6.00$ N.

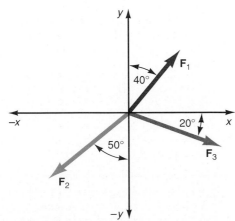

†34. Show that for three nonparallel vectors all in the same plane, any one of them can be represented as a linear sum of the other two.

†35. A unit vector is a vector that has a magnitude of one unit and is in a specified direction. If a unit vector **i** is defined to be in the x-direction, and a unit vector **j** is defined to be in the y-direction, show that any vector **a** can be written in the form

$$\mathbf{a} = a_x\mathbf{i} + a_y\mathbf{j}$$

†36. Prove that $|\mathbf{a} + \mathbf{b}| \leq |\mathbf{a}| + |\mathbf{b}|$.

37. An airplane flies due east at 200 mph straight from city A to city B a distance of 200 mi. A wind of 40 mph from the northwest is blowing. If the pilot doesn't make any corrections, where will the plane be in 1 hr?

38. Given vectors **a** and **b**, where $a = 50$, $\theta_1 = 33°$, $b = 80$, and $\theta_2 = 128°$, find (a) **a** + **b**, (b) **a** − **b**, (c) **a** − 2**b**, (d) 3**a** + **b**, (e) 2**a** − **b**, and (f) 2**b** − **a**.

39. In the accompanying figure the tension T in the cable is 200 N. Find the vertical component T_y and the horizontal component T_x of this tension.

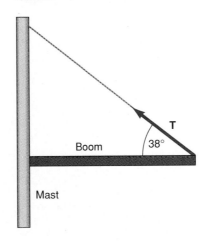

†40. In the accompanying diagram w_1 is 5 lb and w_2 is 3 lb. Find the angle θ such that the component of w_1 parallel to the incline is equal to w_2.

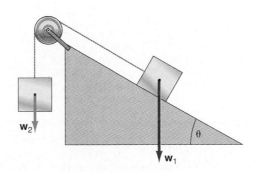

†41. In the accompanying diagram $w_1 = 2$ N, $w_2 = 5$ N, and $\theta = 65°$. Find the angle ϕ such that the components of the two forces parallel to the inclines are equal.

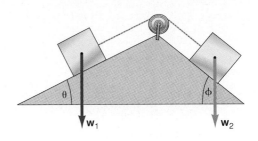

†42. In the accompanying diagram $w = 50$ lb, and $\theta = 10°$. What must be the value of F such that w will be held in place? What happens if the angle is doubled to 20°?

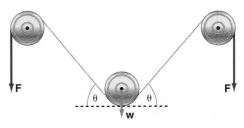

†43. In projectile motion in two dimensions the projectile is located by the displacement vector $\mathbf{r}_1$ at the time t_1 and by the displacement vector $\mathbf{r}_2$ at t_2, as shown in the diagram. If $r_1 = 20$ m, $\theta_1 = 60°$, $r_2 = 25$ m, and $\theta_2 = 25°$, find the magnitude and direction of the vector $\mathbf{r}_2 - \mathbf{r}_1$.

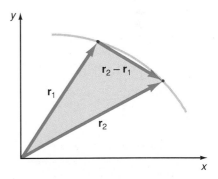

Interactive Tutorials

⊟ 44. A 50.0-N force is directed at an angle of 50° above the horizontal. Resolve this force into vertical and horizontal components.

⊟ 45. Find the resultant of any number of force vectors (up to five vectors).

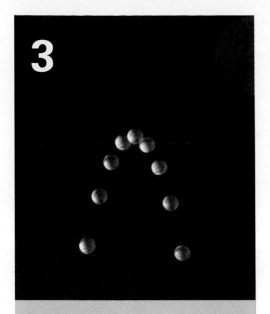

3

Kinematics—The Study of Motion

Captured motion.

My purpose is to set forth a very new science dealing with a very ancient subject. There is, in nature, perhaps nothing older than motion, concerning which the books written by Philosophers are neither few nor small; nevertheless I have discovered by experiment some properties of it which are worth knowing and which have not hitherto been either observed or demonstrated . . . and what I consider more important, there has been opened up to this vast and most excellent science, of which my work is merely the beginning, ways and means by which other minds more acute than mine will explore its remote corners.

Galileo Galilei
Dialogues Concerning Two New Sciences

3.1 Introduction

Kinematics *is defined as that branch of mechanics that studies the motion of a body without regard to the cause of that motion.* In our everyday life we constantly observe objects in motion. For example, an object falls from the table, a car moves along the highway, or a plane flies through the air. In this process of motion, we observe that at one time the object is located at one position in space and then at a later time it has been displaced to some new position. Motion thus entails a movement from one position to another position. Therefore, to describe the motion of a body logically, we need to start by defining the position of a body. To do this we need a reference system. Thus, we introduce a coordinate system, as shown in figure 3.2. The body is located at the point 0 at the time $t = 0$. The point 0, the origin of the coordinate system, is the reference position. We measure the displacement of the moving body from there. After an elapse of time t_1 the object will have moved from 0 and will be found along the x-axis at position 1, a distance x_1 away from 0.

A little later in time, at $t = t_2$, the object will be located at point 2, a distance x_2 away from 0. (As an example, the moving body might be a car on the street. The reference point 0 might be a lamp post on the street, while points 1 and 2 might be telephone poles.) Let us now consider the motion between points 1 and 2.

The **average velocity** of the body in motion between the points 1 and 2 is defined as the displacement of the moving body divided by the time it takes for that displacement. That is,

$$v_{\mathrm{avg}} = \frac{\text{displacement}}{\text{time for displacement}} \tag{3.1}$$

where v_{avg} is the notation used for the average velocity. For this description of one-dimensional motion, it is not necessary to use boldface vector notation. However, a positive value of v implies a velocity in the positive x- or y-direction, while a negative value of v implies a velocity in the negative x- or y-direction. A positive value of x implies a displacement in the positive x-direction, while a negative value of x implies a displacement in the negative x-direction. A positive value of y implies a displacement in the positive y-direction, while a negative value of y implies a displacement in the negative y-direction. The more general case, the velocity of a moving body in two dimensions, is treated in section 3.10.

Figure 3.1

Galileo Galilei.

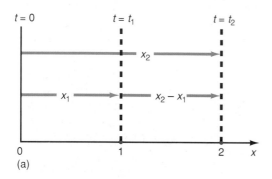

(a)

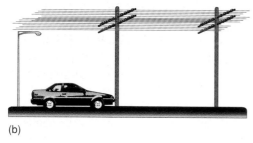

(b)

Figure 3.2

The position of an object at two different times.

From figure 3.2, we can see that during the time interval $t_2 - t_1$, the displacement or change in position of the body is simply $x_2 - x_1$. Therefore, the average velocity of the body in motion between points 1 and 2 is

$$v_{avg} = \frac{x_2 - x_1}{t_2 - t_1} \tag{3.2}$$

Note here that in the example of the car and the telephone poles, t_1 is the time on a clock when the car passes the first telephone pole, position 1, and t_2 is the time on the same clock when the car passes the second telephone pole, position 2.

A convenient notation to describe this change in position with the change in time is the *delta notation*. Delta (the Greek letter Δ) is used as a symbolic way of writing "change in," that is,

$$\Delta x = (\text{change in } x) = x_2 - x_1 \tag{3.3}$$

and

$$\Delta t = (\text{change in } t) = t_2 - t_1 \tag{3.4}$$

Using this delta notation we can write the average velocity as

$$v_{avg} = \frac{x_2 - x_1}{t_2 - t_1} = \frac{\Delta x}{\Delta t} \tag{3.5}$$

Example 3.1

Finding the average velocity using the Δ notation. A car passes telephone pole number 1, located 50.0 ft down the street from the corner lamp post, at a time $t_1 = 8.00$ s. It then passes telephone pole number 2, located 250 ft from the lamp post, at a time of $t_2 = 16.0$ s. What was the average velocity of the car between the positions 1 and 2?

Solution

The average velocity of the car, found from equation 3.5, is

$$v_{avg} = \frac{\Delta x}{\Delta t} = \frac{x_2 - x_1}{t_2 - t_1} = \frac{250 \text{ ft} - 50.0 \text{ ft}}{16.0 \text{ s} - 8.00 \text{ s}}$$

$$= \frac{200 \text{ ft}}{8.00 \text{ s}} = 25.0 \text{ ft/s}$$

(Note that according to the convention that we have adopted, the 25 ft/s represents a velocity because the magnitude of the velocity is 25 ft/s and the direction of the velocity vector is in the positive x-direction. If the answer were -25 ft/s the direction would have been in the negative x-direction.)

For convenience, the reference position 0 that is used to describe the motion is occasionally moved to position 1, then $x_1 = 0$, and the displacement is denoted by x, as shown in figure 3.3. The clock is started at this new reference position 1, so $t_1 = 0$ there. We now express the elapsed time for the displacement as t. In this simplified coordinate system the average velocity is

$$v_{avg} = \frac{x}{t} \tag{3.6}$$

Remember, the average velocity is the same physically in both equations 3.5 and 3.6; the numerator is still the displacement of the moving body, and the denominator is still the elapsed time for this displacement. Because the reference point has been changed, the notation appears differently. We use both notations in the description of motion. The particular notation we use depends on the problem.

Mechanics

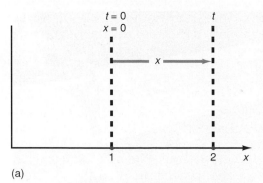

(a)

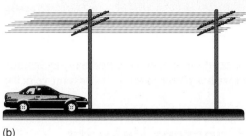

(b)

Figure 3.3

The position of an object determined from a new reference system.

Example 3.2

Changing the reference position. A car passes telephone pole number 1 at $t = 0$ on a watch. It passes a second telephone pole 200 ft down the block 8.00 seconds later. What is the car's average velocity?

Solution

The average velocity, found from equation 3.6, is

$$v_{avg} = \frac{x}{t} = \frac{200 \text{ ft}}{8.00 \text{ s}} = 25.0 \text{ ft/s}$$

Also note that this is the same problem solved in example 3.1; only the reference position for the measurement of the motion has been changed.

◼

Before we leave this section, we should make a distinction between the average velocity of a body and the average speed of a body. *The **average speed** of a body is the distance that a body moves per unit time. The average velocity of a body is the displacement of a body per unit time.* Because the displacement of a body is a vector quantity, that is, it specifies the distance an object moves *in a specified direction,* its velocity is also a vector quantity. *Thus, velocity is a vector quantity while speed is a scalar quantity.* For example, if a girl runs 100 m in the *x*-direction and turns around and returns to the starting point in a total time of 90 s, her average velocity is zero because her displacement is zero. Her average speed, on the other hand, is the total distance she ran divided by the total time it took, or 200 m/90 s = 2.2 m/s. If she ran 100 m in 45 s in one direction only, let us say the positive *x*-direction, her average speed is 100 m/45 s = 2.2 m/s. Her average velocity is 2.2 m/s in the positive *x*-direction. In this case, the speed is the magnitude of the velocity vector. Speed is always a positive quantity, whereas velocity can be either positive or negative depending on whether the motion is in the positive *x*-direction or the negative *x*-direction, respectively.

Section 3.2 shows how the motion of a body can be studied in more detail in the laboratory.

3.2 Experimental Description of a Moving Body

Following Galileo's advice that motion should be studied by experiment, let us go into the laboratory and describe the motion of a moving body on an air track. An air track is a hollow aluminum track that has a right isosceles triangle for its cross section. Air is forced into the air track by a blower and flows out the sides of the track through many small holes. A glider, having an inverted Y cross section, is placed on the track. The air escaping from the holes in the track provides a cushion of air for the glider to move on, thereby substantially reducing the retarding force of friction on the glider. The setup of an air track in the laboratory is shown in figure 3.4.

We attach a piece of spark-timer tape to the air track to act as a permanent record of the position of the moving glider as a function of time. We connect a spark timer, a device that emits electrical pulses at certain prescribed times, to a wire on the air track. The pulse from the timer moves along this wire, which is parallel to the air track, and then jumps across an air gap as a spark to another wire attached to the moving glider. The resulting pulse proceeds along the wire of the glider down to the bottom of the track to the timer tape. A spark now jumps across an air gap between the glider wire and the air track, and in so doing it burns a hole in the timer tape. This burned hole on the tape, which appears as a dot, is a record of the position

Figure 3.4

Setup of an air track.

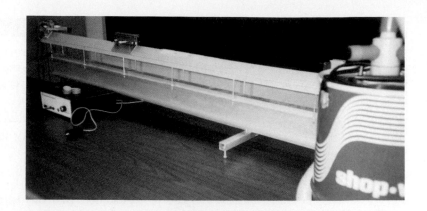

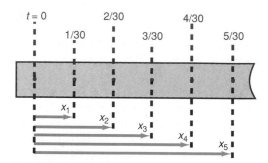

Figure 3.5

Spark-timer paper showing constant velocity.

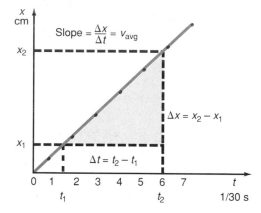

Figure 3.6

Graph of distance versus time for constant velocity.

of the glider at that instant of time. Thus, the combination of a glider, an air track, and a spark timer gives us a record of the position of a moving body at any instant of time. Let us now look at an experiment with a glider moving at constant velocity along the air track.

3.3 A Body Moving at Constant Velocity

To study a body moving at constant velocity we place a glider on a level air track and give it a slight push to initiate its motion along the track. The spark timer is turned on, leaving a permanent record of this motion on a piece of spark-timer tape. The distance traveled by the glider as a function of time is recorded on the spark-timer paper, and appears as in figure 3.5. The spark timer is set to give a spark every $1/30$ of a second. The first dot occurs at the time $t = 0$, and each succeeding dot occurs at a time interval of $1/30$ of a second later. We label the first dot as dot 0, the reference position, and then measure the total distance x from the first dot to each succeeding dot with a meter stick.

The measured data for the total distance traveled by the glider as a function of time are plotted in figure 3.6. Note that the plot is a straight line. If you measure the slope of this line you will observe that it is $\Delta x / \Delta t$, which is the average velocity defined in equation 3.5. Since all the points generate a straight line, which has a constant slope, the velocity of the glider is a constant equal to the slope of this graph. *Whenever a body moves in such a way that it always travels equal distances in equal times, that body is said to be moving with a **constant velocity**.* This can also be observed in figure 3.5 by noting that the dots are equally spaced.

The SI unit for velocity is m/s. The units cm/s and km/hr are also used. In the British engineering system we use ft/s, in./s, and mi/hr. Note that on a graph of the displacement of a moving body versus time, the slope $\Delta x / \Delta t$ always represents a velocity. If the slope is positive, the velocity is positive and the direction of the moving body is toward the right. If the slope is negative, the velocity is negative and the direction of the moving body is toward the left.

Example 3.3

The velocity of a glider on an air track. A glider goes from a position of 12.0 cm at a time of $t = 1/3$ s to a position of 75.0 cm at a time of $t = 5/3$ s. Find the average velocity of the glider during this interval.

Figure 3.7

The tilted air track.

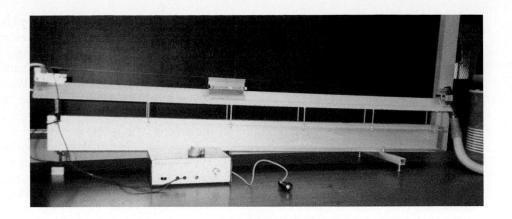

Solution

The average velocity of the glider, found from equation 3.5, is

$$v_{avg} = \frac{\Delta x}{\Delta t} = \frac{x_2 - x_1}{t_2 - t_1}$$

$$= \frac{75.0 \text{ cm} - 12.0 \text{ cm}}{5/3 \text{ s} - 1/3 \text{ s}} = \frac{63.0 \text{ cm}}{4/3 \text{ s}}$$

$$= 47.3 \text{ cm/s}$$

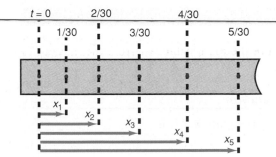

Figure 3.8

Spark-timer tape for accelerated motion.

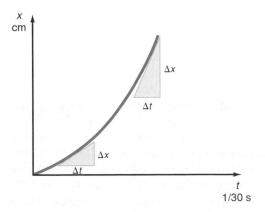

Figure 3.9

Graph of x versus t for constant acceleration.

3.4 A Body Moving at Constant Acceleration

If we tilt the air track at one end it effectively becomes a frictionless inclined plane. We place a glider at the top of the track and then release it from rest. Figure 3.7 is a picture of the glider in its motion on the inclined air track.

The spark timer is turned on, giving a record of the position of the moving glider as a function of time, as illustrated in figure 3.8. The most important feature to immediately note on this record of the motion, is that the dots, representing the positions of the glider, are no longer equally spaced as they were for motion at constant speed, but rather become farther and farther apart as the time increases. The total distance x that the glider moves is again measured as a function of time. If we plot this measured distance x against the time t, we obtain the graph shown in figure 3.9.

The first thing to note in this figure is that the graph of x versus t is not a straight line. However, as you may recall from section 3.3, the slope of the distance versus time graph, $\Delta x/\Delta t$, represents the velocity of the moving body. But in figure 3.9 there are many different slopes to this curve because it is continuously changing with time. Since the slope at any point represents the velocity at that point, we observe that the velocity of the moving body is changing with time. *The change of velocity with time is defined as the **acceleration** of the moving body, and the average acceleration is written as*

$$a_{avg} = \frac{\Delta v}{\Delta t} = \frac{v_2 - v_1}{t_2 - t_1} \tag{3.7}$$

Since the velocity is a vector quantity, acceleration, which is equal to the change in velocity with time, is also a vector quantity. More will be said about this shortly.

Because the velocity is changing continuously, the average velocity for every time interval can be computed from equation 3.5. Thus, subtracting each value of x from the next value of x gives us Δx, the distance the glider moves during one time interval. The average velocity during that interval can then be computed from $v_{avg} = \Delta x/\Delta t$. At the beginning of this interval the actual velocity is less than this

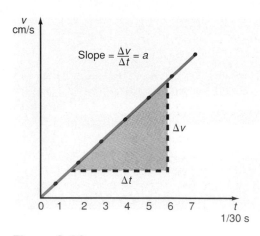

$$\text{Slope} = \frac{\Delta v}{\Delta t} = a$$

Figure 3.10

Graph of v versus t for constant acceleration.

value while at the end of the interval it is greater. Later we will see that *for constant acceleration, the velocity at the center of the time interval is equal to the average velocity for the entire time interval.*

If we plot the velocity at the center of the interval against the time, we obtain the graph in figure 3.10. We can immediately observe that the graph is a straight line. The slope of this line, $\Delta v / \Delta t$, is the experimental acceleration of the glider. Since this graph is a straight line, the slope is a constant; this implies that the acceleration is also a constant. Hence, the acceleration of a body moving down a frictionless inclined plane is a constant. In the case of more general motion, a body can also have its acceleration changing with time. However, most of the accelerated motion discussed in this book is at constant acceleration. The most notable exception is for simple harmonic motion, which we discuss in chapter 11. *Because in constantly accelerated motion the average acceleration is the same as the constant acceleration, the subscript avg will be deleted from the acceleration in all the equations dealing with this type of motion.*

Since acceleration is a change in velocity per unit time, the units for acceleration are velocity divided by the time. In SI units, the acceleration is

$$\frac{m/s}{s}$$

For convenience, this is usually written in the equivalent algebraic form as m/s^2. But we must not forget the physical meaning of a change in velocity of so many m/s every second. Other units used to express acceleration are ft/s^2, cm/s^2, $(km/hr)/s$, mph/s, and $in./s^2$.

Example 3.4

The acceleration of a glider on an air track. A glider's velocity on a tilted air track increases from 17.0 cm/s at the time $t = 1/3$ s to 55.0 cm/s at a time of $t = 7/3$ s. What is the acceleration of the glider?

Solution

The acceleration of the glider, found from equation 3.7, is

$$a = \frac{\Delta v}{\Delta t} = \frac{v_2 - v_1}{t_2 - t_1}$$

$$= \frac{55.0 \text{ cm/s} - 17.0 \text{ cm/s}}{7/3 \text{ s} - 1/3 \text{ s}} = \frac{38.0 \text{ cm/s}}{6/3 \text{ s}}$$

$$= 19.0 \text{ cm/s}^2$$

Before leaving this section we should note that since acceleration is a vector, if the acceleration is a positive quantity, the velocity is increasing with time, and the acceleration vector points toward the right. If the acceleration is a negative quantity, the velocity is decreasing with time, and the acceleration vector points toward the left. When the velocity is positive, indicating that the body is moving in the positive x-direction, and the acceleration is positive, the object is speeding up, or accelerating. However, when the velocity is positive, and the acceleration is negative, the object is slowing down, or decelerating. On the other hand, if the velocity is negative, indicating that the body is moving in the negative x-direction, and the acceleration is negative, the body is speeding up in the negative x-direction. However, when the velocity is negative and the acceleration is positive, the body is slowing down in the negative x-direction. If the acceleration lasts long enough, the body will eventually come to a stop and will then start moving in the positive x-direction. The velocity will then be positive and the body will be speeding up in the positive x-direction.

3.5 The Instantaneous Velocity of a Moving Body

In section 3.4 we observed that the velocity of the glider varies continuously as it "slides" down the frictionless inclined plane. We also stated that the average velocity could be computed from $v_{avg} = \Delta x/\Delta t$. At the beginning of the interval of motion the actual velocity is less than this value while at the end of the interval it is greater. If the interval is made smaller and smaller, the average velocity v_{avg} throughout the interval becomes closer to the actual velocity at the instant the body is at the center of the time interval. Finding the velocity at a particular instant of time leads us to the concept of instantaneous velocity. **Instantaneous velocity is defined as the limit of $\Delta x/\Delta t$ as Δt gets smaller and smaller, eventually approaching zero.** We write this concept mathematically as

$$v = \lim_{\Delta t \to 0} \frac{\Delta x}{\Delta t} \qquad (3.8)$$

As in the case of average velocity in one-dimensional motion, if the limit of $\Delta x/\Delta t$ is a positive quantity, the velocity vector points toward the right. If the limit of $\Delta x/\Delta t$ is a negative quantity, the velocity vector points toward the left.

The concept of instantaneous velocity can be easily understood by performing the following experiment on an air track. First, we tilt the air track to again give an effectively frictionless inclined plane. Then we place a 20-cm length of metal, called a flag, at the top of the glider. A photocell gate, which is a device that can be used to automatically turn a clock on and off, is attached to a clock timer and is placed on the air track. We then allow the glider to slide down the track. When the flag of the glider interrupts the light beam to the photocell, the clock is turned on. When the flag has completely passed through the light beam, the photocell gate turns off the clock. The clock thus records the time for the 20-cm flag to pass through the photocell gate. We find the average velocity of the flag as it moves through the gate from equation 3.5 as $v = \Delta x/\Delta t$. The 20-cm length of the flag is Δx, and Δt is the time interval, as read from the clock.

We repeat the process for a 15-cm, 10-cm, and a 5-cm flag. For each case we measure the time Δt that it takes for the flag to move through the gate. The first thing that we observe is that the time for the flag to move through the gate, Δt, gets smaller for each smaller flag. You might first expect that if Δt approaches 0, the ratio of $\Delta x/\Delta t$ should approach infinity. However, since Δx, the length of the flag, is also getting smaller, the ratio of $\Delta x/\Delta t$ remains finite. If we plot $\Delta x/\Delta t$ as a function of Δt for each flag, we obtain the graph in figure 3.11.

Notice that as Δt approaches 0, $(\Delta t \to 0)$, the plotted line intersects the $\Delta x/\Delta t$ axis. At this point, the distance interval Δx has been reduced from 20 cm to effectively 0 cm. The value of Δt has become progressively smaller so this point represents the limiting value of $\Delta x/\Delta t$ as Δt approaches 0. But this limit is the definition of the instantaneous velocity. Hence, the point where the line intersects the $\Delta x/\Delta t$ axis gives the value of the velocity of the glider at the instant of time that the glider is located at the position of the photocell gate. This limiting process allows us to describe the motion of a moving body in terms of the velocity of the body at any instant of time rather than in terms of the body's average velocity.

Usually we will be more interested in the instantaneous velocity of a moving body than its average velocity. The speedometer of a moving car is a physical example of instantaneous velocity. Whether the car's velocity is constant or changing with time, the instant that the speedometer is observed, the speedometer indicates the speed of the car at that particular instant of time. The instantaneous velocity of the car is that observed value of the speed in the direction that the car is traveling.

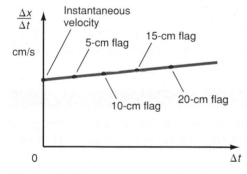

Figure 3.11

Graph of $\Delta x/\Delta t$ versus Δt to obtain the instantaneous velocity of the glider.

(a)

(b)

Figure 3.12

Change in reference system.

3.6 The Kinematic Equations in One Dimension

Because the previous experiments were based on motion at constant acceleration, we can only apply the results of those experiments to motion at a constant acceleration. Let us now compile those results into a set of equations, called the **kinematic equations of linear motion,** that will describe the motion of a moving body. For motion at constant acceleration, the average acceleration is equal to the constant acceleration. Hence, the subscript avg can be deleted from equation 3.7 and that equation now gives the constant acceleration of the moving body as

$$a = \frac{v_2 - v_1}{t_2 - t_1} \tag{3.7}$$

Equation 3.7 indicates that at the time t_1 the body is moving at the velocity v_1, while at the time t_2 the body is moving at the velocity v_2. This motion is represented in figure 3.12(a) for a runner.

Let us change the reference system by starting the clock at the time $t_1 = 0$, as shown in figure 3.12(b). We will now designate the velocity of the moving body at the time 0 as v_0 instead of the v_1 in the previous reference system of figure 3.12(a). Similarly, the time t_2 will correspond to any time t and the velocity v_2 will be denoted by v, the velocity at that time t. Thus, the velocity of the moving body will be v_0 when the time is equal to 0, and v when the time is equal to t. This change of reference system allows us to rewrite equation 3.7 as

$$a = \frac{v - v_0}{t} \tag{3.9}$$

Equation 3.9 is similar to equation 3.7 in that it gives the same definition for acceleration, namely a change in velocity with time, but in a slightly different but equivalent notation. Solving equation 3.9 for v gives the first of the very important kinematic equations, namely,

$$v = v_0 + at \tag{3.10}$$

Equation 3.10 says that the velocity v of the moving object can be found at any instant of time t once the acceleration a and the initial velocity v_0 of the moving body are known.

Example 3.5

Using the kinematic equation for the velocity as a function of time. A car passes a green traffic light while moving at a velocity of 6.00 m/s. It then accelerates at 0.300 m/s² for 15.0 s. What is the car's velocity at 15.0 s?

Solution

The velocity, found from equation 3.10, is

$$v = v_0 + at$$

$$= \left(6.00 \, \frac{m}{s}\right) + \left(0.300 \, \frac{m}{s^2}\right)(15.0 \, s)$$

$$= 10.5 \, m/s$$

The velocity of the car is 10.5 m/s. This means that the car is moving at a speed of 10.5 m/s in the positive x-direction.

In addition to the velocity of the moving body at any time t, we would also like to know the location of the body at that same time. That is, let us obtain an equation for the displacement of the moving body as a function of time. Solving equation 3.6 for the displacement x gives

$$x = v_{avg}t \qquad \text{(3.11)}$$

Hence, the displacement of the moving body is equal to the average velocity of the body times the time it is in motion. For example, if you are driving your car at an average velocity of 50 miles per hour, and you drive for a period of time of two hours, then your displacement is

$$x = 50 \frac{mi}{hr}(2\ hr)$$

$$= 100\ mi$$

You have traveled a total distance of 100 mi from where you started.

Equation 3.11 gives us the displacement of the moving body in terms of its average velocity. The actual velocity during the motion might be greater than or less than the average value. The average velocity does not tell us anything about the body's acceleration. We would like to express the displacement of the body in terms of its acceleration during a particular time interval, and in terms of its initial velocity at the beginning of that time interval.

For example, consider a car in motion along a road between the times $t = 0$ and $t = t$. At the beginning of the time interval the car has an initial velocity v_0, while at the end of the time interval it has the velocity v, as shown in figure 3.13. If the acceleration of the moving body is constant, then the average velocity throughout the entire time interval is

$$v_{avg} = \frac{v_0 + v}{2} \qquad \text{(3.12)}$$

This averaging of velocities for bodies moving at constant acceleration is similar to determining a grade in a course. For example, if you have two test grades in the course, your course grade, the average of the two test grades, is the sum of the test grades divided by 2,

$$\text{Avg. Grade} = \frac{100 + 90}{2} = 95$$

If we substitute this value of the average velocity into equation 3.11, the displacement becomes

$$x = v_{avg}t = \left(\frac{v_0 + v}{2}\right)t \qquad \text{(3.13)}$$

Figure 3.13

A car moving on a road.

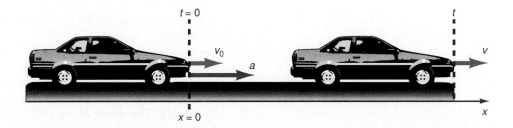

Note that v represents the final value of the velocity at the time t, the end of the time interval. But there already exists an equation for the value of v at the time t, namely equation 3.10. Therefore, substituting equation 3.10 into equation 3.13 gives

$$x = \left[\frac{v_0 + (v_0 + at)}{2} \right] t$$

Simplifying, we get

$$x = \left(\frac{2v_0 + at}{2} \right) t$$

$$= \frac{2v_0 t}{2} + \tfrac{1}{2} at^2$$

$$x = v_0 t + \tfrac{1}{2} at^2 \tag{3.14}$$

Equation 3.14, the second of the kinematic equations, represents the displacement x of the moving body at any instant of time t. In other words, if the original velocity and the constant acceleration of the moving object are known, then we can determine the location of the moving object at any time t. This rather simple equation contains a tremendous amount of information.

Example 3.6

Using the kinematic equation for the displacement as a function of time. A car, initially traveling at 30.0 km/hr, accelerates at the constant rate of 1.50 m/s². How far will the car travel in 15.0 s?

Solution

To express the result in the proper units, km/hr is converted to m/s as

$$v_0 = 30.0 \, \frac{km}{hr} \left(\frac{1 \, hr}{3600 \, s} \right) \left(\frac{1000 \, m}{1 \, km} \right) = 8.33 \, m/s$$

The displacement of the car, found from equation 3.14, is

$$x = v_0 t + \tfrac{1}{2} at^2$$

$$= \left(8.33 \, \frac{m}{s} \right) (15.0 \, s) + \frac{1}{2} \left(1.50 \, \frac{m}{s^2} \right) (15.0 \, s)^2$$

$$= 125 \, m + 169 \, m$$

$$= 294 \, m$$

The first term in the answer, 125 m, represents the distance that the car would travel if there were no acceleration and the car continued to move at the velocity 8.33 m/s for 15.0 s. But there is an acceleration, and the second term shows how much farther the car moves because of that acceleration, namely 169 m. The total displacement of 294 m is the total distance that the car travels because of the two effects.

As a further example of the kinematics of a moving body, consider the car moving along a road at an initial velocity of 60.0 mph = 88.0 ft/s, as shown in figure 3.14. The driver sees a tree fall into the middle of the road 200 ft away. The driver immediately steps on the brakes, and the car starts to decelerate at the constant rate of $a = -18.0 \, ft/s^2$. (As mentioned previously, in one-dimensional motion a negative acceleration means that the acceleration vector is toward the left, in the opposite direction of the motion. If the velocity is positive, a negative value for the acceleration means that the body is slowing down or decelerating.) Will the car come to a stop before hitting the tree?

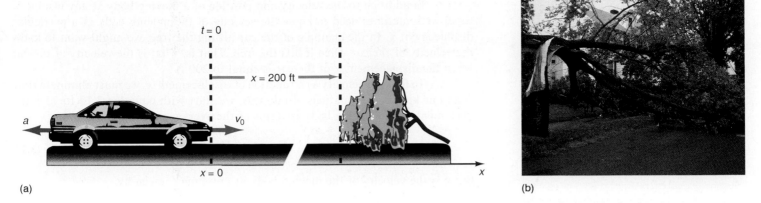

$t = 0$

a v_0

$x = 200$ ft

$x = 0$ x

(a)

(b)

Figure 3.14

A tree falls on the road.

What we need for the solution of this problem is the actual distance the car travels before it can come to a stop while decelerating at the rate of 18.0 ft/s². Before we can find that distance, however, we must know the time it takes for the car to come to a stop. Then we substitute this stopping time into equation 3.14, and the equation tells us how far the car will travel before coming to a stop. (Note that most of the questions that might be asked about the motion of the car can be answered using the kinematic equations 3.10 and 3.14.)

Equation 3.10 tells us the velocity of the car at any instant of time. But when the car comes to rest its velocity is zero. Thus, at the time when the car comes to a stop t_{stop}, the velocity v will be equal to zero. Therefore, equation 3.10 becomes

$$0 = v_0 + at_{stop}$$

Solving for the time for the car to come to a stop, we have

$$t_{stop} = -\frac{v_0}{a} \qquad (3.15)$$

the time interval from the moment the brakes are applied until the car comes to a complete stop. Substituting the values of the initial velocity v_0 and the constant acceleration a into equation 3.15, we have

$$t_{stop} = -\frac{v_0}{a} = \frac{-88.0 \text{ ft/s}}{-18.0 \text{ ft/s}^2} = 4.89 \text{ s}$$

It will take 4.89 s for the car to come to a stop if nothing gets in its way to change its rate of deceleration. Note how the units cancel in the equation until the final unit becomes seconds, that is,

$$\frac{v_0}{a} = \frac{\text{ft/s}}{\text{ft/s}^2} = \frac{1/s}{1/s^2} = \frac{1}{s}\frac{1}{1/s^2} = \frac{1}{1/s} = s$$

Thus, (ft/s)/(ft/s²), comes out to have the unit seconds, which it must since it represents the time for the car to come to a stop.

Now that we know the time for the car to come to a stop, we can substitute that value back into equation 3.14 and find the distance the car will travel in the 4.89 s:

$$x = v_0 t + \tfrac{1}{2} at^2$$

$$= 88.0 \frac{\text{ft}}{\text{s}} (4.89 \text{ s}) + \frac{1}{2}\left(-18.0 \frac{\text{ft}}{\text{s}^2}\right)(4.89 \text{ s})^2$$

$$= 430 \text{ ft} - 215 \text{ ft}$$

$$= 215 \text{ ft}$$

The car will come to a stop in 215 ft. Since the tree is only 200 ft in front of the car, it cannot come to a stop in time and will hit the tree.

In addition to the velocity and position of a moving body at any instant of time, we sometimes need to know the velocity of the moving body at a particular displacement x. In the example of the car hitting the tree, we might want to know the velocity of the car when it hits the tree. That is, what is the velocity of the car when the displacement x of the car is equal to 200 ft?

To find the velocity as a function of displacement x, we must eliminate time from our kinematic equations. To do this, we start with equation 3.13 for the displacement of the moving body in terms of the average velocity,

$$x = v_{avg}t = \left(\frac{v_0 + v}{2}\right)t \tag{3.13}$$

But v is the velocity of the moving body at any time t, given by

$$v = v_0 + at \tag{3.10}$$

Solving for t gives

$$t = \frac{v - v_0}{a}$$

Substituting this value into equation 3.13 gives

$$x = \left(\frac{v_0 + v}{2}\right)t = \left(\frac{v_0 + v}{2}\right)\left(\frac{v - v_0}{a}\right)$$

$$= \left(\frac{v_0 + v}{2a}\right)(v - v_0)$$

$$2ax = v_0v + v^2 - v_0v - v_0^2$$

$$= v^2 - v_0^2$$

Solving for v^2, we obtain *the third kinematic equation,*

$$v^2 = v_0^2 + 2ax \tag{3.16}$$

which is used to determine the velocity v of the moving body at any displacement x.

Let us now go back to the problem of the car moving down the road, with a tree lying in the road 200 ft in front of the car. We already know that the car will hit the tree, but at what velocity will it be going when it hits the tree? That is, what is the velocity of the car at the displacement of 200 feet? Using equation 3.16 with $x = 200$ ft, $v_0 = 60.0$ mph $= 88.0$ ft/s, and $a = -18.0$ ft/s^2, and solving for v gives

$$v^2 = v_0^2 + 2ax$$

$$= (88.0 \text{ ft/s})^2 + 2(-18.0 \text{ ft/s}^2)(200 \text{ ft})$$

$$= 7740 \text{ ft}^2/\text{s}^2 - 7200 \text{ ft}^2/\text{s}^2$$

$$= 540 \text{ ft}^2/\text{s}^2$$

$$v = \frac{23.2 \text{ ft}}{\text{s}}\left(\frac{60.0 \text{ mph}}{88.0 \text{ ft/s}}\right)$$

and finally,

$$v = 15.8 \text{ mph}$$

When the car hits the tree it will be moving at 15.8 mph, so the car may need a new bumper or fender. Equation 3.16 allows us to determine the velocity of the moving body at any displacement x.

A problem similar to that of the car and the tree involves the maximum velocity that a car can move and still have adequate time to stop before hitting something the driver sees on the road in front of the car. Let us again assume that the car decelerates at the same constant rate as before, $a = -18$ ft/s^2, and that the low beam headlights of the car are capable of illuminating a 200 ft distance of the road. Using equation 3.16, which gives the velocity of the car as a function of displacement, let us find the maximum value of v_0 such that v is equal to zero when the car has the displacement x. That is,

$$v^2 = v_0^2 + 2ax$$

$$0 = v_0^2 + 2ax$$

$$v_0 = \sqrt{-2ax}$$

$$= \sqrt{-2(-18.0 \text{ ft/s}^2)(200 \text{ ft})}$$

$$= \sqrt{7200 \text{ ft}^2/\text{s}^2}$$

$$= \frac{84.9 \text{ ft}}{\text{s}} \left(\frac{60.0 \text{ mph}}{88.0 \text{ ft/s}} \right)$$

$$= 57.9 \text{ mph}$$

If the car decelerates at the constant rate of 18.0 ft/s^2 and the low beam headlights are only capable of illuminating a distance of 200 ft, then the maximum safe velocity of the car at night without hitting something is 57.9 miles per hour. For velocities faster than this, the distance it takes to bring the car to a stop is greater than the distance the driver can see with low beam headlights. If you see it, you'll hit it! Of course these results are based on the assumption that the car decelerates at 18 ft/s^2. This number depends on the condition of the brakes and tires and road conditions, and will be different for each car.

In summary, the three kinematic equations,

$$x = v_0 t + \tfrac{1}{2}at^2 \tag{3.14}$$

$$v = v_0 + at \tag{3.10}$$

and

$$v^2 = v_0^2 + 2ax \tag{3.16}$$

are used to describe the motion of an object undergoing constant acceleration. The first equation gives the displacement of the object at any instant of time. The second equation gives the body's velocity at any instant of time. The third equation gives the velocity of the body at any displacement x.

These equations are used for either positive or negative accelerations. Remember the three kinematic equations hold only for constant acceleration. If the acceleration varies with time then more advanced techniques must be used to determine the position and velocity of the moving object.

3.7 **The Freely Falling Body**

Another example of the motion of a body in one dimension is the freely falling body. **A freely falling body** is defined as a body that is moving freely under the influence of gravity, where it is assumed that the effect of air resistance is negligible. The body can have an upward, downward, or even zero initial velocity. The simplest of the freely falling bodies we discuss is the body dropped in the vicinity of the surface of the earth. That is, the first case to be considered is the one with zero initial velocity, $v_0 = 0$. The motion of a body in the vicinity of the surface of the earth with either an upward or downward initial velocity will be considered in section 3.9.

In chapter 4 on Newton's second law of motion, we will see that whenever an unbalanced force acts on an object of mass m, it gives that object an acceleration, a. The gravitational force that the earth exerts on an object causes that object to have an acceleration. This acceleration is called the **acceleration due to gravity** and is denoted by the letter g. Therefore, any time a body is dropped near the surface of the earth, that body, ignoring air friction, experiences an acceleration g. From experiments in the laboratory we know that the value of g near the surface of the earth is constant and is given by

$$g = 9.80 \text{ m/s}^2 = 980 \text{ cm/s}^2 = 32.2 \text{ ft/s}^2$$

Any body that falls with the acceleration due to gravity, g, is called a freely falling body. To simplify the calculations in the British engineering system we will use the value $g = 32.0 \text{ ft/s}^2$.

Originally Aristotle said that a heavier body falls faster than a lighter body and on his authority this statement was accepted as truth for 1800 years. It was not disproved until the end of the sixteenth century when Simon Stevin (Stevinus) of Bruges (1548–1620) dropped balls of very different weights and found that they all fell at the same rate. That is, the balls were all dropped from the same height at the same time and all landed at the ground simultaneously. The argument still persisted that a ball certainly drops faster than a feather, but Galileo Galilei (1564–1642) explained the difference in the motion by saying that it is the air's resistance that slows up the feather. If the air were not present the ball and the feather would accelerate at the same rate.

A standard demonstration of the rate of fall is the penny and the feather demonstration. A long tube containing a penny and a feather is used, as shown in figure 3.15. If we turn the tube upside down, first we observe that the penny falls to the bottom of the tube before the feather. Then we connect the tube to a vacuum pump and evacuate most of the air from the tube. Again we turn the tube upside down, and now the penny and feather do indeed fall at the same rate and reach the bottom of the tube at the same time. Thus, it *is* the air friction that causes the feather to fall at the slower rate.

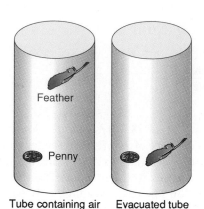

Feather

Penny

Tube containing air   Evacuated tube

Figure 3.15

Free-fall of the penny and the feather.

Another demonstration of a freely falling body, performed by the Apollo astronauts on the surface of the moon, was seen by millions of people on television. One of the astronauts dropped a feather and a hammer simultaneously and millions saw them fall at the same rate, figure 3.16. Remember, there is no atmosphere on the moon.

Therefore, neglecting air friction, all freely falling bodies accelerate downward at the same rate regardless of their mass. Recall that the acceleration of a body was defined as the change in its velocity with respect to time, that is,

$$a = \frac{\Delta v}{\Delta t} \tag{3.7}$$

Hence, a body that undergoes an acceleration due to gravity of 32.0 ft/s^2, has its velocity changing by 32.0 ft/s every second. If we neglect the effects of air friction, every body near the surface of the earth accelerates downward at that rate, whether the body is very large or very small. For all the problems considered in this book, we neglect the effects of air resistance.

Since the acceleration due to gravity is constant near the surface of the earth, we can determine the position and velocity of the freely falling body by using the kinematic equations 3.10, 3.14, and 3.16. However, because the motion is vertical, we designate the displacement by y in the kinematic equations:

$$v = v_0 + at \tag{3.10}$$

$$y = v_0 t + \tfrac{1}{2}at^2 \tag{3.14}$$

$$v^2 = v_0^2 + 2ay \tag{3.16}$$

Figure 3.16

Astronaut David R. Scott holds a geological hammer in his right hand and a feather in his left. The hammer and feather dropped to the lunar surface at the same instant.

Mechanics

Since the first case we consider is a body that is dropped, we will set the initial velocity v_0 equal to zero in the kinematic equations. Also the acceleration of the moving body is now the acceleration due to gravity, therefore we write the acceleration as

$$a = -g \tag{3.17}$$

The minus sign in equation 3.17 is consistent with our previous convention for one-dimensional motion. Motion in the direction of the positive axis is considered positive, while motion in the direction of the negative axis is considered negative. Hence, all quantities in the upward direction (positive y-direction) are considered positive, whether displacements, velocities, or accelerations. And all quantities in the downward direction (negative y-direction) are considered negative, whether displacements, velocities, or accelerations. The minus sign indicates that the direction of the acceleration is down, toward the center of the earth. This notation will be very useful later in describing the motion of projectiles. Therefore, the kinematic equations for a body dropped from rest near the surface of the earth are

$$y = -\tfrac{1}{2}gt^2 \tag{3.18}$$

$$v = -gt \tag{3.19}$$

$$v^2 = -2gy \tag{3.20}$$

Equation 3.18 gives the height or location of the freely falling body at any time, equation 3.19 gives its velocity at any time, and equation 3.20 gives the velocity of the freely falling body at any height y. This sign convention gives a negative value for the displacement y, which means that the zero position of the body is the position from which the body is dropped, and the body's location at any time t will always be below that point. The minus sign on the velocity indicates that the direction of the velocity is downward.

Equations 3.18, 3.19, and 3.20 completely describe the motion of the freely falling body that is dropped from rest. As an example, let us calculate the distance fallen and velocity of a freely falling body as a function of time for the first 5 s of its fall. Because most students are more familiar with the British engineering system of units, we will use this system for the calculations. However, the results of the computations are also written in parentheses in SI units in figure 3.17. At $t = 0$ the body is located at $y = 0$, (top of figure 3.17) and its velocity is zero. We then release the body. Where is it at $t = 1$ s? Using equation 3.18 and British engineering system units, y_1 is the displacement of the body (distance fallen) at the end of 1 s:

$$y_1 = -\tfrac{1}{2}gt^2 = -\tfrac{1}{2}(32 \text{ ft/s}^2)(1 \text{ s})^2 = -16 \text{ ft}$$

The minus sign indicates that the body is 16 ft *below* the starting point. To find the velocity at the end of 1 s, we use equation 3.19:

$$v_1 = -gt = (-32 \text{ ft/s}^2)(1 \text{ s}) = -32 \text{ ft/s}$$

The velocity is 32 ft/s downward at the end of 1 s. The position and velocity at the end of 1 s are shown in figure 3.17. For $t = 2$ s, the displacement and velocity are

$$y_2 = -\tfrac{1}{2}gt^2 = -\tfrac{1}{2}(32 \text{ ft/s}^2)(2 \text{ s})^2 = -64 \text{ ft}$$

$$v = -gt = -(32 \text{ ft/s}^2)(2 \text{ s}) = -64 \text{ ft/s}$$

At the end of 2 s the body has dropped a total distance downward of 64 ft and is moving at a velocity of 64 ft/s downward. For $t = 3$ s we obtain

$$y_3 = -\tfrac{1}{2}gt^2 = -\tfrac{1}{2}(32.0 \text{ ft/s}^2)(3 \text{ s})^2 = -144 \text{ ft}$$

$$v_3 = -gt = -(32.0 \text{ ft/s}^2)(3 \text{ s}) = -96 \text{ ft/s}$$

At the end of 3 s the body has fallen a distance of 144 ft and is moving downward at a velocity of 96 ft/s.

Figure 3.17

The distance and velocity for a freely
falling body.

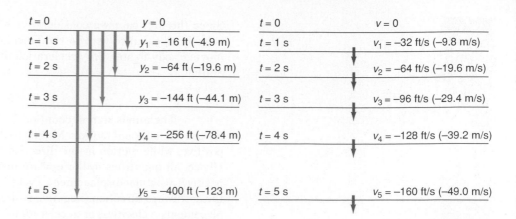

The distance and velocity for $t = 4$ s and $t = 5$ s are found similarly and are shown in figure 3.17. One of the first things to observe in figure 3.17 is that an object falls a relatively large distance in only a few seconds of time. Also note that the object does not fall equal distances in equal times, but rather the distance interval becomes greater for the same time interval as time increases. This is, of course, the result of the t^2 in equation 3.18 and is a characteristic of accelerated motion. Also note that the change in the velocity in any 1-s time interval is 32 ft/s, which is exactly what we meant by saying the acceleration due to gravity is 32 ft/s².

We stated previously that *the average velocity during a time interval is exactly equal to the instantaneous value of the velocity at the exact center of that time interval.* We can see that this is the case by inspecting figure 3.17. For example, if we take the time interval as between $t = 3$ s and $t = 5$ s, then the average velocity between the third and fifth second is

$$v_{35avg} = \frac{v_5 + v_3}{2} = \frac{-160 \text{ ft/s} + (-96 \text{ ft/s})}{2}$$

$$= \frac{-256 \text{ ft/s}}{2}$$

$$= -128 \text{ ft/s} = v_4$$

The average velocity between the time interval of 3 and 5 s, v_{35avg}, is exactly equal to v_4, the velocity at t equals 4 seconds, which is the exact center of the 3–5 time interval, as we can see in figure 3.17. The figure also shows the characteristic of an average velocity. At the beginning of the time interval the actual velocity is less than the average value, while at the end of the time interval the actual velocity is greater than the average value, but right at the center of the time interval the actual velocity is equal to the average velocity. Note that the average velocity occurs at the center of the time interval and not the center of the space interval.

In summary, we can see the enormous power inherent in the kinematic equations. An object was dropped from rest and the kinematic equations completely described the position and velocity of that object at any instant of time. All that information was contained in those equations.

3.8 Determination of Your Reaction Time by a Freely Falling Body

How long a period of time does it take for you to react to something? How can you measure this reaction time? It would be very difficult to use a clock to measure reaction time because it will take some reaction time to turn the clock on and off. However, a freely falling body can be used to measure reaction time. To see how this is accomplished, have one student hold a vertical meter stick near the top, as

(a)

(b)

Figure 3.18

Measurement of reaction time.

shown in figure 3.18(a). The second student then places his or her hand at the zero of the meter stick (the bottom of the stick) with thumb and forefinger extended. The thumb and forefinger should be open about 1 to 2 in. When the first student drops the meter stick, the second student catches it with the thumb and finger [figure 3.18(b)].

As the meter stick is released, it becomes a freely falling body and hence falls a distance y in a time t:

$$y = -\tfrac{1}{2}gt^2$$

The location of the fingers on the meter stick, where the meter stick was caught, represents the distance y that the meter stick has fallen. Solving for the time t we get

$$t = \sqrt{-\frac{2y}{g}} \tag{3.21}$$

Since we have measured y, the distance the meter stick has fallen, and we know the acceleration due to gravity g, we can do the simple calculation in equation 3.21 and determine your reaction time. (Remember that the value of y placed into equation 3.21 will be a negative number and hence we will take the square root of a positive quantity since the square root of a negative number is not defined.)

If you practice catching the meter stick, you will be able to catch it in less time. But eventually you reach a time that, no matter how much you practice, you cannot make smaller. This time is your *minimum reaction time*—the time it takes for your eye to first see the stick drop and then communicate this message to your brain. Your brain then communicates this information through nerves and muscles to your fingers and then you catch the stick. Your *normal reaction time* is most probably the time that you first caught the stick. A normal reaction time to catch the meter stick is about 0.2 to 0.3 seconds.

Note that this is not quite the same reaction time it would take to react to a red light while driving a car, because in that case, part of the communication from the brain would entail lifting your leg from the accelerator, placing it on the brake pedal, and then pressing. The motion of more muscles and mass would consequently take a longer period of time. A normal reaction time in a car is approximately 0.5 s. To obtain a more accurate value of the stopping distance for a car we also need to include the distance that the car moves while the driver reacts to the red light.

3.9 Projectile Motion in One Dimension

A case one step more general than the freely falling body dropped from rest, is the motion of a body that is thrown up or down with an initial velocity v_0 near the surface of the earth. This type of motion is called **projectile motion** in one dimension. Remember, however, that this type of motion still falls into the category of a freely falling body because the object experiences the acceleration g downward throughout its motion. The kinematic equations for projectile motion are

$$y = v_0 t - \tfrac{1}{2}gt^2 \tag{3.22}$$

$$v = v_0 - gt \tag{3.23}$$

$$v^2 = v_0^2 - 2gy \tag{3.24}$$

These three equations completely describe the motion of a projectile in one dimension. Note that these equations are more general than those for the body dropped from rest because they contain the initial velocity v_0. In fact, if v_0 is set equal to zero these equations reduce to the ones studied for the body dropped from rest.

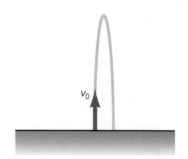

Figure 3.19

Trajectory of a projectile in one dimension.

In the previous cases of motion, we were concerned only with motion in one direction. Here there are two possible directions, up and down. According to our convention the upward direction is positive and the downward direction is negative. Hence, if the projectile is initially thrown upward, v_0 is positive; if the projectile is initially thrown downward, v_0 is negative. Also note that whether the projectile is thrown up or down, the acceleration of gravity always points downward. If it did not, then a ball thrown upward would continue to rise forever and would leave the earth, a result that is contrary to observation.

Let us now consider the motion of a projectile thrown upward. Figure 3.19 shows its path through space, which is called a **trajectory.** The projectile goes straight up, and then straight down. The downward portion of the motion is slightly displaced from the upward portion to clearly show the two different parts of the motion. For example, suppose the projectile is a baseball thrown straight upward with an initial velocity $v_0 = 100$ ft/s. We want to determine

1. The maximum height of the ball.
2. The time it takes for the ball to rise to the top of its trajectory.
3. The total time that the ball is in the air.
4. The velocity of the ball as it strikes the ground.
5. The position and velocity of the ball at any time t, for example, for $t = 4.00$ s.

We are asking for a great deal of information, especially considering that the only data given is the initial position and velocity of the ball. Yet all this information can be obtained using the three kinematic equations 3.22, 3.23, and 3.24. In fact, any time we try to solve any kinematic problem, the first thing is to write down the kinematic equations, because somehow, somewhere, the answers are in those equations. It is just a matter of manipulating them into the right form to obtain the information we want about the motion of the projectile.

Let us now solve the problem of projectile motion in one dimension.

Find the Maximum Height of the Ball

Equation 3.22 tells us the height of the ball at any instant of time. We could find the maximum height if we knew the time for the projectile to rise to the top of its trajectory. But at this point that time is unknown. (In fact, that is question 2.) Equation 3.24 tells us the velocity of the moving body at any height y. The velocity of the ball is positive on the way up, and negative on the way down, so therefore it must have gone through zero somewhere. In fact, the velocity of the ball is zero when the ball is at the very top of its trajectory. If it were greater than zero the ball would continue to rise, if it were less than zero the ball would be on its way down. Therefore, at the top of the trajectory, the position of maximum height, $v = 0$, and equation 3.24,

$$v^2 = v_0^2 - 2gy$$

becomes

$$0 = v_0^2 - 2gy_{max}$$

where y_{max} is the maximum height of the projectile. For any other height y, the velocity is either positive indicating that the ball is on its way up, or negative indicating that it is on the way down. Solving for y_{max}, the maximum height of the ball is

$$2gy_{max} = v_0^2$$

$$y_{max} = \frac{v_0^2}{2g} \tag{3.25}$$

Inserting numbers into equation 3.25, we get

$$y_{max} = \frac{v_0^2}{2g} = \frac{(100 \text{ ft/s})^2}{2(32.0 \text{ ft/s}^2)}$$

$$= 156 \text{ ft}$$

The ball will rise to a maximum height of 156 ft.

Find the Time for the Ball to Rise to the Top of the Trajectory

We have seen that when the projectile is at the top of its trajectory, $v = 0$. Therefore, equation 3.23,

$$v = v_0 - gt$$

becomes

$$0 = v_0 - gt_r$$

where t_r is the time for the projectile to rise to the top of its path. Only at this value of time does the velocity equal zero. At any other time the velocity is either positive or negative, depending on whether the ball is on its way up or down. Solving for t_r we get

$$t_r = \frac{v_0}{g} \tag{3.26}$$

the time for the ball to rise to the top of its trajectory. Inserting numbers into equation 3.26 we obtain

$$t_r = \frac{v_0}{g} = \frac{100 \text{ ft/s}}{32.0 \text{ ft/s}^2}$$

$$= 3.13 \text{ s}$$

It takes 3.13 s for the ball to rise to the top of the trajectory. Notice that the ball has the acceleration $-g$ at the top of the trajectory even though the velocity is zero at that instant. That is, in any kind of motion, we can have a nonzero acceleration even though the velocity is zero. The important thing for an acceleration is the *change in velocity*, not the velocity itself. At the top, the change in velocity is not zero, because the velocity is changing from positive values on the way up, to negative values on the way down.

The time t_r could also have been found using equation 3.22,

$$y = v_0 t - \tfrac{1}{2}gt^2$$

by substituting the maximum height of 156 ft for y. Even though this also gives the correct solution, the algebra and arithmetic are slightly more difficult because a quadratic equation for t would have to be solved.

Find the Total Time that the Object Is in the Air

When t is equal to the total time t_t, that the projectile is in the air, y is equal to zero. That is, during the time from $t = 0$ to $t = t_t$, the projectile goes from the ground to its maximum height and then falls back to the ground. Using equation 3.22, the height of the projectile at any time t,

$$y = v_0 t - \tfrac{1}{2}gt^2$$

with $t = t_t$ and $y = 0$, we get

$$0 = v_0 t_t - \tfrac{1}{2}gt_t^2 \tag{3.27}$$

Solving for t_t we obtain

$$t_t = \frac{2v_0}{g} \qquad (3.28)$$

the total time that the projectile is in the air. Recall from equation 3.26 that the time for the ball to rise to the top of its trajectory is $t_r = v_0/g$. And the total time, equation 3.28, is just twice that value. Therefore, the total time that the projectile is in the air becomes

$$t_t = \frac{2v_0}{g} = 2t_r \qquad (3.29)$$

The total time that the projectile is in the air is twice the time it takes the projectile to rise to the top of its trajectory. Stated in another way, the time for the ball to go up to the top of the trajectory is equal to the time for the ball to come down to the ground.

For this particular problem,

$$t_t = 2t_r = 2(3.13 \text{ s}) = 6.26 \text{ s}$$

The projectile will be in the air for a total of 6.26 s. Also note that equation 3.27 is really a quadratic equation with two roots. One of which we can see by inspection is $t = 0$, which is just the initial moment that the ball is launched.

Find the Velocity of the Ball as It Strikes the Ground

There are two ways to find the velocity of the ball at the ground. The simplest is to use equation 3.24,

$$v^2 = v_0^2 - 2gy$$

noting that the height is equal to zero ($y = 0$) when the ball is back on the ground. Therefore,

$$v_g^2 = v_0^2$$

and

$$v_g = \pm v_0 \qquad (3.30)$$

The two roots represent the velocity at the two times that $y = 0$, namely, when the ball is first thrown up ($t = 0$), with an initial velocity $+v_0$, and when the ball lands ($t = t_t$) with a final velocity of $-v_0$ (the minus sign indicates that the ball is on its way down).

Another way to find the velocity at the ground is to use equation 3.23,

$$v = v_0 - gt$$

which represents the velocity of the projectile at any instant of time. If we let t be the total time that the projectile is in the air (i.e., $t = t_t$), then $v = v_g$, the velocity of the ball at the ground. Thus,

$$v_g = v_0 - gt_t \qquad (3.31)$$

But we have already seen that

$$t_t = \frac{2v_0}{g} \qquad (3.28)$$

Substituting equation 3.28 into equation 3.31 gives

$$v_g = v_0 - \frac{g(2v_0)}{g}$$

Hence,

$$v_g = -v_0$$

The velocity of the ball as it strikes the ground is equal to the negative of the original velocity with which the ball was thrown upward, that is,

$$v_g = -v_0 = -100 \text{ ft/s}$$

Find the Position and Velocity of the Ball at t = 4.00 s

The position of the ball at any time t is given by equation 3.22 as

$$y = v_0 t - \tfrac{1}{2} g t^2$$

Substituting in the values for $t = 4.00$ s gives

$$y_4 = (100 \text{ ft/s})(4.00 \text{ s}) - \tfrac{1}{2}(32.0 \text{ ft/s}^2)(4.00 \text{ s})^2$$

$$= 400 \text{ ft} - 256 \text{ ft}$$

$$= 144 \text{ ft}$$

At $t = 4.00$ s the ball is 144 ft above the ground.

The velocity of the ball at any time is given by equation 3.23 as

$$v = v_0 - gt$$

For $t = 4.00$ s, the velocity becomes

$$v_4 = 100 \text{ ft/s} - (32.0 \text{ ft/s}^2)(4.00 \text{ s})$$
$$= 100 \text{ ft/s} - 128 \text{ ft/s}$$
$$= -28.0 \text{ ft/s}$$

At the end of 4 s the velocity of the ball is -28.0 ft/s, where the negative sign indicates that the ball is on its way down. We could have used equation 3.22 for every value of time and plotted the entire trajectory, as shown in figure 3.20.

There is great beauty and power in these few simple equations, because with them we can completely predict the motion of the projectile for any time, simply by knowing its initial position and velocity. This is a characteristic of the field of physics. First we observe how nature works. Then we make a mathematical model of nature in terms of certain equations. We manipulate these equations until we can make a prediction, and this prediction yields information that we usually have no other way of knowing.

For example, how could you know that the velocity of the ball after 4.00 s would be -28.0 ft/s. In general, there is no way of knowing that. Yet we have actually captured a small piece of nature in our model and have seen how it works.

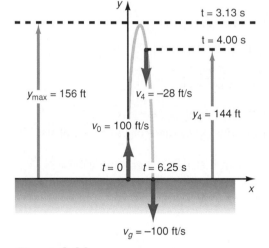

Figure 3.20

Results of projectile motion in one dimension.

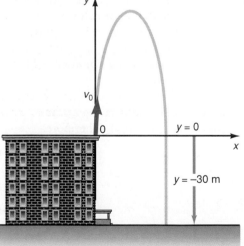

Figure 3.21

A projectile is fired vertically from the top of a building.

Example 3.7

A projectile is fired straight up from the top of a building. A projectile is fired from the top of a building at an initial velocity of 35.0 m/s upward. The top of the building is 30.0 m above the ground. The motion is shown in figure 3.21. Find (a) the maximum height of the projectile, (b) the time for the projectile to reach its maximum height, (c) the velocity of the projectile as it strikes the ground, and (d) the total time that the projectile is in the air.

Solution

We will solve this problem using the techniques just developed.

a. To find the maximum height of the projectile we again note that at the top of the trajectory $v = 0$. Using equation 3.24,

$$v^2 = v_0^2 - 2gy$$

and setting $v = 0$ we obtain

$$0 = v_0^2 - 2gy_{max}$$

Solving for the maximum height,

$$y_{max} = \frac{v_0^2}{2g} = \frac{(35.0 \text{ m/s})^2}{2(9.80 \text{ m/s}^2)}$$

$$= 62.5 \text{ m}$$

The projectile's maximum height is 62.5 m above the roof of the building, or 92.5 m above the ground. Notice that since the initial data was given to us in SI units, we solve the problem in these units and use the value 9.80 m/s² for the acceleration due to gravity.

b. To find the time for the projectile to reach its maximum height we again note that at the maximum height $v = 0$. Substituting $v = 0$ into equation 3.23, we get

$$0 = v_0 - gt_r$$

Solving for the time to rise to the top of the trajectory, we get

$$t_r = \frac{v_0}{g} = \frac{35.0 \text{ m/s}}{9.80 \text{ m/s}^2}$$

$$= 3.57 \text{ s}$$

It takes 3.57 s for the ball to rise from the top of the roof to the top of its trajectory.

c. To find the velocity of the projectile when it strikes the ground, we use equation 3.24. When $y = -30.0$ m the projectile will be on the ground, and its velocity as it strikes the ground is

$$v^2 = v_0^2 - 2gy$$

$$(v_g)^2 = (35.0 \text{ m/s})^2 - 2(9.80 \text{ m/s}^2)(-30.0 \text{ m})$$

$$= 1225 \text{ m}^2/\text{s}^2 + 588 \text{ m}^2/\text{s}^2 = 1813 \text{ m}^2/\text{s}^2$$

$$= -42.6 \text{ m/s}$$

The projectile hits the ground at a velocity of -42.6 m/s. Note that this value is greater than the initial velocity v_0, because the projectile does not hit the roof on its way down, but rather hits the ground 30.0 m below the level of the roof. The acceleration has acted for a longer time, thereby imparting a greater velocity to the projectile.

d. To find the total time that the projectile is in the air we use equation 3.23,

$$v = v_0 - gt$$

But when t is equal to the total time that the projectile is in the air, the velocity is equal to the velocity at the ground (i.e., $v = v_g$). Therefore,

$$v_g = v_0 - gt_t$$

Solving for the total time, we get

$$t_t = \frac{v_0 - v_g}{g}$$

$$= \frac{35.0 \text{ m/s} - (-42.6 \text{ m/s})}{9.80 \text{ m/s}^2}$$

$$= \frac{(35.0 + 42.6)\text{m/s}}{9.80 \text{ m/s}^2}$$

$$= 7.92 \text{ s}$$

The total time that the projectile is in the air is 7.92 s. Note that it is not twice the time for the projectile to rise because the projectile did not return to the level where it started from, but rather to 30.0 m below that level.

3.10 The Kinematic Equations in Vector Form

Up to now we have discussed motion in one dimension only. And although the displacement, velocity, and acceleration of a body are vector quantities, we did not write them in the traditional boldface type, characteristic of vectors. We took into account their vector character by noting that when the displacement, velocity, and acceleration were in the positive x- or y-direction, we considered the quantities positive. When the displacement, velocity, and acceleration were in the negative x- or y-direction, we considered those quantities negative. For two-dimensional motion we must be more general and write the displacement, velocity, and acceleration in boldface type to show their full vector character. Let us now define the kinematic equations in terms of their vector characteristics.

The average velocity of a body is defined as the rate at which the displacement vector changes with time. That is,

$$\mathbf{v_{avg}} = \frac{\Delta \mathbf{r}}{\Delta t} = \frac{\mathbf{r_2} - \mathbf{r_1}}{t_2 - t_1} \tag{3.32}$$

where the letter $\mathbf{r}$ is the displacement vector. The displacement vector $\mathbf{r_1}$ locates the position of the body at the time t_1, while the displacement vector $\mathbf{r_2}$ locates the position of the body at the time t_2. The displacement between the times t_1 and t_2 is just the difference between these vectors, $\mathbf{r_2} - \mathbf{r_1}$, or $\Delta \mathbf{r}$, and is shown in figure 3.22.

We find the instantaneous velocity by taking the limit in equation 3.32 as Δt approaches zero, just as we did in equation 3.8. The magnitude of the instantaneous velocity vector is the instantaneous speed of the body, while the direction of the velocity vector is the direction that the body is moving, which is tangent to the trajectory at that point.

The average acceleration vector is defined as the rate at which the velocity vector changes with time:

$$\mathbf{a} - \frac{\Delta \mathbf{v}}{\Delta t} = \frac{\mathbf{v} - \mathbf{v_0}}{t} \tag{3.33}$$

Since the only cases that we will consider concern motion at constant acceleration, we will not use the subscript avg on $\mathbf{a}$. We find the kinematic equation for the displacement and velocity of the body at any instant of time as in section 3.6, only we write every term except t as a vector:

$$\mathbf{r} = \mathbf{v_0}t + \tfrac{1}{2}\mathbf{a}t^2 \tag{3.34}$$

and

$$\mathbf{v} = \mathbf{v_0} + \mathbf{a}t \tag{3.35}$$

Equation 3.34 represents the vector displacement of the moving body at any time t, while equation 3.35 represents the velocity of the moving body at any time. These vector equations are used to describe the motion of a moving body in two or three directions.

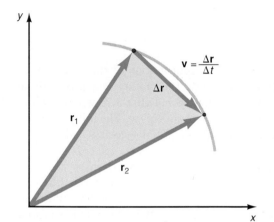

Figure 3.22

The change in the displacement vector.

Figure 3.23

The trajectory of a projectile in two dimensions.

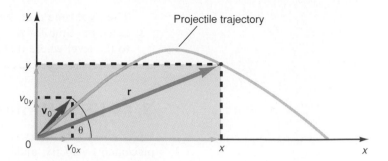

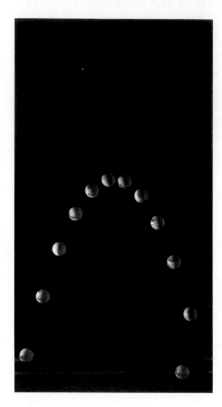

Stroboscopic photograph of a projectile in two dimensions.

3.11 Projectile Motion in Two Dimensions

In the study of kinematics we found that the displacement and velocity of a moving body can be determined if the original velocity $\mathbf{v_0}$ of the body and the acceleration $\mathbf{a}$ acting on it are known. The displacement of the body was given by

$$\mathbf{r} = \mathbf{v_0}t + \tfrac{1}{2}\mathbf{a}t^2 \tag{3.34}$$

while its velocity was given by

$$\mathbf{v} = \mathbf{v_0} + \mathbf{a}t \tag{3.35}$$

These two equations completely describe the resulting motion of the body. As an example of two-dimensional kinematics let us study the motion of a projectile in two dimensions. A projectile is thrown from the point 0 in figure 3.23 with an initial velocity $\mathbf{v_0}$. The trajectory of the projectile is shown in the figure. The initial velocity $\mathbf{v_0}$ has two components: v_{0x}, the x-component of the initial velocity, and v_{0y}, the y-component.

The location of the projectile at any instant of time is given by equation 3.34 and is shown as the displacement vector $\mathbf{r}$ in figure 3.23. We resolve the displacement vector $\mathbf{r}$ into two components: the distance the projectile has moved in the x-direction, we designate as x; the distance (or height) the projectile has moved in the y-direction we designate as y. We can now write the one vector equation 3.34 as two component equations, namely,

$$x = v_{0x}t + \tfrac{1}{2}a_x t^2 \tag{3.36}$$

$$y = v_{0y}t + \tfrac{1}{2}a_y t^2 \tag{3.37}$$

Figure 3.23 shows that v_{0x} is the x-component of the original velocity, given by $v_{0x} = v_0 \cos\theta$, while v_{0y} is the y-component of the original velocity, $v_{0y} = v_0 \sin\theta$. We have resolved the vector acceleration $\mathbf{a}$ into two components a_x and a_y.

In chapter 4 on Newton's laws of motion, we will see that whenever an unbalanced force F acts on a body of mass m, it gives that mass an acceleration a. Because there is no force acting on the projectile in the horizontal x-direction, the acceleration in the x-direction must be zero, that is, $a_x = 0$. Therefore, the x-component of the displacement $\mathbf{r}$ of the projectile, equation 3.36, takes the simple form

$$x = v_{0x}t \tag{3.38}$$

There is, however, a force acting on the projectile in the y-direction, the force of gravity that the earth exerts on any object, directed toward the center of the earth. We define the direction of this gravitational force to be in the negative y-direction. This gravitational force produces a constant acceleration called the acceleration due to gravity g. Hence, the y-component of the acceleration of the projectile is given by $-g$, that is, $a_y = -g$. The y-component of the displacement of the projectile therefore becomes

$$y = v_{0y}t - \tfrac{1}{2}gt^2 \tag{3.39}$$

Using the same arguments, we resolve the velocity **v** at any instant of time, equation 3.35, into the two scalar equations:

$$v_x = v_{0x} \tag{3.40}$$

$$v_y = v_{0y} - gt \tag{3.41}$$

Equation 3.40 does not contain the time t, and therefore the x-component of the velocity v_x is independent of time and is a constant. *Hence, the projectile motion consists of two motions: accelerated motion in the y-direction and motion at constant velocity in the x-direction.*

We can completely describe the motion of the projectile using the four equations, namely,

$$x = v_{0x}t \tag{3.38}$$

$$y = v_{0y}t - \tfrac{1}{2}gt^2 \tag{3.39}$$

$$v_x = v_{0x} \tag{3.40}$$

$$v_y = v_{0y} - gt \tag{3.41}$$

Now let us apply these equations to the projectile motion shown in figure 3.23. We essentially look for the same information that we found for projectile motion in one dimension. Because two-dimensional motion is a superposition of accelerated motion in the y-direction coupled to motion in the x-direction at constant velocity, we can use many of the techniques and much of the information we found in the one-dimensional case.

Let us find (1) the time for the projectile to rise to its maximum height, (2) the total time that the projectile is in the air, (3) the range (or maximum distance in the x-direction) of the projectile, (4) the maximum height of the projectile, (5) the velocity of the projectile as it strikes the ground, and (6) the location and velocity of the projectile at any time t.

To determine this information we use the kinematic equations 3.38 through 3.41.

The Time for the Projectile to Rise to Its Maximum Height

To determine the maximum height of the projectile we use the same reasoning used for the one-dimensional case. As the projectile is moving upward it has some positive vertical velocity v_y. When it is coming down it has some negative vertical velocity $-v_y$. At the very top of the trajectory, $v_y = 0$.

Therefore, at the top of the trajectory, equation 3.41 becomes

$$0 = v_{0y} - gt_r \tag{3.42}$$

Figure 3.24

A punted football is an example of a projectile in two dimensions.

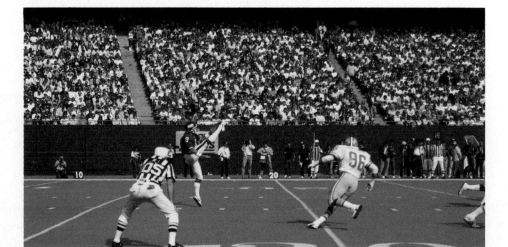

Note that this is very similar to the equation for the one-dimensional case, except for the subscript y on v_0. This is an important distinction between the two motions, because the initial velocity upward v_{0y} is now less than the initial velocity upward v_0 in the one-dimensional case. Solving equation 3.42 for the time to rise to the top of the trajectory t_r, we get

$$t_r = \frac{v_{0y}}{g} \tag{3.43}$$

Since we know v_0 and hence v_{0y}, and because g is a constant, we can immediately compute t_r.

The Total Time the Projectile Is in the Air

To find the total time that the projectile is in the air, we use equation 3.39. When t is the total time t_t, the projectile is back on the ground and the height of the projectile is zero, $y = 0$. Therefore,

$$0 = v_{0y}t_t - \tfrac{1}{2}gt_t^2$$

Solving for the total time that the projectile is in the air, we get

$$t_t = \frac{2v_{0y}}{g} \tag{3.44}$$

But using equation 3.43 for the time to rise, $t_r = v_{0y}/g$, the total time that the projectile is in the air is exactly double this value,

$$t_t = \frac{2v_{0y}}{g} = 2t_r \tag{3.45}$$

which is the same as the one-dimensional case, as expected.

The Range of the Projectile

The **range of a projectile** is defined as the horizontal distance from the point where the projectile is launched to the point where it returns to its launch height. In this case, the range is the maximum distance that the projectile moves in the x-direction before it hits the ground. Because the maximum horizontal distance is the product of the horizontal velocity, which is a constant, and the total time of flight, the range, becomes

$$\text{range} = R = x_{\text{max}} = v_{0x}t_t \tag{3.46}$$

Sometimes it is convenient to express the range in another way. Since $v_{0x} = v_0 \cos\theta$, and the total time in the air is

$$t_t = \frac{2v_{0y}}{g} = \frac{2v_0 \sin\theta}{g}$$

we substitute these values into equation 3.46 to obtain

$$R = \frac{(v_0 \cos\theta)(2v_0 \sin\theta)}{g} = \frac{v_0^2 2 \sin\theta \cos\theta}{g}$$

However, using the well-known trigonometric identity,

$$2 \sin\theta \cos\theta = \sin 2\theta$$

the range of the projectile becomes

$$R = \frac{v_0^2 \sin 2\theta}{g} \tag{3.47}$$

We derived equation 3.47 based on the assumption that the initial and final elevations are the same, and we can use it only in problems where this assumption holds. This formulation of the range is particularly useful when we want to know at what angle a projectile should be fired in order to get the maximum possible range. From equation 3.47 we can see that for a given initial velocity v_0, the maximum range depends on the sine function. Because the sine function varies between -1 and $+1$, the maximum value occurs when $\sin 2\theta = 1$. But this happens when $2\theta = 90°$, hence the maximum range occurs when $\theta = 45°$. We obtain the maximum range of a projectile by firing it at an angle of $45°$.

The Maximum Height of the Projectile

We can find the maximum height of the projectile by substituting the time t_r into equation 3.39 and solving for the maximum height. However, since it is useful to have an equation for vertical velocity as a function of the height, we will use an alternate solution. Equation 3.39 represents the y-component of the displacement of the projectile at any instant of time and equation 3.41 is the y-component of the velocity at any instant of time. If the time is eliminated between these two equations (exactly as it was in section 3.6, for equation 3.16), we obtain the kinematic equation

$$v_y^2 - v_{0y}^2 - 2gy \tag{3.48}$$

which gives the y-component of the velocity of the moving body at any height y.

When the projectile has reached its maximum height, $v_y = 0$. Therefore, equation 3.48 becomes

$$0 = v_{0y}^2 - 2gy_{max}$$

Solving for y_{max} we obtain

$$y_{max} = \frac{v_{0y}^2}{2g} \tag{3.49}$$

the maximum height of the projectile.

The Velocity of the Projectile as It Strikes the Ground

The velocity of the projectile as it hits the ground $\mathbf{v_g}$ can be described in terms of its components, as shown in figure 3.25. The x-component of the velocity at the ground, found from equation 3.40, is

$$v_{xg} = v_x = v_{0x} \tag{3.50}$$

The y-component of the velocity at the ground, found from equation 3.41 with $t = t_t$ is

$$v_{yg} = v_{0y} - gt_t \tag{3.51}$$

$$= v_{0y} - \frac{g(2v_{0y})}{g}$$

$$v_{yg} = -v_{0y} \tag{3.52}$$

The y-component of the velocity of the projectile at the ground is equal to the negative of the y-component of the original velocity. The minus sign just indicates that the projectile is coming down. But this is exactly what we expected from the study of one-dimensional motion. The magnitude of the actual velocity at the ground, found from its two components, is

$$v_g = \sqrt{(v_{xg})^2 + (v_{yg})^2} \tag{3.53}$$

and using equations 3.50 and 3.52, becomes

$$v_g = \sqrt{(v_{0x})^2 + (-v_{0y})^2} = v_0 \tag{3.54}$$

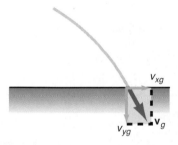

Figure 3.25

The velocity of the projectile at the ground.

The speed of the projectile as it strikes the ground is equal to the original speed of the projectile. The direction that the velocity vector makes with the ground is

$$\theta = \tan^{-1} \frac{v_{yg}}{v_{xg}} = \tan^{-1} -\frac{v_{0y}}{v_{0x}} = -\theta$$

The angle that the velocity vector makes as it hits the ground is the negative of the original angle. That is, if the projectile was fired at an original angle of 30° above the positive x-axis, it will make an angle of 30° below the positive x-axis when it hits the ground.

The Location and Velocity of the Projectile at Any Time t

We find the position and velocity of the projectile at any time t by substituting that value of t into equations 3.38, 3.39, 3.40, and 3.41. Let us look at some examples of projectile motion.

Example 3.8

Projectile motion in two dimensions. A ball is thrown with an initial velocity of 100 ft/s at an angle of 60.0° above the horizontal, as shown in figure 3.26. Find (a) the maximum height of the ball, (b) the time to rise to the top of the trajectory, (c) the total time the ball is in the air, (d) the range of the ball, (e) the velocity of the ball as it strikes the ground, and (f) the position and velocity of the ball at $t = 4$ s.

Solution

The x-component of the initial velocity is

$$v_{0x} = v_0 \cos \theta = (100 \text{ ft/s}) \cos 60° = 50.0 \text{ ft/s}$$

The y-component of the initial velocity is

$$v_{0y} = v_0 \sin \theta = (100 \text{ ft/s}) \sin 60.0° = 86.6 \text{ ft/s}$$

a. The maximum height of the ball, found from equation 3.49, is

$$y_{max} = \frac{v_{0y}^2}{2g} = \frac{(86.6 \text{ ft/s})^2}{2(32.0 \text{ ft/s}^2)} = 117 \text{ ft}$$

b. To find the time to rise to the top of the trajectory, we use equation 3.43,

$$t_r = \frac{v_{0y}}{g} = \frac{86.6 \text{ ft/s}}{32.0 \text{ ft/s}^2} = 2.71 \text{ s}$$

c. To find the total time that the ball is in the air, we use equation 3.45,

$$t_t = 2t_r = 2(2.71 \text{ s}) = 5.42 \text{ s}$$

d. The range of the ball, found from equation 3.47, is

$$R = \frac{v_0^2 \sin 2\theta}{g} = \frac{(100 \text{ ft/s})^2 \sin 120°}{32.0 \text{ ft/s}^2} = 271 \text{ ft}$$

Figure 3.26

Trajectory of thrown ball.

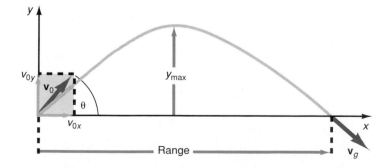

Mechanics

As a check, we can use equation 3.46 to get

$$R = x_{max} = v_{0x}t_t = (50.0 \text{ ft/s})(5.42 \text{ s}) = 271 \text{ ft}$$

e. To find the magnitude of the velocity of the ball at the ground, we use equation 3.53,

$$v_g = \sqrt{(v_{xg})^2 + (v_{yg})^2}$$

where

$$v_{xg} = v_{0x} = 50.0 \text{ ft/s}$$

and

$$v_{yg} = v_{0y} - gt_t = 86.6 \text{ ft/s} - (32 \text{ ft/s}^2)(5.42 \text{ s})$$
$$= 86.6 \text{ ft/s} - 173.4 \text{ ft/s} = -86.8 \text{ ft/s}$$

Hence,

$$v_g = \sqrt{(50.0 \text{ ft/s})^2 + (-86.8 \text{ ft/s})^2}$$
$$= 100 \text{ ft/s}$$

The direction that the velocity vector makes with the ground is

$$\theta = \tan^{-1}\frac{v_{yg}}{v_{xg}} = \tan^{-1}\frac{-86.8 \text{ ft/s}}{50.0 \text{ ft/s}} = -60.0°$$

f. To find the position and velocity of the ball at $t = 4$ s we use the kinematic equations 3.38 through 3.41.

1. $x = v_{0x}t = (50.0 \text{ ft/s})(4 \text{ s}) = 200 \text{ ft}$

2. $y = v_{0y}t - \frac{1}{2}gt^2$

$$= (86.6 \text{ ft/s})(4 \text{ s}) - \frac{1}{2}(32.0 \text{ ft/s}^2)(4 \text{ s})^2$$

$$= 90.4 \text{ ft}$$

The ball is 200 ft down range and is 90.4 ft high.
The components of the velocity at 4 s are

3. $v_x = v_{0x} = 50.0 \text{ ft/s}$

4. $v_y = v_{0y} - gt$

$$= 86.6 \text{ ft/s} - (32.0 \text{ ft/s}^2)(4 \text{ s})$$

$$= -41.4 \text{ ft/s}$$

At the end of 4 s the x-component of the velocity is 50.0 ft/s and the y-component is -41.4 ft/s. To determine the magnitude of the velocity vector at 4 s we have

$$v = \sqrt{(v_x)^2 + (v_y)^2}$$
$$= \sqrt{(50.0 \text{ ft/s})^2 + (-41.4 \text{ ft/s})^2}$$
$$= 64.9 \text{ ft/s}$$

The direction of the velocity vector at 4 s is

$$\theta = \tan^{-1}\frac{v_y}{v_x} = \tan^{-1}\frac{-41.4 \text{ ft/s}}{50.0 \text{ ft/s}} = -39.6°$$

The velocity vector makes an angle of 39.6° below the horizontal at 4 s.

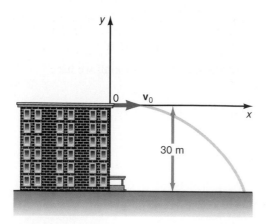

Figure 3.27

Trajectory of projectile thrown horizontally.

Example 3.9

A projectile is fired horizontally from the roof of a building. A projectile is fired horizontally from the roof of a building 30.0 m high at an initial velocity of 20.0 m/s, as shown in figure 3.27. Find (a) the total time the projectile is in the air, (b) where the projectile will hit the ground, and (c) the velocity of the projectile as it hits the ground.

Solution

The x- and y-components of the velocity are

$$v_{0x} = v_0 = 20.0 \text{ m/s}$$

$$v_{0y} = 0$$

a. To find the total time that the projectile is in the air, we use equation 3.39,

$$y = v_{0y}t - \tfrac{1}{2}gt^2$$

However, the initial conditions are that $v_{0y} = 0$. Therefore,

$$y = -\tfrac{1}{2}gt^2$$

Solving for t,

$$t = \sqrt{\frac{-2y}{g}}$$

However, when $t = t_t$, $y = -30.0$ m. Hence,

$$t_t = \sqrt{\frac{-2y}{g}} = \sqrt{\frac{-2(-30.0 \text{ m})}{9.80 \text{ m/s}^2}}$$

$$= 2.47 \text{ s}$$

b. To find where the projectile hits the ground, we use equation 3.38,

$$x = v_{0x}t$$

Now the projectile hits the ground when $t = t_t$, therefore,

$$x = v_{0x}t_t = (20.0 \text{ m/s})(2.47 \text{ s}) = 49.4 \text{ m}$$

The projectile hits the ground at the location $y = -30.0$ m and $x = 49.4$ m.

c. To find the velocity of the projectile at the ground we use equations 3.50, 3.51, and 3.53:

$$v_{xg} = v_{0x} = v_0 = 20.0 \text{ m/s}$$

$$v_{yg} = v_{0y} - gt_t = 0 - (9.80 \text{ m/s}^2)(2.47 \text{ s}) = -24.2 \text{ m/s}$$

$$v_g = \sqrt{v_{xg}^2 + v_{yg}^2}$$

$$= \sqrt{(20.0 \text{ m/s})^2 + (-24.2 \text{ m/s})^2}$$

$$= 31.4 \text{ m/s}$$

The direction that the velocity vector makes with the ground is

$$\theta = \tan^{-1}\frac{v_{yg}}{v_{xg}} = \tan^{-1}\frac{-24.2 \text{ m/s}}{20.0 \text{ m/s}} = -50.4°$$

The velocity vector makes an angle of 50.4° below the horizontal when the projectile hits the ground.

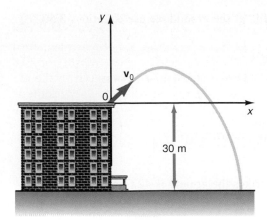

Figure 3.28

Trajectory of a projectile fired from the roof of a building.

Example 3.10

A projectile is fired at an angle from the roof of a building. A projectile is fired at an initial velocity of 35.0 m/s at an angle of 30.0° above the horizontal from the roof of a building 30.0 m high, as shown in figure 3.28. Find (a) the maximum height of the projectile, (b) the time to rise to the top of the trajectory, (c) the total time that the projectile is in the air, (d) the velocity of the projectile at the ground, and (e) the range of the projectile.

Solution

The x- and y-components of the original velocity are

$$v_{0x} = v_0 \cos \theta = (35.0 \text{ m/s}) \cos 30° = 30.3 \text{ m/s}$$

$$v_{0y} = v_0 \sin \theta = (35.0 \text{ m/s}) \sin 30° = 17.5 \text{ m/s}$$

a. To find the maximum height we use equation 3.49:

$$y_{max} = \frac{v_{0y}^2}{2g} = \frac{(17.5 \text{ m/s})^2}{2(9.80 \text{ m/s}^2)}$$

$$= 15.6 \text{ m}$$

above the building. Since the building is 30 m high, the maximum height with respect to the ground is 45.6 m.

b. To find the time to rise to the top of the trajectory we use equation 3.43:

$$t_r = \frac{v_{0y}}{g} = \frac{17.5 \text{ m/s}}{9.80 \text{ m/s}^2} = 1.79 \text{ s}$$

c. To find the total time the projectile is in the air we use equation 3.39:

$$y = v_{0y}t - \tfrac{1}{2}gt^2$$

When $t = t_t$, $y = -30.0$ m. Therefore,

$$-30.0 \text{ m} = (17.5 \text{ m/s})t_t - \tfrac{1}{2}(9.80 \text{ m/s}^2)t_t^2$$

Rearranging the equation, we get

$$4.90 \, t_t^2 - 17.5 \, t_t - 30.0 = 0$$

The units have been temporarily left out of the equation to simplify the following calculations. This is a quadratic equation of the form

$$ax^2 + bx + c = 0$$

with the solution

$$x = \frac{-b \pm \sqrt{b^2 - 4ac}}{2a}$$

In this problem, $x = t_t$, $a = 4.90$, $b = -17.5$, and $c = -30.0$. Therefore,

$$t_t = \frac{+17.5 \pm \sqrt{(17.5)^2 - 4(4.90)(-30.0)}}{2(4.90)}$$

$$= \frac{+17.5 \pm 29.9}{9.80} = 4.84 \text{ s}$$

The total time that the projectile is in the air is 4.84 s. If we had solved the equation for the negative root, we would have found a time of -1.27 s. This corresponds to a time when the height is -30.0 meters but it is a time before the projectile was thrown. If the projectile had been thrown from the ground it would have taken 1.27 seconds to reach the roof.

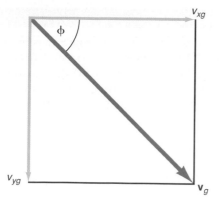

Figure 3.29

Angle of velocity vector as it strikes the ground.

d. To find the velocity of the projectile at the ground we use equations 3.50, 3.51, and 3.53:

$$v_{xg} = v_{0x} = 30.3 \text{ m/s}$$

$$v_{yg} = v_{0y} - gt_t = 17.5 \text{ m/s} - (9.80 \text{ m/s}^2)(4.84 \text{ s})$$

$$= -29.9 \text{ m/s}$$

$$v_g = \sqrt{(v_{xg})^2 + (v_{yg})^2}$$

$$= \sqrt{(30.3 \text{ m/s})^2 + (29.9 \text{ m/s})^2}$$

$$= 42.6 \text{ m/s}$$

The speed of the projectile as it strikes the ground is 42.6 m/s. The angle that the velocity vector makes with the ground, found from figure 3.29, is

$$\tan \phi = \frac{v_{yg}}{v_{xg}}$$

$$\phi = \tan^{-1}\frac{v_{yg}}{v_{xg}} = \tan^{-1}\left(\frac{-29.9}{30.3}\right)$$

$$= -44.6°$$

The velocity vector makes an angle of 44.6° below the horizontal when the projectile hits the ground.

e. To find the range of the projectile we use equation 3.46:

$$x_{max} = v_{0x}t_t = (30.3 \text{ m/s})(4.84 \text{ s})$$

$$= 147 \text{ m}$$

"Have you ever wondered . . . ?"

An Essay on the Application of Physics

Kinematics and Traffic Congestion

Figure 1

Does your expressway look like this?

Have you ever wondered, while sitting in heavy traffic on the expressway, as shown in figure 1, why there is so much traffic congestion? The local radio station tells you there are no accidents on the road, the traffic is heavy because of volume. What does that mean? Why can't cars move freely on the expressway? Why call it an expressway, if you have to move so slowly?

Let us apply some physics to the problem to help understand it. In particular, we will make a simplified model to help analyze the traffic congestion. In this model, we assume that the total length of the expressway L is 10,000 ft (approximately two miles), the length of the car x_c is 10 ft, and the speed of the car v_0 is 55 mph. How many cars of this size can safely fit on this expressway if they are all moving at 55 mph?

First, we need to determine the safe distance required for each car. If the car is moving at 55 mph (80.7 ft/s), and the car is capable of decelerating at -18.0 ft/s^2, the distance required to stop the car is found from equation 3.16,

$$v^2 = v_0^2 + 2ax$$

by noting that $v = 0$ when the car comes to a stop. Solving for the distance x_d that the car moves while decelerating to a stop we get

$$x_d = \frac{-v_0^2}{2a} = \frac{-(80.7 \text{ ft/s})^2}{2(-18 \text{ ft/s}^2)} = 181 \text{ ft}$$

Before the actual deceleration, the car will move, during the reaction time, a distance x_R given by

$$x_R = v_0 t_R = (80.7 \text{ ft/s})(0.500 \text{ s}) = 40.4 \text{ ft}$$

where we assume that it takes the driver 0.500 s to react. The total distance ΔL needed for each car on the expressway to safely come to rest is

$$\Delta L = x_c + x_R + x_d = 10 \text{ ft} + 40.4 \text{ ft} + 181 \text{ ft} = 231 \text{ ft}$$

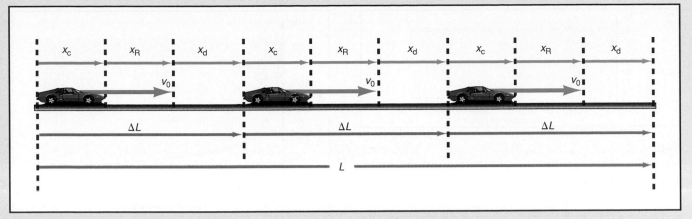

Figure 2

The number of cars on an expressway.

Because it takes a safe distance ΔL for one car to come to rest, N cars will take a distance of $N\Delta L$. The total length of the road L can then hold N cars, each requiring a distance ΔL to stop, as seen in figure 2. Stated mathematically this is

$$L = N\Delta L \qquad \textbf{(H3.1)}$$

Therefore, the number of cars N that can safely fit on this road is

$$N = \frac{L}{\Delta L} = \frac{10{,}000 \text{ ft}}{231 \text{ ft}} = 43 \text{ cars}$$

Thus for a road 10,000 ft long, only 43 cars can fit safely on it when each is moving at 55 mph. If the number of cars on the road doubles, then the safe distance per car ΔL must be halved because the product of N and ΔL must equal L the total length of the road, which is a constant. Rewriting equation H3.1 in the form

$$\Delta L = \frac{L}{N}$$

$$= x_c + x_R + x_d = \frac{L}{N}$$

$$x_c + v_0 t_R - \frac{v_0^2}{2a} = \frac{L}{N} \qquad \textbf{(H3.2)}$$

Notice in equation H3.2 that if the number of cars N increases, the only thing that can change on this fixed length road is the initial velocity v_0 of each car. That is, by increasing the number of cars on the road, the velocity of each car must decrease, in order for each car to move safely. Equation H3.2 can be written in the quadratic form

$$-\frac{v_0^2}{2a} + v_0 t_R + x_c - \frac{L}{N} = 0$$

which can be solved quadratically to yield

$$v_0 = a t_R \mp \sqrt{(a t_R)^2 + 2a(x_c - L/N)} \qquad \textbf{(H3.3)}$$

Equation H3.3 gives the maximum velocity that N cars can safely travel on a road L ft long. (Don't forget that a is a negative number.) Using the same numerical values of a, t_R, x_c, and L as

above, equation H3.3 is plotted in figure 3 to show the safe velocity (in miles per hour) for cars on an expressway as a function of the number of cars on that expressway. Notice from the form of the curve that as the number of cars increases, the safe velocity decreases. As the graph shows, increasing the number of cars on the road to 80, decreases the safe velocity to 38 mph. A further increase in the number of cars on the road to 200, decreases the safe velocity to 20 mph.

Hence, when that radio announcer says, "There is no accident on the road, the heavy traffic comes from volume," he means that by increasing the number of cars on the road, the safe velocity of each car must decrease.

You might wonder if there is some optimum number of cars that a road can handle safely. We can define the capacity C of a road as the number of cars that pass a particular place per unit time. Stated mathematically, this is

$$C = \frac{N}{t} \qquad \textbf{(H3.4)}$$

From the definition of velocity, the time for N cars to pass through a distance L, when moving at the velocity v_0, is

$$t = \frac{L}{v_0}$$

Substituting this into equation H3.4 gives

$$C = \frac{N}{t} = \frac{N}{L/v_0} = \frac{v_0}{L/N}$$

Substituting from equation H3.2 for L/N, the capacity of the road is

$$C = \frac{v_0}{x_c + v_0 t_R - v_0^2/2a} \qquad \textbf{(H3.5)}$$

Using the same values for x_c, t_R, and a as before, equation H3.5 is plotted in figure 4. The number of cars per hour that the road can hold is on the y-axis, and the speed of the cars in miles per hour is on the x-axis. Notice that at a speed of 60 mph, the road can handle 1200 cars per hour. By decreasing the speed of the cars, the number of cars per hour that the road can handle increases. As shown in the figure, if the speed decreases to 40 mph the road can handle about 1600 cars per hour. Notice that the curve peaks at a speed of about 13 mph, allowing about 2300

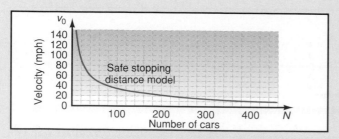

Figure 3

Plot of the velocity of cars (y-axis) as a function of the number of cars on the expressway (x-axis).

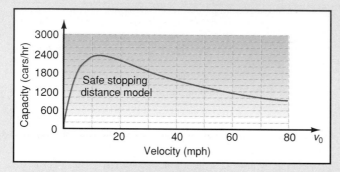

Figure 4

The capacity of a road as a function of the velocity of the cars.

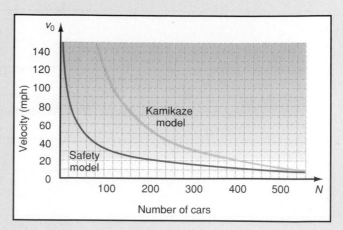

Figure 5

Comparison of traffic with the safe stopping distance model and the kamikaze model.

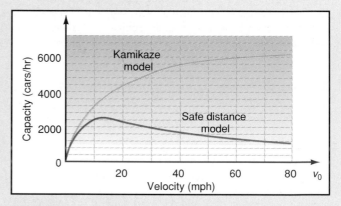

Figure 6

Comparison of the capacity versus velocity for the safe distance model and the kamikaze model.

cars/hour to flow on the expressway. Thus, according to this model, the optimum speed to pass the greatest number of cars per hour is only 13 mph. Hence, even though the road may be called an expressway, if the volume of cars increases significantly, the cars are not going to travel very rapidly. The solution to the problem is to build more lanes to handle the increased volume.

It should also be emphasized that this model is based on safe driving intervals between cars. If an object were to drop from the back of a truck you are following, you would need the safe distance to stop in time to avoid hitting the object. On the other hand, if the car in front of you, also traveling at 55 mph, has to stop, and if both drivers have the same reaction time and both cars decelerate at the same rate, then both cars will need 231 ft to come to a stop. Hence, when both cars come to a stop they will still be separated by the distance of 231 ft. For this reason, in areas of very heavy traffic, many people do not leave the safe distance between them and the car in front. Instead, they get closer and closer to the car in front of them until they are only separated by the reaction distance x_R. I call this the kamikaze model, for obvious reasons. The kamikaze model allows more cars to travel at a greater velocity than are allowed by the safe stopping distance model. The velocity of the cars as a function of the number of cars is found by solving equation H3.2 with the v_0^2 term, which is the term associated with the deceleration distance x_d set equal to zero. The result is shown in figure 5, which compares the safe stopping distance model with the kamikaze model. Notice that many more cars can now fit on the

road. For example, in the safe stopping model, only 40 cars, each traveling at 60 mph, can fit safely on this road. In the kamikaze model about 185 cars can fit on this road, but certainly not safely. There will be only 44 ft between each car, and if you have a slower reaction time than that of the car in front of you, you will almost certainly hit him when he steps on the brakes. This is the reason why there are so many rear-end collisions on expressways. The number of cars on a real expressway falls somewhere between the extremes of these two models. Note that even in the kamikaze model, the velocity of the cars must decrease with volume.

The capacity of the expressway for the kamikaze model is found by setting the v_0^2 term in equation H3.5 to zero. The result is shown in figure 6. Notice that in the kamikaze model the capacity increases with velocity, and there is no optimum speed for the maximum car flow. In practice, the actual capacity of an expressway lies somewhere between these two extremes.

In conclusion, if your expressway is not much of an expressway, it is time to petition your legislators to allocate more money for the widening of the expressway, or maybe it is time to move to a less populated part of the country.

The Language of Physics

Kinematics
The branch of mechanics that describes the motion of a body without regard to the cause of that motion (p. 39).

Average velocity
The average rate at which the displacement vector changes with time. Since a displacement is a vector, the velocity is also a vector (p. 39).

Average speed
The distance that a body moves per unit time. Speed is a scalar quantity (p. 41).

Constant velocity
A body moving in one direction in such a way that it always travels equal distances in equal times (p. 42).

Acceleration
The rate at which the velocity of a moving body changes with time (p. 43).

Instantaneous velocity
The velocity at a particular instant of time. It is defined as the limit of the ratio of the change in the displacement of the body to the change in time, as the time interval approaches zero. The magnitude of the instantaneous velocity is the instantaneous speed of the moving body (p. 45).

Kinematic equations of linear motion
A set of equations that gives the displacement and velocity of the moving body at any instant of time, and the velocity of the moving body at any displacement, if the acceleration of the body is a constant (p. 46).

Freely falling body
Any body that is moving under the influence of gravity only. Hence, any body that is dropped or thrown on the surface of the earth is a freely falling body (p. 51).

Acceleration due to gravity
If air friction is ignored, all objects that are dropped near the surface of the earth, are accelerated toward the center of the earth with an acceleration of 9.80 m/s^2 or 32 ft/s^2 (p. 52).

Projectile motion
The motion of a body thrown or fired with an initial velocity v_0 in a gravitational field (p. 55).

Trajectory
The path through space followed by a projectile (p. 56).

Range of a projectile
The horizontal distance from the point where the projectile is launched to the point where it returns to its launch height (p. 64).

Summary of Important Equations

Average velocity
$$\mathbf{v}_{avg} = \frac{\Delta \mathbf{r}}{\Delta t} = \frac{\mathbf{r}_2 - \mathbf{r}_1}{t_2 - t_1} \tag{3.32}$$

Acceleration
$$\mathbf{a} = \frac{\Delta \mathbf{v}}{\Delta t} = \frac{\mathbf{v} - \mathbf{v}_0}{t} \tag{3.33}$$

Instantaneous velocity in two or more directions, which is a generalization of the instantaneous velocity in one dimension
$$\mathbf{v} = \lim_{\Delta t \to 0} \frac{\Delta \mathbf{r}}{\Delta t}$$
$$v = \lim_{\Delta t \to 0} \frac{\Delta x}{\Delta t} \tag{3.8}$$

Velocity at any time
$$\mathbf{v} = \mathbf{v}_0 + \mathbf{a}t \tag{3.35}$$

Displacement at any time
$$\mathbf{r} = \mathbf{v}_0 t + \frac{1}{2}\mathbf{a}t^2 \tag{3.34}$$

Velocity at any displacement in the x-direction
$$v^2 = v_0^2 + 2ax \tag{3.16}$$

Velocity at any displacement in the y-direction
$$v^2 = v_0^2 + 2ay \tag{3.16}$$

For Projectile Motion

x-displacement
$$x = v_{0x}t \tag{3.38}$$

y-displacement
$$y = v_{0y}t - \frac{1}{2}gt^2 \tag{3.39}$$

x-component of velocity
$$v_x = v_{0x} \tag{3.40}$$

y-component of velocity
$$v_y = v_{0y} - gt \tag{3.41}$$

y-component of velocity at any height y
$$v_y^2 = v_{0y}^2 - 2gy \tag{3.48}$$

Range
$$R = \frac{v_0^2 \sin 2\theta}{g} \tag{3.47}$$

Questions for Chapter 3

1. Discuss the difference between distance and displacement.
2. Discuss the difference between speed and velocity.
3. Discuss the difference between average speed and instantaneous speed.
† 4. Although speed is the magnitude of the instantaneous velocity, is the average speed equal to the magnitude of the average velocity?
5. Why can the kinematic equations be used only for motion at constant acceleration?
6. In dealing with average velocities discuss the statement, "Straight line motion at 60 mph for 1 hr followed by motion in the same direction at 30 mph for 2 hr does not give an average of 45 mph but rather 40 mph."
7. What effect would air resistance have on the velocity of a body that is dropped near the surface of the earth?
8. What is the acceleration of a projectile when its instantaneous vertical velocity is zero at the top of its trajectory?
9. Can an object have zero velocity at the same time that it has an acceleration? Explain and give some examples.
10. Can the velocity of an object be in a different direction than the acceleration? Give some examples.
11. Can you devise a means of using two clocks to measure your reaction time?
†12. A person on a moving train throws a ball straight upward. Describe the motion as seen by a person on the train and by a person on the station platform.
13. You are in free fall, and you let go of your watch. What is the relative velocity of the watch with respect to you?
†14. What kind of motion is indicated by a graph of displacement versus time, if the slope of the curve is (a) horizontal, (b) sloping upward to the right, and (c) sloping downward?
†15. What kind of motion is indicated by a graph of velocity versus time, if the slope of the curve is (a) horizontal, (b) sloping upward at a constant value, (c) sloping upward at a changing rate, (d) sloping downward at a constant value, and (e) sloping downward at a changing rate?

Hints for Problem Solving

To be successful in a physics course it is necessary to be able to solve problems. The following procedure should prove helpful in solving the physics problems assigned. First, as a preliminary step, read the appropriate topic in the textbook. Do not attempt to solve the problems before doing this. Look at the appropriate illustrative problems to see how they are solved. With this background, now read the assigned problem. Now continue with the following procedure.

1. Draw a small picture showing the details of the problem. This is very useful so that you do not lose sight of the problem that you are trying to solve.
2. List all the information that you are given.
3. List all the answers you are expected to find.
4. From the summary of important equations or the text proper, list the equations that are appropriate to this topic.
5. Pick the equation that relates the variables that you are given.
6. Place a check mark ($\checkmark$) over each variable that is given and a question mark (?) over each variable that you are looking for.
7. Solve the equation for the unknown variable.
8. When the answer is obtained check to see if the answer is reasonable.

Let us apply this technique to the following example.

A car is traveling at 30 ft/s when it starts to accelerate at 10 ft/s². Find (a) the velocity and (b) the displacement of the car at the end of 5 s.

1. Draw a picture of the problem.

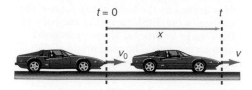

$t = 0$ t

x

v_0 v

2. Given: $v_0 = 30$ ft/s
$$a = 10 \text{ ft/s}^2$$
$$t = 5 \text{ s}$$
3. Find: $v = ?$
$$x = ?$$
4. The problem is one in kinematics and the kinematic equations apply. That is,

$$(1) \quad x = v_0 t + \tfrac{1}{2}at^2$$
$$(2) \quad v = v_0 + at$$
$$(3) \quad v^2 = v_0^2 + 2ax$$

5. Part a of the problem. To solve for the velocity v, we need an equation containing v. Equation 1 does not contain a velocity term v, and hence can not be used to solve for the velocity. Equations 2 and 3, on the other hand, both contain v. Thus, we can use one or possibly both of these equations to solve for the velocity.
6. Write down the equation and place a check mark over the known terms and a question mark over the unknown terms:

$$\overset{?}{} \quad \overset{\checkmark}{} \quad \overset{\checkmark\checkmark}{}$$
$$(2) \quad v = v_0 + at$$

The only unknown in equation 2 is the velocity v and we can now solve for it.

7. The velocity after 5 s, found from equation 2 is

$$v = v_0 + at$$
$$= 30 \text{ ft/s} + (10 \text{ ft/s}^2)(5 \text{ s})$$
$$= 30 \text{ ft/s} + 50 \text{ ft/s}$$
$$= 80 \text{ ft/s}$$

Notice what would happen if we tried to use equation 3 at this time:

$$\overset{?}{} \quad \overset{\checkmark}{} \quad \overset{\checkmark\checkmark?}{}$$
$$(3) \quad v^2 = v_0^2 + 2ax$$

We can not solve for the velocity v from equation 3 because there are two unknowns, both v and x. However, if we had solved part b of the problem for x first, then we could have used this equation.

5. Part b of the problem. To solve for the displacement x, we need an equation containing x. Notice that equation 2 does not contain x, so we can not use it. Equations 1 and 3, on the other hand, do contain x, and we can use either to solve for x.
6. Looking at equation 1, we have

$$\overset{?}{} \quad \overset{\checkmark\checkmark}{} \quad \overset{\checkmark\checkmark\checkmark}{}$$
$$(1) \quad x = v_0 t + \tfrac{1}{2}at^2$$

7. Solving for the only unknown in equation 1, x, we get

$$x = v_0 t + \tfrac{1}{2}at^2$$
$$= (30 \text{ ft/s})(5 \text{ s}) +$$
$$\tfrac{1}{2}(10 \text{ ft/s}^2)(5 \text{ s})^2$$
$$= 150 \text{ ft} + 125 \text{ ft}$$
$$= 275 \text{ ft}$$

Note that at this point we could also have used equation 3 to determine x, because we already found the velocity v in part a of the problem.

Problems for Chapter 3

3.1 Introduction

1. A driver travels 300 mi in 5 hr and 25 min. What is his average speed in (a) mph, (b) ft/s, (c) km/hr, and (d) m/s?
2. A car travels at 40 mph for 2 hr and 60 mph for 3 hr. What is its average speed?
3. A man hears the sound of thunder 5 s after he sees the lightning flash. If the speed of sound in air is 343 m/s, how far away is the lightning? Assume that the speed of light is so large that the lightning was seen essentially at the same time that it was created.

4. The earth-moon distance is 3.84×10^8 m. If it takes 3 days to get to the moon, what is the average speed?
5. Electronic transmission is broadcast at the speed of light, which is 3.00×10^8 m/s. How long would it take for a radio transmission from earth to an astronaut orbiting the planet Mars? Assume that at the time of transmission the distance from earth to Mars is 7.80×10^7 km.
6. In the game of baseball, some excellent fast-ball pitchers have managed to pitch a ball at approximately 100 mph. If the pitcher's mound is 60.5 ft from home

plate, how long does it take the ball to get to home plate? If the pitcher then throws a change-of-pace ball (a slow ball) at 60 mph, how long will it now take the ball to get to the plate?
7. Two students are having a race on a circular track. Student 1 is on the inside track, which has a radius of curvature $r_1 = 250$ m, and is moving at the speed $v_1 = 4.50$ m/s. With what speed must student 2 run to keep up with student 1 if student 2 is on the outside track of radius of curvature $r_2 = 255$ m?

8. A plot of the displacement of a car as a function of time is shown in the diagram. Find the velocity of the car along the paths (a) *O-A*, (b) *A-B*, (c) *B-C*, and (d) *C-D*.

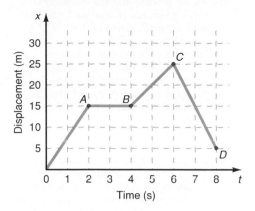

9. A plot of the velocity of a car as a function of time is shown in the diagram. Find the acceleration of the car along the paths (a) *O-A*, (b) *A-B*, (c) *B-C*, and (d) *C-D*.

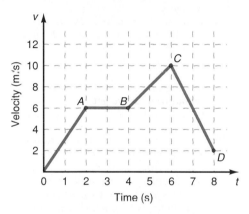

10. If an airplane is traveling at 110 knots, what is its velocity in (a) mph, (b) km/hr, (c) ft/s, and (d) m/s? A knot is a nautical mile per hour, and a nautical mile is equal to 6076 ft.

3.6 The Kinematic Equations in One Dimension

11. A girl who is initially running at 1.00 m/s increases her velocity to 2.50 m/s in 5.00 s. Find her acceleration.
12. A car is traveling at 95.0 km/hr. The driver steps on the brakes and the car comes to a stop in 60.0 m. What is the car's deceleration?
13. A train accelerates from an initial velocity of 20.0 mph to a final velocity of 35.0 mph in 11.8 s. Find its acceleration and the distance the train travels during this time.
14. A train accelerates from an initial velocity of 25.0 km/hr to a final velocity of 65.0 km/hr in 8.50 s. Find its acceleration and the distance the train travels during this time.

15. An airplane travels 1000 ft at a constant acceleration while taking off. If it starts from rest, and takes off in 25.0 s, what is its takeoff velocity?
16. An airplane travels 450 km at a constant acceleration while taking off. If it starts from rest, and takes off in 30.0 s, what is its takeoff velocity?
17. A car starts from rest and acquires a velocity of 20.0 mph in 10.0 s. Where is the car located and what is its velocity at 10.0, 15.0, 20.0, and 25.0 s?
18. A jet airplane goes from rest to a velocity of 250 ft/s in a distance of 400 ft. What is the airplane's average acceleration in ft/s²?

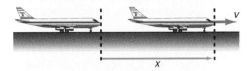

19. An electron in a vacuum tube acquires a velocity of 5.3×10^8 cm/s in a distance of 0.25 cm. Find the acceleration of the electron.
20. A driver traveling at 60.0 mph tries to stop the car and finds that the brakes have failed. The emergency brake is then pulled and the car comes to a stop in 456 ft. Find the car's deceleration.
21. An airplane has a touchdown velocity of 75.0 knots and comes to rest in 400 ft. What is the airplane's average deceleration? How long does it take the plane to stop?
22. A pitcher gives a baseball a horizontal velocity of 110 ft/s by moving his arm through a distance of approximately 3.00 ft. What is the average acceleration of the ball during this throwing process?
23. The speedometer of a car reads 60.0 mph when the brakes are applied. The car comes to rest in 4.55 s. How far does the car travel before coming to rest?
†24. A body with unknown initial velocity moves with constant acceleration. At the end of 8.00 s, it is moving at a velocity of 50.0 m/s and it is 200 m from where it started. Find the body's acceleration and its initial velocity.
†25. A driver traveling at 25.0 mph sees the light turn red at the intersection. If her reaction time is 0.600 s, and the car can decelerate at 18.0 ft/s², find the stopping distance of the car. What would the stopping distance be if the car were moving at 50.0 mph?
26. A driver traveling at 30.0 km/hr sees the light turn red at the intersection. If his reaction time is 0.600 s, and the car can decelerate at 4.50 m/s²,

find the stopping distance of the car. What would the stopping distance be if the car were moving at 90.0 km/hr?
†27. A uniformly accelerating train passes a green light signal at 25.0 km/hr. It passes a second light 125 m farther down the track, 12.0 s later. What is the train's acceleration? What is the train's velocity at the second light?

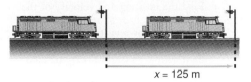

x = 125 m

28. A car accelerates from 50.0 mph to 80.0 mph in 26.9 s. Find its acceleration and the distance the car travels in this time.
†29. A motorcycle starts from rest and accelerates at 4.00 m/s² for 5.00 s. It then moves at constant velocity for 25.0 s, and then decelerates at 2.00 m/s² until it stops. Find the total distance that the motorcycle has moved.
†30. A car starts from rest and accelerates at a constant rate of 3.00 m/s² until it is moving at 18.0 m/s. The car then decreases its acceleration to 0.500 m/s² and continues moving for an additional distance of 250 m. Find the total time taken.

3.7 The Freely Falling Body

31. A passenger, in abandoning a sinking ship, steps over the side. The deck is 15.0 m above the water surface. With what velocity does the passenger hit the water?
32. How long does it take for a stone to fall from a bridge to the water 30.0 m below? With what velocity does the stone hit the water?
33. An automobile traveling at 60.0 mph hits a stone wall. From what height would the car have to fall to acquire the same velocity?
34. A rock is dropped from the top of a building and hits the ground 8.00 s later. How high is the building?
35. A stone is dropped from a bridge 100 ft high. How long will it take for the stone to hit the water below?
36. A ball is dropped from a building 50.0 meters high. How long will it take the ball to hit the ground below?
†37. A girl is standing in an elevator that is moving upward at a velocity of 12.0 ft/s when she drops her handbag. If she was originally holding the bag at a height of 4.00 ft above the elevator floor, how long will it take the bag to hit the floor?

3.9 Projectile Motion in One Dimension

38. A ball is thrown vertically upward with an initial velocity of 130 ft/s. Find its position and velocity at the end of 2, 4, 6, and 8 s and sketch these positions and velocities on a piece of graph paper.

39. A projectile is fired vertically upward with an initial velocity of 40.0 m/s. Find the position and velocity of the projectile at 1, 3, 5, and 7 s.

†40. A ball is thrown vertically upward from the top of a building 40.0 m high with an initial velocity of 25.0 m/s. What is the total time that the ball is in the air?

41. A stone is thrown vertically upward from a bridge 100 ft high at an initial velocity of 50.0 ft/s. How long will it take for the stone to hit the water below?

†42. A stone is thrown vertically downward from a bridge 100 ft high at an initial velocity of −50.0 ft/s. How long will it take for the stone to hit the water below?

43. A rock is thrown vertically downward from a building 40.0 m high at an initial velocity of −15.0 m/s. (a) What is the rock's velocity as it strikes the ground? (b) How long does it take for the rock to hit the ground?

44. A baseball batter fouls a ball vertically upward. The ball is caught right behind home plate at the same height that it was hit. How long was the baseball in flight if it rose a distance of 100 ft? What was the initial velocity of the baseball?

3.11 Projectile Motion in Two Dimensions

45. A projectile is thrown from the top of a building with a horizontal velocity of 15.0 m/s. The projectile lands on the street 85.0 m from the base of the building. How high is the building?

46. To find the velocity of water issuing from the nozzle of a garden hose, the nozzle is held horizontally and the stream is directed against a vertical wall. If the wall is 7.00 m from the nozzle and the water strikes the wall 0.650 m below the horizontal, what is the velocity of the water?

47. A bomb is dropped from an airplane in level flight at a velocity of 970 km/hr. The altitude of the aircraft is 2000 m. At what horizontal distance from the initial position of the aircraft will the bomb land?

†48. A cannon is placed on a hill 20.0 m above level ground. A shell is fired horizontally at a muzzle velocity of 300 m/s. At what horizontal distance from the cannon will the shell land? How long will this take? What will be the shell's velocity as it strikes its target?

49. A shell is fired from a cannon at a velocity of 300 m/s to hit a target 3000 m away. At what angle above the horizontal should the cannon be aimed?

50. In order to hit a target, a marksman finds he must aim 10.0 cm above the target, which is 300 m away. What is the initial speed of the bullet?

51. A golf ball is hit with an initial velocity of 175 ft/s at an angle of 50.0° above the horizontal. (a) How high will the ball go? (b) What is the total time the ball is in the air? (c) How far will the ball travel horizontally before it hits the ground?

52. A projectile is thrown from the ground with an initial velocity of 20.0 m/s at an angle of 40.0° above the horizontal. Find (a) the projectile's maximum height, (b) the time required to reach its maximum height, (c) its velocity at the top of the trajectory, (d) the range of the projectile, and (e) the total time of flight.

Additional Problems

53. A missile has a velocity of 10,000 mph at "burn-out," which occurs 2 min after ignition. Find the average acceleration in (a) ft/s², (b) m/s², and (c) in terms of g, the acceleration due to gravity at the surface of the earth.

54. A block slides down a smooth inclined plane that makes an angle of 25.0° with the horizontal. Find the acceleration of the block. If the plane is 10.0 meters long and the block starts from rest, what is its velocity at the bottom of the plane? How long does it take for the block to get to the bottom?

†55. At the instant that the traffic light turns green, a car starting from rest with an acceleration of 7.00 ft/s² is passed by a truck moving at a constant velocity of 30.0 mph. (a) How long will it take for the car

to overtake the truck? (b) How far from the starting point will the car overtake the truck? (c) At what velocity will the car be moving when it overtakes the truck?

56. At the instant that the traffic light turns green, a car starting from rest with an acceleration of 2.50 m/s² is passed by a truck moving at a constant velocity of 60.0 km/hr. (a) How long will it take for the car to overtake the truck? (b) How far from the starting point will the car overtake the truck? (c) At what velocity will the car be moving when it overtakes the truck?

†57. A boat passes a buoy while moving to the right at a velocity of 8.00 m/s. The boat has a constant acceleration to the left, and 10.0 s later the boat is found to be moving at a velocity of −3.00 m/s. Find (a) the acceleration of the boat, (b) the distance from the buoy when the boat reversed direction, (c) the time for the boat to return to the buoy, and (d) the velocity of the boat when it returns to the buoy.

†58. Two trains are initially at rest on parallel tracks with train 1 50.0 m ahead of train 2. Both trains accelerate simultaneously, train 1 at the rate of 2.00 m/s² and train 2 at the rate of 2.50 m/s². How long will it take train 2 to overtake train 1? How far will train 2 travel before it overtakes train 1?

†59. Repeat problem 58 but with train 1 initially moving at 5.00 m/s and train 2 initially moving at 7.00 m/s.

†60. A policeman driving at 55.0 mph observes a car 200 ft ahead of him speeding at 80.0 mph. If the county line is 1200 ft away from the police car, what must the acceleration of the police car be, in order to catch the speeder before he leaves the county?

61. A policewoman driving at 80.0 km/hr observes a car 50.0 m ahead of her speeding at 120 km/hr. If the county line is 400 m away from the police car, what must the acceleration of the police car be in order to catch the speeder before he leaves the county?

†62. Two trains are approaching each other along a straight and level track. The first train is heading east at 70.0 mph, while the second train is heading west at 45.0 mph. When they are 1.50 miles apart they see each other and start to decelerate. Train 1 decelerates at 5.00 ft/s², while train 2 decelerates at 3.00 ft/s². Will the trains be able to stop or will there be a collision?

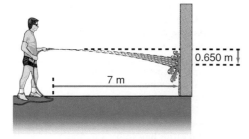

63. Two trains are approaching each other along a straight and level track. The first train is heading south at 125 km/hr, while the second train is heading north at 80.0 km/hr. When they are 2.00 km apart, they see each other and start to decelerate. Train 1 decelerates at 2.00 m/s², while train 2 decelerates at 1.50 m/s². Will the trains be able to stop or will there be a collision?

†64. A boy in an elevator, which is descending at the constant velocity of −5.00 m/s, jumps to a height of 0.500 m above the elevator floor. How far will the elevator descend before the boy returns to the elevator floor?

65. The acceleration due to gravity on the moon is 1.62 m/s². If an astronaut on the moon throws a ball straight upward, with an initial velocity of 25.0 m/s, how high will the ball rise?

†66. A helicopter, at an altitude of 300 m, is rising vertically at 20.0 m/s when a wheel falls off. How high will the wheel go with respect to the ground? How long will it take for the wheel to hit the ground below? At what velocity will the wheel hit the ground?

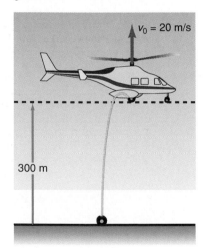

$v_0 = 20$ m/s

300 m

†67. A ball is dropped from the roof of a building 40.0 m high. Simultaneously, another ball is thrown upward from the ground and collides with the first ball at half the distance to the roof. What was the initial velocity of the ball that was thrown upward?

†68. A ball is dropped from the top of a 40.0-m high building. At what initial velocity must a second ball be thrown from the top of the building 2.00 s later, such that both balls arrive at the ground at the same time?

†69. Show that the range of a projectile is the same for either a projection angle of 45.0° +θ or an angle of 45.0° −θ.

70. A projectile hits a target 1.50 km away 10.5 s after it was fired. Find (a) the elevation angle of the gun and (b) the initial velocity of the projectile.

71. A football is kicked with an initial velocity of 70.0 ft/s at an angle of 65.0° above the horizontal. Find (a) how long the ball is in the air, (b) how far down field the ball lands, (c) how high the ball rises, and (d) the velocity of the ball when it strikes the ground.

†72. A baseball is hit at an initial velocity of 110 ft/s at an angle of 45.0° above the horizontal. Will the ball clear a 10.0 ft fence 300 ft from home plate for a home run? If so, by how much will it clear the fence?

†73. A ball is thrown from a bridge 100 m high at an initial velocity of 30.0 m/s at an angle of 50.0° above the horizontal. Find (a) how high the ball goes, (b) the total time the ball is in the air, (c) the maximum horizontal distance that the ball travels, and (d) the velocity of the ball as it strikes the ground.

74. A ball is thrown at an angle of 35.5° below the horizontal at a speed of 22.5 m/s from a building 20.0 m high. (a) How long will it take for the ball to hit the ground below? (b) How far from the building will the ball land?

†75. Using the kinematic equations for the x- and y-components of the displacement, find the equation of the trajectory for two-dimensional projectile motion. Compare this equation with the equation for a parabola expressed in its standard form.

†76. Using the kinematic equations, prove that if two balls are released simultaneously from a table, one with zero velocity and the other with a horizontal velocity v_{0x}, they will both reach the ground at the same time.

Interactive Tutorials

77. A train accelerates from an initial velocity of 20.0 m/s to a final velocity of 35.0 m/s in 11.8 s. Find its acceleration and the distance the train travels in this time.

78. A ball is dropped from a building 50.0 m high. How long will it take the ball to hit the ground below and with what final velocity?

79. A golf ball is hit with an initial velocity $v_0 = 53.0$ m/s at an angle $θ = 50.0°$ above the horizontal. (a) How high will the ball go? (b) What is the total time the ball is in the air? (c) How far will the ball travel horizontally before it hits the ground?

80. Instantaneous velocity. If the equation for the displacement x of a body is known, the average velocity throughout an interval can be computed by the formula

$$v_{avg} = (\Delta x)/(\Delta t)$$

The instantaneous velocity is defined as the limit of the average velocity as Δt approaches zero. That is,

$$v = \lim (\Delta x)/(\Delta t)$$
$$\Delta t \to 0$$

For an acceleration with a displacement given by $x = 0.5\, at^2$, use different values of Δt to see how the average velocity approaches the instantaneous velocity. Compare this to the velocity determined by the equation $v = at$, and determine the percentage error. Plot the average velocity, $(\Delta x)/(\Delta t)$, versus Δt.

81. Free-fall and generalized one-dimensional projectile motion. A projectile is fired from a height y_0 above the ground with an initial velocity v_0 in a vertical direction. Find (a) the time t_r for the projectile to rise to its maximum height, (b) the total time t_t the ball is in the air, (c) the maximum height y_{max} of the projectile, (d) the velocity v_g of the projectile as it strikes the ground, and (e) the location and velocity of the projectile at any time t. (f) Plot a picture of the motion as a function of time.

82. Generalized two-dimensional projectile motion. A projectile is fired from a height y_0 above the horizontal with an initial velocity v_0 at an angle $θ$. Find (a) the time t_r for the projectile to rise to its maximum height; (b) the total time t_t the ball is in the air; (c) the maximum distance the ball travels in the x-direction, x_{max} before it hits the ground; (d) the maximum height y_{max} of the projectile; (e) the velocity v_g of the projectile as it strikes the ground; and (f) the location and velocity of the projectile at any time t. (g) Plot a picture of the trajectory.

4

Newton's Laws of Motion

Artwork to celebrate the 300th anniversary of the publication of Isaac Newton's great work, Philosophical Naturalis Principia Mathematica.

I do not know what I may appear to the world; but to myself I seem to have been only like a boy playing on the sea shore, and diverting myself in now and then finding a smoother pebble or a prettier shell than ordinary, while the great ocean of truth lay all undiscovered before me.

Sir Isaac Newton

4.1 Introduction

Chapter 3 dealt with kinematics, the study of motion. We saw that if the acceleration, initial position, and velocity of a body are known, then the future position and velocity of the moving body can be completely described. But one of the things left out of that discussion, was the cause of the body's acceleration. If a piece of chalk is dropped, it is immediately accelerated downward. The chalk falls because the earth exerts a force of gravity on the chalk pulling it down toward the center of the earth. We will see that any time there is an acceleration, there is always a force present to cause that acceleration. In fact, it is Newton's laws of motion that describe what happens to a body when forces are acting on it. That branch of mechanics concerned with the forces that change or produce the motions of bodies is called **dynamics.**

As an example, suppose you get into your car and accelerate from rest to 50 mph. What causes that acceleration? The acceleration is caused by a force that begins with the car engine. The engine supplies a force, through a series of shafts and gears to the tires, that pushes backward on the road. The road in turn exerts a force on the car to push it forward. Without that force you would never be able to accelerate your car. Similarly, when you step on the brakes, you exert a force through the brake linings, to the wheels and tires of the car to the road. The road exerts a force backward on the car that causes the car to decelerate. All motions are started or stopped by forces.

Before we start our discussion of Newton's laws of motion, let us spend a few moments discussing the life of Sir Isaac Newton, perhaps the greatest scientist who ever lived. Newton was born in the little hamlet of Woolsthorpe in Lincolnshire, England, on Christmas day, 1642. It was about the same time that Galileo Galilei died; it was as though the torch of knowledge had been passed from one generation to another. Newton was born prematurely and was not expected to live; somehow he managed to survive. His father had died three months previously. Isaac grew up with a great curiosity about the things around him. His chief delight was to sit under a tree reading a book. His uncle, a member of Trinity College at Cambridge University, urged that the young Newton be sent to college, and Newton went to Cambridge in June, 1661. He spent the first two years at college learning arithmetic, Euclidean geometry, and trigonometry. He also read and listened to lectures on the Copernican system of astronomy. After that he studied natural philosophy. In 1665 the bubonic plague hit London and Newton returned to his mother's farm at Woolsthorpe. It was there, while observing an apple fall from a tree, that Newton wondered that if the pull of the earth can act through space to pull an apple from

(a)

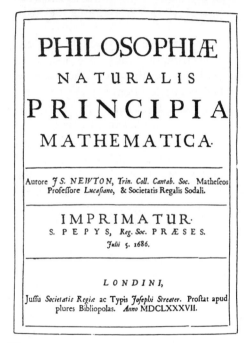

PHILOSOPHIÆ

NATURALIS

PRINCIPIA

MATHEMATICA·

Autore *JS. NEWTON,* Trin. Coll. Cantab. Soc. Mathefeos
Profeffore Lucafiano, & Societatis Regalis Sodali.

IMPRIMATUR·
S. P E P Y S, *Reg. Soc.* P R Æ S E S.
Julii 5. 1686.

L O N D I N I,

Juffu *Societatis Regiæ* ac Typis *Jofephi Streater.* Proftat apud
plures Bibliopolas. *Anno* MDCLXXXVII.

(b)

Figure 4.1

(*a*) Sir Isaac Newton. (*b*) The first page of
Newton's *Principia.*

a tree, could it not also reach out as far as the moon and pull the moon toward the earth? This reasoning became the basis for his law of universal gravitation.

Newton also invented the calculus (he called it fluxions) as a means of solving a problem in gravitation. (We should also note, however, that the German mathematician Gottfried Leibniz also invented the calculus independently of, and simultaneously with, Newton.) Newton's work on mechanics, gravity, and astronomy was published in 1687 as the *Mathematical Principles of Natural Philosophy.* It is commonly referred to as the *Principia,* from its Latin title. Because of its impact on science, it is perhaps one of the most important books ever written. A copy of the first page of the *Principia* is shown in figure 4.1. Newton died in London on March 20, 1727, at the age of 84.

4.2 Newton's First Law of Motion

Newton's first law of motion can be stated as: *A body at rest, will remain at rest and a body in motion at a constant velocity will continue in motion at that constant velocity, unless acted on by some unbalanced external force.* By a **force** we mean a push or a pull that acts on a body. A more sophisticated definition of force will be given after the discussion of Newton's second law.

There are really two statements in the first law. The first statement says that a body at rest will remain at rest unless acted on by some unbalanced force. As an example of this first statement, suppose you placed a book on the desk. That book would remain there forever, unless some unbalanced force moved it. That is, you might exert a force to pick up the book and move it someplace else. But if neither you nor anything else exerts a force on that book, that book will stay there forever. Books, and other inanimate objects, do not just jump up and fly around the room by themselves. A body at rest remains at rest and will stay in that position forever unless acted on by some unbalanced external force. This law is really a simple observation of nature. This is the first part of Newton's first law and it is so basic that it almost seems trivial and unnecessary.

The second part of the statement of Newton's first law is not quite so easy to see. This part states that a body in motion at a constant velocity will continue to move at that constant velocity unless acted on by some unbalanced external force. In fact, at first observation it actually seems to be wrong. For example, if you take this book and give it a shove along the desk, you immediately see that it does not keep on moving forever. In fact, it comes to a stop very quickly. So either Newton's law is wrong or there must be some force acting on the book while it is in motion along the desk. In fact there is a force acting on the book and this force is the force of friction, which tends to oppose the motion of one body sliding on another. (We will go into more details on friction later in this chapter.) But, if instead of trying to slide the book along the desk, we tried to slide it along a sheet of ice (say on a frozen lake), then the book would move a much greater distance before coming to rest. The frictional force acting on the book by the ice is much less than the frictional force that acted on the book by the desk. But there is still a force, regardless of how small, and the book eventually comes to rest. However, we can imagine that in the limiting case where these frictional forces are completely eliminated, an object moving at a constant velocity would continue to move at that same velocity forever, unless it were acted on by a nonzero net force. *The resistance of a body to a change in its motion is called **inertia,** and Newton's first law is also called the law of inertia.*

If you were in outer space and were to take an object and throw it away where no forces acted on it, it would continue to move at a constant velocity. Yet if you take your pen and try to throw it into space, it falls to the floor. Why? Because the force of gravity pulls on it and accelerates it to the ground. It is not free to move in straight line motion but instead follows a parabolic trajectory, as we have seen in the study of projectiles.

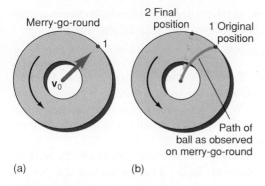

Merry-go-round

2 Final
position   1 Original
position

Path of
ball as observed
on merry-go-round

(a)   (b)

Figure 4.2

A noninertial coordinate system.

Figure 4.3

A merry-go-round is a noninertial
coordinate system.

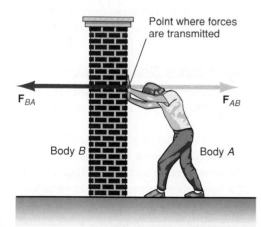

Point where forces
are transmitted

F_{BA} F_{AB}

Body *B*   Body *A*

Figure 4.4

Forces involved when you lean against a
wall.

The first part of Newton's first law—A body at rest, will remain at rest . . .—is really a special case of the second statement—a body in motion at some constant velocity. . . . A body at rest has zero velocity, and will therefore have that same zero velocity forever, unless acted on by some unbalanced external force.

Newton's first law of motion also defines what is called an inertial coordinate system. A coordinate system in which objects experiencing no unbalanced forces remain at rest or continue in uniform motion, is called an inertial coordinate system. *An **inertial coordinate system** (also called an inertial reference system) is a coordinate system that is either at rest or moving at a constant velocity with respect to another coordinate system that is either at rest or also moving at a constant velocity.* In such a coordinate system the first law of motion holds. A good way to understand an inertial coordinate system is to look at a noninertial coordinate system. A rotating coordinate system is an example of a noninertial coordinate system. Suppose you were to stand at rest at the center of a merry-go-round and throw a ball to another student who is on the outside of the rotating merry-go-round at the position 1 in figure 4.2(a). When the ball leaves your hand it is moving at a constant horizontal velocity, v_0. Remember that a velocity is a vector, that is, it has both magnitude and direction. The ball is moving at a constant horizontal speed in a constant direction. The *y*-component of the velocity changes because of gravity, but not the *x*-component. You, being at rest at the center, are in an inertial coordinate system. The person on the rotating merry-go-round is rotating and is in a noninertial coordinate system. As observed by you, at rest at the center of the merry-go-round, the ball moves through space at a constant horizontal velocity. But the person standing on the outside of the merry-go-round sees the ball start out toward her, but then it appears to be deflected to the right of its original path, as seen in figure 4.2(b). Thus, the person on the merry-go-round does not see the ball moving at a constant horizontal velocity, even though you, at the center, do, because she is rotating away from her original position. That student sees the ball changing its direction throughout its flight and the ball appears to be deflected to the right of its path. The person on the rotating merry-go-round is in a noninertial coordinate system and Newton's first law does not hold in such a coordinate system. That is, the ball in motion at a constant horizontal velocity does not appear to continue in motion at that same horizontal velocity. Thus, when Newton's first law is applied it must be done in an inertial coordinate system. In this book nearly all coordinate systems will be either inertial coordinate systems or ones that can be approximated by inertial coordinate systems, hence Newton's first law will be valid. The earth is technically not an inertial coordinate system because of its rotation about its axis and its revolution about the sun. The acceleration caused by the rotation about its axis is only about 1/300 of the acceleration caused by gravity, whereas the acceleration due to its orbital revolution is about 1/1650 of the acceleration due to gravity. Hence, as a first approximation, the earth can usually be used as an inertial coordinate system.

Before discussing the second law, let us first discuss Newton's third law because its discussion is somewhat shorter than the second.

4.3 Newton's Third Law of Motion

Newton stated his third law in the succinct form, "Every action has an equal but opposite reaction." Let us express **Newton's third law of motion** in the form, *if there are two bodies, A and B, and if body A exerts a force on body B, then body B will exert an equal but opposite force on body A.* The first thing to observe in Newton's third law is that two bodies are under consideration, body *A* and body *B*. This contrasts to the first (and second) law, which apply to a single body. As an example of the third law, consider the case of a person leaning against the wall, as shown in figure 4.4. The person is body *A*, the wall is body *B*. The person is exerting a force on the wall, and Newton's third law states that the wall is exerting an equal but opposite force on the person.

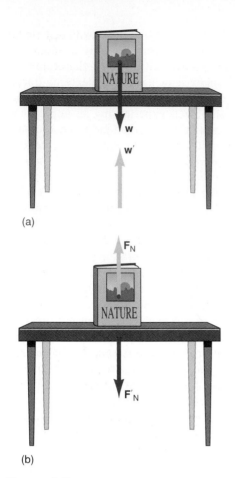

(a)

(b)

Figure 4.5

Newton's third law of motion.

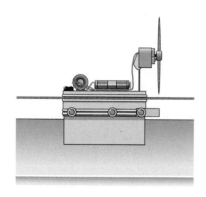

Figure 4.6

Glider and airplane motor.

The key to Newton's third law is that there are two different bodies exerting two equal but opposite forces on each other. Stated mathematically this becomes

$$\mathbf{F}_{AB} = -\mathbf{F}_{BA} \qquad (4.1)$$

where $\mathbf{F}_{AB}$ is the force on body A exerted by body B and $\mathbf{F}_{BA}$ is the force on body B exerted by body A. Equation 4.1 says that all forces in nature exist in pairs. There is no such thing as a single isolated force. We call $\mathbf{F}_{BA}$ the *action force,* whereas we call $\mathbf{F}_{AB}$ the *reaction force* (although either force can be called the action or reaction force). Together these forces are an *action-reaction pair.*

Another example of the application of Newton's third law is a book resting on a table, as seen in figure 4.5. A gravitational force, directed toward the center of the earth, acts on that book. We call the gravitational force on the book its weight **w.** By Newton's third law there is an equal but opposite force **w′** acting on the earth. The forces **w** and **w′** are the action and reaction pair of Newton's third law, and note how they act on two different bodies, the book and the earth. The force **w** acting on the book should cause it to fall toward the earth. However, because the table is in the way, the force down on the book is applied to the table. Hence the book exerts a force down on the table. We label this force on the table, $\mathbf{F}'_N$. By Newton's third law the table exerts an equal but opposite force upward on the book. We call the equal but upward force acting on the book the *normal force,* and designate it as $\mathbf{F}_N$. *When used in this context, normal means perpendicular to the surface.*

If we are interested in the forces acting on the book, they are the gravitational force, which we call the weight **w,** and the normal force $\mathbf{F}_N$. Note however, that these two forces are not an action-reaction pair because they act on the same body, namely the book.

We will discuss Newton's third law in more detail when we consider the law of conservation of momentum in chapter 8.

4.4 Newton's Second Law of Motion

Newton's second law of motion is perhaps the most basic, if not the most important, law of all of physics. We begin our discussion of Newton's second law by noting that whenever an object is dropped, the object is accelerated down toward the earth. We know that there is a force acting on the body, a force called the force of gravity. The force of gravity appears to be the cause of the acceleration downward. We therefore ask the question, *Do all forces cause accelerations? And if so, what is the relation of the acceleration to the causal force?*

Experimental Determination of Newton's Second Law

To investigate the relation between forces and acceleration, we will go into the laboratory and perform an experiment with a propelled glider on an air track, as seen in figure 4.6.[1]

We turn a switch on the glider to apply 0.2 V to the airplane motor mounted on top of the glider. As the propeller turns, it exerts a force on the glider that pulls the glider down the track. We turn on a spark timer, giving a record of the position of the glider as a function of time. From the spark timer tape, we determine the acceleration of the glider as in the experiment on kinematics. We then connect a piece of Mylar tape to the back of the glider and pass it over an air pulley at the end of the track. Weights are hung from the Mylar tape until the force exerted by the weights is equal to the force exerted by the propeller. The glider will then be at rest. In this way, we determine the force exerted by the propeller. This procedure is repeated several times with battery voltages of 0.4, 0.6, 0.8, and 1.0 V. If we plot the acceleration of the glider against the force, we get the result shown in figure 4.7.

1. See Nolan and Bigliani, *Experiments in Physics,* 2d ed., Wm. C. Brown Publishers, Dubuque, Iowa.

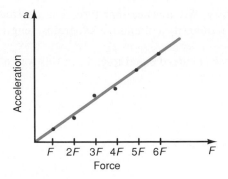

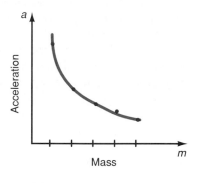

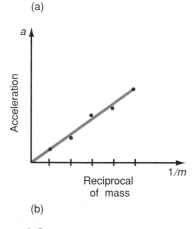

Figure 4.7

Plot of acceleration *a* versus the applied force *F* for a propelled glider.

Figure 4.8

Plot of (*a*) the acceleration *a* versus mass *m* and (*b*) the acceleration *a* versus the reciprocal of the mass (1/*m*) for the propelled gliders.

Whenever a graph of two variables is a straight line, as in figure 4.7, the dependent variable is directly proportional to the independent variable. (See appendix C for a discussion of proportions.) Therefore this graph tells us that the acceleration of the glider is directly proportional to the applied force, that is,

$$a \propto F \qquad (4.2)$$

Thus, not only does a force cause an acceleration of a body but that acceleration is directly proportional to that force, and in the direction of that force. That is, if we double the force, we double the acceleration; if we triple the force, we triple the acceleration; and so forth.

Let us now ask, how is the acceleration affected by the mass of the object being moved? To answer this question we go back to the laboratory and our experiment. This time we connect together two gliders of known mass and place them on the air track. Hence, the mass of the body in motion is increased. We turn on the propeller and the gliders go down the air track with the spark timer again turned on. Then we analyze the spark timer tape to determine the acceleration of the two gliders. We repeat the experiment with three gliders and then with four gliders, all of known mass. We determine the acceleration for each increased mass and plot the acceleration of the gliders versus the mass of the gliders, as shown in figure 4.8(a). The relation between acceleration and mass is not particularly obvious from this graph except that as the mass gets larger, the acceleration gets smaller, which suggests that the acceleration may be related to the reciprocal of the mass. We then plot the acceleration against the reciprocal of the mass in figure 4.8(b), and obtain a straight line.

Again notice the linear relation. This time, however, the acceleration is directly proportional to the reciprocal of the mass. Or saying it another way, the acceleration is inversely proportional to the mass of the moving object. (See appendix C for discussion of inverse proportions.) That is,

$$a \propto \frac{1}{m} \qquad (4.3)$$

Thus, the greater the mass of a body, the smaller will be its acceleration for a given force. *Hence, the mass of a body is a measure of the body's resistance to being put into accelerated motion.* Equations 4.2 and 4.3 can be combined into a single proportionality, namely

$$a \propto \frac{F}{m} \qquad (4.4)$$

The result of this experiment shows that the acceleration of a body is directly proportional to the applied force and inversely proportional to the mass of the moving body. The proportionality in relation 4.4 can be rewritten as an equation if a constant of proportionality *k* is introduced (see the appendix on proportions). Thus,

$$F = kma \qquad (4.5)$$

Let us now define the unit of force in such a way that *k* will be equal to the value one, thereby simplifying the equation. The unit of force in SI units, thus defined, is

$$1 \text{ newton} = 1 \text{ kg} \frac{\text{m}}{\text{s}^2}$$

The abbreviation for a newton is the capital letter N. *A newton is the net amount of force required to give a mass of 1 kg an acceleration of 1 m/s².* Hence, **force is now defined as more than a push or a pull, but rather a force is a quantity that causes a body of mass m to have an acceleration a.** Recall from chapter 1 that the mass of an object is a fundamental quantity. We now see that force is a derived quantity. It is derived from the fundamental quantities of mass in kilograms, length in meters, and time in seconds.

A check on dimensions shows that k is indeed equal to unity in this way of defining force, that is,

$$F = kma$$

$$\text{newton} = (k) \text{ kg m/s}^2$$

$$\text{kg m/s}^2 = (k) \text{ kg m/s}^2$$

$$k = 1$$

Equation 4.5 therefore becomes

$$F = ma \tag{4.6}$$

Equation 4.6 is the mathematical statement of Newton's second law of motion. This is perhaps the most fundamental of all the laws of classical physics. **Newton's second law of motion** can be stated in words as: *If an unbalanced external force F acts on a body of mass m, it will give that body an acceleration a. The acceleration is directly proportional to the applied force and inversely proportional to the mass of the body.* We must understand by Newton's second law that the force F is the resultant external force acting on the body. Sometimes, to be more explicit, Newton's second law is written in the form

$$\Sigma F = ma \tag{4.7}$$

where the greek letter sigma, Σ, means "the sum of." Thus, if there is more than one force acting on a body, it is the resultant unbalanced force that causes the body to be accelerated. For example, if a book is placed on a table as in figure 4.5, the forces acting on the book are the force of gravity pulling the book down toward the earth, while the table exerts a normal force upward on the book. These forces are equal and opposite, so that the resultant unbalanced force acting on the book is zero. Hence, even though forces act on the book, the resultant of these forces is zero and there is no acceleration of the book. It remains on the table at rest.

Newton's second law is the fundamental principle that relates forces to motions, and is the foundation of mechanics. Thus, if an unbalanced force acts on a body, it will give it an acceleration. In particular, the acceleration is found from equation 4.7 to be

$$a = \frac{\Sigma F}{m} \tag{4.8}$$

It is a matter of practice that Σ is usually left out of the equations but do not forget it; it is always implied because it is the resultant force that causes the acceleration.

Once the acceleration of the body is known, its future position and velocity at any time can be determined using the kinematic equations developed in chapter 3, namely,

$$x = v_0 t + \tfrac{1}{2}at^2 \tag{3.14}$$

$$v = v_0 + at \tag{3.10}$$

and

$$v^2 = v_0^2 + 2ax \tag{3.16}$$

provided, of course, that the force, and therefore the acceleration, are constant. When the force and acceleration are not constant, more advanced mathematical techniques are required.

Our determination of Newton's second law has been based on the experimental work performed on the air track. Since the air track is one dimensional, the equations have been written in their one dimensional form. However, recall that acceleration is a vector quantity and therefore force, which is equal to that acceleration times mass, must also be written as a vector quantity. Newton's second law should therefore be written in the more general vector form as

$$\mathbf{F} = m\mathbf{a} \qquad (4.9)$$

The kinematic equations must also be used in their vector form.

Newton's First Law of Motion Is Consistent with His Second Law of Motion

Newton's first law of motion can be shown to be consistent with his second law of motion in the following manner. Let us start with Newton's second law

$$\mathbf{F} = m\mathbf{a} \qquad (4.9)$$

However, the acceleration is defined as the change in velocity with time. Thus,

$$\mathbf{F} = m\mathbf{a} = m\frac{\Delta \mathbf{v}}{\Delta t}$$

If there is no resultant force acting on the body, then $\mathbf{F} = 0$. Hence,

$$0 = m\frac{\Delta \mathbf{v}}{\Delta t}$$

and therefore

$$\Delta \mathbf{v} = 0 \qquad (4.10)$$

which says that there is no change in the velocity of a body if there is no resultant applied force. Another way to see this is to note that

$$\Delta \mathbf{v} = \mathbf{v}_f - \mathbf{v}_0 = 0$$

Hence,

$$\mathbf{v}_f = \mathbf{v}_0 \qquad (4.11)$$

That is, if there is no applied force ($\mathbf{F} = 0$), then the final velocity $\mathbf{v}_f$ is always equal to the original velocity $\mathbf{v}_0$. But that in essence is the first law of motion—a body in motion at a constant velocity will continue in motion at that same constant velocity, unless acted on by some unbalanced external force.

   Also note that the first part of the first law, a body at rest will remain at rest unless acted on by some unbalanced external force, is the special case of $\mathbf{v}_0 = 0$. That is,

$$\mathbf{v}_f = \mathbf{v}_0 = 0$$

indicates that if a body is initially at rest ($\mathbf{v}_0 = 0$), then at any later time its final velocity is still zero ($\mathbf{v}_f = \mathbf{v}_0 = 0$), and the body will remain at rest as long as $\mathbf{F}$ is equal to zero. Thus, the first law, in addition to defining an inertial coordinate system, is also consistent with Newton's second law.

   The ancient Greeks knew that a body at rest under no forces would remain at rest. And they knew that by applying a force to the body they could set it into motion. However, they erroneously assumed that the force had to be exerted continuously in order to keep the body in motion. Galileo was the first to show that this is not true, and Newton showed in his second law that the net force is necessary only to start the body into motion, that is, to accelerate it from rest to a velocity $\mathbf{v}$. Once it is moving at the velocity $\mathbf{v}$, the net force can be removed and the body will continue in motion at that same velocity $\mathbf{v}$.

Figure 4.9

Motion of a block on a smooth horizontal surface.

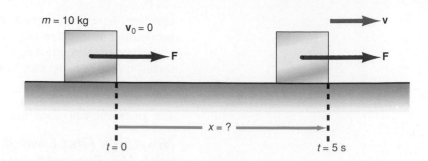

An Example of Newton's Second Law

Example 4.1

Motion of a block on a smooth horizontal surface. A 10.0-kg block is placed on a smooth horizontal table, as shown in figure 4.9. A horizontal force of 6.00 N is applied to the block. Find (a) the acceleration of the block, (b) the position of the block at $t = 5.00$ s, and (c) the velocity of the block at $t = 5.00$ s.

Solution

a. First we draw the forces acting on the block as in the diagram. The statement that the table is smooth implies that there is only a negligible frictional force between the block and the table and it can be ignored. The only unbalanced force[2] acting on the block is the force F, and the acceleration is immediately found from Newton's second law as

$$a = \frac{F}{m} = \frac{6.00 \text{ N}}{10.0 \text{ kg}} = 0.600 \, \frac{\text{kg m/s}^2}{\text{kg}}$$

$$= 0.600 \text{ m/s}^2$$

Note here that this acceleration takes place only as long as the force is applied. If the force is removed, for any reason, then the acceleration becomes zero, and the block continues to move with whatever velocity it had at the time that the force was removed.

b. Now that the acceleration of the block is known, its position at any time can be found using the kinematic equations developed in chapter 3, namely,

$$x = v_0 t + \tfrac{1}{2}at^2 \tag{3.14}$$

But because the block is initially at rest $v_0 = 0$,

$$x = \tfrac{1}{2}at^2 = \tfrac{1}{2}(0.600 \text{ m/s}^2)(5.00 \text{ s})^2$$

$$= 7.50 \text{ m}$$

c. The velocity at the end of 5.00 s, found from equation 3.10, is

$$v = v_0 + at$$

$$= 0 + (0.600 \text{ m/s}^2)(5.00 \text{ s})$$

$$= 3.00 \text{ m/s}$$

In summary, we see that Newton's second law tells us the acceleration imparted to a body because of the forces acting on it. Once this acceleration is known, the position and velocity of the body at any time can be determined by using the kinematic equations.

---

2. Note that there are two other forces acting on the block. One is the weight **w** of the block, which acts downward, and the other is the normal force $\mathbf{F_N}$ that the table exerts upward on the block. However, these forces are balanced and do not cause an acceleration of the block.

Special Case of Newton's Second Law—The Weight of a Body Near the Surface of the Earth

Newton's second law tells us that if an unbalanced force acts on a body of mass m, it will give it an acceleration a. Let the body be a pencil that you hold in your hand. Newton's second law says that if there is an unbalanced force acting on this pencil, it will receive an acceleration. If you let go of the pencil it immediately falls down to the surface of the earth. It is an object in free-fall and, as we have seen, an object in free-fall has an acceleration whose magnitude is g. That is, if Newton's second law is applied to the pencil

$$F = ma$$

But the acceleration a is the acceleration due to gravity, and its magnitude is g. Therefore, Newton's second law can be written as

$$F = mg$$

But this gravitational force pulling an object down toward the earth is called the weight of the body, and its magnitude is w. Hence,

$$F = w$$

and Newton's second law becomes

$$w = mg \tag{4.12}$$

Equation 4.12 thus gives us a relationship between the mass of a body and the weight of a body.

Example 4.2

Finding the weight of a mass. Find the weight of a 1.00-kg mass.

Solution

The weight of a 1.00-kg mass, found from equation 4.12, is

$$w = mg = (1.00 \text{ kg})(9.80 \text{ m/s}^2) = 9.80 \text{ kg m/s}^2$$

$$= 9.80 \text{ N}$$

A mass of 1 kg has a weight of 9.80 N.

In pointing out the distinction between the weight of an object and the mass of an object in chapter 1, we said that a woman on the moon would weigh one-sixth of her weight on the earth. We can now see why. The acceleration due to gravity on the moon g_m is only about one-sixth of the acceleration due to gravity here on the surface of the earth g_E. That is,

$$g_m = \tfrac{1}{6} g_E$$

Hence, the weight of a woman on the moon would be

$$w_m = mg_m = m(\tfrac{1}{6} g_E) = \tfrac{1}{6}(mg_E) = \tfrac{1}{6} w_E$$

The weight of a woman on the moon would be one-sixth of her weight here on the earth. The mass of the woman would be the same on the earth as on the moon, but her weight would be different.

We can see from equations 4.6 and 4.12 that the weight of a body in SI units should be expressed in terms of newtons. And in the scientific community it is. However, the business community does not always follow science. The United

States is now switching over to SI units, but instead of expressing weights in newtons, as defined, the weights of objects are erroneously being expressed in terms of kilograms, a unit of mass.

As an example, if you go to the supermarket and buy a 16 oz can of beans (weight = 1 lb), you will see stamped on the can

NET WT 16 oz (1 lb) 0.453 kg

This is really a mistake, as we now know, because we know that there is a difference between the weight and the mass of a body. To get around this problem, a physics student should realize that in commercial and everyday use, the word "weight" nearly always means mass. So when you buy something that the businessman says weighs 1 kg, he means that it has the weight of a 1-kg mass. We have seen that the weight of a 1-kg mass is 9.80 N. *In this text the word kilogram will always mean mass, and only mass.* If however, you come across any item marked as a weight and expressed in kilograms in your everyday life, you can convert that mass to its proper weight in newtons by simply multiplying the mass by 9.80 m/s².

Example 4.3

Weight and mass at the supermarket. While at the supermarket you buy a bag of potatoes labeled, NET WT 5.00 kg. What is the *correct* weight expressed in newtons?

Solution

We find the weight in newtons by multiplying the mass in kg by 9.80 m/s². Hence,

$$w = (5.00 \text{ kg})(9.80 \text{ m/s}^2) = 49.0 \text{ N}$$

The Mass of an Object in the British Engineering System

In the British engineering system, weight is a fundamental quantity and mass a derived quantity. We can determine the mass from equation 4.12 by solving for *m*:

$$m = \frac{w}{g} \tag{4.13}$$

Thus, the mass of a body in the British engineering system is simply the weight of that body divided by the acceleration due to gravity. The unit of mass in the British engineering system is derived as

$$m = \frac{w}{g} = \frac{\text{lb}}{\text{ft/s}^2}$$

This unit of mass is defined as the slug, where

$$1 \text{ slug} = 1 \frac{\text{lb}}{\text{ft/s}^2}$$

The slug is defined as that amount of mass that receives an acceleration of 1 ft/s² when a force of 1 pound acts on it.

Example 4.4

Mass in the British engineering system. A man weighs 160 lb. What is his mass?

Solution

From equation 4.13 his mass is

$$m = \frac{w}{g} = \frac{160 \text{ lb}}{32.0 \text{ ft/s}^2} = 5.00 \text{ slugs}$$

If we want to use Newton's second law in the British engineering system we must express the mass of the body in slugs.

Example 4.5

Newton's second law in the British engineering system. A force of 1000 lb is used to move a car weighing 3200 lb. What is the acceleration of the car?

Solution

The acceleration is found from Newton's second law as

$$a = \frac{F}{m}$$

but the mass must first be found as

$$m = \frac{w}{g} = \frac{3200 \text{ lb}}{32.0 \text{ ft/s}^2} = 100 \text{ slugs}$$

Hence,

$$a = \frac{F}{m} = \frac{1000 \text{ lb}}{100 \text{ slugs}} = 10.0 \frac{\text{lb}}{\dfrac{\text{lb}}{\text{ft/s}^2}}$$

$$= 10.0 \text{ ft/s}^2$$

Because $m = w/g$ some engineering books even write Newton's second law in the form

$$F = \frac{w}{g} a \qquad (4.14)$$

when dealing with the British engineering system of units. This amounts to the same thing but avoids having to use the unit of slugs directly.

4.5 Applications of Newton's Second Law

A Block on a Frictionless Inclined Plane

Let us find the acceleration of a block that is to slide down a frictionless inclined plane. (The statement that the plane is frictionless means that it is not necessary to take into account the effects of friction on the motion of the block.) The velocity and the displacement of the block at any time can then be found from the kinematic equations. (Note that this problem is equivalent to placing a glider on the tilted air track in the laboratory.) The first thing to do is to draw a diagram of all the forces acting on the block, as shown in figure 4.10. A diagram showing all the forces acting on a body is called a *force diagram* or a *free-body diagram*. Note that all the forces are drawn as if they were acting at the geometrical center of the body. (The reason

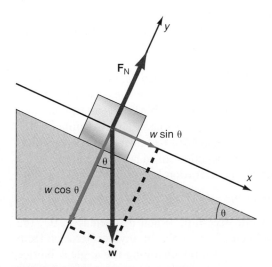

Figure 4.10

A block on a frictionless inclined plane.

for this will be discussed in more detail later when we study the center of mass of a body, but for now we will just say that the body moves as if all the forces were acting at the center of the body.)

The first force we consider is the weight of the body **w,** which acts down toward the center of the earth and is hence perpendicular to the base of the incline. The plane itself exerts a force upward on the block that we denote by the symbol F_N, and call the normal force. (Recall that a normal force is, by definition, a force that is always perpendicular to the surface.)

Let us now introduce a set of axes that are parallel and perpendicular to the plane, as shown in figure 4.10. Thus the parallel axis is the x-axis and lies in the direction of the motion, namely down the plane. The y-axis is perpendicular to the inclined plane, and points upward away from the plane. Take the weight of the block and resolve it into components, one parallel to the plane and one perpendicular to the plane. Recall from chapter 2, on the components of vectors, that if the plane makes an angle θ with the horizontal, then the acute angle between **w** and the perpendicular to the plane is also the angle θ. Hence, the component of **w** parallel to the plane $w_\parallel$ is

$$w_\parallel = w \sin \theta \tag{4.15}$$

whereas the component perpendicular to the plane $w_\perp$ is

$$w_\perp = w \cos \theta \tag{4.16}$$

as can be seen in figure 4.10. One component of the weight, namely $w \cos \theta$, holds the block against the plane, while the other component, $w \sin \theta$, is the force that acts on the block causing the block to accelerate down the plane. To find the acceleration of the block down the plane, we use Newton's second law,

$$F = ma \tag{4.6}$$

The force acting on the block to cause the acceleration is given by equation 4.15. Hence,

$$w \sin \theta = ma \tag{4.17}$$

But by equation 4.12

$$w = mg \tag{4.12}$$

Substituting this into equation 4.17 gives

$$mg \sin \theta = ma$$

Because the mass is contained on both sides of the equation, it divides out, leaving

$$a = g \sin \theta \tag{4.18}$$

as the acceleration of the block down a frictionless inclined plane. An interesting thing about this result is that equation 4.18 does not contain the mass m. That is, the acceleration down the plane is the same, whether the block has a large mass or a small mass. The acceleration is thus independent of mass. This is similar to the case of the freely falling body. There, a body fell at the same acceleration regardless of its mass. Hence, both accelerations are independent of mass. If the angle of the inclined plane is increased to 90°, then the acceleration becomes

$$a = g \sin \theta = g \sin 90° = g (1) = g.$$

Therefore, at $\theta = 90°$ the block goes into free-fall. When θ is equal to 0°, the acceleration is zero. We can use the inclined plane to obtain any acceleration from zero up to the acceleration due to gravity g, by simply changing the angle θ. Notice that the algebraic solution to a problem gives a formula rather than a number for the answer. One of the reasons why algebraic solutions to problems are superior to numerical ones is that we can examine what happens at the extremes (for example at 90° or 0°) to see if they make physical sense, and many times special cases can be considered.

Galileo used the inclined plane extensively to study motion. Since he did not have good devices available to him for measuring time, it was difficult for him to study the velocity and acceleration of a body. By using the inclined plane at relatively small angles of θ, however, he was able to slow down the motion so that he could more easily measure it.

Because we now know the acceleration of the block down the plane, we can determine its velocity and position at any time, or its velocity at any position, using the kinematic equations of chapter 3. However, now the acceleration a is determined from equation 4.18.

Note also in this discussion that if Newton's second law is applied to the perpendicular component we obtain

$$F_\perp = ma_\perp = 0$$

because there is no acceleration perpendicular to the plane. Hence,

$$F_\perp = F_N - w \cos \theta = 0$$

and

$$F_N = w \cos \theta \qquad\qquad (4.19)$$

Example 4.6

A block sliding down a frictionless inclined plane. A 10.0-kg block is placed on a frictionless inclined plane, 5.00 m long, that makes an angle of 30.0° with the horizontal. If the block starts from rest at the top of the plane, what will its velocity be at the bottom of the incline?

Solution

The velocity of the block at the bottom of the plane is found from the kinematic equation

$$v^2 = v_0^2 + 2ax$$

Hence,

$$v = \sqrt{2ax}$$

Before solving for v, we must first determine the acceleration a. Using Newton's second law we obtain

$$a = \frac{F}{m} = \frac{w \sin \theta}{m} = \frac{mg \sin \theta}{m}$$

$$= g \sin \theta = (9.80 \text{ m/s}^2) \sin 30.0°$$

$$= 4.90 \text{ m/s}^2$$

Hence,

$$v = \sqrt{2ax} = \sqrt{2(4.90 \text{ m/s}^2)(5.00 \text{ m})}$$

$$= 7.00 \text{ m/s}$$

The velocity of the block at the bottom of the plane is 7.00 m/s in a direction pointing down the inclined plane.

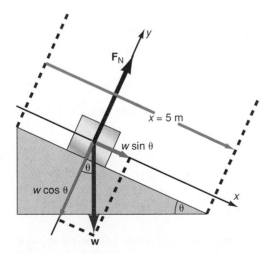

Figure 4.11

Diagram for example 4.6.

It is perhaps appropriate here to discuss the different concepts of mass. In chapter 1, we gave a very simplified definition of mass by saying that mass is a measure of the amount of matter in a body. We picked a certain amount of matter, called it a standard, and gave it the name kilogram. This amount of matter was not placed into motion. It was just the amount of matter in a platinum-iridium cylinder 39 mm in diameter and 39 mm high. The amount of matter in any other body was then compared to this standard kilogram mass. But this comparison was made by placing the different pieces of matter on a balance scale. As pointed out in chapter 1, the balance can be used to show an equality of the amount of matter in a body only because the gravitational force exerts a force downward on each pan of the balance. *Mass determined in this way is actually a measure of the gravitational force on that amount of matter, and hence mass measured on a balance is called gravitational mass.*

In the experimental determination of Newton's second law using the propeller glider, we added additional gliders to the air track to increase the mass that was in motion. The acceleration of the combined gliders was determined as a function of their mass and we observed that *the acceleration was inversely proportional to that mass. Thus, mass used in this way represents the resistance of matter to be placed into motion.* For a person, it would be more difficult to give the same acceleration to a very large mass of matter than to a very small mass of matter. *This characteristic of matter, whereby it resists motion is called inertia. The resistance of a body to be set into motion is called the **inertial mass** of that body.* Hence, in Newton's second law,

$$\mathbf{F} = m\mathbf{a} \qquad (4.9)$$

the mass m stands for the inertial mass of the body. Just as we can determine the gravitational mass of any body in terms of the standard mass of 1 kg using a balance, we can determine the inertial mass of any body in terms of the standard mass of 1 kg using Newton's second law.

As an example, let us go back into the laboratory and use the propelled glider we used early in section 4.4. For a given battery voltage the glider has a constant force acting on the glider. For a glider of mass m_1, the force causes the glider to have an acceleration a_1, which can be represented by Newton's second law as

$$F = m_1 a_1 \qquad (4.20)$$

If a new glider of mass m_2 is used with the same battery setting, and thus the same force F, the glider m_2 will experience the acceleration a_2. We can also represent this by Newton's second law as

$$F = m_2 a_2 \qquad (4.21)$$

Because the force is the same in equations 4.20 and 4.21, the two equations can be set equal to each other giving

$$m_2 a_2 = m_1 a_1$$

Solving for m_2, we get

$$m_2 = \frac{a_1}{a_2} m_1 \qquad (4.22)$$

Thus, the inertial mass of any body can be determined in terms of a mass m_1 and the ratio of the accelerations of the two masses. If the mass m_1 is taken to be the 1-kg mass of matter that we took as our standard, then the mass of any body can be determined inertially in this way. Equation 4.22 defines the inertial mass of a body.

Example 4.7

Finding the inertial mass of a body. A 1.00-kg mass experiences an acceleration of 3.00 m/s² when acted on by a certain force. A second mass experiences an acceleration of 8.00 m/s² when acted on by the same force. What is the value of the second mass?

Solution

The value of the second mass, found from equation 4.22, is

$$m_2 = \frac{a_1}{a_2} m_1$$

$$= \frac{3.00 \text{ m/s}^2}{8.00 \text{ m/s}^2} (1 \text{ kg})$$

$$= 0.375 \text{ kg}$$

Masses measured by the gravitational force can be denoted as m_g, while masses measured by their resistance to motion (i.e., inertial masses) can be represented as m_i. Then, for the motion of a block down the frictionless inclined plane, equation 4.17,

$$w \sin \theta = ma$$

should be changed as follows. The weight of the mass in equation 4.17 is determined in terms of a gravitational mass, and is written as

$$w = m_g g \tag{4.23}$$

whereas the mass in Newton's second law is written in terms of the inertial mass m_i. Hence, equation 4.17 becomes

$$m_g g \sin \theta = m_i a \tag{4.24}$$

It is, however, a fact of experiment that no differences have been found in the two masses even though they are determined differently. That is, experiments performed by Newton could detect no differences between gravitational and inertial masses. Experiments carried out by Roland von Eötvös (1848–1919) in 1890 showed that the relative difference between inertial and gravitational mass is at most 10^{-9}, and Robert H. Dicke found in 1961 the difference could be at most 10^{-11}. That is, the differences between the two masses are

$$m_i - m_g \leq 0.000000001 \text{ kg} \quad \text{(Eötvös)},$$

$$m_i - m_g \leq 0.00000000001 \text{ kg} \quad \text{(Dicke)}.$$

Hence, as best as can be determined,

$$m_i = m_g \tag{4.25}$$

Because of this equivalence between the two different characteristics of mass, the masses on each side of equation 4.24 divide out, giving us the previously found relation, $a = g \sin \theta$. Since a freely falling body is the special case of a body on a 90° inclined plane, the equivalence of these two types of masses is the reason that all objects fall at the same acceleration g near the surface of the earth. This equivalence of gravitational and inertial mass led Einstein to propose it as a general principle called the equivalence principle of which more is said in chapter 30 when general relativity is discussed.

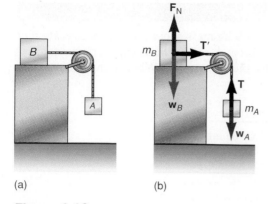

Figure 4.12

Combined motion.

Combined Motion of a Block on a Frictionless Horizontal Plane and a Block Falling Vertically

Let us now find the acceleration of a block, on a smooth horizontal table, that is connected by a cord that passes over a pulley to another block that is hanging over the end of the table, as shown in figure 4.12(a). By a smooth table, we mean there is a negligible frictional force between the block and the table so that the block will move freely over the table. We also assume that the mass of the connecting cord and pulley is negligible and can be ignored in this problem.

We want to find the motion of the blocks. In other words, what is the acceleration of the blocks, and their velocity and position at any time? The two blocks, taken together, are sometimes called a system. To determine the acceleration, we will use Newton's second law. However, before we can do so, we must draw a very careful free-body diagram showing all the forces that are acting on the two blocks, as is done in figure 4.12(b). The forces acting on block A are its weight $\mathbf{w}_A$, pulling it downward, and the tension $\mathbf{T}$ in the cord. It is this tension $\mathbf{T}$ in the cord that restrains block A from falling freely. The forces acting on body B are its weight $\mathbf{w}_B$, the normal force $\mathbf{F_N}$ that the table exerts on block B, and the tension $\mathbf{T}'$ in the cord that acts to pull block B toward the right. Newton's second law, applied to block A, gives

$$\mathbf{F} = m_A \mathbf{a}$$

Here $\mathbf{F}$ is the total resultant force acting on block A and therefore,

$$\mathbf{F} = \mathbf{T} + \mathbf{w}_A = m_A \mathbf{a} \tag{4.26}$$

Equation 4.26 is a vector equation. To simplify its solution, we use our previous convention with vectors in one dimension. That is, the upward direction $(+y)$ is taken as positive and the downward direction $(-y)$ as negative. Therefore, equation 4.26 can be simplified to

$$T - w_A = -m_A a \tag{4.27}$$

However, we can not yet solve equation 4.27 for the acceleration, because the tension T in the cord is unknown. We obviously need more information. We have one equation with two unknowns, the acceleration a and the tension T. *Whenever we want to solve a system of algebraic equations for some unknowns, we must always have as many equations as there are unknowns in order to obtain a solution.* Since there are two unknowns here, we need another equation. We obtain that second equation by applying Newton's second law to block B:

$$\mathbf{F} = m_B \mathbf{a}$$

Here $\mathbf{F}$ is the resultant force on block B and, from figure 4.12(b), we can see that

$$\mathbf{F_N} + \mathbf{w}_B + \mathbf{T}' = m_B \mathbf{a} \tag{4.28}$$

This vector equation is equivalent to the two component equations

$$F_N - w_B = 0 \tag{4.29}$$

and

$$T' = m_B a \tag{4.30}$$

The right-hand side of equation 4.29 is zero, because there is no acceleration of block B perpendicular to the table. It reduces to

$$F_N = w_B$$

That is, the normal force that the table exerts on block B is equal to the weight of block B. The magnitude of the acceleration of block B is also a because block B and block A are tied together by the string and therefore have the same motion.

Equation 4.30 is Newton's second law for the motion of block B to the right. Now we make the assumption that

$$T' = T$$

that is, the magnitude of the tension in the cord pulling on block B is the same as the magnitude of the tension in the cord restraining block A. This is a valid assumption providing the mass of the pulley is very small and friction in the pulley bearing is negligible. The only effect of the pulley is to change the direction of the string and hence the direction of the tension. (In chapter 9 we will again solve this problem, taking the rotational motion of the pulley into account without the assumption of equal tensions.) Therefore, equation 4.30 becomes

$$T = m_B a \tag{4.31}$$

We now have enough information to solve for the acceleration of the system. That is, there are the two equations 4.27 and 4.31 and the two unknowns a and T. By subtracting equation 4.31 from equation 4.27, we eliminate the tension T from both equations:

$$
\begin{aligned}
T - w_A &= -m_A a \tag{4.27}\\
\text{Subtract} \quad T &= m_B a \tag{4.31}\\
\hline
T - T - w_A &= -m_A a - m_B a\\
- w_A &= -m_A a - m_B a\\
w_A &= (m_A + m_B)a
\end{aligned}
$$

Solving for the acceleration a,

$$a = \frac{w_A}{m_A + m_B}$$

To simplify further we note that

$$w_A = m_A g$$

Therefore, the acceleration of the system of two blocks is

$$a = \frac{m_A}{m_A + m_B} g \tag{4.32}$$

To determine the tension T in the cord, we use equations 4.31 and 4.32:

$$T = m_B a = \frac{m_B m_A}{m_A + m_B} g \tag{4.33}$$

Since the acceleration of the system is a constant we can determine the position and velocity of block B in the x-direction at any time using the kinematic equations

$$x = v_0 t + \tfrac{1}{2} a t^2 \tag{3.14}$$

$$v = v_0 + at \tag{3.10}$$

and

$$v^2 = v_0^2 + 2ax \tag{3.16}$$

with the acceleration now given by equation 4.32. We find the position of block A at any time using the same equations, but with x replaced by the displacement y.

Alternate Solution to the Problem There is another way to compute the acceleration of this system that in a sense is a lot easier. But it is an intuitive way of solving the problem. Some students can see the solution right away, others can not. Let us again start with Newton's second law and solve for the acceleration a of the system

$$a = \frac{F}{m} \tag{4.8}$$

Thus, *the acceleration of the system is equal to the total resultant force applied to the system divided by the total mass of the system that is in motion.* The total force that is accelerating the system is the weight w_A. The tension T in the string just transmits the total force from one block to another. The total mass that is in motion is the sum of the two masses, m_A and m_B. Therefore, the acceleration of the system, found from equation 4.8, is

$$a = \frac{w_A}{m_A + m_B}$$

or

$$a = \frac{m_A}{m_A + m_B} g$$

Notice that this is the same acceleration that we determined previously in equation 4.32. The only disadvantage of this second technique is that it does not tell the tension in the cord. Which technique should the student use in the solution of the problem? That depends on the student. If you can see the intuitive approach, and wish to use it, do so. If not, follow the first step-by-step approach.

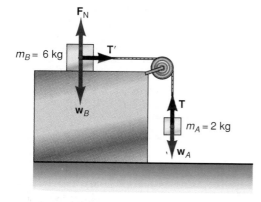

Figure 4.13

Diagram for example 4.8.

Example 4.8

Combined motion of a block moving on a smooth horizontal surface and a mass falling vertically. A 6.00-kg block rests on a smooth table. It is connected by a string of negligible mass to a 2.00-kg block hanging over the end of the table, as shown in figure 4.13. Find (a) the acceleration of each block, (b) the tension in the connecting string, (c) the position of mass A after 0.400 s, and (d) the velocity of mass A at 0.400 s.

Solution

a. To solve the problem, we draw all the forces that are acting on the system and then apply Newton's second law. The magnitude of the acceleration, obtained from equation 4.32, is

$$a = \frac{m_A}{m_A + m_B} g = \frac{2.00 \text{ kg}}{2.00 \text{ kg} + 6.00 \text{ kg}} (9.80 \text{ m/s}^2)$$

$$= 2.45 \text{ m/s}^2$$

b. The tension, found from equation 4.31, is

$$T = m_B a = (6.00 \text{ kg})(2.45 \text{ m/s}^2) = 14.7 \text{ N}$$

c. The position of mass A after 0.400 s is found from the kinematic equation

$$y = v_0 t + \tfrac{1}{2} a t^2$$

Because the block starts from rest, $v_0 = 0$, and the block falls the distance

$$y = \tfrac{1}{2} a t^2 = \tfrac{1}{2}(-2.45 \text{ m/s}^2)(0.400 \text{ s})^2$$

$$= -0.196 \text{ m}$$

d. The velocity of block A is found from the kinematic equation

$$v = v_0 + at$$
$$= 0 + (-2.45 \text{ m/s}^2)(0.400 \text{ s})$$
$$= -0.980 \text{ m/s}$$

The negative sign is used for the acceleration of block A because it accelerated in the negative y-direction. Hence, $y = -0.196$ m indicates that the block is below its starting position. The negative sign on the velocity indicates that block A is moving in the negative y-direction. If we had done the same analysis for block B, the results would have been positive because block B is moving in the positive x-direction.

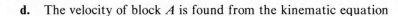

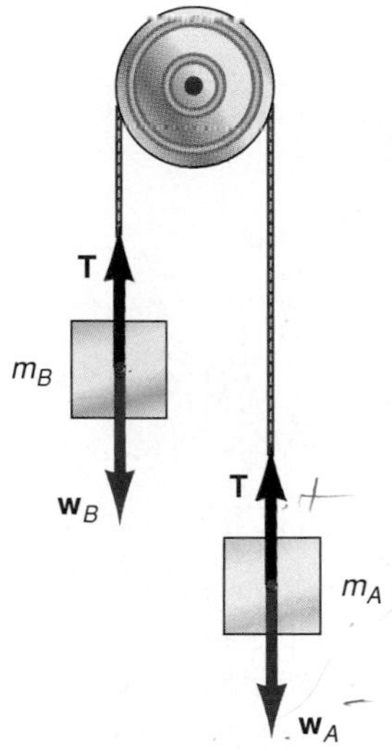

Atwood's Machine

Atwood's machine is a system that consists of a pulley, with a mass m_A on one side, connected by a string of negligible mass to another mass m_B on the other side, as shown in figure 4.14.

We assume that m_A is larger than m_B. When the system is released, the mass m_A will fall downward, pulling the lighter mass m_B, on the other side, upward. We would like to determine the acceleration of the system of two masses. When we know the acceleration we can determine the position and velocity of each of the masses at any time from the kinematic equations.

Let us start by drawing all the forces acting on the masses in figure 4.14 and then apply Newton's second law to each mass. (The assumption that the tension T in the rope is the same for each mass is again utilized. We will solve this problem again in chapter 9, on rotational motion, where the rotating pulley is massive and hence the tensions on both sides of the pulley are not the same.)

For mass A, Newton's second law is

$$\mathbf{F}_A = m_A\mathbf{a}$$

or

$$\mathbf{T} + \mathbf{w}_A = m_A\mathbf{a} \tag{4.34}$$

We can simplify this equation by taking the upward direction as positive and the downward direction as negative, that is,

$$T - w_A = -m_Aa \tag{4.35}$$

We cannot yet solve for the acceleration of the system, because the tension T in the string is unknown. Another equation is needed to eliminate T. We obtain this equation by applying Newton's second law to mass B:

$$\mathbf{F}_B = m_B\mathbf{a}$$

$$\mathbf{T} + \mathbf{w}_B = m_B\mathbf{a} \tag{4.36}$$

Simplifying again by taking the upward direction as positive and the downward direction as negative, we get

$$T - w_B = +m_Ba \tag{4.37}$$

We thus have two equations, 4.35 and 4.37, in the two unknowns of acceleration a and tension T. The tension T is eliminated by subtracting equation 4.37 from equation 4.35. That is,

$$T - w_A = -m_Aa \tag{4.35}$$

Subtract

$$T - w_B = m_Ba \tag{4.37}$$

$$\overline{T - w_A - T + w_B = -m_Aa - m_Ba}$$
$$w_B - w_A = -(m_A + m_B)a$$

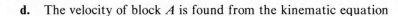

Figure 4.14

Atwood's machine.

Solving for a, we obtain

$$a = \frac{w_A - w_B}{m_A + m_B} = \frac{m_A g - m_B g}{m_A + m_B}$$

Hence, the acceleration of each mass of the system is

$$a = \left(\frac{m_A - m_B}{m_A + m_B}\right) g \qquad (4.38)$$

We find the tension T in the string by substituting equation 4.38 back into equation 4.35:

$$T = w_A - m_A a$$

$$= m_A g - m_A \left(\frac{m_A - m_B}{m_A + m_B}\right) g$$

$$= \left[1 + \left(\frac{m_B - m_A}{m_A + m_B}\right)\right] m_A g$$

$$= \left(\frac{m_A + m_B + m_B - m_A}{m_A + m_B}\right) m_A g \qquad (4.35)$$

Hence,

$$T = \left(\frac{2 m_A m_B}{m_A + m_B}\right) g \qquad (4.39)$$

is the tension in the string of the Atwood's machine.

Special Cases Any formulation in physics should reduce to some simple, recognizable form when certain restrictions are placed on the motion. As an example, suppose a 16-lb (256-oz) bowling ball is placed on one side of Atwood's machine and a small 0.1-oz marble on the other side. What kind of motion would we expect? The bowling ball is so large compared to the marble that the bowling ball should fall like a freely falling body. What does the formulation for the acceleration in equation 4.38 say?

If the bowling ball is m_A and the marble is m_B, then m_A is very much greater than m_B and can be written mathematically as

$$m_A \gg m_B$$

Then,

$$m_A + m_B \approx m_A$$

As an example,

$$\frac{256}{g} + \frac{0.1}{g} = \frac{256.1}{g} \approx \frac{256}{g} = m_A$$

Similarly,

$$m_A - m_B \approx m_A$$

As an example,

$$\frac{256}{g} - \frac{0.1}{g} = \frac{255.9}{g} \approx \frac{256}{g} = m_A$$

Therefore the acceleration of the system, equation 4.38, becomes

$$a = \left(\frac{m_A - m_B}{m_A + m_B}\right) g = \frac{m_A}{m_A} g = g$$

That is, the equation for the acceleration of the system reduces to the acceleration due to gravity, as we would expect if one mass is very much larger than the other.

Another special case is where both masses are equal. That is, if

$$m_A = m_B$$

then the acceleration of the system is

$$a = \left(\frac{m_A - m_B}{m_A + m_B}\right)g = \left(\frac{m_A - m_A}{2m_A}\right)g = 0$$

That is, if both masses are equal there is no acceleration of the system. The system is either at rest or moving at a constant velocity.

Alternate Solution to Atwood's Machine   A simpler solution to Atwood's machine can be obtained directly from Newton's second law by the intuitive approach. The acceleration of the system, found from Newton's second law, is

$$a = \frac{F}{m}$$

where F is the resultant force acting on the system and m is the total mass in motion. The resultant force acting on the system is the difference between the two weights, $w_A - w_B$, and the total mass of the system is the sum of the two masses that are in motion, namely $m_A + m_B$. Thus,

$$a = \frac{F}{m} = \frac{w_A - w_B}{m_A + m_B} = \left(\frac{m_A - m_B}{m_A + m_B}\right)g$$

the same result we found before in equation 4.38.

The Weight of a Person Riding in an Elevator

A scale is placed on the floor of an elevator. A 192-lb person enters the elevator when it is at rest and stands on the scale. What does the scale read when (a) the elevator is at rest, (b) the elevator is accelerating upward at 5.00 ft/s², (c) the acceleration becomes zero and the elevator moves at the constant velocity of 5.00 ft/s upward, (d) the elevator decelerates at 5.00 ft/s² before coming to rest, and (e) the cable breaks and the elevator is in free-fall?

A picture of the person in the elevator showing the forces that are acting is drawn in figure 4.15. The forces acting on the person are his weight **w**, acting down, and the reaction force of the elevator floor acting upward, which we call $\mathbf{F_N}$. Applying Newton's second law we obtain

$$\mathbf{F_N} + \mathbf{w} = m\mathbf{a} \tag{4.40}$$

a.   If the elevator is at rest then $\mathbf{a} = 0$ in equation 4.40. Therefore,

$$\mathbf{F_N} + \mathbf{w} = 0$$

$$\mathbf{F_N} = -\mathbf{w}$$

which shows that the floor of the elevator is exerting a force upward, through the scale, on the person, that is equal and opposite to the force that the person is exerting on the floor. Hence,

$$F_N = w = 192 \text{ lb}$$

We usually think of the operation of a scale in terms of us pressing down on the scale, but we can just as easily think of the scale as pushing upward on us. Thus, the person would read 192 lb on the scale.

b.   The doors of the elevator are now closed and the elevator accelerates upward at a rate of 5.00 ft/s². Newton's second law is again given by equation 4.40. We can write this as a scalar equation if the usual convention of positive for up and negative for down is taken. Hence,

$$F_N - w = ma$$

(a)

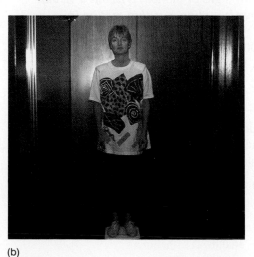

(b)

Figure 4.15

Forces acting on a person in an elevator.

Solving for F_N, we get

$$F_N = w + ma \qquad\qquad (4.41)$$

Substituting the given values into equation 4.41 gives

$$F_N = 192 \text{ lb} + \frac{192 \text{ lb}}{32.0 \text{ ft/s}^2} (5.00 \text{ ft/s}^2)$$

$$= 192 \text{ lb} + 30.0 \text{ lb}$$

$$= 222 \text{ lb}$$

That is, the floor is exerting a force upward on the person of 222 lb. Therefore, the scale would now read 222 lb. Does the person now really weigh 222 lb? Of course not. What the scale is reading is the person's weight plus the additional force of 30.0 lb that is applied to the person, via the scales and floor of the elevator, to cause the person to be accelerated upward along with the elevator. I am sure that all of you have experienced this situation. When you step into an elevator and it accelerates upward you feel as though there is a force acting on you, pushing you down. Your knees feel like they might buckle. It is not that something is pushing you down, but rather that the floor is pushing you up. The floor is pushing upward on you with a force greater than your own weight in order to put you into accelerated motion. That extra force upward on you of 30.0 lb is exactly the force necessary to give you the acceleration of $+5.00$ ft/s^2.

c. The acceleration now stops and the elevator moves upward at the constant velocity of 5.00 ft/s. What does the scale read now?
    Newton's second law is again given by equation 4.40, but since $a = 0$,

$$F_N = w = 192 \text{ lb}$$

Notice that this is the same value as when the elevator was at rest. This is a very interesting phenomenon. *The scale reads the same whether you are at rest or moving at a constant velocity. That is, if you are in motion at a constant velocity, and you have no external references to observe that motion, you cannot tell that you are in motion at all.*

I am sure you also have experienced this while riding an elevator. First you feel the acceleration and then you feel nothing. Your usual reaction is to ask "are we moving, or are we at rest?" You then look for a crack around the elevator door to see if you can see any signs of motion. *Without a visual reference, the only way you can sense a motion is if that motion is accelerated.*

d. The elevator now decelerates at 5.00 ft/s^2. What does the scale read?
    Newton's second law is again given by equation 4.40, and writing it in the simplified form, we have

$$F_N - w = -ma \qquad\qquad (4.42)$$

The minus sign on the right-hand side of equation 4.42 indicates that the acceleration vector is opposite to the direction of the motion because the elevator is decelerating. Solving equation 4.42 for F_N gives

$$F_N = w - ma$$

$$= 192 \text{ lb} - \frac{192 \text{ lb}}{32.0 \text{ ft/s}^2} (5.00 \text{ ft/s}^2)$$

$$= 192 \text{ lb} - 30.0 \text{ lb}$$

$$= 162 \text{ lb}$$

Hence, the force acting on the person is less than the person's weight. The effect is very noticeable when you walk into an elevator and accelerate downward (which is the same as decelerating when the elevator is going upward). You feel as if you are falling. Well, you are falling.

At rest the floor exerts a force upward on a 192-lb person of 192 lb, now it only exerts a force upward of 162 lb. The floor is not exerting enough force to hold the person up. Therefore, the person falls. It is a controlled fall of 5.00 ft/s², but a fall nonetheless. The scale in the elevator now reads 162 lb. The difference in that force and the person's weight is the force that accelerates the person downward.

e. Let us now assume that the cable breaks. What is the acceleration of the system now. Newton's second law is again given by equation 4.40, or in simplified form by

$$F_N - w = -ma \tag{4.42}$$

But if the cable breaks, the elevator becomes a freely falling body with an acceleration g. Therefore, equation 4.42 becomes

$$F_N - w = -mg$$

The force that the elevator exerts upward on the person becomes

$$F_N = w - mg$$

But the weight w is equal to mg. Thus,

$$F_N = w - w = 0$$

or

$$F_N = 0$$

Because the scale reads the force that the floor is pushing upward on the person, the scale now reads zero. This is why *it is sometimes said that in free-fall you are weightless, because in free-fall the scale that reads your weight now reads zero.* This is a somewhat misleading statement because you still have mass, and that mass is still attracted down toward the center of the earth. And in this sense you still have a weight pushing you downward. The difference here is that, while standing on the scale, the scale says that you are weightless, only because the scale itself is also in free-fall. As your feet try to press against the scale to read your weight, the scale falls away from them, and does not permit the pressure of your feet against the scale, and so the scale reads zero. From a reference system outside of the elevator, you would say that the falling person still has weight and that weight is causing that person to accelerate downward at the value g. *However, in the frame of reference of the elevator, not only the person seems weightless, but all weights and gravitational forces on anything around the person seem to have disappeared.* Normally, at the surface of the earth, if a person holds a pen and then lets go, the pen falls. But in the freely falling elevator, if a person lets go of the pen it will not fall to the floor, but will appear to be suspended in space in front of the person as if it were floating. According to the reference frame outside the elevator the pen is accelerating downward at the same rate as the person. But in the elevator, both are falling at the value g and therefore do not move with respect to one another. *In the freely falling reference system of the elevator, the force of gravity and its acceleration appear to have been eliminated.*

4.6 Friction

Whenever we try to slide one body over another body there is a force that opposes that motion. This opposing force is called the force of friction. For example, if this book is placed on the desk and a force is exerted on the book toward the right, there is a force of friction acting on the book toward the left opposing the applied force, as shown in figure 4.16.

Opposed
frictional force

Applied
force

f

F

Figure 4.16

The force of friction.

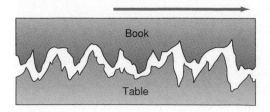

Figure 4.17

The "smooth" surfaces of contact that cause frictional forces.

You push backward on ground

Ground pushes forward on you

Figure 4.18

You can walk because of friction.

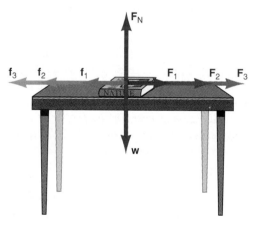

Figure 4.19

The force of static friction.

The basis of this frictional force stems from the fact that the surfaces that slide over each other are really not smooth at all. The top of the desk feels smooth to the hand, and so does the book, but that is because our hands themselves are not particularly smooth. In fact, if we magnified the surface of the book, or the desk, thousands of times, we would see a great irregularity in the supposedly "smooth" surface, as shown in figure 4.17.

As we try to slide the book along the desk these little microscopic chunks of the material get in each others way, and get stuck in the "mountains" and "valleys" of the other material, thereby opposing the tendency of motion. This is why it is difficult to slide one body over another. To get the body into motion we have to break off, or ride over, these microscopic chunks of matter. Because these chunks are microscopic, we do not immediately see the effect of this loss of material. Over a long period of time, however, the effect is very noticeable. As an example, if you observe any step of a stairway, which should be flat and level, you will notice that after a long period of time the middle of the stair is worn from the thousands of times a foot slid on the step in the process of walking up or down the stairs. This effect occurs whether the stairs are made of wood or even marble.

The same wearing process occurs on the soles and heels of shoes, and eventually they must be replaced. In fact the walking process can only take place because there is friction between the shoes and the ground. In the process of walking, in order to step forward, you must press your foot backward on the ground. But because there is friction between your shoe and the ground, there is a frictional force tending to oppose that motion of your shoe backward and therefore the ground pushes forward on your shoe, which allows you to walk forward, as shown in figure 4.18.

If there were no frictional force, your foot would slip backward and you would not be able to walk. This effect can be readily observed by trying to walk on ice. As you push your foot backward, it slips on the ice. You might be able to walk very slowly on the ice because there is some friction between your shoes and the ice. But try to run on the ice and see how difficult it is. If friction were entirely eliminated you could not walk at all.

Force of Static Friction

If this book is placed on the desk, as in figure 4.19, and a small force $\mathbf{F_1}$ is exerted to the right, we observe that the book does not move. There must be a frictional force $\mathbf{f_1}$ to the left that opposes the tendency of motion to the right. That is, $\mathbf{f_1} = -\mathbf{F_1}$.

If we increase the force to the right to $\mathbf{F_2}$, and again observe that the book does not move, the opposing frictional force must also have increased to some new value $\mathbf{f_2}$, where $\mathbf{f_2} = -\mathbf{F_2}$. If we now increase the force to the right to some value $\mathbf{F_3}$, the book just begins to move. The frictional force to the left has increased to some value $\mathbf{f_3}$, where $\mathbf{f_3}$ is infinitesimally less than $\mathbf{F_3}$. The force to the right is now greater than the frictional force to the left and the book starts to move to the right. When the object just begins to move, it has been found experimentally that the frictional force is

$$f_s = \mu_s F_N \tag{4.43}$$

where F_N is the normal or perpendicular force holding the two bodies in contact with each other. As we can see in figure 4.19, the forces acting on the book in the vertical are the weight of the body w, acting downward, and the normal force F_N of the desk, pushing upward on the book. In this case, since the acceleration of the book in the vertical is zero, the normal force F_N is exactly equal to the weight of the book w. (If the desk were tilted, F_N would still be the force holding the two objects together, but it would no longer be equal to w.)

The quantity μ_s in equation 4.43 is called the coefficient of static friction and depends on the materials of the two bodies which are in contact. Coefficients

Table 4.1
Approximate Coefficients of Static and Kinetic Friction
for Various Materials in Contact

Materials in Contact	μ_s	μ_k
Glass on glass	0.95	0.40
Steel on steel (lubricated)	0.15	0.09
Wood on wood	0.50	0.30
Wood on stone	0.50	0.40
Rubber tire on dry concrete	1.00	0.70
Rubber tire on wet concrete	0.70	0.50
Leather on wood	0.50	0.40
Teflon on steel	0.04	0.04
Copper on steel	0.53	0.36

of static friction for various materials are given in table 4.1. It should be noted that these values are approximate and will vary depending on the condition of the rubbing surfaces.

As we have seen, the force of static friction is not always equal to the product of μ_s and F_N, but can be less than that amount, depending on the value of the applied force tending to move the body. Therefore, the **force of static friction** should be written as

$$f_s \leq \mu_s F_N \qquad (4.44)$$

where the symbol $\leq$ means "equal to, or less than." The only time that the equality holds is when the object is just about to go into motion.

Force of Kinetic Friction

Once the object is placed into motion, it is easier to keep it in motion. That is, the force that is necessary to keep the object in motion is much less than the force necessary to start the object into motion. In fact once the object is in motion, we no longer talk of the force of static friction, but rather we talk of the **force of kinetic friction** or sliding friction. For a moving object the frictional force is found experimentally as

$$f_k = \mu_k F_N \qquad (4.45)$$

and is called the force of kinetic friction. The quantity μ_k is called the coefficient of kinetic friction and is also given for various materials in table 4.1. Note from the table that the coefficients of kinetic friction are less than the coefficients of static friction. This means that less force is needed to keep the object in motion, than it is to start it into motion.

We should note here, that these laws of friction are *empirical laws,* and are not exactly like the other laws of physics. For example, with Newton's second law, when we apply an unbalanced external force on a body of mass m, that body is accelerated by an amount given by $a = F/m$, and is always accelerated by that amount. Whereas the frictional forces are different, they are average results. That is, on the average equations 4.44 and 4.45 are correct. At any one given instant of time a force equal to $f_s = \mu_s F_N$, could be exerted on the book of figure 4.19, and yet the book might not move. At still another instance of time a force somewhat less than $f_s = \mu_s F_N$, is exerted and the book does move. Equation 4.43 represents an average result over very many trials. On the average, this equation is correct, but any one individual case may not conform to this result. Hence, this law is not quite as exact as the other laws of physics. In fact, if we return to figure 4.17, we see that it is not so surprising that the frictional laws are only averages, because at any one instant of time there are different interactions between the "mountains" and "valleys" of the two surfaces.

When two substances of the same material are slid over each other, as for example, copper on copper, we get the same kind of results. But if the two surfaces could be made "perfectly smooth," the frictional force would not decrease, but would rather increase. When we get down to the atomic level of each surface that is in contact, the atoms themselves have no way of knowing to which piece of copper they belong, that is, do the atoms belong to the top piece or to the bottom piece. The molecular forces between the atoms of copper would bind the two copper surfaces together.

In most applications of friction in technology, it is usually desirable to minimize the friction as much as possible. Since liquids and gases show much lower frictional effects (liquids and gases possess a quality called *viscosity*—a fluid friction), a layer of oil is usually placed between two metal surfaces as a lubricant, which reduces the friction enormously. The metal now no longer rubs on metal, but rather slides on the layer of the lubricant between the surfaces.

For example, when you put oil in your car, the oil is distributed to the moving parts of the engine. In particular, the oil coats the inside wall of the cylinders in the engine. As the piston moves up and down in the cylinder it slides on this coating of oil, and the friction between the piston and the cylinder is reduced.

Similarly when a glider is placed on an air track, the glider rests on a layer or cushion of air. The air acts as the lubricant, separating the two surfaces of glider and track. Hence, the frictional force between the glider and the air track is so small that in almost all cases it can be neglected in studying the motion of the glider.

When the skates of an ice skater press on the ice, the increased pressure causes a thin layer of the ice to melt. This liquid water acts as a lubricant to decrease the frictional force on the ice skater. Hence the ice skater seems to move effortlessly over the ice, figure 4.20.

Rolling Friction

To reduce friction still further, a wheel or ball of some type is introduced. When something can roll, the frictional force becomes very much less. Many machines in industry are designed with ball bearings, so that the moving object rolls on the ball bearings and friction is greatly reduced.

The whole idea of rolling friction is tied to the concept of the wheel. Some even consider the beginning of civilization as having started with the invention of the wheel, although many never even think of a wheel as something that was invented. The wheel goes so far back into the history of mankind that no one knows for certain when it was first used, but it was an invention. In fact, there were some societies that never discovered the wheel.

The frictional force of a wheel is very small compared with the force of sliding friction, because, theoretically, there is no relative motion between the rim of a wheel and the surface over which it rolls. Because the wheel touches the surface only at a point, there is no sliding friction. The small amount of rolling friction that does occur in practice is caused by the deformation of the wheel as it rolls over the surface, as shown in figure 4.21. Notice that the part of the tire in contact with the ground is actually flat, not circular.

In practice, that portion of the wheel that is deformed does have a tendency to slide along the surface and does produce a frictional force. So the smaller the

Figure 4.20

An ice skater takes advantage of reduced friction.

Figure 4.21

The deformation of a rolling wheel.

deformation, the smaller the frictional force. The harder the substance of the wheel, the less it deforms. For example, with steel on steel there is very little deformation and hence very little friction.

4.7 Applications of Newton's Second Law Taking Friction into Account

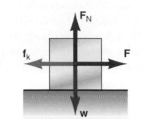

Figure 4.22

A box on a rough floor.

Example 4.9

A box on a rough floor. A 50.0-lb wooden box is at rest on a wooden floor, as shown in figure 4.22. (a) What horizontal force is necessary to start the box into motion? (b) If a force of 20.0 lb is continuously applied once the box is in motion, what will be its acceleration?

Solution

a. Whenever a problem says that a surface is rough, it means that we must take friction into account in the solution of the problem. The minimum force necessary to overcome static friction is found from equation 4.43. Hence, using table 4.1

$$F = f_s = \mu_s F_N$$
$$= \mu_s w = (0.50)(50.0 \text{ lb})$$
$$= 25 \text{ lb}$$

Note that whenever we say that $F = f_s$, we mean that F is an infinitesimal amount greater than f_s, and that it acts for an infinitesimal period of time. If the block is at rest, and $F = f_s$, then the net force acting on the block would be zero, its acceleration would be zero, and the block would therefore remain at rest forever. Thus, F must be an infinitesimal amount greater than f_s for the block to move. Now an infinitesimal quantity is, as the name implies, an extremely small quantity, so for all practical considerations we can assume that the force F plus the infinitesimal quantity, is just equal to the force F in all our equations. This is a standard technique that we will use throughout the study of physics. We will forget about the infinitesimal quantity and just say that the applied force is equal to the force to be overcome. But remember that there really must be that infinitesimal amount more, if the motion is to start.

b. Newton's second law applied to the box is

$$F - f_k = ma \qquad \textbf{(4.46)}$$

The force of kinetic friction, found from equation 4.45 and table 4.1, is

$$f_k = \mu_k F_N = \mu_k w$$
$$= (0.30)(50.0 \text{ lb})$$
$$= 15 \text{ lb}$$

The acceleration of the block, found from equation 4.46, is

$$a = \frac{F - f_k}{m} = \frac{F - f_k}{w/g}$$
$$= \frac{20.0 \text{ lb} - 15 \text{ lb}}{50.0 \text{ lb}/32.0 \text{ ft/s}^2}$$
$$= 3.2 \text{ ft/s}^2$$

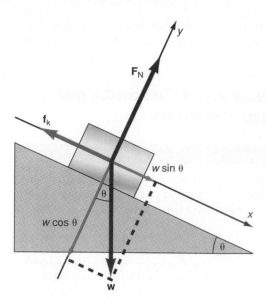

Figure 4.23

Block on an inclined plane with friction.

Example 4.10

A block on a rough inclined plane. Find the acceleration of a block on an inclined plane, as shown in figure 4.23, taking friction into account.

Solution

The problem is very similar to the one solved in figure 4.10, which was for a frictionless plane. We draw all the forces and their components as before, but now we introduce the frictional force. Because the frictional force always opposes the sliding motion, and $w \sin \theta$ acts to move the block down the plane, the frictional force f_k in opposing that motion must be pointed up the plane, as shown in figure 4.23. The block is given a slight push to overcome any force of static friction. To determine the acceleration, we use Newton's second law,

$$\mathbf{F} = m\mathbf{a} \qquad (4.9)$$

However, we can write this as two component equations, one parallel to the inclined plane and the other perpendicular to it.

Components Parallel to the Plane: Taking the direction down the plane as positive, Newton's second law becomes

$$w \sin \theta - f_k = ma \qquad (4.47)$$

Notice that this is very similar to the equation for the frictionless plane, except for the term f_k, the force of friction that is slowing down this motion.

Components Perpendicular to the Plane: Newton's second law for the perpendicular components is

$$F_N - w \cos \theta = 0 \qquad (4.48)$$

The right-hand side is zero because there is no acceleration perpendicular to the plane. That is, the block does not jump off the plane or crash through the plane so there is no acceleration perpendicular to the plane. The only acceleration is the one parallel to the plane, which was just found.

The frictional force f_k, given by equation 4.45, is

$$f_k = \mu_k F_N$$

where F_N is the normal force holding the block in contact with the plane. When the block was on a horizontal surface F_N was equal to the weight w. But now it is not. Now F_N, found from equation 4.48, is

$$F_N = w \cos \theta \qquad (4.49)$$

That is, because the plane is tilted, the force holding the block in contact with the plane is now $w \cos \theta$ rather than just w. Therefore, the frictional force becomes

$$f_k = \mu_k F_N = \mu_k w \cos \theta \qquad (4.50)$$

Substituting equation 4.50 back into Newton's second law, equation 4.47, we get

$$w \sin \theta - \mu_k w \cos \theta = ma$$

but since $w = mg$ this becomes

$$mg \sin \theta - \mu_k mg \cos \theta = ma$$

Since the mass m is in every term of the equation it can be divided out, and the acceleration of the block down the plane becomes

$$a = g \sin \theta - \mu_k g \cos \theta \qquad (4.51)$$

Note that the acceleration is independent of the mass m, since it canceled out of the equation. Also note that this equation reduces to the result for a frictionless plane, equation 4.18, when there is no friction, that is, when $\mu_k = 0$.

In this example, if $\mu_k = 0.300$ and $\theta = 30.0°$, the acceleration becomes

$$a = g \sin \theta - \mu_k g \cos \theta$$
$$= (9.80 \text{ m/s}^2)\sin 30.0° - (0.300)(9.80 \text{ m/s}^2)\cos 30.0°$$
$$= 4.90 \text{ m/s}^2 - 2.55 \text{ m/s}^2$$
$$= 2.35 \text{ m/s}^2$$

Notice the difference between the acceleration when there is no friction (4.90 m/s²) and when there is (2.35 m/s²). The block was certainly slowed down by friction.

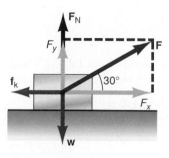

Figure 4.24

Pulling a block on a rough floor.

Example 4.11

Pulling a block on a rough floor. What force is necessary to pull a 50.0-lb wooden box at a constant speed over a wooden floor by a rope that makes an angle θ of 30° above the horizontal, as shown in figure 4.24?

Solution

Let us start by drawing all the forces that are acting on the box in figure 4.24. We break down the applied force into its components F_x and F_y. If Newton's second law is applied to the horizontal components, we obtain

$$F_x - f_k = ma_x \tag{4.52}$$

However, since the box is to move at constant speed, the acceleration a_x must be zero. Therefore,

$$F_x - f_k = 0$$

or

$$F \cos \theta - f_k = 0 \tag{4.53}$$

but

$$f_k = \mu_k F_N$$

where F_N is the normal force holding the box in contact with the floor. Before we can continue with our solution we must determine F_N.

If Newton's second law is applied to the vertical forces we have

$$F_y + F_N - w = ma_y \tag{4.54}$$

but because there is no acceleration in the vertical direction, a_y is equal to zero. Therefore,

$$F_y + F_N - w = 0$$

Solving for F_N we have

$$F_N = w - F_y$$

or

$$F_N = w - F \sin \theta \tag{4.55}$$

Note that F_N is not simply equal to w, as it was in example 4.9, but rather to $w - F \sin \theta$. The y-component of the applied force has the effect of lifting part of the weight from the floor. Hence, the force holding the box in contact with the floor is less than its weight. The frictional force therefore becomes

$$f_k = \mu_k F_N = \mu_k(w - F \sin \theta) \qquad (4.56)$$

and substituting this back into equation 4.53, we obtain

$$F \cos \theta - \mu_k(w - F \sin \theta) = 0$$

or

$$F \cos \theta + \mu_k F \sin \theta - \mu_k w = 0$$

Factoring out the force F,

$$F(\cos \theta + \mu_k \sin \theta) = \mu_k w$$

and finally, solving for the force necessary to move the block at a constant speed, we get

$$F = \frac{\mu_k w}{\cos \theta + \mu_k \sin \theta} \qquad (4.57)$$

Using the value of $\mu_k = 0.30$ (wood on wood) from table 4.1 and substituting the values for w, θ, and μ_k into equation 4.57, we obtain

$$F = \frac{\mu_k w}{\cos \theta + \mu_k \sin \theta} = \frac{(0.30)(50 \text{ lb})}{\cos 30° + 0.30 \sin 30°}$$

$$= 14.8 \text{ lb}$$

Example 4.12

Combined motion of a block moving on a rough horizontal surface and a mass falling vertically. Find the acceleration of a block, on a "rough" table, connected by a cord passing over a pulley to a second block hanging over the table, as shown in figure 4.25.

Solution

This problem is similar to the problem solved in figure 4.12, only now the effects of friction are taken into account. We still assume that the mass of the string and the pulley are negligible. All the forces acting on the two blocks are drawn in figure 4.25. We apply Newton's second law to block A, obtaining

$$T - w_A = -m_A a \qquad (4.58)$$

Applying it to block B, we obtain

$$T - f_k = m_B a \qquad (4.59)$$

where the force of kinetic friction is

$$f_k = \mu_k F_N = \mu_k w_B \qquad (4.60)$$

Substituting equation 4.60 into equation 4.59, we have

$$T - \mu_k w_B = m_B a \qquad (4.61)$$

We eliminate the tension T in the equations by subtracting equation 4.58 from equation 4.61. Thus,

Subtract
$$\begin{array}{ll} T - \mu_k w_B = m_B a & (4.61) \\ T - w_A = -m_A a & (4.58) \\ \hline T - \mu_k w_B - T + w_A = m_B a + m_A a \\ w_A - \mu_k w_B = (m_B + m_A)a \end{array}$$

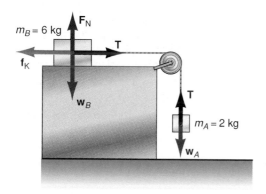

Figure 4.25

Combined motion of a block moving on a rough horizontal surface and a mass falling vertically.

Mechanics

Solving for the acceleration a, we have

$$a = \frac{w_A - \mu_k w_B}{m_A + m_B}$$

But since $w = mg$, this becomes

$$a = \left(\frac{m_A - \mu_k m_B}{m_A + m_B}\right) g \qquad \textbf{(4.62)}$$

the acceleration of the system. Note that if there is no friction, $\mu_k = 0$ and the equation reduces to equation 4.32, the acceleration without friction.

If $m_A = 2.00$ kg, $m_B = 6.00$ kg, and $\mu_k = 0.30$ (wood on wood), then the acceleration of the system is

$$a = \left(\frac{m_A - \mu_k m_B}{m_A + m_B}\right) g = \left[\frac{2.00\text{ kg} - (0.30)6.00\text{ kg}}{2.00\text{ kg} + 6.00\text{ kg}}\right] 9.80 \text{ m/s}^2$$

$$= 0.25 \text{ m/s}^2$$

This is only one-tenth of the acceleration obtained when there was no friction. It is interesting to see what happens if μ_k is equal to 0.40 instead of the value of 0.30 used in this problem. For this new value of μ_k, the acceleration becomes

$$a = \left(\frac{m_A - \mu_k m_B}{m_A + m_B}\right) g = \left[\frac{2.00\text{ kg} - (0.40)6.00\text{ kg}}{2.00\text{ kg} + 6.00\text{ kg}}\right] 9.80 \text{ m/s}^2$$

$$= -0.49 \text{ m/s}^2$$

The negative sign indicates that the acceleration is in the opposite direction of the applied force, which is of course absurd; that is, the block on the table m_B would be moving to the left while block m_A would be moving up. Something is very wrong here. In physics we try to analyze nature and the way it works. But, obviously nature just does not work this way. This is a very good example of trying to use a physics formula when it doesn't apply. Equation 4.62, like all equations, was derived using certain assumptions. If those assumptions hold in the application of the equation, then the equation is valid. If the assumptions do not hold, then the equation is no longer valid. Equation 4.62 was derived from Newton's second law on the basis that block m_B was moving to the right and therefore the force of kinetic friction that opposed that motion would be to the left. For $\mu_k = 0.40$ the force of friction is greater than the tension in the cord and the block does not move at all, that is, the acceleration of the system is zero. In fact if we look carefully at equation 4.62 we see that the acceleration will be zero if

$$m_A - \mu_k m_B = 0$$

which becomes

$$\mu_k m_B = m_A$$

and

$$\mu_k = \frac{m_A}{m_B} \qquad \textbf{(4.63)}$$

Whenever μ_k is equal to or greater than this ratio the acceleration is always zero. Even if we push the block to overcome static friction the kinetic friction is still too great and the block remains at rest. Whenever you solve a problem, always look at the numerical answer and see if it makes sense to you.

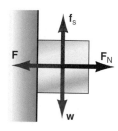

Figure 4.26

Pushing a block up a rough inclined plane.

Example 4.13

Pushing a block up a rough inclined plane. What force F is necessary to push a block up a rough inclined plane at a constant velocity?

Solution

The first thing to note is that if the block is to be pushed up the plane, then the frictional force, which always opposes the sliding motion, must act down the plane. The forces are shown in figure 4.26. Newton's second law for the parallel component becomes

$$-F + w \sin \theta + f_{\mathbf{k}} = 0 \qquad (4.64)$$

The right-hand side of equation 4.64 is 0 because the block is to be moved at constant velocity, that is, $a = 0$. The frictional force $f_{\mathbf{k}}$ is

$$f_{\mathbf{k}} = \mu_{\mathbf{k}} F_{\mathbf{N}} = \mu_{\mathbf{k}} w \cos \theta \qquad (4.65)$$

Hence, equation 4.64 becomes

$$F = w \sin \theta + f_{\mathbf{k}} = w \sin \theta + \mu_{\mathbf{k}} w \cos \theta$$

Finally,

$$F = w(\sin \theta + \mu_{\mathbf{k}} \cos \theta) \qquad (4.66)$$

is the force necessary to push the block up the plane at a constant velocity.

It is appropriate to say something more about this force. If the block is initially at rest on the plane, then there is a force of static friction acting up the plane opposing the tendency of the block to slide down the plane. When the force is exerted to move the block up the plane, then the tendency for the sliding motion is up the plane. Now the force of static friction reverses and acts down the plane. When the applied force F is slightly greater than $w \sin \theta + f_{\mathbf{s}}$, the block will just be put into motion up the plane. Now that the block is in motion, the frictional force to be overcome is the force of kinetic friction, which is less than the force of static friction. The force necessary to move the block up the plane at constant velocity is given by equation 4.66. Because the net force acting on the block is zero, the acceleration of the block is zero. If the block is at rest with a zero net force, then the block would have to remain at rest. However, the block was already set into motion by overcoming the static frictional forces, and since it is in motion, it will continue in that motion as long as the force given by equation 4.66 is applied.

Example 4.14

A book pressed against a rough wall. A 1.00-lb book is held against a wall by pressing it against the wall with a force of 5.00 lb. What must be the minimum coefficient of friction between the book and the wall, such that the book does not slide down the wall? The forces acting on the book are shown in figure 4.27.

Solution

The book has a tendency to slide down the wall because of its weight. Because frictional forces always tend to oppose sliding motion, there is a force of static friction acting upward on the book. If the book is not to fall, then $f_{\mathbf{s}}$ must not be less than the weight of the book w. Therefore, let

$$f_{\mathbf{s}} = w \qquad (4.67)$$

but

$$f_{\mathbf{s}} = \mu_{\mathbf{s}} F_{\mathbf{N}} = \mu_{\mathbf{s}} F \qquad (4.68)$$

Figure 4.27

A book pressed against a rough wall.

Substituting equation 4.68 into equation 4.67, we obtain

$$\mu_s F = w$$

Solving for the coefficient of static friction, we obtain

$$\mu_s = \frac{w}{F} = \frac{1.00\ \text{lb}}{5.00\ \text{lb}} = 0.200$$

Therefore, the minimum coefficient of static friction to hold the book against the wall is $\mu_s = 0.200$.

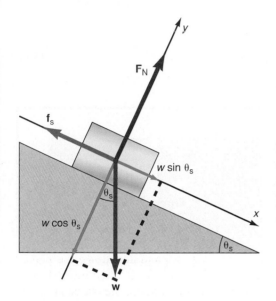

Figure 4.28

Determining the coefficient of static friction.

4.8 Determination of the Coefficients of Friction

If the coefficient of friction for any two materials can not be found in a standardized table, it can always be found experimentally in the laboratory as follows.

Coefficient of Static Friction

To determine the coefficient of static friction, we use an inclined plane whose surface is made up of one of the materials. As an example, let the plane be made of pine wood and the block that is placed on the plane will be made of oak wood. The forces acting on the block are shown in figure 4.28. We increase the angle θ of the plane until the block just begins to slide. We measure this angle where the block starts to slip and call it θ_s, the *angle of repose*.

We assume that the acceleration a of the block is still zero, because the block is just on the verge of slipping. Applying Newton's second law to the block gives

$$w \sin \theta_s - f_s = 0 \qquad (4.69)$$

where

$$f_s = \mu_s F_N = \mu_s w \cos \theta_s \qquad (4.70)$$

Substituting equation 4.70 back into equation 4.69 we have

$$w \sin \theta_s - \mu_s w \cos \theta_s = 0$$
$$w \sin \theta_s = \mu_s w \cos \theta_s$$
$$\mu_s = \frac{\sin \theta_s}{\cos \theta_s}$$

Therefore, the coefficient of static friction is

$$\mu_s = \tan \theta_s \qquad (4.71)$$

That is, the coefficient of static friction μ_s is equal to the tangent of the angle θ_s, found experimentally. With this technique, the coefficient of static friction between any two materials can easily be found.

Coefficient of Kinetic Friction

The coefficient of kinetic friction is found in a similar way. We again place a block on the inclined plane and vary the angle, but now we give the block a slight push to overcome the force of static friction. The block then slides down the plane at a constant velocity. Experimentally, this is slightly more difficult to accomplish because it is difficult to tell when the block is moving at a constant velocity, rather than being accelerated. However, with a little practice we can determine when it is

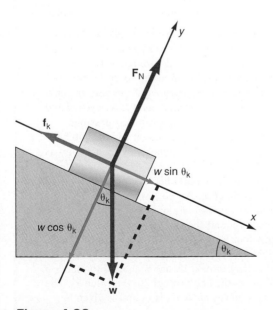

Figure 4.29

Determining the coefficient of kinetic friction.

moving at constant velocity. We measure the angle at which the block moves at constant velocity and call it θ_k. Since there is no acceleration, Newton's second law becomes

$$w \sin \theta_k - f_k = 0 \tag{4.72}$$

but

$$f_k = \mu_k F_N = \mu_k w \cos \theta_k$$
$$w \sin \theta_k - \mu_k w \cos \theta_k = 0$$
$$w \sin \theta_k = \mu_k w \cos \theta_k$$
$$\mu_k = \frac{\sin \theta_k}{\cos \theta_k}$$

Therefore, the coefficient of kinetic friction for the two materials in contact is

$$\mu_k = \tan \theta_k \tag{4.73}$$

"Have you ever wondered . . . ?"

An Essay on the Application of Physics

The Physics of Sports

Have you ever wondered, while watching a baseball game, why the pitcher goes through all those gyrations (figure 1) in order to throw the baseball to the batter? Why can't he throw the ball like all the rest of the players? No one else on the field goes through that big windup. Is there a reason for him to do that?

In order to understand why the pitcher goes through that big windup, let us first analyze the process of throwing a ball, figure 2. From what we already know about Newton's second law, we know you must exert a force on the ball to give it an acceleration. When you hold the ball initially in your hand, with your hand extended behind your head, the ball is at rest and hence has a zero initial velocity, that is, $v_0 = 0$. You now exert the force F on the ball as you move your arm through the distance x_1. The ball is now accelerated by your arm from a zero initial velocity to the final velocity v_1, as it leaves your hand. We find the velocity of the ball from the kinematic equation

$$v_1^2 = v_0^2 + 2ax_1 \tag{H4.1}$$

But since v_0 is equal to zero, the velocity of the ball as it leaves your hand is

$$v_1^2 = 2ax_1$$
$$v_1 = \sqrt{2ax_1} \tag{H4.2}$$

But the acceleration of the ball comes from Newton's second law as

$$a = \frac{F}{m}$$

(a)   (b)   (c)

Figure 1

Look at that form.

Substituting this into the equation for the velocity we get

$$v_1 = \sqrt{2(F/m)x_1} \tag{H4.3}$$

which tells us that the velocity of the ball depends on the mass m of the ball, the force F that your arm exerts on the ball, and the distance x_1 that you move the ball through while you are accelerating it. Since you cannot change the force F that your arm is capable of applying, nor the mass m of the ball, the only way to maximize the velocity v of the ball as it leaves your hand is to increase the value of x to as large a value as possible.

Maximizing the value of x is the reason for the pitcher's long windup. In figure 3, we see the pitcher moving his hand as far backward as possible. In order for the pitcher not to fall down as he leans that far backward, he lifts his left foot forward and upward to maintain his balance. As he lowers his left leg his right arm starts to move forward. As his left foot touches the ground, he lifts his right foot off the ground and swings his body around until his right foot is as far forward as he can make it, while bringing his right arm as far forward as he can, figure 3(b). By going through this long motion he has managed to increase the distance that he moves the ball through, to the value x_2. The velocity of the ball as it leaves his hand is v_2 and is given by

$$v_2 = \sqrt{2(F/m)x_2} \tag{H4.4}$$

Figure 2
The process of throwing a ball.

Figure 3
A pitcher throwing a baseball.

Taking the ratio of these two velocities we obtain

$$\frac{v_2}{v_1} = \frac{\sqrt{2(F/m)x_2}}{\sqrt{2(F/m)x_1}}$$

which simplifies to

$$\frac{v_2}{v_1} = \sqrt{\frac{x_2}{x_1}}$$

The velocity v_2 becomes

$$v_2 = \sqrt{\frac{x_2}{x_1}}\, v_1 \qquad \textbf{(H4.5)}$$

Hence, by going through that long windup, the pitcher has increased the distance to x_2, thereby increasing the value of the

velocity that he can throw the baseball to v_2. For example, for an average person, x_1 is about 50 in., while x_2 is about 125 in. Therefore, the velocity becomes

$$v_2 = \sqrt{\frac{125}{50}}\, v_1$$
$$= 1.58\, v_1$$

Thus, if a pitcher is normally capable of throwing a baseball at a speed of 60 mph, by going through the long windup, the speed of the ball becomes

$$v_2 = 1.58(60 \text{ mph}) = 95 \text{ mph}$$

The long windup has allowed the pitcher to throw the baseball at 95 mph, much faster than the 60 mph that he could normally throw the ball. So this is why the pitcher goes through all those gyrations.

The Language of Physics

Dynamics
That branch of mechanics concerned with the forces that change or produce the motions of bodies. The foundation of dynamics is Newton's laws of motion (p. 79).

Newton's first law of motion
A body at rest will remain at rest, and a body in motion at a constant velocity will continue in motion at that constant velocity, unless acted on by some unbalanced external force. This is sometimes referred to as the law of inertia (p. 80).

Force
The simplest definition of a force is a push or a pull that acts on a body. Force can also be defined in a more general way by Newton's second law, that is, a force is that which causes a mass m to have an acceleration a (p. 80).

Inertia
The characteristic of matter that causes it to resist a change in motion is called inertia (p. 80).

Inertial coordinate system
A coordinate system that is either at rest or moving at a constant velocity with respect to another coordinate system that is either

at rest or also moving at some constant velocity. Newton's first law of motion defines an inertial coordinate system. That is, if a body is at rest or moving at a constant velocity in a coordinate system where there are no unbalanced forces acting on the body, the coordinate system is an inertial coordinate system. Newton's first law must be applied in an inertial coordinate system (p. 81).

Newton's third law of motion
If there are two bodies, A and B, and if body A exerts a force on body B, then body B exerts an equal but opposite force on body A (p. 81).

Newton's second law of motion

If an unbalanced external force **F** acts on a body of mass m, it will give that body an acceleration **a**. The acceleration is directly proportional to the applied force and inversely proportional to the mass of the body. Once the acceleration is determined by Newton's second law, the position and velocity of the body can be determined by the kinematic equations (p. 84).

Inertial mass

The measure of the resistance of a body to a change in its motion is called the inertial mass of the body. The mass of a body in Newton's second law is the inertial mass of the body. The best that can be determined at this time is that the inertial mass of a body is equal to the gravitational mass of the body (p. 92).

Atwood's machine

A simple pulley device that is used to study the acceleration of a system of bodies (p. 97).

Friction

The resistance offered to the relative motion of two bodies in contact. Whenever we try to slide one body over another body, the force that opposes the motion is called the force of friction (p. 101).

Force of static friction

The force that opposes a body at rest from being put into motion (p. 103).

Force of kinetic friction

The force that opposes a body in motion from continuing that motion. The force of kinetic friction is always less than the force of static friction (p. 103).

Summary of Important Equations

Newton's second law
$$F = ma \tag{4.9}$$

The weight of a body
$$w = mg \tag{4.12}$$

Definition of inertial mass
$$m_2 = \frac{a_1}{a_2} m_1 \tag{4.22}$$

Force of static friction
$$f_s \leq \mu_s F_N \tag{4.44}$$

Force of kinetic friction
$$f_k = \mu_k F_N \tag{4.45}$$

Coefficient of static friction
$$\mu_s = \tan \theta_s \tag{4.71}$$

Coefficient of kinetic friction
$$\mu_k = \tan \theta_k \tag{4.73}$$

Questions for Chapter 4

1. A force was originally defined as a push or a pull. Define the concept of force dynamically using Newton's laws of motion.
2. Discuss the difference between the ancient Greek philosophers' requirement of a constantly applied force as a condition for motion with Galileo's and Newton's concept of a force to initiate an acceleration.
3. Is a coordinate system that is accelerated in a straight line an inertial coordinate system? Describe the motion of a projectile in one dimension in a horizontally accelerated system.
4. If you drop an object near the surface of the earth it is accelerated downward to the earth. By Newton's third law, can you also assume that a force is exerted on the earth and the earth should be accelerated upward toward the object? Can you observe such an acceleration? Why or why not?
†5. Discuss an experiment that could be performed on a tilted air track whereby changing the angle of the track would allow you to prove that the acceleration of a body is proportional to the applied force.

Why could you not use this same experiment to show that the acceleration is inversely proportional to the mass?
†6. Discuss the concept of mass as a quantity of matter, a measure of the resistance of matter to being put into motion, and a measure of the gravitational force acting on the mass. Has the original platinum-iridium cylinder, which is stored in Paris, France, and defined as the standard of mass, ever been accelerated so that mass can be defined in terms of its inertial characteristics? Does it have to? Which is the most fundamental definition of mass?
7. From the point of view of the different concepts of mass, discuss why all bodies fall with the same acceleration near the surface of the earth.
8. Discuss why the normal force F_N is not always equal to the weight of the body that is in contact with a surface.
9. In the discussion of Atwood's machine, we assumed that the tension in the string is the same on both sides of the pulley. Can a pulley rotate if the tension is the same on both sides of the pulley?

†10. You are riding in an elevator and the cable breaks. The elevator goes into free fall. The instant before the elevator hits the ground, you jump upward about 3 ft. Will this do you any good? Discuss your motion with respect to the elevator and with respect to the ground. What will happen to you?
†11. Discuss the old saying: "If a horse pulls on a cart with a force F, then by Newton's third law the cart pulls backward on the horse with the same force F, therefore the horse can not move the cart."
12. A football is filled with mercury and taken into space where it is weightless. Will it hurt to kick this football since it is weightless?
†13. A 110-lb lady jumps out of a plane to go skydiving. She extends her body to obtain maximum frictional resistance from the air. After a while, she descends at a constant speed, called her terminal speed. At this time, what is the value of the frictional force of the air?
14. When a baseball player catches a ball he always pulls his glove backward. Why does he do this?

Problems for Chapter 4

In all problems assume that all objects are initially at rest, i.e., $v_0 = 0$, unless otherwise stated.

4.4 Newton's Second Law of Motion

1. What is the mass of a 200-lb person?
2. What is the weight of a 100-kg person at the surface of the earth? What would the person weigh on Mars where $g = 3.84$ m/s²?
3. What horizontal force must be applied to a 15.0-kg body in order to give it an acceleration of 5.00 m/s²?
4. A constant force accelerates a 3200-lb car from 0 to 60.0 mph in 12.0 s. Find (a) the acceleration of the car and (b) the force acting on the car that produces the acceleration.
5. A 3200-lb car is traveling along a highway at 60.0 mph. If the driver immediately applies his brakes and the car comes to rest in a distance of 250 ft, what average force acted on the car during the deceleration?
6. A 910-kg car is traveling along a highway at 88.0 km/hr. If the driver immediately applies his brakes and the car comes to rest in a distance of 70.0 m, what average force acted on the car during the deceleration?
7. A car is traveling at 60.0 mph when it collides with a stone wall. The car comes to rest after the first foot of the car is crushed. What was the average horizontal force acting on a 150-lb driver while the car came to rest? If five cardboard boxes, each 4 ft wide and filled with sand had been placed in front of the wall, and the car moved through all that sand before coming to rest, what would the average force acting on the driver have been then?
8. A rifle bullet of mass 12.0 g has a muzzle velocity of 750 m/s. What is the average force acting on the bullet when the rifle is fired, if the bullet is accelerated over the entire 1.00-m length of the rifle?
9. A car is to tow a 5000-lb truck with a rope. How strong should the rope be so that it will not break when accelerating the truck from rest to 10.0 ft/s in 12.0 s?
10. A force of 200 lb acts on a body that weighs 60.0 lb. (a) What is the mass of the body? (b) What is the acceleration of the body? (c) If the body starts from rest, how fast will it be going after it has moved 10.0 ft?

11. A cable supports an elevator that weighs 8000 N. (a) What is the tension T in the cable when the elevator accelerates upward at 1.50 m/s²? (b) What is the tension when the elevator accelerates downward at 1.50 m/s²?
12. A rope breaks when the tension exceeds 30.0 N. What is the minimum acceleration downward that a 60.0-N load can have without breaking the rope?
13. A 5.00-g bullet is fired at a speed of 100 m/s into a fixed block of wood and it comes to rest after penetrating 6.00 cm into the wood. What is the average force stopping the bullet?
14. A rope breaks when the tension exceeds 100 lb. What is the maximum vertical acceleration that can be given to an 80.0-lb load to lift it with this rope without breaking the rope?
15. What horizontal force must a locomotive exert on a 1000-ton train to increase its speed from 15.0 mph to 30.0 mph in moving 200 ft along a level track?
16. A steady force of 10.0 lb, exerted 30.0° above the horizontal, acts on a 50.0-lb sled on level snow. How far will the sled move in 10.0 s? (Neglect friction.)
17. A steady force of 70.0 N, exerted 43.5° above the horizontal, acts on a 30.0-kg sled on level snow. How far will the sled move in 8.50 s? (Neglect friction.)
18. A helicopter rescues a person at sea by pulling him upward with a cable. If the person weighs 180 lb and is accelerated upward at 1.00 ft/s², what is the tension in the cable?

4.5 Applications of Newton's Second Law

19. A force of 10.0 N acts horizontally on a 20.0-kg mass that is at rest on a smooth table. Find (a) the acceleration, (b) the velocity at 5.00 s, and (c) the position of the body at 5.00 s. (d) If the force is removed at 7.00 s, what is the body's velocity at 7.00, 8.00, 9.00, and 10.0 s?
20. A 50.0-lb box slides down a frictionless inclined plane that makes an angle of 37.0° with the horizontal. (a) What unbalanced force acts on the block? (b) What is the acceleration of the block?

21. A 20.0-kg block slides down a smooth inclined plane. The plane is 10.0 m long and is inclined at an angle of 30.0° with the horizontal. Find (a) the acceleration of the block, and (b) the velocity of the block at the bottom of the plane.
22. A 180-lb person stands on a scale in an elevator. What does the scale read when (a) the elevator is ascending with an acceleration of 4.00 ft/s², (b) it is ascending at a constant velocity of 8.00 ft/s, (c) it decelerates at 4.00 ft/s², (d) it descends at a constant velocity of 8.00 ft/s, and (e) the cable breaks and the elevator is in free-fall?
23. A 90.0-kg person stands on a scale in an elevator. What does the scale read when (a) the elevator is ascending with an acceleration of 1.50 m/s², (b) it is ascending at a constant velocity of 3.00 m/s, (c) it decelerates at 1.50 m/s², (d) it descends at a constant velocity of 3.00 m/s, and (e) the cable breaks and the elevator is in free-fall?
24. A spring scale is attached to the ceiling of an elevator. If a mass of 2.00 kg is placed in the pan of the scale, what will the scale read when (a) the elevator is accelerated upward at 1.50 m/s², (b) it is decelerated at 1.50 m/s², (c) it is moving at constant velocity, and (d) the cable breaks and the elevator is in free-fall?
†25. A block is propelled up a 48.0° frictionless inclined plane with an initial velocity $v_0 = 1.20$ m/s. (a) How far up the plane does the block go before coming to rest? (b) How long does it take to move to that position?
†26. In the diagram m_A is equal to 3.00 kg and m_B is equal to 1.50 kg. The angle of the inclined plane is 38.0°. (a) Find the acceleration of the system of two blocks. (b) Find the tension T_B in the connecting string.

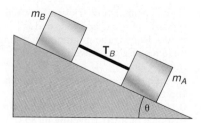

27. The two masses $m_A = 2.00$ kg and $m_B = 20.0$ kg are connected as shown. The table is frictionless. Find (a) the acceleration of the system, (b) the velocity of m_B at $t = 3.00$ s, and (c) the position of m_B at $t = 3.00$ s.

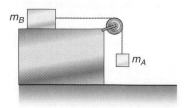

28. A 30.0-g mass and a 50.0-g mass are placed on an Atwood machine. Find (a) the acceleration of the system, (b) the velocity of the 50.0-g block at 4.00 s, (c) the position of the 50.0-g mass at the end of the fourth second, (d) the tension in the connecting string.

†29. Three blocks of mass $m_1 = 100$ g, $m_2 = 200$ g, and $m_3 = 300$ g are connected by strings as shown. (a) What force F is necessary to give the masses a horizontal acceleration of 4 m/s²? Find the tensions T_1 and T_2.

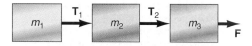

†30. A force of 20.0 lb acts as shown on the two blocks. If the blocks are on a frictionless surface, find the acceleration of each block and the horizontal force exerted on each block.

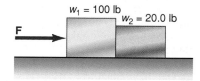

4.7 Applications of Newton's Second Law Taking Friction into Account

31. If the coefficient of friction between the tires of a car and the road is 0.300, what is the minimum stopping distance of a car traveling at 60.0 mph?

32. If the coefficient of friction between the tires of a car and the road is 0.300, what is the minimum stopping distance of a car traveling at 85.0 km/hr?

33. A 200-N container is to be pushed across a rough floor. The coefficient of static friction is 0.500 and the coefficient of kinetic friction is 0.400.

What force is necessary to start the container moving, and what force is necessary to keep it moving at a constant velocity?

34. A 2.00-kg toy accelerates from rest to 3.00 m/s in 8.00 s on a rough surface of $\mu_k = 0.300$. Find the applied force F.

35. A 50.0-lb box is to be moved along a rough floor at a constant velocity. The coefficient of friction is $\mu_k = 0.300$. (a) What force F_1 must you exert if you push downward on the box as shown? (b) What force F_2 must you exert if you pull upward on the box as shown? (c) Which is the better way to move the box?

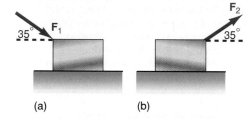

(a)   (b)

36. A 5.00-lb book is held against a rough vertical wall. If the coefficient of static friction between the book and the wall is 0.300, what force perpendicular to the wall is necessary to keep the book from sliding?

37. A block slides along a wooden table with an initial speed of 50.0 cm/s. If the block comes to rest in 150 cm, find the coefficient of kinetic friction between the block and the table.

38. What force must act horizontally on a 20.0-kg mass moving at a constant speed of 4.00 m/s on a rough table of coefficient of kinetic friction of 0.300? If the force is removed, when will the body come to rest? Where will it come to rest?

39. A 10.0-kg package slides down an inclined mail chute 15.0 m long. The top of the chute is 6.00 m above the floor. What is the speed of the package at the bottom of the chute if (a) the chute is frictionless and (b) the coefficient of kinetic friction is 0.300?

40. In order to place a 200-lb air conditioner in a window, a plank is laid between the window and the floor, making an angle of 40.0° with the horizontal. How much force is necessary to push the air conditioner up the plank at a constant speed if the coefficient of kinetic friction between the air conditioner and the plank is 0.300?

41. If a 4.00-kg container has a velocity of 3.00 m/s after sliding down a 2.00-m plane inclined at an angle of 30.0°, what is (a) the force of

friction acting on the container and (b) the coefficient of kinetic friction between the container and the plane?

†42. A 100-lb crate sits on the floor of a truck. If $\mu_s = 0.300$, what is the maximum acceleration of the truck before the crate starts to slip?

43. A skier starts from rest and slides a distance of 200 ft down the ski slope. The slope makes an angle of 35.0° with the horizontal. (a) If the coefficient of friction between the skis and the slope is 0.100, find the speed of the skier at the bottom of the slope. (b) At the bottom of the slope the skier continues to move on level snow. Where does the skier come to a stop?

44. A skier starts from rest and slides a distance of 85.0 m down the ski slope. The slope makes an angle of 23.0° with the horizontal. (a) If the coefficient of friction between the skis and the slope is 0.100, find the speed of the skier at the bottom of the slope. (b) At the bottom of the slope, the skier continues to move on level snow. Where does the skier come to a stop?

†45. A mass of 2.00 kg is pushed up an inclined plane that makes an angle of 50.0° with the horizontal. If the coefficient of kinetic friction between the mass and the plane is 0.400, and a force of 50.0 N is applied parallel to the plane, what is (a) the acceleration of the mass and (b) its velocity after moving 3.00 m up the plane?

46. The two masses $m_A = 20$ kg and $m_B = 20$ kg are connected as shown on a rough table. If the coefficient of friction between block B and the table is 0.45, find (a) the acceleration of each block and (b) the tension in the connecting string.

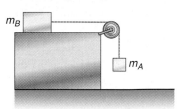

47. To determine the coefficient of static friction, the following system is set up. A mass, $m_B = 2.50$ kg, is placed on a rough horizontal table such as in the diagram for problem 46. When mass m_A is increased to the value of 1.50 kg the system just starts into motion. Determine the coefficient of static friction.

48. To determine the coefficient of kinetic friction, the following system is set up. A mass, $m_B = 2.50$ kg, is placed on a rough horizontal table such as in the diagram for problem 46. Mass m_A has the value of 1.85 kg, and the system goes into accelerated motion with a value a_1. While mass m_A falls to the floor, a distance $x_1 = 30.0$ cm below its starting point, mass m_B will also move through a distance x_1 and will have acquired a velocity v_1 at x_1. When m_A hits the floor, the acceleration a_1 becomes zero. From this point on, the only acceleration m_B experiences is the deceleration a_2 caused by the force of kinetic friction acting on m_B. Mass m_B moves on the rough surface until it comes to rest at the distance $x_2 = 20.0$ cm. From this information, determine the coefficient of kinetic friction.

Additional Problems

†49. Find the force F that is necessary for the system shown to move at constant velocity if $\mu_k = 0.300$ for all surfaces. The masses are $m_A = 6.00$ kg and $m_B = 2.00$ kg.

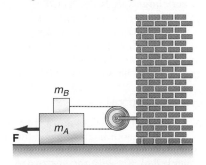

50. A pendulum is placed in a car at rest and hangs vertically. The car then accelerates forward and the pendulum bob is observed to move backward, the string making an angle of 15.0° with the vertical. Find the acceleration of the car.

51. Two gliders are tied together by a string after they are connected together by a compressed spring and placed on an air track. Glider A has a mass of 200 g and the mass of glider B is unknown. The string is now cut and the gliders fly apart. If glider B has an acceleration of 5.00 cm/s² to the right, and the acceleration of glider A to the left is 20.0 cm/s², find the mass of glider B.

52. A mass of 1.87 kg is pushed up a smooth inclined plane with an applied force of 3.50 N parallel to the plane. If the plane makes an angle of 35.8° with the horizontal,

find (a) the acceleration of the mass and (b) its velocity after moving 1.50 m up the plane.

†53. Two blocks $m_1 = 20.0$ kg and $m_2 = 10.0$ kg are connected as shown on a frictionless plane. The angle $\theta = 25.0°$ and $\phi = 35.0°$. Find the acceleration of each block and the tension in the connecting string.

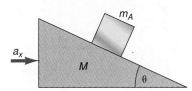

†54. What horizontal acceleration a_x must the inclined block M have in order for the smaller block m_A not to slide down the frictionless inclined plane? What force must be applied to the system to keep the block from sliding down the frictionless plane? $M = 10.0$ kg, $m_A = 1.50$ kg, and $\theta = 43°$.

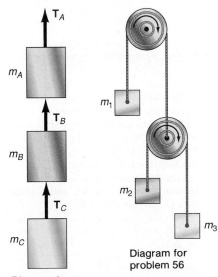

†55. If the acceleration of the system is 3.00 m/s² when it is lifted, and $m_A = 5.00$ kg, $m_B = 3.00$ kg, and $m_C = 2.00$ kg, find the tensions T_A, T_B, and T_C.

†56. Consider the double Atwood's machine as shown. If $m_1 = 50.0$ g, $m_2 = 20.0$ g, and $m_3 = 25.0$ g, what is the acceleration of m_3?

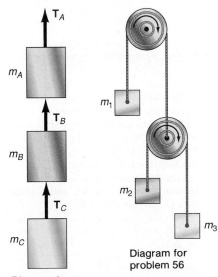

Diagram for problem 55

Diagram for problem 56

†57. Find the tension T_{23} in the string between mass m_2 and m_3, if $m_1 = 10.0$ kg, $m_2 = 2.00$ kg, and $m_3 = 1.00$ kg.

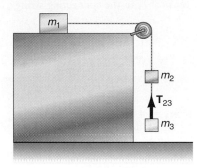

†58. If $m_A = 6.00$ kg, $m_B = 3.00$ kg, and $m_C = 2.00$ kg in the diagram, find the magnitude of the acceleration of the system and the tensions T_A, T_B, and T_C.

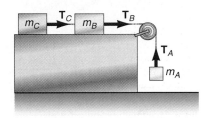

†59. Find (a) the acceleration of mass m_A in the diagram. All surfaces are frictionless. (b) Find the displacement of block A at $t = 0.500$ s. The value of the masses are $m_A = 3.00$ kg and $m_B = 5.00$ kg.

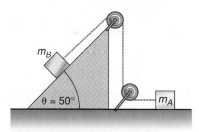

60. A force of 5.00 N acts on a body of mass $m = 2.00$ kg at an angle of 35.0° above the horizontal. If the coefficient of friction between the body and the surface upon which it is resting is 0.250, find the acceleration of the mass.

†61. Derive the formula for the magnitude of the acceleration of the system shown in the diagram.
(a) What problem does this reduce to if $\phi = 90°$?
(b) What problem does this reduce to if both θ and ϕ are equal to 90°?

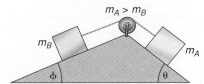

†62. What force is necessary to pull the two masses at constant speed if m_1 = 2.00 kg, m_2 = 5.00 kg, μ_{k_1} = 0.300, and μ_{k_2} = 0.200? What is the tension T_1 in the connecting string?

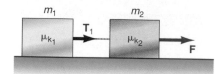

†63. If m_A = 4.00 kg, m_B = 2.00 kg, μ_{k_A} = 0.300, and μ_{k_B} = 0.400, find (a) the acceleration of the system down the plane and (b) the tension in the connecting string.

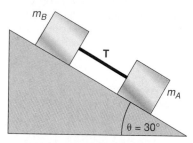

†64. A block m = 0.500 kg slides down a frictionless inclined plane 2.00 m long. It then slides on a rough horizontal table surface of μ_k = 0.300 for 0.500 m. It then leaves the top of the table, which is 1.00 m high. How far from the base of the table does the block land?

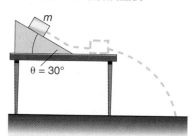

†65. In the diagram m_A = 6.00 kg, m_B = 3.00 kg, m_C = 2.00 kg, μ_{k_C} = 0.400, and μ_{k_B} = 0.300. Find the magnitude of the acceleration of the system and the tension in each string.

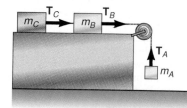

†66. In the diagram m_A = 4.00 kg, m_B = 2.00 kg, m_C = 4.00 kg, and θ = 58°. If all the surfaces are frictionless, find the magnitude of the acceleration of the system.

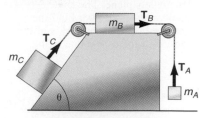

†67. If m_A = 6.00 kg, m_B = 2.00 kg, m_C = 4.00 kg, and the coefficient of kinetic friction for the surfaces are μ_{k_B} = 0.300 and μ_{k_C} = 0.200 find the magnitude of the acceleration of the system shown in the diagram and the tension in each string.

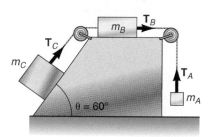

†68. Find (a) the magnitude of the acceleration of the system shown if μ_{k_B} = 0.300, μ_{k_A} = 0.200, m_B = 3.00 kg, and m_A = 5.00 kg, (b) the velocity of block A at 0.500 s.

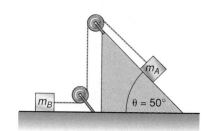

†69. In the diagram, block B rests on a frictionless surface but there is friction between blocks B and C. m_A = 2.00 kg, m_B = 3.00 kg, and m_C = 1 kg. Find (a) the magnitude of the acceleration of the system and (b) the minimum coefficient of friction between blocks C and B such that C will move with B.

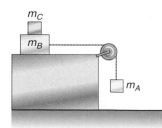

†70. When a body is moving through the air, the effect of air resistance can be taken into account. If the speed of the body is not too great, the force associated with the retarding force of air friction is proportional to the first power of the velocity of the moving body. This retarding force causes the velocity of a falling body at any time t to be

$$v = \frac{mg}{k}(1 - e^{-(k/m)t})$$

where m is the mass of the falling body and k is a constant that depends on the shape of the body. Show that this reduces to the case of a freely falling body if t and k are both small. (*Hint:* expand the term $e^{-(k/m)t}$ in a power series.)

71. Repeat problem 70, but now let the time t be very large (assume it is infinite). What does the velocity of the falling body become now? Discuss this result with Aristotle's statement that heavier objects fall faster than lighter objects. Clearly distinguish between the concepts of velocity and acceleration.

†72. If a body moves through the air at very large speeds the retarding force of friction is proportional to the square of the speed of the body, that is, $f = kv^2$, where k is a constant. Find the equation for the terminal velocity of such a falling body.

Interactive Tutorials

☐ 73. A 20.0-kg block slides down from the top of a smooth inclined plane that is 10.0 m long and is inclined at an angle θ of 30° with the horizontal. Find the acceleration a of the block and its velocity v at the bottom of the plane. Assume the initial velocity v_0 = 0.

☐ 74. Two masses m_A = 40.0 kg and m_B = 30.0 kg are connected by a massless string that hangs over a massless, frictionless pulley in an Atwood's machine arrangement as shown in figure 4.14. Calculate the acceleration a of the system and the tension T in the string.

☐ 75. A mass m_A = 40.0 kg hangs over a table connected by a massless string to a mass m_B = 20.0 kg that is on a rough horizontal table, with a coefficient of friction μ_k = 0.400, that is similar to figure 4.25. Calculate the acceleration a of the system and the tension T in the string.

76. Generalization of problem 61 that also includes friction. Derive the formula for the magnitude of the acceleration of the system shown in the diagram for problem 61. As a general case, assume that the coefficient of kinetic friction between block A and the surface in μ_{k_A} and between block B and the surface is μ_{k_B}. Identify and solve for all the special cases that you can think of.

77. Free fall with friction—variable acceleration—terminal velocity. In the freely falling body studied in chapter 3, we assumed that the resistance of the air could be considered negligible. Let us now remove that constraint. Assume that there is frictional force caused by the motion through the air, and let us further assume that the frictional force is proportional to the square of the velocity of the moving body and is given by

$$f = kv^2$$

Find the displacement, velocity, and acceleration of the falling body and compare it to the displacement, velocity, and acceleration of a freely falling body without friction.

78. The mass of the connecting string is not negligible. In the problem of the combined motion of a block on a frictionless horizontal plane and a block falling vertically, as shown in figure 4.12, it was assumed that the mass of the connecting string was negligible and had no effect on the problem. Let us now remove that constraint. Assume that the string is a massive string. The string has a linear mass density of 0.050 kg/m and is 1.25 m long. Find the acceleration, velocity, and displacement y of the system as a function of time, and compare it to the acceleration, velocity, and displacement of the system with the string of negligible mass.

5

Statics in our lives.

Nature and Nature's laws lay hid in night:
God said, Let Newton be! and all was light.

Alexander Pope

Equilibrium

5.1 The First Condition of Equilibrium

The simplest way to define the **equilibrium** of a body is to say that *a body is in equilibrium if it has no acceleration.* That is, if the acceleration of a body is zero, then it is in equilibrium. Bodies in equilibrium under a system of forces are described as a special case of Newton's second law,

$$\mathbf{F} = m\mathbf{a} \qquad (4.9)$$

where $\mathbf{F}$ is the resultant force acting on the body. As pointed out in chapter 4, to emphasize the point that $\mathbf{F}$ is the resultant force, Newton's second law is sometimes written in the form

$$\Sigma \mathbf{F} = m\mathbf{a}$$

If there are forces acting on a body, but the body is not accelerated (i.e., $\mathbf{a} = 0$), then the body is in equilibrium under these forces and the condition for that body to be in equilibrium is simply

$$\Sigma \mathbf{F} = 0 \qquad (5.1)$$

Equation 5.1 is called the first condition of equilibrium. *The **first condition of equilibrium** states that for a body to be in equilibrium, the vector sum of all the forces acting on that body must be zero.* If the sum of the force vectors are added graphically they will form a closed figure because the resultant vector, which is equal to the sum of all the force vectors, is equal to zero.

Remember that if the acceleration is zero, then there is no change of the velocity with time. Most of the cases considered in this book deal with bodies that are at rest ($v = 0$) under the applications of forces. Occasionally we also consider a body that is moving at a constant velocity (also a case of zero acceleration). At first, we consider only examples where all the forces act through only one point of the body. Forces that act through only one point of the body are called *concurrent forces. That portion of the study of mechanics that deals with bodies in equilibrium is called statics.* When a body is at rest under a series of forces it is sometimes said to be in static equilibrium.

One of the simplest cases of a body in equilibrium is a book resting on the table, as shown in figure 5.1. The forces acting on the book are its weight $\mathbf{w}$, acting downward, and $\mathbf{F}_N$, the normal force that the table exerts upward on the book.

Figure 5.1

A body in equilibrium.

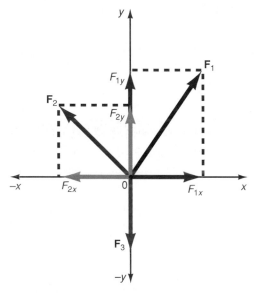

Figure 5.2

Three forces in equilibrium.

Because the book is resting on the table, it has zero acceleration. Hence, the sum of all the forces acting on the book must be zero and the book must be in equilibrium. The sum of all the forces are

$$\Sigma \mathbf{F} = \mathbf{F_N} + \mathbf{w} = 0$$

Taking the upward direction to be positive and the downward direction to be negative, this becomes

$$F_N - w = 0$$

Hence,

$$F_N = w$$

That is, the force that the table exerts upward on the book is exactly equal to the weight of the book acting downward. As we can easily see, this is nothing more than a special case of Newton's second law where the acceleration is zero. That is, forces can act on a body without it being accelerated if these forces balance each other out.

Let us consider another example of a body in equilibrium, as shown in figure 5.2. Suppose three forces $\mathbf{F_1}$, $\mathbf{F_2}$, and $\mathbf{F_3}$ are acting on the body that is located at the point 0, the origin of a Cartesian coordinate system. If the body is in equilibrium, then the vector sum of those forces must add up to zero and the body is not accelerating. Another way to observe that the body is in equilibrium is to look at the components of the forces, which are shown in figure 5.2. From the diagram we can see that if the sum of all the forces in the x-direction is zero, then there will be no acceleration in the x-direction. If the forces in the positive x-direction are taken as positive, and those in the negative x-direction as negative, then the sum of the forces in the x-direction is simply

$$F_{1x} - F_{2x} = 0 \qquad (5.2)$$

Similarly, if the sum of all the forces in the y-direction is zero, there will be no acceleration in the y-direction. As seen in the diagram, this becomes

$$F_{1y} + F_{2y} - F_3 = 0 \qquad (5.3)$$

A generalization of equations 5.2 and 5.3 is

$$\Sigma F_x = 0 \qquad (5.4)$$
$$\Sigma F_y = 0 \qquad (5.5)$$

which is another way of stating the first condition of equilibrium.

The first condition of equilibrium also states that the body is in equilibrium if the sum of all the forces in the x-direction is equal to zero and the sum of all the forces in the y-direction is equal to zero. Equations 5.4 and 5.5 are two component equations that are equivalent to the one vector equation 5.1.

Although only bodies in equilibrium in two dimensions will be treated in this book, if a third dimension were taken into account, an additional equation ($\Sigma \mathbf{F_z} = 0$) would be necessary. Let us now consider some examples of bodies in equilibrium.

Example 5.1

A ball hanging from a vertical rope. A ball is hanging from a rope that is attached to the ceiling, as shown in figure 5.3. Find the tension in the rope. We assume that the mass of the rope is negligible and can be ignored in the problem.

Mechanics

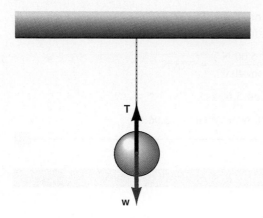

Figure 5.3

Ball hanging from a vertical rope.

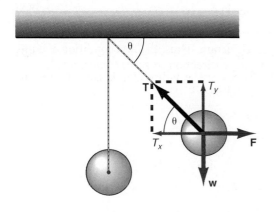

Figure 5.4

Ball pulled to one side.

The first thing that we should observe is that even though there are forces acting on the ball, the ball is at rest. That is, the ball is in *static equilibrium*. Therefore, the first condition of equilibrium must hold, that is,

$$\Sigma F_x = 0 \tag{5.4}$$

$$\Sigma F_y = 0 \tag{5.5}$$

The first step in solving the problem is to draw a diagram showing the forces that are acting on the ball. There is the weight **w,** acting downward in the negative *y*-direction, and the tension **T** in the rope, acting upward in the positive *y*-direction. Note that there are no forces in the *x*-direction so we do not use equation 5.4. The first condition of equilibrium for this problem is

$$\Sigma F_y = 0 \tag{5.5}$$

and, as we can see from the diagram in figure 5.3, this is equivalent to

$$T - w = 0$$

or

$$T = w$$

The tension in the rope is equal to the weight of the ball. If the ball weighs 5 N, then the tension in the rope is 5 N.

Example 5.2

The ball is pulled to one side. A ball hanging from a rope, is pulled to the right by a horizontal force **F** such that the rope makes an angle θ with the ceiling, as shown in figure 5.4. What is the tension in the rope?

Solution

The first thing that we should observe is that the system is at rest. Therefore, the ball is in static equilibrium and the first condition of equilibrium holds. But the tension **T** is neither in the *x*- nor *y*-direction. Before we can use equations 5.4 and 5.5, we must resolve the tension T into its components, T_x and T_y, as shown in figure 5.4. The first condition of equilibrium,

$$\Sigma F_x = 0 \tag{5.4}$$

is applied, which, as we see from figure 5.4 gives

$$\Sigma F_x = F - T_x = 0$$

or

$$F = T_x = T \cos \theta \tag{5.6}$$

Similarly,

$$\Sigma F_y = 0 \tag{5.5}$$

becomes

$$\Sigma F_y = T_y - w = 0$$
$$T_y = T \sin \theta = w \tag{5.7}$$

Note that there are four quantities T, θ, w, and F and only two equations, 5.6 and 5.7. Therefore, if any two of the four quantities are specified, the other two can be determined. *Recall that in order to solve a set of algebraic equations there must always be the same number of equations as unknowns.*

For example, if $w = 5.00$ N and $\theta = 40.0°$, what is the tension T and the force F. We use equation 5.7 to solve for the tension:

$$T = \frac{w}{\sin \theta} = \frac{5.00 \text{ N}}{\sin 40.0°} = 7.78 \text{ N}$$

We determine the force F, from equation 5.6, as

$$F = T \cos \theta = 7.78 \text{ N} \cos 40.0° = 5.96 \text{ N}$$

Example 5.3

Resting in your hammock. A person weighing 150 lb lies in a hammock, as shown in figure 5.5(a). The rope at the person's head makes an angle ϕ of 40.0° with the horizontal, while the rope at the person's feet makes an angle θ of 20.0°. Find the tension in the two ropes.

Solution

All the forces that are acting on the hammock are drawn in figure 5.5(b). The forces are resolved into their components, as shown in figure 5.5(b), where

$$T_{1x} = T_1 \cos \theta$$
$$T_{1y} = T_1 \sin \theta$$
$$T_{2x} = T_2 \cos \phi$$
$$T_{2y} = T_2 \sin \phi \qquad (5.8)$$

The first thing we observe is that the hammock is at rest under the influence of several forces and is therefore in static equilibrium. Thus, the first condition of equilibrium must hold. Setting the forces in the x-direction to zero, equation 5.4,

$$\Sigma F_x = 0$$

gives

$$\Sigma F_x = T_{2x} - T_{1x} = 0$$

and

$$T_{2x} = T_{1x}$$

Figure 5.5

Lying in a hammock.

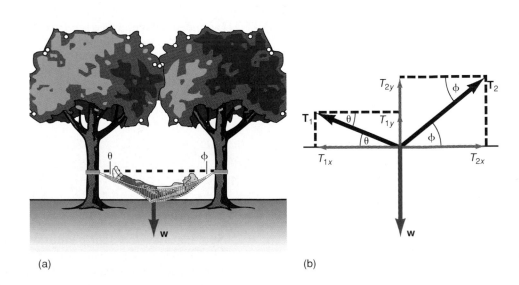

(a)   (b)

Mechanics

Using equations 5.8 for the components, this becomes

$$T_2 \cos \phi = T_1 \cos \theta \tag{5.9}$$

Taking all the forces in the y-direction and setting them equal to zero,

$$\Sigma F_y = 0 \tag{5.5}$$

gives

$$\Sigma F_y = T_{1y} + T_{2y} - w = 0$$

and

$$T_{1y} + T_{2y} = w$$

Using equations 5.8 for the components, this becomes

$$T_1 \sin \theta + T_2 \sin \phi = w \tag{5.10}$$

Equations 5.9 and 5.10 represent the first condition of equilibrium as it applies to this problem. Note that there are five quantities, T_1, T_2, w, θ, and ϕ and only two equations. Therefore, three of these quantities must be specified in order to solve the problem. In this case, θ, ϕ, and w are given and we will determine the tensions T_1 and T_2.

Let us start by solving equation 5.9 for T_2, thus,

$$T_2 = \frac{T_1 \cos \theta}{\cos \phi} \tag{5.11}$$

We cannot use equation 5.11 to solve for T_2 at this point, because T_1 is unknown. Equation 5.11 says that if T_1 is known, then T_2 can be determined. If we substitute this equation for T_2 into equation 5.10, thereby eliminating T_2 from the equations, we can solve for T_1. That is, equation 5.10 becomes

$$T_1 \sin \theta + \left(\frac{T_1 \cos \theta}{\cos \phi} \right) \sin \phi = w$$

Factoring out T_1 we get

$$T_1 \left(\sin \theta + \frac{\cos \theta \sin \phi}{\cos \phi} \right) = w \tag{5.12}$$

Finally, solving equation 5.12 for the tension T_1, we obtain

$$T_1 = \frac{w}{\sin \theta + \cos \theta \tan \phi} \tag{5.13}$$

Note that $\sin \phi / \cos \phi$ in equation 5.12 was replaced by $\tan \phi$, its equivalent, in equation 5.13. Substituting the values of $w = 150$ lb, $\theta = 20.0°$, and $\phi = 40.0°$ into equation 5.13, we find the tension T_1 as

$$\begin{aligned} T_1 &= \frac{w}{\sin \theta + \cos \theta \tan \phi} \\ &= \frac{150 \text{ lb}}{\sin 20.0° + \cos 20.0° \tan 40.0°} \\ &= \frac{150 \text{ lb}}{0.342 + 0.940(0.839)} = \frac{150 \text{ lb}}{1.13} \\ &= 132 \text{ lb} \end{aligned} \tag{5.13}$$

Substituting this value of T_1 into equation 5.11, the tension T_2 in the second rope becomes

$$T_2 = T_1 \frac{\cos \theta}{\cos \phi} = 132 \text{ lb} \frac{\cos 20.0°}{\cos 40.0°}$$

$$= 162 \text{ lb}$$

Note that the tension in each rope is different, that is, T_1 is not equal to T_2. The ropes that are used for this hammock must be capable of withstanding these tensions or they will break.

An interesting special case arises when the angles θ and ϕ are equal. For this case equation 5.11 becomes

$$T_2 = T_1 \frac{\cos \theta}{\cos \phi} = T_1 \frac{\cos \theta}{\cos \theta} = T_1$$

that is,

$$T_2 = T_1$$

For this case, T_1, found from equation 5.13, is

$$T_1 = \frac{w}{\sin \theta + \cos \theta \,(\sin \theta / \cos \theta)}$$

$$= \frac{w}{2 \sin \theta} \qquad (5.14)$$

Thus, when the angle θ is equal to the angle ϕ, the tension in each rope is the same and is given by equation 5.14. Note that if θ were equal to zero in equation 5.14, the tension in the ropes would become infinite. Since this is impossible, the rope must always sag by some amount.

Before leaving this section on the equilibrium of a body let us reiterate that although the problems considered here have been problems where the body is at rest under the action of forces, bodies moving at constant velocity are also in equilibrium. Some of these problems have already been dealt with in chapter 4, that is, examples 4.11 and 4.13 when a block was moving at a constant velocity under the action of several forces, it was a body in equilibrium.

5.2 The Concept of Torque

Let us now consider the familiar seesaw you played on in the local school yard during your childhood. Suppose a 60.0-lb child (w_1) is placed on the left side of a weightless seesaw and another 40.0-lb child (w_2) is placed on the right side, as shown in figure 5.6. The weights of the two children exert forces down on the seesaw, while the support in the middle exerts a force upward, which is exactly equal to the weight of the two children. According to the first condition of equilibrium,

$$\Sigma F_y = 0$$

the body should be in equilibrium. However, we know from experience that if a 60.0-lb child is at the left end, and a 40.0-lb child is at the right end, the 60.0-lb child will move downward, while the 40.0-lb child moves upward. That is, the seesaw rotates in a counterclockwise direction. Even though the first condition of equilibrium holds, the body is not in complete equilibrium because the seesaw has tilted. It is obvious that the first condition of equilibrium is not sufficient to describe equilibrium. The first condition takes care of the problem of *translational equilibrium* (i.e., the body will not accelerate either in the x-direction or the y-direction), but it says nothing about the problem of *rotational equilibrium*.

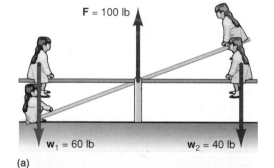

(a)

(b)

Figure 5.6
The seesaw.

Mechanics

In fact, up to this point in almost all our discussions we assumed that all the forces that act on a body all pass through the center of the body. With the seesaw, the forces do not all pass through the center of the body (figure 5.6), but rather act at different locations on the body. Forces acting on a body that do not all pass through one point of the body are called *nonconcurrent forces*. Hence, even though the forces acting on the body cause the body to be in translational equilibrium, the body is still capable of rotating. Therefore, we need to look into the problem of forces acting on a body at a point other than the center of the body; to determine how these off-center forces cause the rotation of the body; and finally to prevent this rotation so that the body will also be in rotational equilibrium. To do this, we need to introduce the concept of torque.

Torque is defined to be the product of the force times the lever arm. The lever arm is defined as the perpendicular distance from the axis of rotation to the line along which the force acts. The line along which the force acts is in the direction of the force vector **F**, and it is sometimes called the *line of action of the force*. The line of action of a force passes through the point of application of the force and is parallel to **F**. This is best seen in figure 5.7. The lever arm appears as $r_\perp$, and the force is denoted by **F**. Note that $r_\perp$ is perpendicular to **F**.

The magnitude of the torque τ (the Greek letter tau) is then defined mathematically as

$$\tau = r_\perp F \tag{5.15}$$

What does this mean physically? Let us consider a very simple example of a torque acting on a body. Let the body be the door to the room. The axes of rotation of the door pass through those hinges that you see at the edge of the door. The distance from the hinge to the door knob is the lever arm $r_\perp$, as shown in figure 5.8. If we exert a force on the door knob by pulling outward, perpendicular to the door, then we have created a torque that acts on the door and is given by equation 5.15. What happens to the door? It opens, just as we would expect. We have caused a rotational motion of the door by applying a torque. Therefore, *an unbalanced torque acting on a body at rest causes that body to be put into rotational motion.* Torque comes from the Latin word *torquere*, which means to twist. We will see in chapter 9, on

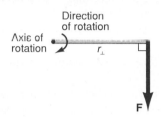

Figure 5.7

Torque defined.

Figure 5.8

An example of a torque applied to a door.

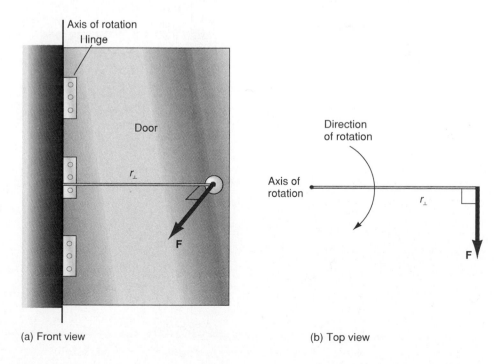

(a) Front view

(b) Top view

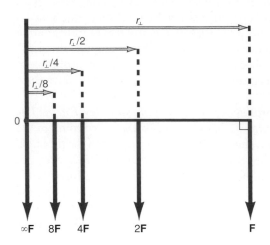

Figure 5.9

If the lever arm decreases, the force must be increased to give the same torque.

rotational motion, that torque is the rotational analogue of force. When an unbalanced force acts on a body, it gives that body a translational acceleration. When an unbalanced torque acts on a body, it gives that body a rotational acceleration.

It is not so much the applied force that opens a door, but rather the applied torque; the product of the force that we apply and the lever arm. A door knob is therefore placed as far away from the hinges as possible to give the maximum lever arm and hence the maximum torque for a given force.

Because the torque is the product of $r_\perp$ and the force F, for a given value of the force, if the distance $r_\perp$ is cut in half, the value of the torque will also be cut in half. If the torque is to remain the same when the lever arm is halved, the force must be doubled, as we easily see in equation 5.15. If a door knob was placed at the center of the door, then twice the original force would be necessary to give the door the same torque. It may even seem strange that some manufacturers of cabinets and furniture place door knobs in the center of cabinet doors because they may have a certain aesthetic value when placed there, but they cause greater exertion by the furniture owner in order to open those doors.

If the door knob was moved to a quarter of the original distance, then four times the original force would have to be exerted in order to supply the necessary torque to open the door. We can see this effect in the diagram of figure 5.9. If the lever arm was finally decreased to zero, then it would take an infinite force to open the door, which is of course impossible. In general, *if a force acts through the axis of rotation of a body, it has no lever arm (i.e., $r_\perp = 0$) and therefore cannot cause a torque to act on the body about that particular axis,* that is, from equation 5.15

$$\tau = r_\perp F = (0)F = 0$$

Instead of exerting a force perpendicular to the door, suppose we exert a force at some other angle θ, as shown in figure 5.10(a), where θ is the angle between the extension of r and the direction of F. Note that in this case r is not a lever arm since it is not perpendicular to F. The definition of a lever arm is the perpendicular distance from the axis of rotation to the line of action of the force. To obtain the lever arm, we extend a line in either the forward or backward direction of the force.

Figure 5.10

If the force is not perpendicular to r.

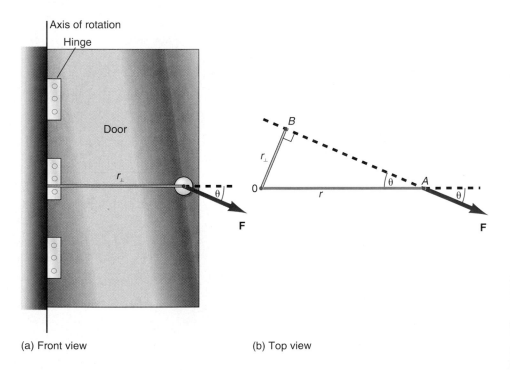

(a) Front view   (b) Top view

Then we drop a perpendicular to this line, as shown in figure 5.10(b). The line extended in the direction of the force vector, and through the point of application of the force, is the line of action of the force. The lever arm, obtained from the figure, is

$$r_\perp = r \sin \theta \qquad (5.16)$$

In general, if the force is not perpendicular to r, the torque equation 5.15 becomes

$$\tau = r_\perp F = rF \sin \theta \qquad (5.17)$$

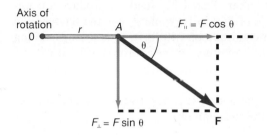

Figure 5.11

The parallel and perpendicular components of a force.

Although this approach to using the lever arm to compute the torque is correct, it may seem somewhat artificial, since the force is really applied at the point A and not the point B in figure 5.10(b). Let us therefore look at the problem from a slightly different point of view, as shown in figure 5.11. Take r, exactly as it is given—the distance from the axis of rotation to the point of application of the force. Then take the force vector **F** and resolve it into two components: one, $F_\parallel$, lies along the direction of r (parallel to r), and the other, $F_\perp$, is perpendicular to r. The component $F_\parallel$ is a force component that goes right through 0, the axis of rotation. But as just shown, if the force goes through the axis of rotation it has no lever arm about that axis and therefore it cannot produce a torque about that axis. Hence, the component of the force parallel to r cannot create a torque about 0.

The component $F_\perp$, on the other hand, does produce a torque, because it is an application of a force that is perpendicular to a distance r. This perpendicular component produces a torque given by

$$\tau = rF_\perp \qquad (5.18)$$

But from figure 5.11 we see that

$$F_\perp = F \sin \theta \qquad (5.19)$$

Thus, the torque becomes

$$\tau = rF_\perp = rF \sin \theta \qquad (5.20)$$

Comparing equation 5.17 to equation 5.20, it is obvious that the results are identical and should be combined into one equation, namely

$$\tau = r_\perp F = rF_\perp = rF \sin \theta \qquad (5.21)$$

Therefore, *the torque acting on a body can be computed either by (a) the product of the force times the lever arm, (b) the product of the perpendicular component of the force times the distance r, or (c) simply the product of r and F times the sine of the angle between F and the extension of r.*

The unit of torque is given by the product of a distance times a force and in SI units, is a m N, (meter newton), while in the British engineering system the unit is a lb ft. According to our definition, we would expect that this unit should be ft lb. However, as we will see in chapter 7, the unit ft lb is used as a unit of energy. In order to avoid using the same unit for two different physical concepts, we will use the notation lb ft for torque in the British engineering system.

5.3 The Second Condition of Equilibrium

Let us now return to the problem of the two children on the seesaw in figure 5.6, which is redrawn schematically in figure 5.12. The entire length of the seesaw is 12.00 ft. From the discussion of torques, it is now obvious that each child produces a torque tending to rotate the seesaw plank. The first child produces a torque about the axis of rotation, sometimes called the *fulcrum,* given by

$$\tau_1 = F_1 r_1 = w_1 r_1 = (60.0 \text{ lb})(6.00 \text{ ft})$$

$$= 360 \text{ lb ft}$$

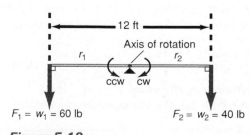

Figure 5.12

The seesaw revisited.

which has a tendency to rotate the seesaw counterclockwise (ccw). *A torque that produces a counterclockwise rotation is sometimes called a counterclockwise torque.* The second child produces a torque about the fulcrum given by

$$\tau_2 = F_2 r_2 = w_2 r_2 = (40.0 \text{ lb})(6.00 \text{ ft})$$

$$= 240 \text{ lb ft}$$

which has a tendency to rotate the seesaw clockwise (cw). *A torque that produces a clockwise rotation is sometimes called a clockwise torque.* These tendencies to rotate the seesaw are opposed to each other. That is, τ_1 tends to produce a counterclockwise rotation with a magnitude of 360 lb ft, while τ_2 has the tendency to produce a clockwise rotation with a magnitude of 240 lb ft. It is a longstanding convention among physicists to designate counterclockwise torques as positive, and clockwise torques as negative. This conforms to the mathematicians' practice of plotting positive angles on an xy plane as measured counterclockwise from the positive x-axis. Hence, τ_1 is a positive torque and τ_2 is a negative torque and the net torque will be the difference between the two, namely

$$\text{net } \tau = \tau_1 - \tau_2 = 360 \text{ lb ft} - 240 \text{ lb ft} = 120 \text{ lb ft}$$

or a net torque τ of 120 lb ft, which will rotate the seesaw counterclockwise.

It is now clear why the seesaw moved. Even though the forces acting on it were balanced, the torques were not. If the torques were balanced then there would be no tendency for the body to rotate, and the seesaw would also be in rotational equilibrium. That is, *the necessary condition for the body to be in rotational equilibrium is that the torques clockwise must be equal to the torques counterclockwise.* That is,

$$\tau_{cw} = \tau_{ccw} \tag{5.22}$$

For this case

$$w_1 r_1 = w_2 r_2 \tag{5.23}$$

We can now solve equation 5.23 for the position r_1 of the first child such that the torques are equal. That is,

$$r_1 = \frac{w_2 r_2}{w_1} = \left(\frac{40.0 \text{ lb}}{60.0 \text{ lb}}\right)(6.00 \text{ ft}) = 4.00 \text{ ft}$$

If the 60.0-lb child moves in toward the axis of rotation by 2.00 ft (4.00 ft from axis), then the torque counterclockwise becomes

$$\tau_1 = \tau_{ccw} = w_1 r_1 = (60.0 \text{ lb})(4.00 \text{ ft}) = 240 \text{ lb ft}$$

which is now equal to the torque τ_2 clockwise. Thus, the torque tending to rotate the seesaw counterclockwise (240 lb ft) is equal to the torque tending to rotate it clockwise (240 lb ft). Hence, the net torque is zero and the seesaw will not rotate. The seesaw is now said to be in rotational equilibrium. This equilibrium condition is shown in figure 5.13.

In general, *for any rigid body acted on by any number of planar torques, the condition for that body to be in rotational equilibrium is that the sum of all the torques clockwise must be equal to the sum of all the torques counterclockwise.* Stated mathematically, this becomes

$$\Sigma \tau_{cw} = \Sigma \tau_{ccw} \tag{5.24}$$

*This condition is called the **second condition of equilibrium.***
If we subtract the term $\Sigma \tau_{cw}$ from both sides of the equation, we obtain

$$\Sigma \tau_{ccw} - \Sigma \tau_{cw} = 0$$

But the net torque is this difference between the counterclockwise and clockwise torques, *so that the second condition for equilibrium can also be written as: for a*

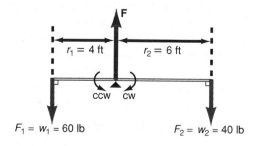

Figure 5.13

The seesaw in equilibrium.

rigid body acted on by any number of torques, the condition for that body to be in rotational equilibrium is that the sum of all the torques acting on that body must be zero, that is,

$$\Sigma \tau = 0 \tag{5.25}$$

The torque is about an axis that is perpendicular to the plane of the paper. Since the plane of the paper is the x,y plane, the torque axis lies along the z-axis. Hence the torque can be represented as a vector that lies along the z-axis. Thus, we can also write equation 5.25 as

$$\Sigma \tau_z = 0$$

In general torques can also be exerted about the x-axis and the y-axis, and for such general cases we have

$$\Sigma \tau_x = 0$$

$$\Sigma \tau_y = 0$$

However, in this text we will restrict ourselves to forces in the x,y plane and torques along the z-axis.

5.4 Equilibrium of a Rigid Body

In general, for a body that is acted on by any number of planar forces, the conditions for that body to be in equilibrium are

$$\Sigma F_x = 0 \tag{5.4}$$

$$\Sigma F_y = 0 \tag{5.5}$$

$$\Sigma \tau_{cw} = \Sigma \tau_{ccw} \tag{5.24}$$

The first condition of equilibrium guarantees that the body will be in translational equilibrium, while the second condition guarantees that the body will be in rotational equilibrium. The solution of various problems of statics reduce to solving the three equations 5.4, 5.5, and 5.25. Section 5.5 is devoted to the solution of various problems of rigid bodies in equilibrium.

5.5 Examples of Rigid Bodies in Equilibrium

Parallel Forces

Two men are carrying a girl on a large plank that is 10.000 m long and weighs 200.0 N. If the girl weighs 445.0 N and sits 3.000 m from one end, how much weight must each man support?

The diagram drawn in figure 5.14(a) shows all the forces that are acting on the plank. We assume that the plank is uniform and the weight of the plank can be located at its center.

The first thing we note is that the body is in equilibrium and therefore the two conditions of equilibrium must hold. The first condition of equilibrium, equation 5.5, applied to figure 5.14 yields,

$$\Sigma F_y = 0$$

$$F_1 + F_2 - w_p - w_g = 0$$

$$F_1 + F_2 = w_p + w_g$$

$$= 200.0 \text{ N} + 445.0 \text{ N}$$

$$F_1 + F_2 = 645.0 \text{ N} \tag{5.26}$$

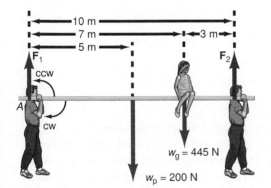

(a) Torques computed about point A

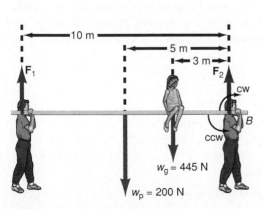

(b) Torques computed about point B

Figure 5.14
A plank in equilibrium under parallel forces.

Since there are no forces in the x-direction, we do not use equation 5.4. The second condition of equilibrium, given by equation 5.24, is

$$\Sigma \tau_{cw} = \Sigma \tau_{ccw}$$

However, before we can compute any torques, we must specify the axis about which the torques will be computed. (In a moment we will see that it does not matter what axis is taken.) For now, let us consider that the axis passes through the point A, where man 1 is holding the plank up with the force F_1. The torques tending to rotate the plank clockwise about axis A are caused by the weight of the plank and the weight of the girl, while the torque tending to rotate the plank counterclockwise about the same axis A is produced by the force F_2 of the second man. Therefore,

$$\Sigma \tau_{cw} = \Sigma \tau_{ccw}$$

$$w_p(5.000 \text{ m}) + w_g(7.000 \text{ m}) = F_2(10.000 \text{ m})$$

Solving for the force F_2 exerted by the second man,

$$F_2 = \frac{w_p(5.000 \text{ m}) + w_g(7.000 \text{ m})}{10.000 \text{ m}}$$

$$= \frac{(200.0 \text{ N})(5.000 \text{ m}) + (445.0 \text{ N})(7.000 \text{ m})}{10.000 \text{ m}}$$

$$= \frac{1000 \text{ m N} + 3115 \text{ m N}}{10.000 \text{ m}}$$

$$F_2 = 411.5 \text{ N} \tag{5.27}$$

Thus, the second man must exert a force upward of 411.5 N. The force that the first man must support, found from equations 5.26 and 5.27, is

$$F_1 + F_2 = 645.0 \text{ N}$$

$$F_1 = 645 \text{ N} - F_2 = 645.0 \text{ N} - 411.5 \text{ N}$$

$$F_1 = 233.5 \text{ N}$$

The first man must exert an upward force of 233.5 N while the second man carries the greater burden of 411.5 N. Note that the force exerted by each man is different. If the girl sat at the center of the plank, then each man would exert the same force.

Let us now see that the same results occur if the torques are computed about any other axis. Let us arbitrarily take the position of the axis to pass through the point B, the location of the force F_2. Since F_2 passes through the axis at point B it cannot produce any torque about that axis because it now has no lever arm. The force F_1 now produces a clockwise torque about the axis through B, while the forces w_p and w_g produce a counterclockwise torque about the axis through B. The solution is

$$\Sigma F_y = 0$$

$$F_1 + F_2 - w_p - w_g = 0$$

$$F_1 + F_2 = w_p + w_g = 645.0 \text{ N}$$

and

$$\Sigma \tau_{cw} = \Sigma \tau_{ccw}$$

$$F_1(10.000 \text{ m}) = w_p(5.000 \text{ m}) + w_g(3.000 \text{ m})$$

Solving for the force F_1,

$$F_1 = \frac{(200.0 \text{ N})(5.000 \text{ m}) + (445.0 \text{ N})(3.000 \text{ m})}{10.000 \text{ m}}$$

$$= \frac{1000 \text{ m N} + 1335 \text{ m N}}{10.000 \text{ m}}$$

$$= 233.5 \text{ N}$$

while the force F_2 is

$$F_2 = 645.0 \text{ N} - F_1$$

$$= 645.0 \text{ N} - 233.5 \text{ N}$$

$$= 411.5 \text{ N}$$

Notice that F_1 and F_2 have the same values as before. As an exercise, take the center of the plank as the point through which the axis passes. Compute the torques about this axis and show that the results are the same.

In general, whenever a rigid body is in equilibrium, every point of that body is in both translational equilibrium and rotational equilibrium, so any point of that body can serve as an axis to compute torques. Even a point outside the body can be used as an axis to compute torques if the body is in equilibrium.

As a general rule, in picking an axis for the computation of torques, try to pick the point that has the largest number of forces acting through it. These forces have no lever arm, and hence produce a zero torque about that axis. This makes the algebra of the problem easier to handle.

The Center of Gravity of a Body

A meter stick of negligible weight has a 10.0-N weight hung from each end. Where, and with what force, should the meter stick be picked up such that it remains horizontal while it moves upward at a constant velocity? This problem is illustrated in figure 5.15.

The meter stick and the two weights constitute a system. If the stick translates with a constant velocity, then the system is in equilibrium under the action of all the forces. The conditions of equilibrium must apply and hence the sum of the forces in the y-direction must equal zero,

$$\Sigma F_y = 0 \tag{5.5}$$

Applying equation 5.5 to this problem gives

$$F - w_1 - w_2 = 0$$

$$F = w_1 + w_2 = 10.0 \text{ N} + 10.0 \text{ N} = 20.0 \text{ N}$$

Therefore, a force of 20 N must be exerted in order to lift the stick. But where should this force be applied? In general, the exact position is unknown so we assume that it can be lifted at some point that is a distance x from the left end of the stick. If this is the correct position, then the body is also in rotational equilibrium and the second condition of equilibrium must also apply. Hence, the sum of the torques clockwise must be set equal to the sum of the torques counterclockwise,

$$\Sigma \tau_{cw} = \Sigma \tau_{ccw} \tag{5.24}$$

Taking the left end of the meter stick as the axis of rotation, the second condition, equation 5.24, becomes

$$w_2 l = Fx \tag{5.28}$$

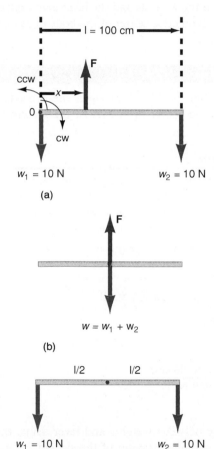

(a)

(b)

(c)

Figure 5.15

The center of gravity of a meter stick.

Since we already found F from the first condition, and w_2 and l are known, we can solve for x, the point where the stick should be lifted:

$$x = \frac{w_2 l}{F} = \frac{(10.0 \text{ N})(100 \text{ cm})}{20.0 \text{ N}}$$

$$= 50.0 \text{ cm}$$

The meter stick should be lifted at its exact geometrical center.

The net effect of these forces can be seen in figure 5.15(b). The force up F is equal to the weight down W. The torque clockwise is balanced by the counterclockwise torque, and there is no tendency for rotation. The stick, with its equal weights at both ends, acts as though all the weights were concentrated at the geometrical center of the stick. *This point that behaves as if all the weight of the body acts through it, is called the center of gravity (cg) of the body.* Hence the center of gravity of the system, in this case a meter stick and two equal weights hanging at the ends, is located at the geometrical center of the meter stick.

The center of gravity is located at the center of the stick because of the symmetry of the problem. The torque clockwise about the center of the stick is w_2 times $1/2$, while the torque counterclockwise about the center of the stick is w_1 times $1/2$, as seen in figure 5.15(c). Because the weights w_1 and w_2 are equal, and the lever arms $(1/2)$ are equal, the torque clockwise is equal to the torque counterclockwise. Whenever such symmetry between the weights and the lever arms exists, the center of gravity is always located at the geometric center of the body or system of bodies.

Example 5.4

The center of gravity when there is no symmetry. If weight w_2 in the preceding discussion is changed to 20.0 N, where will the center of gravity of the system be located?

Solution

The first condition of equilibrium yields

$$\Sigma F_y = 0$$

$$F - w_1 - w_2 = 0$$

$$F = w_1 + w_2 = 10.0 \text{ N} + 20.0 \text{ N}$$

$$= 30.0 \text{ N}$$

The second condition of equilibrium again yields equation 5.28,

$$w_2 l = Fx$$

The location of the center of gravity becomes

$$x = \frac{w_2 l}{F} = \frac{(20.0 \text{ N})(100 \text{ cm})}{30.0 \text{ N}}$$

$$= 66.7 \text{ cm}$$

Thus, when there is no longer the symmetry between weights and lever arms, the center of gravity is no longer located at the geometric center of the system.

General Definition of the Center of Gravity

In the previous section we assumed that the weight of the meter stick was negligible compared to the weights w_1 and w_2. Suppose the weights w_1 and w_2 are eliminated and we want to pick up the meter stick all by itself. The weight of the meter stick can no longer be ignored. But how can the weight of the meter stick be handled?

In the previous problem w_1 and w_2 were discrete weights. Here, the weight of the meter stick is distributed throughout the entire length of the stick. How can the center of gravity of a continuous mass distribution be determined instead of a discrete mass distribution? From the symmetry of the uniform meter stick, we expect that the center of gravity should be located at the geometric center of the 100-cm meter stick, that is, at the point $x = 50$ cm. At this center point, half the mass of the stick is to the left of center, while the other half of the mass is to the right of center. The half of the mass on the left side creates a torque counterclockwise about the center of the stick, while the half of the mass on the right side creates a torque clockwise. Thus, the uniform meter stick has the same symmetry as the stick with two equal weights acting at its ends, and thus must have its center of gravity located at the geometrical center of the meter stick, the 50-cm mark.

To find a general equation for the center of gravity of a body, let us find the equation for the center of gravity of the uniform meter stick shown in figure 5.16. The meter stick is divided up into 10 equal parts, each of length 10 cm. Because the meter stick is uniform, each 10-cm portion contains 1/10 of the total weight of the meter stick, W. Let us call each small weight w_i, where the i is a subscript that identifies which w is being considered.

Because of the symmetry of the uniform mass distribution, each small weight w_i acts at the center of each 10-cm portion. The center of each ith portion, denoted by x_i, is shown in the figure. If a force F is exerted upward at the center of gravity x_{cg}, the meter stick should be balanced. If we apply the first condition of equilibrium to the stick we obtain

$$\Sigma F_y = 0$$

$$F - w_1 - w_2 - w_3 - \cdots - w_{10} = 0$$

$$F = w_1 + w_2 + w_3 + \cdots + w_{10} \tag{5.5}$$

A shorthand notation for this sum can be written as

$$w_1 + w_2 + w_3 + \cdots + w_{10} = \sum_{i=1}^{n} w_i$$

The Greek letter Σ again means "sum of," and when placed in front of w_i it means "the sum of each w_i." The notation $i = 1$ to n, means that we will sum up some n w_i's. In this case, $n = 10$. Using this notation, the first condition of equilibrium becomes

$$F = \sum_{i=1}^{n} w_i = W \tag{5.29}$$

The sum of all these w_i's is equal to the total weight of the meter stick W.

The second condition of equilibrium,

$$\Sigma \tau_{cw} = \Sigma \tau_{ccw} \tag{5.24}$$

Figure 5.16

The weight distribution of a uniform meter stick.

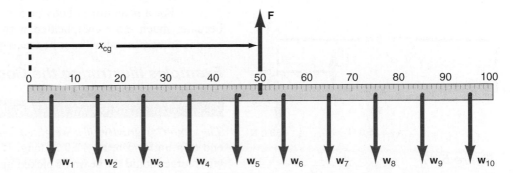

when applied to the meter stick, with the axis taken at the zero of the meter stick, yields

$$(w_1x_1 + w_2x_2 + w_3x_3 + \cdots + w_{10}x_{10}) = Fx_{cg}$$

In the shorthand notation this becomes

$$\sum_{i=1}^{n} w_ix_i = Fx_{cg}$$

Solving for x_{cg}, we have

$$x_{cg} = \frac{\sum_{i=1}^{n} w_ix_i}{F} \tag{5.30}$$

Using equation 5.29, *the general expression for the x-coordinate of the center of gravity of a body is given by*

$$x_{cg} = \frac{\sum_{i=1}^{n} w_ix_i}{W} \tag{5.31}$$

Applying equation 5.31 to the uniform meter stick we have

$$x_{cg} = \frac{\sum w_ix_i}{W} = \frac{w_1x_1 + w_2x_2 + \cdots + w_{10}x_{10}}{W}$$

but since $w_1 = w_2 = w_3 = w_4 = \cdots = w_{10} = W/10$, it can be factored out giving

$$x_{cg} = \frac{W/10}{W}(x_1 + x_2 + x_3 + \cdots + x_{10})$$

$$= 1/10 (5 + 15 + 25 + 45 + \cdots + 95)$$

$$= 500/10$$

$$= 50 \text{ cm}$$

The center of gravity of the uniform meter stick is located at its geometrical center, just as expected from symmetry considerations. The assumption that the weight of a body can be located at its geometrical center, provided that its mass is uniformly distributed, has already been used throughout this book. Now we have seen that this was a correct assumption.

To find the center of gravity of a two-dimensional body, the x-coordinate of the cg is found from equation 5.31, while the y-coordinate, found in an analogous manner, is

$$y_{cg} = \frac{\sum_{i=1}^{n} w_iy_i}{W} \tag{5.32}$$

For a nonuniform body or one with a nonsymmetrical shape, the problem becomes much more complicated with the sums in equations 5.31 and 5.32 becoming integrals and will not be treated in this book.

Examples Illustrating the Concept of the Center of Gravity

Example 5.5

The center of gravity of a weighted beam. A weight of 50.0 N is hung from one end of a uniform beam 12.0 m long. If the beam weighs 25.0 N, where and with what force should the beam be picked up so that it remains horizontal? The problem is illustrated in figure 5.17.

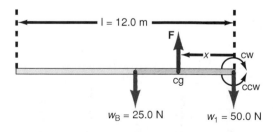

Figure 5.17

The center of gravity of a weighted beam.

Because the beam is uniform, the weight of the beam w_B is located at the geometric center of the beam. Let us assume that the center of gravity of the system of beam and weight is located at a distance x from the right side of the beam. The body is in equilibrium, and the equations of equilibrium become

$$\Sigma F_y = 0 \qquad (5.5)$$

$$F - w_B - w_1 = 0$$

$$F = w_B + w_1$$

$$= 25.0 \text{ N} + 50.0 \text{ N} = 75.0 \text{ N}$$

Taking the right end of the beam as the axis about which the torques are computed, we have

$$\Sigma \tau_{cw} = \Sigma \tau_{ccw} \qquad (5.24)$$

The force F will cause a torque clockwise about the right end, while the force w_B will cause a counterclockwise torque. Hence,

$$Fx = w_B \frac{l}{2}$$

Thus, the center of gravity of the system is located at

$$x_{cg} = \frac{w_B \, l/2}{F}$$

$$= \frac{(25.0 \text{ N})(6.0 \text{ m})}{75.0 \text{ N}} = 2.0 \text{ m}$$

Therefore, we should pick up the beam 2.0 m from the right hand side with a force of 75.0 N.

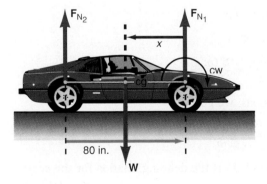

F_{N_2} F_{N_1}

x

cw

cg

80 in.

W

Figure 5.18

The center of gravity of an automobile.

Example 5.6

The center of gravity of an automobile. The front wheels of an automobile, when run onto a platform scale, are found to support 1800 lb, while the rear wheels can support 1500 lb. The auto has an 80.0-in. wheel base (distance from the front axle to the rear axle). Locate the center of gravity of the car. The car is shown in figure 5.18.

Solution

If the car pushes down on the scales with forces w_1 and w_2, then the scale exerts normal forces upward of F_{N_1} and F_{N_2}, respectively, on the car. The total weight of the car is W and can be located at the center of gravity of the car. Since the location of this cg is unknown, let us assume that it is at a distance x from the front wheels. Because the car is obviously in equilibrium, the conditions of equilibrium are applied. Thus,

$$\Sigma F_y = 0 \qquad (5.5)$$

From figure 5.18, we see that this is

$$F_{N_1} + F_{N_2} - W = 0$$

$$F_{N_1} + F_{N_2} = W$$

Solving for W, the weight of the car, we get

$$W = 1800 \text{ lb} + 1500 \text{ lb} = 3300 \text{ lb}$$

The second condition of equilibrium, using the front axle of the car as the axis, gives

$$\Sigma \tau_{cw} = \Sigma \tau_{ccw} \qquad (5.24)$$

The force F_{N_2} will cause a clockwise torque about the front axle, while W will cause a counterclockwise torque. Hence,

$$F_{N_2}(80.0 \text{ in.}) = Wx_{cg}$$

Solving for the center of gravity, we get

$$x_{cg} = F_{N_2}\frac{(80.0 \text{ in.})}{W}$$

$$= \frac{(1500 \text{ lb})(80.0 \text{ in.})}{3300 \text{ lb}}$$

$$= 36.4 \text{ in.}$$

That is, the cg of the car is located 36.4 in. behind the front axle of the car.

Center of Mass

The center of mass (cm) of a body or system of bodies is defined as that point that moves in the same way that a single particle of the same mass would move when acted on by the same forces. Hence, the point reacts as if all the mass of the body were concentrated at that point. All the external forces can be considered to act at the center of mass when the body undergoes any translational acceleration. The general motion of any rigid body can be resolved into the translational motion of the center of mass and the rotation about the center of mass. *On the surface of the earth, where g, the acceleration due to gravity, is relatively uniform, the center of mass (cm) of the body will coincide with the center of gravity (cg) of the body.* To see this, take equation 5.31 and note that

$$w_i = m_i g$$

Substituting this into equation 5.31 we get

$$x_{cg} = \frac{\Sigma w_i x_i}{\Sigma w_i} = \frac{\Sigma (m_i g)x_i}{\Sigma (m_i g)}$$

Factoring the g outside of the summations, we get

$$x_{cg} = \frac{g \Sigma m_i x_i}{g \Sigma m_i} \qquad (5.33)$$

The right-hand side of equation 5.33 is the defining relation for the center of mass of a body, and we will write it as

$$x_{cm} = \frac{\Sigma m_i x_i}{\Sigma m_i} = \frac{\Sigma m_i x_i}{M} \qquad (5.34)$$

where M is the total mass of the body. *Equation 5.34 represents the x-coordinate of the center of mass of the body.* We obtain a similar equation for the y-coordinate by replacing the letter x with the letter y in equation 5.34:

$$y_{cm} = \frac{\Sigma m_i y_i}{\Sigma m_i} = \frac{\Sigma m_i y_i}{M} \qquad (5.35)$$

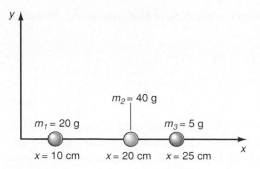

Figure 5.19

The center of mass.

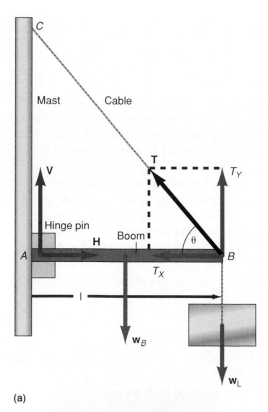

(a)

(b)

Figure 5.20

The crane boom.

Example 5.7

Finding the center of mass. Three masses, $m_1 = 20.0$ g, $m_2 = 40.0$ g, and $m_3 = 5.00$ g are located on the x-axis at 10.0, 20.0, and 25.0 cm, respectively, as shown in figure 5.19. Find the center of mass of the system of three masses.

Solution

The center of mass is found from equation 5.34 with $n = 3$. Thus,

$$x_{cm} = \frac{\Sigma m_i x_i}{\Sigma m_i} = \frac{m_1 x_1 + m_2 x_2 + m_3 x_3}{m_1 + m_2 + m_3}$$

$$= \frac{(20.0 \text{ g})(10.0 \text{ cm}) + (40.0 \text{ g})(20.0 \text{ cm}) + (5.00 \text{ g})(25.0 \text{ cm})}{20.0 \text{ g} + 40.0 \text{ g} + 5.00 \text{ g}}$$

$$= \frac{1125 \text{ g cm}}{65.0 \text{ g}}$$

$$= 17.3 \text{ cm}$$

The center of mass of the three masses is at 17.3 cm.

The Crane Boom

A large uniform boom is connected to the mast by a hinge pin at the point A in figure 5.20. A load w_L is to be supported at the other end B. A cable is also tied to B and connected to the mast at C to give additional support to the boom. We want to determine all the forces that are acting on the boom in order to make sure that the boom, hinge pin, and cable are capable of withstanding these forces when the boom is carrying the load w_L.

First, what are the forces acting on the boom? Because the boom is uniform, its weight $\mathbf{w}_B$ can be situated at its center of gravity, which coincides with its geometrical center. There is a tension $\mathbf{T}$ in the cable acting at an angle θ to the boom. At the hinge pin, there are two forces acting. The first, denoted by $\mathbf{V}$, is a vertical force acting on the end of the boom. If this force were not acting on the boom at this end point, this end of the boom would fall down. That is, the pin with this associated force $\mathbf{V}$ is holding the boom up.

Second, there is also a horizontal force $\mathbf{H}$ acting on the boom toward the right. The horizontal component of the tension $\mathbf{T}$ pushes the boom into the mast. The force $\mathbf{H}$ is the reaction force that the mast exerts on the boom. If there were no force $\mathbf{H}$, the boom would go right through the mast. The vector sum of these two forces, $\mathbf{V}$ and $\mathbf{H}$, is sometimes written as a single contact force at the location of the hinge pin. However, since we want to have the forces in the x- and y-directions, we will leave the forces in the vertical and horizontal directions. The tension $\mathbf{T}$ in the cable also has a vertical component T_y, which helps to hold up the load and the boom.

Let us now determine the forces V, H, and T acting on the system when $\theta = 30.0°$, $w_B = 270$ N, $w_L = 900$ N, and the length of the boom, $l = 6.00$ m. The first thing to do to solve this problem is to observe that the body, the boom, is at rest under the action of several different forces, and must therefore be in equilibrium. Hence, the first and second conditions of equilibrium must apply:

$$\Sigma F_y = 0 \tag{5.5}$$

$$\Sigma F_x = 0 \tag{5.4}$$

$$\Sigma \tau_{cw} = \Sigma \tau_{ccw} \tag{5.24}$$

Using figure 5.20, we observe which forces are acting in the y-direction. Equation 5.5 becomes

$$\Sigma F_y = V + T_y - w_B - w_L = 0$$

or

$$V + T_y = w_B + w_L \tag{5.36}$$

Note from figure 5.20 that $T_y = T \sin \theta$. The right-hand side of equation 5.36 is known, because w_B and w_L are known. But the left-hand side contains the two unknowns, V and T, so we can not proceed any further with this equation at this time.

Let us now consider the second of the equilibrium equations, namely equation 5.4. Using figure 5.20, notice that the force in the positive x-direction is H, while the force in the negative x-direction is T_x. Thus, the equilibrium equation 5.4 becomes

$$\Sigma F_x = H - T_x = 0$$

or

$$H = T_x = T \cos \theta \tag{5.37}$$

There are two unknowns in this equation, namely H and T. At this point, we have two equations with the three unknowns V, H, and T. We need another equation to determine the solution of the problem. This equation comes from the second condition of equilibrium, equation 5.24. In order to compute the torques, we must first pick an axis of rotation. Remember, any point can be picked for the axis to pass through. For convenience we pick the point A in figure 5.20, where the forces V and H are acting, for the axis of rotation to pass through. The forces w_B and w_L are the forces that produce the clockwise torques about the axis at A, while T_y produces the counterclockwise torque. Therefore, equation 5.24 becomes

$$w_B(l/2) + w_L(l) = T_y(l) = T \sin \theta \, (l) \tag{5.38}$$

After dividing term by term by the length l, we can solve equation 5.38 for T. Thus,

$$T \sin \theta = (w_B/2) + w_L$$

The tension in the cable is therefore

$$T = \frac{(w_B/2) + w_L}{\sin \theta} \tag{5.39}$$

Substituting the values of w_B, w_L, and θ, into equation 5.39 we get

$$T = \frac{(270 \text{ N}/2) + 900 \text{ N}}{\sin 30.0°}$$

or

$$T = 2070 \text{ N}$$

The tension in the cable is 2070 N. We can find the second unknown force H by substituting this value of T into equation 5.37:

$$H = T \cos \theta = (2070 \text{ N})\cos 30.0°$$

and

$$H = 1790 \text{ N}$$

The horizontal force exerted on the boom by the hinge pin is 1790 N. We find the final unknown force V by substituting T into equation 5.36, and solving for V, we get

$$V = w_B + w_L - T \sin \theta \qquad (5.40)$$

$$= 270 \text{ N} + 900 \text{ N} - (2070 \text{ N})\sin 30.0°$$

$$= 135 \text{ N}$$

The hinge pin exerts a force of 135 N on the boom in the vertical direction. To summarize, the forces acting on the boom are $V = 135$ N, $H = 1790$ N, and $T = 2070$ N. The reason we are concerned with the value of these forces, is that the boom is designed to carry a particular load. If the boom system is not capable of withstanding these forces the boom will collapse. For example, we just found the tension in the cable to be 2070 N. Is the cable that will be used in the system capable of withstanding a tension of 2070 N? If it is not, the cable will break, the boom will collapse, and the load will fall down. On the other hand, is the hinge pin capable of taking a vertical stress of 135 N and a horizontal stress of 1790 N? If it is not designed to withstand these forces, the pin will be sheared and again the entire system will collapse. Also note that this is not a very well designed boom system in that the hinge pin must be able to withstand only 135 N in the vertical while the horizontal force is 1790 N. In designing a real system the cable could be moved to a much higher position on the mast thereby increasing the angle θ, reducing the component T_x, and hence decreasing the force component H.

There are many variations of the boom problem. Some have the boom placed at an angle to the horizontal. Others have the cable at any angle, and connected to almost any position on the boom. But the procedure for the solution is still the same. The boom is an object in equilibrium and equations 5.4, 5.5, and 5.24 must apply. Variations on the boom problem presented here are included in the problems at the end of the chapter.

The Ladder

A ladder of length L is placed against a wall, as shown in figure 5.21. A person, of weight $\mathbf{w_P}$, ascends the ladder until the person is located a distance d from the top of the ladder. We want to determine all the forces that are acting on the ladder. We assume that the ladder is uniform. Hence, the weight of the ladder $\mathbf{w_L}$ can be located at its geometrical center, that is, at $L/2$. There are two forces acting on the bottom of the ladder, $\mathbf{V}$ and $\mathbf{H}$. The vertical force $\mathbf{V}$ represents the reaction force that the ground exerts on the ladder. That is, since the ladder pushes against the ground, the ground must exert an equal but opposite force upward on the ladder.

With the ladder in this tilted position, there is a tendency for the ladder to slip to the left at the ground. If there is a tendency for the ladder to be in motion to the left, then there must be a frictional force tending to oppose that motion, and therefore that frictional force must act toward the right. We call this horizontal frictional force $\mathbf{H}$. At the top of the ladder there is a force $\mathbf{F}$ on the ladder that acts normal to the wall. This force is the force that the wall exerts on the ladder and is the reaction force to the force that the ladder exerts on the wall. There is also a tendency for the ladder to slide down the wall and therefore there should also be a frictional force on the ladder acting upward at the wall. To solve the general case where there is friction at the wall is extremely difficult. We simplify the problem by assuming that the wall is smooth and hence there is no frictional force acting on the top of the ladder. Thus, whatever results that are obtained in this problem are an approximation to reality.

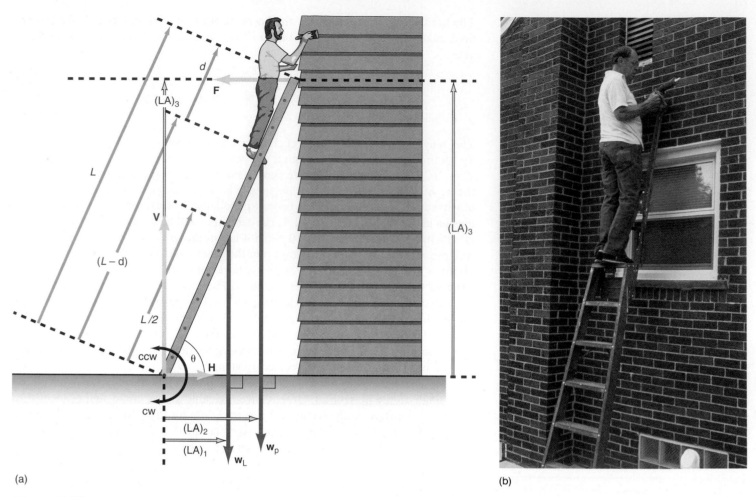

(a)

(b)

Figure 5.21

The ladder.

Since the ladder is at rest under the action of several forces it must be in static equilibrium. Hence, the first and second conditions of equilibrium must apply. Namely,

$$\Sigma F_y = 0 \tag{5.5}$$

$$\Sigma F_x = 0 \tag{5.4}$$

$$\Sigma \tau_{cw} = \Sigma \tau_{ccw} \tag{5.24}$$

Figure 5.21 shows that the force upward is V, while the forces downward are w_L and w_p. Substituting these values into equation 5.5 gives

$$\Sigma F_y = V - w_L - w_p = 0$$

or

$$V = w_L + w_p \tag{5.41}$$

The figure also shows that the force to the right is H, while the force to the left is F. Equation 5.4 therefore becomes

$$\Sigma F_x = H - F = 0$$

or

$$H = F \tag{5.42}$$

It is important that you see how equations 5.41 and 5.42 are obtained from figure 5.21. This is the part that really deals with the physics of the problem. Once all the equations are obtained, their solution is really a matter of simple mathematics.

Mechanics

Before we can compute any of the torques for the second condition of equilibrium, we must pick an axis of rotation. As already pointed out, we can pick any axis to compute the torques. We pick the base of the ladder as the axis of rotation. The forces V and H go through this axis and, therefore, V and H produce no torques about this axis, because they have no lever arms. Observe from the figure that the weights w_L and w_p are the forces that produce clockwise torques, while F is the force that produces the counterclockwise torque. Recall, that torque is the product of the force times the lever arm, where the lever arm is the perpendicular distance from the axis of rotation to the direction or line of action of the force. Note from figure 5.21 that the distance from the axis of rotation to the center of gravity of the ladder does not make a 90° angle with the force w_L, and therefore $L/2$ cannot be a lever arm. If we drop a perpendicular from the axis of rotation to the line showing the direction of the vector $\mathbf{w_L}$, we obtain the lever arm (LA) given by

$$(LA)_1 = (L/2) \cos \theta$$

Thus, the torque clockwise produced by w_L is

$$\tau_{1cw} = w_L(L/2) \cos \theta \tag{5.43}$$

Similarly, the lever arm associated with the weight of the person is

$$(LA)_2 = (L - d) \cos \theta$$

Hence, the second torque clockwise is

$$\tau_{2cw} = w_p(L - d) \cos \theta \tag{5.44}$$

The counterclockwise torque is caused by the force F. However, the ladder does not make an angle of 90° with the force F, and the length L from the axis of rotation to the wall, is not a lever arm. We obtain the lever arm associated with the force F by dropping a perpendicular from the axis of rotation to the direction of the force vector $\mathbf{F}$, as shown in figure 5.21. Note that in order for the force vector to intersect the lever arm, the line from the force had to be extended until it did intersect the lever arm. We call this extended line the *line of action of the force*. This lever arm $(LA)_3$ is equal to the height on the wall where the ladder touches the wall, and is found by the trigonometry of the figure as

$$(LA)_3 = L \sin \theta$$

Hence, the counterclockwise torque produced by F is

$$\tau_{ccw} = FL \sin \theta \tag{5.45}$$

Substituting equations 5.43, 5.44, and 5.45 into equation 5.24 for the second condition of equilibrium, yields

$$w_L(L/2)\cos \theta + w_p(L - d)\cos \theta = FL \sin \theta \tag{5.46}$$

The physics of the problem is now complete. It only remains to solve the three equations 5.41, 5.42, and 5.46 mathematically. There are three equations with the three unknowns V, H, and F.

As a typical problem, let us assume that the following data are given: $\theta = 60.0°$, $w_L = 40.0$ lb, $w_p = 160$ lb, $L = 20.0$ ft, and $d = 5.00$ ft. Equation 5.46, solved for the force F, gives

$$F = \frac{w_L(L/2)\cos \theta + w_p(L - d)\cos \theta}{L \sin \theta} \tag{5.47}$$

Substituting the values just given, we have

$$F = \frac{40.0 \text{ lb}(10.0 \text{ ft})\cos 60.0° + 160 \text{ lb}(15.0 \text{ ft}) \cos 60.0°}{20.00 \text{ ft} \sin 60.0°}$$

$$= \frac{200 \text{ lb ft} + 1200 \text{ lb ft}}{17.32 \text{ ft}}$$

$$= 80.8 \text{ lb}$$

However, since $H = F$ from equation 5.42, we have

$$H = 80.8 \text{ lb}$$

Solving for V from equation 5.41 we obtain

$$V = w_L + w_p = 40.0 \text{ lb} + 160 \text{ lb}$$

$$= 200 \text{ lb}$$

Thus, we have found the three forces F, V, and H acting on the ladder.

As a variation of this problem, we might ask, "What is the minimum value of the coefficient of friction between the ladder and the ground, such that the ladder will not slip out at the ground?" Recall from setting up this problem, that H is indeed a frictional force, opposing the tendency of the bottom of the ladder to slip out, and as such is given by

$$H = f_s = \mu_s F_N \tag{5.48}$$

But the normal force F_N that the ground exerts on the ladder, seen from figure 5.21, is the vertical force V. Hence,

$$H = \mu_s V$$

The coefficient of friction between the ground and the ladder is therefore

$$\mu_s = \frac{H}{V}$$

For this particular example, the minimum coefficient of friction is

$$\mu_s = \frac{80.8 \text{ lb}}{200 \text{ lb}}$$

$$\mu_s = 0.404$$

If μ_s is not equal to, or greater than 0.404, then the necessary frictional force H is absent and the ladder will slide out at the ground.

Applications of the Theory of Equilibrium to the Health Sciences

Example 5.8

A weight lifter's dumbbell curls. A weight lifter is lifting a dumbbell that weighs 75.0 lb, as shown in figure 5.22(a) The biceps muscle exerts a force F_M upward on the forearm at a point approximately 2.00 in. from the elbow joint. The forearm weighs approximately 15.0 lb and its center of gravity is located approximately 7.30 in. from the elbow joint. The upper arm exerts a force at the elbow joint that we denote by F_J. The dumbbell is located approximately 14.5 in. from the elbow. What force must be exerted by the biceps muscle in order to lift the dumbbell?

Figure 5.22

The arm lifting a weight.

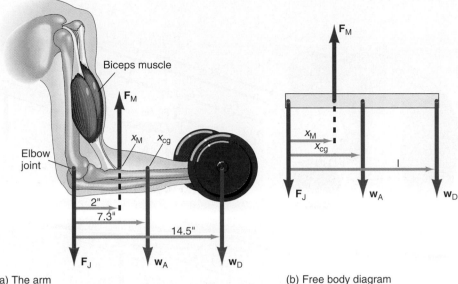

Biceps muscle

Elbow joint

(a) The arm

(b) Free body diagram

Solution

The free body diagram for the arm is shown in figure 5.22(b). The first condition of equilibrium gives

$$\Sigma F_y = F_M - F_J - w_A - w_D = 0$$

$$F_M = F_J + w_A + w_D$$

$$F_M = F_J + 15.0 \text{ lb} + 75.0 \text{ lb} \tag{5.49}$$

$$F_M = F_J + 90.0 \text{ lb} \tag{5.50}$$

Taking the elbow joint as the axis, the second condition of equilibrium gives

$$\Sigma \tau_{cw} = \Sigma \tau_{ccw}$$

$$w_A x_{cg} + w_D l = F_M x_M \tag{5.51}$$

The force exerted by the biceps muscle becomes

$$F_M = \frac{w_A x_{cg} + w_D l}{x_M} \tag{5.52}$$

$$= \frac{(15.0 \text{ lb})(7.30 \text{ in.}) + (75.0 \text{ lb})(14.5 \text{ in.})}{2.00 \text{ in.}}$$

$$= 599 \text{ lb}$$

Thus, the biceps muscle exerts the relatively large force of 599 lb in lifting the 75.0 lb dumbbell. We can now find the force at the joint, from equation 5.50, as

$$F_J = F_M - 90.0 \text{ lb}$$

$$= 599 \text{ lb} - 90 \text{ lb} = 509 \text{ lb}$$

Example 5.9

A weight lifter's bend over rowing. A weight lifter bends over at an angle of 50.0° to the horizontal, as shown in figure 5.23(a). He holds a barbell that weighs 150 lb, w_B. The spina muscle in his back supplies the force F_M to hold the spine of his back in this position. The length of the man's spine is approximately 27.0 in. The spina muscle acts approximately 18.0 in. from the base of the spine and makes an angle

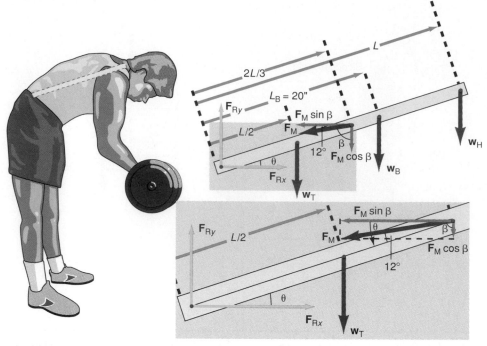

(a) Weight lifter

(b) Free body diagram

(c)

Figure 5.23

Forces on the spinal column.

of 12.0° with the spine, as shown. The man's head weighs about 14.0 lb, w_H, and this force acts at the top of the spinal column, as shown. The torso of the man weighs about 80 lb and this is denoted by w_T, and is located at the center of gravity of the torso, which is taken as 13.5 in. At the base of the spinal column is the fifth lumbar vertebra, which acts as the axis about which the body bends. A reaction force F_R acts on this fifth lumbar vertebra, as shown in the figure. Determine the reaction force F_R and the muscular force F_M on the spine.

Solution

A free body diagram of all the forces is shown in figure 5.23(b). Note that the angle β is

$$\beta = 90° - \theta + 12° = 90° - 50° + 12° = 52°$$

The first condition of equilibrium yields

$$\Sigma F_y = 0$$

$$F_R \sin \theta - w_T - w_B - w_H - F_M \cos \beta = 0$$

or

$$F_R \sin \theta = w_T + w_B + w_H + F_M \cos \beta \qquad (5.53)$$

and

$$\Sigma F_x = 0$$

$$F_R \cos \theta - F_M \sin \beta = 0$$

or

$$F_R \cos \theta = F_M \sin \beta \qquad (5.54)$$

The second condition of equilibrium gives

$$\Sigma \, \tau_{cw} = \Sigma \, \tau_{ccw}$$

$$w_T(L/2)\cos \theta + F_M \cos \beta(2L/3)\cos \theta + w_B L_B \cos \theta$$
$$+ \, w_H L \cos \theta = F_M \sin \beta(2L/3)\sin \theta \qquad (5.55)$$

Solving for F_M, the force exerted by the muscles, gives

$$F_M = \frac{w_T(L/2)\cos \theta + w_B L_B \cos \theta + w_H L \cos \theta}{\sin \beta(2L/3)\sin \theta - \cos \beta(2L/3)\cos \theta} \qquad (5.56)$$

$$= \frac{(80 \text{ lb})(13.5 \text{ in.})(\cos 50°) + (150 \text{ lb})(20 \text{ in.})(\cos 50°) + (14 \text{ lb})(27 \text{ in.})(\cos 50°)}{(\sin 52°)(18 \text{ in.})(\sin 50°) - (\cos 52°)(18 \text{ in.})(\cos 50°)}$$

$$= 764 \text{ lb}$$

The reaction force F_R on the base of the spine, found from equation 5.54, is

$$F_R = \frac{F_M \sin \beta}{\cos \theta}$$

$$= \frac{(764 \text{ lb})\sin 52°}{\cos 50°} = 937 \text{ lb}$$

Thus in lifting a 150 lb barbell there is a force on the spinal disk at the base of the spine of 937 lb.

An Essay on the Application of Physics

Traction

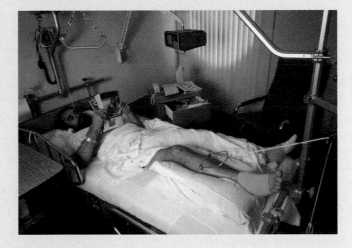

H ave you ever wondered, while visiting Uncle Johnny in the hospital, what they were doing to that poor man in the other bed (figure 1)? As you can see in figure 2, they have him connected to all kinds of pulleys, ropes, and weights. It looks like some kind of medieval torture rack, where they are stretching the man until he tells all he knows. Or perhaps the man is a little short for his weight and they are just trying to stretch him to normal size.

Of course it is none of these things, but the idea of stretching is correct. Actually the man in the other bed is in traction. Traction is essentially a process of exerting a force on a skeletal structure in order to hold a bone in a prescribed position. Traction is used in the treatment of fractures and is a direct application of a body in equilibrium under a number of forces. The object of traction is to exert sufficient force to keep the two sections of the fractured bone in alignment and just touching while they heal. The traction process thus prevents muscle contraction that might cause misalignment at the fracture. The traction force can be exerted through a splint or by a steel pin passed directly through the bone.

Figure 1

A man in traction.

An example of one type of traction, shown in figure 2, is known as Russell traction and is used in the treatment of a fracture of the femur.

Let us analyze the problem from the point of view of equilibrium. First note that almost all of the forces on the bone are transmitted by the ropes that pass around the pulleys. The characteristic of all the systems with pulleys and ropes that are used

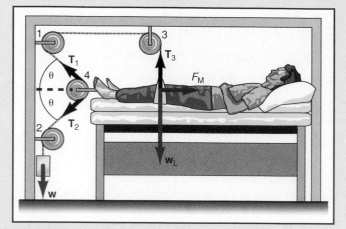

Figure 2
Russell traction.

in traction is that the tension in the taut connecting rope is everywhere the same. Thus, the forces exerted on the bone are the tensions T_1, T_2, T_3, the weight of the leg w_L, and the force exerted by the muscles F_M. The first condition of equilibrium applied to the leg yields

$$\Sigma F_y = 0$$

$$= T_1 \sin \theta + T_3 - T_2 \sin \theta - w_L = 0 \quad \text{(5H.1)}$$

The function of the pulleys is to change the direction of the force, but the tension in the rope is everywhere the same. But the ten-

sion T is supplied by the weight w that is hung from the end of the bed and is thus equal to the weight w. Hence,

$$T_1 = T_2 = T_3 = w \quad \text{(5H.2)}$$

Equation 5H.1 now becomes

$$w \sin \theta + w - w \sin \theta - w_L = 0$$

or

$$w = w_L \quad \text{(5H.3)}$$

Thus the weight w hung from the bottom of the bed must be equal to the weight of the leg w_L.

The second equation of the first condition of equilibrium is

$$\Sigma F_x = 0$$

$$F_M - T_1 \cos \theta - T_2 \cos \theta = 0 \quad \text{(5H.4)}$$

Using equation 5H.2 this becomes

$$F_M - w \cos \theta - w \cos \theta = 0$$

$$F_M = w \cos \theta + w \cos \theta$$

Thus,

$$F_M = 2w \cos \theta \quad \text{(5H.5)}$$

which says that by varying the angle θ, the force to overcome muscle contraction can be varied to any value desired. In this analysis, the force exerted to overcome the muscle contraction lies along the axis of the bone. Variations of this technique can be used if we want to have the traction force exerted at any angle because of the nature of the medical problem.

The Language of Physics

Statics
That portion of the study of mechanics that deals with bodies in equilibrium (p. 121).

Equilibrium
A body is said to be in equilibrium under the action of several forces if the body has zero translational acceleration and no rotational motion (p. 121).

The first condition of equilibrium
For a body to be in equilibrium the vector sum of all the forces acting on the body must be zero. This can also be stated as: a body is in equilibrium if the sum of all the forces in the x-direction is equal to zero and the sum of all the forces in the y-direction is equal to zero (p. 121).

Torque
Torque is defined as the product of the force times the lever arm. Whenever an unbalanced torque acts on a body at rest, it will put that body into rotational motion (p. 127).

Lever arm
The lever arm is defined as the perpendicular distance from the axis of rotation to the direction or line of action of the force. If the force acts through the axis of rotation of the body, it has a zero lever arm and cannot cause a torque to act on the body (p. 127).

The second condition of equilibrium
In order for a body to be in rotational equilibrium, the sum of the torques acting on the body must be equal to zero. This can also be stated as: the necessary condition for a body to be in rotational equilibrium is that the sum of all the torques clockwise must be equal to the sum of all the torques counterclockwise (p. 130).

Center of gravity (cg)
The point that behaves as though the entire weight of the body is located at that point. For a body with a uniform mass distribution located in a uniform

gravitational field, the center of gravity is located at the geometrical center of the body (p. 134).

Center of mass (cm)
The point of a body at which all the mass of the body is assumed to be concentrated. For a body with a uniform mass distribution, the center of mass coincides with the geometrical center of the body. When external forces act on a body to put the body into translational motion, all the forces can be considered to act at the center of mass of the body. For a body in a uniform gravitational field, the center of gravity coincides with the center of mass of the body (p. 138).

Summary of Important Equations

First condition of equilibrium
$$\Sigma \mathbf{F} = 0 \qquad (5.1)$$

First condition of equilibrium
$$\Sigma F_x = 0 \qquad (5.4)$$
$$\Sigma F_y = 0 \qquad (5.5)$$

Torque
$$\tau = r_\perp F = rF_\perp \qquad (5.21)$$
$$= rF \sin \theta$$

Second condition of equilibrium
$$\Sigma \tau = 0 \qquad (5.25)$$

Second condition of equilibrium
$$\Sigma \tau_{cw} = \Sigma \tau_{ccw} \qquad (5.24)$$

Center of gravity
$$x_{cg} = \frac{\Sigma w_i x_i}{W} \qquad (5.31)$$

$$y_{cg} = \frac{\Sigma w_i y_i}{W} \qquad (5.32)$$

Center of mass
$$x_{cm} = \frac{\Sigma m_i x_i}{\Sigma m_i} = \frac{\Sigma m_i x_i}{M} \qquad (5.34)$$

$$y_{cm} = \frac{\Sigma m_i y_i}{\Sigma m_i} = \frac{\Sigma m_i y_i}{M} \qquad (5.35)$$

Questions for Chapter 5

1. Why can a body moving at constant velocity be considered as a body in equilibrium?
2. Why cannot an accelerated body be considered as in equilibrium?
3. Why can a point outside the body in equilibrium be considered as an axis to compute torques?
4. What is the difference between the center of mass of a body and its center of gravity?
5. A ladder is resting against a wall and a person climbs up the ladder. Is the ladder more likely to slip out at the bottom as the person climbs closer to the top of the ladder? Explain.
6. When flying an airplane a pilot frequently changes from the fuel tank in the right wing to the fuel tank in the left wing. Why does he do this?

7. Where would you expect the center of gravity of a sphere to be located? A cylinder?
† 8. When lifting heavy objects why is it said that you should bend your knees and lift with your legs instead of your back? Explain.
9. A short box and a tall box are sitting on the floor of a truck. If the truck makes a sudden stop, which box is more likely to tumble over? Why?
†10. A person is sitting at the end of a row boat that is at rest in the middle of the lake. If the person gets up and walks toward the front of the boat, what will happen to the boat? Explain in terms of the center of mass of the system.

11. Is it possible for the center of gravity of a body to lie outside of the body? (*Hint:* consider a doughnut.)
†12. Why does an obese person have more trouble with lower back problems than a thin person?
13. Describe how a lever works in terms of the concept of torque.
†14. Describe how you could determine the center of gravity of an irregular body such as a plate, experimentally.
†15. Engineers often talk about the moment of a force acting on a body. Is there any difference between the concept of a torque acting on a body and the moment of a force acting on a body?

Problems for Chapter 5

5.1 The First Condition of Equilibrium

1. In a laboratory experiment on a force table, three forces are in equilibrium. One force of 0.300 N acts at an angle of 40.0°. A second force of 0.800 N acts at an angle of 120°. What is the magnitude and direction of the force that causes equilibrium?
2. Two ropes each 10.0 ft long are attached to the ceiling at two points located 15.0 ft apart. The ropes are tied together in a knot at their lower end and a load of 70.0 lb is hung on the knot. What is the tension in each rope?

3. What force must be applied parallel to the plane to make the block move up the frictionless plane at constant speed?

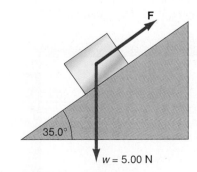

$w = 5.00$ N

4. Two ropes are attached to the ceiling as shown, making angles of 40.0° and 20.0°. A weight of 100 N is hung from the knot. What is the tension in each rope?

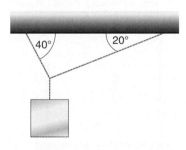

5. Find the force F, parallel to the frictionless plane, that will allow the system to move at constant speed.

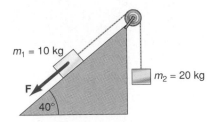

6. A weightless rope is stretched horizontally between two poles 25.0 ft apart. Spiderman, who weighs 160 lb, balances himself at the center of the rope, and the rope is observed to sag 0.500 ft at the center. Find the tension in each part of the rope.

7. A weightless rope is stretched horizontally between two poles 25.0 ft apart. Spiderman, who weighs 160 lb, balances himself 5.00 ft from one end, and the rope is observed to sag 0.300 ft there. What is the tension in each part of the rope?

8. A force of 15.0 N is applied to a 15.0-N block on a rough inclined plane that makes an angle of 52.0° with the horizontal. The force is parallel to the plane. The block moves up the plane at constant velocity. Find the coefficient of kinetic friction between the block and the plane.

9. With what force must a 5.00-N eraser be pressed against a blackboard for it to be in static equilibrium? The coefficient of static friction between the board and the eraser is 0.250.

10. A traffic light, weighing 150 lb is hung from the center of a cable of negligible weight that is stretched horizontally between two poles that are 60.0 ft apart. The cable is observed to sag 2.00 ft. What is the tension in the cable?

11. A traffic light that weighs 600 N is hung from the cable as shown. What is the tension in each cable? Assume the cable to be massless.

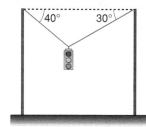

12. Your car is stuck in a snow drift. You attach one end of a 50.0-ft rope to the front of the car and attach the other end to a nearby tree, as shown in the figure. If you can exert a force of 150 lb on the center of the rope, thereby displacing it 3.00 ft to the side, what will be the force exerted on the car?

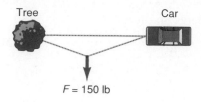

$F = 150$ lb

†13. What force is indicated on the scale in part a and part b of the diagram if $m_1 = m_2 = 20.0$ kg?

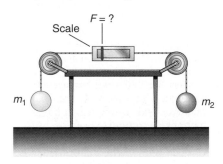

(a)

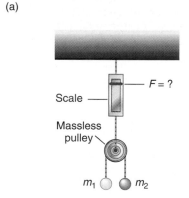

(b)

†14. Find the tension in each cord of the figure, if the block weighs 100 N.

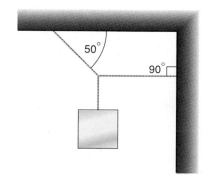

5.2 The Concept of Torque

15. A force of 10.0 lb is applied to a door knob perpendicular to a 30.0-in. door. What torque is produced to open the door?

16. A horizontal force of 10.0 lb is applied at an angle of 40.0° to the door knob of a 30.0-in. door. What torque is produced to open the door?

17. A horizontal force of 50.0 N is applied at an angle of 28.5° to a door knob of a 75.0-cm door. What torque is produced to open the door?

18. A door knob is placed in the center of a 30.0-in. door. If a force of 10.0 lb is exerted perpendicular to the door at the knob, what torque is produced to open the door?

19. Compute the net torque acting on the pulley in the diagram if the radius of the pulley is 0.250 m and the tensions are $T_1 = 50.0$ N and $T_2 = 30.0$ N.

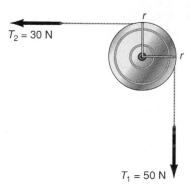

$T_2 = 30$ N

$T_1 = 50$ N

20. Find the torque produced by the bicycle pedal in the diagram if the force $F = 25.0$ lb, the radius of the crank $r = 7.00$ inches, and angle $\theta = 37.0°$.

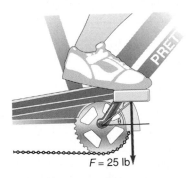

$F = 25$ lb

5.5 Examples of Rigid Bodies in Equilibrium

Parallel Forces

21. Two men are carrying a 25.0-ft telephone pole that weighs 200 lb. If the center of gravity of the pole is 10.0 ft from the right end, and the men lift the pole at the ends, how much weight must each man support?

22. Two men are carrying a 9.00-m telephone pole that has a mass of 115 kg. If the center of gravity of the pole is 3.00 m from the right end, and the men lift the pole at the ends, how much weight must each man support?

23. A uniform board that is 5.00 m long and weighs 450 N is supported by two wooden horses, 0.500 m from each end. If a 800-N person stands on the board 2.00 m from the right end, what force will be exerted on each wooden horse?

24. A 300-N boy and a 250-N girl sit at opposite ends of a 4.00-m seesaw. Where should another 250-N girl sit in order to balance the seesaw?

25. A uniform beam 10.0 ft long and weighing 15.0 lb carries a load of 20.0 lb at one end and 45.0 lb at the other end. It is held horizontal, while resting on a wooden horse 4.00 ft from the heavier load. What torque must be applied to keep it at rest in this position?

26. A uniform beam 3.50 m long and weighing 90.0 N carries a load of 110 N at one end and 225 N at the other end. It is held horizontal, while resting on a wooden horse 1.50 m from the heavier load. What torque must be applied to keep it at rest in this position?

27. A uniform pole 5.00 m long and weighing 100 N is to be carried at its ends by a man and his son. Where should a 250-N load be hung on the pole, such that the father will carry twice the load of his son?

28. A meter stick is hung from two scales that are located at the 20.0- and 70.0-cm marks of the meter stick. Weights of 2.00 N are placed at the 10.0- and 40.0-cm marks, while a weight of 1.00 N is placed at the 90.0-cm mark. The weight of the uniform meter stick is 1.50 N. Determine the scale readings at A and B in the diagram.

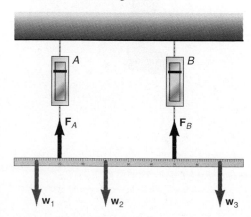

Center of Gravity of a Body

29. A tapered pole 10.0 ft long weighs 25.0 lb. The pole balances at its midpoint when a 5.00-lb weight hangs from the slimmer end. Where is the center of gravity of the pole?

†30. A loaded wheelbarrow that weighs 75.0 lb has its center of gravity 2.00 ft from the front wheel axis. If the distance from the wheel axis to the end of the handles is 6.00 ft, how much of the weight of the wheelbarrow is supported by each arm?

†31. Find the center of gravity of the carpenters square shown in the diagram.

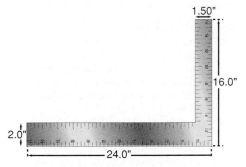

1.50"

16.0"

2.0"

24.0"

32. The front and rear axles of a 3000-lb car are 8.00 ft apart. If the center of gravity of the car is located 3.00 ft behind the front axle, find the load supported by the front and rear wheels of the car.

33. The front and rear axles of a 1110-kg car are 2.50 m apart. If the center of gravity of the car is located 1.15 m behind the front axle, find the load supported by the front and rear wheels of the car.

34. A very bright but lonesome child decides to make a seesaw for one. The child has a large plank, and a wooden horse to act as a fulcrum. Where should the child place the fulcrum, such that the plank will balance, when the child is sitting on the end? The child weighs 60.0 lb and the plank weighs 40.0 lb and is 10.0 ft long. (*Hint:* find the center of gravity of the system.)

Center of Mass

35. Four masses of 20.0, 40.0, 60.0, and 80.0 g are located at the respective distances of 10.0, 20.0, 30.0, and 40.0 cm from an origin. Find the center of mass of the system.

36. Three masses of 15.0, 45.0, and 25.0 g are located on the *x*-axis at 10.0, 25.0, and 45.0 cm. Two masses of 25.0 and 33.0 g are located on the *y*-axis at 35.0 and 50.0 cm, respectively. Find the center of mass of the system.

†37. A 1.00-kg circular metal plate of radius 0.500 m has attached to it a smaller circular plate of the same material of 0.100 m radius, as shown in the diagram. Find the center of mass of the combination with respect to the center of the large plate.

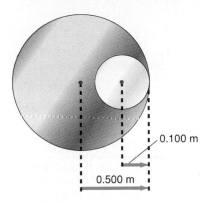

0.100 m

0.500 m

†38. This is the same problem as 37 except the smaller circle of material is removed from the larger plate. Where is the center of mass now?

Crane Boom Problems

39. A horizontal uniform boom that weighs 200 N and is 5.00 m long supports a load w_L of 1000 N, as shown in the figure. Find all the forces acting on the boom.

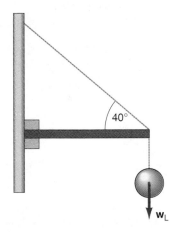

40°

w_L

40. A horizontal, uniform boom 4.00 m long that weighs 200 N supports a load w_L of 1000 N. A guy wire that helps to support the boom, is attached 1.00 m in from the end of the boom. Find all the forces acting on the boom.

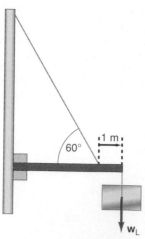

Diagram for problem 40.

41. A horizontal, uniform boom 4.50 m long that weighs 250 N supports a 3500 N load w_L. A guy wire that helps to support the boom is attached 1.0 m in from the end of the boom, as in the diagram for problem 40. If the maximum tension that the cable can withstand is 3500 N, how far out on the boom can a 95.0-kg repairman walk without the cable breaking?

42. A uniform beam 4.00 m long that weighs 200 N is supported, as shown in the figure. The boom lifts a load w_L of 1000 N. Find all the forces acting on the boom.

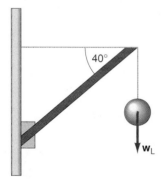

†43. A uniform beam 4.00 m long that weighs 200 N is supported, as shown in the figure. The boom lifts a load w_L of 1000 N. Find all the forces acting on the boom.

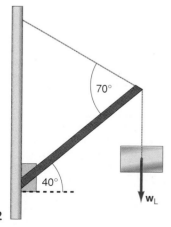

152

44. An 80.0-lb sign is hung on a uniform steel pole that weighs 25.0 lb, as shown in the figure. Find all the forces acting on the boom.

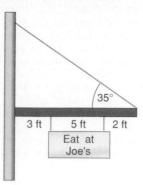

Ladder Problems

45. A uniform ladder 6.00 m long weighing 120 N leans against a frictionless wall. The base of the ladder is 1.00 m away from the wall. Find all the forces acting on the ladder.

46. A uniform ladder 6.00 m long weighing 120 N leans against a frictionless wall. A girl weighing 400 N climbs three-fourths of the way up the ladder. If the base of the ladder makes an angle of 75.0° with the ground, find all the forces acting on the ladder. Compute all torques about the base of the ladder.

47. Repeat problem 46, but compute all torques about the top of the ladder. Is there any difference in the results of the problem?

48. A uniform ladder 15.0 ft long weighing 25.0 lb leans against a frictionless wall. If the base of the ladder makes an angle of 40.0° with the ground, what is the minimum coefficient of friction between the ladder and the ground such that the ladder will not slip out?

†49. A uniform ladder 20.0 ft long weighing 35.0 lb leans against a frictionless wall. The base of the ladder makes an angle of 60.0° with the ground. If the coefficient of friction between the ladder and the ground is 0.300, how high can a 160-lb man climb the ladder before the ladder starts to slip?

50. A uniform ladder 5.50 m long with a mass of 12.5 kg leans against a frictionless wall. The base of the ladder makes an angle of 48.0° with the ground. If the coefficient of friction between the ladder and the ground is 0.300, how high can a 82.3-kg man climb the ladder before the ladder starts to slip?

Applications to the Health Sciences

51. A weight lifter is lifting a dumbbell as in the example shown in figure 5.22 only now the forearm makes an angle of 30.0° with the horizontal. Using the same data as for that problem find the force F_M exerted by the biceps muscle and the reaction force at the elbow joint F_J. Assume that the force F_M remains perpendicular to the arm.

52. Consider the weight lifter in the example shown in figure 5.23. Determine the forces F_M and F_R if the angle $\theta = 00.0°$.

†53. The weight of the upper body of the person in the accompanying diagram acts downward about 8.00 cm in front of the fifth lumbar vertebra. This weight produces a torque about the fifth lumbar vertebra. To counterbalance this torque the muscles in the lower back exert a force F_M that produces a counter torque. These muscles exert their force about 5.00 cm behind the fifth lumbar vertebra. If the person weighs 180 lb find the force exerted by the lower back muscles F_M and the reaction force F_R that the sacrum exerts upward on the fifth lumbar vertebra. The weight of the upper portion of the body is about 65% of the total body weight.

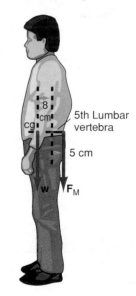

†54. Consider the same situation as in problem 53 except that the person is overweight. The center of gravity with the additional weight is now located 15.0 cm in front of the fifth lumbar vertebra instead of the previous 8.00 cm. Hence a greater torque will be exerted by this additional weight. The distance of the lower back muscles is only slightly greater at 6.00 cm. If the person weighs 240 lb find the force F_R on the fifth lumbar vertebra and the force F_M exerted by the lower back muscles.

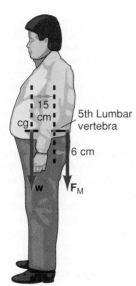

†55. A 668-N person stands evenly on the balls of both feet. The Achilles tendon, which is located at the back of the ankle, provides a tension T_A to help balance the weight of the body as seen in the diagram. The distance from the ball of the foot to the Achilles tendon is approximately 18.0 cm. The tibia leg bone pushes down on the foot with a force F_T. The distance from the tibia to the ball of the foot is about 14.0 cm. The ground exerts a reaction force F_N upward on the ball of the foot that is equal to half of the body weight. Draw a free body diagram of the forces acting and determine the force exerted by the Achilles tendon and the tibia.

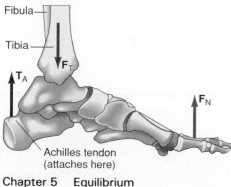

Additional Problems

†56. If w weighs 100 N, find (a) the tension in ropes 1, 2, and 3 and (b) the tension in ropes 4, 5, and 6. The angle $\theta = 52.0°$ and the angle $\phi = 33.0°$.

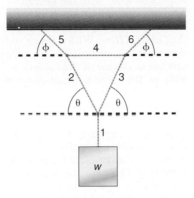

†57. Block A rests on a table and is connected to another block B by a rope that is also connected to a wall. If $M_A = 15.0$ kg and $\mu_s = 0.200$, what must be the value of M_B to start the system into motion?

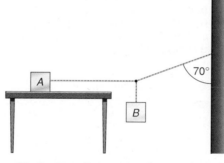

58. In the pulley system shown, what force F is necessary to keep the system in equilibrium?

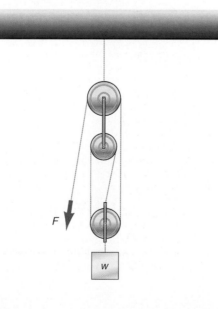

†59. A sling is used to support a leg as shown in the diagram. The leg is elevated at an angle of 20.0°. The bed exerts a reaction force **R** on the thigh as shown. The weight of the thigh, leg, and ankle are given by w_T, w_L, and w_A, respectively, and the locations of these weights are as shown. The sling is located 27.0 in. from the point O in the diagram. A free body diagram is shown in part b of the diagram. Find the weight **w** that is necessary to put the leg into equilibrium.

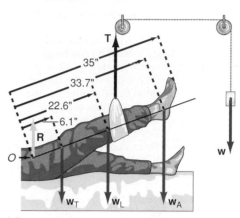

(a)

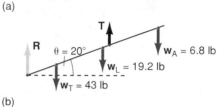

(b)

†60. Find the tensions T_1, T, and T_2 in the figure if $w_1 = 500$ N and $w_2 = 300$ N. The angle $\theta = 35.0°$ and the angle $\phi = 25.0°$.

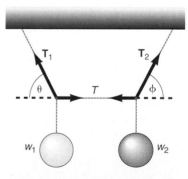

†61. The steering wheel of an auto has a diameter of 18.0 in. The axle that it is connected to has a diameter of 2.00 in. If a force of 25.0 lb is exerted on the rim of the wheel, (a) what is the torque exerted on the steering wheel, (b) what is the torque exerted on the axle, and (c) what force is exerted on the rim of the axle?

62. One type of simple machine is called a wheel and axle. A wheel of radius 35.0 cm is connected to an axle of 2.00 cm radius. A force of $F_{in} = 10.0$ N is applied tangentially to the wheel. What force F_{out} is exerted on the axle? The ratio of the output force F_{out} to the input force F_{in} is called the ideal mechanical advantage (IMA) of the system. Find the IMA of this system.

†63. A box 1.00 m on a side rests on a floor next to a small piece of wood that is fixed to the floor. The box weighs 500 N. At what height h should a force of 400 N be applied so as to just tip the box?

†64. A 200-N door, 0.760 m wide and 2.00 m long, is hung by two hinges. The top hinge is located 0.230 m down from the top, while the bottom hinge is located 0.330 m up from the bottom. Assume that the center of gravity of the door is at its geometrical center. Find the horizontal force exerted by each hinge on the door.

†65. A uniform ladder 6.00 m long weighing 100 N leans against a frictionless wall. If the coefficient of friction between the ladder and the ground is 0.400, what is the smallest angle θ that the ladder can make with the ground before the ladder starts to slip?

†66. If an 800-N man wants to climb a distance of 5.00 m up the ladder of problem 65, what angle θ should the ladder make with the ground such that the ladder will not slip?

†67. A uniform ladder 20.0 ft long weighing 30.0 lb leans against a rough wall, that is, a wall where there is a frictional force between the top of the ladder and the wall. The coefficient of static friction is 0.400. If the base of the ladder makes an angle θ of 40.0° with the ground when the ladder begins to slip down the wall, find all the forces acting on the ladder. (*Hint:* With a rough wall there will be a vertical force f_s acting upward at the top of the ladder. In general, this force is unknown but we do know that it must be less than $\mu_s F_N$. At the moment the ladder starts to slip, this frictional force is known and is given by the equation of static friction, namely, $f_s = \mu_s F_N = \mu_s F$. Although there are now four unknowns, there are also four equations to solve for them.)

†68. A 225-lb person stands three-quarters of the way up a stepladder. The step side weighs 20.0 lb, is 6.00 ft long, and is uniform. The rear side weighs 10.0 lb, is also uniform, and is also 6.00 ft long. A hinge connects the front and back of the ladder at the top. A weightless tie rod, 1.50 ft in length, is connected 2.00 ft from the top of the ladder. Find the forces exerted by the floor on the ladder and the tension in the tie rod.

Interactive Tutorials

69. Two ropes are attached to the ceiling, making angles $\theta = 20.0°$ and $\phi = 40.0°$, suspending a mass $m = 50.0$ kg. Calculate the tensions T_1 and T_2 in each rope.

70. A uniform beam of length $L = 10.0$ m and mass $m = 5.00$ kg is held up at each end by a force F_A (at 0.00 m) and force F_B (at 10.0 m). If a weight $W = 400$ N is placed at the position $x = 8.00$ m, calculate forces F_A and F_B.

71. The crane boom. A uniform boom of weight $w_B = 250$ N and length $l = 8.00$ m is connected to the mast by a hinge pin at the point A in figure 5.20. A load $w_L = 1200$ N is supported at the other end. A cable is connected at the end of the boom making an angle $\theta = 55.0°$, as shown in the diagram. Find the tension T in the cable and the vertical V and horizontal H forces that the hinge pin exerts on the boom.

72. A uniform ladder of weight $w_1 = 100$ N and length $L = 20.0$ m leans against a frictionless wall at a base angle $\theta = 60.0°$. A person weighing $w_p = 150$ N climbs the ladder a distance $d = 6.00$ m from the base of the ladder. Calculate the horizontal H and vertical V forces acting on the ladder, and the force F exerted by the wall on the top of the ladder.

On the moon.

That's one small step for man, one giant leap for mankind.

Neil Armstrong, as he stepped on the surface of the moon July 20, 1969

6 Uniform Circular Motion, Gravitation, and Satellites

6.1 Uniform Circular Motion

Uniform circular motion is defined as motion in a circle at constant speed. Motion in a circle with changing speeds will be discussed in chapter 9. A car moving in a circle at the constant speed of 20 km/hr is an example of a body in uniform circular motion. At every point on that circle the car would be moving at 20 km/hr. This type of motion is shown in figure 6.1. At the time t_0, the car is located at the point A and is moving with the velocity $\mathbf{v_0}$, which is tangent to the circle at that point. At a later time t, the car will have moved through the angle θ, and will be located at the point B. At the point B the car has a velocity $\mathbf{v}$, which is tangent to the circle at B. (The velocity is always tangent to the circle, because at any instant the tangent specifies the direction of motion.) The lengths of the two vectors, $\mathbf{v}$ and $\mathbf{v_0}$, are the same because the magnitude of any vector is represented as the length of that vector. The magnitude of the velocity is the speed, which is a constant for uniform circular motion.

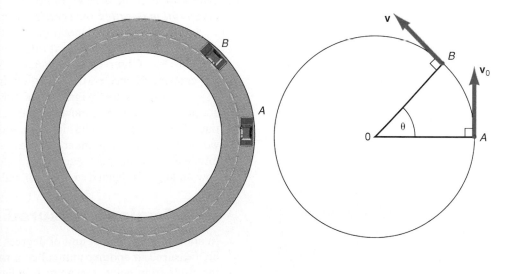

(a)   (b)

Figure 6.1
Uniform circular motion.

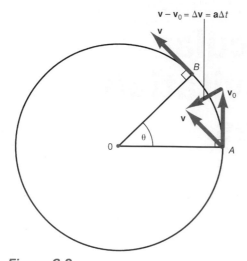

Figure 6.2

The direction of the centripetal acceleration.

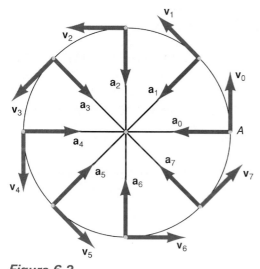

Figure 6.3

The centripetal acceleration always points toward the center of the circle.

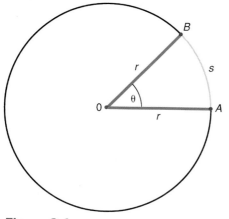

Figure 6.4

Definition of an angle expressed in radians.

The first thing we observe in figure 6.1 is that the direction of the velocity vector has changed in going from the point *A* to the point *B*. Recall from chapter 3, on kinematics, that the acceleration is defined as the change in velocity with time, that is,

$$\mathbf{a} = \frac{\Delta \mathbf{v}}{\Delta t} \tag{6.1}$$

Even though the speed is a constant in uniform circular motion, the direction is always changing with time. Hence, the velocity is changing with time, and there must be an acceleration. *Thus, motion in a circle at constant speed is accelerated motion.* We must now determine the direction of this acceleration and its magnitude.

6.2 Centripetal Acceleration and its Direction

To determine the direction of the centripetal acceleration, let us start by moving the vector **v**, located at the point *B* in figure 6.1, parallel to itself to the point *A*, as shown in figure 6.2. The difference between the two velocity vectors is $\mathbf{v} - \mathbf{v_0}$ and points approximately toward the center of the circle in the direction shown. But this difference between the velocity vectors is the change in the velocity vector $\Delta \mathbf{v}$, that is,

$$\Delta \mathbf{v} = \mathbf{v} - \mathbf{v_0} \tag{6.2}$$

But from equation 6.1

$$\Delta \mathbf{v} = \mathbf{a} \Delta t \tag{6.3}$$

This is a vector equation, and whatever direction the left-hand side of the equation has, the right-hand side must have the same direction. Therefore, the vector $\Delta \mathbf{v}$ points in the same direction as the acceleration vector **a.** Observe from figure 6.2 that $\Delta \mathbf{v}$ points approximately toward the center of the circle. (If the angle θ, between the points *A* and *B*, were made very small, then $\Delta \mathbf{v}$ would point exactly at the center of the circle.) Thus, since $\Delta \mathbf{v}$ points toward the center, the acceleration vector must also point toward the center of the circle. *This is the characteristic of uniform circular motion. Even though the body is moving at constant speed, there is an acceleration and the acceleration vector points toward the center of the circle. This acceleration is called the* **centripetal acceleration.**

The word centripetal means "center seeking" or seeking the center. If this circular motion were shown at intervals of 45°, we would obtain the picture shown in figure 6.3. Observe in figure 6.3 that no matter where the body is on the circle, the centripetal acceleration always points toward the center of the circle.

What is the magnitude of this acceleration? The problem of calculating accelerations of objects moving in circles at constant speed was first solved by Christian Huygens (1629–1695) in 1673, and his solution is effectively the same one that we use today. The argument is basically a geometric one. However, before the magnitude of the centripetal acceleration can be determined, we need first to determine how an angle is defined in terms of radian measure.

6.3 Angles Measured in Radians

In addition to the usual unit of degrees used to measure an angle, an angle can also be measured in another unit called a **radian.** As the body moves along the arc *s* of the circle from point *A* to point *B* in figure 6.4, it sweeps out an angle θ in the time *t*. This angle θ, measured in radians (abbreviated rad), is defined as the ratio of the arc length *s* traversed to the radius of the circle *r*. That is,

$$\theta = \frac{s}{r} = \frac{\text{arc length}}{\text{radius}} \tag{6.4}$$

Thus an angle of 1 radian is an angle swept out such that the distance s, traversed along the arc, is equal to the radius of the circle:

$$\theta = \frac{s}{r} = \frac{r}{r} = 1 \text{ rad}$$

Notice that a radian is a dimensionless quantity. If s is measured in meters and r is measured in meters, then the ratio yields units of meters over meters and the units will thus cancel.

For an entire rotation around the circle, that is, for one revolution, the arc subtended is the circumference of the circle, $2\pi r$. Therefore, an angle of one revolution, measured in radians, becomes

$$\theta = \frac{s}{r} = \frac{2\pi r}{r} = 2\pi \text{ rad}$$

That is, one revolution is equal to 2π rad. The relationship between an angle measured in degrees, and one measured in radians can be found from the fact that one revolution is also equal to 360 degrees. Thus,

$$1 \text{ rev} = 2\pi \text{ rad} = 360°$$

and solving for a radian, we get

$$1 \text{ rad} = \frac{360°}{2\pi} = 57.296°$$

Similarly,

$$1 \text{ degree} = 0.01745 \text{ rad}$$

For ease in calculations on an electronic calculator it is helpful to use the conversion factor

$$\pi \text{ rad} = 180°$$

In almost all problems in circular motion the angles will be measured in radians.

The relationship between the arc length s and the angle θ, measured in radians, for circular motion, found from equation 6.4, is

$$s = r\theta \tag{6.5}$$

6.4 The Magnitude of the Centripetal Acceleration

Having determined the relation between the arc length s and the angle θ swept out, we can now determine the magnitude of the centripetal acceleration. In moving at the constant speed v, along the arc of the circle from A to B in figure 6.2, the body has traveled the distance

$$s = vt \tag{6.6}$$

But in this same time t, the angle θ has been swept out in moving the distance s along the arc. If the distance s moved along the arc from equations 6.5 and 6.6 are equated, we have

$$r\theta = vt$$

Solving for θ, we obtain

$$\theta = \frac{vt}{r} \tag{6.7}$$

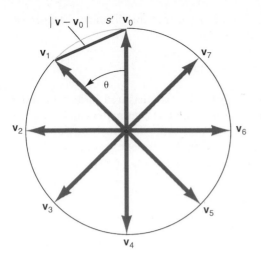

Figure 6.5

The velocity circle.

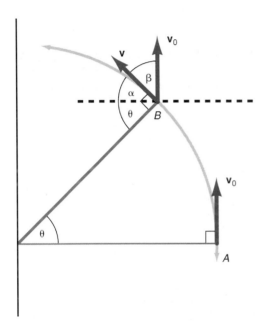

Figure 6.6

The angle between the velocity vectors **v** and **v₀** is the same as the angle θ swept out in moving from point A to point B.

This is the angle θ swept out in the uniform circular motion, in terms of the speed v, time t, and the radius r of the circle. We will return to equation 6.7 in a moment, but first let us look at the way that these velocity vectors are changing with time.

As we see in figure 6.3, the velocity vector **v** points in a different direction at every instant of time. Let us slide each velocity vector in figure 6.3 parallel to itself to a common point. If we draw a curve connecting the tips of each velocity vector, we obtain the circle shown in figure 6.5. That is, since the magnitude of the velocity vector is a constant, a circle of radius v is generated. As the object moves from A to B and sweeps out the angle θ in figure 6.2, the velocity vector also moves through the same angle θ, figure 6.5. To prove this, notice that the velocity vectors **v₀** at A and **v** at B are each tangent to the circle there, figure 6.6. In moving through the angle θ in going from A to B, the velocity vector turns through this same angle θ. This is easily seen in figure 6.6. The angle α is

$$\alpha = \frac{\pi}{2} - \theta \tag{6.8}$$

while the angle β is

$$\beta = \frac{\pi}{2} - \alpha \tag{6.9}$$

Substituting equation 6.8 into equation 6.9 gives

$$\beta = \frac{\pi}{2} - \left(\frac{\pi}{2} - \theta\right)$$

Hence, β, the angle between **v** and **v₀** in figure 6.6, is

$$\beta = \theta$$

Thus, the angle between the velocity vectors **v** and **v₀** is the same as the angle θ swept out in moving from point A to point B.

Therefore, in moving along the velocity circle in figure 6.5, an amount of arc s' is swept out with the angle θ. This velocity circle has a radius of v, the constant speed in the circle. Using equation 6.4, as it applies to the velocity circle, we have

$$\theta = \frac{\text{arc length}}{\text{radius}} = \frac{s'}{v} \tag{6.10}$$

If the angle θ is relatively small, then the arc of the circle s' is approximately equal to the chord of the circle $|\mathbf{v} - \mathbf{v_0}|$ in figure 6.5.[1] That is,

$$\text{arc} \approx \text{chord}$$
$$s' = |\mathbf{v} - \mathbf{v_0}|$$

But

$$\mathbf{v} - \mathbf{v_0} = \mathbf{a}t$$

hence,

$$s' = at$$

Substituting this result into equation 6.10 gives

$$\theta = \frac{at}{v} \tag{6.11}$$

Thus we have obtained a second relation for the angle θ swept out, expressed now in terms of acceleration, speed, and time. Returning to equation 6.7, which gave us

---

1. Note that $|\mathbf{v} - \mathbf{v_0}|$ is the magnitude of the difference in the velocity vectors and is the straight line between the tip of the velocity vector **v₀** and the tip of the velocity vector **v**, and as such, is equal to the chord of the circle in figure 6.5.

the angle θ swept out as the moving body went from point A to point B along the circular path, and comparing it to equation 6.11, which gives the angle θ swept out in the velocity circle. Because both angles θ are equal, equation 6.11 can now be equated to equation 6.7, giving

$$\theta = \theta$$

$$\frac{at}{v} = \frac{vt}{r}$$

Solving for the acceleration we obtain

$$a = \frac{v^2}{r}$$

Placing a subscript c on the acceleration to remind us that this is the centripetal acceleration, we then have

$$a_c = \frac{v^2}{r} \qquad (6.12)$$

Therefore, *for the uniform circular motion of an object moving at constant speed v in a circle of radius r, the object undergoes an acceleration a_c, pointed toward the center of the circle, and having the magnitude given by equation 6.12.*

Example 6.1

Find the centripetal acceleration. An object moves in a circle of 10.0-m radius, at a constant speed of 5.00 m/s. What is its centripetal acceleration?

Solution

The centripetal acceleration, found from equation 6.12, is

$$a_c = \frac{v^2}{r} \qquad (6.12)$$

$$= \frac{(5.00 \text{ m/s})^2}{10.0 \text{ m}}$$

$$= 2.50 \text{ m/s}^2$$

Example 6.2

The special case of the centripetal acceleration equal to the gravitational acceleration. At what uniform speed should a body move in a circular path of 10.0 m radius such that the acceleration experienced will be the same as the acceleration of gravity?

Solution

We find the velocity of the moving body in terms of the centripetal acceleration by solving equation 6.12 for v:

$$v = \sqrt{ra_c}$$

To have the body experience the same acceleration as the acceleration of gravity, we set $a_c = g$ and get

$$v = \sqrt{ra_c} = \sqrt{rg}$$

$$= \sqrt{(10.0 \text{ m})(9.80 \text{ m/s}^2)}$$

$$= 9.90 \text{ m/s}$$

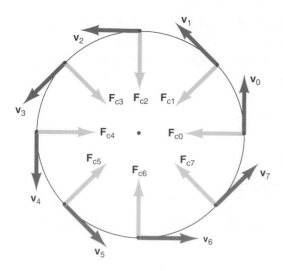

Figure 6.7

The centripetal force always points toward the center of the circle.

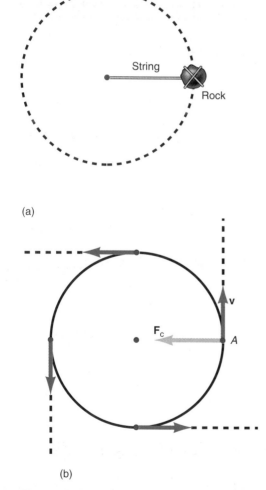

(a)

(b)

Figure 6.8

The string supplies the centripetal force on a rock moving in a circle.

6.5 The Centripetal Force

We have just seen that an object in uniform circular motion experiences a centripetal acceleration. However, because of Newton's second law of motion, there must be a force acting on the object to give it the necessary centripetal acceleration. Applying Newton's second law to the body in uniform circular motion we have

$$F = ma = ma_c = \frac{mv^2}{r} \qquad (6.13)$$

A subscript c is placed on the force to remind us that this is the centripetal force, and equation 6.13 becomes

$$F_c = \frac{mv^2}{r} \qquad (6.14)$$

The force, given by equation 6.14, that causes an object to move in a circle at constant speed is called the **centripetal force.** Because the centripetal acceleration is pointed toward the center of the circle, then from Newton's second law in vector form, we see that

$$\mathbf{F_c} = m\mathbf{a_c} \qquad (6.15)$$

Hence, the centripetal force must also point toward the center of the circle. Therefore, *when an object moves in uniform circular motion there must always be a centripetal force acting on the object toward the center of the circle as seen in figure 6.7.*

We should note here that we need to physically supply the force to cause the body to go into uniform circular motion. The centripetal force is the amount of force necessary to put the body into uniform circular motion, but it is not a real physical force in itself that is applied to the body. It is the amount of force necessary, but something must supply that force, such as a tension, a weight, gravity, and the like. As an example, consider the motion of a rock, tied to a string of negligible mass, and whirled in a horizontal circle, at constant speed v. At every instant of time there must be a centripetal force acting on the rock to pull it toward the center of the circle, if the rock is to move in the circle. This force is supplied by your hand, and transmitted to the rock, by the string. It is evident that such a force must be acting by the following consideration. Consider the object at point *A* in figure 6.8 moving with a velocity **v** at a time *t*. By Newton's first law, a body in motion at a constant velocity will continue in motion at that same constant velocity, unless acted on by some unbalanced external force. Therefore, if there were no centripetal force acting on the object, the object would continue to move at its same constant velocity and would fly off in a direction tangent to the circle. In fact, if you were to cut the string, while the rock is in motion, you would indeed observe the rock flying off tangentially to the original circle. (Cutting the string removes the centripetal force.)

Example 6.3

Finding the centripetal force. A 500-g rock attached to a string is whirled in a horizontal circle at the constant speed of 10.0 m/s. The length of the string is 1.00 m. Neglecting the effects of gravity, find (a) the centripetal acceleration of the rock and (b) the centripetal force acting on the rock.

Solution

a. The centripetal acceleration, found from equation 6.12, is

$$a_c = \frac{v^2}{r} = \frac{(10.0 \text{ m/s})^2}{1.00 \text{ m}} = 100 \frac{\text{m}^2/\text{s}^2}{\text{m}}$$

$$= 100 \text{ m/s}^2$$

b. The centripetal force, which is supplied by the tension in the string, found from equation 6.14, is

$$F_c = \frac{mv^2}{r} = \frac{(0.500 \text{ kg})(10.0 \text{ m/s})^2}{1.00 \text{ m}} = 50.0 \frac{\text{kg m}^2}{\text{m s}^2}$$

$$= 50.0 \text{ N}$$

Notice how the units combine so that the final unit is a newton, the unit of force.

6.6 The Centrifugal Force

In the preceding example of the rock revolving in a horizontal circle, there was a centripetal force acting on the rock by the string. But by Newton's third law, if body *A* exerts a force on body *B*, then body *B* exerts an equal but opposite force on body *A*. Thus, if the string (body *A*) exerts a force on the rock (body *B*), then the rock (body *B*) must exert an equal but opposite force on the string (body *A*). *This reaction force to the centripetal force is called the* **centrifugal force.** Note that the centrifugal force does not act on the same body as does the centripetal force. The centripetal force acts on the rock, the centrifugal force acts on the string. The centrifugal force is shown in figure 6.9 as the dashed line that goes around the rock to emphasize that the force does not act on the rock but on the string.

If we wish to describe the motion of the rock, then we must use the centripetal force, because it is the centripetal force that acts on the rock and is necessary for the rock to move in a circle. The reaction force is the centrifugal force. But *the centrifugal force does not act on the rock, which is the object in motion.*

The word centrifugal means to fly from the center, and hence the centrifugal force acts away from the center. This has been the cause of a great deal of confusion. Many people mistakenly believe that the centrifugal force acts outward on the rock, keeping it out on the end of the string. We can show that this reasoning is incorrect by merely cutting the string. If there really were a centrifugal force acting outward on the rock, then the moment the string is cut the rock should fly radially away from the center of the circle, as in figure 6.10(a). It is a matter of observation that the rock does not fly away radially but rather flies away tangentially as predicted by Newton's first law.

A similar example is furnished by a car wheel when it goes through a puddle of water, as in figure 6.10(b). Water droplets adhere to the wheel. The water droplet is held to the wheel by the adhesive forces between the water molecules and the tire. As the wheel turns, the drop of water wants to move in a straight line as it is governed by Newton's first law but the adhesive force keeps the drop attached to the

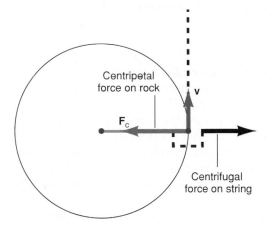

Figure 6.9

The centrifugal force is the reaction force on the string.

Figure 6.10

There is no radial force outward acting on the rock.

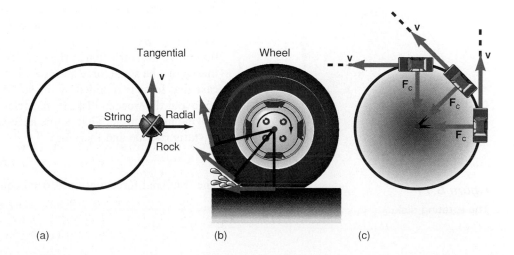

(a)   (b)   (c)

wheel. That is, the adhesive force is supplying the necessary centripetal force. As the wheel spins faster, v increases and the centripetal force necessary to keep the droplet attached to the wheel also increases ($F_c = mv^2/r$). If the wheel spins fast enough, the adhesive force is no longer large enough to supply the necessary centripetal force and the water droplet on the rotating wheel flies away tangentially from the wheel according to Newton's first law.

Another example illustrating the difference between centripetal force and centrifugal force is supplied by a car when it goes around a turn, as in figure 6.10(c). Suppose you are in the passenger seat as the driver makes a left turn. Your first impression as you go through the turn is that you feel a force pushing you outward against the right side of the car. We might assume that there is a centrifugal force acting on you and you can feel that centrifugal force pushing you outward toward the right. This however is not a correct assumption. Instead what is really happening is that at the instant the driver turns the wheels, a frictional force between the wheels and the pavement acts on the car to deviate it from its straight line motion, and deflects it toward the left. You were originally moving in a straight line at an initial velocity **v**. By Newton's first law, you want to continue in that same straight ahead motion. But now the car has turned and starts to push inward on you to change your motion from the straight ahead motion, to a motion that curves toward the left. It is the right side of the car, the floor, and the seat that is supplying, through friction, the necessary centripetal force on you to turn your straight ahead motion into circular motion. There is no centrifugal force pushing you toward the right, but rather the car, through friction, is supplying the centripetal force on you to push you to the left.

Other mistaken beliefs about the centrifugal force will be mentioned as we proceed. However, in almost all of the physical problems that you will encounter, you can forget entirely about the centrifugal force, because it will not be acting on the body in motion. Only in a noninertial coordinate system, such as a rotating coordinate system, do "fictitious" forces such as the centrifugal force need to be introduced. However, in this book we will limit ourselves to inertial coordinate systems.

6.7 Examples of Centripetal Force

The Rotating Disk in the Amusement Park

Amusement parks furnish many examples of the application of centripetal force and circular motion. In one such park there is a large, horizontal, highly polished wooden disk, very close to a highly polished wooden floor. While the disk is at rest, children come and sit down on it. Then the disk starts to rotate faster and faster until the children slide off the disk onto the floor.

Let us analyze this circular motion. In particular let us determine the maximum velocity that the child can move and still continue to move in the circular path. At any instant of time, the child has some tangential velocity **v**, as seen in figure 6.11. By Newton's first law, the child has the tendency to continue moving in that tangential direction at the velocity **v**. However, if the child is to move in a circle, there must be some force acting on the child toward the center of the circle. In this case that force is supplied by the static friction between the seat of the pants of the child and the wooden disk. If that frictional force is present, the child will continue moving in the circle. That is, the necessary centripetal force is supplied by the force of static friction and therefore

$$F_c = f_s \qquad (6.16)$$

The frictional force, obtained from equation 4.44, is

$$f_s \leq \mu_s F_N$$

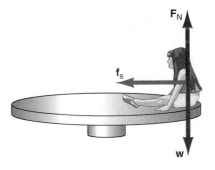

(a) Side view

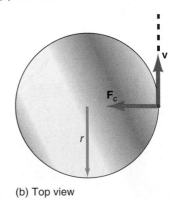

(b) Top view

Figure 6.11

The rotating disk.

Recall that the frictional force is usually less than the product $\mu_s F_N$, and is only equal at the moment that the body is about to slip. In this example, we are finding the maximum velocity of the child and that occurs when the child is about to slip off the disk. Hence, we will use the equality sign for the frictional force in equation 4.44. Using the centripetal force from equation 6.14 and the frictional force from equation 4.44, we obtain

$$\frac{mv^2}{r} = \mu_s F_N \tag{6.17}$$

As seen from figure 6.11, $F_N = w = mg$. Therefore, equation 6.17 becomes

$$\frac{mv^2}{r} = \mu_s mg \tag{6.18}$$

The first thing that we observe in equation 6.18 is that the mass m of the child is on both sides of the equation and divides out. Thus, whatever happens to the child, it will happen to a big massive child or a very small one. When equation 6.18 is solved for v, we get

$$v = \sqrt{\mu_s rg} \tag{6.19}$$

This is the maximum speed that the child can move and still stay in the circular path. For a speed greater than this, the frictional force will not be great enough to supply the necessary centripetal force. Depending on the nature of the children's clothing, μ_s will, in general, be different for each child, and therefore each child will have a different maximum value of v allowable. If the disk's speed is slowly increased until v is greater than that given by equation 6.19, there is not enough frictional force to supply the necessary centripetal force, and the children gleefully slide tangentially from the disk in all directions across the highly polished floor.

Example 6.4

The rotating disk. A child is sitting 1.50 m from the center of a highly polished, wooden, rotating disk. The coefficient of static friction between the disk and the child is 0.30. What is the maximum tangential speed that the child can have before slipping off the disk?

Solution

The maximum speed, obtained from equation 6.19, is

$$v = \sqrt{\mu_s rg}$$
$$= \sqrt{(0.30)(1.50 \text{ m})(9.80 \text{ m/s}^2)}$$
$$= 2.1 \text{ m/s}$$

The Rotating Circular Room in the Amusement Park

In another amusement park there is a ride that consists of a large circular room. (It looks as if you were on the inside of a very large barrel.) Everyone enters the room and stands against the wall. The door closes, and the entire room starts to rotate. As the speed increases each person feels as if he or she is being pressed up against the wall. Eventually, as everyone is pinned against the wall, the floor of the room drops out about 4 or 5 ft, leaving all the children apparently hanging on the wall. After several minutes of motion, the rotation slows down and the children eventually slide down the wall to the lowered floor and the ride ends.

Let us analyze the motion, in particular let us find the value of μ_s, the minimum value of the coefficient of static friction such that the child will not slide down the wall. The room is shown in figure 6.12. As the room reaches its operational

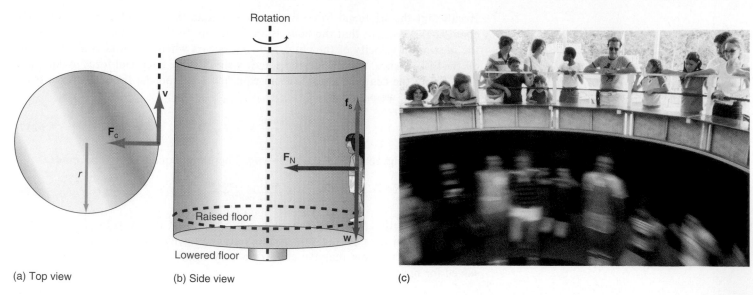

Rotation

F_c

r

(a) Top view

v

f_s

F_N

Raised floor

w

Lowered floor

(b) Side view

(c)

Figure 6.12

Circular room in an amusement park.

speed, the child, at any instant, has a velocity **v** that is tangential to the room, as in figure 6.12(a). By Newton's first law, the child should continue in this straight line motion, but the wall of the room exerts a normal force on the child toward the center of the room, causing the child to deviate from the straight line motion and into the circular motion of the wall of the room. This normal force of the wall on the child, toward the center of the room, supplies the necessary centripetal force. When the floor drops out, the weight **w** of the child is acting downward and would cause the child to slide down the wall. But the frictional force f_s between the wall and the child's clothing opposes the weight, as seen in figure 6.12(b). The frictional force f_s is again given by equation 4.44. We are looking for the minimum value of μ_s that will just keep the child pinned against the wall. That is, the child will be just on the verge of slipping down the wall. Hence, we use the equality sign in equation 4.44. Thus, the frictional force is

$$f_s = \mu_s F_N \tag{6.20}$$

where F_N is the normal force holding the two objects in contact. The centripetal force F_c is supplied by the normal force F_N, that is,

$$F_c = F_N = \frac{mv^2}{r} \tag{6.21}$$

Therefore, the greater the value of the normal force, the greater will be the frictional force.

The child does not slide down the wall because the frictional force f_s is equal to the weight of the child:

$$f_s = w \tag{6.22}$$

Substituting equation 6.20 into 6.22 gives

$$\mu_s F_N = w \tag{6.23}$$

Substituting the normal force F_N from equation 6.21 into equation 6.23 gives

$$\mu_s \frac{mv^2}{r} = w$$

But the weight w of the child is equal to mg. Thus,

$$\mu_s \frac{mv^2}{r} = mg \tag{6.24}$$

Notice that the mass m is contained on both sides of equation 6.24 and can be cancelled. Hence,

$$\mu_s \frac{v^2}{r} = g \qquad (6.25)$$

We can solve equation 6.25 for μ_s, the minimum value of the coefficient of static friction such that the child will not slide down the wall:

$$\mu_s = \frac{rg}{v^2} \qquad (6.26)$$

Example 6.5

The rotating room. The radius r of the rotating room is 15 ft, and the speed v of the child is 40 ft/s. Find the minimum value of μ_s to keep the child pinned against the wall.

Solution

The minimum value of μ_s, found from equation 6.26, is

$$\mu_s = \frac{rg}{v^2} = \frac{(15 \text{ ft})(32 \text{ ft/s}^2)}{(40 \text{ ft/s})^2} = 0.30$$

As long as μ_s is greater than 0.30, the child will be held against the wall.

As the ride comes to an end, the speed v decreases, thereby decreasing the centripetal force, which is supplied by the normal force F_N. The frictional force, $f_s = \mu_s F_N$, also decreases and is no longer capable of holding up the weight w of the child, and the child slides slowly down the wall.

Again, we should note that there is no centrifugal force acting on the child pushing the child against the wall. It is the wall that is pushing against the child with the centripetal force that is supplied by the normal force. There are many variations of this ride in different amusement parks, where you are held tight against a rotating wall. The analysis will be similar.

A Car Making a Turn on a Level Road

Consider a car making a turn at a corner. The portion of the turn can be approximated by an arc of a circle of radius r, as shown in figure 6.13. If the car makes the turn at a constant speed v, then during that turn, the car is going through uniform circular motion and there must be some centripetal force acting on the car. The necessary centripetal force is supplied by the frictional force between the tires of the car and the pavement.

Let us analyze the motion, in particular let us find the minimum coefficient of static friction that must be present between the tires of the car and the pavement in order for the car to make the turn without skidding. As the steering wheel of the car is turned, the tires turn into the direction of the turn. But the tire also wants to continue in straight line motion by Newton's first law. Because all real tires are slightly deformed, part of the tire in contact with the road is actually flat. Hence, the portion of the tire in contact with the ground has a tendency to slip and there is therefore a frictional force that opposes this motion. Hence, the force that allows the car to go into that circular path is the static frictional force f_s between the flat portion of the tire and the road. The problem is therefore very similar to the rotating disk discussed previously. The frictional force f_s is again given by equation 4.44. We are looking for the minimum value of μ_s that will just keep the car moving in the

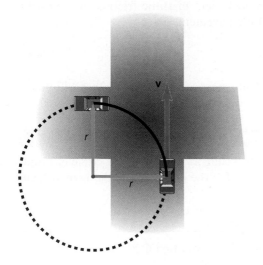

(a) Top view

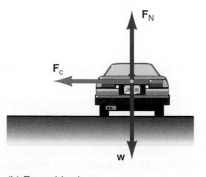

(b) Rear side view

Figure 6.13

A car making a turn on a level road.

circle. That is, the car will be just on the verge of slipping. Hence, we use the equality sign in equation 4.44. Because the centripetal force is supplied by the frictional force, we equate them as

$$F_c = f_s \qquad (6.27)$$

We obtain the centripetal force from equation 6.14 and the frictional force from equation 6.20. Hence,

$$\frac{mv^2}{r} = \mu_s F_N$$

But as we can see from figure 6.13, the normal force F_N is equal to the weight w, thus,

$$\frac{mv^2}{r} = \mu_s w$$

The weight of the car $w = mg$, therefore,

$$\frac{mv^2}{r} = \mu_s mg \qquad (6.28)$$

Notice that the mass m is on each side of equation 6.28 and can be divided out. Solving equation 6.28 for the minimum coefficient of static friction that must be present between the tires of the car and the pavement, gives

$$\mu_s = \frac{v^2}{rg} \qquad (6.29)$$

Because equation 6.29 is independent of the mass of the car, the effect will be the same for a large massive car or a small one.

<hr>

Example 6.6

Making a turn on a level road. A car is traveling at 20.0 mph in a circle of radius $r = 200$ ft. Find the minimum value of μ_s for the car to make the turn without skidding.

Solution

The minimum coefficient of friction, found from equation 6.29, is

$$\mu_s = \frac{v^2}{rg}$$

$$= \frac{[(20.0 \text{ mph})(88.0 \text{ ft/s})/(60.0 \text{ mph})]^2}{(200 \text{ ft})(32.0 \text{ ft/s}^2)}$$

$$= 0.134$$

The minimum value of the coefficient of static friction between the tires and the road must be 0.134.

For all values of μ_s, equal to or greater than this value, the car can make the turn without skidding. From table 4.1, the coefficient of friction for a tire on concrete is much greater than this, and there will be no problem in making the turn. However, if there is snow or freezing rain, then the coefficient of static friction between the tires and the snow or ice will be much lower. If it is lower than the value of 0.134 just determined, then the car will skid out in the turn. That is, there will no longer be enough frictional force to supply the necessary centripetal force.

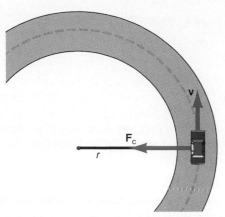

(a) Top view

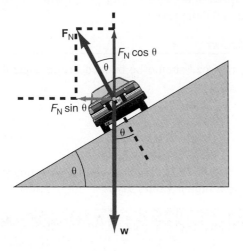

(b) Rear view

(c)

Figure 6.14

A car making a turn on a banked road.

If you ever go into a skid what should you do? The standard procedure is to turn the wheels of the car into the direction of the skid. You are then no longer trying to make the turn, and therefore you no longer need the centripetal force. You will stop skidding and proceed in the direction that was originally tangent to the circle. By tapping the brakes, you then slow down so that you can again try to make the turn. At a slower speed you may now be able to make the turn. As an example, if the speed of the car in example 6.6 is reduced from 20 mph to 10 mph, that is, in half, then from equation 6.29 the minimum value of μ_s would be cut by a fourth. Therefore, $\mu_s = 0.034$. The car should then be able to make the turn.

Even on a hot sunny day with excellent road conditions there could be a problem in making the original turn, if the car is going too fast.

Example 6.7

Making a level turn while driving too fast. If the car in example 6.6 tried to make the turn at a speed of 60 mph, that is, three times faster than before, what would the value of μ_s have to be?

Solution

The minimum coefficient of friction, found from equation 6.29, is

$$\mu_s = \frac{(3v_0)^2}{rg} = 9 \frac{(v_0^2)}{rg} = 9 \, \mu_{s0} = 1.21$$

That is, by increasing the speed by a factor of three, the necessary value of μ_s has been increased by a factor of 9 to the value of 1.21.

From the possible values of μ_s given in table 4.1, we cannot get such high values of μ_s. Therefore, the car will definitely go into a skid. When the original road was designed, it could have been made into a more gentle curve with a much larger value of the radius of curvature r, thereby reducing the minimum value of μ_s needed. This would certainly help, but there are practical constraints on how large we can make r.

A Car Making a Turn on a Banked Road

On large highways that handle cars at high speeds, the roads are usually banked to make the turns easier. By banking the road, a component of the reaction force of the road points into the center of curvature of the road, and that component will supply the necessary centripetal force to move the car in the circle. The car on the banked road is shown in figure 6.14. The road is banked at an angle θ. The forces acting on the car are the weight **w**, acting downward, and the reaction force of the road $\mathbf{F_N}$, acting upward on the car, perpendicular to the road. We resolve the force $\mathbf{F_N}$ into vertical and horizontal components. Using the value of θ as shown, the vertical component is $F_N \cos \theta$, while the horizontal component is $F_N \sin \theta$. As we can see from the figure, the horizontal component points toward the center of the circle. Hence, the necessary centripetal force is supplied by the component $F_N \sin \theta$. That is,

$$F_c = F_N \sin \theta \tag{6.30}$$

The vertical component is equal to the weight of the car, that is,

$$w = F_N \cos \theta \tag{6.31}$$

The problem can be simplified by eliminating F_N by dividing equation 6.30 by equation 6.31:

$$\frac{F_N \sin \theta}{F_N \cos \theta} = \frac{F_c}{w} = \frac{mv^2/r}{mg}$$

and finally, using the fact that $\sin\theta/\cos\theta = \tan\theta$, we have

$$\tan\theta = \frac{v^2}{rg} \qquad (6.32)$$

Solving for θ, the angle of bank, we get

$$\theta = \tan^{-1}\frac{v^2}{rg} \qquad (6.33)$$

which says that if the road is banked by this angle θ, then the necessary centripetal force for any car to go into the circular path will be supplied by the horizontal component of the reaction force of the road.

Example 6.8

Making a turn on a banked road. The car from example 6.7 is to manipulate a turn with a radius of curvature of 200 ft at a speed of 60 mph = 88 ft/s. At what angle should the road be banked for the car to make the turn?

Solution

To have the necessary centripetal force, the road should be banked at the angle θ given by equation 6.33 as

$$\theta = \tan^{-1}\frac{v^2}{rg} = \tan^{-1}\left[\frac{(88\text{ ft/s})^2}{(200\text{ ft})(32\text{ ft/s}^2)}\right]$$
$$= 50°$$

This angle is a little large for practical purposes. A reasonable compromise might be to increase the radius of curvature r, to a higher value, say $r = 600$ ft, then,

$$\theta = \tan^{-1}\frac{v^2}{rg} = \tan^{-1}\left[\frac{(88\text{ ft/s})^2}{(600\text{ ft})(32\text{ ft/s}^2)}\right]$$
$$= 22°$$

a more reasonable angle of bank.

The design of the road becomes a trade-off between the angle of bank and the radius of curvature, but the necessary centripetal force is supplied by the horizontal component of the reaction force of the road on the car.

An Airplane Making a Turn

During straight and level flight, the following forces act on the aircraft shown in figure 6.15: **T** is the thrust on the aircraft pulling it forward into the air, **w** is the weight of the aircraft acting downward, **L** is the lift on the aircraft that causes the

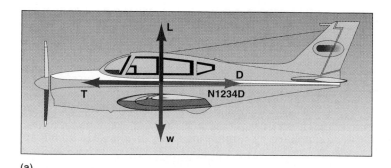

(a)

(b)

Figure 6.15

Forces acting on an aircraft in flight.

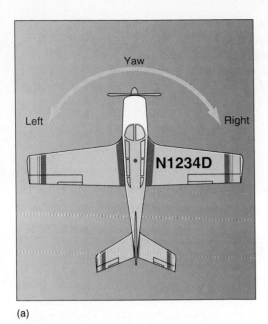

(a)

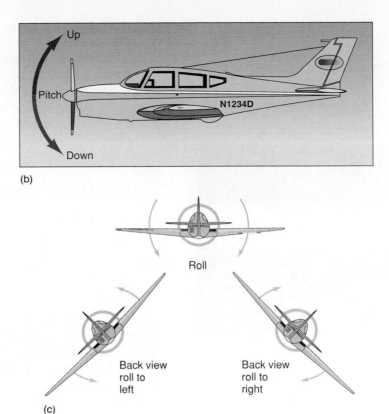

(b)

(c)

Figure 6.16

(a) Yaw of an aircraft. (b) Pitch of an aircraft. (c) Roll of an aircraft.

(a)

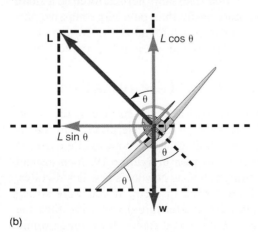

(b)

Figure 6.17

Forces acting on the aircraft during a turn.

plane to ascend, and **D** is the drag on the aircraft that tends to slow down the aircraft. The drag is opposite to the thrust. Lift and drag are just the vertical and horizontal components of the fluid force of the air on the aircraft. In normal straight and level flight, the aircraft is in equilibrium under all these forces. The lift overcomes the weight and holds the plane up; the thrust overcomes the frictional drag forces, allowing the plane to fly at a constant speed. An aircraft has three ways of changing the direction of its motion.

Yaw Control: By applying a force on the rudder pedals with his feet, the pilot can make the plane turn to the right or left, as shown in figure 6.16(a).

Pitch Control: By pushing the stick forward, the pilot can make the plane dive; by pulling the stick backward, the pilot can make the plane climb, as shown in figure 6.16(b).

Roll Control: By pushing the stick to the right, the pilot makes the plane roll to the right; by pushing the stick to the left, the pilot makes the plane roll to the left, as shown in figure 6.16(c).

To make a turn to the right or left, therefore, the pilot could simply use the rudder pedals and yaw the aircraft to the right or left. However, this is not an efficient way to turn an aircraft. As the aircraft yaws, it exposes a larger portion of its fuselage to the air, causing a great deal of friction. This increased drag causes the plane to slow down. To make the most efficient turn, a pilot performs a coordinated turn. In a coordinated turn, the pilot yaws, rolls, and pitches the aircraft simultaneously. The attitude of the aircraft is as shown in figure 6.17. In level flight the forces acting on the plane in the vertical are the lift **L** and the weight **w.** If the aircraft was originally in equilibrium in level flight, then $L = w$. Because of the turn, however, only a component of the lift is in the vertical, that is,

$$L \cos \theta = w$$

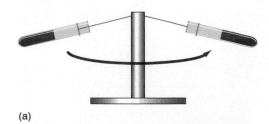

(a)

(b)

Figure 6.18

The centrifuge.

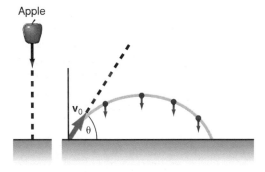

Apple

v_0

θ

Figure 6.19

Motion of an apple or a projectile.

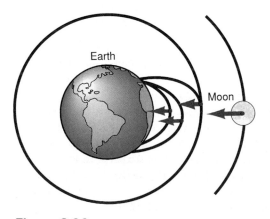

Earth

Moon

Figure 6.20

The same force acting on a projectile acts
on the moon.

Therefore, the aircraft loses altitude in a turn, unless the pilot pulls back on the stick, pitching the nose of the aircraft upward. This new attitude of the aircraft increases the angle of attack of the wings, thereby increasing the lift L of the aircraft. In this way the turn can be made at a constant altitude.

The second component of the lift, $L \sin \theta$, supplies the necessary centripetal force for the aircraft to make its turn. That is,

$$L \sin \theta = F_c = \frac{mv^2}{r} \tag{6.34}$$

while

$$L \cos \theta = w = mg \tag{6.35}$$

Dividing equation 6.34 by equation 6.35 gives

$$\frac{L \sin \theta}{L \cos \theta} = \frac{mv^2/r}{mg}$$

$$\tan \theta = \frac{v^2}{rg} \tag{6.36}$$

That is, for an aircraft traveling at a speed v, and trying to make a turn of radius of curvature r, the pilot must bank or roll the aircraft to the angle θ given by equation 6.36. Note that this is the same equation found for the banking of a road. A similar analysis would show that when a bicycle makes a turn on a level road, the rider leans into the turn by the same angle θ given by equation 6.36, to obtain the necessary centripetal force to make the turn.

The Centrifuge

The **centrifuge** is a device for separating particles of different densities in a liquid. The liquid is placed in a test tube and the test tube in the centrifuge, as shown in figure 6.18. The centrifuge spins at a high speed. The more massive particles in the mixture separate to the bottom of the test tube while the particles of smaller mass separate to the top. There is no centrifugal force acting on these particles to separate them as is often stated in chemistry, biology, and medical books. Instead, each particle at any instant has a tangential velocity v and wants to continue at that same velocity by Newton's first law. The centripetal force necessary to move the particles in a circle is given by equation 6.14 ($F_c = mv^2/r$). The normal force of the bottom of the glass tube on the particles supplies the necessary centripetal force on the particles to cause them to go into circular motion. The same normal force on a small mass causes it to go into circular motion more easily than on a large massive particle. The result is that the more massive particles are found at the bottom of the test tube, while the particles of smaller mass are found at the top of the test tube.

6.8 Newton's Law of Universal Gravitation

Newton observed that an object, an apple, released near the surface of the earth, was accelerated toward the earth. Since the cause of an acceleration is an unbalanced force, there must, therefore, be a force pulling objects toward the earth. If you throw a projectile at an initial velocity v_0, as seen in figure 6.19, then instead of that object moving off into space in a straight line as Newton's first law dictates, it is continually acted on by a force pulling it back to earth. If you were strong enough to throw the projectile with greater and greater initial velocities, then the projectile paths would be as shown in figure 6.20. The distance down range would become greater and greater until at some initial velocity, the projectile would not hit the earth at all, but would go right around it in an orbit. But at any point along

NEWTON'S SYSTEM OF THE WORLD   551

force, as it is a force which is directed towards some centre; and as it regards more particularly a body in that centre, we call it circumsolar, circumterrestrial, circumjovial; and so in respect of other central bodies.

[3.] *The action of centripetal forces.*

That by means of centripetal forces the planets may be retained in certain orbits, we may easily understand, if we consider the motions of projectiles (pp. 2–4); for a stone that is projected is by the pressure of its own weight forced out of the rectilinear path, which by the initial projection alone it should have pursued, and made to describe a curved line in the air; and

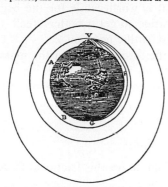

through that crooked way is at last brought down to the ground; and the greater the velocity is with which it is projected, the farther it goes before it falls to the earth. We may therefore suppose the velocity to be so increased, that it would describe an arc of 1, 2, 5, 10, 100, 1000 miles before it arrived at the earth, till at last, exceeding the limits of the earth, it should pass into space without touching it.

Figure 6.21

A page from Newton's *Principia*.

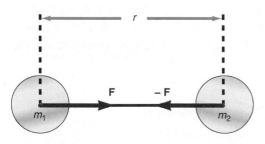

Figure 6.22

Newton's law of universal gravitation.

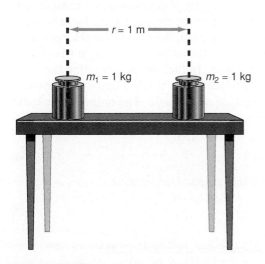

Figure 6.23

Two 1-kg masses sitting on a table.

its path the projectile would still have a force acting on it pulling it down toward the surface of the earth just as it had in figure 6.19. Figure 6.21 shows a page from the translated version of Newton's *Principia* showing these ideas.

Newton was led to the conclusion that the same force that causes the apple to fall to the earth also causes the moon to be pulled to the earth. Thus, the moon moves in its orbit about the earth because it is pulled toward the earth. It is falling toward the earth. But if there is a force between the moon and the earth, why not a force between the sun and the earth? Or for that matter why not a force between the sun and all the planets? Newton proposed that the same gravitational force that acts on objects near the surface of the earth also acts on all the heavenly bodies. This was a revolutionary hypothesis at that time, for no one knew why the planets revolved around the sun. Following this line of reasoning to its natural conclusion, Newton proposed that there was a force of gravitation between each and every mass in the universe.

Newton's law of universal gravitation was stated as follows: between every two masses in the universe there is a force of attraction between them that is directly proportional to the product of their masses, and inversely proportional to the square of the distance separating them. If the two masses are as shown in figure 6.22 with r the distance between the centers of the two masses, then the force of attraction is

$$F = \frac{Gm_1m_2}{r^2} \qquad (6.37)$$

where G is a constant, called the universal gravitational constant, given by

$$G = 6.67 \times 10^{-11} \frac{\text{N m}^2}{\text{kg}^2}$$

We assume here that the radii of the masses are relatively small compared to the distance separating them so that the distance separating the masses is drawn to the center of the masses. Such masses are sometimes treated as *point masses* or *particles*. Spherical masses are usually treated like particles.

Note here that the numerical value of the constant G was determined by a celebrated experiment by Henry Cavendish (1731–1810) over 100 years after Newton's statement of the law of gravitation. Cavendish used a torsion balance with known masses. The force between the masses was measured and G was then calculated.

6.9 Gravitational Force between Two 1-Kg Masses

Newton's law of universal gravitation says that there is a force between any two masses in the universe. Let us set up a little experiment to test this law. Let us take two standard 1-kg masses and place them on the desk, so that they are 1 m apart, as shown in figure 6.23. According to Newton's theory of gravitation, there is a force between these masses, and according to Newton's second law, they should be accelerated toward each other. However, we observe that the two masses stay right where they are. They do not move together! Is Newton's law of universal gravitation correct or isn't it?

Let us compute the gravitational force between these two 1-kg masses. We assume that the gravitational force acts at the center of each of the 1-kg masses. By equation 6.37, we have

$$F = \frac{Gm_1m_2}{r^2} = \frac{6.67 \times 10^{-11} \text{ N m}^2}{\text{kg}^2} \frac{(1 \text{ kg})(1 \text{ kg})}{(1 \text{ m})^2}$$

and therefore the force acting between these two 1-kg masses is

$$F = 6.67 \times 10^{-11} \text{ N}$$

This is, of course, a very small force. In fact, if this is written in ordinary decimal notation we have

$$F = 0.0000000000667 \text{ N}$$

A very, very small force indeed. (Sometimes it is worth while for the beginning student to write numbers in this ordinary notation to get a better "feel" for the meaning of the numbers that are expressed in scientific notation.)

If we redraw figure 6.23 showing all the forces acting on the masses, we get figure 6.24. The gravitational force on mass m_2 is trying to pull it toward the left. But if the body tends to slide toward the left, there is a force of static friction that acts to oppose that tendency and acts toward the right. The frictional force that must be overcome is

$$f_s = \mu_s F_{N2} = \mu_s w_2 = \mu_s m_2 g$$

Assuming a reasonable value of $\mu_s = 0.50$ we obtain for this frictional force,

$$f_s = \mu_s m_2 g = (0.50)(1.00 \text{ kg})(9.80 \text{ m/s}^2) = 4.90 \text{ N}$$

Hence, to initiate the movement of the 1-kg mass across the table, a force greater than 4.90 N is needed. As you can see, the gravitational force (6.67×10^{-11} N) is nowhere near this value, and is thus not great enough to overcome the force of static friction. Hence you do not, in general, observe different masses attracting each other. That is, two chairs do not slide across the room and collide due to the gravitational force between them.

If these small 1-kg masses were taken somewhere out in space, where there is no frictional force opposing the gravitational force, the two masses would be pulled together. It will take a relatively long time for the masses to come together because the force, and hence the acceleration is small, but they will come together within a few days.

6.10 Gravitational Force between a 1-Kg Mass and the Earth

The reason why the computed gravitational force between the two 1-kg masses was so small is because G, the universal gravitational constant, is very small compared to the masses involved. If, instead of considering two 1-kg masses, we consider one mass to be the 1-kg mass and the second mass to be the earth, then the force between them is very noticeable. If you let go of the 1-kg mass, the gravitational force acting on it immediately pulls it toward the surface of the earth. The cause of the greater force in this case is the larger mass of the earth. In fact, let us determine the gravitational force on a 1-kg mass near the surface of the earth. Figure 6.25 shows a mass m_1 of 1 kg a distance h above the surface of the earth. The radius of the earth r_e is $r_e = 6.371 \times 10^6$ m, and its mass m_e is $m_e = 5.977 \times 10^{24}$ kg. The separation distance between m_1 and m_e is

$$r = r_e + h \approx r_e \qquad (6.38)$$

since $r_e \gg h$. The gravitational force acting on that 1-kg mass is

$$F_g = \frac{G m_e m_1}{r_e^2} \qquad (6.39)$$

$$= \left(\frac{6.67 \times 10^{-11} \text{ N m}^2}{\text{kg}^2} \right) \frac{(5.977 \times 10^{24} \text{ kg})(1.00 \text{ kg})}{(6.371 \times 10^6 \text{ m})^2}$$

$$= 9.82 \text{ N}$$

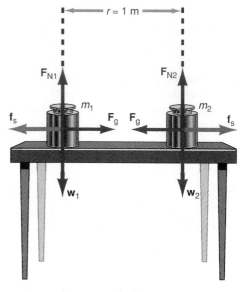

Figure 6.24

Gravitational force on two 1-kg masses.

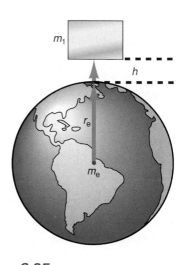

Figure 6.25

Gravitational force on a 1-kg mass near the surface of the earth.

Mechanics

This number should be rather familiar. Recall that the weight of a 1.00-kg mass was determined from Newton's second law as

$$w = mg = (1.00 \text{ kg})(9.80 \text{ m/s}^2) = 9.80 \text{ N}$$

(The standard value of $g = 9.80 \text{ m/s}^2$ has been used. We will see shortly that g can actually vary between 9.78 m/s^2 at the equator to 9.83 m/s^2 at the pole. Also the radius of the earth r_e used in equation 6.39 is the mean value of r_e. The actual value of r_e varies slightly with latitude.)

The point to notice here is that the weight of a body is in fact the gravitational force acting on that body by the earth and pulling it down toward the center of the earth. Thus, the weight of a body is actually determined by Newton's law of universal gravitation. This points up even further the difference between the mass and the weight of a body.

6.11 The Acceleration Due to Gravity and Newton's Law of Universal Gravitation

Newton's second law states that when an external unbalanced force acts on an object, it will give that object an acceleration a, that is,

$$F = ma$$

But if the force acting on a body near the surface of the earth is the gravitational force, then that body experiences the acceleration g of a freely falling body, as shown in section 3.7. That is, the force acting on the object is called its weight, and it experiences the acceleration g. Thus, Newton's second law becomes

$$w = mg \qquad \qquad \textbf{(6.40)}$$

But as just shown, the weight of a body is equal to the gravitational force acting on that body and therefore,

$$w = F_g \qquad \qquad \textbf{(6.41)}$$

Using equations 6.40 and 6.39 we get

$$mg = \frac{Gm_e m}{r_e^2} \qquad \qquad \textbf{(6.42)}$$

Solving for g, we obtain

$$g = \frac{Gm_e}{r_e^2} \qquad \qquad \textbf{(6.43)}$$

That is, *the acceleration due to gravity, g, which in chapter 3 was accepted as an experimental fact, can be deduced from theoretical considerations of Newton's second law and his law of universal gravitation.*

Example 6.9

Determine the acceleration of gravity g at the surface of the earth.

Solution

The value of g, determined from equation 6.43, is

$$g = \frac{Gm_e}{r_e^2} = \left(\frac{6.67 \times 10^{-11} \text{ N m}^2}{\text{kg}^2} \right) \frac{(5.977 \times 10^{24} \text{ kg})}{(6.371 \times 10^6 \text{ m})^2}$$

$$= 9.82 \text{ m/s}^2$$

Newton introduced his law of universal gravitation, and a by-product of it is a theoretical explanation of the acceleration due to gravity g. This is an example of the beauty and simplicity of physics. There is no way that we could have predicted the relation of equation 6.43 from purely experimental grounds. Yet Newton's second law and his law of universal gravitation, in combination, have made that prediction.

6.12 Variation of the Acceleration Due to Gravity

Figure 6.26

The earth is an oblate spheroid.

We can see from equation 6.43 why the acceleration due to gravity g is very nearly a constant. G is a constant and m_e is a constant, but r_e is not exactly a constant. The earth is not, in fact, a perfect sphere. It is, rather, an oblate spheroid. That is, the radius of the earth at the equator r_{ee} is slightly greater than the radius of the earth at the poles r_{ep}, as seen in figure 6.26. The diagram is, of course, exaggerated to show this difference. The actual values of r_{ee} and r_{ep} are

$$r_{ee} = 6.378 \times 10^6 \text{ m}$$
$$r_{ep} = 6.356 \times 10^6 \text{ m}$$

with the mean radius

$$r_e = 6.371 \times 10^6 \text{ m}$$

The variation in the radius of the earth is thus quite small. However, the variation, although small, does contribute to the observed variation in the acceleration due to gravity on the earth from a low of 9.78 m/s² at the equator to a high of 9.83 m/s² at the North Pole, as seen in table 6.1. This analysis also assumes that the earth is not rotating. A more sophisticated analysis takes into account the variation in g caused by the centripetal acceleration, which varies with latitude on the surface of the earth. The standard value of g, adopted for most calculations in physics, is

$$g = 9.80 \text{ m/s}^2$$

the value at 45° north latitude at the surface of the earth.

At greater heights, g also varies slightly from that given in equation 6.43 because of the approximation

$$r_e \approx r_e + h$$

that was made for that equation. Although this approximation is, in general, quite good for most locations, if you are on the top of a mountain, such as Pikes Peak, this higher altitude (large value of h) will give you a slightly smaller value of g, as we can see in table 6.1.

Table 6.1 Different Values of g on the Earth	
Location	**Value of g in m/s²**
Equator at sea level	9.78
New York City	9.80
45° N latitude (standard)	9.80
North Pole	9.83
Pikes Peak—elevation 4293 m	9.79
Denver, Colorado—elevation 1638 m	9.80

Mechanics

Again it is quite remarkable that these slight variations in the observed experimental values of g on the surface of this earth can be explained and predicted by Newton's law of universal gravitation, with slight corrections for the radius of the earth, the centripetal acceleration (which is a function of latitude), and the height of the location above mean sea level. There are also slight local variations in g due to the nonhomogeneous nature of the mass distribution of the earth. These variations in g due to different mass distributions are used in geophysical explorations. One of the many scientific experiments performed on the moon was a mapping of the acceleration due to gravity on the moon to disclose the possible locations of different mineral deposits.

6.13 Acceleration Due to Gravity on the Moon and on Other Planets

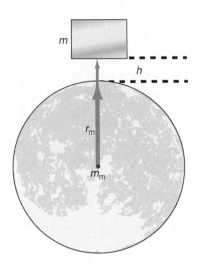

Figure 6.27

A mass placed close to the surface of the moon.

Equation 6.43 was derived on the basis of the gravitational force of the earth acting on a mass near the surface of the earth. The result is perfectly general however. If, for example, an object were placed close to the surface of the moon, as shown in figure 6.27, the force on that mass would be its lunar weight, which is just the gravitational force of the moon acting on it. Therefore the weight of an object on the moon is

$$w_{\mathrm{m}} = F_{\mathrm{g}} \qquad (6.44)$$

This becomes

$$mg_{\mathrm{m}} = \frac{Gm_{\mathrm{m}}m}{r_{\mathrm{m}}^2} \qquad (6.45)$$

where g_{m} is the acceleration due to gravity on the moon and m_{m} and r_{m} are the mass and radius of the moon, respectively. Hence, the acceleration due to gravity on the moon is

$$g_{\mathrm{m}} = \frac{Gm_{\mathrm{m}}}{r_{\mathrm{m}}^2} \qquad (6.46)$$

Equation 6.46 is identical to equation 6.43 except for the subscripts. Therefore, we can use equation 6.43 to determine the acceleration due to gravity on any of the planets, simply by using the mass of that planet and the radius of that planet in equation 6.43.

Example 6.10

Determine the acceleration due to gravity on the moon.

Solution

The acceleration due to gravity on the moon, found by solving equation 6.46, is

$$g_{\mathrm{m}} = \frac{Gm_{\mathrm{m}}}{r_{\mathrm{m}}^2} = \left(\frac{6.67 \times 10^{-11} \text{ N m}^2}{\text{kg}^2} \right) \frac{(7.34 \times 10^{22} \text{ kg})}{(1.738 \times 10^6 \text{ m})^2}$$

$$= 1.62 \text{ m/s}^2 = 0.165 \, g_{\mathrm{e}} \approx \tfrac{1}{6} \, g_{\mathrm{e}}$$

The acceleration due to gravity on the moon is approximately $1/6$ the acceleration due to gravity on the earth.

Planet	Mean Orbital Radius (m)	Mass (kg)	Mean Radius of Planet (m)	g at Equator (m/s²)	(g_e)
Mercury	5.80×10^{10}	3.24×10^{23}	2.340×10^6	3.95	0.4
Venus	1.08×10^{11}	4.86×10^{24}	6.10×10^6	8.72	0.89
Earth	1.50×10^{11}	5.97×10^{24}	6.371×10^6	9.78	1.00
Mars	2.28×10^{11}	6.40×10^{23}	3.32×10^6	3.84	0.39
Jupiter	7.80×10^{11}	1.89×10^{27}	69.8×10^6	23.16	2.37
Saturn	1.43×10^{12}	5.67×10^{26}	58.2×10^6	8.77	0.90
Uranus	2.88×10^{12}	8.67×10^{25}	23.8×10^6	9.46	0.97
Neptune	4.52×10^{12}	1.05×10^{26}	22.4×10^6	13.66	1.40
Pluto	5.91×10^{12}	6.6×10^{23}	2.9×10^6	5.23	0.53
Earth's Moon	3.84×10^8	7.34×10^{22}	1.738×10^6	1.62	1/6

Table 6.2
Some Characteristics of the Planets and the Moon

The Earth and its satellite.

Because the weight of an object is

$$w = mg$$

the weight of an object on the moon is

$$w_m = mg_m = m(1/6\ g_e) = 1/6\ (mg_e) = 1/6\ w_e$$

which is 1/6 of the weight that it had on the earth. That is, if you weigh 180 lb on earth, you will only weigh 30 lb on the moon.

Table 6.2 is a list of the masses, radii, and values of g on the various planets. Note that the most massive planet is Jupiter, and it has an acceleration due to gravity of

$$g_J = 2.37\ g_e$$

Therefore, the weight of an object on Jupiter will be

$$w_J = 2.37\ w_e$$

If you weighed 180 lb on earth, you would weigh 427 lb on Jupiter.

6.14 Satellite Motion

Consider the motion of a satellite around its parent body. This could be the motion of the earth around the sun, the motion of any planet around the sun, the motion of the moon around the earth, the motion of any other moon around its planet, or the motion of an artificial satellite around the earth, the moon, another planet, and so forth.

Let us start with the analysis of the motion of an artificial satellite in a circular orbit around the earth. Perhaps the first person to ever conceive of the possibility of an artificial earth satellite was Sir Isaac Newton, when he wrote in 1686 in his *Principia:*

But if we now imagine bodies to be projected in the directions of lines parallel to the horizon from greater heights, as of 5, 10, 100, 1000, or more miles or rather as many semi-diameters of the earth, those bodies according

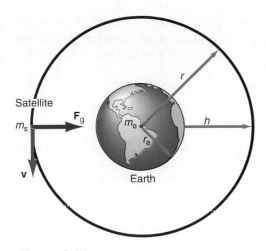

Figure 6.28

Satellite motion.

to their different velocity, and the different force of gravity in different heights, will describe arcs either concentric with the earth, or variously eccentric, and go on revolving through the heavens in those orbits just as the planets do in their orbits.[2]

For the satellite to be in motion in a circular orbit, there must be a centripetal force acting on the satellite to force it into the circular motion. This centripetal force acting on the satellite, is supplied by the force of gravity of the earth. Let us assume that the satellite is in orbit a distance h above the surface of the earth, as shown in figure 6.28. Because the centripetal force is supplied by the gravitational force, we have

$$F_c = F_g \tag{6.47}$$

or

$$\frac{m_s v^2}{r} = \frac{G m_e m_s}{r^2} \tag{6.48}$$

Solving for the speed of the satellite in the circular orbit, we obtain

$$v = \sqrt{\frac{G m_e}{r}} \tag{6.49}$$

The first thing that we note is that m_s, the mass of the satellite, divided out of equation 6.48. This means that the speed of the satellite is independent of its mass. That is, the speed is the same, whether it is a very large massive satellite or a very small one.

Equation 6.49 represents the speed that a satellite must have if it is to remain in a circular orbit, at a distance r from the center of the earth. Because the satellite is at a height h above the surface of the earth, the orbital radius r is

$$r = r_e + h \tag{6.50}$$

Equation 6.49 also says that the speed depends only on the radius of the orbit r. For large values of r, the required speed will be relatively small; whereas for small values of r, the speed v must be much larger. If the actual speed of a satellite at a distance r is less than the value of v, given by equation 6.49, then the satellite will move closer toward the earth. If it gets close enough to the earth's atmosphere, the air friction will slow the satellite down even further, eventually causing it to spiral into the earth. The increased air friction will then cause it to burn up and crash. If the actual speed at the distance r is greater than that given by equation 6.49, then the satellite will go farther out into space, eventually going into either an elliptical, parabolic, or hyperbolic orbit, depending on the speed v.

How do we get the satellite into this circular orbit? The satellite is placed in the orbit by a rocket. The rocket is launched vertically from the earth, and at a predetermined altitude it pitches over, so as to approach the desired circular orbit tangentially. The engines are usually turned off and the rocket coasts on a projectile trajectory to the orbital intercept point I in figure 6.29. Let the velocity of the rocket on the ascent trajectory at the point of intercept be $\mathbf{v_A}$. The velocity necessary for the satellite to be in circular orbit at the height h is $\mathbf{v}$ and its speed is given by equation 6.49. Therefore, the rocket must undergo a change in velocity $\Delta\mathbf{v}$ to match its ascent velocity to the velocity necessary for the circular orbit. That is,

$$\Delta\mathbf{v} = \mathbf{v} - \mathbf{v_A} \tag{6.51}$$

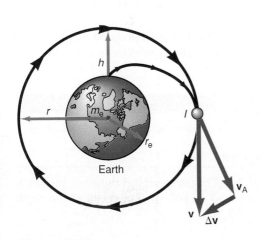

Figure 6.29

Placing a satellite in a circular orbit.

2. Quoted from Sir Isaac Newton's *Mathematical Principles of Natural Philosophy*, p. 552. Translated by A. Motte. University of California Press, 1960.

This change in velocity is of course produced by the thrust of the rocket engines. How long should these engines be turned on to get this necessary change in speed Δv? As a first approximation we take Newton's second law in the form

$$F = ma = m\frac{\Delta v}{\Delta t}$$

Solving Newton's second law for Δt, gives

$$\Delta t = \frac{m}{F}\Delta v \qquad (6.52)$$

where F is the thrust of the rocket engine, Δv is the necessary speed change, determined from equation 6.51; and m is the mass of the space craft at this instant of time. Therefore, equation 6.52 tells the astronaut the length of time to "burn" his engines. At the end of this time the engines are shut off, and the spacecraft has the necessary orbital speed to stay in its circular orbit.

This is, of course, a greatly simplified version of orbital insertion, for we need not only the magnitude of Δv but also its direction. An attitude control system is necessary to determine the proper direction for the $\Delta \mathbf{v}$. Also it is important to note that using equation 6.52 is only an approximation, because as the spacecraft burns its rocket propellant, its mass m is continually changing. This example points out a deficiency in using Newton's second law in the form $F = ma$, because this form assumes that the mass under consideration is a constant. In chapter 8, on momentum, we will write Newton's second law in another form that allows for the case of variable mass.

We should also note here that the orbits of all the planets around the sun are ellipses rather than circles. But, in general, the amount of ellipticity is relatively small, and as a first approximation it is quite often assumed that their orbits are circular. For this approximation, we can use equation 6.49, with the appropriate change in subscripts, to determine the approximate speed of any of the planets. For precise astronomical work, the elliptical orbit must be used. Extremely precise experimental determinations of the orbits of the planets were made by the Danish astronomer Tycho Brahe (1546–1609). Johannes Kepler (1571–1630) analyzed this work and expressed the result in what are now called **Kepler's laws of planetary motion.** Kepler's laws are

1. The orbit of each planet is an ellipse with the sun at one focus.
2. The speed of the planet varies in such a way that the line joining the planet and the sun sweeps out equal areas in equal times.
3. The cube of the semimajor axes of the elliptical orbit is proportional to the square of the time for the planet to make a complete revolution about the sun.

Example 6.11

Determine the speed of the earth in its orbit about the sun, shown in figure 6.30.

Solution

The mass of the sun is $m_{sun} = 1.99 \times 10^{30}$ kg. The mean orbital radius of the earth around the sun, found in table 6.2, is $r_{es} = 1.50 \times 10^{11}$ m. The speed of the earth around the sun v_{es}, found from equation 6.49, is

$$v_{es} = \sqrt{\frac{Gm_s}{r_{es}}}$$

$$= \sqrt{\left(\frac{6.67 \times 10^{-11} \text{ N m}^2}{\text{kg}^2}\right)\frac{1.99 \times 10^{30} \text{ kg}}{1.50 \times 10^{11} \text{ m}}}$$

$$= 2.97 \times 10^4 \text{ m/s} = 29.7 \text{ km/s} = 66,600 \text{ mi/hr}$$

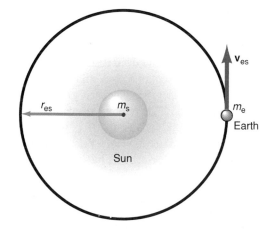

Figure 6.30

The speed of the earth in its orbit about the sun.

That is, as you sit and read this book, you are speeding through space at 66,600 mph. This is a little over 18 miles each second. The mean orbital speed of any of the planets or satellites can be determined in the same way.

6.15 The Geosynchronous Satellite

An interesting example of satellite motion is the geosynchronous satellite. The **geosynchronous satellite** is a satellite whose orbital motion is synchronized with the rotation of the earth. In this way the geosynchronous satellite is always over the same point on the equator as the earth turns. The geosynchronous satellite is obviously very useful for world communication, weather observations, and military use.

What should the orbital radius of such a satellite be, in order to stay over the same point on the earth's surface? The speed necessary for the circular orbit, given by equation 6.49, is

$$v = \sqrt{\frac{Gm_e}{r}}$$

But this speed must be equal to the average speed of the satellite in one day, namely

$$v = \frac{s}{t} = \frac{2\pi r}{\tau} \tag{6.53}$$

where τ is the period of revolution of the satellite that is equal to one day. That is, the satellite must move in one complete orbit in a time of exactly one day. Because the earth rotates in one day and the satellite will revolve around the earth in one day, the satellite at A' will always stay over the same point on the earth A, as in figure 6.31(a). That is, the satellite is at A', which is directly above the point A on the earth. As the earth rotates, A' is always directly above A. Setting equation 6.53 equal to equation 6.49 for the speed of the satellite, we have

$$\frac{2\pi r}{\tau} = \sqrt{\frac{Gm_e}{r}} \tag{6.54}$$

Squaring both sides of equation 6.54 gives

$$\frac{4\pi^2 r^2}{\tau^2} = \frac{Gm_e}{r}$$

or

$$r^3 = \frac{Gm_e \tau^2}{4\pi^2}$$

Solving for r, gives the required orbital radius of

$$r = \left(\frac{Gm_e \tau^2}{4\pi^2}\right)^{1/3} \tag{6.55}$$

Substituting the values for the earth into equation 6.55 gives

$$r = \left\{\frac{(6.67 \times 10^{-11} \text{ Nm}^2/\text{kg}^2)(5.97 \times 10^{24} \text{ kg})[24 \text{ hr}(3600 \text{ s/hr})]^2}{4(3.14159)^2}\right\}^{1/3}$$

$$r = 4.22 \times 10^7 \text{ m} = 4.22 \times 10^4 \text{ km} = 26,200 \text{ miles}$$

the orbital radius, measured from the center of the earth, for a geosynchronous satellite. A satellite at this height will always stay directly above a particular point on the surface of the earth.

A satellite communication system can be set up by placing several geosynchronous satellites in orbits over different points on the surface of the earth. As an

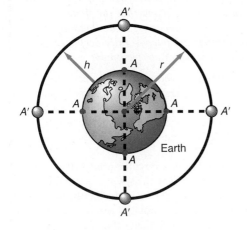

(a)

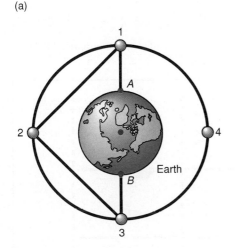

(b)

Figure 6.31

The geosynchronous satellite.

example, suppose four geosynchronous satellites were placed in orbit, as shown in figure 6.31(b). Let us say that we want to communicate, by radio or television, between the points *A* and *B,* which are on opposite sides of the earth. The communication would first be sent from point *A* to geosynchronous satellite 1, which would retransmit the communication to geosynchronous satellite 2. This satellite would then transmit to geosynchronous satellite 3, which would then transmit to the point *B* on the opposite side of the earth. Since these geosynchronous satellites appear to hover over one place on earth, continuous communication with any place on the surface of the earth can be attained.

"Have you ever wondered . . . ?"

An Essay on the Application of Physics

Space Travel

Earth is the cradle of man, but man was never made to stay in a cradle forever.

K. Tsiolkovsky

Have you ever wondered what it would be like to go to the moon or perhaps to another planet or to travel anywhere in outer space? But how can you get there? How can you travel into space?

Man has long had a fascination with the possibility of space travel. Jules Verne's novel, *From the Earth to the Moon,* was first published in 1868. In it he describes a trip to the moon inside a gigantic cannon shell. It is interesting to note that he says

Now as the Moon is never in the zenith, or directly overhead, in countries further than 28° from the equator, to decide on the exact spot for casting the Columbiad became a question that required some nice consultation. [And then a little further on] The 28th parallel of north latitude, as every school boy knows, strikes the American continent a little below Cape Canaveral. (pp. 66 and 68)

As I am sure we all know, Cape Canaveral is the site of the present Kennedy Space Center, the launch site for the Apollo mission to the moon. The first astronauts landed on the moon on July 20, 1969, just over a hundred years after the publication of Verne's novel. (Actually Jules Verne did not pick Cape Canaveral as the launch site, but rather Tampa, Florida, a relatively short distance away, because of its "position and easiness of approach, both by sea and land.")[3]

The idea of space travel left the realm of science fiction by the work of three men, Konstantin Tsiolkovsky, a Russian; Robert Goddard, an American; and Hermann Oberth, a German. Tsiolkovsky's first paper, "Free Space," was published in 1883. In his *Dreams of Earth and Sky,* 1895, he wrote of an artificial earth satellite. Goddard's first paper, "A Method of Reaching Extreme Altitudes," was written in 1919. The extreme altitudes he was referring to was the moon. Goddard launched the first liquid-

[3]*From the Earth to the Moon* in *The Space Novels of Jules Verne,* p. 69, Dover Publications, N.Y.

fueled rocket in history on March 16, 1926. Meanwhile, Oberth published his work, *The Rocket into Inter Planetary Space,* in 1923, which culminated with the German V-2 rocket in World War II. Another analysis of the problems associated with space flight was published in 1925 by Walter Hohmann in *Die Erreichbarkeit der Himmelskorper (The Attainability of the Heavenly Bodies).* In the preface, Hohmann says,

The present work will contribute to the recognition that space travel is to be taken seriously and that the final successful solution of the problem cannot be doubted, if existing technical possibilities are purposefully perfected as shown by conservative mathematical treatment.

Hohmann's original work had been written 10 years previous to its publication. In this work, Hohmann shows how to get to the Moon, Venus, and Mars. His simple approach to reach these heavenly bodies is by use of the cotangential ellipse. Before this approach is described, let us first say a word about the elliptical orbit.

Just as the speed of a satellite in a circular orbit is determined by equation 6.49, we can determine the speed of a satellite in an elliptical orbit. The mathematical derivation is slightly more complicated and will not be given here, but the result is quite simple. The speed of a satellite in an elliptical orbit is given by

$$v = \sqrt{GM\left(\frac{2}{r} - \frac{1}{a}\right)} \qquad \textbf{(6H.1)}$$

where *a* is a constant of the orbit called the semimajor axis of the ellipse and is shown in figure 1.

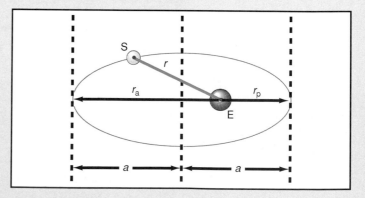

Figure 1

An elliptical orbit.

Let us assume that this is an elliptical satellite orbit about the earth. The earth is located at the focus of the ellipse, labeled E in figure 1 and r is the distance from the center of the earth to the satellite S, at any instant of time. The first thing we observe about elliptical motion is that the speed v is not a constant as it is for circular orbital motion. The speed varies with the location r in the orbit as we see in equation 6H.1. When the satellite is at its closest approach to the earth, $r = r_p$, the satellite is said to be at its perigee position. From equation 6H.1 we see that because this is the smallest value of r in the orbit, this corresponds to the greatest speed of the satellite. Hence, the satellite moves at its greatest speed when it is closest to the earth. When the satellite is at its farthest position from the earth, $r = r_a$, the satellite is said to be at its apogee position. Because this is the largest value of r in the orbit, it is the largest value of r in equation 6H.1. Since r is in the denominator of equation 6H.1, the largest value of r corresponds to the smallest value of v. Hence, the satellite moves at its slowest speed when it is the farthest distance from the earth. Thus, the motion in the orbit is not uniform, it speeds up as the satellite approaches the earth and slows down as the satellite recedes away from the earth.

We can express the semimajor axis of the ellipse in terms of the perigee and apogee distances by observing from figure 1 that

$$2a = r_a + r_p$$

or

$$a = \frac{r_a + r_p}{2} \qquad \text{(6H.2)}$$

For the special case where the ellipse degenerates into a circle, $r_a = r_p = r$, the radius of the circular orbit and then

$$a = \frac{r_a + r_p}{2} = \frac{r + r}{2} = \frac{2r}{2} = r$$

The equation for the speed, equation 6H.1 then becomes

$$v = \sqrt{GM\left(\frac{2}{r} - \frac{1}{r}\right)}$$

$$= \sqrt{\frac{GM}{r}} \qquad \text{(6H.3)}$$

But equation 6H.3 is the equation for the speed of a satellite in a circular orbit, equation 6.49. Hence, the elliptical orbit is the more general orbit, with the circular orbit as a special case.

Example 6H.1

The earth is at its closest position to the sun, its perihelion, on about January 3 when it is approximately 1.47×10^{11} m away from the sun. The earth reaches its aphelion distance, its greatest distance, on July 4, when it is about 1.53×10^{11} m away from the sun. Find the speed of the earth at its perihelion and aphelion position in its orbit.

Solution

The semimajor axis of the elliptical orbit, found from equation 6H.2, is

$$a = \frac{r_a + r_p}{2}$$
$$= \frac{1.53 \times 10^{11} \text{ m} + 1.47 \times 10^{11} \text{ m}}{2}$$
$$= 1.50 \times 10^{11} \text{ m} \qquad \text{(6H.2)}$$

The speed of the earth at perihelion is found from equation 6H.1 with $r = r_p$, the perihelion distance,

$$v = \sqrt{GM_s\left(\frac{2}{r} - \frac{1}{a}\right)}$$

Hence,

$$v_p = \sqrt{GM_s\left(\frac{2}{r_p} - \frac{1}{a}\right)}$$
$$= \sqrt{(6.67 \times 10^{-11} \text{ Nm}^2/\text{kg}^2)(1.99 \times 10^{30} \text{ kg})(2/1.47 \times 10^{11} \text{ m} - 1/1.50 \times 10^{11} \text{ m})}$$
$$= 3.03 \times 10^4 \text{ m/s}$$

The speed of the earth at aphelion is found from equation 6H.1 with $r = r_a = 1.53 \times 10^{11}$ m,

$$v = \sqrt{GM_s\left(\frac{2}{r} - \frac{1}{a}\right)}$$

$$v_a = \sqrt{GM_s\left(\frac{2}{r_a} - \frac{1}{a}\right)}$$
$$= \sqrt{(6.67 \times 10^{-11} \text{ Nm}^2/\text{kg}^2)(1.99 \times 10^{30} \text{ kg})(2/1.53 \times 10^{11} \text{ m} - 1/1.50 \times 10^{11} \text{ m})}$$
$$= 2.92 \times 10^4 \text{ m/s}$$

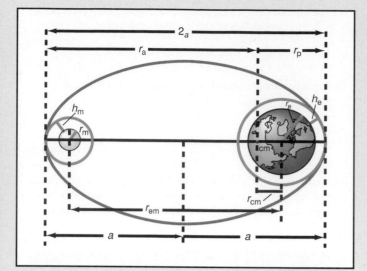

Figure 2

The Hohmann transfer orbit.

It is thus easy to see why the earth, in its orbit about the sun, is sometimes approximated as a circular orbit. The aphelion distance, perihelion distance, and the mean distance are very close, that is, 1.53×10^{11} m, 1.47×10^{11} m, and 1.50×10^{11} m, respectively. Also the speed of the earth at aphelion, perihelion, and in a circular orbit is 2.92×10^4 m/s, 3.03×10^4 m/s, and 2.97×10^4 m/s, respectively, which are also very close. The error in using the circular approximation rather than the elliptical analysis is no more than about 2%.

The simplest approach to space flight to the moon or to a planet is by use of the *Hohmann transfer ellipse*. Let us assume that the spacecraft is launched from the surface of the earth on an ascent trajectory. It is then desired to place the spacecraft in a circular parking orbit about the earth. If the circular parking orbit is to be at a height h_e above the surface of the earth then the necessary speed for the spacecraft, given by equation 6.49, is

$$v_{oe} = \sqrt{\frac{GM_e}{r_e + h_e}} \qquad \text{(6H.4)}$$

Knowing the speed of the spacecraft on the ascent trajectory from an on-board inertial navigational system, equation 6.51 is then used to determine the necessary "delta v," Δv, to get into this orbit. The engines are then turned on for the value of Δt, determined by equation 6.52, and the spacecraft is thus inserted into the circular parking orbit about the earth.

Before descending to the surface of the moon, it would be desirable to first go into a circular lunar parking orbit. To get to this circular lunar parking orbit, a "cotangential ellipse," the Hohmann transfer ellipse, is placed onto the two parking orbits, such that the focus of the ellipse is placed at the center of mass of the earth-moon system, and the ellipse is tangential to each parking orbit, as seen in figure 2. All positions in the orbit are measured from the center of mass of the earth-moon system. The semimajor axis a, of this ellipse, found from figure 2, is

$$a = \frac{r_{em} + r_m + h_m + r_e + h_e}{2} \qquad \text{(6H.5)}$$

where

r_{em} is the distance from the center of the earth to the center of the moon.
r_m is the radius of the moon.
h_m is the height of the spacecraft above the surface of the moon.
r_e is the radius of the earth.
h_m is the height of the spacecraft above the surface of the earth when it is in its circular parking orbit.

The insertion of the spacecraft into the transfer ellipse occurs at the perigee position of the elliptical orbit, which from figure 2 is

$$r_p = r_{cm} + r_e + h_e \qquad \text{(6H.6)}$$

where r_{cm} is the distance from the center of earth to the center of mass of the earth-moon system. The necessary speed to get into this cotangential ellipse at the perigee position, found from equation 6H.1, is

$$v_{TEp} = \sqrt{GM_e\left(\frac{2}{r_p} - \frac{1}{a}\right)} \qquad \text{(6H.7)}$$

where a and r_p are found from equations 6H.5 and 6H.6, respectively. The notation v_{TEp} stands for the speed in the transfer ellipse at perigee.

Because the speed of the spacecraft in the earth parking orbit is known, equation 6H.4, and the necessary speed for the transfer orbit is known, equation 6H.7, the necessary change in speed (Δv_I) of the spacecraft is just the difference between these speeds. Hence, the required Δv for insertion into the transfer ellipse is given by

$$\Delta v_I = v_{TEp} - v_{oe} \qquad \text{(6H.8)}$$

The spacecraft engines must be turned on to supply this necessary change in speed (Δv). When this Δv_I is achieved, the spacecraft engines are turned off and the spacecraft coasts toward the moon. If the engines are not turned on again, then the spacecraft would coast to the moon, reach it, and would then continue back toward the earth on the second half of the transfer ellipse. Thus, if there were some type of malfunction on the spacecraft, it would automatically return to earth.

Assuming there is no failure, the astronauts on board the spacecraft would like to change from their transfer orbit into the circular lunar parking orbit. The speed of the spacecraft on the transfer ellipse is given by equation 6H.1, with $r = r_a$ the apogee distance, as

$$v_{TEa} = \sqrt{GM_e\left(\frac{2}{r_a} - \frac{1}{a}\right)} \qquad \text{(6H.9)}$$

The apogee distance r_a, found from figure 2, is

$$r_a = r_{em} + r_m + h_m - r_{cm} \qquad \text{(6H.10)}$$

The necessary speed that the spacecraft must have to enter a circular lunar parking orbit v_{om} is found from modifying equation 6.49 to

$$v_{om} = \sqrt{\frac{GM_m}{r_m + h_m}} \qquad \text{(6H.11)}$$

where M_m is the mass of the moon, r_m is the radius of the moon, and h_m is the height of the spacecraft above the surface of the moon in its circular lunar parking orbit. The necessary change in speed to transfer from the Hohmann ellipse to the circular lunar parking orbit is obtained by subtracting equation 6H.11 from equation 6H.9. Thus the necessary Δv is

$$\Delta v_{II} = v_{TEa} - v_{om} \qquad \text{(6H.12)}$$

The spacecraft engines are turned on to obtain this necessary change in speed. When the engines are shut off the spacecraft will have the speed v_{om}, and will stay in the circular lunar parking orbit until the astronauts are ready to descend to the lunar surface. The process is repeated for the return to earth.

The Hohmann transfer is the simplest of the transfer orbits and is also the orbit of minimum energy. However, it has the disadvantage of having a large flight time. In the very early stages of the Apollo program, the Hohmann transfer ellipse was considered for the lunar transfer orbit. However, because of its long flight time, it was discarded for a hyperbolic transfer orbit that had been perfected by the Jet Propulsion Laboratories in California on its *Ranger, Surveyor,* and *Lunar Orbiter* unmanned spacecrafts to the moon. The hyperbolic orbit requires a great deal more energy, but its flight time is relatively small. The procedure for a trip on a hyperbolic orbit is similar to the elliptical orbit, only another equation is necessary for the speed of the spacecraft in the hyperbolic orbit. The principle however is the

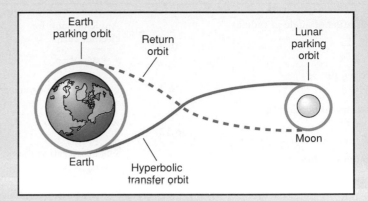

Figure 3

A hyperbolic transfer orbit.

same. Determine the current speed in the particular orbit, then determine the speed that is necessary for the other orbit. The difference between the two of them is the necessary Δv. The spacecraft engines are turned on until this value of Δv is obtained. A typical orbital picture for this type of transfer is shown in figure 3.

Unmanned satellites have since traveled to Mars, Venus, Saturn, Jupiter, Uranus, and Neptune. And what about manned trips to these planets? On July 20, 1989, the twentieth anniversary of the first landing on the moon, the president of the United States, George Bush, announced to the world that the United States will begin planning a manned trip to the planet Mars and eventually to an exploration of our entire solar system. Man is thus getting ready to leave his cradle.

The Language of Physics

Uniform circular motion
Motion in a circle at constant speed. Because the velocity vector changes in direction with time, this type of motion is accelerated motion (p. 155).

Centripetal acceleration
When a body moves in uniform circular motion, the acceleration is called centripetal acceleration. The direction of the centripetal acceleration is toward the center of the circle (p. 156).

Radian
A unit that is used to measure an angle. It is defined as the ratio of the arc length subtended to the radius of the circle, where 2π radians equals 360° (p. 156).

Centripetal force
The force that is necessary to cause an object to move in a circle at constant speed. The centripetal force acts toward the center of the circle (p. 160).

Centrifugal force
The reaction force to the centripetal force. The reaction force does not act on the same body as the centripetal force. That is, if a string were tied to a rock and the rock were swung in a horizontal circle at constant speed, the centripetal force would act on the rock while the centrifugal force would act on the string (p. 161).

Centrifuge
A device for separating particles of different densities in a liquid. The centrifuge spins at a high speed. The more massive particles in the mixture will separate to the bottom of the test tube while the particles of smaller mass will separate to the top (p. 170).

Newton's law of universal gravitation
Between every two masses in the universe there is a force of attraction that is directly proportional to the product of their masses

and inversely proportional to the square of the distance separating them (p. 171).

Kepler's laws of planetary motion
(1) The orbit of each planet is an ellipse with the sun at one focus. (2) The speed of the planet varies in such a way that the line joining the planet and the sun sweeps out equal areas in equal times. (3) The cube of the semimajor axes of the elliptical orbit is proportional to the square of the time for the planet to make a complete revolution about the sun (p. 178).

Geosynchronous satellite
A satellite whose orbital motion is synchronized with the rotation of the earth. In this way the satellite is always over the same point on the equator as the earth turns (p. 179).

Summary of Important Equations

Definition of angle in radians

$$\theta = \frac{s}{r} \qquad (6.4)$$

Arc length

$$s = r\theta \qquad (6.5)$$

Centripetal acceleration

$$a_c = \frac{v^2}{r} \qquad (6.12)$$

Centripetal force

$$F_c = ma_c = \frac{mv^2}{r} \qquad (6.14)$$

Angle of bank for circular turn

$$\theta = \tan^{-1}\frac{v^2}{rg} \qquad (6.33)$$

Newton's law of universal gravitation

$$F = \frac{Gm_1 m_2}{r^2} \qquad (6.37)$$

The acceleration due to gravity on earth

$$g_e = \frac{Gm_e}{r^2} \qquad (6.43)$$

The acceleration due to gravity on the moon

$$g_m = \frac{Gm_m}{r_m^2} \qquad (6.46)$$

Speed of a satellite in a circular orbit

$$v = \sqrt{Gm_e/r} \qquad (6.49)$$

Questions for Chapter 6

1. If a car is moving in uniform circular motion at a speed of 5.00 m/s and has a centripetal acceleration of 2.50 m/s², will the speed of the car increase at 2.50 m/s every second?

2. Does it make any sense to say that a car in uniform circular motion is moving with a velocity that is tangent to a circle and yet the acceleration is perpendicular to the tangent? Should not the acceleration be tangential because that is the direction that the car is moving?

3. If a car is moving in uniform circular motion, and the acceleration is toward the center of that circle, why does the car not move into the center of the circle?

4. Answer the student's question, "If an object moving in uniform circular motion is accelerated motion, why doesn't the speed change with time?"

5. Reply to the student's statement, "I know there is a centrifugal force acting on me when I move in circular motion in my car because I can feel the force pushing me against the side of the car."

† 6. Is it possible to change to a noninertial coordinate system, say a coordinate system that is fixed to the rotating body, to study uniform circular motion? In this rotating coordinate system is there a centrifugal force?

7. If you take a pail of water and turn it upside down all the water will spill out. But if you take the pail of water, attach a rope to the handle, and turn it rapidly in a vertical circle the water will not spill out when it is upside down at the top of the path. Why is this?

† 8. In high-performance jet aircraft the pilot must wear a pressure suit that exerts pressure on the abdomen and upper thighs of the pilot when the pilot pulls out of a steep dive. Why is this necessary?

9. If the force of gravity acting on a body is directly proportional to its mass, why does a massive body fall at the same rate as a less massive body?

10. Why does the earth bulge at the equator and not at the poles?

11. If the acceleration of gravity varies from place to place on the surface of the earth, how does this affect records made in the Olympics in such sports as shot put, javelin throwing, high jump, and the like?

12. What is wrong with applying Newton's second law in the form $F = ma$ to satellite motion? Does this same problem occur in the motion of an airplane?

†13. How can you use Kepler's second law to explain that the earth moves faster in its motion about the sun when it is closer to the sun?

14. Could you place a synchronous satellite in a polar orbit about the earth? At 45° latitude?

15. Explain how you can use a Hohmann transfer orbit to allow one satellite in an earth orbit to rendezvous with another satellite in a different earth orbit.

†16. A satellite is in a circular orbit. Explain what happens to the orbit if the engines are momentarily turned on to exert a thrust (a) in the direction of the velocity, (b) opposite to the velocity, (c) toward the earth, and (d) away from the earth.

17. A projectile fired close to the earth falls toward the earth and eventually crashes to the earth. The moon in its orbit about the earth is also falling toward the earth. Why doesn't it crash into the earth?

†18. The gravitational force on the earth caused by the sun is greater than the gravitational force on the earth caused by the moon. Why then does the moon have a greater effect on the tides than the sun?

†19. How was the universal gravitational constant G determined experimentally?

20. A string is tied to a rock and then the rock is put into motion in a vertical circle. Is this an example of uniform circular motion?

Problems for Chapter 6

6.3 Angles Measured in Radians

1. Express the following angles in radians: (a) 360°, (b) 270°, (c) 180°, (d) 90°, (e) 60°, (f) 30°, and (g) 1 rev.
2. Express the following angles in degrees: (a) 2π rad, (b) π rad, (c) 1 rad, and (d) 0.500 rad.
3. A record player turns at 33-1/3 rpm. What distance along the arc has a point on the edge moved in 1.00 min if the record has a diameter of 10.0 in.?

6.4 and 6.5 The Centripetal Acceleration and the Centripetal Force

4. A 4.00-kg stone is whirled at the end of a 2.00-m rope in a horizontal circle at a speed of 15.0 m/s. Ignoring the gravitational effects (a) calculate the centripetal acceleration and (b) calculate the centripetal force.
5. An automatic washing machine, in the spin cycle, is spinning wet clothes at the outer edge at 8.00 m/s. The diameter of the drum is 0.450 m. Find the acceleration of a piece of clothing in this spin cycle.
6. A 3200-lb car moving at 60.0 mph goes around a curve of 1000-ft radius. What is the centripetal acceleration? What is the centripetal force on the car?
7. A 1500-kg car moving at 86.0 km/hr goes around a curve of 325-m radius. What is the centripetal acceleration? What is the centripetal force on the car?
8. An electron is moving at a speed of 2.00×10^6 m/s in a circle of radius 0.0500 m. What is the force on the electron?
9. Find the centripetal force on a 70.0-lb girl on a merry-go-round that turns through one revolution in 40.0 s. The radius of the merry-go-round is 10.0 ft.

6.7 Examples of Centripetal Force

10. A boy sits on the edge of a polished wooden disk. The disk has a radius of 3.00 m and the coefficient of friction between his pants and the disk is 0.300. What is the maximum speed of the disk at the moment the boy slides off?

11. A 3200-lb car begins to skid when traveling at 60.0 mph around a level curve of 500-ft radius. Find the centripetal acceleration and the coefficient of friction between the tires and the road.
12. A 1200-kg car begins to skid when traveling at 80.0 km/hr around a level curve of 125-m radius. Find the centripetal acceleration and the coefficient of friction between the tires and the road.
13. At what angle should a bobsled turn be banked if the sled, moving at 26.0 m/s, is to round a turn of radius 100 m?
14. A motorcyclist goes around a curve of 400-ft radius at a speed of 60.0 mph, without leaning into the turn. (a) What must the coefficient of friction between the tires and the road be in order to supply the necessary centripetal force? (b) If the road is iced and the motorcyclist can not depend on friction, at what angle from the vertical should the motorcyclist lean to supply the necessary centripetal force?
15. A motorcyclist goes around a curve of 100-m radius at a speed of 95.0 km/hr, without leaning into the turn. (a) What must the coefficient of friction between the tires and the road be in order to supply the necessary centripetal force? (b) If the road is iced and the motorcyclist can not depend on friction, at what angle from the vertical should the motorcyclist lean to supply the necessary centripetal force?
16. At what angle should a highway be banked for cars traveling at a speed of 60.0 mph, if the radius of the road is 1000 ft and no frictional forces are involved?
17. At what angle should a highway be banked for cars traveling at a speed of 100 km/hr, if the radius of the road is 400 m and no frictional forces are involved?
18. An airplane is flying in a circle with a speed of 200 knots. (A knot is a nautical mile per hour.) The aircraft weighs 2000 lb and is banked at an angle of 30.0°. Find the radius of the turn in feet.
19. An airplane is flying in a circle with a speed of 650 km/hr. At what angle with the horizon should a pilot make a turn of radius of 8.00 km such that a component of the lift of the aircraft supplies the necessary centripetal force for the turn?

6.8 Newton's Law of Universal Gravitation

20. Two large metal spheres are separated by a distance of 2.00 m from center to center. If each sphere has a mass of 5000 kg, what is the gravitational force between them?
21. A 5.00-kg mass is 1.00 m from a 10.0-kg mass. (a) What is the gravitational force that the 5.00-kg mass exerts on the 10.0-kg mass? (b) What is the gravitational force that the 10.0-kg mass exerts on the 5.00-kg mass? (c) If both masses are free to move, what will their initial acceleration be?
22. Three point masses of 10.0 kg, 20.0 kg, and 30.0 kg are located on a line at 10.0 cm, 50.0 cm, and 80.0 cm, respectively. Find the resultant gravitational force on (a) the 10.0-kg mass, (b) the 20.0-kg mass, and (c) the 30.0-kg mass.
23. Pete meets Eileen for the first time and is immediately attracted to her. If Pete has a mass of 75.0 kg and Eileen has a mass of 50.0 kg and they are separated by a distance of 3.00 m, is their attraction purely physical?

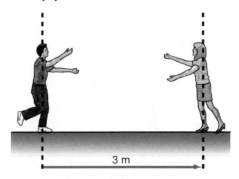

24. What is the gravitational force between a proton and an electron in a hydrogen atom if they are separated by a distance of 5.29×10^{-11} m?

6.11–6.13 The Acceleration Due to Gravity

25. What is the value of g at a distance from the center of the earth of (a) 1 earth radius, (b) 2 earth radii, (c) 10 earth radii, and (d) at the distance of the moon?

26. What is the weight of a body, in terms of its weight at the surface of the earth, at a distance from the center of the earth of (a) 1 earth radius, (b) 2 earth radii, (c) 10 earth radii, and (d) at the distance of the moon? How can an object in a satellite, at say 2 earth radii, be considered to be weightless?

27. Calculate the acceleration of gravity on the surface of Mars. What would a man who weighs 180 lb on earth weigh on Mars?

†28. It is the year 2020 and a base has been established on Mars. An enterprising businessman decides to buy coffee on earth at $5.00/lb and sell it on Mars for $10.00/lb. How much does he make or lose per lb when he sells it on Mars? Ignore the cost of transportation from earth to Mars.

29. The sun's radius is 110 times that of the earth, and its mass is 333,000 times as large. What would be the weight of a 1.00-kg object at the surface of the sun, assuming that it does not melt or evaporate there?

6.14 Satellite Motion

30. What is the velocity of the moon around the earth in a circular orbit? What is the time for one revolution?

31. Calculate the velocity of the earth in an approximate circular orbit about the sun. Calculate the time for one revolution.

32. A satellite orbits the earth 700 mi above the surface. Find its speed and its period of revolution.

33. Calculate the speed of a satellite orbiting 100 km above the surface of Mars. What is its period?

†34. An Apollo space capsule orbited the moon in a circular orbit at a height of 112 km above the surface. The time for one complete orbit, the period T, was 120 min. Find the mass of the moon.

†35. A satellite orbits the earth in a circular orbit in 130 min. What is the distance of the satellite to the center of the earth? What is its height above the surface? What is its speed?

Additional Problems

†36. A rock attached to a string hangs from the roof of a moving train. If the train is traveling at 50.0 mph around a level curve of 500-ft radius, find the angle that the string makes with the vertical.

37. Find the centripetal force due to the rotation of the earth acting on a 100 kg person at (a) the equator, (b) 45.0° north latitude, and (c) the north pole.

38. Find the resultant vector acceleration caused by the acceleration due to gravity and the centripetal acceleration for a person located at (a) the equator, (b) 45.0° north latitude, and (c) the north pole.

†39. A pilot who weighs 180 lb pulls out of a vertical dive at 400 mph along an arc of a circle of 5280-ft radius. Find the centripetal acceleration, centripetal force, and the net normal force on the pilot at the bottom of the dive.

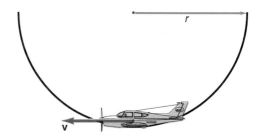

40. A 90-kg pilot pulls out of a vertical dive at 685 km/hr along an arc of a circle of 1500-m radius. Find the centripetal acceleration, centripetal force, and the net force on the pilot at the bottom of the dive.

†41. What is the minimum speed of an airplane in making a vertical loop such that an object in the plane will not fall during the peak of the loop? The radius of the loop is 1000 ft.

†42. A rope is attached to a pail of water and the pail is then rotated in a vertical circle of 80.0-cm radius. What must the minimum speed of the pail of water be such that the water will not spill out?

43. A mass is attached to a string and is swung in a vertical circle. At a particular instant the mass is moving at a speed v, and its velocity vector makes an angle θ with the horizontal. Show that the normal component of the acceleration is given by

$$T + w \sin\theta = mv^2/r$$

and the tangential component of the acceleration is given by

$$a_T = -g \cos\theta$$

Hence show why this motion in a vertical circle is not uniform circular motion.

†44. A 10.0-N ball attached to a string 1.00 m long moves in a horizontal circle. The string makes an angle of 60.0° with the vertical. (a) Find the tension in the string. (b) Find the component of the tension that supplies the necessary centripetal force. (c) Find the speed of the ball.

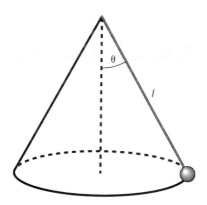

45. A mass $m_A = 35.0$ g is on a smooth horizontal table. It is connected by a string that passes through the center of the table to a mass $m_B = 25.0$ g. At what uniform speed should m_A move in a circle of radius $r = 40.0$ cm such that mass m_B remains motionless?

†46. Three point masses of 30.0 kg, 50.0 kg, and 70.0 kg are located at the vertices of an equilateral triangle 1.00 m on a side. Find the resultant gravitational force on each mass.

†47. Four metal spheres are located at the corners of a square of sides of 0.300 m. If each sphere has a mass of 10.0 kg, find the force on the sphere in the lower right-hand corner.

48. What is the gravitational force between the earth and the moon? If a steel cable can withstand a force of 7.50×10^4 N/cm², what must the diameter of a steel cable be to sustain the equivalent force?

†49. At what speed would the earth have to rotate such that the centripetal force at the equator would be equal to the weight of a body there? If the earth rotated at this velocity, how long would a day be? If a 200-lb man stood on a weighing scale there, what would the scales read?

†50. What would the mass of the earth have to be in order that the gravitational force is inadequate to supply the necessary centripetal force to keep a person on the surface of the earth at the equator? What density would this correspond to? Compare this to the actual density of the earth.

†51. Compute the gravitational force of the sun on the earth. Then compute the gravitational force of the moon on the earth. Which do you think would have a greater effect on the tides, the sun or the moon? Which has the greatest effect?

†52. Find the force exerted on 1.00 kg of water by the moon when (a) the 1.00 kg is on the side nearest the moon and (b) when the 1.00 kg is on the side farthest from the moon. Would this account for tides?

†53. By how much does (a) the sun and (b) the moon change the value of g at the surface of the earth?

54. How much greater would the range of a projectile be on the moon than on the earth?

†55. Find the point between the earth and the moon where the gravitational forces of earth and moon are equal. Would this be a good place to put a satellite?

†56. An earth satellite is in a circular orbit 110 mi above the earth. The period, the time for one orbit, is 88.0 min. Determine the velocity of the satellite and the acceleration of gravity in the satellite at the satellite altitude.

†57. Show that Kepler's third law, which shows the relationship between the period of motion and the radius of the orbit, can be found for circular orbits by equating the centripetal force to the gravitational force, and obtaining

$$T^2 = \frac{4\pi^2 r^3}{Gm}$$

†58. Using Kepler's third law from problem 57, find the mass of the sun. If the radius of the sun is 7.00×10^8 m, find its density.

†59. The speed of the earth around the sun was found, using dynamical principles in the example 6.11 of section 6.14, to be 66,600 mph. Show that this result is consistent with a purely kinematical calculation of the speed of the earth about the sun.

†60. A better approximation for equation 6.52, the "burn time" for the rocket engines, can be obtained if the rate at which the rocket fuel burns, is a known constant. The rate at which the fuel burns is then given by $\Delta m/\Delta t = K$. Hence, the mass at any time during the burn will be given by $(m_0 - K\Delta t)$, where m_0 is the initial mass of the rocket ship before the engines are turned on. Show that for this approximation the time of burn becomes

$$\Delta t = \frac{m_0 \Delta v}{F + K\Delta v}$$

†61. If a spacecraft is to transfer from a 200 nautical mile earth parking orbit to an 80.0 nautical mile lunar parking orbit by a Hohmann transfer ellipse, find (a) the location of the center of mass of the earth-moon system, (b) the perigee distance of the transfer ellipse, (c) the apogee distance, (d) the semimajor axis of the ellipse, (e) the speed of the spacecraft in the earth circular

parking orbit, (f) the speed necessary for insertion into the Hohmann transfer ellipse, (g) the necessary Δv for this insertion, (h) the speed of the spacecraft in a circular lunar parking orbit, (i) the speed of the spacecraft on the Hohmann transfer at time of lunar insertion, and (j) the necessary Δv for insertion into the lunar parking orbit.

Interactive Tutorials

62. Two masses $m_1 = 5.10 \times 10^{21}$ kg and $m_2 = 3.00 \times 10^{14}$ kg are separated by a distance $r = 4.30 \times 10^5$ m. Calculate their gravitational force of attraction.

63. Planet X has mass $m_p = 3.10 \times 10^{25}$ kg and a radius $r_p = 5.40 \times 10^7$ m. Calculate the acceleration due to gravity g at distances of 1–10 planet radii from the planet's surface.

64. Find the angle of bank for a car making a turn on a banked road.

65. Find the speed of a satellite in a circular orbit about its parent body.

66. You are to plan a trip to the planet Mars using the Hohmann transfer ellipse described in the "Have you ever wondered . . . ?" section. The spacecraft is to transfer from a 925-km earth circular parking orbit to a 185-km circular parking orbit around Mars. Find (a) the center of mass of the Earth-Sun-Mars system, (b) the perigee distance of the transfer ellipse, (c) the apogee distance of the transfer ellipse, (d) the semimajor axis of the ellipse, (e) the speed of the spacecraft in the earth parking orbit, (f) the speed necessary for insertion into the Hohmann transfer ellipse, (g) the necessary Δv for insertion into the transfer ellipse, (h) the necessary speed in the Mars circular parking orbit, (i) the speed of the spacecraft in the transfer ellipse at Mars, and (j) the necessary Δv for insertion into the Mars parking orbit.

Energy and Its Conservation

An ellipsoid roller coaster where potential energy is converted to kinetic energy.

The fundamental principle of natural philosophy is to attempt to reduce the apparently complex physical phenomena to some simple fundamental ideas and relations.

Einstein and Infeld

7.1 Energy

The fundamental concept that connects all of the apparently diverse areas of natural phenomena such as mechanics, heat, sound, light, electricity, magnetism, chemistry, and others, is the concept of energy. Energy can be subdivided into well-defined forms, such as (1) mechanical energy, (2) heat energy, (3) electrical energy, (4) chemical energy, and (5) atomic energy. In any process that occurs in nature, energy may be transformed from one form to another. The history of technology is one of a continuing process of transforming one type of energy into another. Some examples include the light bulb, generator, motor, microphone, and loudspeakers.

In its simplest form, *energy can be defined as the ability of a body or system of bodies to perform work. A **system** is an aggregate of two or more particles that is treated as an individual unit*. In order to describe the energy of a body or a system, we must first define the concept of work.

7.2 Work

Almost everyone has an intuitive grasp for the concept of work. However, we need a precise definition of the concept of work so let us define it as follows. Let us exert a force **F** on the block in figure 7.1, causing it to be displaced a distance x along the table. The ***work*** W *done in displacing the body a distance x along the table is defined as the product of the force acting on the body, in the direction of the displacement, times the displacement x of the body*. Mathematically this is

$$W = Fx \tag{7.1}$$

We will always use a capital W to designate the work done, in order to distinguish it from the weight of a body, for which we use the lower case w. The important thing to observe here is that there must be a displacement x if work is to be done. If you push as hard as you can against the wall with your hands, then from the point of view of physics, you do no work on the wall as long as the wall has not moved through a displacement x. This may not appeal to you intuitively because after pushing against that wall for a while, you will become tired and will feel that you certainly did do work. But again, from the point of view of physics, no work on the wall is accomplished because there is no displacement of the wall. In order to do work on an object, you must exert a force F on that object and move that object from one place to another. If that object is not moved, no work is done.

From the point of view of expending energy in pushing against the immovable wall, your body used chemical energy in its tissues and muscles to hold

189

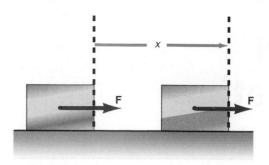

Figure 7.1

The concept of work.

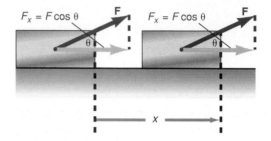

Figure 7.2

Work done when the force is not in the direction of the displacement.

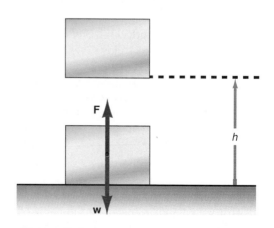

Figure 7.3

Work done in lifting a box.

your hands against the wall. As the body uses this energy, it becomes tired and that energy must eventually be replaced by eating. We will consider the energy used by the body in sustaining the force chemical energy. But, in terms of mechanical energy, no work is done in pressing your hands against an immovable wall. Hence, work as it is used here, is *mechanical work*.

In order to be consistent with the definition of work stated above, if the force acting on the body is not parallel to the displacement, as in figure 7.2, then the work done is the product of the force in the direction of the displacement, times the displacement. That is, the *x*-component of the force,

$$F_x = F \cos \theta$$

is the component of the force in the direction of the displacement. Therefore, the work done on the body is

$$W = (F \cos \theta)x$$

which is usually written as

$$W = Fx \cos \theta \qquad (7.2)$$

This is the general equation used to find the work done on a body. If the force is in the same direction as the displacement, then the angle θ equals zero. But $\cos 0° = 1$, and equation 7.2 reduces to equation 7.1, where the force was in the direction of the displacement.

Units of Work

Since the unit of force in SI units is a newton, and the unit of length is a meter, the SI unit of work is defined as 1 newton meter, which we call 1 joule, that is,

$$1 \text{ joule} = 1 \text{ newton meter}$$

Abbreviated, this is

$$1 \text{ J} = 1 \text{ N m}$$

One joule of work is done when a force of one newton acts on a body, moving it through a distance of one meter. The unit joule is named after James Prescott Joule (1818–1889), a British physicist. *Since energy is the ability to do work, the units of work will also be the units of energy.* In the British engineering system, the force is expressed in pounds and the distance in feet. Hence, the unit of work is defined as

$$1 \text{ unit of work} = 1 \text{ ft lb}$$

One foot-pound is the work done when a force of one pound acts on a body moving it through a distance of one foot. Unlike SI units, the unit of work in the British engineering system is not given a special name. (Remember that the unit lb ft was used for torque to distinguish it from the unit of ft lb for energy. The concepts are very distinct, in that torque was equal to a force times a perpendicular distance, while work is equal to a force times a parallel distance.)

Example 7.1

Work done in lifting a box. What is the minimum amount of work that is necessary to lift a 1.00-lb box to a height of 5.00 ft (figure 7.3)?

We find the work done by noting that F is the force that is necessary to lift the block and h is the distance that the block is lifted. Since the force is in the same direction as the displacement, θ is equal to zero in equation 7.2. Thus,

$$W = Fx \cos \theta = Fh \cos 0°$$
$$= Fh = (1.00 \text{ lb})(5.00 \text{ ft})$$
$$= 5.00 \text{ ft lb}$$

Note here that if a force of only 1.00 lb is exerted to lift the block, then the block will be in equilibrium and will not be lifted from the table at all. If, however, a force that is just infinitesimally greater than w is exerted for just an infinitesimal period of time, then this will be enough to set w into motion. Once the block is moving, then a force F, equal to w, will keep it moving upward at a constant velocity, regardless of how small that velocity may be. In all such cases where forces are exerted to lift objects, such that $F = w$, we will tacitly assume that some additional force was applied for an infinitesimal period of time, to start the motion.

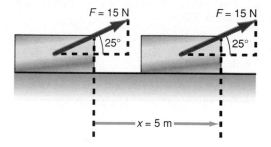

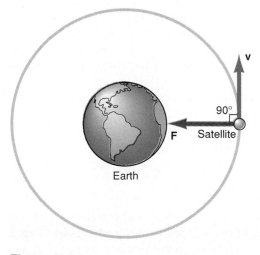

Figure 7.4

Work done when pulling a box.

Figure 7.5

The work done to keep a satellite in orbit.

Example 7.2

When the force is not in the same direction as the displacement. A force of 15.0 N acting at an angle of 25.0° to the horizontal is used to pull a box a distance of 5.00 m across a floor (figure 7.4). How much work is done?

Solution

The work done, found by using equation 7.2, is

$$W = Fx \cos \theta = (15.0 \text{ N})(5.00 \text{ m})(\cos 25.0°)$$
$$= 68.0 \text{ N m} = 68.0 \text{ J}$$

Example 7.3

Work done keeping a satellite in orbit. Find the work done to keep a satellite in a circular orbit about the earth.

Solution

A satellite in a circular orbit about the earth has a gravitational force acting on it that is perpendicular to the orbit, as seen in figure 7.5. The displacement of the satellite in its orbit is perpendicular to that gravitational force. Note that if the displacement is perpendicular to the direction of the applied force, then θ is equal to 90°, and cos 90° = 0. Hence, the work done on the satellite by gravity, found from equation 7.2, is

$$W = Fx \cos \theta = Fx \cos 90° = 0$$

Therefore, no work is done by gravity on the satellite as it moves in its orbit. Work had to be done to get the satellite into the orbit, but once there, no additional work is required to keep it moving in that orbit. *In general, whenever the applied force is perpendicular to the displacement, no work is done by that applied force.*

Figure 7.6

Work done in stopping a car.

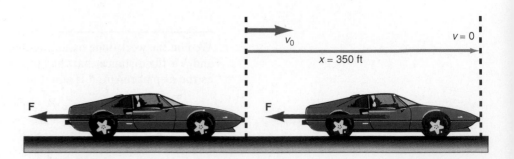

| Example 7.4 |

Work done in stopping a car. A force of 867 lb is applied to a car to bring it to rest in a distance of 350 ft, as shown in figure 7.6. How much work is done in stopping the car?

Solution

To determine the work done in bringing the car to rest, note that the applied force is opposite to the displacement of the car. Therefore, θ is equal to 180° in equation 7.2. Hence, the work done, found from equation 7.2, is

$$W = Fx \cos \theta = (867 \text{ lb})(350 \text{ ft}) \cos 180°$$

$$= -3.03 \times 10^5 \text{ ft lb}$$

Notice that $\cos 180° = -1$, and hence, the work done is negative. In general, whenever the force is opposite to the displacement, the work will always be negative.

7.3 Power

When you walk up a flight of stairs, you do work because you are lifting your body up those stairs. You know, however, that there is quite a difference between walking up those stairs slowly and running up them very rapidly. The work that is done is the same in either case because the net result is that you lifted up the same weight w to the same height h. But you know that if you ran up the stairs you would be more tired than if you walked up them slowly. There is, therefore, a difference in the rate at which work is done.

    Power is defined as the time rate of doing work. We express this mathematically as

$$\text{Power} = \frac{\text{work done}}{\text{time}}$$

$$P = \frac{W}{t} \tag{7.3}$$

When you ran up the stairs rapidly, the time t was small, and therefore the power P, which is the work divided by that small time, was relatively large. Whereas, when you walked up the stairs slowly, t was much larger, and therefore the power P was smaller than before. Hence, when you go up the stairs rapidly you expend more power than when you go slowly.

Units of Power

In SI units, the unit of power is defined as a watt, that is,

$$1 \text{ watt} = 1 \frac{\text{joule}}{\text{second}}$$

which we abbreviate as

$$1 \text{ W} = 1 \frac{\text{J}}{\text{s}}$$

One watt of power is expended when one joule of work is done each second. The watt is named in honor of James Watt (1736–1819), a Scottish engineer who perfected the steam engine. The kilowatt, a unit with which you may already be more familiar, is a thousand watts:

$$1 \text{ kw} = 1000 \text{ W}$$

Another unit with which you may also be familiar is the kilowatt-hour (kwh), but this is not a unit of power, but energy, as can be seen from equation 7.3. Since

$$P = \frac{W}{t}$$

then

$$W = Pt = (\text{kilowatt})(\text{hour})$$

Your monthly electric bill is usually expressed in kilowatt-hours, which is the amount of electric energy you have used for that month. It is the number of kilowatts of power that you used times the number of hours that you used them. To convert kilowatt-hours to joules note

$$1 \text{ kwh} = (1000 \text{ J/s})(1 \text{ hr})(3600 \text{ s/hr}) = 3.6 \times 10^6 \text{ J}$$

To determine the units of power in the British engineering system, we have

$$P = \frac{W}{t} = \frac{\text{ft lb}}{\text{s}}$$

and although this would be the logical unit to express power in the British engineering system, it is not the unit used. Instead, the unit of power in the British engineering system is the horsepower. The horsepower is defined as

$$1 \text{ horsepower} = 1 \text{ hp} = 550 \frac{\text{ft lb}}{\text{s}}$$

The story goes that when Watt had perfected the steam engine, he approached the coal mine owners and leading industrialists of the time, and tried to convince them that they should use his steam engine to take the coal out of the mines instead of the workhorses being used. Supposedly, they wanted to know how good his engine really was. For instance, how many horses could one of his engines replace? At that time, Watt did not know the answer, so he went down into the mines and measured the average amount of work that each horse could do per minute. Each workhorse could do, on the average, 33,000 ft lb of work per minute or 550 ft lb per second. Watt then figured out how many ft lb per second his steam engine could do, and then rated his engine in terms of horsepower. He was then in a position to tell the mine owners how many horses could be replaced by his engine. Even today, in the United States, most engine power is still rated in terms of horsepower. As an example, your car engine may be rated at 200 hp, or the electric motor in your washing machine may be rated at 3/4 hp. A handy conversion factor from watts to horsepower is

$$1 \text{ hp} = 745.7 \text{ W}$$

It seems strange in this day and age of high technology to continue rating engines in terms of horses. It is another reason for abandoning the British engineering system of units.

Example 7.5

Power expended. A person pulls a block with a force of 15.0 N at an angle of 25.0° with the horizontal. If the block is moved 5.00 m in the horizontal direction in 5.00 s, how much power is expended?

Solution

The power expended, found from equations 7.3 and 7.2, is

$$P = \frac{W}{t} = \frac{Fx \cos \theta}{t}$$

$$= \frac{(15.0 \text{ N})(5.00 \text{ m})\cos 25.0°}{5.00 \text{ s}} = 13.6 \frac{\text{N m}}{\text{s}} = 13.6 \text{ W}$$

When a constant force acts on a body in the direction of the body's motion, we can also express the power as

$$P = W = \frac{Fx}{t} = F\frac{x}{t}$$

but

$$\frac{x}{t} = v$$

the velocity of the moving body. Therefore,

$$P = Fv \tag{7.4}$$

is the power expended by a force F, acting on a body that is moving at the velocity v.

Example 7.6

Power to move your car. An applied force of 5500 N keeps a car moving at 95 km/hr. How much power is expended by the car?

Solution

The power expended by the car, found from equation 7.4, is

$$P = Fv = (5500 \text{ N})\left(95 \frac{\text{km}}{\text{hr}}\right)\left(\frac{1 \text{ hr}}{3600 \text{ s}}\right)\left(\frac{1000 \text{ m}}{1 \text{ km}}\right)$$

$$= 1.45 \times 10^5 \text{ N m/s} = 1.45 \times 10^5 \text{ J/s}$$

$$= 1.45 \times 10^5 \text{ W}$$

7.4 Gravitational Potential Energy

Gravitational potential energy is defined as the energy that a body possesses by virtue of its position. If the block shown in figure 7.7, were lifted to a height h above the table, then that block would have potential energy in that raised position. That is, in the raised position, the block has the ability to do work whenever it is allowed to fall. The most obvious example of gravitational potential energy is a waterfall (figure 7.8). Water at the top of the falls has potential energy. When the water falls to the bottom, it can be used to turn turbines and thus do work. A similar example is a pile driver. A pile driver is basically a large weight that is raised above a pile

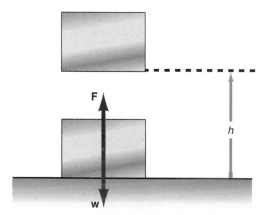

Figure 7.7

Gravitational potential energy.

Figure 7.8

Water at the top of the falls has potential energy.

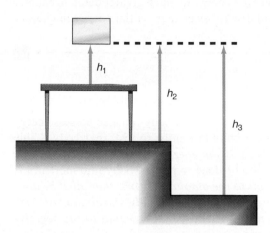

Figure 7.9

Reference plane for potential energy.

Figure 7.10

Changing potential energy.

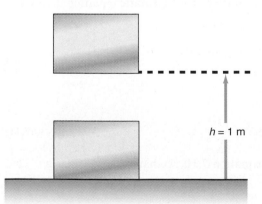

Figure 7.11

The potential energy of a block.

that is to be driven into the ground. In the raised position, the driver has potential energy. When the weight is released, it falls and hits the pile and does work by driving the pile into the ground.

Therefore, whenever an object in the gravitational field of the earth is placed in a position above some reference plane, then that object will have potential energy because it has the ability to do work.

As in all the concepts studied in physics, we want to make this concept of potential energy quantitative. That is, how much potential energy does a body have in the raised position? How should potential energy be measured?

Because work must be done on a body to put the body into the position where it has potential energy, the work done is used as the measure of this potential energy. That is, the potential energy of a body is equal to the work done to put the body into the particular position. Thus, the potential energy (PE) is

$$PE = \text{Work done to put body into position} \qquad (7.5)$$

We can now compute the potential energy of the block in figure 7.7 as

$$PE = \text{Work done}$$

$$PE = W = Fh = wh \qquad (7.6)$$

The applied force F necessary to lift the weight is set equal to the weight w of the block. And since $w = mg$, the potential energy of the block becomes

$$PE = mgh \qquad (7.7)$$

We should emphasize here that the potential energy of a body is referenced to a particular plane, as in figure 7.9. If we raise the block a height h_1 above the table, then with respect to the table it has a potential energy

$$PE_1 = mgh_1$$

While at the same position, it has the potential energy

$$PE_2 = mgh_2$$

with respect to the floor, and

$$PE_3 = mgh_3$$

with respect to the ground outside the room. All three potential energies are different because the block can do three different amounts of work depending on whether it falls to the table, the floor, or the ground. Therefore, it is very important that when the potential energy of a body is stated, it is stated with respect to a particular reference plane. We should also note that it is possible for the potential energy to be negative with respect to a reference plane. That is, if the body is not located above the plane but instead is found below it, it will have negative potential energy with respect to that plane. In such a position the body can not fall to the reference plane and do work, but instead work must be done on the body to move the body up to the reference plane.

<table>
<tr><td>Example 7.7</td></tr>
</table>

The potential energy. A mass of 1.00 kg is raised to a height of 1.00 m above the floor (figure 7.11). What is its potential energy with respect to the floor?

Solution

The potential energy, found from equation 7.7, is

$$PE = mgh = (1.00 \text{ kg})(9.80 \text{ m/s}^2)(1.00 \text{ m})$$

$$= 9.80 \text{ J}$$

In addition to gravitational potential energy, a body can have elastic potential energy and electrical potential energy. An example of elastic potential energy is a compressed spring. When the spring is compressed, the spring has potential energy because when it is released, it has the ability to do work as it expands to its normal position. Its potential energy is equal to the work that is done to compress it. We will discuss the spring and its potential energy in much greater detail in chapter 11 on simple harmonic motion. We will discuss electric potential energy in chapter 19 on electric fields.

7.5 Kinetic Energy

In addition to having energy by virtue of its position, a body can also possess energy by virtue of its motion. When we bring a body in motion to rest, that body is able to do work. *The **kinetic energy** of a body is the energy that a body possesses by virtue of its motion.* Because work had to be done to place a body into motion, *the kinetic energy of a moving body is equal to the amount of work that must be done to bring a body from rest into that state of motion. Conversely, the amount of work that you must do in order to bring a moving body to rest is equal to the negative of the kinetic energy of the body.* That is,

$$\text{Kinetic energy (KE)} = \text{Work done to put body into motion}$$

$$= -\text{Work done to bring body to a stop} \qquad (7.8)$$

The work done to put a body at rest into motion is positive and hence the kinetic energy is positive, and the body has gained energy. The work done to bring a body in motion to a stop is negative, and hence the change in its kinetic energy is negative. This means that the body has lost energy as it goes from a velocity v to a zero velocity.

Consider a block at rest on the frictionless table as shown in figure 7.12. A constant net force F is applied to the block to put it into motion. When it is a distance x away, it is moving at a speed v. What is its kinetic energy at this point? The kinetic energy, found from equation 7.8, is

$$\text{KE} = \text{Work done} = W = Fx \qquad (7.9)$$

But by Newton's second law, the force acting on the body gives the body an acceleration. That is, $F = ma$, and substituting this into equation 7.9 we have

$$\text{KE} = Fx = max \qquad (7.10)$$

But for a body moving at constant acceleration, the kinematic equation 3.16 was

$$v^2 = v_0^2 + 2ax$$

Since the block started from rest, $v_0 = 0$, giving us

$$v^2 = 2ax$$

Solving for the term ax,

$$ax = \frac{v^2}{2} \qquad (7.11)$$

Substituting equation 7.11 back into equation 7.10, we have

$$\text{KE} = m(ax) = \frac{mv^2}{2}$$

or

$$\text{KE} = \tfrac{1}{2}mv^2 \qquad (7.12)$$

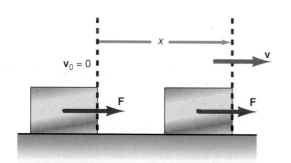

$v_0 = 0$

x

v

F F

Figure 7.12

The kinetic energy of a body.

Equation 7.12 is the classical expression for the kinetic energy of a body in motion at speed v.

Example 7.8

Kinetic energy. Let the block of figure 7.12 have a mass $m = 2.00$ kg and let it be moving at a speed of 5.00 m/s when $x = 5.00$ m. What is its kinetic energy at $x = 5.00$ m?

Solution

Using equation 7.12 for the kinetic energy we obtain

$$KE = \tfrac{1}{2}mv^2 = \tfrac{1}{2}(2.00 \text{ kg})(5.00 \text{ m/s})^2$$
$$= 25.0 \text{ kg m}^2/\text{s}^2 = 25.0(\text{kg m/s}^2)\text{m} = 25.0 \text{ N m}$$
$$= 25.0 \text{ J}$$

Example 7.9

Kinetic energy in the British system of units. A block of mass 10.0 slug is moving at a speed of 5.00 ft/s. Find its kinetic energy.

Solution

The kinetic energy, found from equation 7.12, is

$$KE = \tfrac{1}{2}mv^2 = \tfrac{1}{2}(10.0 \text{ slug})(5.00 \text{ ft/s})^2$$
$$= \left(125 \frac{\text{slug ft}^2}{\text{s}^2}\right)\left(\frac{1}{\text{slug}}\right)\left(\frac{\text{lb}}{\text{ft/s}^2}\right)$$
$$= 125 \text{ ft lb}$$

Example 7.10

The effect of doubling the speed on the kinetic energy. If a car doubles its speed, what happens to its kinetic energy?

Solution

Let us assume that the car of mass m is originally moving at a speed v_0. Its original kinetic energy is

$$(KE)_0 = \tfrac{1}{2}mv_0^2$$

If the speed is doubled, then $v = 2v_0$ and its kinetic energy is

$$KE = \tfrac{1}{2}mv^2 = \tfrac{1}{2}m(2v_0)^2 = \tfrac{1}{2}m4v_0^2$$
$$= 4(\tfrac{1}{2}mv_0^2) = 4KE_0$$

That is, doubling the speed results in quadrupling the kinetic energy. Increasing the speed by a factor of 4 increases the kinetic energy by a factor of 16. This is why automobile accidents at high speeds cause so much damage.

Before we leave this section, we should note that in our derivation of the kinetic energy, work was done to bring an object from rest into motion. The work done on the body to place it into motion was equal to the acquired kinetic energy of the body. If an object is already in motion when the constant force is applied to it, the work done is equal to the change in kinetic energy of the body. That is, equation 7.9 can be written as

$$\text{Work done} = W = Fx$$

$$W = Fx = max$$

but if the block is already in motion at an initial velocity v_0 when the force was applied,

$$v^2 = v_0^2 + 2ax$$

$$ax = \frac{v^2 - v_0^2}{2}$$

Hence,

$$W = Fx = max = m\left(\frac{v^2 - v_0^2}{2}\right)$$

$$= m\frac{v^2}{2} - m\frac{v_0^2}{2}$$

$$= \text{KE}_f - \text{KE}_i = \Delta\text{KE}$$

7.6 The Conservation of Energy

When we say that something is conserved, we mean that that quantity is a constant and does not change with time. It is a somewhat surprising aspect of nature that when a body is in motion, its position is changing with time, its velocity is changing with time, yet certain characteristics of that motion still remain constant. One of the quantities that remain constant during motion is the total energy of the body. The analysis of systems whose energy is conserved leads us to the law of conservation of energy.

   *In any **closed system**, that is, an isolated system, the total energy of the system remains a constant. This is the **law of conservation of energy**.* There may be a transfer of energy from one form to another, but the total energy remains the same.

   As an example of the conservation of energy applied to a mechanical system without friction, let us go back and look at the motion of a projectile in one dimension. Assume that a ball is thrown straight upward with an initial velocity v_0. The ball rises to some maximum height and then descends to the ground, as shown in figure 7.13. At the point 1, a height h_1 above the ground, the ball has a potential energy given by

$$\text{PE}_1 = mgh_1 \tag{7.13}$$

At this same point it is moving at a velocity v_1 and thus has a kinetic energy given by

$$\text{KE}_1 = \tfrac{1}{2}mv_1^2 \tag{7.14}$$

The total energy of the ball at point 1 is the sum of its potential energy and its kinetic energy. Hence, using equations 7.13 and 7.14, we get

$$E_1 = \text{PE}_1 + \text{KE}_1 \tag{7.15}$$

$$E_1 = mgh_1 + \tfrac{1}{2}mv_1^2 \tag{7.16}$$

When the ball reaches point 2 it has a new potential energy because it is higher up, at the height h_2. Hence, its potential energy is

$$\text{PE}_2 = mgh_2$$

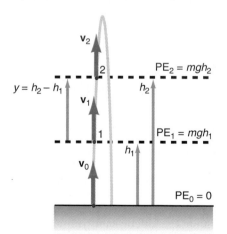

Figure 7.13

The conservation of energy and projectile motion.

As the ball rises, it slows down. Hence, it has a smaller velocity v_2 at point 2 than it had at point 1. Its kinetic energy is now

$$KE_2 = \tfrac{1}{2}mv_2^2$$

The total energy of the ball at position 2 is the sum of its potential energy and its kinetic energy:

$$E_2 = PE_2 + KE_2 \qquad (7.17)$$

$$E_2 = mgh_2 + \tfrac{1}{2}mv_2^2 \qquad (7.18)$$

Let us now look at the difference in the total energy of the ball between when it is at position 2 and when it is at position 1. The change in the total energy of the ball between position 2 and position 1 is

$$\Delta E - E_2 \quad E_1 \qquad (7.19)$$

Using equations 7.16 and 7.18, this becomes

$$\Delta E = mgh_2 + \tfrac{1}{2}mv_2^2 - mgh_1 - \tfrac{1}{2}mv_1^2$$

Simplifying,

$$\Delta E = mg(h_2 - h_1) + \tfrac{1}{2}m(v_2^2 - v_1^2) \qquad (7.20)$$

Let us return, for the moment, to the third of the kinematic equations for projectile motion developed in chapter 3, namely

$$v^2 = v_0^2 - 2gy \qquad (3.24)$$

Recall that v was the velocity of the ball at a height y above the ground, and v_0 was the initial velocity at the ground. We can apply equation 3.24 to the present situation by noting that v_2 is the velocity of the ball at a height $h_2 - h_1 = y$, above the level where the velocity was v_1. Hence, we can rewrite equation 3.24 as

$$v_2^2 = v_1^2 - 2gy$$

Rearranging terms, this becomes

$$v_2^2 - v_1^2 = -2gy \qquad (7.21)$$

If we substitute equation 7.21 into equation 7.20, we get

$$\Delta E = mg(h_2 - h_1) + \tfrac{1}{2}m(-2gy)$$

But, as we can see from figure 7.13, $h_2 - h_1 = y$. Hence,

$$\Delta E = mgy - mgy$$

or

$$\Delta E = 0 \qquad (7.22)$$

which tells us that there is no change in the total energy of the ball between the arbitrary levels 1 and 2. But, since $\Delta E = E_2 - E_1$ from equation 7.19, equation 7.22 is also equivalent to

$$\Delta E = E_2 - E_1 = 0 \qquad (7.23)$$

Therefore,

$$E_2 = E_1 = \text{constant} \qquad (7.24)$$

That is, the total energy of the ball at position 2 is equal to the total energy of the ball at position 1. *Equations 7.22, 7.23, and 7.24 are equivalent statements of the law of conservation of energy. There is no change in the total energy of the ball throughout its entire flight. Or similarly, the total energy of the ball remains the same throughout its entire flight, that is, it is a constant.*

We can glean even more information from these equations by combining equations 7.15, 7.17, and 7.23 into

$$\Delta E = E_2 - E_1 = PE_2 + KE_2 - PE_1 - KE_1 = 0$$

$$PE_2 - PE_1 + KE_2 - KE_1 = 0 \qquad (7.25)$$

But,

$$PE_2 - PE_1 = \Delta PE \qquad (7.26)$$

is the change in the potential energy of the ball, and

$$KE_2 - KE_1 = \Delta KE \qquad (7.27)$$

is the change in the kinetic energy of the ball. Substituting equations 7.26 and 7.27 back into equation 7.25 gives

$$\Delta PE + \Delta KE = 0 \qquad (7.28)$$

or

$$\Delta PE = -\Delta KE \qquad (7.29)$$

Equation 7.29 says that the change in potential energy of the ball will always be equal to the change in the kinetic energy of the ball. Hence, if the velocity decreases between level 1 and level 2, ΔKE will be negative. When this is multiplied by the minus sign in equation 7.29, we obtain a positive number. Hence, there is a positive increase in the potential energy ΔPE. *Thus, the amount of kinetic energy of the ball lost between levels 1 and 2 will be equal to the gain in potential energy of the ball between the same two levels. Thus, energy can be transformed between kinetic energy and potential energy but, the total energy will always remain a constant.* The energy described here is mechanical energy. But the law of conservation of energy is, in fact, more general and applies to all forms of energy, not only mechanical energy. We will say more about this later.

This transformation of energy between kinetic and potential is illustrated in figure 7.14. When the ball is launched at the ground with an initial velocity v_0, all the energy is kinetic, as seen on the bar graph. When the ball reaches position 1, it is at a height h_1 above the ground and hence has a potential energy associated with that height. But since the ball has slowed down to v_1, its kinetic energy has decreased. But the sum of the kinetic energy and the potential energy is still the same constant energy, E_{tot}. The ball has lost kinetic energy but its potential energy has increased by the same amount lost. That is, energy was transformed from kinetic energy to potential energy. At position 2 the kinetic energy has decreased even further but the potential energy has increased correspondingly. At position 3, the ball is at the top of its trajectory. Its velocity is zero, hence its kinetic energy at the top is also zero. The total energy of the ball is all potential. At position 4, the ball

Figure 7.14

Bar graph of energy during projectile motion.

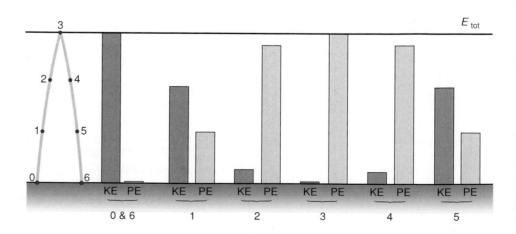

has started down. Its kinetic energy is small but nonzero, and its potential energy is starting to decrease. At position 5, the ball is moving much faster and the kinetic energy has increased accordingly. The potential energy has decreased to account for the increase in the kinetic energy. At position 6, the ball is back on the ground, and hence has no potential energy. All of the energy has been converted back into kinetic energy. As we can observe from the bar graph, the total energy remained constant throughout the flight.

Example 7.11

Conservation of energy and projectile motion. A 0.140-kg ball is thrown upward with an initial velocity of 35.0 m/s. Find (a) the total energy of the ball, (b) the maximum height of the ball, and (c) the kinetic energy and velocity of the ball at 30.0 m.

Solution

a. The total energy of the ball is equal to the initial kinetic energy of the ball, that is,

$$E_{tot} = KE_i = \tfrac{1}{2}mv^2$$

$$= \tfrac{1}{2}(0.140 \text{ kg})(35.0 \text{ m/s})^2$$

$$= 85.8 \text{ J}$$

b. At the top of the trajectory the velocity of the ball is equal to zero and hence its kinetic energy is also zero there. Thus, the total energy at the top of the trajectory is all in the form of potential energy. Therefore,

$$E_{tot} = PE = mgh$$

and the maximum height is

$$h = \frac{E_{tot}}{mg}$$

$$= \frac{85.8 \text{ J}}{(0.140 \text{ kg})(9.80 \text{ m/s}^2)}$$

$$= 62.5 \text{ m}$$

c. The total energy of the ball at 30 m is equal to the total energy of the ball initially. That is,

$$E_{30} = PE_{30} + KE_{30} = E_{tot}$$

The kinetic energy of the ball at 30.0 m is

$$KE_{30} = E_{tot} - PE_{30} = E_{tot} - mgh_{30}$$

$$= 85.8 \text{ J} - (0.140 \text{ kg})(9.80 \text{ m/s}^2)(30.0 \text{ m})$$

$$= 44.6 \text{ J}$$

The velocity of the ball at 30 m is found from

$$\tfrac{1}{2}mv^2 = KE_{30}$$

$$v = \sqrt{\frac{2KE_{30}}{m}}$$

$$= \sqrt{\frac{2(44.6 \text{ J})}{0.140 \text{ kg}}}$$

$$= 25.2 \text{ m/s}$$

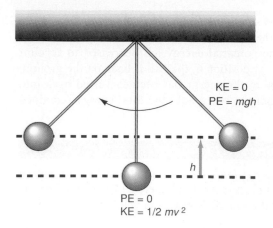

KE = 0
PE = mgh

PE = 0
KE = 1/2 mv^2

h

Figure 7.15

The simple pendulum.

Another example of this transformation of energy back and forth between kinetic and potential is given by the pendulum. The simple pendulum, as shown in figure 7.15, is a string, one end of which is attached to the ceiling, the other to a bob. The pendulum is pulled to the right so that it is a height h above its starting point. All its energy is in the form of potential energy. When it is released, it falls toward the center. As its height h decreases, it loses potential energy, but its velocity increases, increasing its kinetic energy. At the center position h is zero, hence its potential energy is zero. All its energy is now kinetic, and the bob is moving at its greatest velocity. Because of the inertia of the bob it keeps moving toward the left. As it does, it starts to rise, gaining potential energy. This gain in potential energy is of course accompanied by a corresponding loss in kinetic energy, until the bob is all the way to the left. At that time its velocity and hence kinetic energy is zero and, since it is again at the height h, all its energy is potential and equal to the potential energy at the start.

We can find the maximum velocity, which occurs at the bottom of the swing, by equating the total energy at the bottom of the swing to the total energy at the top of the swing:

$$E_{bottom} = E_{top}$$

$$KE_{bottom} = PE_{top} \quad (7.30)$$

$$\tfrac{1}{2}mv^2 = mgh$$

$$v = \sqrt{2gh} \quad (7.31)$$

Thus, the velocity at the bottom of the swing is independent of the mass of the bob and depends only on the height.

Let us now consider the following important example showing the relationship between work, potential energy, and kinetic energy.

Example 7.12

When the work done is not equal to the potential energy. A 5.00-kg block is lifted vertically through a height of 5.00 m by a force of 60.0 N. Find (a) the work done in lifting the block, (b) the potential energy of the block at 5.00 m, (c) the kinetic energy of the block at 5.00 m, (d) the velocity of the block at 5.00 m.

Solution

a. The work done in lifting the block, found from equation 7.1, is

$$W = Fy = (60.0 \text{ N})(5.00 \text{ m}) = 300 \text{ J}$$

b. The potential energy of the block at 5.00 m, found from equation 7.7, is

$$PE = mgh = (5.00 \text{ kg})(9.80 \text{ m/s}^2)(5.00 \text{ m}) = 245 \text{ J}$$

It is important to notice something here. We defined the potential energy as the work done to move the body into its particular position. Yet in this problem the work done to lift the block is 300 J, while the PE is only 245 J. The numbers are not the same. It seems as though something is wrong. Looking at the problem more carefully, however, we see that everything is okay. In the defining relation for the potential energy, we assumed that the work done to raise the block to the height h is done at a constant velocity, approximately a zero velocity. (Remember the force up F was just equal to the weight of the block). In this problem, the weight of the block is

$$w = mg = (5.00 \text{ kg})(9.80 \text{ m/s}^2) = 49.0 \text{ N}$$

Since the force exerted upward of 60.0 N is greater than the weight of the block, 49.0 N, the block is accelerated upward and arrives at the height of

5.00 m with a nonzero velocity and hence kinetic energy. Thus, the work done has raised the mass and changed its velocity so that the block arrives at the 5.00-m height with both a potential energy and a kinetic energy.

c. The kinetic energy is found by the law of conservation of energy, equation 7.15,

$$E_{tot} = KE + PE$$

Hence, the kinetic energy is

$$KE = E_{tot} - PE$$

The total energy of the block is equal to the total amount of work done on the block, namely 300 J, and as shown, the potential energy of the block is 245 J. Hence, the kinetic energy of the block at a height of 5.00 m is

$$KE = E_{tot} - PE = 300 \text{ J} - 245 \text{ J} = 55 \text{ J}$$

d. The velocity of the block at 5.00 m, found from equation 7.12 for the kinetic energy of the block, is

$$KE = \tfrac{1}{2}mv^2$$
$$v = \sqrt{2KE/m} = \sqrt{2(55 \text{ J}/5.00 \text{ kg})}$$
$$= 4.69 \text{ m/s}$$

7.7 Further Analysis of the Conservation of Energy

There are many rather difficult problems in physics that are greatly simplified and easily solved by the principle of conservation of energy. In fact, in advanced physics courses, most of the analysis is done by energy methods. Let us consider the following simple example. A block starts from rest at the top of the frictionless plane, as seen in figure 7.16. What is the speed of the block at the bottom of the plane?

Let us first solve this problem by Newton's second law. The force acting on the block down the plane is $w \sin \theta$, which is a constant. Newton's second law gives

$$F = ma$$
$$w \sin \theta = ma$$
$$mg \sin \theta = ma$$

Hence, the acceleration down the plane is

$$a = g \sin \theta \qquad (7.32)$$

which is a constant. The speed of the block at the bottom of the plane is found from the kinematic formula,

$$v^2 = v_0^2 + 2ax$$
$$v = \sqrt{2ax}$$

or, since $a = g \sin \theta$,

$$v = \sqrt{2g \sin \theta\, x} \qquad (7.33)$$

but

$$x \sin \theta = h$$

Therefore,

$$v = \sqrt{2gh} \qquad (7.34)$$

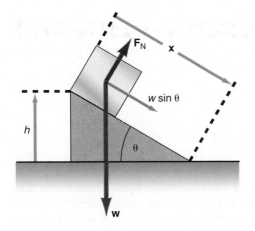

Figure 7.16

A block on an inclined plane.

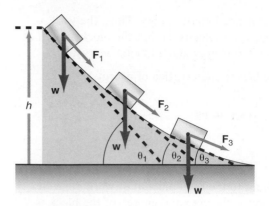

Figure 7.17

A block on a curved surface.

The problem is, of course, quite simple because the force acting on the block is a constant and hence the acceleration is a constant. The kinematic equations were derived on the basis of a constant acceleration and can be used only when the acceleration is a constant. What happens if the forces and accelerations are not constant? As an example, consider the motion of a block that starts from rest at the top of a frictionless *curved* surface, as shown in figure 7.17. The weight w acting downward is always the same, but at each position, the angle the block makes with the horizontal is different. Therefore, the force is different at every position on the surface, and hence the acceleration is different at every point. Thus, the simple techniques developed so far can not be used. (The calculus would be needed for the solution of this case of variable acceleration.)

Let us now look at the same problem from the point of view of energy. The law of conservation of energy says that the total energy of the system is a constant. Therefore, the total energy at the top must equal the total energy at the bottom, that is,

$$E_{top} = E_{bot}$$

The total energy at the top is all potential because the block starts from rest ($v_0 = 0$, hence KE = 0), while at the bottom all the energy is kinetic because at the bottom $h = 0$ and hence PE = 0. Therefore,

$$PE_{top} = KE_{bot}$$
$$mgh = \tfrac{1}{2}mv^2$$
$$v = \sqrt{2gh} \tag{7.35}$$

the speed of the block at the bottom of the plane. We have just solved a very difficult problem, but by using the law of conservation of energy, its solution is very simple.

A very interesting thing to observe here is that the speed of the block down a frictionless inclined plane of height h, equation 7.34, is the same as the speed of a block down the frictionless curved surface of height h, equation 7.35. In fact, if the block were dropped over the top of the inclined plane (or curved surface) so that it fell freely to the ground, its speed at the bottom would be found as

$$E_{top} = E_{bot}$$
$$mgh = \tfrac{1}{2}mv^2$$
$$v = \sqrt{2gh}$$

which is the same speed obtained for the other two cases. This is a characteristic of the law of conservation of energy. The speed of the moving object at the bottom is the same regardless of the path followed by the moving object to get to the final position. This is a consequence of the fact that the same amount of energy was used to place the block at the top of the plane for all three cases, and therefore that same amount of energy is obtained when the block returns to the bottom of the plane.

The energy that the block has at the top of the plane is equal to the work done on the block to place the block at the top of the plane. If the block in figure 7.18 is lifted vertically to the top of the plane, the work done is

$$W = Fh = wh = mgh \tag{7.36}$$

If the block is pushed up the frictionless plane at a constant speed, then the work done is

$$W = Fx = w \sin \theta \, x$$
$$W = mgx \sin \theta \tag{7.37}$$

but

$$x \sin \theta = h$$

Figure 7.18

A conservative system.

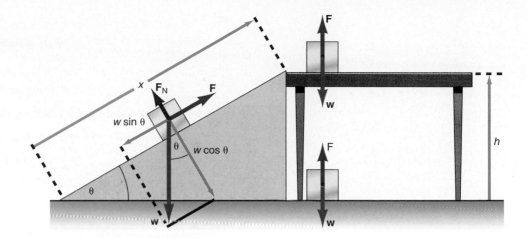

and hence, the work done in pushing the block up the plane is

$$W = mgh \qquad (7.38)$$

which is the identical amount of work just found in lifting the block vertically into the same position. Therefore, the energy at the top is independent of the path taken to get to the top. *Systems for which the energy is the same regardless of the path taken to get to that position are called conservative systems.* **Conservative systems** are systems for which the energy is conserved, that is, the energy remains constant throughout the motion. A conservative system is a system in which the difference in energy is the same regardless of the path taken between two different positions. In a conservative system the total mechanical energy is conserved.

For a better understanding of a conservative system it is worthwhile to consider a nonconservative system. The nonconservative system that we will examine is an inclined plane on which friction is present, as shown in figure 7.19. Let us compute the work done in moving the block up the plane at a constant speed. The force F, exerted up the plane, is

$$F = w \sin \theta + f_k \qquad (7.39)$$

where

$$f_k = \mu_k F_N = \mu_k w \cos \theta \qquad (7.40)$$

Figure 7.19

A nonconservative system.

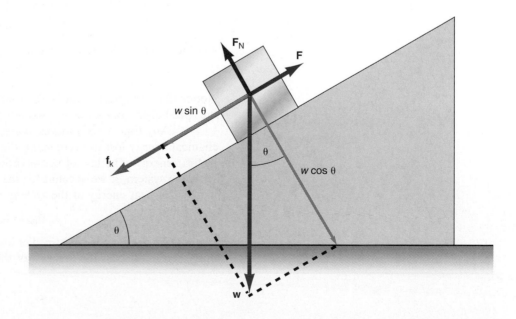

or

$$F = w \sin \theta + \mu_k w \cos \theta$$

or

$$F = mg \sin \theta + \mu_k mg \cos \theta \qquad (7.41)$$

The work done in sliding the block up the plane is

$$W_s = Fx = (mg \sin \theta + \mu_k mg \cos \theta)x$$

$$= mgx \sin \theta + \mu_k mgx \cos \theta \qquad (7.42)$$

but

$$x \sin \theta = h$$

Therefore,

$$W_s = mgh + \mu_k mgx \cos \theta \qquad (7.43)$$

That is, the work done in sliding the block up the plane against friction is greater than the amount of work necessary to lift the block to the top of the plane. The work done in lifting it is

$$W_L = mgh$$

But there appears to be a contradiction here. Since both blocks end up at the same height h above the ground, they should have the same energy mgh. This seems to be a violation of the law of conservation of energy. The problem is that an inclined plane with friction is not a conservative system. Energy is expended by the person exerting the force, to overcome the friction of the inclined plane. The amount of energy lost is found from equation 7.43 as

$$E_{lost} = \mu_k mgx \cos \theta \qquad (7.44)$$

This energy that is lost in overcoming friction shows up as heat energy in the block and the plane. At the top of the plane, both blocks will have the same potential energy. But we must do more work to slide the block up the frictional plane than in lifting it straight upward to the top.

If we now let the block slide down the plane, the same amount of energy, equation 7.44, is lost in overcoming friction as it slides down. Therefore, the total energy of the block at the bottom of the plane is less than in the frictionless case and therefore its speed is also less. That is, the total energy at the bottom is now

$$\tfrac{1}{2}mv^2 = mgh - \mu_k mgx \cos \theta \qquad (7.45)$$

and the speed at the bottom is now

$$v = \sqrt{2gh - 2\mu_k gx \cos \theta} \qquad (7.46)$$

Notice that the speed of the block down the rough plane, equation 7.46, is less than the speed of the block down a smooth plane, equation 7.34.

Any time a body moves against friction, there is always an amount of mechanical energy lost in overcoming this friction. This lost energy always shows up as heat energy. The law of conservation of energy, therefore, holds for a nonconservative system, if we account for the lost mechanical energy of the system as an increase in heat energy of the system, that is,

$$E_{tot} = KE + PE + Q \qquad (7.47)$$

where Q is the heat energy gained or lost during the process. We will say more about this when we discuss the first law of thermodynamics in chapter 17.

Example 7.13

Losing kinetic energy to friction. A 1.50-kg block slides along a smooth horizontal surface at 2.00 m/s. It then encounters a rough horizontal surface. The coefficient of kinetic friction between the block and the rough surface is $\mu_k = 0.400$. How far will the block move along the rough surface before coming to rest?

Solution

When the block slides along the smooth surface it has a total energy that is equal to its kinetic energy. When the block slides over the rough surface it slows down and loses its kinetic energy. Its kinetic energy is equal to the work done on the block by friction as it is slowed to a stop. Therefore,

$$KE = W_f$$
$$\tfrac{1}{2}mv^2 = f_k x = \mu_k F_N x = \mu_k w x = \mu_k mg x$$

Solving for x, the distance the block moves as it comes to a stop, we get

$$\mu_k mg x = \tfrac{1}{2}mv^2$$
$$x = \frac{\tfrac{1}{2}v^2}{\mu_k g}$$
$$= \frac{\tfrac{1}{2}(2 \text{ m/s})^2}{(0.400)(9.80 \text{ m/s}^2)}$$
$$= 0.510 \text{ m}$$

"Have you ever wondered . . . ?"

An Essay on the Application of Physics

The Great Pyramids

Have you ever wondered how the great pyramids of Egypt were built? The largest, Cheops, located 10 mi outside of the city of Cairo, figure 1, is about 400 ft high and contains more than 2½ million blocks of limestone and granite weighing between 2 and 70 ton, apiece. Yet these pyramids were built over 4000 years ago. How did these ancient people ever raise these large stones to such great heights with the very limited equipment available to them?

It is usually supposed that the pyramids were built using the principle of the mechanical advantage obtained by the inclined plane. The first level of stones for the pyramid were assembled on the flat surface, as in figure 2(a). Then an incline was built out of sand and pressed against the pyramid, as in figure 2(b). Another level of stones were then put into place. As each succeeding level was made, more sand was added to the incline in order to reach the next level. The process continued with additional sand added to the incline for each new level of stones. When the final stones were at the top, the sand was removed leaving the pyramids as seen today.

Figure 1
The great pyramid of Cheops.[1]

The advantage gained by using the inclined plane can be explained as follows. An ideal frictionless inclined plane is shown in figure 3. A stone that has the weight w_s is to be lifted from the ground to the height h. If it is lifted straight up, the work that must be done to lift the stone to the height h, is

$$W_1 = F_A h = w_s h \qquad \text{(H7.1)}$$

where F_A is the applied force to lift the stone and w_s is the weight of the stone.

If the same stone is on an inclined plane, then the component of the weight of the stone, $w_s \sin \theta$, acts down the plane and hence

a force, $F = w_s \sin \theta$, must be exerted on the stone in order to push the stone up the plane. The work done pushing the stone a distance L up the plane is

$$W_2 = FL \qquad \text{(H7.2)}$$

Whether the stone is lifted to the top of the plane directly, or pushed up the inclined plane to the top, the stone ends up at the top and the work done in pushing the stone up the plane is equal to the work done in lifting the stone to the height h. Therefore,

$$W_2 = W_1 \qquad \text{(H7.3)}$$

$$FL = w_s h \qquad \text{(H7.4)}$$

Hence, the force F that must be exerted to push the block up the inclined plane is

$$F = \frac{h}{L} w_s \qquad \text{(H7.5)}$$

If the length of the incline L is twice as large as the height h (i.e., $L = 2h$), then the force necessary to push the stone up the incline is

$$F = \frac{h}{L} w_s = \frac{h}{2h} w_s = \frac{w_s}{2}$$

Therefore, if the length of the incline is twice the length of the height, the force necessary to push the stone up the incline is only half the weight of the stone. If the length of the incline is increased to $L = 10h$, then the force F is

$$F = \frac{h}{L} w_s = \frac{h}{10h} w_s = \frac{w_s}{10}$$

That is, by increasing the length of the incline to ten times the height, the force that we must exert to push the stone up the incline is only 1/10 of the weight of the stone. Thus by making L very large, the force that we must exert to push the stone up the inclined plane is made relatively small. If $L = 100h$, then the force necessary would only be one-hundredth of the weight of the stone.

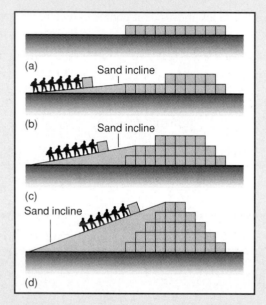

Figure 2

The construction of the pyramids.

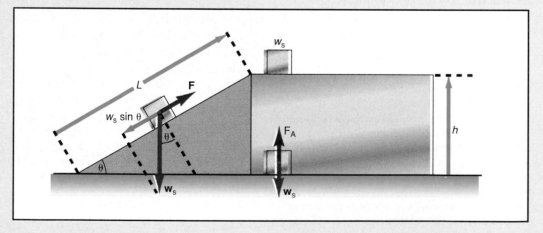

Figure 3

The inclined plane.

Figure 4
Aerial view of the pyramid of Dashur.

The inclined plane is called a simple machine. With it, we have amplified our ability to move a very heavy stone to the top of the hill. This amplification is called the *ideal mechanical advantage* (IMA) of the inclined plane and is defined as

$$\text{Ideal mechanical advantage} = \frac{\text{Force out}}{\text{Force in}} \qquad \textbf{(H7.6)}$$

or

$$\text{IMA} = \frac{F_{\text{out}}}{F_{\text{in}}} \qquad \textbf{(H7.7)}$$

The force that we get out of the machine, in this example, is the weight of the stone w_s, which ends up at the top of the incline, while the force into the machine is equal to the force F that is exerted on the stone in pushing it up the incline. Thus, the ideal mechanical advantage is

$$\text{IMA} = \frac{w_s}{F} \qquad \textbf{(H7.8)}$$

Using equation H7.4 this becomes

$$\text{IMA} = \frac{w_s}{F} = \frac{L}{h} \qquad \textbf{(H7.9)}$$

Hence if $L = 10h$, the IMA is

$$\text{IMA} = 10\frac{h}{h} = 10$$

and the amplification of the force is 10.

The angle θ of the inclined plane, found from the geometry of figure 3, is

$$\sin \theta = \frac{h}{L} \qquad \textbf{(H7.10)}$$

Thus, by making θ very small, a slight incline, a very small force could be applied to move the very massive stones of the pyramid into position. The inclined plane does not give us something for nothing, however. The work done in lifting the stone or pushing the stone is the same. Hence, the smaller force F must be exerted for a very large distance L to do the same work as lifting the very massive stone to the relatively short height h. However, if we are limited by the force F that we can exert, as were the ancient Egyptians, then the inclined plane gives us a decided advantage. An aerial view of the pyramid of Dashur is shown in figure 4. Notice the ramp under the sands leading to the pyramid.[1]

1. This picture is taken from *Secrets of the Great Pyramids* by Peter Tompkins, Harper Colophon Books, 1978.

The Language of Physics

Energy
The ability of a body or system of bodies to perform work (p. 189).

System
An aggregate of two or more particles that is treated as an individual unit (p. 189).

Work
The product of the force acting on a body in the direction of the displacement, times the displacement of the body (p. 189).

Power
The time rate of doing work (p. 192).

Gravitational potential energy
The energy that a body possesses by virtue of its position in a gravitational field. The potential energy is equal to the work that must be done to put the body into that particular position (p. 194).

Kinetic energy
The energy that a body possesses by virtue of its motion. The kinetic energy is equal to the work that must be done to bring the body from rest into that state of motion (p. 196).

Closed system
An isolated system that is not affected by any external influences (p. 198).

Law of conservation of energy
In any closed system, the total energy of the system remains a constant. To say that energy is conserved means that the energy is a constant (p. 198).

Conservative system
A system in which the difference in energy is the same regardless of the path taken between two different positions. In a conservative system the total mechanical energy is conserved (p. 205).

Summary of Important Equations

Work done
$$W = Fx \qquad (7.1)$$

Work done in general
$$W = Fx \cos \theta \qquad (7.2)$$

Power
$$P = W/t \qquad (7.3)$$

Power of moving system
$$P = Fv \qquad (7.4)$$

Gravitational potential energy
$$PE = mgh \qquad (7.7)$$

Kinetic energy
$$KE = \tfrac{1}{2}mv^2 \qquad (7.12)$$

Total mechanical energy
$$E_{tot} = KE + PE$$

Conservation of
mechanical energy
$$\Delta E = E_2 - E_1 = 0 \qquad (7.23)$$

$$E_2 = E_1 = \text{constant} \qquad (7.24)$$

Questions for Chapter 7

1. If the force acting on a body is perpendicular to the displacement, how much work is done in moving the body?
2. A person is carrying a heavy suitcase while walking along a horizontal corridor. Does the person do work (a) against gravity (b) against friction?
3. A car is moving at 90 km/hr when it is braked to a stop. Where does all the kinetic energy of the moving car go?
†4. A rowboat moves in a northerly direction upstream at 3 mph relative to the water. If the current moves south at 3 mph relative to the bank, is any work being done?
†5. For a person to lose weight, is it more effective to exercise or to cut down on the intake of food?
6. If you lift a body to a height h with a force that is greater than the weight of a body, where does the extra energy go?
7. Potential energy is energy that a body possesses by virtue of its position, while kinetic energy is energy that a body possesses by virtue of its speed. Could there be an energy that a body possesses by virtue of its acceleration? Discuss.
8. For a conservative system, what is $\Delta E / \Delta t$?
9. Describe the transformation of energy in a pendulum as it moves back and forth.
10. If positive work is done putting a body into motion, is the work done in bringing a moving body to rest negative work? Explain.

Problems for Chapter 7

7.2 Work

1. A 500-lb box is raised through a height of 15.0 ft. How much work is done in lifting the box at a constant velocity?
2. How much work is done if (a) a force of 150 N is used to lift a 10.0-kg mass to a height of 5.00 m and (b) a force of 150 N, parallel to the surface, is used to pull a 10.0-kg mass, 5.00 m on a horizontal surface?
3. A force of 8.00 N is used to pull a sled through a distance of 100 m. If the force makes an angle of 40.0° with the horizontal, how much work is done?
4. A person pushes a lawn mower with a force of 50.0 N at an angle of 35.0° below the horizontal. If the mower is moved through a distance of 25.0 m, how much work is done?
5. A consumer's gas bill indicates that they have used a total of 37 therms of gas for a 30-day period. Express this energy in joules. A therm is a unit of energy equal to 100,000 Btu and a Btu (British thermal unit) is a unit of energy equal to 778 ft lb.

6. A 670-kg man lifts a 200-kg mass to a height of 1.00 m above the floor and then carries it through a horizontal distance of 10.0 m. How much work is done (a) against gravity in lifting the mass, (b) against gravity in carrying it through the horizontal distance, and (c) against friction in carrying it through the horizontal distance?
7. Calculate the work done in (a) pushing a 4.00-kg block up a frictionless inclined plane 10.0 m long that makes an angle of 30.0° with the horizontal and (b) lifting the block vertically from the ground to the top of the plane, 5.00 m high. (c) Compare the force used in parts a and b.

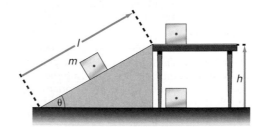

8. A football player weighs 200 lb and does a chin-up by pulling himself up by his arms to an additional height of 30.0 in. above the floor. If he does a total of 20 chin-ups, how much work does he do?
9. A 110-kg football player does a chin-up by pulling himself up by his arms an additional height of 50.0 cm above the floor. If he does a total of 25 chin-ups, how much work does he do?

7.3 Power

10. A consumer's electric bill indicates that they have used a total of 793 kwh of electricity for a 30-day period. Express this energy in (a) joules and (b) ft lb. (c) What is the average power used per hour?
11. A 150-lb person climbs a rope at a constant velocity of 2.00 ft/s in a period of time of 10.0 s. (a) How much power does the person expend? (b) How much work is done?
12. You are designing an elevator that must be capable of lifting a load (elevator plus passengers) of 4000 lb to a height of 12 floors (120 ft) in

1 min. What horsepower motor should you require if half of the power is used to overcome friction?

13. A locomotive pulls a train at a velocity of 40.0 mph with a force of 15,000 lb. What power is exerted by the locomotive?

14. A locomotive pulls a train at a velocity of 88.0 km/hr with a force of 55,000 N. What power is exerted by the locomotive?

7.4 Gravitational Potential Energy

15. Find the potential energy of a 7.00-kg mass that is raised 2.00 m above the desk. If the desk is 1.00 m high, what is the potential energy of the mass with respect to the floor?

16. A 5.00-kg block is at the top of an inclined plane that is 4.00 m long and makes an angle of 35.0° with the horizontal. Find the potential energy of the block.

17. A 15.0-kg sledge hammer is 2.00 m high. How much work can it do when it falls to the ground?

18. A pile driver lifts a 500-lb hammer 10.0 ft before dropping it on a pile. If the pile is driven 4.00 in. into the ground when hit by the hammer, what is the average force exerted on the pile?

7.5 Kinetic Energy

19. What is the kinetic energy of the earth as it travels at a velocity of 30.0 km/s in its orbit about the sun?

20. Compute the kinetic energy of a 3200-lb auto traveling at (a) 15.0 mph, (b) 30.0 mph, and (c) 60.0 mph.

21. Compare the kinetic energy of a 1200-kg auto traveling at (a) 30.0 km/hr, (b) 60.0 km/hr, and (c) 120 km/hr.

22. If an electron in a hydrogen atom has a velocity of 2.19×10^6 m/s, what is its kinetic energy?

23. A 1000-lb airplane traveling at 150 mph is 3000 ft above the terrain. What is its kinetic energy and its potential energy?

24. A 700-kg airplane traveling at 320 km/hr is 1500 m above the terrain. What is its kinetic energy and its potential energy?

25. A 10.0-g bullet, traveling at a velocity of 900 m/s hits and is embedded 2.00 cm into a large piece of oak wood that is fixed at rest. What is the kinetic energy of the bullet? What is the average force stopping the bullet?

26. A little league baseball player throws a baseball (0.15 kg) at a speed of 8.94 m/s. (a) How much work must be done to catch this baseball? (b) If the catcher moves his glove backward by 2.00 cm while catching the ball, what is the average force exerted on his glove by the ball? (c) What is the average force if the distance is 20.0 cm? Is there an advantage in moving the glove backward?

7.6 The Conservation of Energy

27. A 2.00-kg block is pushed along a horizontal frictionless table a distance of 3.00 m, by a horizontal force of 12.0 N. Find (a) how much work is done by the force, (b) the final kinetic energy of the block, and (c) the final velocity of the block. (d) Using Newton's second law, find the acceleration and then the final velocity.

28. A 2.75-kg block is placed at the top of a 40.0° frictionless inclined plane that is 40.0 cm high. Find (a) the work done in lifting the block to the top of the plane, (b) the potential energy at the top of the plane, (c) the kinetic energy when the block slides down to the bottom of the plane, (d) the velocity of the block at the bottom of the plane, and (e) the work done in sliding down the plane.

29. A projectile is fired vertically with an initial velocity of 200 ft/s. Using the law of conservation of energy, find how high the projectile rises.

30. A 3.00-kg block is lifted vertically through a height of 6.00 m by a force of 40.0 N. Find (a) the work done in lifting the block, (b) the potential energy of the block at 6.00 m, (c) the kinetic energy of the block at 6.00 m, and (d) the velocity of the block at 6.00 m.

31. Apply the law of conservation of energy to an Atwood's machine and find the velocity of block A as it hits the ground. $m_B = 40.0$ g, $m_A = 50.0$ g, $h_B = 0.500$ m, and $h_A = 1.00$ m.

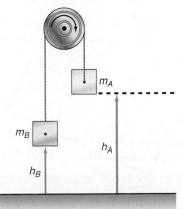

†32. Determine the velocity of block 2 when the height of block 1 is equal to $h_0/4$. $m_2 = 35.0$ g, $m_1 = 20.0$ g, $h_0 = 1.50$ m, and $h_2 = 2.00$ m.

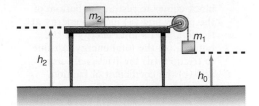

33. A 250-g bob is attached to a string 1.00 m long to make a pendulum. If the pendulum bob is pulled to the right, such that the string makes an angle of 15.0° with the vertical, what is (a) the maximum potential energy, (b) the maximum kinetic energy, and (c) the maximum velocity of the bob and where does it occur?

34. A 45.0-kg girl is on a swing that is 2.00 m long. If the swing is pulled to the right, such that the rope makes an angle of 30.0° with the vertical, what is (a) the maximum potential energy of the girl, (b) her maximum kinetic energy, and (c) the maximum velocity of the swing and where does it occur?

7.7 Further Analysis of the Conservation of Energy

35. A 3.56-kg mass moving at a speed of 3.25 m/s enters a region where the coefficient of kinetic friction is 0.500. How far will the block move before it comes to rest?

36. A 5.00-kg mass is placed at the top of a 35.0° rough inclined plane that is 30.0 cm high. The coefficient of kinetic friction between the mass and the plane is 0.400. Find (a) the potential energy at the top of the plane, (b) the work done against friction as it slides down the plane, (c) the kinetic energy of the mass at the bottom of the plane, and (d) the velocity of the mass at the bottom of the plane.

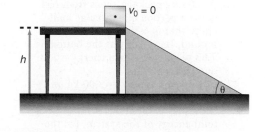

37. A 100-g block is pushed down a rough inclined plane with an initial velocity of 1.50 m/s. The plane is 2.00 m long and makes an angle of 35.0° with the horizontal. If the block comes to rest at the bottom of the plane, find (a) its total energy at the top, (b) its total energy at the bottom, (c) the total energy lost due to friction, (d) the frictional force, and (e) the coefficient of friction.

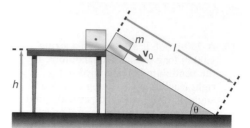

38. A 1.00-kg block is pushed along a rough horizontal floor with a horizontal force of 5.00 N for a distance of 5.00 m. If the block is moving at a constant velocity of 4.00 m/s, find (a) the work done on the block by the force, (b) the kinetic energy of the block, and (c) the energy lost to friction.

39. A 500-lb box is pushed along a rough floor by a horizontal force. The block moves at constant velocity for a distance of 15.0 ft. If the coefficient of friction between the box and the floor is 0.30, how much work is done in moving the box?

40. A 10.0-lb package slides from rest down a portion of a circular mail chute that is 20.0 ft above the ground. Its velocity at the bottom is 20.0 ft/s. How much energy is lost due to friction?

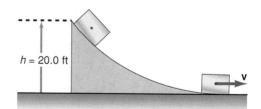

41. A 6.68-kg package slides from rest down a portion of a circular mail chute that is 4.58 m above the ground. Its velocity at the bottom is 7.63 m/s. How much energy is lost due to friction?

42. In the diagram m_2 = 3.00 kg, m_1 = 5.00 kg, h_2 = 1.00 m, h_1 = 0.750 m, and μ_k = 0.400. Find (a) the initial total energy of the system, (b) the work done against friction as m_2 slides on the rough surface, (c) the

velocity v_1 of mass m_1 as it hits the ground, and (d) the kinetic energy of m_1 as it hits the ground.

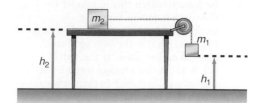

†43. A 5.00-kg body is placed at the top of the track, position A, 2.00 m above the base of the track, as shown in the diagram. (a) Find the total energy of the block. (b) The block is allowed to slide from rest down the frictionless track to the position B. Find the velocity of the body at B. (c) The block then moves over the level rough surface of μ_k = 0.300. How far will the block move before coming to rest?

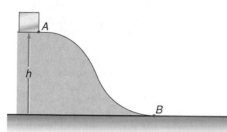

44. A 0.500-kg ball is dropped from a height of 3.00 m. Upon hitting the ground it rebounds to a height of 1.50 m. (a) How much mechanical energy is lost in the rebound, and what happens to this energy? (b) What is the velocity just before and just after hitting the ground?

Additional Problems

†45. The concept of work can be used to describe the action of a lever. Using the principle of work in equals work out, show that

$$F_{out} = \frac{r_{in}}{r_{out}} F_{in}$$

Show how this can be expressed in terms of a mechanical advantage.

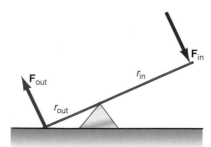

†46. Show how the inclined plane can be considered as a simple machine by comparing the work done in sliding an object up the plane with the work done in lifting the block to the top of the plane. How does the inclined plane supply a mechanical advantage?

47. A force acting on a 300-g mass causes it to move at a constant speed over a rough surface. The coefficient of kinetic friction is 0.350. Find the work required to move the mass a distance of 2.00 m.

48. A 5.00-kg projectile is fired at an angle of 58.0° above the horizontal with the initial velocity of 30.0 m/s. Find (a) the total energy of the projectile, (b) the total energy in the vertical direction, (c) the total energy in the horizontal direction, (d) the total energy at the top of the trajectory, (e) the potential energy at the top of the trajectory, (f) the maximum height of the projectile, (g) the kinetic energy at the top of the trajectory, and (h) the velocity of the projectile as it hits the ground.

49. It takes 20,000 W to keep a 1600-kg car moving at a constant speed of 60.0 km/hr on a level road. How much power is required to keep the car moving at the same speed up a hill inclined at an angle of 22.0° with the horizontal?

50. John consumes 5000 kcal/day. His metabolic efficiency is 70.0%. If his normal activity utilizes 2000 kcal/ day, how many hours will John have to exercise to work off the excess calories by (a) walking, which uses 3.80 kcal/hr; (b) swimming, which uses 8.00 kcal/hr; and (c) running, which uses 11.0 kcal/hr?

51. A 2.50-kg mass is at rest at the bottom of a 5.00-m-long rough inclined plane that makes an angle of 25.0° with the horizontal. When a constant force is applied up the plane and parallel to it, it causes the mass to arrive at the top of the incline at a speed of 0.855 m/s. Find (a) the total energy of the mass when it is at the top of the incline, (b) the work done against friction, and (c) the magnitude of the applied force. The coefficient of friction between the mass and the plane is 0.350.

†52. A 2.00-kg block is placed at the position A on the track that is 3.00 m above the ground. Paths A-B and C-D of the track are frictionless, while section B-C is rough with a coefficient of kinetic friction of 0.350 and a length of 1.50 m. Find (a) the total energy of the block at A, (b) the velocity of the block at B, (c) the energy lost along path B-C, and (d) how high the block rises along path C-D.

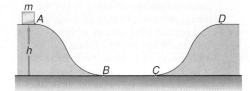

53. A mass $m = 3.50$ kg is launched with an initial velocity $v_0 = 1.50$ m/s from the position A at a height $h = 3.80$ m above the reference plane in the diagram for problem 52. Paths A-B and C-D of the track are frictionless, while path B-C is rough with a coefficient of kinetic friction of 0.300 and a length of 3.00 m. Find (a) the number of oscillations the block makes before coming to rest along the path B-C and (b) where the block comes to rest on path B-C.

54. A ball starts from rest at position A at the top of the track. Find (a) the total energy at A, (b) the total energy at B, (c) the velocity of the ball at B, and (d) the velocity of the ball at C.

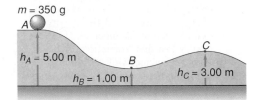

55. A 20.0-kg mass is at rest on a rough horizontal surface. It is then accelerated by a net constant force of 8.6 N. After the mass has moved 1.5 m from rest, the force is removed and the mass comes to rest in 2.00 m. Using energy methods find the coefficient of kinetic friction.

56. In an Atwood's machine $m_B = 30.0$ g, $m_A = 50.0$ g, $h_B = 0.400$ m, and $h_A = 0.800$ m. The machine starts from rest and mass m_A acquires a

velocity of 1.25 m/s as it strikes the ground. Find the energy lost due to friction in the bearings of the pulley.

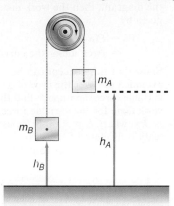

†57. What is the total energy of the Atwood's machine in the position shown in the diagram? If the blocks are released and m_1 falls through a distance of 1.00 m, what is the kinetic and potential energy of each block, and what are their velocities?

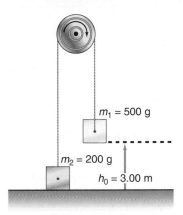

†58. The gravitational potential energy of a mass m with respect to infinity is given by

$$PE = -\frac{Gm_E m}{r}$$

where G is the universal gravitational constant, m_E is the mass of the earth, and r is the distance from the center of the earth to the mass m. Find the escape velocity of a spaceship from the earth. (The escape velocity is the necessary velocity to remove a body from the gravitational attraction of the earth.)

†59. Modify problem 58 and find the escape velocity for (a) the moon, (b) Mars, and (c) Jupiter.

†60. The entire Atwood's machine shown is allowed to go into free-fall. Find the velocity of m_1 and m_2 when the entire system has fallen 1.00 m.

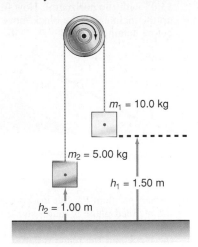

†61. A 1.50-kg block moves along a smooth horizontal surface at 2.00 m/s. The horizontal surface is at a height h_0 above the ground. The block then slides down a rough hill, 20.0 m long, that makes an angle of 30.0° with the horizontal. The coefficient of kinetic friction between the block and the hill is 0.600. How far down the hill will the block move before coming to rest?

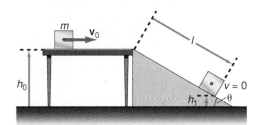

†62. At what point above the ground must a car be released such that when it rolls down the track and into the circular loop it will be going fast enough to make it completely around the loop? The radius of the circular loop is R.

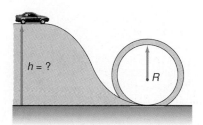

†63. A 1.50-kg block moves along a smooth horizontal surface at 2.00 m/s. It then encounters a smooth inclined plane that makes an angle of 53.0° with the horizontal. How far up the incline will the block move before coming to rest?

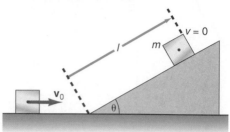

†64. Repeat problem 63, but in this case the inclined plane is rough and the coefficient of kinetic friction between the block and the plane is 0.400.

†65. In the diagram mass m_1 is located at the top of a rough inclined plane that has a length $l_1 = 0.500$ m. $m_1 = 0.500$ kg, $m_2 = 0.200$ kg, $\mu_{k_1} = 0.500$, $\mu_{k_2} = 0.300$, $\theta = 50.0°$, and $\phi = 50.0°$. (a) Find the total energy of the system in the position shown. (b) The system is released from rest. Find the work done for block 1 to overcome friction as it slides down the plane. (c) Find the work done for block 2 to overcome friction as it slides up the plane. (d) Find the potential energy of block 2 when it arrives at the top of the plane. (e) Find the velocity of block 1 as it reaches the bottom of the plane. (f) Find the kinetic energy of each block at the end of their travel.

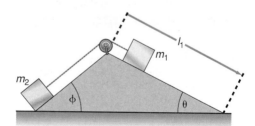

†66. If a constant force acting on a body is plotted against the displacement of the body from x_1 to x_2, as shown in the diagram, then the work done is given by

$$W = F(x_2 - x_1)$$
$$= \text{Area under the curve}$$

Show that this concept can be extended to cover the case of a variable force, and hence find the work done for the variable force, $F = kx$, where $k = 2.00$ N/m as the body is displaced from x_1 to x_2. Draw a graph showing your results.

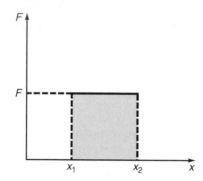

Interactive Tutorials

◪ 67. A projectile of mass $m = 100$ kg is fired vertically upward at a velocity $v_0 = 50.0$ m/s. Calculate its potential energy PE (relative to the ground), its kinetic energy KE, and its total energy E_{tot} for the first 10.0 s of flight.

◪ 68. Consider the general motion in an Atwood's machine such as the one shown in the diagram of problem 31; $m_A = 0.650$ kg and is at a height $h_A = 2.55$ m above the reference plane and mass $m_B = 0.420$ kg is at a height $h_B = 0.400$ m. If the system starts from rest, find (a) the initial potential energy of mass A, (b) the initial potential energy of mass B, and (c) the total energy of the system. When m_A has fallen a distance $y_A = 0.75$ m, find (d) the potential energy of mass A, (e) the potential energy of mass B, (f) the speed of each mass at that point, (g) the kinetic energy of mass A, and (h) the kinetic energy of mass B. (i) When mass A hits the ground, find the speed of each mass.

◪ 69. Consider the general motion in the combined system shown in the diagram of problem 42; $m_1 = 0.750$ kg and is at a height $h_1 = 1.85$ m above the reference plane and mass $m_2 = 0.285$ kg is at a height $h_2 = 2.25$ m, $\mu_k = 0.450$. If the system starts from rest, find (a) the initial potential energy of mass 1, (b) the initial potential energy of mass 2, and (c) the total energy of the system. When m_1 has fallen a distance $y_1 = 0.35$ m, find (d) the potential energy of mass 1, (e) the potential energy of mass 2, (f) the energy lost due to friction as mass 2 slides on the rough surface, (g) the speed of each mass at that point, (h) the kinetic energy of mass 1, and (i) the kinetic energy of mass 2. (j) When mass 1 hits the ground, find the speed of each mass.

◪ 70. Consider the general case of motion shown in the diagram with mass m_A initially located at the top of a rough inclined plane of length l_A, and mass m_B is at the bottom of the second plane; x_A is the distance from the mass A to the bottom of the plane. Let $m_A = 0.750$ kg, $m_B = 0.250$ kg, $l_A = 0.550$ m, $\theta = 40.0°$, $\phi = 30.0°$, $\mu_{k_A} = 0.400$, $\mu_{k_B} = 0.300$, and $x_A = 0.200$ m. When $x_A = 0.200$ m, find (a) the initial total energy of the system, (b) the distance block B has moved, (c) the potential energy of mass A, (d) the potential energy of mass B, (e) the energy lost due to friction for block A, (f) the energy lost due to friction for block B, (g) the velocity of each block, (h) the kinetic energy of mass A, and (i) the kinetic energy of mass B.

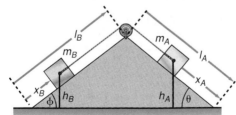

Momentum and Its Conservation

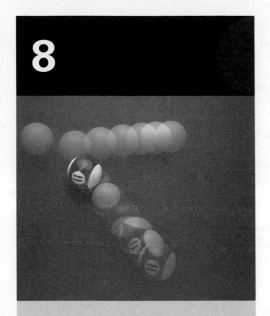

Conservation of momentum in a collision.

The quantity of motion is the measure of the same, arising from the velocity and quantity conjointly.

Isaac Newton, *Principia*

8.1 Momentum

In dealing with some problems in mechanics, we find that in many cases, it is exceedingly difficult, if not impossible, to determine the forces that are acting on a body, and/or for how long the forces are acting. These difficulties can be overcome, however, by using the concept of momentum.

    *The **linear momentum** of a body is defined as the product of the mass of the body in motion times its velocity.* That is,

$$\mathbf{p} = m\mathbf{v} \tag{8.1}$$

Because velocity is a vector, linear momentum is also a vector, and points in the same direction as the velocity vector. We use the word linear here to indicate that the momentum of the body is along a line, in order to distinguish it from the concept of angular momentum. Angular momentum applies to bodies in rotational motion and will be discussed in chapter 9. In this book, whenever the word momentum is used by itself it will mean linear momentum.

    This definition of momentum may at first seem rather arbitrary. Why not define it in terms of v^2, or v^3? We will see that this definition is not arbitrary at all. Let us consider Newton's second law

$$\mathbf{F} = m\mathbf{a} = m\frac{\Delta\mathbf{v}}{\Delta t}$$

However, since $\Delta\mathbf{v} = \mathbf{v}_f - \mathbf{v}_i$, we can write this as

$$\mathbf{F} = m\left(\frac{\mathbf{v}_f - \mathbf{v}_i}{\Delta t}\right)$$

$$\mathbf{F} = \frac{m\mathbf{v}_f - m\mathbf{v}_i}{\Delta t} \tag{8.2}$$

But $m\mathbf{v}_f = \mathbf{p}_f$, the final value of the momentum, and $m\mathbf{v}_i = \mathbf{p}_i$, the initial value of the momentum. Substituting this into equation 8.2, we get

$$\mathbf{F} = \frac{\mathbf{p}_f - \mathbf{p}_i}{\Delta t} \tag{8.3}$$

However, the final value of any quantity, minus the initial value of that quantity, is equal to the change of that quantity and is denoted by the delta Δ symbol. Hence,

$$\mathbf{p}_f - \mathbf{p}_i = \Delta\mathbf{p} \tag{8.4}$$

the change in the momentum. Therefore, Newton's second law becomes

$$F = \frac{\Delta p}{\Delta t} \qquad (8.5)$$

Newton's second law in terms of momentum *can be stated as: When a resultant applied force* ***F*** *acts on a body, it causes the linear momentum of that body to change with time.*

The interesting thing we note here is that this is essentially the form in which Newton expressed his second law. Newton did not use the word momentum, however, but rather the expression, "quantity of motion," which is what today would be called momentum. Thus, defining momentum as $p = mv$ is not arbitrary at all. In fact, Newton's second law in terms of the time rate of change of momentum is more basic than the form $F = ma$. In the form $F = ma$, we assume that the mass of the body remains constant. But suppose the mass does not remain constant? As an example, consider an airplane in flight. As it burns fuel its mass decreases with time. At any one instant, Newton's second law in the form $F = ma$, certainly holds and the aircraft's acceleration is

$$a = \frac{F}{m}$$

But only a short time later the mass of the aircraft is no longer m, and therefore the acceleration changes. Another example of a changing mass system is a rocket. Newton's second law in the form $F = ma$ does not properly describe the motion because the mass is constantly changing. Also when objects move at speeds approaching the speed of light, the theory of relativity predicts that the mass of the body does not remain a constant, but rather it increases. In all these variable mass systems, Newton's second law in the form $F = \Delta p/\Delta t$ is still valid, even though $F = ma$ is not.

8.2 The Law of Conservation of Momentum

A very interesting result, and one of extreme importance, is found by considering the behavior of mechanical systems containing two or more particles. Recall from chapter 7 that a system is an aggregate of two or more particles that is treated as an individual unit. Newton's second law, in the form of equation 8.5, can be applied to the entire system if **F** is the total force acting on the system and **p** is the total momentum of the system. Forces acting on a system can be divided into two categories: external forces and internal forces. ***External forces*** *are forces that originate outside the system and act on the system.* ***Internal forces*** *are forces that originate within the system and act on the particles within the system.* The net force acting on and within the system is equal to the sum of the external forces and the internal forces. If the total external force **F** acting on the system is zero then, since

$$F = \frac{\Delta p}{\Delta t} \qquad (8.5)$$

this implies that

$$\frac{\Delta p}{\Delta t} = 0$$

or

$$\Delta p = 0 \qquad (8.6)$$

But

$$\Delta p = p_f - p_i$$

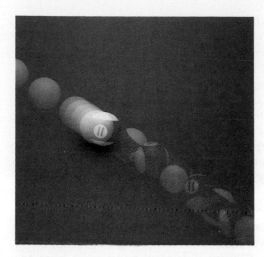

Figure 8.1

Collision of billiard balls in an example of conservation of momentum.

Therefore,

$$\mathbf{p_f} - \mathbf{p_i} = 0$$

and

$$\mathbf{p_f} = \mathbf{p_i} \tag{8.7}$$

*Equation 8.7 is called the **law of conservation of linear momentum**. It says that if the total external force acting on a system is equal to zero, then the final value of the total momentum of the system is equal to the initial value of the total momentum of the system.* That is, the total momentum is a constant, or as usually stated, the total momentum is conserved.

As an example of the law of conservation of momentum let us consider the head-on collision of two billiard balls. The collision is shown in a stroboscopic picture in figure 8.1 and schematically in figure 8.2. Initially the ball of mass m_1 is moving to the right with an initial velocity $\mathbf{v_{1i}}$, while the second ball of mass m_2 is moving to the left with an initial velocity $\mathbf{v_{2i}}$.

At impact, the two balls collide, and ball 1 exerts a force $\mathbf{F_{21}}$ on ball 2, toward the right. But by Newton's third law, ball 2 exerts an equal but opposite force on ball 1, namely $\mathbf{F_{12}}$. (The notation, $\mathbf{F}_{ij}$, means that this is the force on ball i, caused by ball j.) If the system is defined as consisting of the two balls that are enclosed within the green region of figure 8.2, then the net force on the system of the two balls is equal to the forces on ball 1 plus the forces on ball 2, plus any external forces acting on these balls. The forces $\mathbf{F_{12}}$ and $\mathbf{F_{21}}$ are internal forces in that they act completely within the system.

It is assumed in this problem that there are no external horizontal forces acting on either of the balls. Hence, the net force on the system is

$$\text{Net } \mathbf{F} = \mathbf{F_{12}} + \mathbf{F_{21}}$$

But by Newton's third law

$$\mathbf{F_{21}} = -\mathbf{F_{12}}$$

Therefore, the net force becomes

$$\text{Net } \mathbf{F} = \mathbf{F_{12}} + (-\mathbf{F_{12}}) = 0 \tag{8.8}$$

That is, the net force acting on the system of the two balls during impact is zero, and equation 8.7, the law of conservation of momentum, must hold. The

Figure 8.2

Example of conservation of momentum.

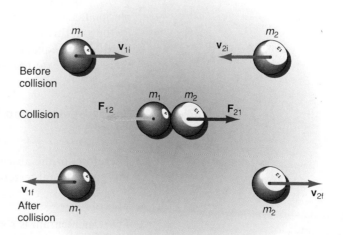

total momentum of the system after the collision must be equal to the total momentum of the system before the collision. Although the momentum of the individual bodies within the system may change, the total momentum will not. After the collision, ball m_1 moves to the left with a final velocity v_{1f}, and ball m_2 moves off to the right with a final velocity v_{2f}.

We will go into more detail on collisions in section 8.5. The important thing to observe here, is what takes place during impact. First, we are no longer considering the motion of a single body, but rather the motion of two bodies. The two bodies are the system. Even though there is a force on ball 1 and ball 2, these forces are internal forces, and the internal forces can not exert a net force on the system, only an external force can do that. *Whenever a system exists without external forces—a system that we call a closed system—the net force on the system is always zero and the law of conservation of momentum always holds.*

The law of conservation of momentum is a consequence of Newton's third law. Recall that because of the third law, all forces in nature exist in pairs; there is no such thing as a single isolated force. Because all internal forces act in pairs, the net force on an isolated system must always be zero, and the system's momentum must always be conserved. Therefore, all systems to which the law of conservation of momentum apply, must consist of at least two bodies and could consist of even millions or more, such as the number of atoms in a gas. If the entire universe is considered as a closed system, then it follows that the total momentum of the universe is also a constant.

The law of conservation of momentum, like the law of conservation of energy, is independent of the type of interaction between the interacting bodies, that is, it applies to colliding billiard balls as well as to gravitational, electrical, magnetic, and other similar interactions. It applies on the atomic and nuclear level as well as on the astronomical level. It even applies in cases where Newtonian mechanics fails. *Like the conservation of energy, the conservation of momentum is one of the fundamental laws of physics.*

8.3 **Examples of the Law of Conservation of Momentum**

Firing a Gun or a Cannon

Let us consider the case of firing a bullet from a gun. The bullet and the gun are the system to be analyzed and they are initially at rest in our frame of reference. We also assume that there are no external forces acting on the system. Because there is no motion of the bullet with respect to the gun at this point, the initial total momentum of the system of bullet and gun p_i is zero, as shown in figure 8.3(a).

At the moment the trigger of the gun is pulled, a controlled chemical explosion takes place within the gun, figure 8.3(b). A force F_{BG} is exerted on the bullet by the gun through the gases caused by the exploding gun powder. But by Newton's third law, an equal but opposite force F_{GB} is exerted on the gun by the bullet. Since there are no external forces, the net force on the system of bullet and gun is

$$\text{Net Force} = F_{BG} + F_{GB} \qquad (8.9)$$

But by Newton's third law

$$F_{BG} = -F_{GB}$$

Therefore, in the absence of external forces, the net force on the system of bullet and gun is equal to zero:

$$\text{Net Force} = F_{BG} - F_{GB} = 0 \qquad (8.10)$$

Thus, momentum is conserved and

$$p_f = p_i \qquad (8.11)$$

Figure 8.3

Conservation of momentum in firing a gun.

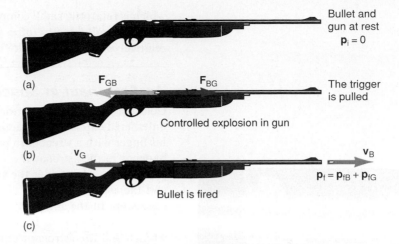

Bullet and gun at rest
$\mathbf{p}_i = 0$

(a)

$\mathbf{F}_{GB}$ $\mathbf{F}_{BG}$

The trigger is pulled

Controlled explosion in gun

(b) $\mathbf{v}_G$

$\mathbf{v}_B$

$\mathbf{p}_f = \mathbf{p}_{fB} + \mathbf{p}_{fG}$

Bullet is fired

(c)

However, because the initial total momentum was zero,

$$\mathbf{p}_i = 0 \qquad (8.12)$$

the total final momentum must also be zero. But because the bullet is moving with a velocity $\mathbf{v}_B$ to the right, and therefore has momentum to the right, the gun must move to the left with the same amount of momentum in order for the final total momentum to be zero, figure 8.3(c). That is, calling $\mathbf{p}_{fB}$ the final momentum of the bullet, and $\mathbf{p}_{fG}$ the final momentum of the gun, the total final momentum is

$$\mathbf{p}_f = \mathbf{p}_{fB} + \mathbf{p}_{fG} = 0$$

$$m_B \mathbf{v}_B + m_G \mathbf{v}_G = 0$$

Solving for the velocity $\mathbf{v}_G$ of the gun, we get

$$\mathbf{v}_G = -\frac{m_B}{m_G} \mathbf{v}_B \qquad (8.13)$$

Because $\mathbf{v}_B$ is the velocity of the bullet to the right, we see that because of the minus sign in equation 8.13, the velocity of the gun must be in the opposite direction, namely to the left. We call $\mathbf{v}_G$ the recoil velocity and its magnitude is

$$v_G = \frac{m_B}{m_G} v_B \qquad (8.14)$$

Even though v_B, the speed of the bullet, is quite large, v_G, the recoil speed of the gun, is relatively small because v_B is multiplied by the ratio of the mass of the bullet m_B to the mass of the gun m_G. Because m_B is relatively small, while m_G is relatively large, the ratio is a small number.

Example 8.1

Recoil of a gun. If the mass of the bullet is 5.00 g, and the mass of the gun is 10.0 kg, and the velocity of the bullet is 300 m/s, find the recoil speed of the gun.

Solution

The recoil speed of the gun, found from equation 8.14, is

$$v_G = \frac{m_B}{m_G} v_B$$

$$= \frac{5.00 \times 10^{-3} \text{ kg}}{10.0 \text{ kg}} 300 \text{ m/s}$$

$$= 0.150 \text{ m/s} = 15.0 \text{ cm/s}$$

Figure 8.4

Recoil of a cannon.

which is relatively small compared to the speed of the bullet. Because it is necessary for this recoil velocity to be relatively small, the mass of the gun must always be relatively large compared to the mass of the bullet.

An Astronaut in Space Throws an Object Away

Consider the case of an astronaut repairing the outside of his spaceship while on an untethered extravehicular activity. While trying to repair the radar antenna he bangs his finger with a wrench. In pain and frustration he throws the wrench away. What happens to the astronaut?

Let us consider the system as an isolated system consisting of the wrench and the astronaut. Let us place a coordinate system, a frame of reference, on the spaceship. In the analysis that follows, we will measure all motion with respect to this reference system. In this frame of reference there is no relative motion of the wrench and the astronaut initially and hence their total initial momentum is zero, as shown in figure 8.5(a).

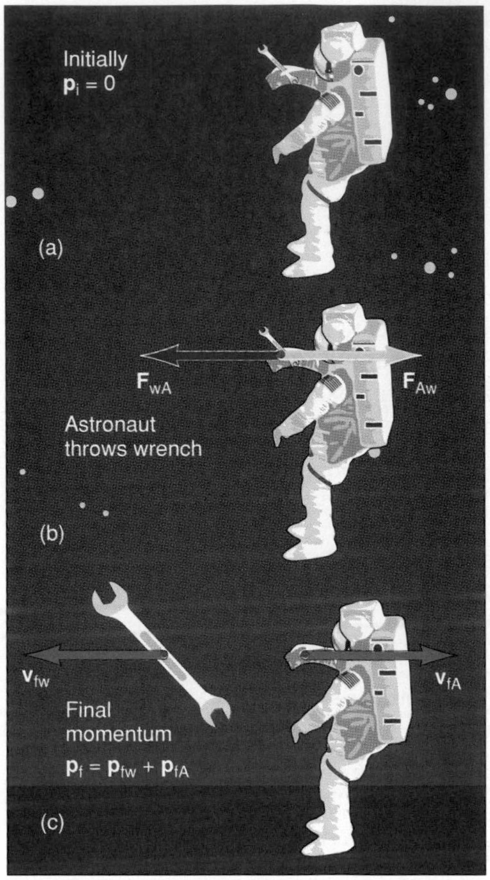

During the throwing process, the astronaut exerts a force $\mathbf{F}_{wA}$ on the wrench. But by Newton's third law, the wrench exerts an equal but opposite force $\mathbf{F}_{Aw}$ on the astronaut, figure 8.5(b). The net force on this isolated system is therefore zero and the law of conservation of momentum must hold. Thus, the final total momentum must equal the initial total momentum, that is,

$$\mathbf{p}_f = \mathbf{p}_i$$

But initially, $\mathbf{p}_i = 0$ in our frame of reference. Also, the final total momentum is the sum of the final momentum of the wrench and the astronaut, figure 8.5(c). Therefore,

$$\mathbf{p}_f = \mathbf{p}_{fw} + \mathbf{p}_{fA} = 0$$
$$m_w \mathbf{v}_{fw} + m_A \mathbf{v}_{fA} = 0$$

Solving for the final velocity of the astronaut, we get

$$\mathbf{v}_{fA} = \frac{-m_w}{m_A} \mathbf{v}_{fw} \tag{8.15}$$

Thus, as the wrench moves toward the left, the astronaut must recoil toward the right. The magnitude of the final velocity of the astronaut is

$$v_{fA} = \frac{m_w}{m_A} v_{fw} \tag{8.16}$$

Figure 8.5

Conservation of momentum and an astronaut.

Example 8.2

The hazards of being an astronaut. An 80.0-kg astronaut throws a 0.250-kg wrench away at a speed of 3.00 m/s. Find (a) the speed of the astronaut as he recoils away from his space station and (b) how far will he be from the space ship in 1 hr?

Solution

a. The recoil speed of the astronaut, found from equation 8.16, is

$$v_{fA} = \frac{m_w}{m_A} v_{fw}$$

$$= \frac{(0.250 \text{ kg})(3.00 \text{ m/s})}{80.0 \text{ kg}}$$

$$= 9.38 \times 10^{-3} \text{ m/s}$$

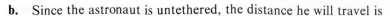

b. Since the astronaut is untethered, the distance he will travel is

$$x_A = v_{fA}t = (9.38 \times 10^{-3} \text{ m/s})(3600 \text{ s})$$

$$= 33.8 \text{ m}$$

The astronaut will have moved a distance of 33.8 m away from his space ship in 1 hr.

A Person on the Surface of the Earth Throws a Rock Away

The result of the previous subsection may at first seem somewhat difficult to believe. An astronaut throws an object away in space and as a consequence of it, the astronaut moves off in the opposite direction. This seems to defy our ordinary experiences, for if a person on the surface of the earth throws an object away, the person does not move backward. What is the difference?

Let an 80.0-kg person throw a 0.250-kg rock away, as shown in figure 8.6. As the person holds the rock, its initial velocity is zero. The person then applies a force to the rock accelerating it from zero velocity to a final velocity v_f. While the rock is leaving the person's hand, the force $\mathbf{F}_{Rp}$ is exerted on the rock by the person. But by Newton's third law, the rock is exerting an equal but opposite force $\mathbf{F}_{pR}$ on the person. But the system that is now being analyzed is not an isolated system, consisting only of the person and the rock. Instead, the system also contains the surface of the earth, because the person is connected to it by friction. The force $\mathbf{F}_{pR}$, acting on the person, is now opposed by the frictional force between the person and the earth and prevents any motion of the person.

As an example, let us assume that in throwing the rock the person's hand moves through a distance x of 1.00 m, as shown in figure 8.6(a), and it leaves the person's hand at a velocity of 3.00 m/s. The acceleration of the rock can be found from the kinematic equation

$$v^2 = v_0^2 + 2a_R x$$

by solving for a_R. Thus,

$$a_R = \frac{v^2}{2x} = \frac{(3.00 \text{ m/s})^2}{2(1.00 \text{ m})} = 4.50 \text{ m/s}^2$$

The force acting on the rock F_{Rp}, found by Newton's second law, is

$$F_{Rp} = m_R a_R = (0.250 \text{ kg})(4.50 \text{ m/s}^2)$$

$$= 1.13 \text{ N}$$

But by Newton's third law this must also be the force exerted on the person by the rock, F_{pR}. That is, there is a force of 1.13 N acting on the person, tending to push that person to the left. But since the person is standing on the surface of the earth there is a frictional force that tends to oppose that motion and is shown in figure 8.6(b). The maximum value of that frictional force is

$$f_s = \mu_s F_N = \mu_s w_p$$

The weight of the person w_p is

$$w_p = mg = (80.0 \text{ kg})(9.80 \text{ m/s}^2) = 784 \text{ N}$$

Assuming a reasonable value of $\mu_s = 0.500$ (leather on wood), we have

$$f_s = \mu_s w_p = (0.500)(784 \text{ N})$$

$$= 392 \text{ N}$$

That is, before the person will recoil from the process of throwing the rock, the recoil force F_{pR}, acting on the person, must be greater than the maximum frictional force of 392 N. We found the actual reaction force on the person to be only

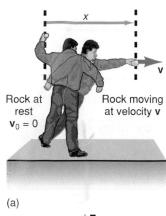

Rock at rest
$\mathbf{v}_0 = 0$

Rock moving at velocity $\mathbf{v}$

(a)

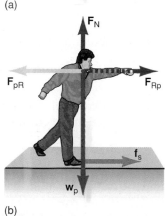

(b)

Figure 8.6

A person throwing a rock on the surface of the earth.

1.13 N, which is no where near the amount necessary to overcome friction. Hence, when a person on the surface of the earth throws an object, the person does not recoil like an astronaut in space.

If friction could be minimized, then the throwing of the object would result in a recoil velocity. For example, if a person threw a rock to the right, while standing in a boat on water, then because the frictional force between the boat and the water is relatively small, the person and the boat would recoil to the left.

In a similar way, if a person is standing at the back of a boat, which is at rest, and then walks toward the front of the boat, the boat will recoil backward to compensate for his forward momentum.

8.4 Impulse

Let us consider Newton's second law in the form of change in momentum as found in equation 8.5,

$$F = \frac{\Delta p}{\Delta t}$$

If both sides of equation 8.5 are multiplied by Δt, we have

$$F\Delta t = \Delta p \tag{8.17}$$

The quantity $F\Delta t$, is called the **impulse**[1] of the force and is given by

$$J = F\Delta t \tag{8.18}$$

The impulse J is a measure of the force that is acting, times the time that force is acting. Equation 8.17 then becomes

$$J = \Delta p \tag{8.19}$$

That is, the impulse acting on a body changes the momentum of that body. Since $\Delta p = p_f - p_i$, equation 8.19 also can be written as

$$J = p_f - p_i \tag{8.20}$$

In many cases, the force F that is exerted is not a constant during the collision process. In that case an average force F_{avg} can be used in the computation of the impulse. That is,

$$F_{avg}\Delta t = \Delta p \tag{8.21}$$

Examples of the use of the concept of impulse can be found in such sports as baseball, golf, tennis, and the like, see figure 8.7. If you participated in such sports, you were most likely told that the "follow through" is extremely important. For example, consider the process of hitting a golf ball. The ball is initially at rest on the tee. As the club hits the ball, the club exerts an average force F_{avg} on the ball. By "following through" with the golf club, as shown in figure 8.8, we mean that the longer the time interval Δt that the club is exerting its force on the ball, the greater is the impulse imparted to the ball and hence the greater will be the change in momentum of the ball. The greater change in momentum implies a greater change in the velocity of the ball and hence the ball will travel a greater distance.

The principle is the same in baseball, tennis, and other similar sports. The better the follow through, the longer the bat or racket is in contact with the ball and the greater the change in momentum the ball will have. Those interested in the application of physics to sports can read the excellent book, *Sport Science* by Peter Brancazio (Simon and Schuster, 1984).

(a)

(b)

Figure 8.7

Physics in sports. When hitting (a) a baseball or (b) a tennis ball, the "follow-through" is very important.

---

1. In some books the letter I is used to denote the impulse. In order to not confuse it with the moment of inertia of a body, also designated by the letter I and treated in detail in chapter 9, we will use the letter J for impulse.

Mechanics

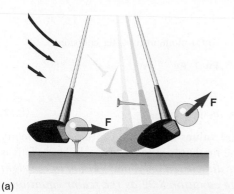

(a)

(b)

Figure 8.8

The effect of "follow through" in hitting a golf ball.

Figure 8.9

A perfectly elastic collision.

8.5 Collisions in One Dimension

We saw in section 8.2 that momentum is always conserved in a collision if the net external force on the system is zero. In physics three different kinds of collisions are usually studied. Momentum is conserved in all of them, but kinetic energy is conserved in only one. These different types of collisions are

1. *A perfectly elastic collision*—a collision in which no kinetic energy is lost, that is, kinetic energy is conserved.
2. *An inelastic collision*—a collision in which some kinetic energy is lost. All real collisions belong to this category.
3. *A perfectly inelastic collision*—a collision in which the two objects stick together during the collision. A great deal of kinetic energy is usually lost in this collision.

In all real collisions in the macroscopic world, some kinetic energy is lost. As an example, consider a collision between two billiard balls. As the balls collide they are temporarily deformed. Some of the kinetic energy of the balls goes into the potential energy of deformation. Ideally, as each ball returns to its original shape, all the potential energy stored by the ball is converted back into the kinetic energy of the ball. In reality, some kinetic energy is lost in the form of heat and sound during the deformation process. The mere fact that we can hear the collision indicates that some of the mechanical energy has been transformed into sound energy. But in many cases, the amount of kinetic energy that is lost is so small that, as a first approximation, it can be neglected. For such cases we assume that no energy is lost during the collision, and the collision is treated as a perfectly elastic collision. The reason why we like to solve perfectly elastic collisions is simply that they are much easier to analyze than inelastic collisions.

Perfectly Elastic Collisions

Between Unequal Masses Consider the collision shown in figure 8.9 between two different masses, m_1 and m_2, having initial velocities $\mathbf{v}_{1i}$ and $\mathbf{v}_{2i}$, respectively. We assume that $\mathbf{v}_{1i}$ is greater than $\mathbf{v}_{2i}$, so that a collision will occur. We can write the law of conservation of momentum as

$$\mathbf{p}_i = \mathbf{p}_f$$

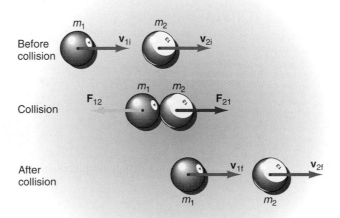

That is,

Total momentum before collision = Total momentum after collision

$$\mathbf{p}_{1i} + \mathbf{p}_{2i} = \mathbf{p}_{1f} + \mathbf{p}_{2f}$$

or

$$m_1\mathbf{v}_{1i} + m_2\mathbf{v}_{2i} = m_1\mathbf{v}_{1f} + m_2\mathbf{v}_{2f} \qquad (8.22)$$

where the subscript i stands for the initial values of the momentum and velocity (before the collision) while f stands for the final values (after the collision). This is a vector equation. *If the collision is in one dimension only, and motion to the right is considered positive, then we can rewrite equation 8.22 as the scalar equation*

$$m_1 v_{1i} + m_2 v_{2i} = m_1 v_{1f} + m_2 v_{2f} \qquad (8.23)$$

Usually we know v_{1i} and v_{2i} and need to find v_{1f} and v_{2f}. In order to solve for these final velocities, we need another equation.

The second equation comes from the law of conservation of energy. Since the collision occurs on a flat surface, which we take as our reference level and assign the height zero, there is no change in potential energy to consider during the collision. Thus, we need only consider the conservation of kinetic energy. The law of conservation of energy, therefore, becomes

$$KE_{BC} = KE_{AC} \qquad (8.24)$$

That is,

Kinetic energy before collision = Kinetic energy after collision $\qquad$ (8.25)

which becomes

$$\tfrac{1}{2}m_1 v_{1i}^2 + \tfrac{1}{2}m_2 v_{2i}^2 = \tfrac{1}{2}m_1 v_{1f}^2 + \tfrac{1}{2}m_2 v_{2f}^2 \qquad (8.26)$$

If the initial values of the speed of the two bodies are known, then we find the final values of the speed by solving equations 8.23 and 8.26 simultaneously. The algebra involved can be quite messy for a direct simultaneous solution. (A simplified solution is given below. However, even the simplified solution is a little long. Those students not interested in the derivation can skip directly to the solution in equation 8.30.)

To simplify the solution, we rewrite equation 8.23, the conservation of momentum, in the form

$$m_1(v_{1i} - v_{1f}) = m_2(v_{2f} - v_{2i}) \qquad (8.27)$$

where the masses have been factored out. Similarly, we factor the masses out in equation 8.26, the conservation of energy, and rewrite it in the form

$$m_1(v_{1i}^2 - v_{1f}^2) = m_2(v_{2f}^2 - v_{2i}^2) \qquad (8.28)$$

We divide equation 8.28 by equation 8.27 to eliminate the mass terms:

$$\frac{m_1(v_{1i}^2 - v_{1f}^2)}{m_1(v_{1i} - v_{1f})} = \frac{m_2(v_{2f}^2 - v_{2i}^2)}{m_2(v_{2f} - v_{2i})}$$

Note that we can rewrite the numerators as products of factors:

$$\frac{(v_{1i} + v_{1f})(v_{1i} - v_{1f})}{v_{1i} - v_{1f}} = \frac{(v_{2i} + v_{2f})(v_{2f} - v_{2i})}{v_{2f} - v_{2i}}$$

which simplifies to

$$v_{1i} + v_{1f} = v_{2i} + v_{2f} \qquad (8.29)$$

Solving for v_{2f} in equation 8.29, we get

$$v_{2f} = v_{1i} + v_{1f} - v_{2i}$$

Substituting this into equation 8.27, we have

$$m_1(v_{1i} - v_{1f}) = m_2[(v_{1i} + v_{1f} - v_{2i}) - v_{2i}]$$

$$m_1 v_{1i} - m_1 v_{1f} = m_2 v_{1i} + m_2 v_{1f} - m_2 v_{2i} - m_2 v_{2i}$$

Collecting terms of v_{1f}, we have

$$-m_1 v_{1f} - m_2 v_{1f} = -2m_2 v_{2i} + m_2 v_{1i} - m_1 v_{1i}$$

Multiplying both sides of the equation by -1, we get

$$+m_1 v_{1f} + m_2 v_{1f} = +2m_2 v_{2i} - m_2 v_{1i} + m_1 v_{1i}$$

Simplifying,

$$(m_1 + m_2)v_{1f} = (m_1 - m_2)v_{1i} + 2m_2 v_{2i}$$

Solving for the final speed of ball 1, we have

$$v_{1f} = \left(\frac{m_1 - m_2}{m_1 + m_2}\right)v_{1i} + \left(\frac{2m_2}{m_1 + m_2}\right)v_{2i} \qquad (8.30)$$

In a similar way, we can solve equation 8.29 for v_{1f}, which we then substitute into equation 8.27. After the same algebraic treatment (which is left as an exercise), the final speed of the second ball becomes

$$v_{2f} = \left(\frac{2m_1}{m_1 + m_2}\right)v_{1i} - \left(\frac{m_1 - m_2}{m_1 + m_2}\right)v_{2i} \qquad (8.31)$$

Equations 8.30 and 8.31 were derived on the assumption that balls 1 and 2 were originally moving with a positive velocity to the right before the collision, and both balls had a positive velocity to the right after the collision. If v_{1f} comes out to be a negative number, ball 1 will have a negative velocity after the collision and will rebound to the left.

If the collision looks like the one depicted in figure 8.2, we can still use equations 8.30 and 8.31. However, ball 2 will be moving to the left, initially, and will thus have a negative velocity v_{2i}. This means that v_{2i} has to be a negative number when placed in these equations. If v_{1f} comes out to be a negative number in the calculations, that means that ball 1 has a negative final velocity and will be moving to the left.

Example 8.3

Perfectly elastic collision, ball 1 catches up with ball 2. Consider the perfectly elastic collision between masses $m_1 = 100$ g and $m_2 = 200$ g. Ball 1 is moving with a velocity v_{1i} of 30.0 cm/s to the right, and ball 2 has a velocity $v_{2i} = 20.0$ cm/s, also to the right, as shown in figure 8.9. Find the final velocities of the two balls.

Solution

The final velocity of the first ball, found from equation 8.30, is

$$v_{1f} = \left(\frac{m_1 - m_2}{m_1 + m_2}\right)v_{1i} + \left(\frac{2m_2}{m_1 + m_2}\right)v_{2i}$$

$$= \left(\frac{100\text{ g} - 200\text{ g}}{100\text{ g} + 200\text{ g}}\right)(30.0\text{ cm/s}) + \frac{2(200\text{ g})}{100\text{ g} + 200\text{ g}}(20.0\text{ cm/s})$$

$$= 16.7\text{ cm/s}$$

Since v_{1f} is a positive quantity, the final velocity of ball 1 is toward the right. The final velocity of the second ball, obtained from equation 8.31, is

$$v_{2f} = \left(\frac{2m_1}{m_1 + m_2}\right)v_{1i} - \left(\frac{m_1 - m_2}{m_1 + m_2}\right)v_{2i}$$

$$= \left[\frac{2(100 \text{ g})}{100 \text{ g} + 200 \text{ g}}\right](30.0 \text{ cm/s}) - \left(\frac{100 \text{ g} - 200 \text{ g}}{100 \text{ g} + 200 \text{ g}}\right)(20.0 \text{ cm/s})$$

$$= 26.7 \text{ cm/s}$$

Since v_{2f} is a positive quantity, the second ball has a positive velocity and is moving toward the right.

Example 8.4

Perfectly elastic collision with masses approaching each other. Consider the perfectly elastic collision between masses $m_1 = 100$ g, $m_2 = 200$ g, with velocity $v_{1i} = 20.0$ cm/s to the right, and velocity $v_{2i} = -30.0$ cm/s to the left, as shown in figure 8.2. Find the final velocities of the two balls.

Solution

The final velocity of ball 1, found from equation 8.30, is

$$v_{1f} = \left(\frac{m_1 - m_2}{m_1 + m_2}\right)v_{1i} + \left(\frac{2m_2}{m_1 + m_2}\right)v_{2i}$$

$$= \left(\frac{100 \text{ g} - 200 \text{ g}}{100 \text{ g} + 200 \text{ g}}\right)(20.0 \text{ cm/s}) + \left[\frac{2(200 \text{ g})}{100 \text{ g} + 200 \text{ g}}\right](-30.0 \text{ cm/s})$$

$$= -46.7 \text{ cm/s}$$

Since v_{1f} is a negative quantity, the final velocity of the first ball is negative, indicating that the first ball moves to the left after the collision. The final velocity of the second ball, found from equation 8.31, is

$$v_{2f} = \left(\frac{2m_1}{m_1 + m_2}\right)v_{1i} - \left(\frac{m_1 - m_2}{m_1 + m_2}\right)v_{2i}$$

$$= \left[\frac{2(100 \text{ g})}{100 \text{ g} + 200 \text{ g}}\right](20.0 \text{ cm/s}) - \left(\frac{100 \text{ g} - 200 \text{ g}}{100 \text{ g} + 200 \text{ g}}\right)(-30.0 \text{ cm/s})$$

$$= 3.33 \text{ cm/s}$$

Since v_{2f} is a positive quantity, the final velocity of ball 2 is positive, and the ball will move toward the right.

Let us now look at a few special types of collisions.

Between Equal Masses If the elastic collision occurs between two equal masses, then the final velocities after the collision are again given by equations 8.30 and 8.31, only with mass m_1 set equal to m_2. That is,

$$v_{1f} = \left(\frac{m_2 - m_2}{m_2 + m_2}\right)v_{1i} + \left(\frac{2m_2}{m_2 + m_2}\right)v_{2i}$$

$$= 0 + \frac{2m_2}{2m_2}v_{2i}$$

$$v_{1f} = v_{2i} \qquad (8.32)$$

and

$$v_{2f} = \left(\frac{2\,m_2}{m_2 + m_2}\right)v_{1i} - \left(\frac{m_2 - m_2}{m_2 + m_2}\right)v_{2i}$$

$$= \frac{2m_2}{2m_2}v_{1i} + 0$$

$$v_{2f} = v_{1i} \tag{8.33}$$

Equations 8.32 and 8.33 tell us that the bodies exchange their velocities during the collision.

Both Masses Equal, One Initially at Rest This is the same case, except that one mass is initially at rest, that is, $v_{2i} = 0$. From equation 8.32 we get

$$v_{1f} = v_{2i} = 0 \tag{8.34}$$

while equation 8.33 remains the same

$$v_{2f} = v_{1i}$$

as before. This is an example of the first body being "stopped cold" while the second one "takes off" with the original velocity of the first ball.

A Ball Thrown against a Wall When you throw a ball against a wall, figure 8.10, you have another example of a collision. Assuming the collision to be elastic, equations 8.30 and 8.31 apply. The wall is initially at rest, so $v_{2i} = 0$. Because the wall is very massive compared to the ball we can say that

$$m_2 \gg m_1$$

which implies that

$$m_1 - m_2 \approx -m_2$$

and

$$m_1 + m_2 \approx m_2$$

Solving equation 8.30 for v_{1f}, we have

$$v_{1f} = \left(\frac{m_1 - m_2}{m_1 + m_2}\right)v_{1i} + \left(\frac{2m_2}{m_1 + m_2}\right)v_{2i}$$

$$= -\frac{m_2}{m_2}v_{1i} + 0$$

Therefore, the final velocity of the ball is

$$v_{1f} = -v_{1i} \tag{8.35}$$

The negative sign indicates that the final velocity of the ball is negative, so the ball rebounds from the wall and is now moving toward the left with the original speed.

The velocity of the wall, found from equation 8.31, is

$$v_{2f} = \left(\frac{2m_1}{m_1 + m_2}\right)v_{1i} - \left(\frac{m_1 - m_2}{m_1 + m_2}\right)v_{2i}$$

$$= \frac{2m_1}{m_2}v_{1i} + 0$$

However, since

$$m_2 \gg m_1$$

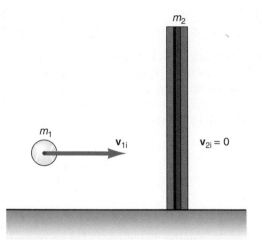

Figure 8.10
A ball bouncing off a wall.

m_2

m_1

$\mathbf{v}_{1i}$

$\mathbf{v}_{2i} = 0$

then

$$\frac{m_1}{m_2} \approx 0$$

Therefore,

$$v_{2f} = 0 \tag{8.36}$$

The ball rebounds from the wall with the same speed that it hit the wall, and the wall, because it is so massive, remains at rest.

Inelastic Collisions

Let us consider for a moment equation 8.29, which we developed earlier in the section, namely

$$v_{1i} + v_{1f} = v_{2f} + v_{2i}$$

If we rearrange this equation by placing all the initial velocities on one side of the equation and all the final velocities on the other, we have

$$v_{1i} - v_{2i} = v_{2f} - v_{1f} \tag{8.37}$$

However, as we can observe from figure 8.9,

$$v_{1i} - v_{2i} = V_A \tag{8.38}$$

that is, the difference in the velocities of the two balls is equal to the *velocity of approach* V_A of the two billiard balls. (The velocity of approach is also called the *relative velocity* between the two balls.) As an example, if ball 1 is moving to the right initially at 10.00 cm/s and ball 2 is moving to the right initially at 5.00 cm/s, then the velocity at which they approach each other is

$$V_A = v_{1i} - v_{2i} = 10.00 \text{ cm/s} - 5.00 \text{ cm/s}$$
$$= 5.00 \text{ cm/s}$$

Similarly,

$$v_{2f} - v_{1f} = V_S \tag{8.39}$$

is the velocity at which the two balls separate. That is, if the final velocity of ball 1 is toward the left at the velocity $v_{1f} = -10.0$ cm/s, and ball 2 is moving to the right at the velocity $v_{2f} = 5.00$ cm/s, then the velocity at which they move away from each other, the *velocity of separation,* is

$$V_S = v_{2f} - v_{1f} = 5.00 \text{ cm/s} - (-10.0 \text{ cm/s})$$
$$= 15.0 \text{ cm/s}$$

Therefore, we can write equation 8.37 as

$$V_A = V_S \tag{8.40}$$

That is, *in a perfectly elastic collision, the velocity of approach of the two bodies is equal to the velocity of separation.*

   *In an inelastic collision, the velocity of separation is not equal to the velocity of approach, and a new parameter, the **coefficient of restitution,** is defined as a measure of the inelastic collision.* That is, we define the coefficient of restitution e as

$$e = \frac{V_S}{V_A} \tag{8.41}$$

and the velocity of separation becomes

$$V_S = eV_A \tag{8.42}$$

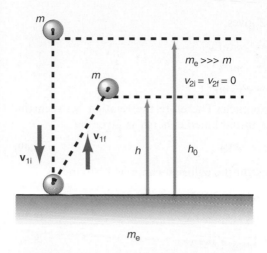

(a)

(b)

Figure 8.11

Imperfectly elastic collision of a ball with the earth.

For a perfectly elastic collision $e = 1$. For a perfectly inelastic collision $e = 0$, which implies $V_S = 0$. Thus, the objects stick together and do not separate at all. *For the inelastic collision*

$$0 < e < 1 \qquad (8.43)$$

Determination of the Coefficient of Restitution If the inelastic collision is between a ball and the earth, as shown in figure 8.11, then, because the earth is so massive, $v_{2i} = v_{2f} = 0$. Equation 8.42 reduces to

$$v_{1f} = ev_{1i} \qquad (8.44)$$

The ball attained its speed v_{1i} by falling from the height h_0, where it had the potential energy

$$PE_0 = mgh_0$$

Immediately before impact its kinetic energy is

$$KE_i = \tfrac{1}{2}mv_{1i}^2$$

And, by the law of the conservation of energy,

$$KE_i = PE_0$$

or

$$\tfrac{1}{2}mv_{1i}^2 = mgh_0$$

Thus, the initial speed before impact with the earth is

$$v_{1i} = \sqrt{2gh_0} \qquad (8.45)$$

After impact, the ball rebounds with a speed v_{1f}, and has a kinetic energy of

$$KE_f = \tfrac{1}{2}mv_{1f}^2$$

which will be less than KE_i because some energy is lost in the collision. After the collision the ball rises to a new height h, as seen in the figure. The final potential energy of the ball is

$$PE_f = mgh$$

However, by the law of conservation of energy

$$KE_f = PE_f$$
$$\tfrac{1}{2}mv_{1f}^2 = mgh$$

Hence, the final speed after the collision is

$$v_{1f} = \sqrt{2gh} \qquad (8.46)$$

We can now find the coefficient of restitution from equations 8.44, 8.45, and 8.46, as

$$e = \frac{v_{1f}}{v_{1i}} = \frac{\sqrt{2gh}}{\sqrt{2gh_0}} = \sqrt{\frac{h}{h_0}} \qquad (8.47)$$

Thus, by measuring the final and initial heights of the ball and taking their ratio, we can find the coefficient of restitution.

The loss of energy in an inelastic collision can easily be found using equation 8.42,

$$V_S = eV_A$$

The kinetic energy after separation is

$$KE_S = \tfrac{1}{2}mV_S^2 \qquad (8.48)$$

Substituting for V_S from equation 8.42 gives,

$$KE_S = \tfrac{1}{2}m(eV_A)^2$$

$$KE_S = \tfrac{1}{2}me^2V_A^2$$

$$KE_S = e^2(\tfrac{1}{2}mV_A^2)$$

But $\tfrac{1}{2}mV_A^2$ is the kinetic energy of approach. Therefore the relation between the kinetic energy after separation and the initial kinetic energy is given by

$$KE_S = e^2KE_A \tag{8.49}$$

The total amount of energy lost in the collision can now be found as

$$\Delta E_{lost} = KE_A - KE_S$$

$$= KE_A - e^2KE_A \tag{8.50}$$

$$\Delta E_{lost} = (1 - e^2)KE_A \tag{8.51}$$

Example 8.5

An imperfectly elastic collision. A 20.0-g racquet ball is dropped from a height of 1.00 m and impacts a tile floor. If the ball rebounds to a height of 76.0 cm, (a) what is the coefficient of restitution, (b) what percentage of the initial energy is lost in the collision, and (c) what is the actual energy lost in the collision?

Solution

a. The coefficient of restitution, found from equation 8.47, is

$$e = \sqrt{\frac{h}{h_0}} = \sqrt{\frac{76.0 \text{ cm}}{100 \text{ cm}}} = 0.872$$

b. The percentage energy lost, found from equation 8.51, is

$$\Delta E_{lost} = (1 - e^2)KE_A$$

$$= (1 - (0.872)^2)KE_A$$

$$= 0.240 \; KE_A$$

$$= 24.0\% \text{ of the initial KE}$$

c. The actual energy lost in the collision with the floor is

$$\Delta E = PE_0 - PE_f$$

$$= mgh_0 - mgh$$

$$= (0.020 \text{ kg})(9.80 \text{ m/s}^2)(1.00 \text{ m}) - (0.020 \text{ kg})(9.80 \text{ m/s}^2)(0.76 \text{ m})$$

$$= 0.047 \text{ J lost}$$

Perfectly Inelastic Collision

Between Unequal Masses In the perfectly inelastic collision, figure 8.12, the two bodies join together during the collision process and move off together as one body after the collision. We assume that v_{1i} is greater than v_{2i}, so a collision will occur. The law of conservation of momentum, when applied to figure 8.12, becomes

$$m_1\mathbf{v}_{1i} + m_2\mathbf{v}_{2i} = (m_1 + m_2)\mathbf{V}_f \tag{8.52}$$

Taking motion to the right as positive, we write this in the scalar form,

$$m_1v_{1i} + m_2v_{2i} = (m_1 + m_2)V_f \tag{8.53}$$

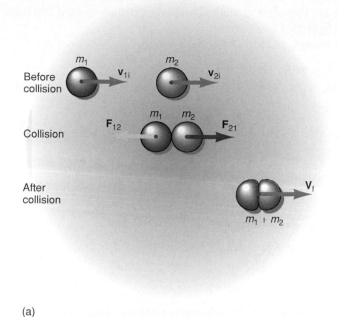

(a)

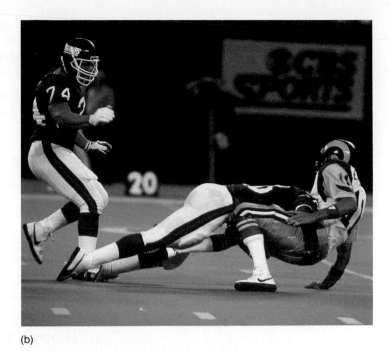

(b)

Figure 8.12

(a) Perfectly inelastic collision. (b) A football player being tackled is also an example of a perfectly inelastic collision.

Solving for the final speed V_f of the combined masses, we get

$$V_f = \left(\frac{m_1}{m_1 + m_2}\right)v_{1i} + \left(\frac{m_2}{m_1 + m_2}\right)v_{2i} \tag{8.54}$$

It is interesting to determine the initial and final values of the kinetic energy of the colliding bodies.

$$KE_i = \tfrac{1}{2}m_1v_{1i}^2 + \tfrac{1}{2}m_2v_{2i}^2 \tag{8.55}$$

$$KE_f = \tfrac{1}{2}(m_1 + m_2)V_f^2 \tag{8.56}$$

Is kinetic energy conserved for this collision?

Example 8.6

A perfectly inelastic collision. A 50.0-g piece of clay moving at a velocity of 5.00 cm/s to the right has a head-on collision with a 100-g piece of clay moving at a velocity of -10.0 cm/s to the left. The two pieces of clay stick together during the impact. Find (a) the final velocity of the clay, (b) the initial kinetic energy, (c) the final kinetic energy, and (d) the amount of energy lost in the collision.

Solution

a. The initial velocity of the first piece of clay is positive, because it is in motion toward the right. The initial velocity of the second piece of clay is negative, because it is in motion toward the left. The final velocity of the clay, given by equation 8.54, is

$$V_f = \left(\frac{m_1}{m_1 + m_2}\right)v_{1i} + \left(\frac{m_2}{m_1 + m_2}\right)v_{2i}$$

$$= \frac{(50.0 \text{ g})(5.00 \text{ cm/s})}{50 \text{ g} + 100 \text{ g}} + \frac{(100 \text{ g})(-10.0 \text{ cm/s})}{50 \text{ g} + 100 \text{ g}}$$

$$= -5.00 \text{ cm/s} = -5.00 \times 10^{-2} \text{ m/s}$$

The minus sign means that the velocity of the combined pieces of clay is negative and they are therefore moving toward the left, not toward the right as we assumed in figure 8.12.

b. The initial kinetic energy, found from equation 8.55, is

$$KE_i = \tfrac{1}{2}m_1v_{1i}^2 + \tfrac{1}{2}m_2v_{2i}^2$$
$$= \tfrac{1}{2}(0.050 \text{ kg})(5.00 \times 10^{-2} \text{ m/s})^2 + \tfrac{1}{2}(0.100 \text{ kg})(-10.0 \times 10^{-2} \text{ m/s})^2$$
$$= 5.63 \times 10^{-4} \text{ J}$$

c. The kinetic energy after the collision, found from equation 8.56, is

$$KE_f = \tfrac{1}{2}(m_1 + m_2)V_f^2$$
$$= \tfrac{1}{2}(0.050 \text{ kg} + 0.100 \text{ kg})(-5.00 \times 10^{-2} \text{ m/s})^2$$
$$= 1.88 \times 10^{-4} \text{ J}$$

d. The mechanical energy lost in the collision is found from

$$\Delta E = KE_i - KE_f$$
$$= 5.63 \times 10^{-4} \text{ J} - 1.88 \times 10^{-4} \text{ J}$$
$$= 3.75 \times 10^{-4} \text{ J}$$

Hence, 3.75×10^{-4} J of energy are lost in the deformation caused by the collision.

†8.6 Collisions in Two Dimensions— Glancing Collisions

In the collisions treated so far, the collisions were head-on collisions, and the forces exerted on the two colliding bodies were on a line in the direction of motion of the two bodies. As an example, consider the collision to be between two billiard balls. For a head-on collision, as in figure 8.13(a), the force on ball 2 caused by ball 1,

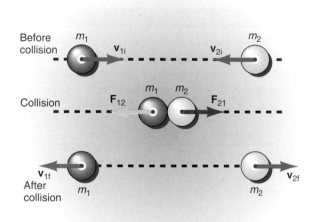

(a) One-dimensional collision (on centers)

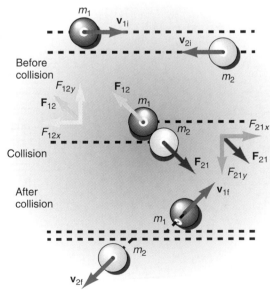

(b) Two-dimensional collision (off centers)

Figure 8.13

Comparison of one-dimensional and two-dimensional collisions.

$\mathbf{F}_{21}$, is in the positive x-direction, while $\mathbf{F}_{12}$, the force on ball 1 caused by ball 2, is in the negative x-direction. After the collision, the two balls move along the original line of action. In a glancing collision, on the other hand, the motion of the centers of mass of each of the two balls do not lie along the same line of action, figure 8.13(b). Hence, when the balls collide, the force exerted on each ball does not lie along the original line of action but is instead a force that is exerted along the line connecting the center of mass of each ball, as shown in the diagram. Thus the force on ball 2 caused by ball 1, $\mathbf{F}_{21}$, is a two-dimensional vector, and so is $\mathbf{F}_{12}$, the force on ball 1 caused by ball 2. As we can see in the diagram, these forces can be decomposed into x- and y-components. Hence, a y-component of force has been exerted on each ball causing it to move out of its original direction of motion. Therefore, after the collision, the two balls move off in the directions indicated. All glancing collisions must be treated as two-dimensional problems. Since the general solution of the two-dimensional collision problem is even more complicated than the one-dimensional problem solved in the last section, we will solve only some special cases of the two-dimensional problem.

Consider the glancing collision between two billiard balls shown in figure 8.14. Ball 1 is moving to the right at the velocity $\mathbf{v}_{1i}$ and ball 2 is at rest ($\mathbf{v}_{2i} = 0$). After the collision, ball 1 is found to be moving at an angle $\theta = 45.0°$ above the horizontal and ball 2 is moving at an angle $\phi = 45.0°$ below the horizontal. Let us find the velocities of both balls after the collision. As in all collisions, the law of conservation of momentum holds, that is,

$$\mathbf{p}_f = \mathbf{p}_i$$
$$m_1\mathbf{v}_{1f} + m_2\mathbf{v}_{2f} = m_1\mathbf{v}_{1i}$$

The last single vector equation is equivalent to the two scalar equations

$$m_1 v_{1f} \cos \theta + m_2 v_{2f} \cos \phi = m_1 v_{1i} \qquad (8.57)$$

$$m_1 v_{1f} \sin \theta - m_2 v_{2f} \sin \phi = 0 \qquad (8.58)$$

Solving equation 8.58 for v_{2f} with $\theta = \phi = 45.0°$, we get

$$m_1 v_{1f} \sin 45.0° = m_2 v_{2f} \sin 45.0°$$

$$v_{2f} = \frac{m_1}{m_2} v_{1f} \qquad (8.59)$$

Figure 8.14

A glancing collision.

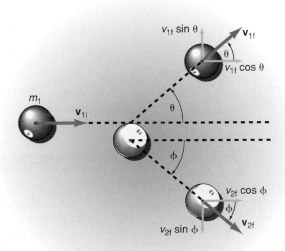

Inserting equation 8.59 into equation 8.57 we can solve for v_{1f} as

$$m_1 v_{1f} \cos 45.0° + m_2\left(\frac{m_1}{m_2} v_{1f}\right) \cos 45.0° = m_1 v_{1i}$$

$$2m_1 v_{1f} \cos 45.0° = m_1 v_{1i}$$

$$v_{1f} = \frac{v_{1i}}{2 \cos 45.0°} \qquad (8.60)$$

Example 8.7

A glancing collision. Billiard ball 1 is moving at a speed of $v_{1i} = 10.0$ cm/s, when it has a glancing collision with an identical billiard ball that is at rest. After the collision, $\theta = \phi = 45.0°$. The mass of the billiard ball is 0.170 kg. (a) Find the speed of ball 1 and 2 after the collision. (b) Is energy conserved in this collision?

Solution

a. The speed of ball 1, found from equation 8.60, is

$$v_{1f} = \frac{v_{1i}}{2 \cos 45.0°}$$

$$= \frac{10.0 \text{ cm/s}}{2 \cos 45.0°}$$

$$= 7.07 \text{ cm/s}$$

and the speed of ball 2, found from equation 8.59, is

$$v_{2f} = \frac{m_1}{m_2} v_{1f}$$

$$= \frac{m_1}{m_1} v_{1f}$$

$$= v_{1f} = 7.07 \text{ cm/s}$$

b. The kinetic energy before the collision is

$$KE_i = \tfrac{1}{2}m_1 v_{1i}^2 = \tfrac{1}{2}(0.170 \text{ kg})(0.100 \text{ m/s})^2$$
$$= 8.50 \times 10^{-4} \text{ J}$$

while the kinetic energy after the collision is

$$KE_f = \tfrac{1}{2}m_1 v_{1f}^2 + \tfrac{1}{2}m_2 v_{2f}^2$$
$$= \tfrac{1}{2}(0.170 \text{ kg})(0.0707 \text{ m/s})^2 + \tfrac{1}{2}(0.170 \text{ kg})(0.0707 \text{ m/s})^2$$
$$= 8.50 \times 10^{-4} \text{ J}$$

Notice that the kinetic energy after the collision is equal to the kinetic energy before the collision. Therefore the collision is perfectly elastic.

Example 8.8

Colliding cars. Two cars collide at an intersection as shown in figure 8.15. Car 1 has a mass of 1200 kg and is moving at a velocity of 95.0 km/hr due east and car 2 has a mass of 1400 kg and is moving at a velocity of 100 km/hr due north. The cars stick together and move off as one at an angle θ as shown in the diagram. Find (a) the angle θ and (b) the final velocity of the combined cars.

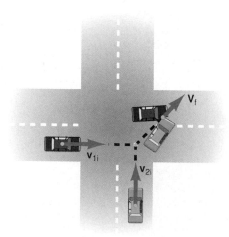

Figure 8.15

A perfectly inelastic glancing collision.

Mechanics

a. This is an example of a perfectly inelastic collision in two dimensions. The law of conservation of momentum yields

$$\mathbf{p_f} = \mathbf{p_i}$$
$$(m_1 + m_2)\mathbf{V_f} = m_1\mathbf{v_{1i}} + m_2\mathbf{v_{2i}} \tag{8.61}$$

Resolving this equation into its x- and y-component equations, we get for the x-component:

$$(m_1 + m_2)V_f \cos\theta = m_1 v_{1i} \tag{8.62}$$

and for the y-component:

$$(m_1 + m_2)V_f \sin\theta = m_2 v_{2i} \tag{8.63}$$

Dividing the y-component equation by the x-component equation we get

$$\frac{(m_1 + m_2)V_f \sin\theta}{(m_1 + m_2)V_f \cos\theta} = \frac{m_2 v_{2i}}{m_1 v_{1i}}$$

$$\frac{\sin\theta}{\cos\theta} = \frac{m_2 v_{2i}}{m_1 v_{1i}}$$

$$\tan\theta = \frac{m_2 v_{2i}}{m_1 v_{1i}}$$

$$\tan\theta = \frac{(1400\text{ kg})(100\text{ km/hr})}{(1200\text{ kg})(95.0\text{ km/hr})}$$

$$\theta = 50.8°$$

b. The combined final speed, found by solving for V_f in equation 8.62, is

$$V_f = \frac{m_1 v_{1i}}{(m_1 + m_2)\cos\theta}$$

$$= \frac{(1200\text{ kg})(95.0\text{ km/hr})}{(1200\text{ kg} + 1400\text{ kg})\cos 50.8°}$$

$$= 69.4\text{ km/hr}$$

The Language of Physics

Linear momentum
The product of the mass of the body in motion times its velocity (p. 215).

Newton's second law in terms of linear momentum
When a resultant applied force acts on a body, it causes the linear momentum of that body to change with time (p. 216).

External forces
Forces that originate outside the system and act on the system (p. 216).

Internal forces
Forces that originate within the system and act on the particles within the system (p. 216).

Law of conservation of linear momentum
If the total external force acting on a system is equal to zero, then the final value of the total momentum of the system is

equal to the initial value of the total momentum of the system. Thus, the total momentum is a constant, or as usually stated, the total momentum is conserved. The law of conservation of momentum is a consequence of Newton's third law (p. 217).

Impulse
The product of the force that is acting and the time that the force is acting. The impulse acting on a body is equal to the change in momentum of the body (p. 222).

Perfectly elastic collision
A collision in which no kinetic energy is lost, that is, the kinetic energy is conserved. Momentum is conserved in all collisions for which there are no external forces. In this type of collision, the velocity of separation of the two bodies is equal to the velocity of approach (p. 223).

Inelastic collision
A collision in which some kinetic energy is lost. The velocity of separation of the two bodies in this type of collision is not equal to the velocity of approach. The coefficient of restitution is a measure of the inelastic collision (p. 223).

Perfectly inelastic collision
A collision in which the two objects stick together during the collision. A great deal of kinetic energy is usually lost in this type of collision (p. 223).

Coefficient of restitution
The measure of the amount of the inelastic collision. It is equal to the ratio of the velocity of separation of the two bodies to the velocity of approach (p. 228).

Summary of Important Equations

Definition of momentum
$$\mathbf{p} = m\mathbf{v} \tag{8.1}$$

Newton's second law in terms of momentum
$$\mathbf{F} = \frac{\Delta \mathbf{p}}{\Delta t} \tag{8.5}$$

Law of conservation of momentum for
$$\mathbf{F}_{net} = 0$$
$$\mathbf{p}_f = \mathbf{p}_i \tag{8.7}$$

Recoil speed of a gun
$$v_G = \frac{m_B}{m_G} v_B \tag{8.14}$$

Impulse
$$\mathbf{J} = \mathbf{F}\Delta t \tag{8.18}$$

Impulse is equal to the change in momentum
$$\mathbf{J} = \Delta \mathbf{p} \tag{8.19}$$

Conservation of momentum in a collision
$$m_1\mathbf{v}_{1i} + m_2\mathbf{v}_{2i} = m_1\mathbf{v}_{1f} + m_2\mathbf{v}_{2f} \tag{8.22}$$

Conservation of momentum in scalar form, both bodies in motion in same direction, and $v_{1i} > v_{2i}$.
$$m_1 v_{1i} + m_2 v_{2i} = m_1 v_{1f} + m_2 v_{2f} \tag{8.23}$$

Conservation of energy in a perfectly elastic collision
$$\tfrac{1}{2} m_1 v_{1i}^2 + \tfrac{1}{2} m_2 v_{2i}^2$$
$$= \tfrac{1}{2} m_1 v_{1f}^2 + \tfrac{1}{2} m_2 v_{2f}^2 \tag{8.26}$$

Final velocity of ball 1 in a perfectly elastic collision
$$v_{1f} = \left(\frac{m_1 - m_2}{m_1 + m_2}\right) v_{1i}$$
$$+ \left(\frac{2m_2}{m_1 + m_2}\right) v_{2i} \tag{8.30}$$

Final velocity of ball 2 in a perfectly elastic collision
$$v_{2f} = \left(\frac{2m_1}{m_1 + m_2}\right) v_{1i}$$
$$- \left(\frac{m_1 - m_2}{m_1 + m_2}\right) v_{2i} \tag{8.31}$$

The velocity of approach
$$v_{1i} - v_{2i} = V_A \tag{8.38}$$

The velocity of separation
$$v_{2f} - v_{1f} = V_S \tag{8.39}$$

For any collision
$$V_S = eV_A \tag{8.42}$$

For a perfectly elastic collision
$$e = 1$$

For an inelastic collision
$$0 < e < 1 \tag{8.43}$$

For a perfectly inelastic collision
$$e = 0$$

Perfectly inelastic collision
$$V_f = \left(\frac{m_1}{m_1 + m_2}\right) v_{1i}$$
$$+ \left(\frac{m_2}{m_1 + m_2}\right) v_{2i} \tag{8.54}$$

Questions for Chapter 8

1. If the velocity of a moving body is doubled, what does this do to the kinetic energy and the momentum of the body?
2. Why is Newton's second law in terms of momentum more appropriate than the form $F = ma$?
3. State and discuss the law of conservation of momentum and show its relation to Newton's third law of motion.
4. Discuss what is meant by an isolated system and how it is related to the law of conservation of momentum.
5. Is it possible to have a collision in which all the kinetic energy is lost? Describe such a collision.

6. An airplane is initially flying at a constant velocity in plane and level flight. If the throttle setting is not changed, explain what happens to the plane as it continues to burn its fuel?
†7. In the early days of rocketry it was assumed by many people that a rocket would not work in outer space because there was no air for the exhaust gases to push against. Explain why the rocket does work in outer space.
8. Discuss the possibility of a fourth type of collision, a super elastic collision, in which the particles have more kinetic energy after the collision than before. As an example,

consider a car colliding with a truck loaded with dynamite.
9. If the net force acting on a body is equal to zero, what happens to the center of mass of the body?
†10. A bird is sitting on a swing in an enclosed bird cage that is resting on a mass balance. If the bird leaves the swing and flies around the cage without touching anything, does the balance show any change in its reading?
11. From the point of view of impulse, explain why an egg thrown against a wall will break, while an egg thrown against a loose vertical sheet will not.

Problems for Chapter 8

8.1 Momentum

1. What is the momentum of a 3200-lb car traveling at 55.0 mph?
2. A 3200-lb car traveling at 55.0 mph collides with a tree and comes to a stop in 0.100 s. What is the change in momentum of the car? What average force acted on the car during impact? What is the impulse?

3. Answer the same questions in problem 2 if the car hit a sand barrier in front of the tree and came to rest in 0.300 s.
4. A 0.150-kg ball is thrown straight upward at an initial velocity of 30.0 m/s. Two seconds later the ball has a velocity of 10.4 m/s. Find (a) the initial momentum of the ball, (b) the momentum of the ball at 2 s, (c) the force acting on the ball, and (d) the weight of the ball.

5. How long must a force of 5.00 N act on a block of 3.00-kg mass in order to give it a velocity of 4.00 m/s?
6. A force of 25.0 N acts on a 10.0-kg mass in the positive x-direction, while another force of 13.5 N acts in the negative x-direction. If the mass is initially at rest, find (a) the time rate of change of momentum, (b) the change in momentum after 1.85 s, and (c) the velocity of the mass at the end of 1.85 s.

8.2 and 8.3 Conservation of Momentum

7. A 10.0-g bullet is fired from a 5.00-kg rifle with a velocity of 300 m/s. What is the recoil velocity of the rifle?

8. In an ice skating show, a 200-lb man at rest pushes a 100-lb woman away from him at a speed of 4.00 ft/s. What happens to the man?

9. A 5000-kg cannon fires a shell of 3.00-kg mass with a velocity of 250 m/s. What is the recoil velocity of the cannon?

10. A cannon of 3.50×10^3 kg fires a shell of 2.50 kg with a muzzle speed of 300 m/s. What is the recoil velocity of the cannon?

11. A 150-lb boy at rest on roller skates throws a 2.00-lb ball horizontally with a speed of 25.0 ft/s. With what speed does the boy recoil?

12. An 80.0-kg astronaut pushes herself away from a 1200-kg space capsule at a velocity of 3.00 m/s. Find the recoil velocity of the space capsule.

13. A 200-lb man is standing in a 500-lb boat. The man walks forward at 3.00 ft/s relative to the water. What is the final velocity of the boat? Neglect any resistive force of the water on the boat.

14. A 78.5-kg man is standing in a 275-kg boat. The man walks forward at 1.25 m/s relative to the water. What is the final velocity of the boat? Neglect any resistive force of the water on the boat.

15. A water hose sprays 2 kg of water against the side of a building in 1 s. If the velocity of the water is 15 m/s, what force is exerted on the wall by the water? (Assume that the water does not bounce off the wall of the building.)

8.4 Impulse

16. A boy kicks a football with an average force of 20.0 lb for a time of 0.200 s. (a) What is the impulse? (b) What is the change in momentum of the football? (c) If the football has a mass of 250 g, what is the velocity of the football as it leaves the kicker's foot?

17. A boy kicks a football with an average force of 66.8 N for a time of 0.185 s. (a) What is the impulse? (b) What is the change in momentum of the football? (c) If the football has mass of 250 g, what is the velocity of the football as it leaves the kicker's foot?

18. A baseball traveling at 150 km/hr is struck by a bat and goes straight back to the pitcher at the same speed. If the baseball has a mass of 200 g, find (a) the change in momentum of the baseball, (b) the impulse imparted to the ball, and (c) the average force acting if the bat was in contact with the ball for 0.100 s.

19. A 10.0-kg hammer strikes a nail at a velocity of 12.5 m/s and comes to rest in a time interval of 0.004 s. Find (a) the impulse imparted to the nail and (b) the average force imparted to the nail.

20. If a gas molecule of mass 5.30×10^{-26} kg and an average speed of 425 m/s collides perpendicularly with a wall of a room and rebounds at the same speed, what is its change of momentum? What impulse is imparted to the wall?

8.5 Collisions in One Dimension

21. Two gliders moving toward each other, one of mass 200 g and the other of 250 g, collide on a frictionless air track. If the first glider has an initial velocity of 25.0 cm/s toward the right and the second of −35.0 cm/s toward the left, find the velocities after the collision if the collision is perfectly elastic.

22. A 250-g glider overtakes and collides with a 200-g glider on an air track. If the 250-g glider is moving at 35.0 cm/s and the second glider at 25.0 cm/s, find the velocities after the collision if the collision is perfectly elastic.

†23. A 200-g ball makes a perfectly elastic collision with an unknown mass that is at rest. If the first ball rebounds with a final speed of $v_{1f} = \frac{1}{2}v_{1i}$, (a) what is the unknown mass, and (b) what is the final velocity of the unknown mass?

24. A 30.0-g ball, m_1, collides perfectly elastically with a 20.0-g ball, m_2. If the initial velocities are $v_{1i} = 50.0$ cm/s to the right and $v_{2i} = -30.0$ cm/s to the left, find the final velocities v_{1f} and v_{2f}. Compute the initial and final momenta. Compute the initial and final kinetic energies.

25. A 150-g ball moving at a velocity of 25.0 cm/s to the right collides with a 250-g ball moving at a velocity of 18.5 cm/s to the left. The collision is imperfectly elastic with a coefficient of restitution of 0.65. Find (a) the velocity of each ball after the collision, (b) the kinetic energy before the collision, (c) the kinetic

energy after the collision, and (d) the percentage of energy lost in the collision.

26. A 2000-lb car traveling at 60.0 mph collides "head-on" with a 20,000-lb truck traveling toward the car at 30.0 mph. The car becomes stuck to the truck during the collision. What is the final velocity of the car and truck?

27. A 1150-kg car traveling at 110 km/hr collides "head-on" with a 9500-kg truck traveling toward the car at 40.0 km/hr. The car becomes stuck to the truck during the collision. What is the final velocity of the car and truck?

28. A 3.00-g bullet is fired at 200 m/s into a wooden block of 10-kg mass that is at rest. If the bullet becomes embedded in the wooden block, find the velocity of the block and bullet after impact.

29. A 20,000-lb freight car traveling at 3.00 ft/s collides with a stationary freight car that weighs 15,000 lb. If the cars couple together find the resultant velocity of the cars after the collision.

30. A 9500-kg freight car traveling at 5.50 km/hr collides with an 8000-kg stationary freight car. If the cars couple together, find the resultant velocity of the cars after the collision.

31. Two gliders are moving toward each other on a frictionless air track. Glider 1 has a mass of 200 g and glider 2 of 250 g. The first glider has an initial speed of 25.0 cm/s while the second has a speed of 35.0 cm/s. If the collision is perfectly inelastic, find (a) the final velocity of the gliders, (b) the kinetic energy before the collision, and (c) the kinetic energy after the collision. (d) How much energy is lost, and where did it go?

8.6 Collisions in Two Dimensions— Glancing Collisions

32. A 230-lb linebacker moving due east at 25 mph tackles a 175-lb halfback moving south at 40 mph. The two stick together during the collision. What is the resultant velocity of the two of them?

33. A 10,000-kg truck enters an intersection heading north at 45 km/hr when it makes a perfectly inelastic collision with a 1000-kg car traveling at 90 km/hr due east. What is the final velocity of the car and truck?

†34. Billiard ball 2 is at rest when it is hit with a glancing collision by ball 1 moving at a velocity of 50.0 cm/s toward the right. After the collision ball 1 moves off at an angle of 35.0° from the original direction while ball 2 moves at an angle of 40.0°, as shown in the diagram. The mass of each billiard ball is 0.017 kg. Find the final velocity of each ball after the collision. Find the kinetic energy before and after the collision. Is the collision elastic?

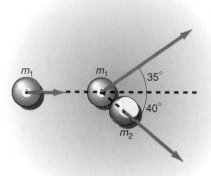

35. A 0.150-kg ball, moving at a speed of 25.0 m/s, makes an elastic collision with a wall at an angle of 40.0°, and rebounds at an angle of 40.0°. Find (a) the change in momentum of the ball and (b) the magnitude and direction of the momentum imparted to the wall. The diagram is a view from the top.

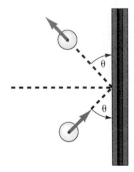

Additional Problems

†36. A 0.250-kg ball is dropped from a height of 1.00 m. It rebounds to a height of 0.750 m. If the ground exerts a force of 300 N on the ball, find the time the ball is in contact with the ground.

37. A 200-g ball is dropped from the top of a building. If the speed of the ball before impact is 40.0 m/s, and right after impact it is 25.0 cm/s, find (a) the momentum of the ball before impact, (b) the momentum of the ball after impact, (c) the kinetic energy of the ball before impact, (d) the kinetic energy of the ball after impact, and (e) the coefficient of restitution of the ball.

†38. A ball is dropped from a height of 1.00 m and rebounds to a height of 0.920 m. Approximately how many bounces will the ball make before losing 90% of its energy?

39. A 60.0-g tennis ball is dropped from a height of 1.00 m. If it rebounds to a height of 0.560 m, (a) what is the coefficient of restitution of the tennis ball and the floor, and (b) how much energy is lost in the collision?

†40. A 25.0-g bullet strikes a 5.00-kg ballistic pendulum that is initially at rest. The pendulum rises to a height of 14.0 cm. What is the initial speed of the bullet?

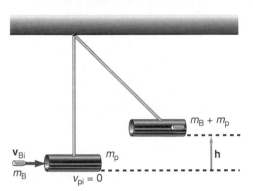

41. A 25.0-g bullet with an initial speed of 400 m/s strikes a 5-kg ballistic pendulum that is initially at rest. (a) What is the speed of the combined bullet-pendulum after the collision? (b) How high will the pendulum rise?

42. An 80-kg caveman, standing on a branch of a tree 5 m high, swings on a vine and catches a 60-kg cavegirl at the bottom of the swing. How high will both of them rise?

†43. A hunter fires an automatic rifle at an attacking lion that weighs 300 lb. If the lion is moving toward the hunter at 10.0 ft/s, and the rifle bullets weigh 2.00 oz each and have a muzzle velocity of 2500 ft/s, how many bullets must the man fire at the lion in order to stop the lion in his tracks?

†44. Two gliders on an air track are connected by a compressed spring and a piece of thread as shown; m_1 = 300 g and m_2 is unknown. If the

connecting string is cut, the gliders separate. Glider 1 experiences the velocity v_1 = 10.0 cm/s, and glider 2 experiences the velocity v_2 = 20.0 cm/s, what is the unknown mass?

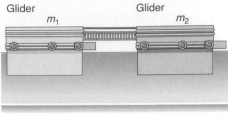

Air track

†45. Two gliders on an air track are connected by a compressed spring and a piece of thread as shown. The masses of the gliders are m_1 = 300 g and m_2 = 250 g. The connecting string is cut and the compressed string causes the two gliders to separate from each other. If glider 1 has moved 35.0 cm from its starting point, where is glider 2 located?

†46. Two balls, m_1 = 100 g and m_2 = 200 g, are suspended near each other as shown. The two balls are initially in contact. Ball 2 is then pulled away so that it makes a 45.0° angle with the vertical and is then released. (a) Find the velocity of ball 2 just before impact and the velocity of each ball after the perfectly elastic impact. (b) How high will each ball rise?

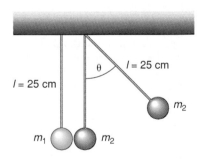

†47. Two swimmers simultaneously dive off opposite ends of a 200-lb boat. If the first swimmer has a weight w_1 = 200 lb and a velocity of 5.00 ft/s toward the right, while the second swimmer has a weight w_2 = 150 lb and a velocity of −3.00 ft/s toward the left, what is the final velocity of the boat?

48. Two swimmers simultaneously dive off opposite ends of a 110-kg boat. If the first swimmer has a mass m_1 = 66.7 kg and a velocity of 1.98 m/s toward the right, while the second swimmer has a mass m_2 = 77.8 kg and a velocity of −7.63 m/s toward the left, what is the final velocity of the boat?

†49. Show that the kinetic energy of a moving body can be expressed in terms of the linear momentum as KE = $p^2/2m$.

†50. A machine gun is mounted on a small train car and fires 100 bullets per minute backward. If the mass of each bullet is 10.0 g and the speed of each bullet as it leaves the gun is 900 m/s, find the average force exerted on the gun. If the mass of the car and machine gun is 225 kg, what is the acceleration of the train car while the gun is firing?

†51. An open toy railroad car of mass 250 g is moving at a constant speed of 30 cm/s when a wooden block of 50 g is dropped into the open car. What is the final speed of the car and block?

†52. Masses m_1 and m_2 are located on the top of the two frictionless inclined planes as shown in the diagram. It is given that $m_1 = 30.0$ g, $m_2 = 50.0$ g, $l_1 = 50.0$ cm, $l_2 = 20.0$ cm, $l = 100$ cm, $\theta_1 = 50.0°$, and $\theta_2 = 25.0°$. Find (a) the speeds v_1 and v_2 at the bottom of each inclined plane, note that ball 1 reaches the bottom of the plane before ball 2; (b) the position between the planes where the masses will collide elastically; (c) the speeds of the two masses after the collision; and (d) the final locations l_1' and l_2' where the two masses will rise up the plane after the collision.

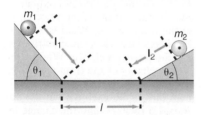

†53. The mass $m_1 = 40.0$ g is initially located at a height $h_1 = 1.00$ m on the frictionless surface shown in the diagram. It is then released from rest and collides with the mass $m_2 = 70.0$ g, which is at rest at the bottom of the surface. After the collision, will the mass m_2 make it over the top of the hill at position B, which is at a height of 0.500 m?

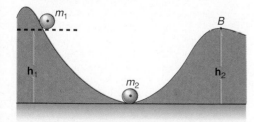

†54. Two balls of mass m_1 and m_2 are placed on a frictionless surface as shown in the diagram. Mass $m_1 = 30.0$ g is at a height $h_1 = 50.0$ cm above the bottom of the bowl, while mass $m_2 = 60.0$ g is at a height of 3/4 h_1. The distance $l = 100$ cm. Assuming that both balls reach the bottom at the same time, find (a) the speed of each ball at the bottom of each surface, (b) the position where the two balls collide, (c) the speed of each ball after the collision, and (d) the height that each ball will rise to after the collision.

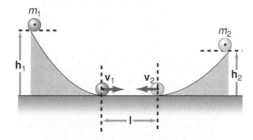

†55. A person is in a small train car that weighs 500 lb and contains 500 lb of rocks. The train is initially at rest. The person starts to throw large rocks, each weighing 100 lb, from the rear of the train at a speed of 5.00 ft/s. (a) If the person throws out 1 rock what will the recoil velocity of the train be? The person then throws out another rock at the same speed. (b) What is the recoil velocity now? (c) The person continues to throw out the rest of the rocks one at a time. What is the final velocity of the train when all the rocks have been thrown out?

†56. A bullet of mass 20.0 g is fired into a block of mass 5.00 kg that is initially at rest. The combined block and bullet moves a distance of 5.00 m over a rough surface of coefficient of kinetic friction of 0.500, before coming to rest. Find the initial velocity of the bullet.

†57. A bullet of mass 20.0 g is fired at an initial velocity of 200 m/s into a 15.0-kg block that is initially at rest. The combined bullet and block move over a rough surface of coefficient of kinetic friction of 0.500. How far will the combined bullet and block move before coming to rest?

58. A 15.0-kg bullet moving at a speed of 250 m/s hits a 2.00-kg block of wood, which is initially at rest. The bullet emerges from the block of wood at 150 m/s. Find (a) the final velocity of the block of wood and (b) the amount of energy lost in the collision.

†59. A 5-kg pendulum bob, at a height of 0.750 m above the floor, swings down to the ground where it hits a 2.15 kg block that is initially at rest. The block then slides up a 30.0° incline. Find how far up the incline the block will slide if (a) the plane is frictionless and (b) if the plane is rough with a value of $\mu_k = 0.450$.

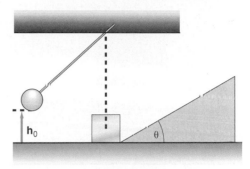

†60. A 0.15-kg baseball is thrown upward at an initial velocity of 35.0 m/s. Two seconds later, a 20.0-g bullet is fired at 250 m/s into the rising baseball. How high will the combined bullet and baseball rise?

†61. A 25-g ball slides down a smooth inclined plane, 0.850 m high, that makes an angle of 35.0° with the horizontal. The ball slides into an open box of 200-g mass and the ball and box slide on a rough surface of $\mu_k = 0.450$. How far will the combined ball and box move before coming to rest?

†62. A 25-g ball slides down a smooth inclined plane, 0.850 m high, that makes an angle of 35.0° with the horizontal. The ball slides into an open box of 200-g mass and the ball and box slide off the end of a table 1.00 m high. How far from the base of the table will the combined ball and box hit the ground?

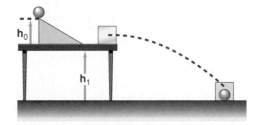

†63. A 1300-kg car collides with a 15,000-kg truck at an intersection and they couple together and move off as one leaving a skid mark 5 m long that makes an angle of 30.0° with the original direction of the car. If $\mu_k = 0.700$, find the initial velocities of the car and truck before the collision.

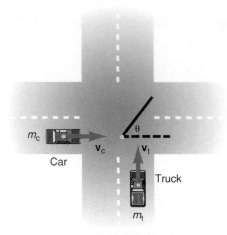

Car

Truck

m_c

v_c

v_t

m_t

θ

64. A bomb of mass $M = 2.50$ kg, moving in the x-direction at a speed of 10.5 m/s, explodes into three pieces. One fragment, $m_1 = 0.850$ kg, flies off at a velocity of 3.5 m/s at an angle of 30.0° above the x-axis. Fragment $m_2 = 0.750$ kg, flies off at an angle of 43.5° below the positive x-axis, and the third fragment flies off at an angle of 150° with respect to the positive x-axis. Find the velocities of m_2 and m_3.

Interactive Tutorials

65. Recoil velocity of a gun. A bullet of mass $m_b = 10.0$ g is fired at a velocity $v_b = 300$ m/s from a rifle of mass $m_r = 5.00$ kg. Calculate the recoil velocity v_r of the rifle. If the bullet is in the barrel of the rifle for $t = 0.004$ s, what is the bullet's acceleration and what force acted on the bullet? Assume the force is a constant.

66. An inelastic collision. A car of mass $m_1 = 1000$ kg is moving at a velocity $v_1 = 50.0$ m/s and collides inelastically with a car of mass $m_2 = 750$ kg moving in the same direction at a velocity of $v_2 = 20.0$ m/s. Calculate (a) the final velocity v_f of both vehicles; (b) the initial momentum p_i; (c) the final momentum p_f; (d) the initial kinetic energy KE_i; (e) the final kinetic energy KE_f of the system; (f) the energy lost in the collision ΔE; and (g) the percentage of the original energy lost in the collision, $\%E_{lost}$.

67. A perfectly elastic collision. A mass, $m_1 = 3.57$ kg, moving at a velocity, $v_1 = 2.55$ m/s, overtakes and collides with a second mass, $m_2 = 1.95$ kg, moving at a velocity $v_2 = 1.35$ m/s. If the collision is perfectly elastic, find (a) the velocities after the collision, (b) the momentum before the collision, (c) the momentum after the collision, (d) the kinetic energy before the collision, and (e) the kinetic energy after the collision.

68. An imperfectly elastic collision. A mass, $m = 2.84$ kg, is dropped from a height $h_0 = 3.42$ m and hits a wooden floor. The mass rebounds to a height $h = 2.34$ m. If the collision is imperfectly elastic, find (a) the velocity of the mass as it hits the floor, v_{1i}; (b) the velocity of the mass after it rebounds from the floor, v_{1i}; (c) the coefficient of restitution, e; (d) the kinetic energy, KE_A, just as the mass approached the floor; (e) the kinetic energy, KE_S, after the separation of the mass from the floor; (f) the actual energy lost in the collision; (g) the percentage of energy lost in the collision; (h) the momentum before the collision; and (i) the momentum after the collision.

69. An imperfectly elastic collision—the bouncing ball. A ball of mass, $m = 1.53$ kg, is dropped from a height $h_0 = 1.50$ m and hits a wooden floor. The collision with the floor is imperfectly elastic and the ball only rebounds to a height $h = 1.12$ m for the first bounce. Find (a) the initial velocity of the ball, v_i, as it hits the floor on its first bounce; (b) the velocity of the ball v_f, after it rebounds from the floor on its first bounce; (c) the coefficient of restitution, e; (d) the initial kinetic energy, KE_i, just as the ball approaches the floor; (e) the final kinetic energy, KE_f, of the ball after the bounce from the floor; (f) the actual energy lost in the bounce, $E_{lost}/bounce$; and (g) the percentage of the initial kinetic energy lost by the ball in the bounce, $\%E_{lost}/bounce$. The ball continues to bounce until it loses all its energy. (h) Find the cumulative total percentage energy lost, % Energy lost, for all the bounces. (i) Plot a graph of the % of Total Energy lost as a function of the number of bounces.

70. A variable mass system. A train car of mass $m_T = 1500$ kg, contains 35 rocks each of mass $m_r = 30$ kg. The train is initially at rest. A man throws out each rock from the rear of the train at a speed $v_r = 8.50$ m/s. (a) When the man throws out one rock, what will the recoil velocity, V_T, of the train be? (b) What is the recoil velocity when the man throws out the second rock? (c) What is the recoil velocity of the train when the nth rock is thrown out? (d) If the man throws out each rock at the rate $R = 1.5$ rocks/s, find the change in the velocity of the train and its acceleration. (e) Draw a graph of the velocity of the train as a function of the number of rocks thrown out of the train. (f) Draw a graph of the acceleration of the train as a function of time.

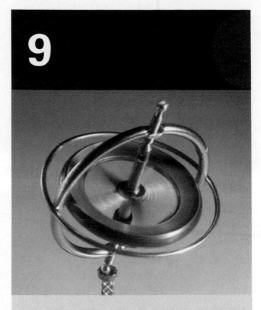

Rotational Motion

Conservation of angular momentum in a gyroscope.

In experimental philosophy we are to look upon propositions inferred by general induction from phenomena as accurately or very nearly true, notwithstanding any contrary hypotheses that may be imagined, till such time as other phenomena occur, by which they may either be made more accurate, or liable to exceptions.

Isaac Newton

9.1 Introduction

Up to now, the main emphasis in the description of the motion of a body dealt with the translational motion of that body. But in addition to translating, a body can also rotate about some axis, called the *axis of rotation.* Therefore, for a complete description of the motion of a body we also need to describe any rotational motion the body might have. As a matter of fact, the most general motion of a rigid body is composed of the translation of the center of mass of the body plus its rotation about the center of mass.

In the analysis of rotational motion, we will see a great similarity to translational motion. In fact, this chapter can serve as a review of all the mechanics discussed so far.

9.2 Rotational Kinematics

In the study of translational kinematics the first concept we defined was the position of an object. The position of the body was defined as the displacement x from a reference point. In a similar way, the position of a point on a rotating body is defined by the **angular displacement** θ from some reference line that connects the point to the axis of rotation, as shown in figure 9.1. That is, if the point was originally at P, and a little later it is at the point P', then the body has rotated through the angular displacement θ. The angular displacement can be represented as a vector that is perpendicular to the plane of the motion. If the angular displacement is a positive quantity, the rotation of the body is counterclockwise and the angular displacement vector points upward. If the angular displacement is a negative quantity, the rotation of the body is clockwise and the angular displacement vector points downward. The magnitude of the angular displacement is the angle θ itself. We measure the angle θ in radians, which we defined in section 6.3. The linear distance between the points P and P' is given by the arc length s, and is related to the angular displacement by

$$s = r\theta \tag{6.5}$$

The average velocity of a translating body was defined as the displacement of the body divided by the time for that displacement:

$$v_{avg} = \frac{\Delta x}{\Delta t}$$

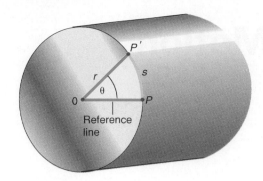

Figure 9.1

The angular displacement.

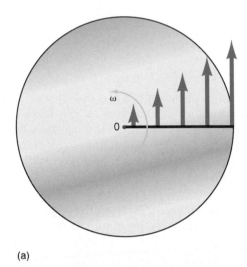

(a)

(b)

Figure 9.2

The linear velocity varies with the distance r from the center of the rotating body.

In the same way, *the average **angular velocity** of a rotating body is defined as the angular displacement of the body about the axis of rotation divided by the time for that displacement:*

$$\omega_{avg} = \frac{\Delta\theta}{\Delta t} \tag{9.1}$$

The units for angular velocity are radians/second, abbreviated as rad/s. It is important to remember that the radian is a dimensionless quantity, and can be added or deleted from an equation whenever it is convenient. The angular velocity, like the angular displacement, can also be represented as a vector quantity that is also perpendicular to the plane of the motion. It is positive and points upward for counterclockwise rotations and is negative and points downward for clockwise rotations.

The similarities between translational and rotational motion can be seen in table 9.1. The relation between the linear velocity of a point on the rotating body and the angular velocity of the body is found by dividing both sides of equation 6.5 by t, that is,

$$\frac{s}{t} = \frac{r\theta}{t}$$

but $s/t = v$ and $\theta/t = \omega$. Therefore,

$$v = r\omega \tag{9.2}$$

Equation 9.2 says that for a body rotating at an angular velocity ω, the farther the distance r that the body is from the axis of rotation, the greater is its linear velocity. This can be seen in figure 9.2(a). You may recall when you were a child and went on the merry-go-round, you usually wanted to ride on the outside horses because they moved the fastest. You can now see why. They are at the greatest distance from the axis of rotation and hence have the greatest linear velocity. The linear velocity of a point on the rotating body can also be called the *tangential velocity* because the point is moving along the tangential direction at any instant.

Table 9.1

Comparison of Translational and Rotational Motion

Translational Motion	Rotational Motion
$v_{avg} = \dfrac{\Delta x}{\Delta t} = \dfrac{x}{t}$	$\omega_{avg} = \dfrac{\Delta\theta}{\Delta t} = \dfrac{\theta}{t}$
$a = \dfrac{\Delta v}{\Delta t} = \dfrac{v - v_0}{t}$	$\alpha = \dfrac{\Delta\omega}{\Delta t} = \dfrac{\omega - \omega_0}{t}$
$v = v_0 + at$	$\omega = \omega_0 + \alpha t$
$x = v_0 t + \frac{1}{2}at^2$	$\theta = \omega_0 t + \frac{1}{2}\alpha t^2$
$v^2 = v_0^2 + 2ax$	$\omega^2 = \omega_0^2 + 2\alpha\theta$
$KE = \frac{1}{2}mv^2$	$KE = \frac{1}{2}I\omega^2$
$F = ma$	$\tau = I\alpha$
$p = mv$	$L = I\omega$
$F = \dfrac{\Delta p}{\Delta t}$	$\tau = \dfrac{\Delta L}{\Delta t}$
$p_f = p_i$	$L_f = L_i$

Mechanics

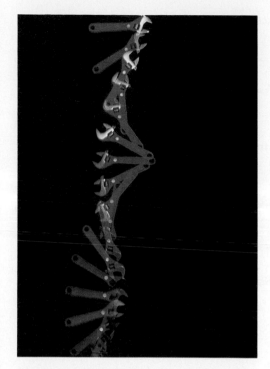

The most general motion of a rigid body is composed of the translation of the center of mass of the body plus its rotation about the center of mass.

Another example of the relation between the tangential velocity and the angular velocity is seen in the old fashioned "whip" that you formed by holding hands while you were ice skating or roller skating, figure 9.2(b). The person at the inside end of the "whip" hardly moved at all ($r = 0$), but the person at the far end of the whip (maximum r) moved at very high speeds.

Example 9.1

The angular velocity of a wheel. A wheel of radius 15.0 cm starts from rest and turns through 2.00 rev in 3.00 s. (a) What is its average angular velocity? (b) What is the tangential velocity of a point on the rim of the wheel?

Solution

a. The average angular velocity, found from equation 9.1, is

$$\omega_{\text{avg}} = \frac{\theta}{t}$$

$$= \frac{(2.00 \text{ rev})(2\pi \text{ rad})}{3.00 \text{ s} (1 \text{ rev})}$$

$$= 4.19 \text{ rad/s}$$

Note that we accomplished the conversion from revolutions to radians using the identity that one revolution is equal to 2π radians.

b. The tangential velocity of a point on the rim of the wheel, found from equation 9.2, is

$$v = r\omega$$

$$= (0.150 \text{ m})(4.19 \text{ rad/s})$$

$$= 0.628 \text{ m/s}$$

In the study of kinematics we defined the average translational acceleration of a body in equations 3.7 and 3.9 as the change in the velocity of the body with time, that is

$$a = \frac{\Delta v}{\Delta t} = \frac{v - v_0}{t}$$

Because we considered only problems where motion was at constant acceleration, the average acceleration was equal to the instantaneous acceleration. In the same way, *we now define the average **angular acceleration** α of the rotating body as the change in the angular velocity of the body with time,* that is,

$$\alpha = \frac{\Delta \omega}{\Delta t} = \frac{\omega - \omega_0}{t} \tag{9.3}$$

Again, since the only problems that we will consider concern motion at constant angular acceleration, the average angular acceleration is equal to the angular acceleration at any instant of time. We should note that angular acceleration, like angular velocity, can also be represented as a vector that lies along the axis of rotation of the rotating body. If the angular velocity vector is increasing with time, α is positive, and points upward from the plane of the rotation. If the angular velocity is decreasing with time, α is negative, and points downward into the plane of the rotation. The units for angular acceleration are radians/second per second, abbreviated as rad/s².

From the definition of the acceleration, equation 3.9, the first of the kinematic equations became

$$v = v_0 + at \tag{3.10}$$

the velocity of the moving body at any instant of time. Similarly, if equation 9.3 is solved for ω, we have

$$\omega = \omega_0 + \alpha t \tag{9.4}$$

the first of the **kinematical equations for rotational motion.** *Equation 9.4 gives the angular velocity of the rotating body at any instant of time for a constant acceleration, α.*

Example 9.2

The angular acceleration of a cylinder. A cylinder rotating at 10.0 rad/s is accelerated to 50.0 rad/s in 10.0 s. What is the angular acceleration of the cylinder?

Solution

The angular acceleration, found from equation 9.4, is

$$\alpha = \frac{\omega - \omega_0}{t} = \frac{50.0 \text{ rad/s} - 10.0 \text{ rad/s}}{10.0 \text{ s}}$$

$$= 4.00 \text{ rad/s}^2$$

Example 9.3

The angular velocity of a crankshaft. A crankshaft rotating at 10.0 rad/s undergoes an angular acceleration of 0.500 rad/s². What is the angular velocity of the shaft after 10.0 s?

Solution

The angular velocity, found from equation 9.4, is

$$\omega = \omega_0 + \alpha t$$

$$= 10.0 \text{ rad/s} + (0.500 \text{ rad/s}^2)(10.0 \text{ s})$$

$$= 15.0 \text{ rad/s}$$

The relationship between the magnitude of the tangential acceleration of a point on the rim of the rotating body and the angular acceleration is found by dividing both sides of equation 9.2 by t, that is,

$$v = r\omega \tag{9.2}$$

$$\frac{v}{t} = \frac{r\omega}{t}$$

but $v/t = a$ and $\omega/t = \alpha$, therefore,

$$a = r\alpha \tag{9.5}$$

Equation 9.5 gives the relationship between the magnitude of the tangential acceleration and the angular acceleration.

The next kinematic derivation was the equation giving the position of the moving body as a function of time. Recall that the average velocity was substituted in the equation $x = v_{avg}t$ to yield the kinematic equation for the position of the moving body as a function of time as

$$x = v_0 t + \tfrac{1}{2}at^2$$

Similarly, to find the angular displacement of a rotating body at any instant of time, we use equation 9.1 in the form

$$\theta = \omega_{avg}t \tag{9.6}$$

But for a body rotating at constant angular acceleration, the average angular velocity is

$$\omega_{avg} = \frac{\omega + \omega_0}{2} \tag{9.7}$$

where ω_0 is the initial angular velocity and ω is the final angular velocity at some time t. Substituting 9.7 into 9.6 gives

$$\theta = \left(\frac{\omega + \omega_0}{2}\right)t \tag{9.8}$$

Substituting equation 9.4 for the angular velocity ω into equation 9.8 gives

$$\theta = \left[\frac{(\omega_0 + \alpha t) + \omega_0}{2}\right]t$$

Rearranging terms we get

$$\theta = \omega_0 t + \tfrac{1}{2}\alpha t^2 \tag{9.9}$$

Equation 9.9 gives the angular displacement of the rotating body as a function of time for constant angular acceleration. This is the second kinematic equation for rotational motion.

Example 9.4

The angular displacement of a wheel. A wheel rotating at 15.0 rad/s undergoes an angular acceleration of 10.0 rad/s². Through what angle has the wheel turned when $t = 5.00$ s?

Solution

The angular displacement, found from equation 9.9, is

$$\theta = \omega_0 t + \tfrac{1}{2}\alpha t^2$$
$$= (15.0 \text{ rad/s})(5.00 \text{ s}) + \tfrac{1}{2}(10.0 \text{ rad/s}^2)(5.00 \text{ s})^2$$
$$= 200 \text{ rad}$$

We obtained the third translational kinematic equation,

$$v^2 = v_0^2 + 2ax \tag{3.16}$$

from the first two translational kinematic equations by eliminating the time t between them. We can find a similar equation for the angular velocity as a function of the angular displacement by eliminating the time between equations 9.4 and 9.9 and we suggest that the student do this as an exercise. We will obtain the third kinematic equation for rotational motion in a slightly different manner, however. Let us start with

$$v^2 = v_0^2 + 2ax \tag{3.16}$$

But we know that a relationship exists between the translational variables and the rotational variables. Those relationships are

$$s = r\theta \tag{6.5}$$

$$v = r\omega \tag{9.2}$$

$$a = r\alpha \tag{9.5}$$

For the rotating body, we replace the linear distance x by the distance s along the arc of the circle. If we substitute the above equations into equation 3.16, we get

$$v^2 = v_0^2 + 2as$$
$$(r\omega)^2 = (r\omega_0)^2 + 2(r\alpha)(r\theta)$$
$$r^2\omega^2 = r^2\omega_0^2 + 2r^2\alpha\theta$$

Dividing each term by r^2, we obtain

$$\omega^2 = \omega_0^2 + 2\alpha\theta \tag{9.10}$$

Equation 9.10 represents the angular velocity of the rotating body at any angular displacement θ for constant angular acceleration α.

Example 9.5

The angular velocity at a particular angular displacement. A wheel, initially rotating at 10.0 rad/s, undergoes an angular acceleration of 5.00 rad/s². What is the angular velocity when the wheel has turned through an angle of 50.0 rad?

Solution

The angular velocity, found from equation 9.10, is

$$\omega^2 = \omega_0^2 + 2\alpha\theta$$
$$= (10.0 \text{ rad/s})^2 + 2(5.00 \text{ rad/s}^2)(50.0 \text{ rad})$$
$$= 100 \text{ rad}^2/\text{s}^2 + 500 \text{ rad}^2/\text{s}^2 = 600 \text{ rad}^2/\text{s}^2$$
$$\omega = 24.5 \text{ rad/s}$$

Note in table 9.1 the similarity in the translational and rotational equations. Everywhere there is an x in the translational equations, there is a θ in the rotational equations. Everywhere there is a v in the translational equations, there is an ω in the rotational equations. And finally, everywhere there is an a in the translational equations, there is an α in the rotational equations. We will see additional analogues as we proceed in the discussion of rotational motion.

Another way to express the magnitude of the centripetal acceleration discussed in chapter 6,

$$a_c = \frac{v^2}{r} \tag{6.12}$$

is to use

$$v = r\omega \tag{9.2}$$

to obtain

$$a_c = \frac{\omega^2 r^2}{r}$$

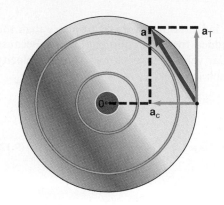

Figure 9.3

The total acceleration of a point on a rotating body is equal to the vector sum of the tangential acceleration and the centripetal acceleration.

Hence, we can represent the magnitude of the centripetal acceleration in terms of the angular velocity as

$$a_c = \omega^2 r \qquad (9.11)$$

For nonuniform circular motion, the resultant acceleration of a point on a rim of a rotating body becomes the vector sum of the tangential acceleration and the centripetal acceleration, as seen in figure 9.3.

Example 9.6

The total acceleration of a point on a rotating body. A cylinder 35.0 cm in diameter is at rest initially. It is then given an angular acceleration of 0.0400 rad/s². Find (a) the angular velocity at 7.00 s, (b) the centripetal acceleration of a point at the edge of the cylinder at 7.00 s, (c) the tangential acceleration at the edge of the cylinder at 7.00 s, and (d) the resultant acceleration of a point at the edge of the cylinder at 7.00 s.

Solution

a. The angular velocity at 7.00 s, found from equation 9.4, is

$$\omega = \omega_0 + \alpha t$$
$$= 0 + (0.0400 \text{ rad/s}^2)(7.00 \text{ s})$$
$$= 0.280 \text{ rad/s}$$

b. The centripetal acceleration, found from equation 9.11, is

$$a_c = \omega^2 r$$
$$= (0.280 \text{ rad/s})^2(17.5 \text{ cm})$$
$$= 1.37 \text{ cm/s}^2$$

c. The tangential acceleration, found from equation 9.5, is

$$a_T = r\alpha = (17.5 \text{ cm})(0.0400 \text{ rad/s}^2)$$
$$= 0.700 \text{ cm/s}^2$$

d. The resultant acceleration at 7.00 s, found from figure 9.3, is

$$a = \sqrt{(a_c)^2 + (a_T)^2}$$
$$= \sqrt{(1.37 \text{ cm/s}^2)^2 + (0.700 \text{ cm/s}^2)^2}$$
$$= 1.54 \text{ cm/s}^2$$

■

9.3 The Kinetic Energy of Rotation

Let us consider the motion of four point masses m_1, m_2, m_3, and m_4 located at distances r_1, r_2, r_3, and r_4, respectively, rotating at an angular speed ω about an axis through the point 0, as shown in figure 9.4.

Let us assume that the masses are connected to the center of rotation by rigid, massless rods. (A massless rod is one whose mass is so small compared to the mass at the end of the rod that we can neglect it in the analysis.) Let us determine the total kinetic energy of these rotating masses. The total energy is equal to the sum of the kinetic energy of each mass. That is,

$$KE_{total} = KE_1 + KE_2 + KE_3 + KE_4 + \cdots$$

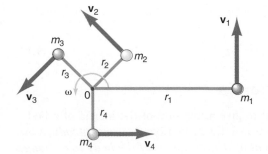

Figure 9.4

Rotational kinetic energy.

The plus sign and dots after the last term indicate that if there are more than the four masses considered, another term is added for each additional mass. Because each mass is rotating with the same angular velocity ω, each has a linear velocity v, as shown. Since the kinetic energy of each mass is

$$KE = \tfrac{1}{2}mv^2$$

the total kinetic energy is

$$KE_{tot} = \tfrac{1}{2}m_1v_1^2 + \tfrac{1}{2}m_2v_2^2 + \tfrac{1}{2}m_3v_3^2 + \tfrac{1}{2}m_4v_4^2 + \cdots \qquad (9.12)$$

but from equation 9.2,

$$v = r\omega$$

hence, for each mass

$$\left.\begin{aligned}
v_1 &= r_1\omega \\
v_2 &= r_2\omega \\
v_3 &= r_3\omega \\
v_4 &= r_4\omega
\end{aligned}\right\} \qquad (9.13)$$

Substituting equations 9.13 back into equation 9.12, gives

$$KE_{tot} = \tfrac{1}{2}m_1(r_1\omega)^2 + \tfrac{1}{2}m_2(r_2\omega)^2 + \tfrac{1}{2}m_3(r_3\omega)^2 + \tfrac{1}{2}m_4(r_4\omega)^2 + \cdots$$
$$= \tfrac{1}{2}m_1r_1^2\omega^2 + \tfrac{1}{2}m_2r_2^2\omega^2 + \tfrac{1}{2}m_3r_3^2\omega^2 + \tfrac{1}{2}m_4r_4^2\omega^2 + \cdots$$

Note that there is a 1/2 and an ω^2 in every term, so let us factor them out:

$$KE_{tot} = \tfrac{1}{2}(m_1r_1^2 + m_2r_2^2 + m_3r_3^2 + m_4r_4^2 + \cdots)\omega^2$$

Looking at the form of the equation for the translational kinetic energy ($\tfrac{1}{2}mv^2$), and remembering all the symmetry in the translational-rotational equations, it is reasonable to expect that the equation for the rotational kinetic energy might have an analogous form. That symmetry is maintained by defining the term in parentheses as the moment of inertia, the rotational analogue of the mass m. That is, the moment of inertia about the axis of rotation for these four masses is

$$I = m_1r_1^2 + m_2r_2^2 + m_3r_3^2 + m_4r_4^2 \qquad (9.14)$$

We will discuss the concept of the moment of inertia in more detail in section 9.4. For now, we see that the equation for the total energy of the four rotating masses is

$$KE_{tot} = \tfrac{1}{2}I\omega^2$$

And finally let us note that the total kinetic energy of the rotating masses can simply be called the **kinetic energy of rotation.** Therefore, the kinetic energy of rotation about a specified axis is

$$KE_{rot} = \tfrac{1}{2}I\omega^2 \qquad (9.15)$$

9.4 The Moment of Inertia

The concept of mass m was introduced to give a measure of the inertia of a body, that is, its resistance to a change in its translational motion. *Now we introduce the moment of inertia to give a measurement of the resistance of the body to a change in its rotational motion.* For example, the larger the moment of inertia of a body, the more difficult it is to put that body into rotational motion. Conversely, the larger the moment of inertia of a body, the more difficult it is to stop its rotational motion.

For the particular configuration studied in section 9.3, the moment of inertia about the axis of rotation was defined as

$$I = m_1 r_1^2 + m_2 r_2^2 + m_3 r_3^2 + m_4 r_4^2 \qquad (9.14)$$

For any number of masses, this definition can be generalized to

$$I = \sum_{i=1}^{n} m_i r_i^2 \qquad (9.16)$$

where the Greek letter sigma, Σ, means the "sum of," as used before. The subscript i on m and r means that the index i takes on the values from 1 up to the number n. So when $n = 4$ the identical result found in equation 9.14 is obtained.

For the very special case of *the moment of inertia of a single mass m rotating about an axis,* equation 9.16 reduces to ($i = n = 1$),

$$I = mr^2 \qquad (9.17)$$

Thus, the significant feature for rotational motion is not the mass of the rotating body, but rather the square of the distance of that body from the axis of rotation. A small mass m, at a great distance r from the axis of rotation, has a greater moment of inertia than a large mass, very close to the axis of rotation.

For continuous mass distributions, the moments of inertia are given in figure 9.5. More extensive tables of moments of inertia are found in various handbooks, such as the *Handbook of Chemistry and Physics* (published by the Chemical Rubber Co. Press, Cleveland, Ohio), if the need for them arises.

It is important to note here that when we ask for the moment of inertia of a body, we must specify about what axis the rotation will occur. Because r is different for each axis, and since I varies as r^2, I is also different for each axis. As an example,

Figure 9.5

Moments of inertia for various mass distributions.

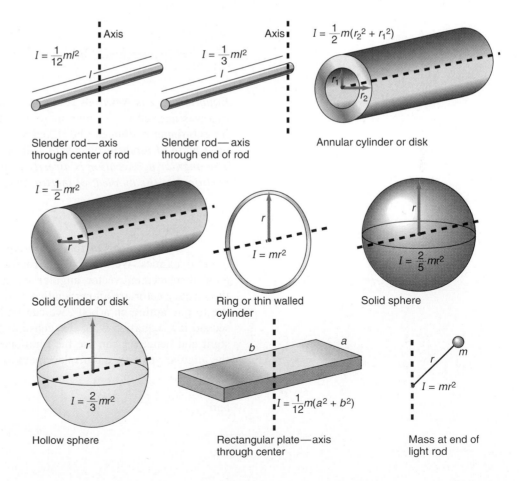

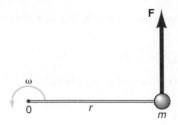

Figure 9.6

Torque causes a body to rotate.

consider the slender rod in figure 9.5. When the axis is taken through the center of the rod, as shown, $I = \frac{1}{12} ml^2$, while if the axis of rotation is at the end of the rod, then $I = \frac{1}{3} ml^2$. The unit for the moment of inertia is kg m² and has no special name.

9.5 Newton's Laws for Rotational Motion

Let us consider a single mass m connected by a rigid rod, of negligible mass, to an axis passing through the point 0, as shown in figure 9.6. Let us apply a tangential force F, in the plane of the page, to the body of mass m. The force acting on the constrained body causes a torque, given by

$$\tau = rF \tag{9.18}$$

This torque causes the body to rotate about the axis through 0. The force F acting on the mass m causes a tangential acceleration given by Newton's second law as

$$F = ma \tag{9.19}$$

If we substitute equation 9.19 into 9.18, we have

$$\tau = rma \tag{9.20}$$

But the tangential acceleration a is related to the angular acceleration by

$$a = r\alpha \tag{9.5}$$

Substituting this into equation 9.20, gives

$$\tau = rm(r\alpha) = mr^2\alpha \tag{9.21}$$

But, as already seen, the moment of inertia of a single mass rotating about an axis is

$$I = mr^2 \tag{9.17}$$

Therefore, equation 9.21 becomes

$$\tau = I\alpha \tag{9.22}$$

Equation 9.22 is Newton's second law for rotational motion. Although this equation was derived for a single mass, it is true in general, and **Newton's second law for rotational motion** can be stated as: *When an unbalanced external torque acts on a body of moment of inertia I, it gives that body an angular acceleration, α. The angular acceleration is directly proportional to the torque and inversely proportional to the moment of inertia,* that is,

$$\alpha = \frac{\tau}{I} \tag{9.23}$$

The problems of rotational dynamics are very similar to those in translational dynamics. We will consider rotational motion only in the *x-y* plane. The angular displacement vector, angular velocity vector, angular acceleration vector, and the torque vector are all perpendicular to the plane of the rotation. By determining the torque acting on a body, we can find the angular acceleration from Newton's second law, equation 9.23. For constant torque, the angular acceleration is a constant and hence we can use the rotational kinematic equations. Therefore, we find the angular velocity and displacement at any time from the kinematic equations

$$\omega = \omega_0 + \alpha t \tag{9.4}$$

and

$$\theta = \omega_0 t + \frac{1}{2}\alpha t^2 \tag{9.9}$$

To determine Newton's first law for rotational motion, we note that

$$\tau = I\alpha = I\frac{\Delta\omega}{\Delta t}$$

and if there is no external torque (i.e., if $\tau = 0$), then

$$\Delta\omega = 0$$

$$\omega_f - \omega_i = 0$$

or

$$\omega_f = \omega_i \qquad\qquad (9.24)$$

That is, equation 9.24 says that if there is no external torque acting on a body, then a body rotating at an initial angular velocity ω_i will continue to rotate at that same angular velocity forever.

Stated in more formal terms, *Newton's first law for rotational motion* is *A body in motion at a constant angular velocity will continue in motion at that same angular velocity, unless acted on by some unbalanced external torque.*

One of the most obvious examples of Newton's first law for rotational motion is the earth itself. Somehow, someway in its creation, the earth was given an initial angular velocity ω_i of 7.27×10^{-5} rad/s. Since there is no external torque acting on the earth it continues to rotate at this same angular velocity.

For completeness, we can state **Newton's third law of rotational motion** as *If body A and body B have the same common axis of rotation, and if body A exerts a torque on body B, then body B exerts an equal but opposite torque on body A.* That is, if body A exerts a torque on body B that tends to rotate body B in a clockwise direction, then body B will exert a torque on body A that will tend to rotate body A in a counterclockwise direction. An application of this principle is found in a helicopter (see figure 9.7). As the main rotor blades above the helicopter turn counterclockwise, the helicopter itself would start to turn clockwise. To prevent this rotation of the helicopter, a second but smaller set of rotor blades are located at the side and end of the helicopter to furnish a countertorque to prevent the helicopter from turning.

9.6 Rotational Dynamics

Now let us look at some examples of the use of Newton's laws in solving problems in rotational motion.

Figure 9.7

Newton's third law for rotational motion and the helicopter.

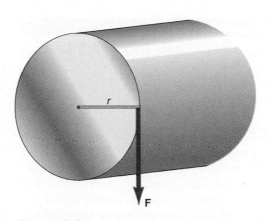

Figure 9.8

Rotational motion of a cylinder.

Example 9.7

Rotational dynamics of a cylinder. Consider a solid cylinder of mass $m = 3.00$ kg and radius $r = 0.500$ m, which is free to rotate about an axis through its center, as shown in figure 9.8. The cylinder is initially at rest when a constant force of 8.00 N is applied tangentially to the cylinder. Find (a) the moment of inertia of the cylinder, (b) the torque acting on the cylinder, (c) the angular acceleration of the cylinder, (d) its angular velocity after 10.0 s, and (e) its angular displacement after 10.0 s.

Solution

a. The equation for the moment of inertia of a cylinder about its main axis, found in figure 9.5, is

$$I = \tfrac{1}{2}mr^2$$

$$= \tfrac{1}{2}(3.00 \text{ kg})(0.500 \text{ m})^2$$

$$= 0.375 \text{ kg m}^2$$

b. The torque acting on the cylinder is the product of the force times the lever arm. From figure 9.8, we see that the lever arm is just the radius of the cylinder. Therefore,

$$\tau = rF \tag{9.18}$$

$$= (0.500 \text{ m})(8.00 \text{ N})$$

$$= 4.00 \text{ m N}$$

c. The angular acceleration of the cylinder, determined by Newton's second law, is

$$\alpha = \frac{\tau}{I} \tag{9.23}$$

$$= \frac{4.00 \text{ m N}}{0.375 \text{ kg m}^2} = \frac{10.7 \text{ m kg m/s}^2}{\text{kg m}^2}$$

$$= 10.7 \text{ rad/s}^2$$

Note that in the solution all the units cancel out except the s² in the denominator. We then introduced the unit radian in the numerator to give us the desired unit for angular acceleration, namely, rad/s². Recall that the radian is a unit that can be multiplied by or divided into an equation at will, because the radian is a dimensionless quantity. It was defined as the ratio of the arc length to the radius of the circle,

$$\theta = \frac{s}{r} = \frac{\text{meter}}{\text{meter}} = 1 = \text{radian}$$

d. To determine the angular velocity of the rotating cylinder we use the kinematic equation for the angular velocity, namely

$$\omega = \omega_0 + \alpha t \tag{9.4}$$

$$= 0 + \left(10.7 \frac{\text{rad}}{\text{s}^2}\right)(10.0 \text{ s})$$

$$= 107 \text{ rad/s}$$

e. The angular displacement, found by the kinematic equation, is

$$\theta = \omega_0 t + \tfrac{1}{2}\alpha t^2 \tag{9.9}$$

$$= 0 + \tfrac{1}{2}(10.7 \text{ rad/s}^2)(10.0 \text{ s})^2$$

$$= 535 \text{ rad}$$

Example 9.8

Combined translational and rotational motion of a sphere rolling down an inclined plane. A solid sphere of 1.00 kg mass rolls down an inclined plane, as shown in figure 9.9. Find (a) the acceleration of the sphere, (b) its velocity at the bottom of the 1.00 m long plane, and (c) the frictional force acting on the sphere.

Solution

a. First, we draw all the forces acting on the sphere. The component of the weight acting down the plane, $w \sin \theta$, is shown acting through the center of mass of the sphere. Because of this force there is a tendency for the sphere to slide down the plane. A force of static friction opposes this motion and is directed up the plane, as shown. This frictional force can not be shown as acting at the center of the body as was done in problems with "blocks" sliding on the inclined plane. It is this frictional force acting at the point of

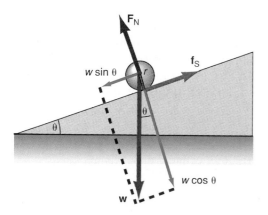

Figure 9.9

A sphere rolling down an inclined plane.

Mechanics

contact of the sphere that creates the necessary torque to rotate the sphere so that it rolls down the plane. The motion is therefore composed of two motions, the translation of the center of mass of the sphere, and the rotation about the center of mass of the sphere. Applying Newton's second law for the translational motion of the center of mass of the sphere gives

$$F = ma$$

$$w \sin \theta - f_s = ma \tag{9.25}$$

Applying the second law for the rotation of the sphere about its center of mass gives

$$\tau = I\alpha \tag{9.22}$$

But the torque is the product of the frictional force f_s and the radius of the sphere. Therefore,

$$f_s r = I\alpha \tag{9.26}$$

Now we eliminate the frictional force f_s between the two equations 9.25 and 9.26. That is, from 9.26,

$$f_s = \frac{I\alpha}{r}$$

Substituting this into equation 9.25, we get

$$w \sin \theta - \frac{I\alpha}{r} = ma$$

The moment of inertia of a solid sphere, found from figure 9.5, is

$$I = \tfrac{2}{5}mr^2 \tag{9.27}$$

Therefore,

$$ma = w \sin \theta - (\tfrac{2}{5}mr^2)\frac{\alpha}{r}$$

$$= w \sin \theta - \tfrac{2}{5}mr\alpha$$

But recall that

$$a = r\alpha \tag{9.5}$$

Therefore,

$$ma = w \sin \theta - \tfrac{2}{5}ma$$

$$ma + \tfrac{2}{5}ma = mg \sin \theta$$

$$\tfrac{7}{5}a = g \sin \theta$$

Solving for the acceleration of the sphere, we get

$$a = \tfrac{5}{7}g \sin \theta \tag{9.28}$$

$$a = \tfrac{5}{7}(9.80 \text{ m/s}^2)\sin 30.0°$$

$$= 3.50 \text{ m/s}^2$$

b. The velocity of the center of mass of the sphere at the bottom of the plane is found from the kinematic equation

$$v^2 = v_0^2 + 2ax \qquad (3.16)$$

Because the sphere starts from rest, $v_0 = 0$. Therefore,

$$v = \sqrt{2ax}$$
$$= \sqrt{(2)(3.50 \text{ m/s}^2)(1.00 \text{ m})}$$
$$= 2.65 \text{ m/s}$$

c. The frictional force can be determined from equation 9.25, that is,

$$w \sin \theta - f_s = ma$$
$$f_s = w \sin \theta - ma$$
$$= mg \sin \theta - m(\tfrac{5}{7}g \sin \theta)$$
$$= \tfrac{2}{7}mg \sin \theta$$
$$= \tfrac{2}{7}(1.00 \text{ kg})(9.80 \text{ m/s}^2)\sin 30.0°$$
$$= 1.40 \text{ N}$$

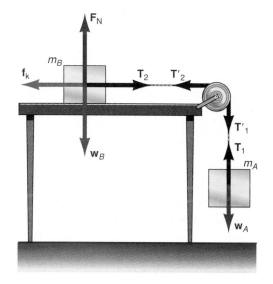

Figure 9.10

Combined motion taking the rotational motion of the pulley into account.

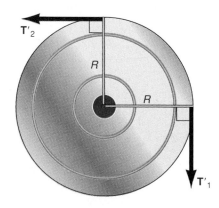

Figure 9.11

Forces acting on the pulley.

As we can see the general motion of a rigid body can become quite complicated. We will see in section 9.8 how these problems can be simplified by the use of the law of conservation of energy.

†Combined Translational and Rotational Motion Treated by Newton's Second Law

It is appropriate here to return to some of the problems discussed in chapter 4, in which we assumed that the tension in the rope on both sides of a pulley are equal. Let us analyze these problems taking the rotational motion of the pulley into account. Consider the problem of a block moving on a rough horizontal surface, as shown in figure 9.10. What is the acceleration of each block in the system?

Applying Newton's second law to block A, we obtain

$$T_1 - w_A = -m_A a \qquad (9.29)$$

Applying the second law to block B, we get

$$T_2 - f_k = m_B a \qquad (9.30)$$

We find the frictional force f_k from

$$f_k = \mu_k F_N = \mu_k w_B$$

Substituting this into equation 9.30, gives

$$T_2 - \mu_k w_B = m_B a \qquad (9.31)$$

It was at this point in chapter 4 that we made the assumption that the tension $T_1 = T_2$, and then determined the acceleration of each block of the system. Let us now look a little more closely at the assumption of the equality of tensions. The string exerts a force T_1 upward on weight w_A, but by Newton's third law the weight w_A exerts a force down on the string, call it T_1'. Figure 9.11 shows the pulley with the appropriate tensions in the string. The force T_1' acting on the pulley causes a torque

$$\tau_1 = T_1'R$$

which tends to rotate the pulley clockwise. The radius of the pulley is R.

Similarly, the string exerts a tension force T_2 on mass m_B. But by Newton's third law, block B exerts a force on the string, which we call T_2'. The force T_2' causes a counterclockwise torque about the axis of the pulley, given by

$$\tau_2 = T_2'R$$

Because the motion of the system causes the pulley to rotate in a clockwise direction, the net torque on the pulley is equal to the difference in these two torques, namely,

$$\tau = \tau_1 - \tau_2$$

$$\tau = T_1'R - T_2'R \tag{9.32}$$

But by Newton's second law for rotational motion,

$$\tau = I\alpha \tag{9.22}$$

Substituting equation 9.22 into equation 9.32, gives

$$I\alpha = T_1'R - T_2'R$$

$$I\alpha = (T_1' - T_2')R \tag{9.33}$$

From figure 9.10 and Newton's third law, we have

$$\left. \begin{array}{l} T_1' = T_1 \\ T_2' = T_2 \end{array} \right\} \tag{9.34}$$

Substituting equations 9.34 into equation 9.33, gives

$$I\alpha = (T_1 - T_2)R \tag{9.35}$$

However, the angular acceleration $\alpha = a/R$. Therefore, equation 9.35 becomes

$$I\frac{a}{R} = (T_1 - T_2)R \tag{9.36}$$

The pulley resembles a disk, whose moment of inertia, found from figure 9.5, is $I_{\text{Disk}} = \frac{1}{2}MR^2$, where M is the mass of the pulley and R is the radius of the pulley. Substituting this result into equation 9.36, gives

$$(\tfrac{1}{2}MR^2)\frac{a}{R} = (T_1 - T_2)R$$

Simplifying,

$$\tfrac{1}{2}Ma = (T_1 - T_2) \tag{9.37}$$

There are now three equations 9.29, 9.31, and 9.37 in terms of the three unknowns a, T_1, and T_2. Solving equation 9.29 for T_1, gives

$$T_1 = w_A - m_A a \tag{9.38}$$

Solving equation 9.31 for T_2, gives

$$T_2 = \mu_k w_B + m_B a \tag{9.39}$$

Subtracting equation 9.39 from equation 9.38, we get

$$T_1 - T_2 = w_A - m_A a - \mu_k w_B - m_B a$$

Substituting for $T_1 - T_2$ from equation 9.37, gives

$$\tfrac{1}{2}Ma = w_A - m_A a - \mu_k w_B - m_B a$$

Gathering the terms with a in them to the left-hand side of the equation, we get

$$\tfrac{1}{2}Ma + m_A a + m_B a = w_A - \mu_k w_B$$

Factoring out the a, and writing each weight w as mg, we get

$$a(\tfrac{1}{2}M + m_A + m_B) = m_A g - \mu_k m_B g$$

Solving for the acceleration of the system, we have

$$a = \frac{(m_A - \mu_k m_B)g}{m_A + m_B + M/2} \tag{9.40}$$

It is immediately apparent in equation 9.40 that the acceleration of the system depends on the mass M of the pulley. If this mass is very small compared to the masses m_A and m_B (i.e., $M \approx 0$), then equation 9.40 would reduce to the simpler problem already found in equation 4.62.

Example 9.9

Combined translational and rotational motion. If $m_A = 2.00$ kg, $m_B = 6.00$ kg, $\mu_k = 0.300$, and $M = 8.00$ kg in figure 9.10, find the acceleration of each block of the system.

Solution

The acceleration of each block in the system, found from equation 9.40, is

$$a = \frac{(m_A - \mu_k m_B)g}{m_A + m_B + M/2}$$

$$= \frac{[2.00 \text{ kg} - (0.300)(6.00 \text{ kg})](9.80 \text{ m/s}^2)}{2.00 \text{ kg} + 6.00 \text{ kg} + 8.00 \text{ kg}/2}$$

$$= 0.163 \text{ m/s}^2$$

If we compare this example with example 4.12 in chapter 4, we see a relatively large difference in the acceleration of the system by assuming M to be negligible.

Example 9.10

The effect of a smaller pulley. Let us repeat example 9.9, but now let us use a much smaller plastic pulley, with $M = 25$ g. Find the acceleration of each block of the system.

Solution

The acceleration of each block, again found from equation 9.40, is

$$a = \frac{(m_A - \mu_k m_B)g}{m_A + m_B + M/2}$$

$$= \frac{[2.00 \text{ kg} - (0.300)(6.00 \text{ kg})](9.80 \text{ m/s}^2)}{2.00 \text{ kg} + 6.00 \text{ kg} + 0.025 \text{ kg}/2}$$

$$= 0.244 \text{ m/s}^2$$

which agrees very closely to the value found in example 4.12, of chapter 4, when the effect of the pulley was assumed to be negligible.

Mechanics

Example 9.11

The tension in the strings. If the radius of the pulley is 5.00 cm, find the tension in the strings of examples 9.9 and 9.10.

Solution

For example 9.9, the tension T_1, found from equation 9.38, is

$$T_1 = w_A - m_A a = m_A g - m_A a = m_A(g - a)$$
$$= (2.00 \text{ kg})(9.80 \text{ m/s}^2 - 0.163 \text{ m/s}^2)$$
$$= 19.3 \text{ N}$$

Tension T_2, found from equation 9.39, is

$$T_2 = \mu_k w_B + m_B a$$
$$= \mu_k m_B g + m_B a$$
$$= (0.300)(6.00 \text{ kg})(9.80 \text{ m/s}^2) + (6.00 \text{ kg})(0.163 \text{ m/s}^2)$$
$$= 17.6 \text{ N} + 0.978 \text{ N}$$
$$= 18.6 \text{ N}$$

Thus the tensions in the strings on both sides of the pulley are unequal. It is this difference in the tensions that causes the torque,

$$\tau = R(T_1 - T_2)$$
$$= (0.05 \text{ m})(19.3 \text{ N} - 18.6 \text{ N})$$
$$= 3.50 \times 10^{-2} \text{ m N}$$

on the pulley. This torque gives the pulley its angular acceleration.

For example 9.10, the tension T_1 is again found from equation 9.38, only now the acceleration of the system is 0.244 m/s². Thus,

$$T_1 = w_A - m_A a = m_A g - m_A a = m_A(g - a)$$
$$= (2.00 \text{ kg})(9.80 \text{ m/s}^2 - 0.244 \text{ m/s}^2)$$
$$= 19.1 \text{ N}$$

Tension T_2, found from equation 9.39, is

$$T_2 = \mu_k w_B + m_B a$$
$$= \mu_k m_B g + m_B a$$
$$= (0.300)(6.00 \text{ kg})(9.80 \text{ m/s}^2) + (6.00 \text{ kg})(0.244 \text{ m/s}^2)$$
$$= 17.6 \text{ N} + 1.46 \text{ N}$$
$$= 19.1 \text{ N}$$

Hence in this case, where the pulley has a small mass, the tensions are equal, at least to three significant figures, and there is no resultant torque to cause the pulley to rotate. The two tensions must be different to cause a net torque to rotate the pulley.

†Atwood's Machine

Let us reconsider the Atwood's machine problem solved in chapter 4, only this time we no longer assume the tensions on each side of the pulley to be equal, figure 9.12. We apply Newton's second law to mass m_A to obtain

$$T_1 - w_A = -m_A a \tag{9.41}$$

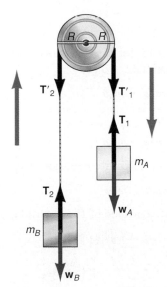

Figure 9.12

Atwood's machine with the rotational motion of the pulley taken into account.

Applying the second law to mass m_B, we obtain

$$T_2 - w_B = m_B a \qquad \text{(9.42)}$$

Let us now consider the pulley. The tension T_1' causes a clockwise torque,

$$\tau_1 = T_1' R$$

whereas the tension T_2' causes the counterclockwise torque,

$$\tau_2 = T_2' R$$

The net torque acting on the pulley is

$$\tau = \tau_1 - \tau_2 = T_1' R - T_2' R$$

But by Newton's second law for rotational motion

$$\tau = I\alpha$$

Therefore,

$$I\alpha = (T_1' - T_2')R \qquad \text{(9.43)}$$

However, by Newton's third law of motion

$$T_2' = T_2$$
$$T_1' = T_1$$

Hence, the second law, equation 9.43, becomes

$$I\alpha = (T_1 - T_2)R \qquad \text{(9.44)}$$

The moment of inertia of the pulley, found in figure 9.5, is

$$I_{\text{disk}} = \tfrac{1}{2}MR^2$$

and the angular acceleration is given by $\alpha = a/R$. Substituting these two values into equation 9.44, gives

$$\tfrac{1}{2}MR^2(a/R) = (T_1 - T_2)R$$

or

$$\tfrac{1}{2}Ma = (T_1 - T_2) \qquad \text{(9.45)}$$

There are now three equations, 9.41, 9.42, and 9.45, for the three unknowns, a, T_1, and T_2. Subtracting equation 9.41 from equation 9.42, we get

$$T_2 - w_B - T_1 + w_A = m_B a + m_A a$$

or

$$T_2 - T_1 = w_B - w_A + m_B a + m_A a \qquad \text{(9.46)}$$

Substituting the value for $T_2 - T_1$ from equation 9.45 into equation 9.46, we get

$$-\tfrac{1}{2}Ma = w_B - w_A + m_B a + m_A a$$

Gathering the terms with the acceleration a onto one side of the equation,

$$\tfrac{1}{2}Ma + m_B a + m_A a = w_A - w_B$$

Factoring out the a, we get,

$$a(\tfrac{1}{2}M + m_B + m_A) = w_A - w_B$$

Expressing the weights as $w = mg$, and solving for the acceleration of each mass of the system, we get

$$a = \frac{(m_A - m_B)g}{m_A + m_B + M/2} \qquad (9.47)$$

Equation 9.47 is the acceleration of each mass in Atwood's machine, when the rotational motion of the pulley is taken into account. If the mass of the pulley M is very small, then equation 9.47 reduces to the simplified solution in equation 4.38.

Example 9.12

Combined motion in an Atwood's machine. In an Atwood's machine, $m_B = 30.0$ g, $m_A = 50.0$ g, and the mass M of the pulley is 2.00 kg. Find the acceleration of each mass.

Solution

The acceleration, found from equation 9.47, is

$$a = \frac{(m_A - m_B)g}{m_A + m_B + M/2}$$

$$= \frac{(50.0 \text{ g} - 30.0 \text{ g})(9.80 \text{ m/s}^2)}{[50.0 \text{ g} + 30.0 \text{ g} + (2000 \text{ g})/2]}$$

$$= 0.181 \text{ m/s}^2$$

If the pulley were made of light plastic and, therefore, had a negligible mass, then the acceleration of the system would have been, $a = 2.45 \text{ m/s}^2$, which is a very significant difference.

Example 9.13

Velocity in an Atwood's machine. If block m_A of example 9.12, located a distance $h_A = 2.00$ m above the floor, falls from rest, find its velocity as it hits the floor.

Solution

Because m_A falls at the constant acceleration given by equation 9.47, the kinematic equation can be used to find its velocity at the floor. Thus,

$$v^2 = v_0^2 + 2ay$$

$$v = \sqrt{2ay} = \sqrt{2ah_A}$$

$$= \sqrt{2(0.181 \text{ m/s}^2)(2.00 \text{ m})}$$

$$= 0.850 \text{ m/s}$$

9.7 Angular Momentum and Its Conservation

Just as the linear momentum of a body was defined as the product of its mass and its linear velocity, $p = mv$, the **angular momentum** of a rotating body is defined as the product of its moment of inertia and its angular velocity. That is, the angular momentum L, with respect to a given axis, is defined as

$$L = I\omega \qquad (9.48)$$

As the concept of momentum led to an alternative form of Newton's second law for the translational case, angular momentum also leads to an alternative form for the rotational case, as shown below.

Translational Case	Rotational Case
$F = ma = m\dfrac{\Delta v}{\Delta t}$	$\tau = I\alpha = I\dfrac{\Delta \omega}{\Delta t}$
$F = m\left(\dfrac{v_f - v_i}{\Delta t}\right)$	$\tau = I\left(\dfrac{\omega_f - \omega_i}{\Delta t}\right)$
$F = \dfrac{mv_f - mv_i}{\Delta t} = \dfrac{p_f - p_i}{\Delta t}$	$\tau = \dfrac{I\omega_f - I\omega_i}{\Delta t} = \dfrac{L_f - L_i}{\Delta t}$
$F = \dfrac{\Delta p}{\Delta t}$	$\tau = \dfrac{\Delta L}{\Delta t}$

Thus, we can write Newton's second law in terms of angular momentum as

$$\tau = \frac{\Delta L}{\Delta t} \tag{9.49}$$

If we apply equation 9.49 to a system of bodies, the total torque τ arises from two sources, external torques and internal torques. Because of Newton's third law for rotational motion, the internal torques will add to zero and equation 9.49 becomes

$$\tau_{ext} = \frac{\Delta L}{\Delta t} \tag{9.50}$$

If the total external torque acting on the system is zero, then

$$0 = \frac{\Delta L}{\Delta t}$$

$$\Delta L = 0 \tag{9.51}$$

$$L_f - L_i = 0$$

Therefore,

$$L_f = L_i \tag{9.52}$$

*Equations 9.51 and 9.52 are a statement of the **law of conservation of angular momentum.** They say: if the total external torque acting on a system is zero, then there is no change in the angular momentum of the system, and the final angular momentum is equal to the initial angular momentum.*

Let us now consider some examples of the conservation of angular momentum.

The Rotating Earth

Because there is no external torque acting on the earth, $\tau = 0$, and there is conservation of angular momentum. Hence,

$$L_f = L_i \tag{9.52}$$

But since the angular momentum is the product of the moment of inertia and the angular velocity, this becomes

$$I_f \omega_f = I_i \omega_i \tag{9.53}$$

Figure 9.13

Because there is no torque acting on the earth, its angular momentum is conserved, and it will continue to spin with the same angular velocity forever.

Mechanics

However, the moment of inertia of the earth does not change with time and thus, $I_f = I_i$. Therefore,

$$\omega_f = \omega_i$$

That is, the angular velocity of the earth is a constant and will continue to spin forever with the same angular velocity unless it is acted on by some external torque. We also assume that the moment of inertia of the earth does not change.

The Spinning Ice Skater

The familiar picture of the spinning ice skater, as shown in figure 9.14, gives another example of the conservation of angular momentum. As the skater (body A) pushes against the ice (body B), thereby creating a torque, the ice (body B) pushes back on the skater (body A), creating a torque on her. The net torque on the skater and the ice is therefore zero and angular momentum is conserved.

Because the earth is so massive there will be no measurable change in the angular momentum of the earth and we need consider only the skater. The skater first starts spinning relatively slowly with her hands outstretched. We assume that any friction between the skater and the ice is negligible. As the skater draws her arms to her sides, she starts to spin very rapidly. Let us analyze the motion by the law of conservation of angular momentum. The conservation of angular momentum gives

$$L_f = L_i \tag{9.52}$$

or

$$I_f\omega_f = I_i\omega_i \tag{9.53}$$

For simplicity of calculation, let us assume that the skater is holding a set of dumbbells in her hands so that her moment of inertia can be considered to come only from the dumbbells. (That is, we assume that the moment of inertia of the girl's hands and arms can be considered negligible compared to the dumbbells in order to simplify the calculation.) The skater's initial moment of inertia is

$$I_i = mr_i^2$$

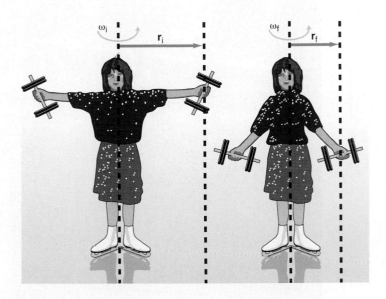

Figure 9.14

The spinning ice skater.

where m is the mass of the dumbbells and r_i the distance from the center of the body (the axis of rotation) to the outstretched dumbbells. When the skater pulls her hand down to her side the new moment of inertia is

$$I_f = mr_f^2$$

where r_f is now the distance from the axis of rotation to the dumbbell, as seen in figure 9.14(b). As we can immediately see from the figure, r_f is less than r_i, therefore I_f is less than I_i. But if the moment of inertia is changing, what happens to the skater as a consequence of the conservation of angular momentum? The angular momentum must remain the same, as given by equation 9.53. The final angular momentum must be equal to the initial angular momentum, which is equal to the product of $I_i \omega_i$, which remains a constant. Thus, the final angular momentum $I_f \omega_f$ must equal that same constant. But if I_f has decreased, the only way to maintain the equality is to have the final value of the angular velocity ω_f increase. And this is, in fact, exactly what happens. As the girl's arms are dropped to her side, the spinning increases. When the skater wishes to come out of the spin, she merely raises her arms to the original outstretched position, her moment of inertia increases and her angular velocity decreases.

A Man Diving from a Diving Board

When a man pushes down on a diving board, the board reacts by pushing back on him, as in figure 9.15. As the man leans forward at the start of the dive, the reaction force on him causes a torque to set him into rotational motion, about an axis through his center of mass, with a relatively small angular velocity ω_i. As the man leaves the board there is no longer a torque acting on him, and his angular momentum must be conserved. His initial moment of inertia is I_i, and he is spinning at an an-

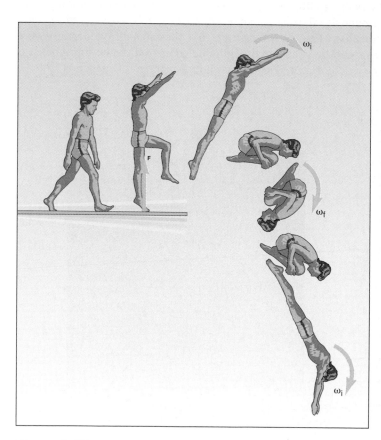

Figure 9.15

A man diving from a diving board.

Mechanics

gular velocity ω_i. If he now bends his knees and pulls his legs and arms up toward himself to form a ball, his moment of inertia decreases to a value I_f. But by the conservation of angular momentum

$$I_f\omega_f = I_i\omega_i \qquad (9.53)$$

Since I_f has decreased, his angular velocity ω_f must increase to maintain the equality of the conservation of momentum. The man now rotates relatively rapidly for one or two turns. He then stretches his body out to its original configuration with the larger value of the moment of inertia. His angular velocity then decreases to the relatively low value ω_i that he started with. If he has timed his dive properly, his outstretched body will enter the water head first at the end of his dive. The force of gravity acts on the man throughout the motion and causes the center of mass of the man to follow the parabolic trajectory associated with any projectile. Thus, the center of mass of the man is moving under the force of gravity while the man rotates around his center of mass. A trapeze artist uses the same general techniques when she rotates her body while moving through the air from one trapeze to another.

Rotational Collision (an Idealized Clutch)

Consider two disks rotating independently, as shown in figure 9.16(a). The original angular momentum of the two rotating disks is the sum of the angular momentum of each disk, that is,

$$L_i = L_{1i} + L_{2i} \qquad (9.54)$$

The initial angular momentum of disk 1 is

$$L_{1i} = I_1\omega_{1i}$$

and the initial angular momentum of disk 2 is

$$L_{2i} = I_2\omega_{2i}$$

Hence, the total initial angular momentum is

$$L_i = I_1\omega_{1i} + I_2\omega_{2i} \qquad (9.55)$$

The two disks are now forced together along their axes. Initially there may be some slipping of the disks but very quickly the two disks are coupled together by friction and spin as one, with one final angular velocity ω_f, as shown in figure 9.16(b). During the coupling process disk 1 exerted a torque on disk 2, while by Newton's third law, disk 2 exerted an equal but opposite torque on disk 1. Therefore, the net torque is zero and angular momentum must be conserved; that is, the final value of the angular momentum must equal the initial value:

$$L_f = L_i \qquad (9.52)$$

The final value of the angular momentum is the sum of the angular momentum of each disk:

$$L_f = L_{1f} + L_{2f}$$

The final value of the angular momentum of disk 1 is

$$L_{1f} = I_1\omega_f$$

while for disk 2, we have

$$L_{2f} = I_2\omega_f$$

Note that both disks have the same final angular velocity, since they are coupled together. The final momentum is therefore

$$L_f = I_1\omega_f + I_2\omega_f = (I_1 + I_2)\omega_f \qquad (9.56)$$

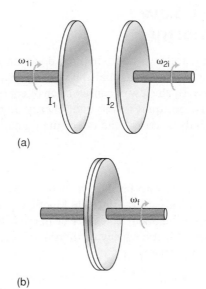

(a)

(b)

Figure 9.16

Rotational collision—the clutch.

Substituting equations 9.55 and 9.56 into the conservation of angular momentum, equation 9.52, we get

$$(I_1 + I_2)\omega_f = I_1\omega_{1i} + I_2\omega_{2i} \qquad (9.57)$$

Solving for the final angular velocity of the coupled disks, we have

$$\omega_f = \frac{I_1\omega_{1i} + I_2\omega_{2i}}{I_1 + I_2} \qquad (9.58)$$

This idealized device is the basis of a clutch. For a real clutch, the first spinning disk could be attached to the shaft of a motor, while the second disk could be connected through a set of gears to the wheels of the vehicle. When disk 2 is coupled to disk 1, the wheels of the vehicle turn. When the disks are separated, the wheels are disengaged.

9.8 Combined Translational and Rotational Motion Treated by the Law of Conservation of Energy

Let us now consider the motion of a ball that rolls, without slipping, down an inclined plane, as shown in figure 9.17. In particular, let us find the velocity of the ball at the bottom of the one meter long incline. By the law of conservation of energy, the total energy at the top of the plane must be equal to the total energy at the bottom of the plane. Because the ball is initially at rest at the top of the plane, all the energy at the top is potential energy:

$$E_{top} = PE_{top} = mgh$$

At the bottom of the plane the potential energy is zero because $h = 0$. Since the body is translating at the bottom of the incline, it has a translational kinetic energy of its center of mass of $\frac{1}{2}mv^2$. But it is also rotating about its center of mass at the bottom of the plane, and therefore it also has a kinetic energy of rotation of $\frac{1}{2}I\omega^2$. Therefore the total energy at the bottom of the plane is

$$E_{bot} = KE_{trans} + KE_{rot}$$
$$E_{bot} = \tfrac{1}{2}mv^2 + \tfrac{1}{2}I\omega^2 \qquad (9.59)$$

Equating the total energy at the bottom to the total energy at the top, we have

$$E_{bot} = E_{top}$$
$$\tfrac{1}{2}mv^2 + \tfrac{1}{2}I\omega^2 = mgh \qquad (9.60)$$

The moment of inertia for the ball is the same as a solid sphere,

$$I = \tfrac{2}{5}mr^2 \qquad (9.61)$$

The angular velocity ω of the rotating ball is related to the linear velocity of a point on the surface of the ball by

$$\omega = \frac{v}{r} \qquad (9.62)$$

The distance that a point on the edge of the ball moves along the incline is the same as the distance that the center of mass of the ball moves along the incline. Hence,

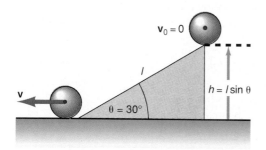

Figure 9.17

Combined translational and rotational motion.

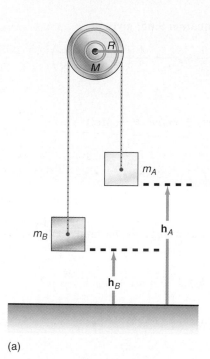

(a)

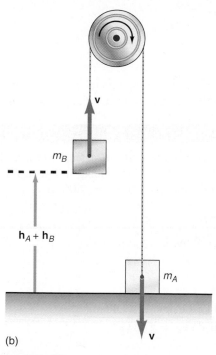

(b)

Figure 9.18

Atwood's machine revisited.

the velocity of the edge of the ball is equal to the velocity of the center of mass of the ball. Substituting equations 9.61 and 9.62 into equation 9.60, we have

$$\frac{1}{2}mv^2 + \frac{1}{2}\left(\frac{2}{5}mr^2\right)\left(\frac{v}{r}\right)^2 = mgh$$

$$\frac{1}{2}mv^2 + \frac{1}{2}\frac{2}{5}mr^2\frac{v^2}{r^2} = mgh$$

Simplifying,

$$\frac{v^2}{2} + \frac{v^2}{5} = gh$$

$$\frac{5v^2 + 2v^2}{10} = gh$$

$$\frac{7v^2}{10} = gh$$

$$v = \sqrt{\left(\frac{10}{7}\right)gh} \qquad (9.63)$$

the velocity of the ball at the bottom of the plane. The height h, found from the trigonometry of the triangle in figure 9.17, is

$$h = l \sin \theta = (1 \text{ m})\sin 30.0° = 0.500 \text{ m}$$

Therefore the velocity at the bottom of the plane is

$$v = \sqrt{(10/7)(9.80 \text{ m/s}^2)(0.500 \text{ m})}$$

$$= 2.65 \text{ m/s}$$

Note that this is the same result obtained in example 9.8 in section 9.6. The energy approach is obviously much easier.

As another example of the combined translational and rotational motion of a rigid body, let us consider the Atwood's machine shown in figure 9.18(a). Using the law of conservation of energy let us find the velocity of the mass m_A as it hits the ground.

The total energy of the system in the configuration shown consists only of the potential energy of the two masses m_A and m_B, that is,

$$E_{tot} = m_A g h_A + m_B g h_B \qquad (9.64)$$

When the system is released, m_A loses potential energy as it falls but gains kinetic energy due to its motion. Mass, m_B gains potential energy as it rises and also acquires a kinetic energy. The pulley, when set into rotational motion, also has kinetic energy of rotation. The total energy of the system as m_A strikes the ground, found from figure 9.18(b), is

$$E_{tot} = PE_B + KE_A + KE_B + KE_{pulley}$$

$$E_{tot} = m_B g(h_A + h_B) + \tfrac{1}{2}m_A v^2 + \tfrac{1}{2}m_B v^2 + \tfrac{1}{2}I\omega^2 \qquad (9.65)$$

The speed of masses A and B are equal because they are tied together by the string. The moment of inertia of the pulley (disk), found from figure 9.5, is

$$I_{disk} = \tfrac{1}{2}MR^2 \qquad (9.66)$$

Also, the angular velocity ω of the disk is related to the tangential velocity of the string as it passes over the pulley by

$$\omega = \frac{v}{R} \qquad (9.67)$$

Substituting equations 9.66 and 9.67 into equation 9.65, gives

$$E_{tot} = m_Bg(h_A + h_B) + \frac{1}{2}m_Av^2 + \frac{1}{2}m_Bv^2 + \frac{1}{2}\left(\frac{1}{2}MR^2\right)\left(\frac{v}{R}\right)^2$$

Simplifying,

$$E_{tot} = m_Bg(h_A + h_B) + \frac{1}{2}(m_A + m_B)v^2 + \frac{1}{4}Mv^2$$

or

$$E_{tot} = m_Bg(h_A + h_B) + \frac{1}{2}\left(m_A + m_B + \frac{M}{2}\right)v^2 \qquad (9.68)$$

By the law of conservation of energy, we equate the total energy in the initial configuration, equation 9.64, to the total energy in the final configuration, equation 9.68, obtaining

$$m_Agh_A + m_Bgh_B = m_Bg(h_A + h_B) + \frac{1}{2}\left(m_A + m_B + \frac{M}{2}\right)v^2$$

$$\frac{1}{2}\left(m_A + m_B + \frac{M}{2}\right)v^2 = m_Agh_A + m_Bgh_B - m_Bgh_A - m_Bgh_B,$$

$$\frac{1}{2}\left(m_A + m_B + \frac{M}{2}\right)v^2 = (m_A - m_B)gh_A$$

$$v^2 = \frac{(m_A - m_B)gh_A}{\frac{1}{2}(m_A + m_B + M/2)}$$

Solving for v, we get

$$v = \sqrt{\frac{(m_A - m_B)gh_A}{\frac{1}{2}(m_A + m_B + M/2)}} \qquad (9.69)$$

Example 9.14

Conservation of energy and combined translational and rotational motion. If $m_B = 30.0$ g, $m_A = 50.0$ g, and the mass of the pulley M is 2.00 kg in figure 9.18, find the velocity of mass m_A as it falls through the distance $h_A = 2.00$ m.

Solution

The velocity of block *B*, found from equation 9.69, is

$$v = \sqrt{\frac{(m_A - m_B)gh_A}{\frac{1}{2}(m_A + m_B + M/2)}}$$

$$= \sqrt{\frac{(0.0500 \text{ kg} - 0.0300 \text{ kg})(9.80 \text{ m/s}^2)(2.00 \text{ m})}{\frac{1}{2}(0.0300 \text{ kg} + 0.0500 \text{ kg} + 2.00 \text{ kg}/2)}}$$

$$= 0.850 \text{ m/s}$$

Note that this is the same result obtained by treating the Atwood's machine by Newton's laws of motion rather than the energy technique.

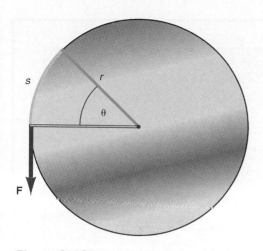

Figure 9.19

Work in rotational motion.

9.9 Work in Rotational Motion

The work done in translating a body from one position to another was found in chapter 7 as

$$W = Fx \tag{7.1}$$

where F is the force in the direction of the displacement and x is the magnitude of the displacement. We can find the work done in causing a body to rotate from equation 7.1 and figure 9.19. In figure 9.19, a string is wrapped around the disk and pulled with a constant force F, causing the disk to rotate through the angle θ. The rim of the disk moves through the distance s. The work done by the force is

$$W = Fx$$

But $x = s$ and $s = r\theta$. Therefore,

$$W = Fr\theta \tag{9.70}$$

But F times r is equal to the torque τ acting on the disk, that is

$$Fr = \tau \tag{9.71}$$

Substituting equation 9.71 into equation 9.70 gives the work done to rotate the disk as

$$W = \tau\theta \tag{9.72}$$

The power expended in rotating the disk for a time t is

$$P = \frac{W}{t} = \tau\frac{\theta}{t} \tag{9.73}$$

but $\theta/t = \omega$, the angular velocity. Therefore,

$$P = \tau\omega \tag{9.74}$$

Example 9.15

Work done in rotational motion. A constant force of 5.00 N is applied to a string that is wrapped around a disk of 0.500-m radius. If the wheel rotates through an angle of 2.00 rev, how much work is done?

Solution

The work done, given by equation 9.72, is

$$W = \tau\theta = rF\theta$$

$$= (0.500 \text{ m})(5.00 \text{ N})(2.00 \text{ rev})\left(\frac{2\pi \text{ rad}}{\text{rev}}\right)$$

$$= 31.4 \text{ J}$$

An Essay on the Application of Physics

*Attitude Control of Airplanes
and Spaceships*

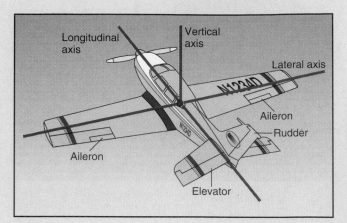

Figure 1

Control surfaces.

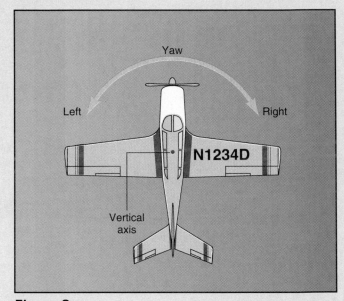

Figure 2

Aircraft yaw.

H ave you ever wondered how an airplane or space vehicle is able to change its direction of flight? A plane or spacecraft can turn, climb, and dive. But how does it do this?

Attitude Control of Aircraft

An aircraft changes its attitude by the use of control surfaces, figure 1. As we saw in section 6.7, an airplane has three ways of changing the direction of its motion. They are yaw, pitch, and roll. Yaw is a rotation about the vertical axis of the aircraft. The control surface to yaw the aircraft is the rudder, which is located at the rear of the vertical stabilizer. Pitch is a rotation about the lateral axis of the aircraft. The control surface to pitch the aircraft is the elevator, which is located at the rear of the horizontal stabilizer. Roll is a rotation about the longitudinal axis of the aircraft. The control surfaces to roll the aircraft are the ailerons, which are located on the trailing edge of the wings.

1. **Yaw Control:** Yaw is a rotation of the aircraft about a vertical axis that passes through the center of gravity of the aircraft, as shown in figure 2. The aircraft can yaw to the right or left, as seen from the position of the pilot in the aircraft.

   Before the pilot presses either rudder pedal in the cockpit, the rudder is aligned with the vertical stabilizer and the air streams past the rudder exerting no unbalanced forces on it. When the pilot presses the right rudder pedal the rudder moves toward the right, as seen from above and behind the aircraft, figure 3(a). In this position the air stream exerts a normal force **F** on the rudder surface, as shown in the figure. If we draw the line r from the center of gravity of the aircraft to the point of application of the force, we see that this force produces a

torque about the vertical axis. We find the lever arm for this torque by dropping a perpendicular from the axis of rotation to the line of action of the force. As seen in the figure, the lever arm is $r \sin \psi$. Hence, the torque is

$$\tau = Fr \sin \psi \qquad (9\text{H}.1)$$

This torque produces a clockwise torque about the center of gravity causing the aircraft to rotate (yaw) to the right. The greater the angle ψ, the greater the torque acting on the aircraft, and hence the greater will be its yaw. When the pilot moves the rudder pedals back to the neutral position, the force acting on the rudder is reduced to zero, the torque $\tau = 0$, and there is no further yaw of the aircraft.

   When the pilot presses the left rudder pedal, the rudder moves toward the left, as seen from above and

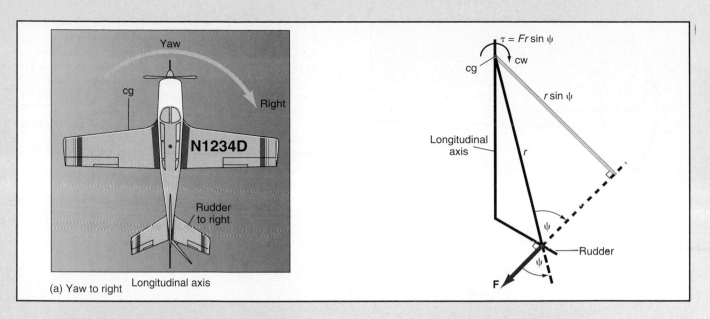

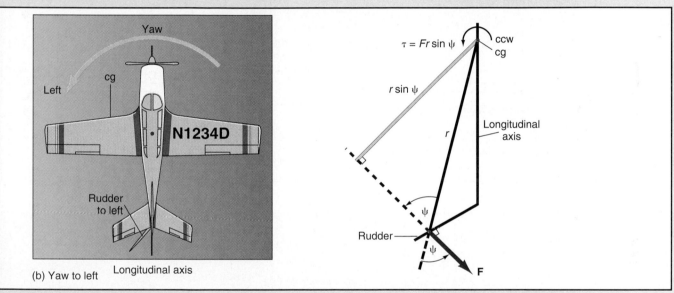

Figure 3

Dynamics of aircraft yaw.

behind the aircraft, figure 3(b). For this case the force of the air on the rudder produces a counterclockwise torque that causes the aircraft to rotate to the left, as seen in the diagram. Thus the rudder is a control surface that produces a torque on the aircraft that causes it to rotate either clockwise or counterclockwise about the vertical axis.

2. **Pitch Control:** Pitch is a rotation of the aircraft about a lateral axis that passes through the center of gravity of the aircraft, figure 4. In straight and level flight, the thrust vector of the aircraft lies along the longitudinal axis of the aircraft and thus the aircraft moves straight ahead.

When the pilot pulls the "stick" backward, the elevator is pushed upward, figure 5(a). The air that hits the elevator exerts a normal force **F** on the elevator, as seen in the diagram. If we draw the line r from the center of gravity of the aircraft to the point of application of the force, we see that this force produces a clockwise torque about the lateral axis of the aircraft. We find the lever arm for this torque by dropping a perpendicular from the axis of rotation to the line of action of the force. As seen in the figure, the lever arm is $r \sin \theta$. Hence the torque acting on the aircraft is given by

$$\tau = Fr \sin \theta \qquad \textbf{(9H.2)}$$

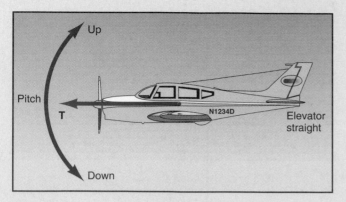

Figure 4

Aircraft pitch.

This torque causes the aircraft to rotate (pitch) about the lateral axis, such that the tail goes downward and the nose goes upward, figure 5(a). The thrust vector of the aircraft is no longer horizontal but now makes a positive angle with the horizontal, and hence the plane climbs. The farther back the pilot pulls on the stick the greater the torque and hence the steeper the climb.

When the pilot pushes the stick forward, the elevator is pushed downward, figure 5(b). The air that hits the elevator exerts a normal force **F** on the elevator, as shown. We find the lever arm for this torque by dropping a perpendicular from the axis of rotation to the line of action of the force, as shown. The resulting counterclockwise torque pushes the tail up and the nose down. The thrust vector now falls below the horizontal and the plane dives. The farther forward the pilot pushes

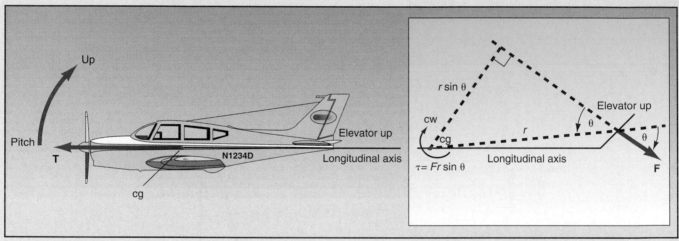

(a) Pitch up

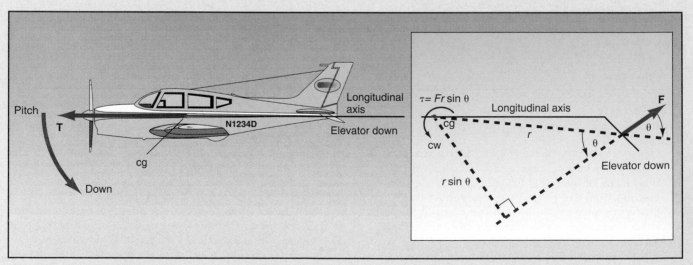

(b) Pitch down

Figure 5

Dynamics of aircraft pitch.

Mechanics

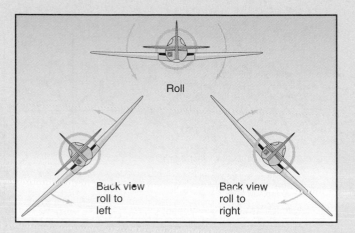

Figure 6

Aircraft roll.

the stick, the greater the torque and hence the steeper the dive. In this way the pilot can make the aircraft climb or dive.

3. **Roll Control:** Roll is a rotation of the aircraft about the longitudinal axis of the aircraft. When the pilot pushes the stick to the left, the plane will roll to the left; when he pushes the stick to the right, the plane will roll to the right, figure 6.

When the pilot pushes the stick to the left, the right aileron is pushed downward and the left aileron is pushed upward, figure 7(c). The wind blowing over the wings exerts a force on the ailerons as shown in figure 7(a,b). The force acting on the raised left aileron pushes the left wing downward, while the force acting on the lowered

right aileron pushes the right wing upward. The ailerons act similar to the elevator in that they produce a torque about the lateral axis of the aircraft. However, with one aileron up and one down the torques they produce to pitch the aircraft are equal and opposite and hence have no effect on pitching the aircraft. However, the force up on the right wing and the force down on the left wing cause a counterclockwise torque about the longitudinal axis, as viewed from the rear of the aircraft (the view that is seen by the pilot). Therefore the aircraft rolls to the left, figure 7(c). When the aircraft has rolled to the required bank angle, the pilot places the stick back to the neutral position and the aircraft stays at this angle of bank. To bring the aircraft back to level flight the pilot must push the stick to the right. The aircraft now rolls to the right until the aircraft is level. Then the pilot places the stick in the neutral position.

To roll the aircraft to the right the pilot pushes the stick to the right. The right aileron now goes up and the left aileron now goes down, figure 7(d). The force down on the right wing and the force up on the left wing causes a clockwise torque about the longitudinal axis. Thus, the aircraft rotates (rolls) to the right.

The force exerted on a control surface by the air creates the necessary torque to rotate the aircraft in any specified direction.

Attitude Control of Space Vehicles

An aircraft will not work in space because there is no air to exert the necessary lift on the wings of the aircraft. Nor can rudders, elevators, or ailerons work in space because there is no air to exert forces on the control surfaces to change the attitude of the vehicle.

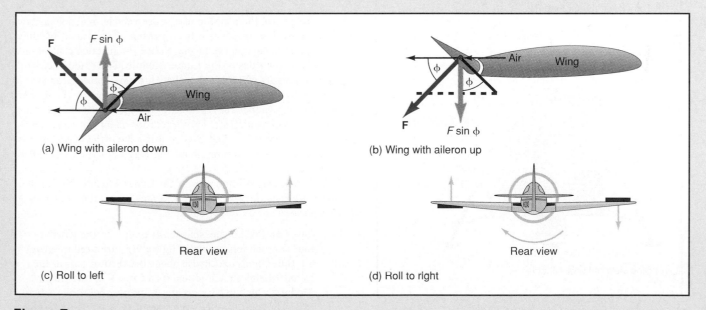

(a) Wing with aileron down

(b) Wing with aileron up

(c) Roll to left

(d) Roll to right

Figure 7

Dynamics of aircraft roll.

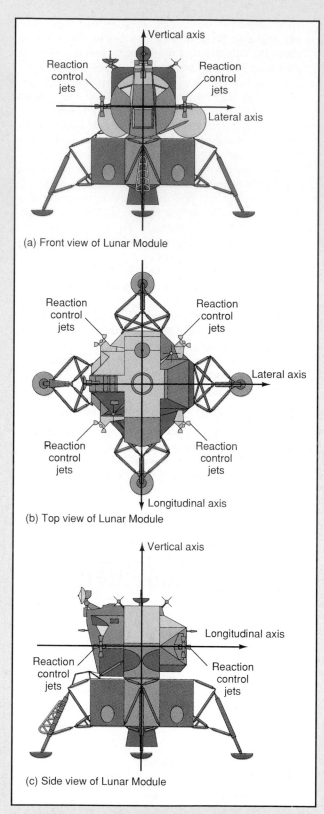

(a) Front view of Lunar Module

(b) Top view of Lunar Module

(c) Side view of Lunar Module

Figure 8

The Lunar Module.

Source: NASA.

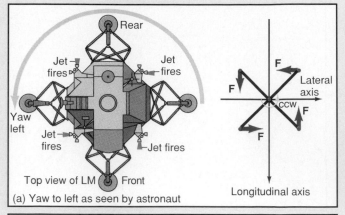

(a) Yaw to left as seen by astronaut

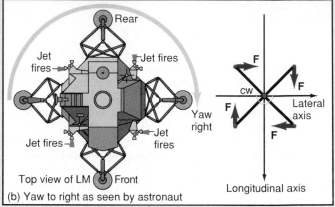

(b) Yaw to right as seen by astronaut

Figure 9

Dynamics of Lunar Module yaw.

Source (left): NASA.

To control the attitude of a space vehicle, reaction control jets are used. Figure 8 is a line drawing of the Lunar Module (LM) that landed on the moon. Notice the reaction control jets located on the sides of the Lunar Module. The reaction control system consists of 16 small rocket thrusters placed around the vehicle to control the translation and rotation of the Lunar Module. Also notice that the axes of the spacecraft are the same as the axes of an aircraft. Thus a rotation about the vertical axis of the spacecraft is called yaw, rotation about the lateral axis is called pitch, and rotation about the longitudinal axis is called roll.

Figure 8(b) is a top view of the Lunar Module. Notice that there are four thruster assemblies, each containing four rocket jets, located on the Lunar Module.

1. **Yaw Control:** For the spacecraft to yaw to the left the four reaction jets shown in figure 9(a) are fired to create a torque counterclockwise about the vertical axis of the Lunar Module. Each jet exerts a force **F** on the Lunar Module, which in turn creates a torque about the vertical axis. The total torque is the sum of the four torques. For the spacecraft to yaw to the right the four reaction jets shown in figure 9(b) are fired to create a torque clockwise

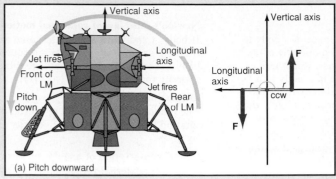

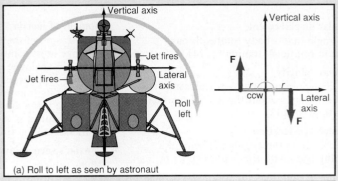

(a) Pitch downward

(a) Roll to left as seen by astronaut

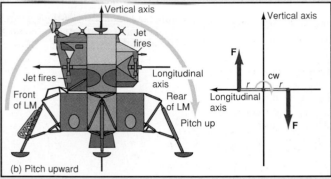

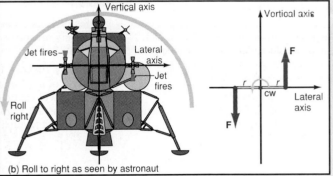

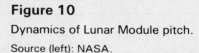

(b) Pitch upward

(b) Roll to right as seen by astronaut

Figure 10

Dynamics of Lunar Module pitch.

Source (left): NASA.

Figure 11

Dynamics of Lunar Module roll.

Source (left): NASA.

about the vertical axis of the Lunar Module. Notice that a different set of jets are used to yaw to the right than to yaw to the left.

2. **Pitch Control:** For the Lunar Module to pitch downward, the two reaction jets on each side of the Lunar Module (total of 4 jets) shown in figure 10(a) are fired to create a torque counterclockwise about the lateral axis of the Lunar Module. For the spacecraft to pitch upward the two reaction jets on each side of the Lunar Module (total of 4 jets) shown in figure 10(b) are fired to create a torque clockwise about the lateral axis of the Lunar Module.

3. **Roll Control:** For the spacecraft to roll to the left the two reaction jets on each side of the Lunar Module (total of 4

jets) shown in figure 11(a) are fired to create a torque counterclockwise about the longitudinal axis of the Lunar Module. (Don't forget that left and right are defined from the position of the pilot. Figure 11 shows the Lunar Module from a front view, and hence left and right appear to be reversed.)

A roll to the right is accomplished by firing the two reaction jets on each side of the Lunar Module (total of 4 jets) shown in figure 11(b) to create a torque clockwise about the longitudinal axis of the Lunar Module.

Thus the Lunar Module, and any spacecraft for that matter, can control its attitude by supplying torques for its rotation by the suitable firing of the different reaction control jets.

The Language of Physics

Angular displacement
The angle that a body rotates through while in rotational motion (p. 241).

Angular velocity
The change in the angular displacement of a rotating body about the axis of rotation with time (p. 242).

Angular acceleration
The change in the angular velocity of a rotating body with time (p. 243).

Kinematic equations for rotational motion
A set of equations that give the angular displacement and angular velocity of a rotating body at any instant of time, and the angular velocity at a particular angular displacement, if the angular acceleration of the body is constant (p. 244).

Kinetic energy of rotation
The energy that a body possesses by virtue of its rotational motion (p. 248).

Moment of inertia
The measure of the resistance of a body to a change in its rotational motion. It is the rotational analogue of mass, which is a measure of the resistance of a body to a change in its translational motion. The larger the moment of inertia of a body the more difficult it is to put that body into rotational motion (p. 248).

Newton's second law for rotational motion
When an unbalanced external torque acts on a body, it gives that body an angular acceleration. The angular acceleration is directly proportional to the torque and inversely proportional to the moment of inertia (p. 250).

Newton's first law for rotational motion
A body in motion at a constant angular velocity will continue in motion at that same constant angular velocity unless acted upon by some unbalanced external torque (p. 251).

Newton's third law of rotational motion
If body A and body B have the same axis of rotation, and if body A exerts a torque on body B, then body B exerts an equal but opposite torque on body A (p. 251).

Angular momentum
The product of the moment of inertia of a rotating body and its angular velocity (p. 259).

Law of conservation of angular momentum
If the total external torque acting on a system is zero, then there is no change in the angular momentum of the system, and the final angular momentum is equal to the initial angular momentum (p. 260).

Summary of Important Equations

Angular velocity
$$\omega = \frac{\Delta\theta}{\Delta t} = \frac{\theta}{t} \tag{9.1}$$

Angular acceleration
$$\alpha = \frac{\Delta\omega}{\Delta t} = \frac{\omega - \omega_0}{t} \tag{9.3}$$

Kinematic equations
$$\omega = \omega_0 + \alpha t \tag{9.4}$$
$$\theta = \omega_0 t + \tfrac{1}{2}\alpha t^2 \tag{9.9}$$
$$\omega^2 = \omega_0^2 + 2\alpha\theta \tag{9.10}$$

Relations between translational and rotational variables
$$s = r\theta \tag{6.5}$$
$$v = r\omega \tag{9.2}$$
$$a = r\alpha \tag{9.5}$$

Centripetal acceleration
$$a_c = \omega^2 r \tag{9.11}$$

Kinetic energy of rotation
$$KE_{rot} = \tfrac{1}{2}I\omega^2 \tag{9.15}$$

Moment of inertia
$$I = \sum_{i=1}^{n} m_i r_i^2 \tag{9.16}$$

Moment of inertia for a single mass
$$I = mr^2 \tag{9.17}$$

Newton's second law for rotational motion
$$\tau = I\alpha \tag{9.22}$$

Angular momentum
$$L = I\omega \tag{9.48}$$

Newton's second law in terms of momentum
$$\tau = \frac{\Delta L}{\Delta t} \tag{9.49}$$

Law of conservation of angular momentum (no external torques)
$$L_f = L_i \tag{9.52}$$

Work done in rotational motion
$$W = \tau\theta \tag{9.72}$$

Power expended in rotational motion
$$P = \tau\omega \tag{9.74}$$

Questions for Chapter 9

1. Discuss the similarity between the equations for translational motion and the equations for rotational motion.
2. When moving in circular motion at a constant angular velocity, why does the body at the greatest distance from the axis of rotation move faster than the body closest to the axis of rotation?
3. It is easy to observe the angular velocity of the second hand of a clock. Why is it more difficult to observe the angular velocity of the minute and hour hands of the clock?
†4. If a cylinder, a ball, and a ring are placed at the top of an inclined plane and then allowed to roll down the plane, in what order will they arrive at the bottom of the plane? Why?
†5. How would you go about approximating the rotational kinetic energy of our galaxy?
6. Which would be more difficult to put into rotational motion, a large sphere or a small sphere? Why?
7. Why must the axis of rotation be specified when giving the moment of inertia of an object?

†8. If two balls collide such that the force transmitted lies along a line connecting the center of mass of each body, can either ball be put into rotational motion? If the balls collide in a glancing collision in which there is also friction between the two surfaces as they collide, can either ball be put into rotational motion? Draw a diagram of the collision in both cases and discuss both possibilities.

†9. As long as there are no external torques acting on the earth, the earth will continue to spin forever at its present angular velocity. Discuss the possibility of small perturbative torques that might act on the earth and what effect they might have.

†10. Can the angular displacement, angular velocity, and angular acceleration be treated as vectors? Consider a rotation of your book through an angular displacement of 90° about the x-axis, then a rotation through an angular displacement of 90° about the y-axis, and finally a rotation through an angular displacement of 90° about the z-axis. Would you get the same result if you changed the order of the rotations to the y-, x-, and then z-axis? What happens if the rotations are infinitesimal?

†11. If the instantaneous angular velocity can be considered as a vector, should the angular momentum also be considered as a vector? If so, what direction would it have? What would the change in the direction of the angular momentum look like?

†12. It is said that if you throw a cat, upside down, into the air, it will always land on its feet. Discuss this possibility from the point of view of the cat moving his legs and tail and thus changing his moment of inertia and hence his angular velocity.

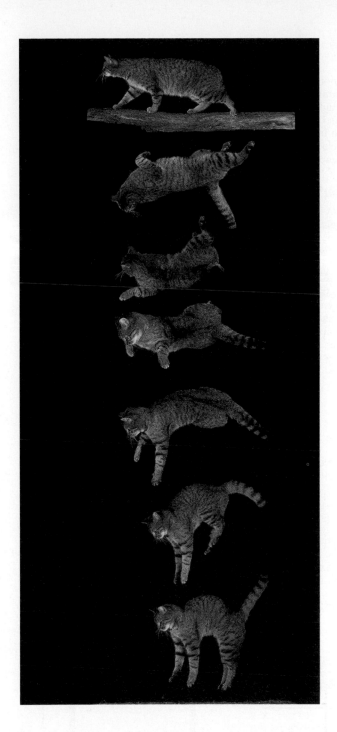

Problems for Chapter 9

9.2 Rotational Kinematics

1. Express the following angular velocities of a phonograph turntable in terms of rad/s. (a) 33 1/3 rpm (revolutions per minute), (b) 45 rpm, and (c) 78 rpm.

2. Determine the angular velocity of the following hands of a clock: (a) the second hand, (b) the minute hand, and (c) the hour hand.

3. A cylinder 15.0 cm in diameter rotates at 1000 rpm. (a) What is its angular velocity in rad/s? (b) What is the tangential velocity of a point on the rim of the cylinder?

4. A circular saw blade rotating at 3600 rpm is reduced to 3450 rpm in 2.00 s. What is the angular acceleration of the blade?

5. A circular saw blade rotating at 3600 rpm is braked to a stop in 6 s. What is the angular acceleration? How many revolutions did the blade make before coming to a stop?

6. A wheel 50.0 cm in diameter is rotating at an initial angular velocity of 6.00 rad/s. It is given an acceleration of 2.00 rad/s². Find (a) the angular velocity at 5.00 s, (b) the angular displacement at 5.00 s, (c) the tangential velocity of a point on the rim at 5.00 s, (d) the tangential acceleration of a point on the rim, (e) the centripetal acceleration of a point on the rim, and (f) the resulting acceleration of a point on the rim.

9.3 The Kinetic Energy of Rotation

7. Find the kinetic energy of a 2.00-kg cylinder, 25.0 cm in diameter, if it is rotating about its longitudinal axis at an angular velocity of 0.550 rad/s.

8. A 3.00-kg ball, 15.0 cm in diameter, rotates at an angular velocity of 3.45 rad/s. Find its kinetic energy.

9.4 The Moment of Inertia

9. Calculate the moment of inertia of a 0.500-kg meterstick about an axis through its center, and perpendicular to its length.

10. Compute the moment of inertia through its center of a 16.0-lb bowling ball of radius 4.00 in.

11. Find the moment of inertia for the system of point masses shown for (a) rotation about the y-axis and (b) for rotation about the x-axis. Given are $m_1 = 2.00$ kg, $m_2 = 3.50$ kg, $r_1 = 0.750$ m, and $r_2 = 0.873$ m.

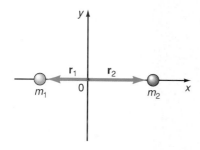

†12. Find the moment of inertia for the system shown for rotation about (a) the y-axis, (b) the x-axis, and (c) an axis going through masses m_2 and m_4. Assume $m_1 = 0.532$ kg, $m_2 = 0.425$ kg, $m_3 = 0.879$ kg, and $m_4 = 0.235$ kg.

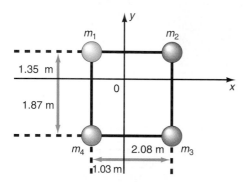

9.5 Newton's Laws for Rotational Motion and 9.6 Rotational Dynamics

13. A solid wheel of mass 5.00 kg and radius 0.350 m is set in motion by a constant force of 6.00 N applied tangentially. Determine the angular acceleration of the wheel.

14. A torque of 5.00 m N is applied to a body. Of this torque, 2.00 m N of it is used to overcome friction in the bearings. The body has a resultant angular acceleration of 5.00 rad/s². (a) When the applied torque is removed, what is the angular acceleration of the body? (b) If the angular velocity of the body was 100 rad/s when the applied torque was removed how long will it take the body to come to rest?

15. A mass of 200 g is attached to a wheel by a string wrapped around the wheel. The wheel has a mass of 1.00 kg. Find the acceleration of the mass. Assume that the moment of inertia of the wheel is the same as a disk.

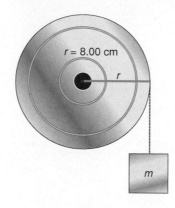

16. A mass m_A of 10.0 kg is attached to another mass m_B of 4.00 kg by a string that passes over a pulley of mass $M = 1.00$ kg. The coefficient of kinetic friction between block B and the table is 0.400. Find (a) the acceleration of each block of the system, (b) the tensions in the cords, and (c) the velocity of block A as it hits the floor 0.800 m below its starting point.

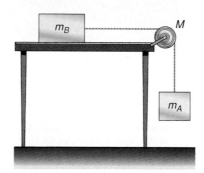

17. A mass $m_A = 100$ g, and another mass $m_B = 200$ g are attached to an Atwood's machine that has a pulley mass $M = 1.00$ kg. (a) Find the acceleration of each block of the system. (b) Find the velocity of mass B as it hits the floor 1.50 meters below its starting point.

9.7 Angular Momentum and Its Conservation

†18. A 75-kg student stands at the edge of a large disk of 150-kg mass that is rotating freely at an angular velocity of 0.800 rad/s. The disk has a radius of $R = 3.00$ m. (a) Find the initial moment of inertia of the disk and student and its kinetic energy. The student now walks toward the center of the disk. Find the moment of inertia, the angular velocity, and the kinetic energy when the student is at (b) $3R/4$, (c) $R/2$, and (d) $R/4$.

19. Two disks are to be made into an idealized clutch. Disk 1 has a mass of 3.00 kg and a radius of 20.0 cm, while disk 2 has a mass of 1.00 kg and a radius of 20.0 cm. If disk 2 is originally at rest and disk 1 is rotating at 2000 rpm, what is the final angular velocity of the coupled disks?

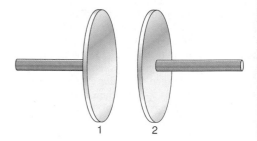

†20. Two beads are fixed on a thin long wire on the x-axis at $r_1 = 0.700$ m and $r_2 = 0.800$ m, as shown in the diagram. Assume $m_1 = 85.0$ g and $m_2 = 63.0$ g. The combination is spinning about the y-axis at an angular velocity of 4.00 rad/s. A catch is then released allowing the beads to move freely to the stops at the end of the wire, which is 1.00 m from the origin. Find (a) the initial moment of inertia of the system, (b) the initial angular momentum of the system, (c) the initial kinetic energy of the system, (d) the final angular momentum of the system, (e) the final angular velocity of the system, and (f) the final kinetic energy of the system.

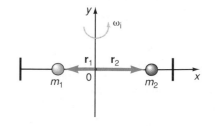

9.8 Combined Translational and Rotational Motion Treated by the Law of Conservation of Energy

21. Find the velocity of (a) a cylinder and (b) a ring at the bottom of an inclined plane that is 2.00 m high. The cylinder and ring start from rest and roll down the plane.

22. Compute the velocity of a cylinder at the bottom of a plane 1.5 m high if (a) it slides without rotating on a frictionless plane and (b) it rotates on a rough plane.

23. A 1.50-kg ball, 10.0 cm in radius, is rolling on a table at a velocity of 0.500 m/s. (a) What is its angular velocity about its center of mass? (b) What is the translational kinetic energy of its center of mass? (c) What is its rotational kinetic energy about its center of mass? (d) What is its total kinetic energy?

24. Using the law of conservation of energy for the Atwood's machine shown, find the velocity of m_A at the ground, if $m_A = 20.0$ g, $m_B = 10.0$ g, $M = 1.00$ kg, and $r = 15.0$ cm.

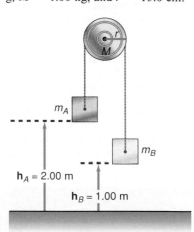

$h_A = 2.00$ m

$h_B = 1.00$ m

9.9 Work in Rotational Motion

25. A constant force of 2.50 N acts tangentially on a cylinder of 12.5-cm radius and the cylinder rotates through an angle of 5.00 rev. How much work is done in rotating the cylinder?

26. An engine operating at 1800 rpm develops 200 hp, what is the torque developed?

Additional Problems

27. Determine (a) the angular velocity of the earth, (b) its moment of inertia, and (c) its kinetic energy of rotation. (d) Compare this with its kinetic energy of translation. (e) Find the angular momentum of the earth.

28. The earth rotates once in a day. If the earth could collapse into a smaller sphere, what would be the radius of that sphere that would give a point on the equator a linear velocity equal to the velocity of light $c = 3.00 \times 10^8$ m/s? Use the initial angular velocity of the earth and the moment of inertia determined in problem 27.

29. A disk of 10.0-cm radius, having a mass of 100 g, is set into motion by a constant tangential force of 2.00 N. Determine (a) the moment of inertia of the disk, (b) the torque applied to the disk, (c) the angular acceleration of the disk, (d) the angular velocity at 2.00 s, (e) the angular displacement at 2.00 s, (f) the kinetic energy at 2.00 s, and (g) the angular momentum at 2.00 s.

30. A 3.50-kg solid disk of 25.5 cm diameter has a cylindrical hole of 3.00-cm radius cut into it. The hole is 1.00 cm in from the edge of the solid disk. Find (a) the initial moment of inertia of the disk about an axis perpendicular to the disk before the hole was cut into it and (b) the moment of inertia of the solid disk with the hole in it. State the assumptions you use in solving the problem.

†31. Due to slight effects caused by tidal friction between the water and the land and the nonsphericity of the sun, there is a slight angular deceleration of the earth. The length of a day will increase by approximately 1.5×10^{-3} s in a century. (a) What will be the angular velocity of the earth after one century? (b) What will be the change in the angular velocity of the earth per century? (c) As a first approximation, is it reasonable to assume that there are no external torques acting on the earth and the angular velocity of the earth is a constant?

32. A string of length 1.50 m with a small bob at one end is connected to a horizontal disk of negligible radius at the other end. The disk is put into rotational motion and is now rotating at an angular velocity $\omega = 5.00$ rad/s. Find the angle that the string makes with the vertical.

33. A constant force of 5.00 N acts on a disk of 3.00-kg mass and diameter of 50.0 cm for 10.0 s. Determine (a) the angular acceleration, (b) the angular velocity after 10.0 s, and (c) the kinetic energy after 10.0 s. (d) Compute the work done to cause the disk to rotate and compare with your answer to part c.

†34. A 5.00-kg block is at rest at the top of the inclined plane shown in the diagram. The plane makes an angle of 32.5° with the horizontal. A string is attached to the block and tied around the disk, which has a mass of 2.00 kg and a radius of 8.00 cm. Find the acceleration of the block down the plane if (a) the plane is frictionless, and (b) the plane is rough with a value of $\mu_k = 0.54$.

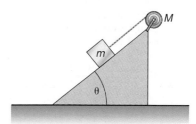

35. A large cylinder has a radius of 12.5 cm and it is pressed against a smaller cylinder of radius 4.50 cm such that the two axes of the cylinders are parallel. When the larger cylinder rotates about its axis, it causes the smaller cylinder to rotate about its axis. The larger cylinder accelerates from rest to a constant angular velocity of 20 rad/s. Find (a) the tangential velocity of a point on the surface of the large cylinder, (b) the tangential velocity of a point on the surface of the smaller cylinder, and (c) the angular velocity of the smaller cylinder. Can you think of this setup as a kind of mechanical advantage?

†36. A small disk of $r_1 = 5.00$-cm radius is attached to a larger disk of $r_2 = 15.00$-cm radius such that they have a common axis of rotation, as shown in the diagram. The small disk has a mass $M_1 = 0.250$ kg and the large disk has a mass $M_2 = 0.850$ kg. A string is wrapped around the small disk and a force is applied to the string causing a constant tangential force of 2.00 N to be applied to the disk. Find (a) the applied torque, (b) the moment of inertia of the system, (c) the angular acceleration of the system, and (d) the angular velocity at 4.00 s.

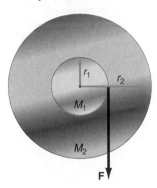

†37. Repeat problem 36 with the string wrapped around the large disk instead of the small disk.

†38. A small disk of mass $M_1 = 50.0$ g is connected to a larger disk of mass $M_2 = 200.0$ g such that they have a common axis of rotation. The small disk has a radius $r_1 = 10.0$ cm, while the large disk has a radius of $r_2 = 30.0$ cm. A mass $m_1 = 25.0$ g is connected to a string that is wrapped around the small disk, while a mass $m_2 = 35.0$ g is connected to a string and wrapped around the large disk, as shown in the diagram. Find (a) the moment of inertia of each disk, (b) the moment of inertia of the combined disks, (c) the net torque acting on the disks, (d) the angular acceleration of the disks, (e) the angular velocity of the disks at 4.00 s, (f) the kinetic energy of the disks at 4.00 s, and (g) the angular momentum of the disks at 4.00 s.

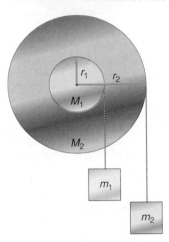

†39. One end of a string is wrapped around a pulley and the other end is connected to the ceiling, which is 3.00 m above the floor. The mass of the pulley is 200 g and has a radius of 10.0 cm. The pulley is released from rest and is allowed to fall. Find (a) the initial total energy of the system, and (b) the velocity of the pulley just before it hits the floor.

40. This is essentially the same problem as problem 39 but is to be treated by Newton's second law for rotational motion. Find the angular acceleration of the cylinder and the tension in the string.

†41. A 1.5-kg disk of 0.500-m radius is rotating freely at an angular velocity of 2.00 rad/s. Small 5-g balls of clay are dropped onto the disk at 3/4 of the radius at a rate of 4 per second. Find the angular velocity of the disk at 10.0 s.

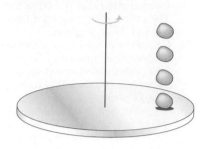

†42. A space station is to be built in orbit in the shape of a large wheel of outside radius 100.0 m and inside radius of 97.0 m. The satellite is to rotate such that it will have a centripetal acceleration exactly equal to the acceleration of gravity g on earth. The astronauts will then be able to walk about and work on the rim of the wheel in an environment similar to earth. (a) At what angular velocity must the wheel rotate to simulate the earth's gravity? (b) If the mass of the spaceship is 40,000 kg, what is its approximate moment of inertia? (c) How much energy will be necessary to rotate the space station? (d) If it takes 20.0 rev to bring the space station up to its operating angular velocity, what torque must be applied in the form of gas jets attached to the outer rim of the wheel?

†43. At the instant that ball 1 is released from rest at the top of a rough inclined plane a second ball (2) moves past it on the horizontal surface below at a constant velocity of 2.30 m/s. The plane makes an angle $\theta = 35.0°$ with the horizontal and the height of the plane is 0.500 m. Using Newton's second law for combined translational and rotational motion find (a) the acceleration a of ball 1 down the plane, (b) the velocity of ball 1 at the base of the incline, (c) the time it takes for ball 1 to reach the bottom of the plane, (d) the distance that ball 2 has moved in this time, and (e) at what horizontal distance from the base of the incline will ball 1 overtake ball 2.

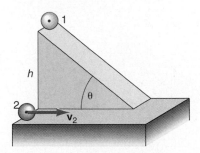

Interactive Tutorials

44. A cylinder of mass $m = 4.00$ kg and radius $r = 2.00$ m is rotating at an angular velocity $\omega = 3600$ rpm. Calculate (a) its angular velocity ω in rad/s, (b) its moment of inertia I, (c) its rotational kinetic energy KE_{rot}, and (d) its angular momentum L.

45. A mass $m = 2.00$ kg is attached by a string that is wrapped around a frictionless solid cylinder of mass $M = 8.00$ kg and radius $R = 0.700$ m that is free to rotate. Calculate (a) the acceleration a of the mass m and (b) the tension T in the string.

46. A cylinder of mass $m = 2.35$ kg and radius $r = 0.345$ m is initially rotating at an angular velocity $\omega_0 = 1.55$ rad/s when a constant force $F = 9.25$ N is applied tangentially to the cylinder as in figure 9.8. Find

(a) the moment of inertia I of the cylinder, (b) the torque τ acting on the cylinder, (c) the angular acceleration α of the cylinder, (d) the angular velocity ω of the cylinder at $t = 4.55$ s, and (e) the angular displacement θ at $t = 4.55$ s.

47. The moment of inertia of a continuous mass distribution. A meterstick, $m = 0.149$ kg, lies on the x-axis with the zero of the meterstick at the origin of the coordinate system. Determine the moment of inertia of the meterstick about an axis that passes through the zero of the meterstick and perpendicular to it. Assume that the meterstick can be divided into $N = 10$ equal parts.

48. This is a generalization of Interactive Tutorial problem 76 of chapter 4 but it also takes the rotational motion of the pulley into account. Derive the formula for the magnitude of the acceleration of the system shown in the diagram for problem 61 of chapter 4. The pulley has a mass M and the radius R. As a general case assume that the coefficient of kinetic friction between block A and the surface is μ_{kA} and between block B and the surface is μ_{kB}. Solve for all the special cases that you can think of. In all the cases, consider different values for the mass M of the pulley and see the effect it has on the results of the problem.

49. Consider the general motion in an Atwood's machine such as the one shown in figure 9.18. Mass $m_A = 0.650$ kg and is at a height $h_A = 2.55$ m above the reference plane and mass $m_B = 0.420$ kg is at a height $h_B = 0.400$ m. The pulley has a mass of $M = 2.00$ kg and a radius $R = 0.100$ m. If the system starts from rest, find (a) the initial potential energy of mass A, (b) the initial potential energy of mass B, and (c) the total energy of the system. When mass m_A has fallen a distance $y_A = 0.750$ m, find (d) the potential energy of mass A, (e) the potential energy of mass B, (f) the speed of

each mass at that point, (g) the kinetic energy of mass A, (h) the kinetic energy of mass B, (i) the moment of inertia of the pulley (assume it to be a disk), (j) the angular velocity ω of the pulley, and (k) the rotational kinetic energy of the pulley. (l) When mass A hits the ground, find the speed of each mass and the angular velocity of the pulley.

50. Consider the general motion in the combined system shown in the diagram of problem 16. Mass $m_A = 0.750$ kg and is at a height $h_A = 1.85$ m above the reference plane and mass $m_B = 0.285$ kg is at a height $h_B = 2.25$ m, $\mu_k = 0.450$. The pulley has a mass $M = 1.85$ kg and a radius $R = 0.0800$ m. If the system starts from rest, find (a) the initial potential energy of mass A, (b) the initial potential energy of mass B, and (c) the total energy of the system. When m_A has fallen a distance $y_A = 0.35$ m, find (d) the potential energy of mass A, (e) the potential energy of mass B, (f) the energy lost due to friction as mass B slides on the rough surface, (g) the speed of each mass at that point, (h) the kinetic energy of mass A, (i) the kinetic energy of mass B, (j) the moment of inertia of the pulley (assumed to be a disk), (k) the angular velocity ω of the pulley, and (l) the rotational kinetic energy of the pulley. (m) When mass A hits the ground, find the speed of each mass.

51. A disk of mass $M = 3.55$ kg and a radius $R = 1.25$ m is rotating freely at an initial angular velocity $\omega_i = 1.45$ rad/s. Small balls of clay of mass $m_b = 0.025$ kg are dropped onto the rotating disk at the radius $r = 0.85$ m at the rate of $n = 5$ ball/s. Find (a) the initial moment of inertia of the disk, (b) the initial angular momentum of the disk, and (c) the angular velocity ω at $t = 6.00$s. (d) Plot the angular velocity ω as a function of the number of balls dropped.

Vibratory Motion, Wave Motion, and Fluids

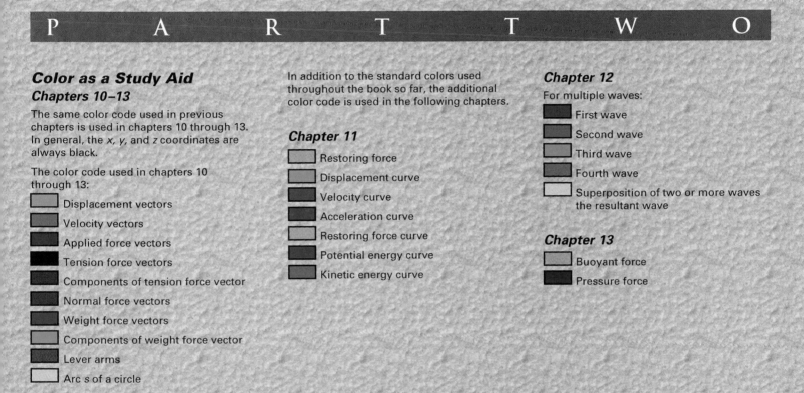

P A R T   T W O

Color as a Study Aid
Chapters 10–13

The same color code used in previous chapters is used in chapters 10 through 13. In general, the x, y, and z coordinates are always black.

The color code used in chapters 10 through 13:

- Displacement vectors
- Velocity vectors
- Applied force vectors
- Tension force vectors
- Components of tension force vector
- Normal force vectors
- Weight force vectors
- Components of weight force vector
- Lever arms
- Arc s of a circle

In addition to the standard colors used throughout the book so far, the additional color code is used in the following chapters.

Chapter 11

- Restoring force
- Displacement curve
- Velocity curve
- Acceleration curve
- Restoring force curve
- Potential energy curve
- Kinetic energy curve

Chapter 12
For multiple waves:

- First wave
- Second wave
- Third wave
- Fourth wave
- Superposition of two or more waves the resultant wave

Chapter 13

- Buoyant force
- Pressure force

10

Elasticity

Distribution of stress shown by photoelasticity. The greatest stress occurs where there are the most color changes.

If I have seen further than other men, it is because I stood on the shoulders of giants.

Isaac Newton

10.1 The Atomic Nature of Elasticity

Elasticity is that property of a body by which it experiences a change in size or shape whenever a deforming force acts on the body. When the force is removed the body returns to its original size and shape. Most people are familiar with the stretching of a rubber band. All materials, however, have this same elastic property, but in most materials it is not so pronounced.

The explanation of the elastic property of solids is found in an atomic description of a solid. Most solids are composed of a very large number of atoms or molecules arranged in a fixed pattern called the **lattice structure of a solid** and shown schematically in figure 10.1(a). These atoms or molecules are held in their positions by electrical forces. The electrical force between the molecules is attractive and tends to pull the molecules together. Thus, the solid resists being pulled apart. Any one molecule in figure 10.1(a) has an attractive force pulling it to the right and an equal attractive force pulling it to the left. There are also equal attractive forces pulling the molecule up and down, and in and out. A repulsive force between the molecules also tends to repel the molecules if they get too close together. This is why solids are difficult to compress. To explain this repulsive force we would need to invoke the Pauli exclusion principle of quantum mechanics (which we discuss in section 32.8). Here we simply refer to all these forces as molecular forces.

The net result of all these molecular forces is that each molecule is in a position of equilibrium. If we try to pull one side of a solid material to the right, let us say, then we are in effect pulling all these molecules slightly away from their equilibrium position. The displacement of any one molecule from its equilibrium position is quite small, but since there are billions of molecules, the total molecular displacements are directly measurable as a change in length of the material. When the applied force is removed, the attractive molecular forces pull all the molecules back to their original positions, and the material returns to its original length.

If we now exert a force on the material in order to compress it, we cause the molecules to be again displaced from their equilibrium position, but this time they are pushed closer together. The repulsive molecular force prevents them from getting too close together, but the total molecular displacement is directly measurable as a reduction in size of the original material. When the compressive force is removed, the repulsive molecular force causes the atoms to return to their equilibrium position and the solid returns to its original size. *Hence, the elastic properties of matter are a manifestation of the molecular forces that hold solids together.* Figure 10.1(b) shows a typical lattice structure of atoms in a solar cell analyzed with a scanning tunneling microscope.

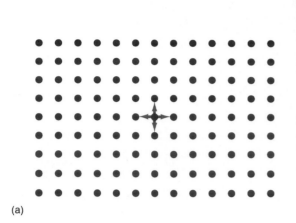

(a)

(b)

Figure 10.1

(a) Lattice structure of a solid. (b) Actual pictures of atoms in a solar cell.

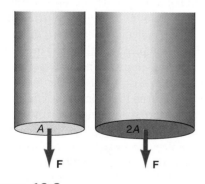

Figure 10.2

Stretching an object.

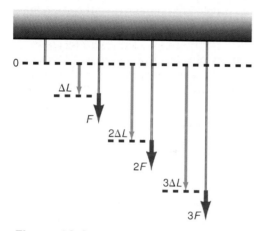

Figure 10.3

The deformation is inversely proportional to the cross-sectional area of the wire.

10.2 Hooke's Law—Stress and Strain

If we apply a force to a rubber band, we find that the rubber band stretches. Similarly, if we attach a wire to a support, as shown in figure 10.2, and sequentially apply forces of magnitude F, $2F$, and $3F$ to the wire, we find that the wire stretches by an amount ΔL, $2\Delta L$, and $3\Delta L$, respectively. (Note that the amount of stretching is greatly exaggerated in the diagram for illustrative purposes.) The deformation, ΔL, is directly proportional to the magnitude of the applied force F and is written mathematically as

$$\Delta L \propto F \qquad (10.1)$$

This aspect of elasticity is true for all solids. It would be tempting to use equation 10.1 as it stands to formulate a theory of elasticity, but with a little thought it becomes obvious that although it is correct in its description, it is incomplete.

Let us consider two wires, one of cross-sectional area A, and another with twice that area, namely $2A$, as shown in figure 10.3. When we apply a force $\mathbf{F}$ to the first wire, that force is distributed over all the atoms in that cross-sectional area A. If we subject the second wire to the same applied force $\mathbf{F}$, then this same force is distributed over twice as many atoms in the area $2A$ as it was in the area A. Equivalently we can say that each atom receives only half the force in the area $2A$ that it received in the area A. Hence, the total stretching of the $2A$ wire is only $1/2$ of what it was in wire A. Thus, the elongation of the wire ΔL is inversely proportional to the cross-sectional area A of the wire, and this is written

$$\Delta L \propto \frac{1}{A} \qquad (10.2)$$

Note also that the original length of the wire must have something to do with the amount of stretch of the wire. For if a force of magnitude F is applied to two wires of the same cross-sectional area, but one has length L_0 and the other has length $2L_0$, the same force is transmitted to every molecule in the length of the wire. But because there are twice as many molecules to stretch apart in the wire having length $2L_0$, there is twice the deformation, or $2\Delta L$, as shown in figure 10.4. We write this as the proportion

$$\Delta L \propto L_0 \qquad (10.3)$$

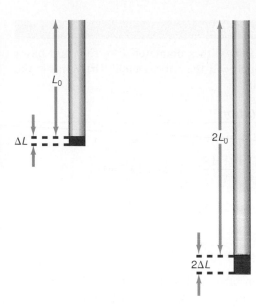

Figure 10.4

The deformation is directly proportional to the original length of the wire.

The results of equations 10.1, 10.2 and 10.3 are, of course, also deduced experimentally. *The deformation ΔL of the wire is thus directly proportional to the magnitude of the applied force F (equation 10.1), inversely proportional to the cross-sectional area A (equation 10.2), and directly proportional to the original length of the wire L_0 (equation 10.3).* These results can be incorporated into the one proportionality:

$$\Delta L \propto \frac{FL_0}{A}$$

which we rewrite in the form

$$\frac{F}{A} \propto \frac{\Delta L}{L_0} \qquad \qquad (10.4)$$

The ratio of the magnitude of the applied force to the cross-sectional area of the wire is called the **stress** *acting on the wire, while the ratio of the change in length to the original length of the wire is called the* **strain** *of the wire.* Equation 10.4 is a statement of **Hooke's law** *of elasticity, which says that in an elastic body the stress is directly proportional to the strain,* that is,

$$\text{stress} \propto \text{strain} \qquad \qquad (10.5)$$

The stress is what is applied to the body, while the resulting effect is called the strain.

To make an equality out of this proportion, we must introduce a constant of proportionality (see appendix C on proportionalities). This constant depends on the type of material used, since the molecules, and hence the molecular forces of each material, are different. This constant, called **Young's modulus of elasticity** is denoted by the letter Y. Equation 10.4 thus becomes

$$\frac{F}{A} = Y\frac{\Delta L}{L_0} \qquad \qquad (10.6)$$

The value of Y for various materials is given in table 10.1.

Table 10.1

Some Elastic Constants

Substance	Young's Modulus		Shear Modulus		Bulk Modulus		Elastic Limit		Ultimate Tensile Stress	
	$N/m^2 \times 10^{10}$	$lb/in.^2 \times 10^6$	$N/m^2 \times 10^{10}$	$lb/in.^2 \times 10^6$	$N/m^2 \times 10^{10}$	$lb/in.^2 \times 10^6$	$N/m^2 \times 10^{10}$	$lb/in.^2 \times 10^6$	$N/m^2 \times 10^8$	$lb/in.^2 \times 10^4$
Aluminum	7.0	10	3	3.8	7	10	1.4	1.9	1.4	3.1
Bone	1.5		8.0						1.30	
Brass	9.1	14	3.6	5.1	6	8.5	3.5	5.5	4.5	6.5
Copper	11.0	17	4.2	6.0	14	17	1.6	2.2	4.1	4.9
Iron	9.1	13	7.0		10	15	1.7	2.5	3.2	4.6
Lead	1.6	2.3	0.56		0.77	1.1			0.2	0.29
Steel	21	30	8.4	12	16	24	2.4	3.7	4.8	7.0

Example 10.1

Stretching a wire. A steel wire 1.00 m long with a diameter $d = 1.00$ mm has a 10.0-kg mass hung from it. (a) How much will the wire stretch? (b) What is the stress on the wire? (c) What is the strain?

Solution

a. The cross-sectional area of the wire is given by

$$A = \frac{\pi d^2}{4} = \frac{\pi(1.00 \times 10^{-3} \text{ m})^2}{4} = 7.85 \times 10^{-7} \text{ m}^2$$

We assume that the cross-sectional area of the wire does not change during the stretching process. The force stretching the wire is the weight of the 10.0-kg mass, that is,

$$F = mg = (10.0 \text{ kg})(9.80 \text{ m/s}^2) = 98.0 \text{ N}$$

Young's modulus for steel is found in table 10.1 as $Y = 21 \times 10^{10} \text{ N/m}^2$. The elongation of the wire, found from modifying equation 10.6, is

$$\Delta L = \frac{FL_0}{AY}$$

$$= \frac{(98.0 \text{ N})(1.00 \text{ m})}{(7.85 \times 10^{-7} \text{ m}^2)(21.0 \times 10^{10} \text{ N/m}^2)}$$

$$= 0.594 \times 10^{-3} \text{ m} = 0.594 \text{ mm}$$

b. The stress acting on the wire is

$$\frac{F}{A} = \frac{98.0 \text{ N}}{7.85 \times 10^{-7} \text{ m}^2} = 1.25 \times 10^8 \text{ N/m}^2$$

c. The strain of the wire is

$$\frac{\Delta L}{L_0} = \frac{0.594 \times 10^{-3} \text{ m}}{1.00 \text{ m}} = 0.594 \times 10^{-3}$$

The applied stress on the wire cannot be increased indefinitely if the wire is to remain elastic. Eventually a point is reached where the stress becomes so great that the atoms are pulled permanently away from their equilibrium position in the lattice structure. This point is called the **elastic limit** of the material and is shown in figure 10.5. When the stress exceeds the elastic limit the material does not return to its original size or shape when the stress is removed. The entire lattice structure of the material has been altered.

If the stress is increased beyond the elastic limit, eventually the ultimate stress point is reached. This is the highest point on the stress-strain curve and represents the greatest stress that the material can bear. Brittle materials break suddenly at this point, while some ductile materials can be stretched a little more due to a decrease in the cross-sectional area of the material. But they too break shortly thereafter at the breaking point. *Hooke's law is only valid below the elastic limit,* and it is only that region that will concern us.

Although we have been discussing the stretching of an elastic body, a body is also elastic under compression. If a large load is placed on a column, then the column is compressed, that is, it shrinks by an amount ΔL. When the load is removed the column returns to its original length.

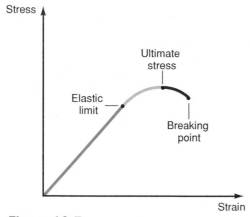

Figure 10.5

Stress-strain relationship.

Example 10.2

Compressing a steel column. A 100,000-lb load is placed on top of a steel column 10.0 ft long and 4.00 in. in diameter. By how much is the column compressed?

Solution

The cross-sectional area of the column is

$$A = \frac{\pi d^2}{4} = \frac{\pi(4.00 \text{ in.})^2}{4} = 12.6 \text{ in.}^2$$

The change in length of the column, found from equation 10.6, is

$$\Delta L = \frac{FL_0}{AY}$$

$$= \frac{(100,000 \text{ lb})(10.0 \text{ ft})}{(12.6 \text{ in.}^2)(30 \times 10^6 \text{ lb/in.}^2)} \left(\frac{12 \text{ in.}}{1 \text{ ft}}\right)$$

$$= 3.18 \times 10^{-2} \text{ in.} = 0.032 \text{ in.}$$

Note that the compression is quite small (0.032 in.) considering the very large load (100,000 lb). This is indicative of the very strong molecular forces in the lattice structure of the solid.

Example 10.3

Exceeding the ultimate compressive strength. A human bone is subjected to a compressive force of 5.00×10^5 N/m². The bone is 25.0 cm long and has an approximate area of 4.00 cm². If the ultimate compressive strength for a bone is 1.70×10^8 N/m², will the bone be compressed or will it break under this force?

Solution

The stress acting on the bone is found from

$$\frac{F}{A} = \frac{5.00 \times 10^5 \text{ N}}{4.00 \times 10^{-4} \text{ m}^2} = 12.5 \times 10^8 \text{ N/m}^2$$

Since this stress exceeds the ultimate compressive stress of a bone, 1.70×10^8 N/m², the bone will break.

10.3 Hooke's Law for a Spring

A simpler formulation of Hooke's law is sometimes useful and can be found from equation 10.6 by a slight rearrangement of terms. That is, solving equation 10.6 for F gives

$$F = \frac{AY}{L_0}\Delta L$$

Because A, Y, and L_0 are all constants, the term AY/L_0 can be set equal to a new constant k, namely

$$k = \frac{AY}{L_0} \tag{10.7}$$

We call k a force constant or a spring constant. Then,

$$F = k\Delta L \tag{10.8}$$

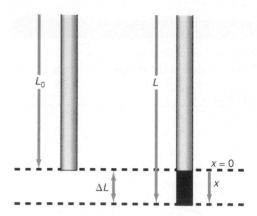

Figure 10.6

Changing the reference system.

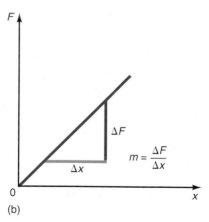

(a)

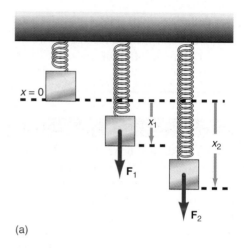

(b)

Figure 10.7

Experimental determination of a spring constant.

The change in length ΔL of the material is simply the final length L minus the original length L_0. We can introduce a new reference system to measure the elongation, by calling the location of the end of the material in its unstretched position, $x = 0$. Then we measure the stretch by the value of the displacement x from the unstretched position, as seen in figure 10.6. Thus, $\Delta L = x$, in the new reference system, and we can write equation 10.8 as

$$F = kx \qquad \textbf{(10.9)}$$

Equation 10.9 is a simplified form of Hooke's law that we use in vibratory motion containing springs. For a helical spring, we can not obtain the spring constant from equation 10.7 because the geometry of a spring is not the same as a simple straight wire. However, we can find k experimentally by adding various weights to a spring and measuring the associated elongation x, as seen in figure 10.7(a). A plot of the magnitude of the applied force F versus the elongation x gives a straight line that goes through the origin, as in figure 10.7(b). Because Hooke's law for the spring, equation 10.9, is an equation of the form of a straight line passing through the origin, that is,

$$y = mx$$

the slope m of the straight line is the spring constant k. In this way, we can determine experimentally the spring constant for any spring.

Example 10.4

The elongation of a spring. A spring with a force constant of 50.0 N/m is loaded with a 0.500-kg mass. Find the elongation of the spring.

Solution

The elongation of the spring, found from Hooke's law, equation 10.9, is

$$x = \frac{F}{k} = \frac{mg}{k}$$

$$= \frac{(0.500 \text{ kg})(9.80 \text{ m/s}^2)}{50.0 \text{ N/m}}$$

$$= 0.098 \text{ m}$$

10.4 Elasticity of Shape—Shear

In addition to being stretched or compressed, a body can be deformed by changing the shape of the body. If the body returns to its original shape when the distorting stress is removed, the body exhibits the property of elasticity of shape, sometimes called **shear**.

As an example, consider the cube fixed to the surface in figure 10.8(a). A tangential force $\mathbf{F_t}$ is applied at the top of the cube, a distance h above the bottom. The magnitude of this force F_t times the height h of the cube would normally cause a torque to act on the cube to rotate it. However, since the cube is not free to rotate, the body instead becomes deformed and changes its shape, as shown in figure 10.8(b). The normal lattice structure is shown in figure 10.8(c), and the deformed lattice structure in figure 10.8(d). The tangential force applied to the body causes the layers of atoms to be displaced sideways; one layer of the lattice structure slides over another.

Vibratory Motion, Wave Motion, and Fluids

Figure 10.8

Elasticity of shear.

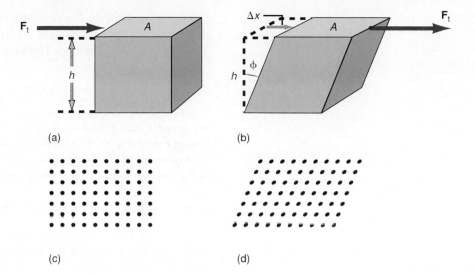

(a)   (b)

(c)   (d)

The tangential force thus causes a change in the shape of the body that is measured by the angle ϕ, called the *angle of shear*. We can also relate ϕ to the linear change from the original position of the body by noting from figure 10.8(b) that

$$\tan \phi = \frac{\Delta x}{h}$$

Because the deformations are usually quite small, as a first approximation the $\tan \phi$ can be replaced by the angle ϕ itself, expressed in radians. Thus,

$$\phi = \frac{\Delta x}{h} \tag{10.10}$$

Equation 10.10 represents the **shearing strain** *of the body.*

The tangential force $\mathbf{F_t}$ causes a deformation ϕ of the body and we find experimentally that

$$\phi \propto F_t \tag{10.11}$$

That is, the angle of shear is directly proportional to the magnitude of the applied tangential force F_t. We also find the deformation of the cube experimentally to be inversely proportional to the area of the top of the cube. With a larger area, the distorting force is spread over more molecules and hence the corresponding deformation is less. Thus,

$$\phi \propto \frac{1}{A} \tag{10.12}$$

Equations 10.11 and 10.12 can be combined into the single equation

$$\phi \propto \frac{F_t}{A} \tag{10.13}$$

Note that F_t/A has the dimensions of a stress and it is now defined as the **shearing stress**:

$$\text{Shearing stress} = \frac{F_t}{A} \tag{10.14}$$

Since ϕ is the shearing strain, equation 10.13 shows the familiar proportionality that stress is directly proportional to the strain. Introducing a constant of proportionality S, called the **shear modulus,** Hooke's law for the elasticity of shear is given by

$$\frac{F_t}{A} = S\phi \qquad (10.15)$$

Values of S for various materials are given in table 10.1. The larger the value of S, the greater the resistance to shear. Note that the shear modulus is smaller than Young's modulus Y. This implies that it is easier to slide layers of molecules over each other than it is to compress or stretch them. The shear modulus is also known as the *torsion modulus* and the *modulus of rigidity.*

Example 10.5

Elasticity of shear. A sheet of copper 0.750 m long, 1.00 m high, and 0.500 cm thick is acted on by a tangential force of 50,000 N, as shown in figure 10.9. The value of S for copper is 4.20×10^{10} N/m². Find (a) the shearing stress, (b) the shearing strain, and (c) the linear displacement Δx.

Solution

a. The area that the tangential force is acting over is

$$A = bt = (0.750 \text{ m})(5.00 \times 10^{-3} \text{ m})$$

$$= 3.75 \times 10^{-3} \text{ m}^2$$

where b is the length of the base and t is the thickness of the copper wire shown in figure 10.9. The shearing stress is

$$\frac{F_t}{A} = \frac{50,000 \text{ N}}{3.75 \times 10^{-3} \text{ m}^2} = 1.33 \times 10^7 \text{ N/m}^2$$

b. The shearing strain, found from equation 10.15, is

$$\phi = \frac{F_t}{AS} = \frac{1.33 \times 10^7 \text{ N/m}^2}{4.20 \times 10^{10} \text{ N/m}^2}$$

$$= 3.17 \times 10^{-4} \text{ rad}$$

c. The linear displacement Δx, found from equation 10.10, is

$$\Delta x = h\phi = (1.00 \text{ m})(3.17 \times 10^{-4} \text{ rad})$$

$$= 3.17 \times 10^{-4} \text{ m} = 0.317 \text{ mm}$$

Figure 10.9

An example of shear.

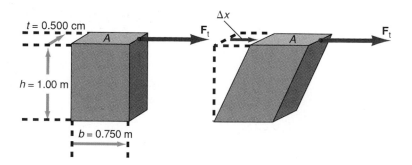

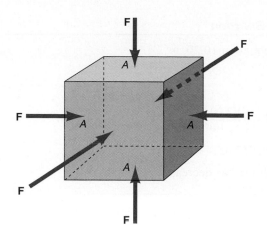

Figure 10.10

Volume elasticity.

10.5 Elasticity of Volume

If a uniform force is exerted on all sides of an object, as in figure 10.10, such as a block under water, each side of the block is compressed. Thus, the entire volume of the block decreases. The compressional stress is defined as

$$\text{stress} = \frac{F}{A} \tag{10.16}$$

where F is the magnitude of the normal force acting on the cross-sectional area A of the block. The strain is measured by the change in volume per unit volume, that is,

$$\text{strain} = \frac{\Delta V}{V_0} \tag{10.17}$$

Since the stress is directly proportional to the strain, by Hooke's law, we have

$$\frac{F}{A} \propto \frac{\Delta V}{V_0} \tag{10.18}$$

To obtain an equality, we introduce a constant of proportionality B, called the **bulk modulus,** and Hooke's law for **elasticity of volume** becomes

$$\frac{F}{A} = -B\frac{\Delta V}{V_0} \tag{10.19}$$

The minus sign is introduced in equation 10.19 because an increase in the stress (F/A) causes a decrease in the volume, leaving ΔV negative. The bulk modulus is a measure of how difficult it is to compress a substance. The reciprocal of the bulk modulus B, called the *compressibility k,* is a measure of how easy it is to compress the substance. The bulk modulus B is used for solids, while the compressibility k is usually used for liquids.

Quite often the body to be compressed is immersed in a liquid. In dealing with liquids and gases it is convenient to deal with the pressure exerted by the liquid or gas. We will see in detail in chapter 13 that pressure is defined as the force that is acting over a unit area of the body, that is,

$$p = \frac{F}{A}$$

For the case of volume elasticity, the stress F/A, acting on the body by the fluid, can be replaced by the pressure of the fluid itself. Thus, Hooke's law for volume elasticity can also be written as

$$p = -B\frac{\Delta V}{V_0} \tag{10.20}$$

Example 10.6

Elasticity of volume. A solid copper sphere of 0.500-m³ volume is placed 100 ft below the ocean surface where the pressure is 3.00×10^5 N/m². What is the change in volume of the sphere? The bulk modulus for copper is 14×10^{10} N/m².

The change in volume, found from equation 10.20, is

$$\Delta V = -\frac{V_0}{B} p$$

$$= -\frac{(0.500 \text{ m}^3)(3.00 \times 10^5 \text{ N/m}^2)}{14 \times 10^{10} \text{ N/m}^2}$$

$$= -1.1 \times 10^{-6} \text{ m}^3$$

The minus sign indicates that the volume has decreased.

The Language of Physics

Elasticity
That property of a body by which it experiences a change in size or shape whenever a deforming force acts on the body. The elastic properties of matter are a manifestation of the molecular forces that hold solids together (p. 283).

Lattice structure of a solid
A regular, periodically repeated, three-dimensional array of the atoms or molecules comprising the solid (p. 283).

Stress
For a body that can be either stretched or compressed, the stress is the ratio of the applied force acting on a body to the cross-sectional area of the body (p. 285).

Strain
For a body that can be either stretched or compressed, the ratio of the change in length to the original length of the body is called the strain (p. 285).

Hooke's law
In an elastic body, the stress is directly proportional to the strain (p. 285).

Young's modulus of elasticity
The proportionality constant in Hooke's law. It is equal to the ratio of the stress to the strain (p. 285).

Elastic limit
The point where the stress on a body becomes so great that the atoms of the body are pulled permanently away from their equilibrium position in the lattice structure. When the stress exceeds the elastic limit, the material will not return to its original size or shape when the stress is removed. Hooke's law is no longer valid above the elastic limit (p. 286).

Shear
That elastic property of a body that causes the shape of the body to be changed when a stress is applied. When the stress is removed the body returns to its original shape (p. 288).

Shearing strain
The angle of shear, which is a measure of how much the body's shape has been deformed (p. 289).

Shearing stress
The ratio of the tangential force acting on the body to the area of the body over which the tangential force acts (p. 289).

Shear modulus
The constant of proportionality in Hooke's law for shear. It is equal to the ratio of the shearing stress to the shearing strain (p. 290).

Bulk modulus
The constant of proportionality in Hooke's law for volume elasticity. It is equal to the ratio of the compressional stress to the strain. The strain for this case is equal to the change in volume per unit volume (p. 291).

Elasticity of volume
When a uniform force is exerted on all sides of an object, each side of the object becomes compressed. Hence, the entire volume of the body decreases. When the force is removed the body returns to its original volume (p. 291).

Summary of Important Equations

Hooke's law in general
stress $\propto$ strain   **(10.5)**

Hooke's law for stretching
or compression
$$\frac{F}{A} = Y\frac{\Delta L}{L_0}$$ **(10.6)**

Hooke's law for a spring
$$F = kx$$ **(10.9)**

Hooke's law for shear
$$\frac{F_t}{A} = S\phi$$ **(10.15)**

Hooke's law for volume
elasticity
$$\frac{F}{A} = -B\frac{\Delta V}{V_0}$$ **(10.19)**

Hooke's law for volume
elasticity
$$p = -B\frac{\Delta V}{V_0}$$ **(10.20)**

Questions for Chapter 10

1. Why is concrete often reinforced with steel?
†2. An amorphous solid such as glass does not have the simple lattice structure shown in figure 10.1. What effect does this have on the elastic properties of glass?
3. Discuss the assumption that the diameter of a wire does not change when under stress.
4. Compare the elastic constants of a human bone with the elastic constants of other materials listed in table 10.1. From this standpoint discuss the bone as a structural element.

5. Why are there no Young's moduli for liquids or gases?
6. Describe the elastic properties of a cube of jello.
7. If you doubled the diameter of a human bone, what would happen to the maximum compressive force that the bone could withstand without breaking?
†8. In the profession of Orthodontics, a dentist uses braces to realign teeth. Discuss this process from the point of view of stress and strain.

†9. Discuss Hooke's law as it applies to the bending of a beam that is fixed at one end and has a load placed at the other end.

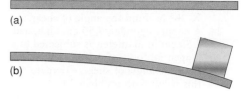

(a)

(b)

†10. How do the elastic properties of a material affect the vibration of that material?

Problems for Chapter 10

10.2 Hooke's Law—Stress and Strain

1. An aluminum wire has a diameter of 0.850 mm and is subjected to a force of 1000 N. Find the stress acting on the wire.
2. A copper wire experiences a stress of 5.00×10^3 N/m². If the diameter of the wire is 0.750 mm, find the force acting on the wire.
3. A brass wire 0.750 cm long is stretched by 0.001 cm. Find the strain of the wire.
4. A steel wire, 1.00 m long, has a diameter of 1.50 mm. If a mass of 3.00 kg is hung from the wire, by how much will it stretch?
5. A load of 50,000 lb is placed on an aluminum column 4.00 in. in diameter. If the column was originally 4.00 ft high find the amount that the column has shrunk.
6. A mass of 25,000 kg is placed on a steel column, 3.00 m high and 15.0 cm in diameter. Find the decrease in length of the column under this compression.

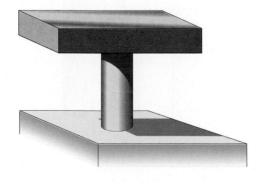

7. An aluminum wire, 1.50 m long, has a diameter of 0.750 mm. If a force of 60.0 N is suspended from the wire, find (a) the stress on the wire, (b) the elongation of the wire, and (c) the strain of the wire.
8. A copper wire, 1.00 m long, has a diameter of 0.750 mm. When an unknown weight is suspended from the wire it stretches 0.200 mm. What was the load placed on the wire?
9. A steel wire is 1.00 m long and has a diameter of 0.75 mm. Find the maximum value of a mass that can be suspended from the wire before exceeding the elastic limit of the wire.
10. A steel wire is 1.00 m long and has a 10.0-kg mass suspended from it. What is the minimum diameter of the wire such that the load will not exceed the elastic limit of the wire?
11. Find the maximum load that can be applied to a brass wire, 0.750 mm in diameter, without exceeding the elastic limit of the wire.
12. Find the maximum change in length of a 1.00-m brass wire, of 0.800 mm diameter, such that the elastic limit of the wire is not exceeded.
13. If the thigh bone is about 25.0 cm in length and about 4.00 cm in diameter determine the maximum compression of the bone before it will break. The ultimate compressive strength of bone is 1.70×10^8 N/m².
14. If the ultimate tensile strength of glass is 7.00×10^7 N/m², find the maximum weight that can be placed on a glass cylinder of 0.100 m² area, 25.0 cm long, if the glass is not to break.

15. A human bone is 2.00 cm in diameter. Find the maximum compression force the bone can withstand without fracture. The ultimate compressive strength of bone is 1.70×10^8 N/m².
16. A copper rod, 0.400 cm in diameter, supports a load of 150 kg suspended from one end. Will the rod return to its initial length when the load is removed or has this load exceeded the elastic limit of the rod?

10.3 Hooke's Law for a Spring

17. A coil spring stretches 4.00 cm when a mass of 500 g is suspended from it. What is the force constant of the spring?
18. A coil spring stretches by 2.00 cm when an unknown load is placed on the spring. If the spring has a force constant of 3.5 N/m, find the value of the unknown force.
19. A coil spring stretches by 2.50 cm when a mass of 750 g is suspended from it. (a) Find the force constant of the spring. (b) How much will the spring stretch if 800 g is suspended from it?
20. A horizontal spring stretches 20.0 cm when a force of 10.0 N is applied to the spring. By how much will it stretch if a 30.0-N force is now applied to the spring? If the same spring is placed in the vertical and a weight of 10.0 N is hung from the spring, will the results change?
21. A coil spring stretches by 4.50 cm when a mass of 250 g is suspended from it. What force is necessary to stretch the spring an additional 2.50 cm?

10.4 Elasticity of Shape—Shear

22. A brass cube, 5.00 cm on a side, is subjected to a tangential force. If the angle of shear is measured in radians to be 0.010 rad, what is the magnitude of the tangential force?

23. A copper block, 7.50 cm on a side, is subjected to a tangential force of 3.5 $\times$ 10^3 N. Find the angle of shear.

24. A copper cylinder, 7.50 cm high, and 7.50 cm in diameter, is subjected to a tangential force of 3.5 $\times$ 10^3 N. Find the angle of shear. Compare this result with problem 23.

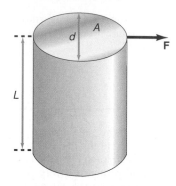

25. An annular copper cylinder, 7.50 cm high, inner radius of 2.00 cm and outer radius of 3.75 cm, is subjected to a tangential force of 3.5 $\times$ 10^3 N. Find the angle of shear. Compare this result with problems 23 and 24.

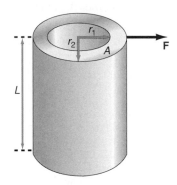

10.5 Elasticity of Volume

26. A cube of lead 15.0 cm on a side is subjected to a uniform pressure of 5.00 $\times$ 10^5 N/m^2. By how much does the volume of the cube change?

27. A liter of glycerine contracts 0.21 cm^3 when subject to a pressure of 9.8 $\times$ 10^5 N/m^2. Calculate the bulk modulus of glycerine.

28. A pressure of 1.013 $\times$ 10^7 N/m^2 is applied to a volume of 15.0 m^3 of water. If the bulk modulus of water is 0.020 $\times$ 10^{10} N/m^2, by how much will the water be compressed?

29. Repeat problem 28, only this time use glycerine that has a bulk modulus of 0.45 $\times$ 10^{10} N/m^2.

30. Normal atmospheric pressure is 1.013 $\times$ 10^5 N/m^2. How many atmospheres of pressure must be applied to a volume of water to compress it to 1.00% of its original volume? The bulk modulus of water is 0.020 $\times$ 10^{10} N/m^2.

31. Find the ratio of the density of water at the bottom of a 50.0-m lake to the density of water at the surface of the lake. The pressure at the bottom of the lake is 4.90 $\times$ 10^5 N/m^2. (*Hint:* the volume of the water will be decreased by the pressure of the water above it.) The bulk modulus for water is 0.21 $\times$ 10^{10} N/m^2.

Additional Problems

32. A lead block 50.0 cm long, 10.0 cm wide, and 10.0 cm thick, has a force of 200,000 N placed on it. Find the stress, the strain, and the change in length if (a) the block is standing upright, and (b) the block is lying flat.

33. An aluminum cylinder must support a load of 450,000 N. The cylinder is 5.00 m high. If the maximum allowable stress is 1.4 $\times$ 10^8, what must be the minimum radius of the cylinder in order for the cylinder to support the load? What will be the length of the cylinder when under load?

34. This is essentially the same problem as 33, but now the cylinder is made of steel. Find the minimum radius of the steel cylinder that is necessary to support the load and compare it to the radius of the aluminum cylinder. The maximum allowable stress for steel is 2.4 $\times$ 10^{10} N/m^2.

35. How many 1.00-kg masses may be hung from a 1.00-m steel wire, 0.750 mm in diameter, without exceeding the elastic limit of the wire?

36. A solid copper cylinder 1.50 m long and 10.0 cm in diameter, has a mass of 5000 kg placed on its top. Find the compression of the cylinder.

37. This is the same problem as 36, except that the cylinder is an annular cylinder with an inner radius of 3.50 cm and outer radius of 5.00 cm. Find the compression of the cylinder and compare with problem 36.

38. This is the same problem as problem 36 except the body is an I-beam with the dimensions shown in the diagram. Find the compression of the I-beam and compare to problems 36 and 37. The crossbar width is 2.00 cm.

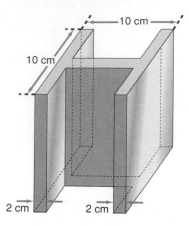

†39. Two pieces of metal rod, 2.00 cm thick, are to be connected together by riveting a steel plate to them as shown in the diagram. Two rivets, each 1.00 cm in diameter, are used. What is the maximum force that can be applied to the metal rod without exceeding a shearing stress of 8.4 $\times$ 10^8 N/m^2.

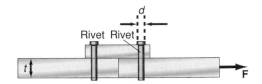

†40. A copper and steel wire are welded together at their ends as shown. The original length of each wire is 50.0 cm and each has a diameter of 0.780 mm. A mass of 10.0 kg is suspended from the combined wire. By how much will the combined wire stretch?

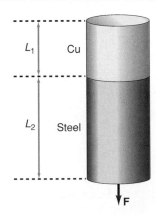

†41. A copper and steel wire each 50.0 cm in length and 0.780 mm in diameter are connected in parallel to a load of 98.0 N, as shown in the diagram. If the strain is the same for each wire, find (a) the force on wire 1, (b) the force on wire 2, and (c) the total displacement of the load.

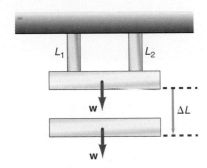

†42. Repeat problem 41 with the diameter of wire 1 equal to 1.00 mm and the diameter of wire 2 equal to 1.50 mm.

†43. Two steel wires of diameters 1.50 mm and 1.00 mm, and each 50.0 cm long, are welded together in series as shown in the diagram. If a weight of 98.0 N is suspended from the bottom of the combined wire, by how much will the combined wire stretch?

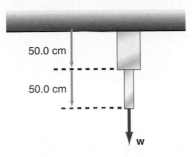

†44. Two springs are connected in parallel as shown in the diagram. The spring constants are $k_1 = 5.00$ N/m and $k_2 = 3.00$ N/m. A force of 10.0 N is applied as shown. If the strain is the same in each spring, find (a) the displacement of mass m, (b) the force on spring 1, and (c) the force on spring 2.

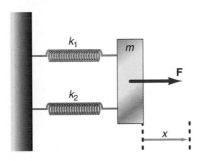

†45. Two springs are connected in series as shown in the diagram. The spring constants are $k_1 = 5.00$ N/m and $k_2 = 3.00$ N/m. A force of 10.0 N is applied as shown. Find (a) the displacement of mass m, (b) the displacement of spring 1, and (c) the displacement of spring 2.

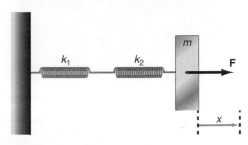

Interactive Tutorials

📟 46. Young's modulus for a wire is $Y = 2.10 \times 10^{11}$ N/m². The wire has an initial length of $L_0 = 0.700$ m and a diameter $d = 0.310$ mm. A force $F = 1.00$ N is applied in steps from 1.00 to 10.0 N. Calculate the wire's change in length ΔL with increasing load F, and graph the result.

11

A simple pendulum.

We are to admit no more causes of natural things than such as are both true and sufficient to explain their appearances.

Isaac Newton

Simple Harmonic Motion

11.1 Introduction to Periodic Motion

Periodic motion is any motion that repeats itself in equal intervals of time. The uniformly rotating earth represents a periodic motion that repeats itself every 24 hours. The motion of the earth around the sun is periodic, repeating itself every 12 months. A vibrating spring and a pendulum also exhibit periodic motion. The period of the motion is defined as the time for the motion to repeat itself. A special type of periodic motion is simple harmonic motion and we now proceed to investigate it.

11.2 Simple Harmonic Motion

An example of simple harmonic motion is the vibration of a mass m, attached to a spring of negligible mass, as the mass slides on a frictionless surface, as shown in figure 11.1. We say that the mass, in the unstretched position, figure 11.1(a), is in its equilibrium position. If an applied force $\mathbf{F_A}$ acts on the mass, the mass will be displaced to the right of its equilibrium position a distance x, figure 11.1(b). The distance that the spring stretches, obtained from Hooke's law, is

$$F_A = kx$$

*The **displacement** x is defined as the distance the body moves from its equilibrium position.* Because F_A is a force that pulls the mass to the right, it is also the force that pulls the spring to the right. By Newton's third law there is an equal but opposite elastic force exerted by the spring on the mass pulling the mass toward the left. Since this force tends to restore the mass to its original position, we call it the restoring force F_R. Because the restoring force is opposite to the applied force, it is given by

$$F_R = -F_A = -kx \tag{11.1}$$

When the applied force F_A is removed, the elastic restoring force F_R is then the only force acting on the mass m, figure 11.1(c), and it tries to restore m to its equilibrium position. We can then find the acceleration of the mass from Newton's second law as

$$ma = F_R$$
$$= -kx$$

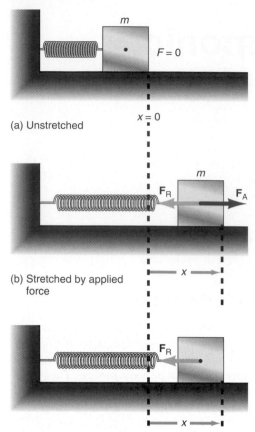

(a) Unstretched

(b) Stretched by applied force

(c) Applied force removed

Figure 11.1

The vibrating spring.

Thus,

$$a = -\frac{k}{m}x \qquad \text{(11.2)}$$

Equation 11.2 is the defining equation for simple harmonic motion. **Simple harmonic motion** *is motion in which the acceleration of a body is directly proportional to its displacement from the equilibrium position but in the opposite direction.* A vibrating system that executes simple harmonic motion is sometimes called a *harmonic oscillator. Because the acceleration is directly proportional to the displacement x in simple harmonic motion, the acceleration of the system is not constant but varies with x.* At large displacements, the acceleration is large, at small displacements the acceleration is small. Describing the vibratory motion of the mass *m* requires some new techniques because the kinematic equations derived in chapter 3 were based on the assumption that the acceleration of the system was a constant. As we can see from equation 11.2, this assumption is no longer valid. We need to derive a new set of kinematic equations to describe simple harmonic motion, and we will do so in section 11.3. However, let us first look at the motion from a physical point of view. The mass *m* in figure 11.2(a) is pulled a distance $x = A$ to the right, and is then released. The maximum restoring force on *m* acts to the left at this position because

$$F_{Rmax} = -kx_{max} = -kA$$

*The maximum displacement A is called the **amplitude** of the motion.* At this position the mass experiences its maximum acceleration to the left. From equation 11.2 we obtain

$$a = -\frac{k}{m}A$$

The mass continues to move toward the left while the acceleration continuously decreases. At the equilibrium position, figure 11.2(b), $x = 0$ and hence, from equation 11.2, the acceleration is also zero. However, because the mass has inertia it moves past the equilibrium position to negative values of *x,* thereby compressing the spring. The restoring force F_R now points to the right, since for negative values of *x,*

$$F_R = -k(-x) = kx$$

The force acting toward the right causes the mass to slow down, eventually coming to rest at $x = -A$. At this point, figure 11.2(c), there is a maximum restoring force pointing toward the right

$$F_{Rmax} = -k(-A)_{max} = kA$$

and hence a maximum acceleration

$$a_{max} = -\frac{k}{m}(-A) = \frac{k}{m}A$$

also to the right. The mass moves to the right while the force F_R and the acceleration *a* decreases with *x* until *x* is again equal to zero, figure 11.2(d). Then F_R and *a* are also zero. Because of the inertia of the mass, it moves past the equilibrium position to positive values of *x.* The restoring force again acts toward the left, slowing down the mass. When the displacement *x* is equal to *A,* figure 11.2(e), the mass momentarily comes to rest and then the motion repeats itself. *One complete motion from $x = A$ and back to $x = A$ is called a **cycle** or an oscillation. The **period** T is the*

Figure 11.2

Detailed motion of the vibrating spring.

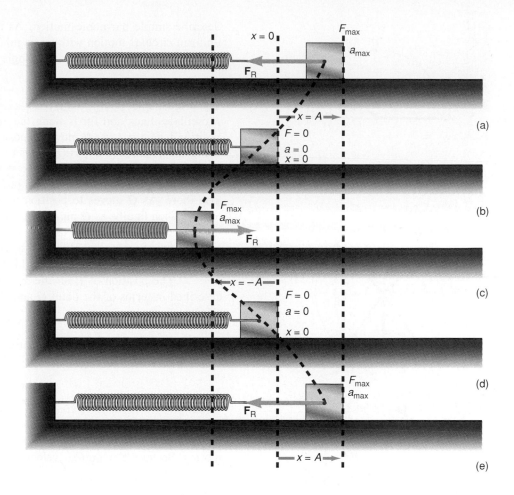

*time for one complete oscillation, and the **frequency** f is the number of complete oscillations or cycles made in unit time.* The period and the frequency are reciprocal to each other, that is,

$$f = \frac{1}{T} \tag{11.3}$$

The unit for a period is the second, while the unit for frequency, called a hertz, is one cycle per second. The hertz is abbreviated, Hz. Also note that a cycle is a number not a dimensional quantity and can be dropped from the computations whenever doing so is useful.

11.3 Analysis of Simple Harmonic Motion— The Reference Circle

As pointed out in section 11.2, the acceleration of the mass in the vibrating spring system is not a constant, but rather varies with the displacement x. Hence, the kinematic equations of chapter 3 can not be used to describe the motion. (We derived those equations on the assumption that the acceleration was constant.) Thus, a new set of equations must be derived to describe simple harmonic motion.

Simple harmonic motion is related to the uniform circular motion studied in chapter 6. An analysis of uniform circular motion gives us a set of equations to

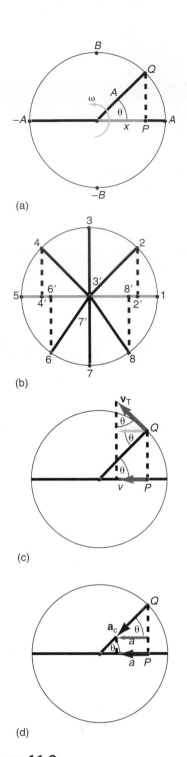

(a)

(b)

(c)

(d)

Figure 11.3

Simple harmonic motion and the reference circle.

describe simple harmonic motion. As an example, consider a point Q moving in uniform circular motion with an angular velocity ω, as shown in figure 11.3(a). At a particular instant of time t, the angle θ that Q has turned through is

$$\theta = \omega t \tag{11.4}$$

The projection of point Q onto the x-axis gives the point P. As Q rotates in the circle, P oscillates back and forth along the x-axis, figure 11.3(b). That is, when Q is at position 1, P is at 1. As Q moves to position 2 on the circle, P moves to the left along the x-axis to position 2'. As Q moves to position 3, P moves on the x-axis to position 3', which is of course the value of $x = 0$. As Q moves to position 4 on the circle, P moves along the negative x-axis to position 4'. When Q arrives at position 5, P is also there. As Q moves to position 6 on the circle, P moves to position 6' on the x-axis. Then finally, as Q moves through positions 7, 8, and 1, P moves through 7', 8', and 1, respectively. The oscillatory motion of point P on the x-axis corresponds to the simple harmonic motion of a body m moving under the influence of an elastic restoring force, as shown in figure 11.2.

The position of P on the x-axis and hence the position of the mass m is described in terms of the point Q and the angle θ found in figure 11.3(a) as

$$x = A \cos \theta \tag{11.5}$$

Here A is the amplitude of the vibratory motion and using the value of θ from equation 11.4 we have

$$x = A \cos \omega t \tag{11.6}$$

Equation 11.6 is the first kinematic equation for simple harmonic motion; it gives the displacement of the vibrating body at any instant of time t. The angular velocity ω of point Q in the **reference circle** is related to the frequency of the simple harmonic motion. Because the angular velocity was defined as

$$\omega = \frac{\theta}{t} \tag{11.7}$$

then, for a complete rotation of point Q, θ rotates through an angle of 2π rad. But this occurs in exactly the time for P to execute one complete vibration. We call this time for one complete vibration the period T. Hence, we can also write the angular velocity, equation 11.7, as

$$\omega = \frac{\theta}{t} = \frac{2\pi}{T} \tag{11.8}$$

Since the frequency f is the reciprocal of the period T (equation 11.3) we can write equation 11.8 as

$$\omega = 2\pi f \tag{11.9}$$

*Thus, the **angular velocity** of the uniform circular motion in the reference circle is related to the frequency of the vibrating system.* Because of this relation between the angular velocity and the frequency of the system, we usually call the angular velocity ω the *angular frequency of the vibrating system.* We can substitute equation 11.9 into equation 11.6 to give *another form for the first kinematic equation of simple harmonic motion, namely*

$$x = A \cos(2\pi f t) \tag{11.10}$$

We can find the velocity of the mass m attached to the end of the spring in figure 11.2 with the help of the reference circle in figure 11.3(c). The point Q moves

Vibratory Motion, Wave Motion, and Fluids

with the tangential velocity V_T. The x-component of this velocity is the velocity of the point P and hence the velocity of the mass m. From figure 11.3(c) we can see that

$$v = -V_T \sin \theta \qquad (11.11)$$

The minus sign indicates that the velocity of P is toward the left at this position. The linear velocity V_T of the point Q is related to the angular velocity ω by equation 9.2 of chapter 9, that is

$$v = r\omega$$

For the reference circle, $v = V_T$ and r is the amplitude A. Hence, the tangential velocity V_T is given by

$$V_T = \omega A \qquad (11.12)$$

Using equations 11.11, 11.12, and 11.4, the velocity of point P becomes

$$v = -\omega A \sin \omega t \qquad (11.13)$$

Equation 11.13 is the second of the kinematic equations for simple harmonic motion and it gives the speed of the vibrating mass at any time t.

A third kinematic equation for simple harmonic motion giving the speed of the vibrating body as a function of displacement can be found from equation 11.13 by using the trigonometric identity

$$\sin^2\theta + \cos^2\theta = 1$$

or

$$\sin \theta = \pm \sqrt{1 - \cos^2\theta}$$

From figure 11.3(a) or equation 11.5, we have

$$\cos \theta = \frac{x}{A}$$

Hence,

$$\sin \theta = \pm \sqrt{1 - \frac{x^2}{A^2}} \qquad (11.14)$$

Substituting equation 11.14 back into equation 11.13, we get

$$v = \pm \omega A \sqrt{1 - \frac{x^2}{A^2}}$$

or

$$v = \pm \omega \sqrt{A^2 - x^2} \qquad (11.15)$$

Equation 11.15 is the third of the kinematic equations for simple harmonic motion and it gives the velocity of the moving body at any displacement x. The $\pm$ sign in equation 11.15 indicates the direction of the vibrating body. If the body is moving to the right, then the positive sign $(+)$ is used. If the body is moving to the left, then the negative sign $(-)$ is used.

Finally, we can find the acceleration of the vibrating body using the reference circle in figure 11.3(d). The point Q in uniform circular motion experiences a centripetal acceleration $\mathbf{a}_c$ pointing toward the center of the circle in figure 11.3(d). The x-component of the centripetal acceleration is the acceleration of the vibrating body at the point P. That is,

$$a = -a_c \cos \theta \qquad (11.16)$$

The minus sign again indicates that the acceleration is toward the left. But recall from chapter 6 that the magnitude of the centripetal acceleration is

$$a_c = \frac{v^2}{r} \qquad (6.12)$$

where v represents the tangential speed of the rotating object, which in the present case is V_T, and r is the radius of the circle, which in the present case is the radius of the reference circle A. Thus,

$$a_c = \frac{V_T^2}{A}$$

But we saw in equation 11.12 that $V_T = \omega A$, therefore

$$a_c = \omega^2 A$$

The acceleration of the mass m, equation 11.16, thus becomes

$$a = -\omega^2 A \cos \omega t \qquad (11.17)$$

Equation 11.17 is the fourth of the kinematic equations for simple harmonic motion. It gives the acceleration of the vibrating body at any time t. This equation has no counterpart in chapter 3, because there the acceleration was always a constant. Also, since $F = ma$ by Newton's second law, the force acting on the mass m, becomes

$$F = -m\omega^2 A \cos \omega t \qquad (11.18)$$

Thus, the force acting on the mass m is a variable force.

Equations 11.6 and 11.17 can be combined into the simple equation

$$a = -\omega^2 x \qquad (11.19)$$

If equation 11.19 is compared with equation 11.2,

$$a = -\frac{k}{m} x$$

we see that the acceleration of the mass at P, equation 11.19, is directly proportional to the displacement x and in the opposite direction. But this is the definition of simple harmonic motion as stated in equation 11.2. Hence, the projection of a point at Q, in uniform circular motion, onto the x-axis does indeed represent simple harmonic motion. Thus, the kinematic equations developed to describe the motion of the point P, also describe the motion of a mass attached to a vibrating spring.

An important relation between the characteristics of the spring and the vibratory motion can be easily deduced from equations 11.2 and 11.19. That is, because both equations represent the acceleration of the vibrating body they can be equated to each other, giving

$$\omega^2 = \frac{k}{m}$$

or

$$\omega = \sqrt{\frac{k}{m}} \qquad (11.20)$$

The value of ω in the kinematic equations is expressed in terms of the force constant k of the spring and the mass m attached to the spring. The physics of simple harmonic motion is thus connected to the angular frequency ω of the vibration.

In summary, the kinematic equations for simple harmonic motion are

$$x = A \cos \omega t \qquad (11.6)$$

$$v = -\omega A \sin \omega t \qquad (11.13)$$

$$v = \pm \omega \sqrt{A^2 - x^2} \qquad (11.15)$$

$$a = -\omega^2 A \cos \omega t \qquad (11.17)$$

$$F = -m\omega^2 A \cos \omega t \qquad (11.18)$$

where, from equations 11.9 and 11.20, we have

$$\omega = 2\pi f = \sqrt{\frac{k}{m}}$$

A plot of the displacement, velocity, and acceleration of the vibrating body as a function of time are shown in figure 11.4. We can see that the mathematical description follows the physical description in figure 11.2. When $x = A$, the maximum displacement, the velocity v is zero, while the acceleration is at its maximum value of $-\omega^2 A$. The minus sign indicates that the acceleration is toward the left. The force is at its maximum value of $-m\omega^2 A$, where the minus sign shows that the restoring force is pulling the mass back toward its equilibrium position. At the equilibrium position $x = 0$, $a = 0$, and $F = 0$, but v has its maximum velocity of $-\omega A$ toward the left. As x goes to negative values, the force and the acceleration become positive, slowing down the motion to the left, and hence v starts to decrease. At $x = -A$ the velocity is zero and the force and acceleration take on their maximum values toward the right, tending to restore the mass to its equilibrium position. As

Figure 11.4

Displacement, velocity, and acceleration in simple harmonic motion.

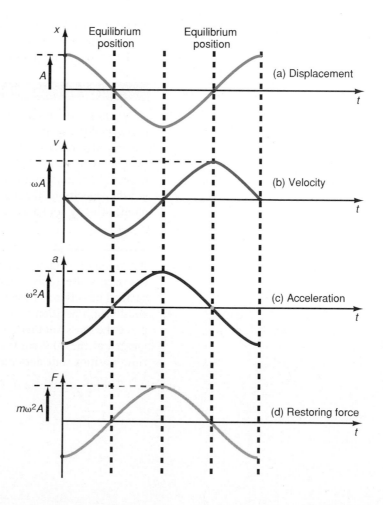

x becomes less negative, the velocity to the right increases, until it picks up its maximum value of ωA at $x = 0$, the equilibrium position, where F and a are both zero. Because of this large velocity, the mass passes the equilibrium position in its motion toward the right. However, as soon as x becomes positive, the force and the acceleration become negative thereby slowing down the mass until its velocity becomes zero at the maximum displacement A. One entire cycle has been completed, and the motion starts over again. (We should emphasize here that in this vibratory motion there are two places where the velocity is instantaneously zero, $x = A$ and $x = -A$, even though the instantaneous acceleration is nonzero there.)

Sometimes the vibratory motion is so rapid that the actual displacement, velocity, and acceleration at every instant of time are not as important as the gross motion, which can be described in terms of the frequency or period of the motion. We can find the frequency of the vibrating mass in terms of the spring constant k and the vibrating mass m by setting equation 11.9 equal to equation 11.20. Thus,

$$\omega = 2\pi f = \sqrt{\frac{k}{m}}$$

Solving for the frequency f, we obtain

$$f = \frac{1}{2\pi} \sqrt{\frac{k}{m}} \tag{11.21}$$

Equation 11.21 gives the frequency of the vibration. Because the period of the vibrating motion is the reciprocal of the frequency, we get for the period

$$T = 2\pi \sqrt{\frac{m}{k}} \tag{11.22}$$

Equation 11.22 gives the period of the simple harmonic motion in terms of the mass m in motion and the spring constant k. Notice that for a particular value of m and k, the period of the motion remains a constant throughout the motion.

Example 11.1

An example of simple harmonic motion. A mass of 0.300 kg is placed on a vertical spring and the spring stretches by 10.0 cm. It is then pulled down an additional 5.00 cm and then released. Find (a) the spring constant k, (b) the angular frequency ω, (c) the frequency f, (d) the period T, (e) the maximum velocity of the vibrating mass, (f) the maximum acceleration of the mass, (g) the maximum restoring force, (h) the velocity of the mass at $x = 2.00$ cm, and (i) the equation of the displacement, velocity, and acceleration at any time t.

Solution

Although the original analysis dealt with a mass on a horizontal frictionless surface, the results also apply to a mass attached to a spring that is allowed to vibrate in the vertical direction. The constant force of gravity on the 0.300-kg mass displaces the equilibrium position to $x = 10.0$ cm. When the additional force is applied to displace the mass another 5.00 cm, the mass oscillates about the equilibrium position, located at the 10.0-cm mark. Thus, the force of gravity only displaces the equilibrium position, but does not otherwise influence the result of the dynamic motion.

a. The spring constant, found from Hooke's law, is

$$k = \frac{F_A}{x} = \frac{mg}{x}$$

$$= \frac{(0.300 \text{ kg})(9.80 \text{ m/s}^2)}{0.100 \text{ m}}$$

$$= 29.4 \text{ N/m}$$

b. The angular frequency ω, found from equation 11.20, is

$$\omega = \sqrt{\frac{k}{m}}$$

$$= \sqrt{\frac{29.4 \text{ N/m}}{0.300 \text{ kg}}}$$

$$= 9.90 \text{ rad/s}$$

c. The frequency of the motion, found from equation 11.9, is

$$f = \frac{\omega}{2\pi}$$

$$= \frac{9.90 \text{ rad/s}}{2\pi \text{ rad}}$$

$$= 1.58 \frac{\text{cycles}}{\text{s}} = 1.58 \text{ Hz}$$

d. We could find the period from equation 11.22 but since we already know the frequency f, it is easier to compute T from equation 11.3. Thus,

$$T = \frac{1}{f} = \frac{1}{1.58 \text{ cycles/s}} = 0.633 \text{ s}$$

e. The maximum velocity, found from equation 11.13, is

$$v_{\text{max}} = \omega A = (9.90 \text{ rad/s})(5.00 \times 10^{-2} \text{ m})$$

$$= 0.495 \text{ m/s}$$

f. The maximum acceleration, found from equation 11.17, is

$$a_{\text{max}} = \omega^2 A = (9.90 \text{ rad/s})^2(5.00 \times 10^{-2} \text{ m})$$

$$= 4.90 \text{ m/s}^2$$

g. The maximum restoring force, found from Hooke's law, is

$$F_{\text{max}} = k x_{\text{max}} = kA$$

$$= (29.4 \text{ N/m})(5.00 \times 10^{-2} \text{ m})$$

$$= 1.47 \text{ N}$$

h. The velocity of the mass at $x = 2.00$ cm, found from equation 11.15, is

$$v = \pm\omega\sqrt{A^2 - x^2}$$

$$= \pm(9.90 \text{ rad/s})\sqrt{(5.00 \times 10^{-2} \text{ m})^2 - (2.00 \times 10^{-2} \text{ m})^2}$$

$$= \pm 0.454 \text{ m/s}$$

where v is positive when moving to the right and negative when moving to the left.

i. The equation of the displacement at any instant of time, found from equation 11.6, is

$$x = A \cos \omega t$$

$$= (5.00 \times 10^{-2} \text{ m}) \cos(9.90 \text{ rad/s})t$$

The equation of the velocity at any instant of time, found from equation 11.13, is

$$v = -\omega A \sin \omega t$$

$$= -(9.90 \text{ rad/s})(5.00 \times 10^{-2} \text{ m})\sin(9.90 \text{ rad/s})t$$

$$= -(0.495 \text{ m/s})\sin(9.90 \text{ rad/s})t$$

The equation of the acceleration at any time, found from equation 11.17, is

$$a = -\omega^2 A \cos \omega t$$

$$= -(9.90 \text{ rad/s})^2(5.00 \times 10^{-2} \text{ m})\cos(9.90 \text{ rad/s})t$$

$$= -(4.90 \text{ m/s}^2)\cos(9.90 \text{ rad/s})t$$

11.4 The Potential Energy of a Spring

In chapter 7 we defined the gravitational potential energy of a body as the energy that a body possesses by virtue of its position in a gravitational field. A body can also have elastic potential energy. For example, *a compressed spring has potential energy because it has the ability to do work as it expands to its equilibrium configuration. Similarly, a stretched spring must also contain potential energy because it has the ability to do work as it returns to its equilibrium position.* Because work must be done on a body to put the body into the configuration where it has the elastic potential energy, this work is used as the measure of the potential energy. Thus, the **potential energy of a spring** is equal to the work that you, the external agent, must do to compress (or stretch) the spring to its present configuration. We defined the potential energy as

$$\text{PE} = W = Fx \qquad (11.23)$$

However, we can not use equation 11.23 in its present form to determine the potential energy of a spring. Recall that the work defined in this way, in chapter 7, was for a constant force. We have seen in this chapter that the force necessary to compress or stretch a spring is not a constant but is rather a variable force depending on the value of x, $(F = -kx)$. We can still solve the problem, however, by using the average value of the force between the value at the equilibrium position and the value at the position x. That is, because the restoring force is directly proportional to the displacement, the average force exerted in moving the mass m from $x = 0$ to the value x in figure 11.5(a) is

$$F_{\text{avg}} = \frac{F_0 + F}{2}$$

Thus, we find the potential energy in this configuration by using the average force, that is,

$$\text{PE} = W = F_{\text{avg}}x$$

$$= W = \left(\frac{F_0 + F}{2}\right)x$$

$$= \left(\frac{0 + kx}{2}\right)x$$

Hence,

$$\text{PE} = \tfrac{1}{2}kx^2 \qquad (11.24)$$

Because of the x^2 in equation 11.24, the potential energy of a spring is always positive, whether x is positive or negative. The zero of potential energy is defined at the equilibrium position, $x = 0$.

Note that equation 11.24 could also be derived by plotting the force F acting on the spring versus the displacement x of the spring, as shown in figure 11.5(b). Because the work is equal to the product of the force F and the displacement x, the work is also equal to the area under the curve in figure 11.5(b). The area of that triangle is $\tfrac{1}{2}(x)(F) = \tfrac{1}{2}(x)(kx) = \tfrac{1}{2}kx^2$. (For the more general problem where the

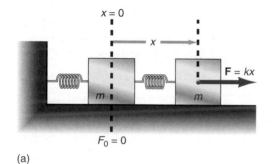

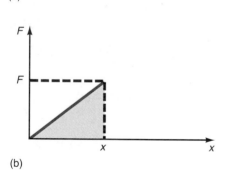

Figure 11.5
The potential energy of a spring.

Vibratory Motion, Wave Motion, and Fluids

force is not a linear function of the displacement x, if the force is plotted versus the displacement x, the work done, and hence the potential energy, will still be equal to the area under the curve.)

Example 11.2

The potential energy of a spring. A spring, with a spring constant of 29.4 N/m, is stretched 5.00 cm. How much potential energy does the spring possess?

Solution

The potential energy of the spring, found from equation 11.24, is

$$PE = \tfrac{1}{2}kx^2$$
$$= \tfrac{1}{2}(29.4 \text{ N/m})(5.00 \times 10^{-2} \text{ m})^2$$
$$= 3.68 \times 10^{-2} \text{ J}$$

11.5 Conservation of Energy and the Vibrating Spring

The vibrating spring system of figure 11.2 can also be described in terms of the law of conservation of energy. When the spring is stretched to its maximum displacement A, work is done on the spring, and hence the spring contains potential energy. The mass m attached to the spring also has that potential energy. The total energy of the system is equal to the potential energy at the maximum displacement because at that point, $v = 0$, and therefore the kinetic energy is equal to zero, that is,

$$E_{\text{tot}} = PE = \tfrac{1}{2}kA^2 \tag{11.25}$$

When the spring is released, the mass moves to a smaller displacement x, and is moving at a speed v. At this arbitrary position x, the mass will have both potential energy and kinetic energy. The law of conservation of energy then yields

$$E_{\text{tot}} = PE + KE$$
$$E_{\text{tot}} = \tfrac{1}{2}kx^2 + \tfrac{1}{2}mv^2 \tag{11.26}$$

But the total energy imparted to the mass m is given by equation 11.25. Hence, the law of conservation of energy gives

$$E_{\text{tot}} = E_{\text{tot}}$$
$$\tfrac{1}{2}kA^2 = \tfrac{1}{2}kx^2 + \tfrac{1}{2}mv^2 \tag{11.27}$$

We can also use equation 11.27 to find the velocity of the moving body at any displacement x. Thus,

$$\frac{1}{2}mv^2 = \frac{1}{2}kA^2 - \frac{1}{2}kx^2$$

$$v^2 = \frac{k}{m}(A^2 - x^2)$$

$$v = \pm\sqrt{\frac{k}{m}(A^2 - x^2)} \tag{11.28}$$

We should note that this is the same equation for the velocity as derived earlier (equation 11.15). It is informative to replace the values of x and v from their respective equations 11.6 and 11.13 into equation 11.26. Thus,

$$E_{\text{tot}} = \tfrac{1}{2}k(A\cos\omega t)^2 + \tfrac{1}{2}m(-\omega A\sin\omega t)^2$$

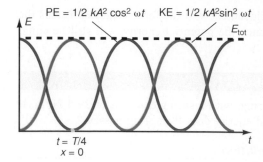

$PE = 1/2\ kA^2 \cos^2 \omega t$ $KE = 1/2\ kA^2 \sin^2 \omega t$

Figure 11.6

Conservation of energy and simple harmonic motion.

or

$$E_{tot} = \tfrac{1}{2}kA^2 \cos^2\omega t + \tfrac{1}{2}m\omega^2 A^2 \sin^2\omega t$$

but since

$$\omega^2 = \frac{k}{m}$$

$$E_{tot} = \frac{1}{2}kA^2 \cos^2\omega t + \frac{1}{2}m\frac{k}{m}A^2 \sin^2\omega t$$

$$= \frac{1}{2}kA^2 \cos^2\omega t + \frac{1}{2}kA^2 \sin^2\omega t \qquad (11.29)$$

These terms are plotted in figure 11.6.

The total energy of the vibrating system is a constant and this is shown as the horizontal line, E_{tot}. At $t = 0$ the total energy of the system is potential energy (v is zero, hence the kinetic energy is zero). As the time increases the potential energy decreases and the kinetic energy increases, as shown. However, the total energy remains the same. From equation 11.24 and figure 11.6, we see that at $x = 0$ the potential energy is zero and hence all the energy is kinetic. This occurs when $t = T/4$. The maximum velocity of the mass m occurs here and is easily found by equating the maximum kinetic energy to the total energy, that is,

$$\frac{1}{2}mv^2_{max} = \frac{1}{2}kA^2$$

$$v_{max} = \sqrt{\frac{k}{m}}\,A = \omega A \qquad (11.30)$$

When the oscillating mass reaches $x = A$, the kinetic energy becomes zero since

$$\tfrac{1}{2}kA^2 = \tfrac{1}{2}kA^2 + \tfrac{1}{2}mv^2$$

$$\tfrac{1}{2}mv^2 = \tfrac{1}{2}kA^2 - \tfrac{1}{2}kA^2 = 0$$

$$= KE = 0$$

As the oscillation continues there is a constant interchange of energy between potential energy and kinetic energy but the total energy of the system remains a constant.

Example 11.3

Conservation of energy applied to a spring. A horizontal spring has a spring constant of 29.4 N/m. A mass of 300 g is attached to the spring and displaced 5.00 cm. The mass is then released. Find (a) the total energy of the system, (b) the maximum velocity of the system, and (c) the potential energy and kinetic energy for $x = 2.00$ cm.

Solution

a. The total energy of the system is

$$E_{tot} = \tfrac{1}{2}kA^2$$

$$= \tfrac{1}{2}(29.4\ \text{N/m})(5.00 \times 10^{-2}\ \text{m})^2$$

$$= 3.68 \times 10^{-2}\ \text{J}$$

b. The maximum velocity occurs when $x = 0$ and the potential energy is zero. Therefore, using equation 11.30,

$$v_{max} = \sqrt{\frac{k}{m}}\,A$$

$$= \sqrt{\frac{29.4 \text{ N/m}}{3.00 \times 10^{-1} \text{ kg}}}\,(5.00 \times 10^{-2} \text{ m})$$

$$= 0.495 \text{ m/s}$$

c. The potential energy at 2.00 cm is

$$PE = \tfrac{1}{2}kx^2 = \tfrac{1}{2}(29.4 \text{ N/m})(2.00 \times 10^{-2} \text{ m})^2$$

$$= 5.88 \times 10^{-3} \text{ J}$$

The kinetic energy at 2.00 cm is

$$KE = \frac{1}{2}mv^2 = \frac{1}{2}m\frac{k}{m}(A^2 - x^2)$$

$$= \frac{1}{2}(29.4 \text{ N/m})[(5.00 \times 10^{-2} \text{ m})^2 - (2.00 \times 10^{-2} \text{ m})^2]$$

$$= 3.09 \times 10^{-2} \text{ J}$$

Note that the sum of the potential energy and the kinetic energy is equal to the same value for the total energy found in part a.

11.6 The Simple Pendulum

Another example of periodic motion is a pendulum. A **simple pendulum** is a bob that is attached to a string and allowed to oscillate, as shown in figure 11.7(a). The bob oscillates because there is a restoring force, given by

$$\text{Restoring force} = -mg \sin \theta \tag{11.31}$$

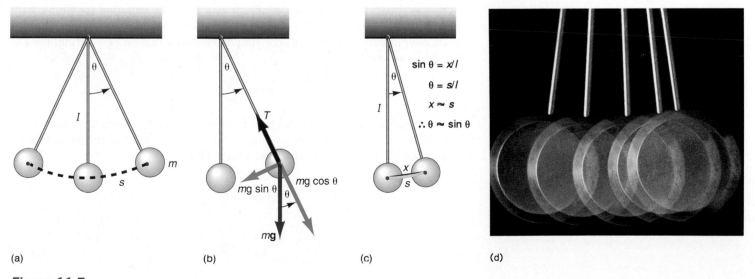

(a) (b) (c) (d)

Figure 11.7

The simple pendulum.

This restoring force is just the component of the weight of the bob that is perpendicular to the string, as shown in figure 11.7(b). If Newton's second law, $F = ma$, is applied to the motion of the pendulum bob, we get

$$-mg \sin \theta = ma$$

The tangential acceleration of the pendulum bob is thus

$$a = -g \sin \theta \qquad (11.32)$$

Note that although this pendulum motion is periodic, it is not, in general, simple harmonic motion because the acceleration is not directly proportional to the displacement of the pendulum bob from its equilibrium position. However, if the angle θ of the simple pendulum is small, then the sine of θ can be replaced by the angle θ itself, expressed in radians. (The discrepancy in using θ rather than the $\sin \theta$ is less than 0.2% for angles less than 10 degrees.) That is, for small angles

$$\sin \theta \approx \theta$$

The acceleration of the bob is then

$$a = -g\theta \qquad (11.33)$$

From figure 11.7 and the definition of an angle in radians ($\theta =$ arc length/radius), we have

$$\theta = \frac{s}{l}$$

where s is the actual path length followed by the bob. Thus

$$a = -g \frac{s}{l} \qquad (11.34)$$

The path length s is curved, but if the angle θ is small, the arc length s is approximately equal to the chord x, figure 11.7(c). Hence,

$$a = -\frac{g}{l} x \qquad (11.35)$$

which is an equation having the same form as that of the equation for simple harmonic motion. Therefore, if the angle of oscillation θ is small, the pendulum will execute simple harmonic motion. For simple harmonic motion of a spring, the acceleration was found to be

$$a = -\frac{k}{m} x \qquad (11.2)$$

We can now use the equations developed for the vibrating spring to describe the motion of the pendulum. We find an equivalent spring constant of the pendulum by setting equation 11.2 equal to equation 11.35. That is

$$\frac{k}{m} = \frac{g}{l}$$

or

$$k_P = \frac{mg}{l} \qquad (11.36)$$

Equation 11.36 states that the motion of a pendulum can be described by the equations developed for the vibrating spring by using the equivalent spring constant of the pendulum k_p. Thus, the period of motion of the pendulum, found from equation 11.22, is

$$T_p = 2\pi \sqrt{\frac{m}{k_p}}$$

$$= 2\pi \sqrt{\frac{m}{mg/l}}$$

$$T_p = 2\pi \sqrt{\frac{l}{g}} \tag{11.37}$$

The period of motion of the pendulum is independent of the mass m of the bob but is directly proportional to the square root of the length of the string. If the angle θ is equal to 15° on either side of the central position, then the true period differs from that given by equation 11.37 by less than 0.5%.

The pendulum can be used as a very simple device to measure the acceleration of gravity at a particular location. We measure the length l of the pendulum and then set the pendulum into motion. We measure the period by a clock and obtain the acceleration of gravity from equation 11.37 as

$$g = \frac{4\pi^2}{T_p^2} l \tag{11.38}$$

Example 11.4

The period of a pendulum. Find the period of a simple pendulum 1.50 m long.

Solution

The period, found from equation 11.37, is

$$T_p = 2\pi \sqrt{\frac{l}{g}}$$

$$= 2\pi \sqrt{\frac{1.50 \text{ m}}{9.80 \text{ m/s}^2}}$$

$$= 2.46 \text{ s}$$

Example 11.5

The length of a pendulum. Find the length of a simple pendulum whose period is 1.00 s.

Solution

The length of the pendulum, found from equation 11.37, is

$$l = \frac{T_p^2}{4\pi^2} g$$

$$= \frac{(1.00 \text{ s})^2}{4\pi^2} (9.80 \text{ m/s}^2)$$

$$= 0.248 \text{ m}$$

Example 11.6

The pendulum and the acceleration of gravity. A pendulum 1.50 m long is observed to have a period of 2.47 s at a certain location. Find the acceleration of gravity there.

Solution

The acceleration of gravity, found from equation 11.38, is

$$g = \frac{4\pi^2}{T_p^2} l$$

$$= \frac{4\pi^2}{(2.47 \text{ s})^2}(1.50 \text{ m})$$

$$= 9.71 \text{ m/s}^2$$

We can also use a pendulum to measure an acceleration. If a pendulum is placed on board a rocket ship in interstellar space and the rocket ship is accelerated at 9.80 m/s², the pendulum oscillates with the same period as it would at rest on the surface of the earth. An enclosed person or thing on the rocket ship could not distinguish between the acceleration of the rocket ship at 9.80 m/s² and the acceleration of gravity of 9.80 m/s² on the earth. (This is an example of Einstein's principle of equivalence in general relativity.) An oscillating pendulum of measured length *l* can be placed in an elevator and the period *T* measured. We can use equation 11.38 to measure the resultant acceleration experienced by the pendulum in the elevator.

11.7 Springs in Parallel and in Series

Sometimes more than one spring is used in a vibrating system. The motion of the system will depend on the way the springs are connected. As an example, suppose there are three massless springs with spring constants k_1, k_2, and k_3. These springs can be connected in parallel, as shown in figure 11.8(a), or in series, as in figure 11.8(b). The period of motion of either configuration can be found by using an equivalent spring constant k_E.

Springs in Parallel

If the total force pulling the mass *m* a distance *x* to the right is F_{tot}, this force will distribute itself among the three springs such that there will be a force F_1 on spring 1, a force F_2 on spring 2, and a force F_3 on spring 3. If the displacement of each spring is equal to *x*, then the springs are said to be in parallel. Then we can write the total force as

$$F_{tot} = F_1 + F_2 + F_3 \tag{11.39}$$

However, since we assumed that each spring was displaced the same distance *x*, Hooke's law for each spring is

$$F_1 = k_1 x$$
$$F_2 = k_2 x$$
$$F_3 = k_3 x \tag{11.40}$$

Substituting equation 11.40 into equation 11.39 gives

$$F_{tot} = k_1 x + k_2 x + k_3 x$$
$$= (k_1 + k_2 + k_3)x$$

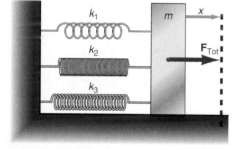

(a) Springs in parallel

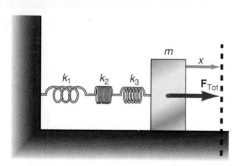

(b) Springs in series

Figure 11.8

Springs in parallel and in series.

We now define an equivalent spring constant k_E for springs connected in parallel as

$$k_E = k_1 + k_2 + k_3 \qquad (11.41)$$

Hooke's law for the combination of springs is given by

$$F_{tot} = k_E x \qquad (11.42)$$

The springs in parallel will execute a simple harmonic motion whose period, found from equation 11.22, is

$$T = 2\pi \sqrt{\frac{m}{k_E}} = 2\pi \sqrt{\frac{m}{k_1 + k_2 + k_3}} \qquad (11.43)$$

Springs in Series

If the same springs are connected in series, as in figure 11.8(b), the total force F_{tot} displaces the mass m a distance x to the right. But in this configuration, each spring stretches a different amount. Thus, the total displacement x is the sum of the displacements of each spring, that is,

$$x = x_1 + x_2 + x_3 \qquad (11.44)$$

The displacement of each spring, found from Hooke's law, is

$$x_1 = \frac{F_1}{k_1}$$

$$x_2 = \frac{F_2}{k_2}$$

$$x_3 = \frac{F_3}{k_3} \qquad (11.45)$$

Substituting these values of the displacement into equation 11.44, yields

$$x = \frac{F_1}{k_1} + \frac{F_2}{k_2} + \frac{F_3}{k_3} \qquad (11.46)$$

But because the springs are in series the total applied force is transmitted equally from spring to spring. Hence,

$$F_{tot} = F_1 = F_2 = F_3 \qquad (11.47)$$

Substituting equation 11.47 into equation 11.46, gives

$$x = \frac{F_{tot}}{k_1} + \frac{F_{tot}}{k_2} + \frac{F_{tot}}{k_3}$$

and

$$x = \left(\frac{1}{k_1} + \frac{1}{k_2} + \frac{1}{k_3} \right) F_{tot} \qquad (11.48)$$

We now define the equivalent spring constant for springs connected in series as

$$\frac{1}{k_E} = \frac{1}{k_1} + \frac{1}{k_2} + \frac{1}{k_3} \qquad (11.49)$$

Thus, the total displacement, equation 11.48, becomes

$$x = \frac{F_{tot}}{k_E} \qquad (11.50)$$

and Hooke's law becomes

$$F_{tot} = k_E x \qquad (11.51)$$

where k_E is given by equation 11.49. Hence, the combination of springs in series executes simple harmonic motion and the period of that motion, given by equation 11.22, is

$$T = 2\pi \sqrt{\frac{m}{k_E}} = 2\pi \sqrt{m\left(\frac{1}{k_1} + \frac{1}{k_2} + \frac{1}{k_3}\right)} \qquad (11.52)$$

Example 11.7

Springs in parallel. Three springs with force constants $k_1 = 10.0$ N/m, $k_2 = 12.5$ N/m, and $k_3 = 15.0$ N/m are connected in parallel to a mass of 0.500 kg. The mass is then pulled to the right and released. Find the period of the motion.

Solution

The period of the motion, found from equation 11.43, is

$$T = 2\pi \sqrt{\frac{m}{k_1 + k_2 + k_3}}$$

$$= 2\pi \sqrt{\frac{0.500 \text{ kg}}{10.0 \text{ N/m} + 12.5 \text{ N/m} + 15.0 \text{ N/m}}}$$

$$= 0.726 \text{ s}$$

Example 11.8

Springs in series. The same three springs as in example 11.7 are now connected in series. Find the period of the motion.

Solution

The period, found from equation 11.52, is

$$T = 2\pi \sqrt{m\left(\frac{1}{k_1} + \frac{1}{k_2} + \frac{1}{k_3}\right)}$$

$$= 2\pi \sqrt{(0.500 \text{ kg})\left(\frac{1}{10.0 \text{ N/m}} + \frac{1}{12.5 \text{ N/m}} + \frac{1}{15.0 \text{ N/m}}\right)}$$

$$= 1.56 \text{ s}$$

The Language of Physics

Periodic motion
Motion that repeats itself in equal intervals of time (p. 297).

Displacement
The distance a vibrating body moves from its equilibrium position (p. 297).

Simple harmonic motion
Periodic motion in which the acceleration of a body is directly proportional to its displacement from the equilibrium position but in the opposite direction. Because the acceleration is directly proportional to the

displacement, the acceleration of the body is not constant. The kinematic equations developed in chapter 3 are no longer valid to describe this type of motion (p. 298).

Amplitude
The maximum displacement of the vibrating body (p. 298).

Cycle
One complete oscillation or vibratory motion (p. 298).

Period
The time for the vibrating body to complete one cycle (p. 298).

Frequency
The number of complete cycles or oscillations in unit time. The frequency is the reciprocal of the period (p. 299).

Reference circle
A body executing uniform circular motion does so in a circle. The projection of the position of the rotating body onto the x- or y-axis is equivalent to simple harmonic motion along that axis. Thus, vibratory motion is related to motion in a circle, the reference circle (p. 300).

Vibratory Motion, Wave Motion, and Fluids

Angular velocity
The angular velocity of the uniform circular motion in the reference circle is related to the frequency of the vibrating system. Hence, the angular velocity is called the angular frequency of the vibrating system (p. 300).

Potential energy of a spring
The energy that a body possesses by virtue of its configuration. A compressed spring has potential energy because it has the ability to do work as it expands to its equilibrium configuration. A stretched spring can also do work as it returns to its equilibrium configuration (p. 306).

Simple pendulum
A bob that is attached to a string and allowed to oscillate to and fro under the action of gravity. If the angle of the pendulum is small the pendulum will oscillate in simple harmonic motion (p. 309).

Summary of Important Equations

Restoring force in a spring
$$F_R = -kx \tag{11.1}$$

Defining relation for simple harmonic motion
$$a = -\frac{k}{m}x \tag{11.2}$$

Frequency
$$f = \frac{1}{T} \tag{11.3}$$

Displacement in simple harmonic motion
$$x = A \cos \omega t \tag{11.6}$$

Angular frequency
$$\omega = 2\pi f \tag{11.9}$$

Velocity as a function of time in simple harmonic motion
$$v = -\omega A \sin \omega t \tag{11.13}$$

Velocity as a function of displacement
$$v = \pm \omega \sqrt{A^2 - x^2} \tag{11.15}$$

Acceleration as a function of time
$$a = -\omega^2 A \cos \omega t \tag{11.17}$$

Angular frequency of a spring
$$\omega = \sqrt{\frac{k}{m}} \tag{11.20}$$

Frequency in simple harmonic motion
$$f = \frac{1}{2\pi}\sqrt{\frac{k}{m}} \tag{11.21}$$

Period in simple harmonic motion
$$T = 2\pi\sqrt{\frac{m}{k}} \tag{11.22}$$

Potential energy of a spring
$$PE = \frac{1}{2}kx^2 \tag{11.24}$$

Conservation of energy for a vibrating spring
$$\frac{1}{2}kA^2 = \frac{1}{2}kx^2 + \frac{1}{2}mv^2 \tag{11.27}$$

Period of motion of a simple pendulum
$$T_p = 2\pi\sqrt{\frac{l}{g}} \tag{11.37}$$

Equivalent spring constant for springs in parallel
$$k_E = k_1 + k_2 + k_3 \tag{11.41}$$

Period of motion for springs in parallel
$$T = 2\pi\sqrt{\frac{m}{k_1 + k_2 + k_3}} \tag{11.43}$$

Equivalent spring constant for springs in series
$$\frac{1}{k_E} = \frac{1}{k_1} + \frac{1}{k_2} + \frac{1}{k_3} \tag{11.49}$$

Period of motion for springs in series
$$T = 2\pi\sqrt{m\left(\frac{1}{k_1} + \frac{1}{k_2} + \frac{1}{k_3}\right)} \tag{11.52}$$

Questions for Chapter 11

1. Can the periodic motion of the earth be considered to be an example of simple harmonic motion?
2. Can the kinematic equations derived in chapter 3 be used to describe simple harmonic motion?
3. In the simple harmonic motion of a mass attached to a spring, the velocity of the mass is equal to zero when the acceleration has its maximum value. How is this possible? Can you think of other examples in which a body has zero velocity with a nonzero acceleration?
4. What is the characteristic of the restoring force that makes simple harmonic motion possible?
5. Discuss the significance of the reference circle in the analysis of simple harmonic motion.

6. How can a mass that is undergoing a one-dimensional translational simple harmonic motion have anything to do with an angular velocity or an angular frequency, which is a characteristic of two or more dimensions?
7. How is the angular frequency related to the physical characteristics of the spring and the vibrating mass in simple harmonic motion?
†8. In the entire derivation of the equations for simple harmonic motion we have assumed that the springs are massless and friction can be neglected. Discuss these assumptions. Describe qualitatively what you would expect to happen to the motion if the springs are not small enough to be considered massless.

†9. Describe how a geological survey for iron might be undertaken on the moon using a simple pendulum.
†10. How could a simple pendulum be used to make an accelerometer?
†11. Discuss the assumption that the displacement of each spring is the same when the springs are in parallel. Under what conditions is this assumption valid and when would it be invalid?

Problems for Chapter 11

11.2 Simple Harmonic Motion and 11.3 Analysis of Simple Harmonic Motion

1. A mass of 0.200 kg is attached to a spring of spring constant 30.0 N/m. If the mass executes simple harmonic motion, what will be its frequency?
2. A 30.0-g mass is attached to a vertical spring and it stretches 10.0 cm. It is then stretched an additional 5.00 cm and released. Find its period of motion and its frequency.
3. A 0.200-kg mass on a spring executes simple harmonic motion at a frequency f. What mass is necessary for the system to vibrate at a frequency of $2f$?
4. A simple harmonic oscillator has a frequency of 2.00 Hz and an amplitude of 10.0 cm. What is its maximum acceleration? What is its acceleration at $t = 0.25$ s?
5. A ball attached to a string travels in uniform circular motion in a horizontal circle of 50.0 cm radius in 1.00 s. Sunlight shining on the ball throws its shadow on a wall. Find the velocity of the shadow at (a) the end of its path and (b) the center of its path.
6. A 50.0-g mass is attached to a spring of force constant 10.0 N/m. The spring is stretched 20.0 cm and then released. Find the displacement, velocity, and acceleration of the mass at 0.200 s.
7. A 25.0-g mass is attached to a vertical spring and it stretches 15.0 cm. It is then stretched an additional 10.0 cm and then released. What is the maximum velocity of the mass? What is its maximum acceleration?
8. The displacement of a body in simple harmonic motion is given by $x = (0.15$ m$)\cos[(5.00$ rad/s$)t]$. Find (a) the amplitude of the motion, (b) the angular frequency, (c) the frequency, (d) the period, and (e) the displacement at 3.00 s.
9. A 500-g mass is hung from a coiled spring and it stretches 10.0 cm. It is then stretched an additional 15.0 cm and released. Find (a) the frequency of vibration, (b) the period, and (c) the velocity and acceleration at a displacement of 10.0 cm.

10. A mass of 0.200 kg is placed on a vertical spring and the spring stretches by 15.0 cm. It is then pulled down an additional 10.0 cm and then released. Find (a) the spring constant, (b) the angular frequency, (c) the frequency, (d) the period, (e) the maximum velocity of the mass, (f) the maximum acceleration of the mass, (g) the maximum restoring force, and (h) the equation of the displacement, velocity, and acceleration at any time t.

11.5 Conservation of Energy and the Vibrating Spring

11. A simple harmonic oscillator has a spring constant of 5.00 N/m. If the amplitude of the motion is 15.0 cm, find the total energy of the oscillator.
12. A body is executing simple harmonic motion. At what displacement is the potential energy equal to the kinetic energy?
13. A 20.0-g mass is attached to a horizontal spring on a smooth table. The spring constant is 3.00 N/m. The spring is then stretched 15.0 cm and then released. What is the total energy of the motion? What is the potential and kinetic energy when $x = 5.00$ cm?
14. A body is executing simple harmonic motion. At what displacement is the speed v equal to one-half the maximum speed?

11.6 The Simple Pendulum

15. Find the period and frequency of a simple pendulum 0.75 m long.
16. If a pendulum has a length L and a period T, what will be the period when (a) L is doubled and (b) L is halved?
17. Find the frequency of a child's swing whose ropes have a length of 3.25 m.
18. What is the period of a 0.500-m pendulum on the moon where $g_m = (1/6)g_e$?
19. What is the period of a pendulum 0.750 m long on a spaceship (a) accelerating at 4.90 m/s² and (b) moving at constant velocity?
20. What is the period of a pendulum in free-fall?
21. A pendulum has a period of 0.750 s at the equator at sea level. The pendulum is carried to another place on the earth and the period is now found to be 0.748 s. Find the acceleration due to gravity at this location.

11.7 Springs in Parallel and in Series

†22. Springs with spring constants of 5.00 N/m and 10.0 N/m are connected in parallel to a 5.00-kg mass. Find (a) the equivalent spring constant and (b) the period of the motion.
†23. Springs with spring constants 5.00 N/m and 10.0 N/m are connected in series to a 5.00-kg mass. Find (a) the equivalent spring constant and (b) the period of the motion.

Additional Problems

24. A 500-g mass is attached to a vertical spring of spring constant 30.0 N/m. How far should the spring be stretched in order to give the mass an upward acceleration of 3.00 m/s²?
25. A ball is caused to move in a horizontal circle of 40.0-cm radius in uniform circular motion at a speed of 25.0 cm/s. Its projection on the wall moves in simple harmonic motion. Find the velocity and acceleration of the shadow of the ball at (a) the end of its motion, (b) the center of its motion, and (c) halfway between the center and the end of the motion.
†26. The motion of the piston in the engine of an automobile is approximately simple harmonic. If the stroke of the piston (twice the amplitude) is equal to 8.00 in. and the engine turns at 1800 rpm, find (a) the acceleration at $x = A$ and (b) the speed of the piston at the midpoint of the stroke.
†27. A 535-g mass is dropped from a height of 25.0 cm above an uncompressed spring of $k = 20.0$ N/m. By how much will the spring be compressed?
28. A simple pendulum is used to operate an electrical device. When the pendulum bob sweeps through the midpoint of its swing, it causes an electrical spark to be given off. Find the length of the pendulum that will give a spark rate of 30.0 sparks per minute.
†29. The general solution for the period of a simple pendulum, without making the assumption of small angles of swing, is given by

$$T = 2\pi \sqrt{\frac{l}{g}} \left[1 + \frac{(\frac{1}{2})^2 \sin^2\theta}{2} + \frac{(\frac{1}{2})^2(\frac{3}{4})^2 \sin^4\theta}{2} + \ldots \right]$$

Find the period of a 1.00-m pendulum for $\theta = 10.0°$, $30.0°$, and $50.0°$ and compare with the period obtained with the small angle approximation. Determine the percentage error in each case by using the small angle approximation.

30. A pendulum clock on the earth has a period of 1.00 s. Will this clock run slow or fast, and by how much if taken to (a) Mars, (b) Moon, and (c) Venus?

†31. A pendulum bob, 355 g, is raised to a height of 12.5 cm before it is released. At the bottom of its path it makes a perfectly elastic collision with a 500-g mass that is connected to a horizontal spring of spring constant 15.8 N/m, that is at rest on a smooth surface. How far will the spring be compressed?

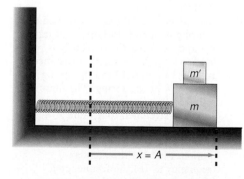

†32. A 500-g block is in simple harmonic motion as shown in the diagram. A mass m' is added to the top of the block when the block is at its maximum extension. How much mass should be added to change the frequency by 25%?

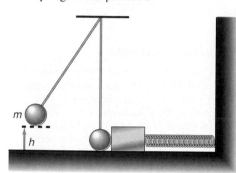

†33. A pendulum clock keeps correct time at a location at sea level where the acceleration of gravity is equal to 9.80 m/s². The clock is then taken up to the top of a mountain and the clock loses 3.00 s per day. How high is the mountain?

†34. Three people, who together weigh 422 lb, get into a car and the car is observed to move 2.00 in. closer to the ground. What is the spring constant of the car springs?

†35. In the accompanying diagram, the mass m is pulled down a distance of 9.50 cm from its equilibrium position and is then released. The mass then executes simple harmonic motion. Find (a) the total potential energy of the mass with respect to the ground when the mass is located at positions 1, 2, and 3; (b) the total energy of the mass at positions 1, 2, and 3; and (c) the speed of the mass at position 2. Assume $m = 55.6$ g, $k = 25.0$ N/m, $h_0 = 50.0$ cm.

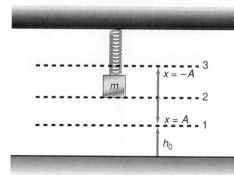

†36. A 20.0-g ball rests on top of a vertical spring gun whose spring constant is 20 N/m. The spring is compressed 10.0 cm and the gun is then fired. Find how high the ball rises in its vertical trajectory.

†37. A toy spring gun is used to fire a ball as a projectile. Find the value of the spring constant, such that when the spring is compressed 10.0 cm, and the gun is fired at an angle of 62.5°, the range of the projectile will be 1.50 m. The mass of the ball is 25.2 g.

†38. In the simple pendulum shown in the diagram, find the tension in the string when the height of the pendulum is (a) h, (b) $h/2$, and (c) $h = 0$. The mass $m = 500$ g, the initial height $h = 15.0$ cm, and the length of the pendulum $l = 1.00$ m.

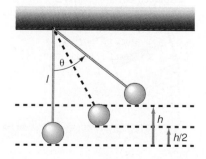

†39. A mass is attached to a horizontal spring. The mass is given an initial amplitude of 10.0 cm on a rough surface and is then released to oscillate in simple harmonic motion. If 10.0% of the energy is lost per cycle due to the friction of the mass moving over the rough surface, find the maximum displacement of the mass after 1, 2, 4, 6, and 8 complete oscillations.

†40. Find the maximum amplitude of vibration after 2 periods for a 85.0-g mass executing simple harmonic motion on a rough horizontal surface of $\mu_k = 0.350$. The spring constant is 24.0 N/m and the initial amplitude is 20.0 cm.

41. A 240-g mass slides down a circular chute without friction and collides with a horizontal spring, as shown in the diagram. If the original position of the mass is 25.0 cm above the table top and the spring has a spring constant of 18 N/m, find the maximum distance that the spring will be compressed.

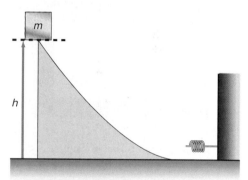

†42. A 235-g block slides down a frictionless inclined plane, 1.25 m long, that makes an angle of 34.0° with the horizontal. At the bottom of the plane the block slides along a rough horizontal surface 1.50 m long. The coefficient of kinetic friction between the block and the rough horizontal surface is 0.45. The block then collides with a horizontal spring of $k = 20.0$ N/m. Find the maximum compression of the spring.

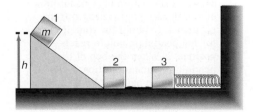

†43. A 335-g disk that is free to rotate about its axis is connected to a spring that is stretched 35.0 cm. The spring constant is 10.0 N/m. When the disk is released it rolls without slipping as it moves toward the equilibrium position. Find the speed of the disk when the displacement of the spring is equal to -17.5 cm.

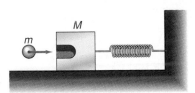

†44. A 25.0-g ball moving at a velocity of 200 cm/s to the right makes an inelastic collision with a 200-g block that is initially at rest on a frictionless surface. There is a hole in the large block for the small ball to fit into. If $k = 10$ N/m, find the maximum compression of the spring.

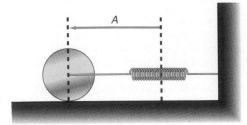

†45. Show that the period of simple harmonic motion for the mass shown is equivalent to the period for two springs in parallel.

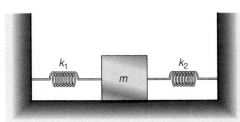

†46. A nail is placed in the wall at a distance of $l/2$ from the top, as shown in the diagram. A pendulum of length 85.0 cm is released from position 1. (a) Find the time it takes for the pendulum bob to reach position 2. When the bob of the pendulum reaches position 2, the string hits the nail. (b) Find the time it takes for the pendulum bob to reach position 3.

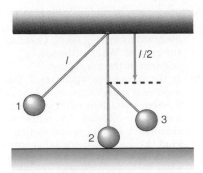

†47. A spring is attached to the top of an Atwood's machine as shown. The spring is stretched to $A = 10$ cm before being released. Find the velocity of m_2 when $x = -A/2$. Assume $m_1 = 28.0$ g, $m_2 = 43.0$ g, and $k = 10.0$ N/m.

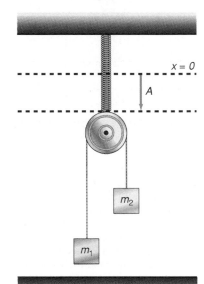

†48. A 280-g block is connected to a spring on a rough inclined plane that makes an angle of 35.5° with the horizontal. The block is pulled down the plane a distance $A = 20.0$ cm, and is then released. The spring constant is 40.0 N/m and the coefficient of kinetic friction is 0.100. Find the speed of the block when the displacement $x = -A/2$.

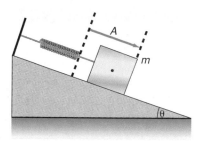

49. The rotational analog of simple harmonic motion, is angular simple harmonic motion, wherein a body rotates periodically clockwise and then counterclockwise. Hooke's law for rotational motion is given by

$$\tau = -C\theta$$

where τ is the torque acting on the body, θ is the angular displacement, and C is a constant, like the spring constant. Use Newton's second law for rotational motion to show

$$\alpha = \frac{C}{I}\theta$$

Use the analogy between the linear result, $a = -\omega^2 x$, to show that the frequency of vibration of an object executing angular simple harmonic motion is given by

$$f = \frac{1}{2\pi}\sqrt{\frac{C}{I}}$$

Interactive Tutorials

⊟ 50. Calculate the period T of a simple pendulum located on a planet having a gravitational acceleration of $g = 9.80$ m/s², if its length $l = 1.00$ m is increased from 1 to 10 m in steps of 1.00 m. Plot the results as the period T versus the length l.

⊟ 51. The displacement x of a body undergoing simple harmonic motion is given by the formula $x = A\cos\omega t$, where A is the amplitude of the vibration, ω is the angular frequency in rad/s, and t is the time in seconds. Plot the displacement x as a function of t for an amplitude $A = 0.150$ m and an angular frequency $\omega = 5.00$ rad/s as t increases from 0 to 2 s in 0.10 s increments.

⊟ 52. A mass $m = 0.500$ kg is attached to a spring on a smooth horizontal table. An applied force $F_A = 4.00$ N is used to stretch the spring a distance $x_0 = 0.150$ m. (a) Find the spring constant k of the spring. The mass is returned to its equilibrium position and then stretched to a value $A = 0.15$ m and then released. The mass then executes simple harmonic motion. Find (b) the angular frequency ω, (c) the frequency f, (d) the period T, (e) the maximum velocity v_{max} of the vibrating mass, (f) the maximum acceleration a_{max} of the vibrating mass, (g) the maximum restoring force F_{Rmax}, and (h) the velocity of the mass at the displacement $x = 0.08$ m. (i) Plot the displacement x, velocity v, acceleration a, and the restoring force F_R at any time t.

53. A mass $m = 0.350$ kg is attached to a horizontal spring. The mass is then pulled a distance $x = A = 0.200$ m from its equilibrium position and when released the mass executes simple harmonic motion. Find (a) the total energy E_{tot} of the mass when it is at its maximum displacement A from its equilibrium position. When the mass is at the displacement $x = 0.120$ m find, (b) its potential energy PE, (c) its kinetic energy KE, and (d) its speed v. (e) Plot the total energy, potential energy, and kinetic energy of the mass as a function of the displacement x. The spring constant $k = 35.5$ N/m.

54. A mass $m = 0.350$ kg is attached to a vertical spring. The mass is at a height $h_0 = 1.50$ m from the floor. The mass is then pulled down a distance $A = 0.220$ m from its equilibrium position and when released executes simple harmonic motion. Find (a) the total energy of the mass when it is at its maximum displacement A below its equilibrium position, (b) the gravitational potential energy when it is at the displacement $x = 0.120$ m, (c) the elastic potential energy when it is at the same displacement x, (d) the kinetic energy at the displacement x, and (e) the speed of the mass when it is at the displacement x. The spring constant $k = 35.5$ N/m.

12 **Wave Motion**

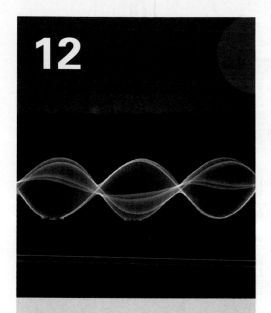

Standing transverse waves in a vibrating string.

Query 17. If a stone be thrown into stagnating water, the waves excited thereby continue some time to arise in the place where the stone fell into the water, and are propagated from thence in concentric circles upon the surface of the water to great distances.

Isaac Newton

12.1 **Introduction**

Everyone has observed that when a rock is thrown into a pond of water, waves are produced that move out from the point of the disturbance in a series of concentric circles. *The **wave** is a propagation of the disturbance through the medium without any net displacement of the medium.* In this case the rock hitting the water initiates the disturbance and the water is the medium through which the wave travels. Of the many possible kinds of waves, the simplest to understand, and the one that we will analyze, is the wave that is generated by an object executing simple harmonic motion. As an example, consider the mass m executing simple harmonic motion in figure 12.1. Attached to the right of m is a very long spring. The spring is so long that it is not necessary to consider what happens to the spring at its far end at this time. When the mass m is pushed out to the position $x = A$, the portion of the spring immediately to the right of A is compressed. This *compression* exerts a force on the portion of the spring immediately to its right, thereby compressing it. It in turn compresses part of the spring to its immediate right. The process continues with the compression moving along the spring, as shown in figure 12.1. As the mass m moves in simple harmonic motion to the displacement $x = -A$, the spring immediately to its right becomes elongated. We call the elongation of the spring a *rarefaction;* it is the converse of a compression. As the mass m returns to its equilibrium position, the rarefaction moves down the length of the spring. The combination of a compression and rarefaction comprise part of a longitudinal wave. *A **longitudinal wave** is a wave in which the particles of the medium oscillate in simple harmonic motion parallel to the direction of the wave propagation.* The compressions and rarefactions propagate down the spring, as shown in figure 12.1(f). The mass m in simple harmonic motion generated the wave and the wave moves to the right with a velocity v. Every portion of the medium, in this case the spring, executes simple harmonic motion around its equilibrium position. The medium oscillates back and forth with motion parallel to the wave velocity. Sound is an example of a longitudinal wave.

Another type of wave, and one easier to visualize, is a transverse wave. *A **transverse wave** is a wave in which the particles of the medium execute simple harmonic motion in a direction perpendicular to its direction of propagation.* A transverse wave can be generated by a mass having simple harmonic motion in the vertical direction, as shown in figure 12.2. A horizontal string is connected to the mass as shown. As the mass executes simple harmonic motion in the vertical direction, the end of the string does likewise. As the end moves up and down, it causes the particle next to it to follow suit. It, in turn, causes the next particle to move. Each particle

Figure 12.1

Generation of a longitudinal wave.

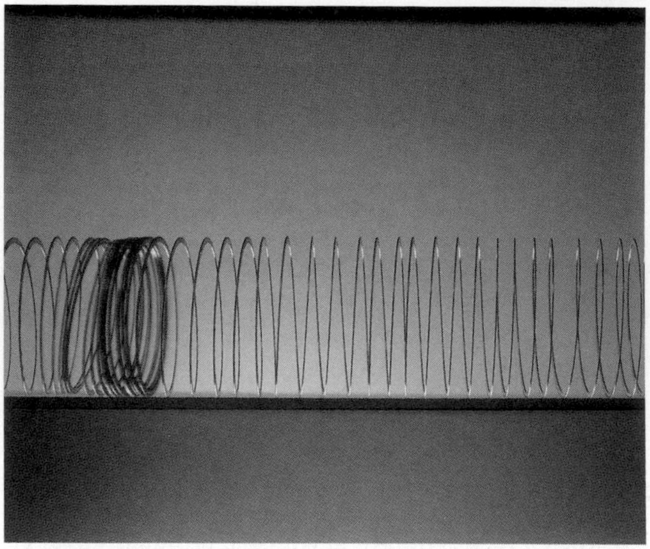

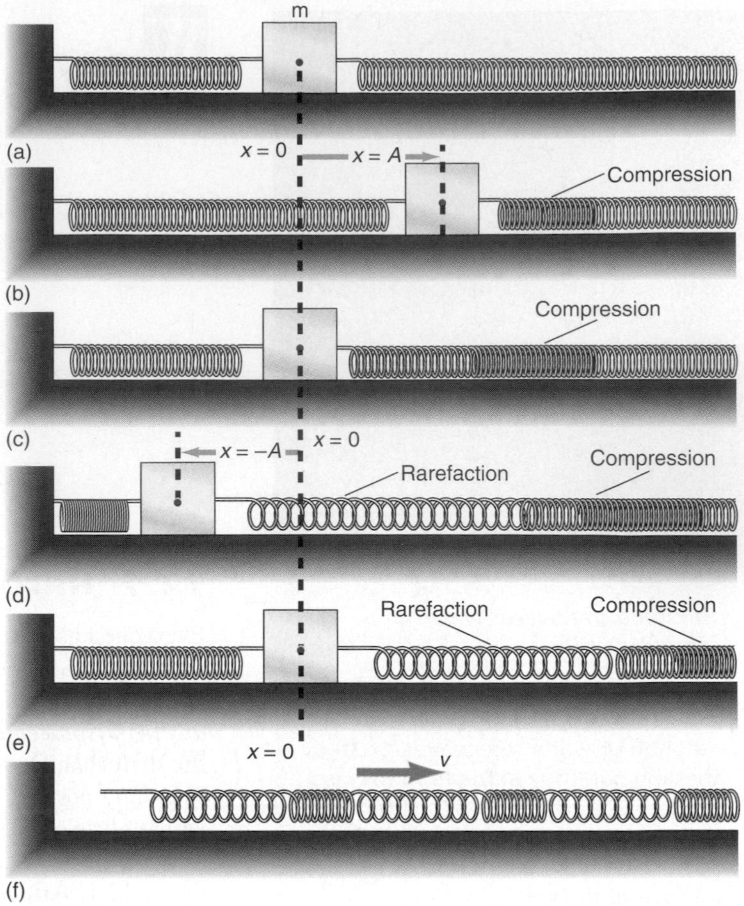

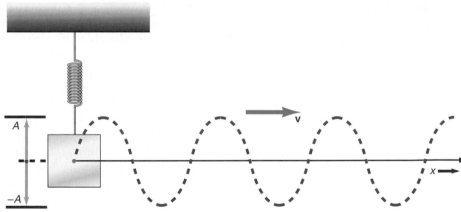

(a)
(b)
(c)
(d)
(e)
(f)

Figure 12.2

A transverse wave.

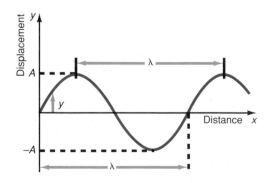

Figure 12.3

Characteristics of a simple wave.

transmits the motion to the next particle along the entire length of the string. The resulting wave propagates in the horizontal direction with a velocity v, while any one particle of the string executes simple harmonic motion in the vertical direction. The particle of the string is moving perpendicular to the direction of wave propagation, and is not moving in the direction of the wave.

Using figure 12.3, let us now define the characteristics of a transverse wave moving in a horizontal direction.

*The **displacement** of any particle of the wave is the displacement of that particle from its equilibrium position* and is measured by the vertical distance y.

*The **amplitude** of the wave is the maximum value of the displacement* and is denoted by A in figure 12.3.

The **wavelength** of a wave is the distance, in the direction of propagation, in which the wave repeats itself and is denoted by λ.

The **period** T of a wave is the time it takes for one complete wave to pass a particular point.

The **frequency** f of a wave is defined as the number of waves passing a particular point per second.

It is obvious from the definitions that the frequency is the reciprocal of the period, that is,

$$f = \frac{1}{T} \tag{12.1}$$

The speed of propagation of the wave is the distance the wave travels in unit time. Because a wave of one wavelength passes a point in a time of one period, its speed of propagation is

$$v = \frac{\text{distance traveled}}{\text{time}} = \frac{\lambda}{T} \tag{12.2}$$

Using equation 12.1, this becomes

$$v = \lambda f \tag{12.3}$$

Equation 12.3 is the fundamental equation of wave propagation. It relates the speed of the wave to its wavelength and frequency.

Example 12.1

Wavelength of sound. The human ear can hear sounds from a low of 20.0 Hz up to a maximum frequency of about 20,000 Hz. If the speed of sound in air at a temperature of 0 °C is 331 m/s, find the wavelengths associated with these frequencies.

Solution

The wavelength of a sound wave, determined from equation 12.3, is

$$\lambda = \frac{v}{f}$$

$$= \frac{331 \text{ m/s}}{20.0 \text{ cycles/s}}$$

$$= 16.6 \text{ m}$$

and

$$\lambda = \frac{v}{f} = \frac{331 \text{ m/s}}{20,000 \text{ cycles/s}}$$

$$= 0.0166 \text{ m}$$

The type of waves we consider in this chapter are called mechanical waves. *The wave causes a transfer of energy from one point in the medium to another point in the medium without the actual transfer of matter between these points.* Another type of wave, called an electromagnetic wave, is capable of traveling through empty space without the benefit of a medium. This type of wave is extremely unusual in this respect and we will treat it in more detail in chapters 25 and 29.

12.2 Mathematical Representation of a Wave

The simple wave shown in figure 12.3 is a picture of a transverse wave in a string at a particular time, let us say at $t = 0$. The wave can be described as a sine wave and can be expressed mathematically as

$$y = A \sin x \tag{12.4}$$

The value of y represents the displacement of the string at every position x along the string, and A is the maximum displacement, and is called the amplitude of the wave. Equation 12.4 is plotted in figure 12.4. We see that the wave repeats itself for $x = 360° = 2\pi$ rad. Also plotted in figure 12.4 is $y = A \sin 2x$ and $y = A \sin 3x$. Notice from the figure that $y = A \sin 2x$ repeats itself twice in the same interval of 2π that $y = A \sin x$ repeats itself only once. Also note that $y = A \sin 3x$ repeats itself three times in that same interval of 2π. The wave $y = A \sin kx$ would repeat itself k times in the interval of 2π. We call the space interval in which $y = A \sin x$ repeats itself its wavelength, denoted by λ_1. Thus, when $x = \lambda_1 = 2\pi$, the wave starts to repeat itself. The wave represented by $y = A \sin 2x$ repeats itself for $2x = 2\pi$, and hence its wavelength is

$$\lambda_2 = x = \frac{2\pi}{2} = \pi$$

The wave $y = A \sin 3x$ repeats itself when $3x = 2\pi$, hence its wavelength is

$$\lambda_3 = x = \frac{2\pi}{3}$$

Using this notation any wave can be represented as

$$y = A \sin kx \tag{12.5}$$

where k is a number, called the *wave number*. The wave repeats itself whenever

$$kx = 2\pi \tag{12.6}$$

Because the value of x for a wave to repeat itself is its wavelength λ, equation 12.6 can be written as

$$k\lambda = 2\pi \tag{12.7}$$

Figure 12.4

Plot of $A \sin x$, $A \sin 2x$, and $A \sin 3x$.

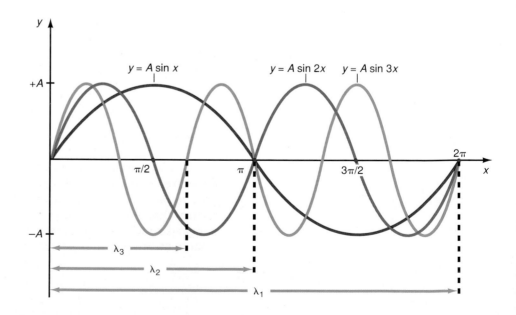

We can obtain the wavelength λ from equation 12.7 as

$$\lambda = \frac{2\pi}{k} \tag{12.8}$$

Note that equation 12.8 gives the wavelengths in figure 12.4 by letting k have the values 1, 2, 3, and so forth, that is,

$$\lambda_1 = \frac{2\pi}{1}$$

$$\lambda_2 = \frac{2\pi}{2}$$

$$\lambda_3 = \frac{2\pi}{3}$$

$$\lambda_4 = \frac{2\pi}{4}$$

We observe that the wave number k is the number of waves contained in the interval of 2π. We can express the wave number k in terms of the wavelength λ by rearranging equation 12.8 into the form

$$k = \frac{2\pi}{\lambda} \tag{12.9}$$

Note that in order for the units to be consistent, the wave number must have units of m^{-1}. The quantity x in equation 12.5 represents the location of any point on the string and is measured in meters. The quantity kx in equation 12.5 has the units $(m^{-1}m = 1)$ and is thus a dimensionless quantity and represents an angle measured in radians. Also note that the wave number k is a different quantity than the spring constant k, discussed in chapter 11.

Equation 12.5 represents a snapshot of the wave at $t = 0$. That is, it gives the displacement of every particle of the string at time $t = 0$. As time passes, this wave, and every point on it, moves. Since each particle of the string executes simple harmonic motion in the vertical, we can look at the particle located at the point $x = 0$ and see how that particle moves up and down with time. Because the particle executes simple harmonic motion in the vertical, it is reasonable to represent the displacement of the particle of the string at any time t as

$$y = A \sin \omega t \tag{12.10}$$

just as a simple harmonic motion on the x-axis was represented as $x = A \cos \omega t$ in chapter 11. The quantity t is the time and is measured in seconds, whereas the quantity ω is an angular velocity or an angular frequency and is measured in radians per second. Hence the quantity ωt represents an angle measured in radians. The displacement y repeats itself when $t = T$, the period of the wave. Since the sine function repeats itself when the argument is equal to 2π, we have

$$\omega T = 2\pi \tag{12.11}$$

The period of the wave is thus

$$T = \frac{2\pi}{\omega}$$

but the period of the wave is the reciprocal of the frequency. Therefore,

$$T = \frac{1}{f} = \frac{2\pi}{\omega}$$

Solving for the angular frequency ω, in terms of the frequency f, we get

$$\omega = 2\pi f \tag{12.12}$$

Notice that the wave is periodic in both space and time. The space period is represented by the wavelength λ, and the time by the period T.

Equation 12.5 represents every point on the string at $t = 0$, while equation 12.10 represents the point $x = 0$ for every time t. Obviously the general equation for a wave must represent every point x of the wave at every time t. We can arrange this by combining equations 12.5 and 12.10 into the one equation for a wave given by

$$y = A \sin(kx - \omega t) \tag{12.13}$$

The reason for the minus sign for ωt is explained below. We can find the relation between the wave number k and the angular frequency ω by combining equations 12.7 and 12.11 as

$$k\lambda = 2\pi \tag{12.7}$$

$$\omega T = 2\pi \tag{12.11}$$

Thus,

$$\omega T = k\lambda$$

and

$$\omega = \frac{k\lambda}{T}$$

However, the wavelength λ, divided by the period T is equal to the velocity of propagation of the wave v, equation 12.2. Therefore, the angular frequency becomes

$$\omega = kv \tag{12.14}$$

Now we can write equation 12.13 as

$$y = A \sin(kx - kvt) \tag{12.15}$$

or

$$y = A \sin k(x - vt) \tag{12.16}$$

The minus sign before the velocity v determines the direction of propagation of the wave. As an example, consider the wave

$$y_1 = A \sin k(x - vt) \tag{12.17}$$

We will now see that this is the equation of a wave traveling to the right with a speed v at any time t. A little later in time, Δt, the wave has moved a distance Δx to the right such that the same point of the wave now has the coordinates $x + \Delta x$ and $t + \Delta t$, figure 12.5(a). Then we represent the wave as

$$y_2 = A \sin k[(x + \Delta x) - v(t + \Delta t)]$$

or

$$y_2 = A \sin k[(x - vt) + \Delta x - v\Delta t] \tag{12.18}$$

If this equation for y_2 is to represent the same wave as y_1, then y_2 must be equal to y_1. It is clear from equations 12.18 and 12.17 that if

$$v = \frac{\Delta x}{\Delta t} \tag{12.19}$$

the velocity of the wave to the right, then

$$\Delta x - v\Delta t = \Delta x - \frac{\Delta x}{\Delta t}\Delta t = 0$$

Figure 12.5

A traveling wave.

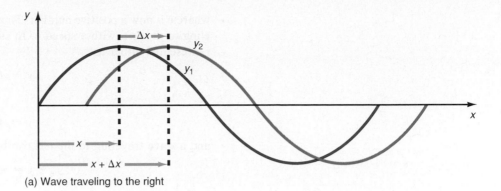

(a) Wave traveling to the right

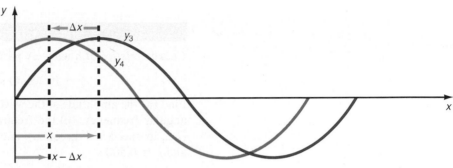

(b) Wave traveling to the left

and y_2 is equal to y_1. Because the term $\Delta x - v\Delta t$ is indeed equal to zero, y_2 is the same wave as y_1 only displaced a distance Δx to the right in the time Δt. Thus, equation 12.17 represents a wave traveling to the right with a velocity of propagation v.

A wave traveling to the left is depicted in figure 12.5(b) and we will begin by representing it as

$$y_3 = A \sin k(x - vt) \qquad (12.20)$$

In a time Δt, the wave y_3 moves a distance $-\Delta x$ to the left. The coordinates (x,t) of a point on y_3 now has the coordinates $x - \Delta x$ and $t + \Delta t$ for the same point on y_4. We can now write the new wave as

$$y_4 = A \sin k[(x - \Delta x) - v(t + \Delta t)]$$

or

$$y_4 = A \sin k[(x - vt) + (-\Delta x - v\Delta t)] \qquad (12.21)$$

The wave y_4 represents the same wave as y_3, providing $-\Delta x - v\Delta t = 0$ in equation 12.21. If $v = -\Delta x/\Delta t$, the velocity of the wave to the left, then

$$-\Delta x - v\Delta t = -\Delta x - \left(\frac{-\Delta x}{\Delta t}\right)\Delta t = 0$$

Thus, $-\Delta x - v\Delta t$ is indeed equal to zero, and wave y_4 represents the same wave as y_3 only it is displaced a distance $-\Delta x$ to the left in the time Δt. Instead of writing the equation 12.20 as a wave to the left with v a negative number, it is easier to write the equation for the wave to the left as

$$y = A \sin k(x + vt) \qquad (12.22)$$

where v is now a positive number. Therefore, equation 12.22 represents a wave traveling to the left, with a speed v. In summary, a wave traveling to the right can be represented either as

$$y = A \sin k(x - vt) \qquad (12.23)$$

or

$$y = A \sin(kx - \omega t) \qquad (12.24)$$

and a wave traveling to the left can be represented as either

$$y = A \sin k(x + vt) \qquad (12.25)$$

or

$$y = A \sin(kx + \omega t) \qquad (12.26)$$

Example 12.2

Characteristics of a wave. A particular wave is given by

$$y = (0.200 \text{ m}) \sin[(0.500 \text{ m}^{-1})x - (8.20 \text{ rad/s})t]$$

Find (a) the amplitude of the wave, (b) the wave number, (c) the wavelength, (d) the angular frequency, (e) the frequency, (f) the period, (g) the velocity of the wave (i.e., its speed and direction), and (h) the displacement of the wave at $x = 10.0$ m and $t = 0.500$ s.

Solution

The characteristics of the wave are determined by writing the wave in the standard form

$$y = A \sin(kx - \omega t)$$

a. The amplitude A is determined by inspection of both equations as $A = 0.200$ m.
b. The wave number k is found from inspection to be $k = 0.500 \text{ m}^{-1}$ or a half a wave in an interval of 2π.
c. The wavelength λ, found from equation 12.8, is

$$\lambda = \frac{2\pi}{k} = \frac{2\pi}{0.500 \text{ m}^{-1}}$$

$$= 12.6 \text{ m}$$

d. The angular frequency ω, found by inspection, is

$$\omega = 8.20 \text{ rad/s}$$

e. The frequency f of the wave, found from equation 12.12, is

$$f = \frac{\omega}{2\pi} = \frac{8.20 \text{ rad/s}}{2\pi \text{ rad}} = 1.31 \text{ cycles/s} = 1.31 \text{ Hz}$$

f. The period of the wave is the reciprocal of the frequency, thus

$$T = \frac{1}{f} = \frac{1}{1.31 \text{ Hz}} = 0.766 \text{ s}$$

g. The speed of the wave, found from equation 12.14, is

$$v = \frac{\omega}{k} = \frac{8.20 \text{ rad/s}}{0.500 \text{ m}^{-1}} = 16.4 \text{ m/s}$$

We could also have determined this by

$$v = f\lambda = \left(1.31\ \frac{1}{s}\right)(12.6\ m) = 16.4\ m/s$$

The direction of the wave is to the right because the sign in front of ω is negative.

h. The displacement of the wave at $x = 10.0$ m and $t = 0.500$ s is

$$y = (0.200\ m)\sin[(0.500\ m^{-1})(10.0\ m) - (8.20\ rad/s)(0.500\ s)]$$
$$= (0.200\ m)\sin[0.900\ rad] = (0.200\ m)(0.783)$$
$$= 0.157\ m$$

12.3 The Speed of a Transverse Wave on a String

Let us consider the motion of a transverse wave on a string, as shown in figure 12.6. The wave is moving to the right with a velocity v. Let us observe the wave by moving with the wave at the same velocity v. In this reference frame, the wave appears stationary, while the particles composing the string appear to be moving through the wave to the left. One such particle is shown at the top of the wave of figure 12.6 moving to the left at the velocity v. If we consider only a small portion of the top of the wave, we can approximate it by an arc of a circle of radius R, as shown. If the angle θ is small, the length of the string considered is small and the mass m of this small portion of the string can be approximated by a mass moving in uniform circular motion. Hence, there must be a centripetal force acting on this small portion of the string and its magnitude is given by

$$F_c = \frac{mv^2}{R} \tag{12.27}$$

This centripetal force is supplied by the tension in the string. In figure 12.6, the tensions on the right and left side of m are resolved into components. There is a force $T\cos\theta$ acting to the right of m and a force $T\cos\theta$ acting to the left. These components are equal and opposite and cancel each other out, thus exerting a zero

Figure 12.6

Velocity of a transverse wave.

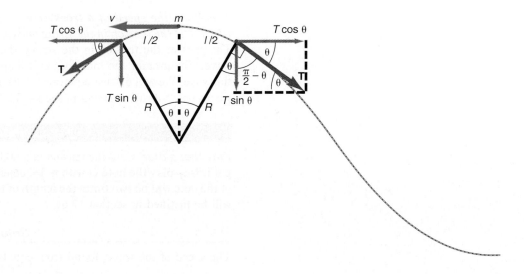

net force in the horizontal direction. The components $T \sin \theta$ on the right and left side of m act downward on m and thus supply the necessary centripetal force for m to be in uniform circular motion. Thus,

$$F_c = 2T \sin \theta$$

Since we assume that θ is small, the $\sin \theta$ can be replaced by the angle θ itself, expressed in radians. Thus,

$$F_c = 2T \theta \qquad (12.28)$$

The small portion of the string l approximates an arc of a circle and the arc of a circle is given by $s = R\theta$. Therefore,

$$s = \frac{l}{2} + \frac{l}{2} = R(\theta + \theta) = 2\theta R$$

and

$$\theta = \frac{l}{2R}$$

The centripetal force, equation 12.28, becomes

$$F_c = 2T\frac{l}{2R}$$

and

$$F_c = \frac{Tl}{R} \qquad (12.29)$$

Equating the centripetal force in equations 12.27 and 12.29 we get

$$\frac{mv^2}{R} = \frac{Tl}{R}$$

$$v^2 = \frac{Tl}{m} = \frac{T}{m/l}$$

Solving for the speed of the wave we get

$$v = \sqrt{\frac{T}{m/l}} \qquad (12.30)$$

Therefore, the speed of a transverse wave in a string is given by equation 12.30, where T is the tension in the string and m/l is the mass per unit length of the string. The greater the tension in the string, the greater the speed of propagation of the wave. The greater the mass per unit length of the string, the smaller the speed of the wave. We will discuss equation 12.30 in more detail when dealing with traveling waves on a vibrating string in section 12.6.

Example 12.3

Play that guitar. Find the tension in a 60.0-cm guitar string that has a mass of 1.40 g if it is to play the note G with a frequency of 396 Hz. Assume that the wavelength of the note will be two times the length of the string, or $\lambda = 120$ cm (this assumption will be justified in section 12.6).

Solution

The speed of the wave, found from equation 12.3, is

$$v = \lambda f = (1.20 \text{ m})(396 \text{ cycles/s})$$

$$= 475 \text{ m/s}$$

The mass density of the string is

$$\frac{m}{l} = \frac{1.40 \times 10^{-3} \text{ kg}}{0.600 \text{ m}} = 2.33 \times 10^{-3} \text{ kg/m}$$

The tension that the guitar string must have in order to play this note, found from equation 12.30, is

$$T = v^2 \frac{m}{l} = (475 \text{ m/s})^2 (2.33 \times 10^{-3} \text{ kg/m})$$

$$= 526 \text{ N}$$

Example 12.4

Sounds flat to me. If the tension in the guitar string of example 12.3 was 450 N, would the guitar play that note flat or sharp?

Solution

The mass density of the string is 2.33×10^{-3} kg/m. With a tension of 450 N, the speed of the wave is

$$v = \sqrt{\frac{T}{m/l}} = \sqrt{\frac{450 \text{ N}}{2.33 \times 10^{-3} \text{ kg/m}}}$$

$$= 439 \text{ m/s}$$

The frequency of the wave is then

$$f = \frac{v}{\lambda} = \frac{439 \text{ m/s}}{1.20 \text{ m}} = 366 \text{ Hz}$$

The string now plays a note at too low a frequency and the note is flat by 396 Hz − 366 Hz = 30 Hz.

12.4 Reflection of a Wave at a Boundary

In the analysis of the vibrating string we assumed that the string was infinitely long so that it was not necessary to consider what happens when the wave gets to the end of the string. Now we need to rectify this omission by considering the **reflection of a wave at a boundary.** To simplify the discussion let us deal with a single pulse rather than the continuous waves dealt with in the preceding sections.

First, let us consider how a pulse is generated. Take a piece of string fixed at one end and hold the other end in your hand, as shown in figure 12.7(a). The string and hand are at rest. If the hand is moved up rapidly, the string near the hand will also be pulled up. This is shown in figure 12.7(b) with arrows pointing upward representing the force upward on the particles of the string. Each particle that moves upward exerts a force on the particle immediately to its right by the tension in the string. In this way, the force upward is passed from particle to particle along the string. In figure 12.7(c), the hand is quickly moved downward pulling the end of the string down with it. The force acting on the string downward is shown by the arrows pointing downward in figure 12.7(c). Note that the arrows pointing upward caused by the force upward in figure 12.7(b) are still upward and moving toward the right. In figure 12.7(d), the hand has returned to the equilibrium position and is at rest. However, the motion of the hand upward and downward has created a pulse that is moving along the string with a velocity of propagation v. The arrows upward represent the force pulling the string upward in advance of the center of the pulse, while the arrows downward represent the force pulling the string down-

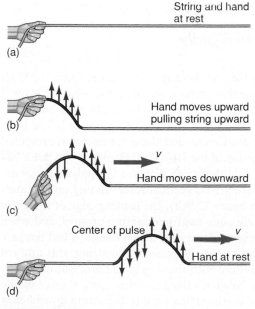

Figure 12.7

Creation of a pulse on a string.

String and hand at rest (a)

Hand moves upward pulling string upward (b)

Hand moves downward (c)

Center of pulse Hand at rest (d)

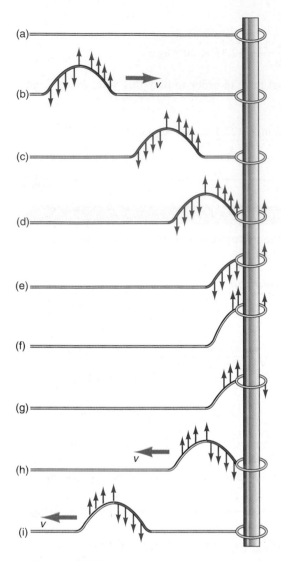

Figure 12.8

Reflection of a pulse on a string that is free to move in the transverse direction.

ward, behind the center of the pulse, back to its rest position. As the pulse propagates so will these forces. Let us now consider what happens to this pulse as it comes to a boundary, in this case, the end of the string.

The End of the String Is Not Fixed Rigidly but Is Allowed to Move

Let us first consider the case of a pulse propagating along a string that is free to move at its end point. This is shown in figure 12.8. The end of the stationary string, figure 12.8(a), is attached to a ring that is free to move in the vertical direction on a frictionless pole. A pulse is now sent down the string in figure 12.8(b). Let us consider the forces on the string that come from the pulse, that is, we will ignore the gravitational forces on the string. The arrows up and down on the pulse represent the forces upward and downward, respectively, on the particles of the string. The pulse propagates to the right in figure 12.8(c) and reaches the ring in figure 12.8(d). The force upward that has been propagating to the right causes the ring on the end of the string to move upward. The ring now rises to the height of the pulse as the center of the pulse arrives at the ring, figure 12.8(e). Although there is no additional force upward on the ring, the ring continues to move upward because of its momentum.

We can also consider this from an energy standpoint. At the location of the top of the pulse, the ring has a kinetic energy upward. The ring continues upward until this kinetic energy is lost. As the ring moves upward it now pulls the string up with it, eventually overcoming the forces downward at the rear of the pulse, until the string to the left of the ring has a net force upward acting on it, figure 12.8(f). This upward force is now propagated along the string to the left by pulling each adjacent particle to its left upward. Because the ring pulled upward on the string, by Newton's third law the string also pulls downward on the ring, and the ring eventually starts downward, figure 12.8(g). As the ring moves downward it exerts a force downward on the string, as shown by the arrows in figure 12.8(h). The forces upward and downward propagate to the left as the pulse shown in figure 12.8(i).

The net result of the interaction of the pulse to the right with the movable ring is a reflected pulse of the same size and shape that now moves to the left with the same speed of propagation. The incoming pulse was right side up, and the reflected pulse is also right side up. The movable ring at the end of the string acts like the hand, moving up and down to create the pulse in figure 12.7.

The End of the String Is Fixed Rigidly and Not Allowed to Move

A pulse is sent down a string that has the end fixed to a wall, as in figure 12.9(a). Let us consider the forces on the string that come from the pulse, that is, we will ignore the gravitational forces on the string. The front portion of the pulse has forces that are acting upward and are represented by arrows pointing upward. The back portion of the pulse has forces acting downward, and these forces are represented by arrows pointing downward. Any portion of the string in advance of the pulse has no force in the vertical direction acting on the string because the pulse has not arrived yet. Hence, in figure 12.9(a), there are no vertical forces acting on that part of the string that is tied to the wall. In figure 12.9(b), the leading edge of the pulse has just arrived at the wall. This leading edge has forces acting upward, and when they make contact with the wall they exert a force upward on the wall. But because of the enormous mass of the wall compared to the mass of the string, this upward force can not move the wall upward as it did with the ring in the previous case. The end of the string remains fixed. But by Newton's third law this upward force on the wall causes a reaction force downward on the string pulling the string down below the equilibrium position of the string, figure 12.9(c). This initiates the beginning of a pulse that moves to the left. At this point the picture becomes rather complicated,

Vibratory Motion, Wave Motion, and Fluids

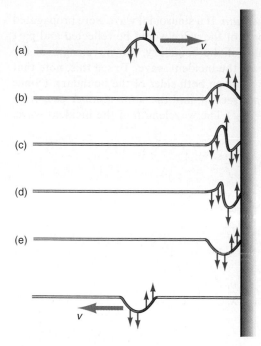

Figure 12.9

A reflected pulse on a string with a fixed end.

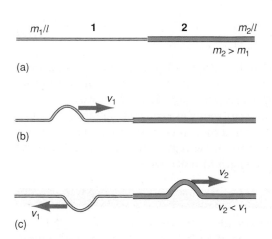

Figure 12.10

A pulse goes from a less dense medium to a more dense medium.

because while the back portion of the original pulse is still moving toward the right, the front portion has become reflected and is moving toward the left, figure 12.9(d). The resulting pulse becomes a superposition of these two pulses, one traveling toward the right, the other toward the left. The forces acting on each particle of the string become the sum of the forces caused by each pulse. When the back portion of the original pulse reaches the wall it exerts a force downward on the wall. By Newton's third law the wall now exerts a reaction force upward on the string that pulls the string of the rear of the pulse upward to its equilibrium position, figure 12.9(e). The pulse has now been completely reflected by the wall and moves to the left with the same speed v, figure 12.9(f). Note, however, that in reflecting the pulse, the reaction force of the wall on the string has caused the reflected pulse to be inverted or turned upside down. *Hence, a wave or pulse that is reflected from a fixed end is inverted.* The reflected pulse is said to be 180° out of phase with the incident pulse. This was not the case for the string whose end was free to move.

Reflection and Transmission of a Wave at the Boundary of Two Different Media

When an incident wave impinges upon a boundary separating two different media, part of the incident wave is transmitted into the second medium while part is reflected back into the first medium. We can easily see this effect by connecting together two strings of different mass densities. Let us consider two different cases of the **reflection and transmission of a wave at the boundary of two different media.**

Case 1: *The Wave Goes from the Less Dense Medium to the More Dense Medium*

Consider the string in figure 12.10(a). The left-hand side of the combined string is a light string of mass density m_1/l, while the right-hand side is a heavier string of mass density m_2/l, where m_2 is greater than m_1. The combined string is pulled tight so that both strings have the same tension T. A pulse is now sent down the lighter string at a velocity v_1 to the right in figure 12.10(b). As the pulse hits the boundary between the two strings, the upward force in the leading edge of the pulse on string 1 exerts an upward force on string 2. Because string 2 is much more massive than string 1, the boundary acts like the fixed end in figure 12.9, and the reaction force of the massive string causes an inverted reflected pulse to travel back along string 1, as shown in figure 12.10(c). Because the massive string is not infinite, like a rigid wall, the forces of the incident pulse pass through to the massive string, thus also transmitting a pulse along string 2, as shown in figure 12.10(c). Since string 2 is more massive than string 1, the transmitted force can not displace the massive string elements of string 2 as much as in string 1. Hence the amplitude of the transmitted pulse is less than the amplitude of the incident pulse.

Because the tension in each string is the same, the speed of the pulses in medium 1 and 2 are

$$v_1 = \sqrt{\frac{T}{m_1/l}} \tag{12.31}$$

$$v_2 = \sqrt{\frac{T}{m_2/l}} \tag{12.32}$$

Because the tension T in each string is the same, they can be equated to find the speed of the pulse in medium 2 as

$$v_2 = \sqrt{\frac{m_1/l}{m_2/l}}\, v_1 \tag{12.33}$$

However, because m_2 is greater than m_1, equation 12.33 implies that v_2 will be less than v_1. That is, the speed v_2 of the transmitted pulse will be less than v_1, the speed of the incident and reflected pulses. *Thus, the pulse slows down in going from the less dense medium to the more dense medium.* If a sinusoidal wave were propagated along the string instead of the pulse, part of the wave would be reflected and part would be transmitted. However, because of the boundary, the wavelength of the transmitted wave would be different from the incident wave. To see this, note that the frequency of the wave must be the same on both sides of the boundary. (Since the frequency is the number of waves per second, and the same number pass from medium 1 into medium 2, we have $f_1 = f_2$.) The wavelength of the incident wave, found from equation 12.3, is

$$\lambda_1 = \frac{v_1}{f}$$

whereas the wavelength of the transmitted wave λ_2 is

$$\lambda_2 = \frac{v_2}{f}$$

Because the frequency is the same,

$$f = \frac{v_1}{\lambda_1}$$

and

$$f = \frac{v_2}{\lambda_2}$$

they can be equated giving

$$\frac{v_1}{\lambda_1} = \frac{v_2}{\lambda_2}$$

Thus, the wavelength of the transmitted wave λ_2 is

$$\lambda_2 = \frac{v_2}{v_1} \lambda_1 \tag{12.34}$$

Since v_2 is less than v_1, equation 12.34 tells us that λ_2 is less than λ_1. *Hence, when a wave goes from a less dense medium to a more dense medium, the wavelength of the transmitted wave is less than the wavelength of the incident wave.*

Although these results were derived from waves on a string, they are quite general. In chapter 27 we will see that when a light wave goes from a region of low density such as a vacuum or air, into a more dense region, such as glass, the speed of the light wave decreases and its wavelength also decreases.

Example 12.5

A wave going from a less dense to a more dense medium. One end of a 60.0-cm steel wire of mass 1.40 g is welded to the end of a 60.0-cm steel wire of 6.00 g mass. The combined wires are placed under uniform tension. (a) If a wave propagates down the lighter wire at a speed of 475 m/s, at what speed will it be transmitted along the heavier wire? (b) If the wavelength on the first wire is 1.20 m, what is the wavelength on the second wire?

Solution

a. The mass per unit length of each wire is

$$\frac{m_1}{l} = \frac{1.40 \times 10^{-3} \text{ kg}}{0.600 \text{ m}} = 2.20 \times 10^{-3} \text{ kg/m}$$

$$\frac{m_2}{l} = \frac{6.00 \times 10^{-3} \text{ kg}}{0.600 \text{ m}} = 1.00 \times 10^{-2} \text{ kg/m}$$

Vibratory Motion, Wave Motion, and Fluids

The speed of the transmitted wave, found from equation 12.33, is

$$v_2 = \sqrt{\frac{m_1/l}{m_2/l}}\, v_1$$

$$= \sqrt{\frac{2.20 \times 10^{-3} \text{ kg/m}}{1.00 \times 10^{-2} \text{ kg/m}}} \, (475 \text{ m/s})$$

$$= 223 \text{ m/s}$$

b. The wavelength of the transmitted wave, found from equation 12.34, is

$$\lambda_2 = \frac{v_2}{v_1} \lambda_1$$

$$= \left(\frac{223 \text{ m/s}}{475 \text{ m/s}}\right)(1.20 \text{ m})$$

$$= 0.563 \text{ m}$$

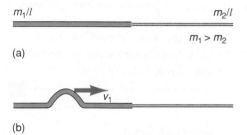

m_1/l m_2/l

$m_1 > m_2$

(a)

(b)

v_1 v_2

$v_2 > v_1$

(c)

Figure 12.11

A pulse goes from a more dense medium to a less dense medium.

Case 2: *A Wave Goes from a More Dense Medium to a Less Dense Medium*

Consider the string in figure 12.11(a). The left-hand side of the combined string is a heavy string of mass density m_1/l, whereas the right-hand side is a light string of mass density m_2/l, where m_2 is now less than m_1. A pulse is sent down the string in figure 12.11(b). When the pulse hits the boundary the boundary acts like the free end of the string in figure 12.8 because of the low mass of the second string. A pulse is reflected along the string that is erect or right side up, figure 12.11(c). However, the forces of the incident pulse are transmitted very easily to the lighter second string and a transmitted pulse also appears in string 2, figure 12.11(c). Because the tension is the same in both strings, a similar analysis to case 1 shows that *when a wave goes from a more dense medium to a less dense medium, the transmitted wave moves faster than the incident wave and has a longer wavelength.*

Example 12.6

A wave going from a more dense medium to a less dense medium. The first half of a combined string has a linear mass density of 0.100 kg/m, whereas the second half has a linear mass density of 0.0500 kg/m. A sinusoidal wave of wavelength 1.20 m is sent along string 1. If the combined string is under a tension of 10.0 N, find (a) the speed of the incident wave in string 1, (b) the speed of the transmitted wave in string 2, (c) the wavelength of the transmitted wave, and (d) the speed and wavelength of the reflected wave.

Solution

a. The speed of the incident wave in string 1, found from equation 12.31, is

$$v_1 = \sqrt{\frac{T}{m_1/l}}$$

$$= \sqrt{\frac{10.0 \text{ N}}{0.100 \text{ kg/m}}}$$

$$= 10.0 \text{ m/s}$$

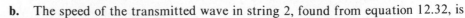

b. The speed of the transmitted wave in string 2, found from equation 12.32, is

$$v_2 = \sqrt{\frac{T}{m_2/l}}$$

$$= \sqrt{\frac{10.0 \text{ N}}{0.0500 \text{ kg/m}}}$$

$$= 14.1 \text{ m/s}$$

c. The wavelength of the transmitted wave, found from equation 12.34, is

$$\lambda_2 = \frac{v_2}{v_1}\lambda_1$$

$$= \left(\frac{14.1 \text{ m/s}}{10.0 \text{ m/s}}\right)(1.20 \text{ m})$$

$$= 1.69 \text{ m}$$

d. The speed and wavelength of the reflected wave are the same as the incident wave because the reflected wave is in the same medium as the incident wave. Note that the mass of string 1 is greater than string 2 and the speed of the wave in medium 2 is greater than the speed of the wave in medium 1 (i.e., $v_2 > v_1$). Also note that the wavelength of the transmitted wave is greater than the wavelength of the incident wave (i.e., $\lambda_2 > \lambda_1$).

The superposition of two waves in water.

12.5 The Principle of Superposition

Up to this point in our discussion we have considered only one wave passing through a medium at a time. What happens if two or more waves pass through the same medium at the same time? To solve the problem of multiple waves we use the principle of superposition. This principle is based on the vector addition of the displacement associated with each wave. *The **principle of superposition** states that whenever two or more wave disturbances pass a particular point in a medium, the resultant displacement of the point of the medium is the sum of the displacements of each individual wave disturbance.* For example, if the two waves

$$y_1 = A_1 \sin(k_1 x - \omega_1 t)$$

$$y_2 = A_2 \sin(k_2 x - \omega_2 t)$$

are acting in a medium at the same time, the resultant wave is given by

$$y = y_1 + y_2 \tag{12.35}$$

or

$$y_1 = A_1 \sin(k_1 x - \omega_1 t) + A_2 \sin(k_2 x - \omega_2 t) \tag{12.36}$$

The superposition principle holds as long as the resultant displacement of the medium does not exceed its elastic limit. Sometimes the two waves are said to interfere with each other, or cause *interference*.

Example 12.7

Superposition. The following two waves interfere with each other:

$$y_1 = (5.00 \text{ m})\sin[(0.800 \text{ m}^{-1})x - (6.00 \text{ rad/s})t]$$

$$y_2 = (10.00 \text{ m})\sin[(0.900 \text{ m}^{-1})x - (3.00 \text{ rad/s})t]$$

Find the resultant displacement when $x = 5.00$ m and $t = 1.10$ s.

The resultant displacement found by the superposition principle, equation 12.35, is

$$y = y_1 + y_2$$

where

$$y_1 = (5.00 \text{ m})\sin[(0.800 \text{ m}^{-1})(5.00 \text{ m}) - (6.00 \text{ rad/s})(1.10 \text{ s})]$$
$$= (5.00 \text{ m})\sin(4.00 \text{ rad} - 6.6 \text{ rad})$$
$$= (5.00 \text{ m})\sin(-2.60 \text{ rad})$$
$$= (5.00 \text{ m})(-0.5155) = -2.58 \text{ m}$$

and

$$y_2 = (10.00 \text{ m})\sin[(0.900 \text{ m}^{-1})(5.00 \text{ m}) - (3.00 \text{ rad/s})(1.10 \text{ s})]$$
$$= (10.00 \text{ m})\sin(4.50 \text{ rad} - 3.30 \text{ rad})$$
$$= (10.00 \text{ m})\sin(1.20 \text{ rad})$$
$$= (10.00 \text{ m})(0.932) = 9.32 \text{ m}$$

Hence, the resultant displacement is

$$y = y_1 + y_2 = -2.58 \text{ m} + 9.32 \text{ m}$$
$$= 6.74 \text{ m}$$

Note that this is the resultant displacement only for the values of $x = 5.00$ m and $t = 1.10$ s. We can find the entire resultant wave for any value of the time by substituting a series of values of x into the equation for that value of t. Then we determine the resultant displacement y for each value of x. A graph of the resultant displacement y versus x gives a snapshot of the resultant wave at that value of time t. The process can be repeated for various values of t, and the sequence of the graphs will show how that resultant wave travels with time.

It is possible that when dealing with two or more waves the waves may not be in phase with each other. *Two waves are in phase if they reach their maximum amplitudes at the same time, are zero at the same time, and have their minimum amplitudes at the same time.* An example of two waves in phase is shown in figure 12.12(a). An example of two waves that are out of phase with each other is shown in figure 12.12(b). Note that the second wave does not have its maximum, zero, and minimum displacements at the same place as the first wave. Instead these positions are translated to the right of their position in wave y_1. We say that wave 2 is out of phase with wave 1 by an angle ϕ, where ϕ is measured in radians. The equation for the first wave is

$$y_1 = A_1 \sin(kx - \omega t) \tag{12.37}$$

whereas the equation for the wave displaced to the right is

$$y_2 = A_2 \sin(kx - \omega t - \phi) \tag{12.38}$$

*The angle ϕ is called the **phase angle** and is a measure of how far wave 2 is displaced in the horizontal from wave 1.* Just as the minus sign on $-\omega t$ indicated a wave traveling to the right, the minus sign on ϕ indicates a wave displaced to the right. The second wave lags the first wave by the phase angle ϕ. That is, wave 2 has its maximum, zero, and minimum displacements after wave 1 does, and the amount of lag is given by the phase angle ϕ. If the wave was displaced to the left, the equation for the wave would be

$$y_2 = A_2 \sin(kx - \omega t + \phi) \tag{12.39}$$

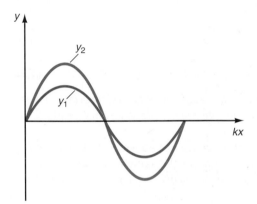

(a) Waves in phase

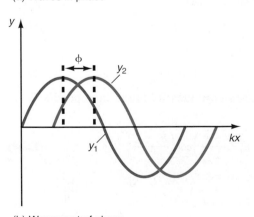

(b) Waves out of phase

Figure 12.12

Phase of a wave.

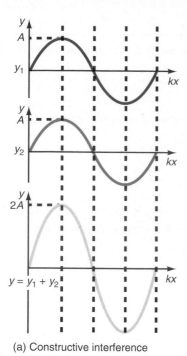

(a) Constructive interference

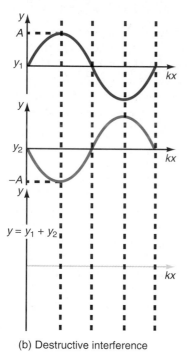

(b) Destructive interference

Figure 12.13

Interference of waves.

An important special case of the addition of waves is shown in figure 12.13. *When two waves are in phase with each other, $\phi = 0$ in equation 12.38, and the waves are said to exhibit* **constructive interference,** *figure 12.13(a). That is,*

$$y_1 = A \sin(kx - \omega t)$$

$$y_2 = A \sin(kx - \omega t)$$

and the resultant wave is

$$y = y_1 + y_2 = 2A \sin(kx - \omega t) \tag{12.40}$$

That is, the resultant amplitude has doubled. If the two waves are 180°, or π rad, out of phase with each other, then y_2 is

$$y_2 = A_2 \sin(kx - \omega t - \pi)$$

Setting $kx - \omega t = B$ and $\pi = C$, we can use the formula for the sine of the difference between two angles, which is found in appendix B. That is,

$$\sin(B - C) = \sin B \cos C - \cos B \sin C \tag{12.41}$$

Thus,

$$\sin[(kx - \omega t) - \pi] = \sin(kx - \omega t)\cos \pi - \cos(kx - \omega t)\sin \pi \tag{12.42}$$

But $\sin \pi = 0$, and the last term drops out. And because the $\cos \pi = -1$, equation 12.42 becomes

$$\sin[(kx - \omega t) - \pi] = -\sin(kx - \omega t)$$

Therefore we can write the second wave as

$$y_2 = -A \sin(kx - \omega t) \tag{12.43}$$

The superposition principle now yields

$$y = y_1 + y_2 = A \sin(kx - \omega t) - A \sin(kx - \omega t) = 0 \tag{12.44}$$

Thus, if the waves are 180° out of phase the resultant wave is zero everywhere. This is shown in figure 12.13(b) and is called **destructive interference.** Wave 2 has completely canceled out the effects of wave 1.

A more general solution for the interference of two waves of the same frequency, same wave number, same amplitude, and in the same direction but out of phase with each other by a phase angle ϕ, can be easily determined by the superposition principle. Let the two waves be

$$y_1 = A \sin(kx - \omega t) \tag{12.45}$$

$$y_2 = A \sin(kx - \omega t - \phi) \tag{12.46}$$

The resultant wave is

$$y = y_1 + y_2 = A \sin(kx - \omega t) + A \sin(kx - \omega t - \phi) \tag{12.47}$$

To simplify this result, we use the trigonometric identity found in appendix B for the sum of two sine functions, namely

$$\sin B + \sin C = 2 \sin\left(\frac{B + C}{2}\right)\cos\left(\frac{B - C}{2}\right) \tag{12.48}$$

For this problem

$$B = kx - \omega t$$

and

$$C = kx - \omega t - \phi$$

Thus,

$$\sin(kx - \omega t) + \sin(kx - \omega t - \phi) =$$

$$2 \sin\left(\frac{kx - \omega t + kx - \omega t - \phi}{2}\right) \cos\left(\frac{kx - \omega t - kx + \omega t + \phi}{2}\right)$$

$$= 2 \sin\left(kx - \omega t - \frac{\phi}{2}\right) \cos\left(\frac{\phi}{2}\right) \qquad \textbf{(12.49)}$$

Substituting equation 12.49 into equation 12.47 we obtain for the resultant wave

$$y = 2A \cos\left(\frac{\phi}{2}\right) \sin\left(kx - \omega t - \frac{\phi}{2}\right) \qquad \textbf{(12.50)}$$

Equation 12.50 is a more general result than found before and contains constructive and destructive interference as special cases. For example, if $\phi = 0$ the two waves are in phase and since $\cos 0 = 1$, equation 12.50 becomes

$$y = 2A \sin(kx - \omega t)$$

which is identical to equation 12.40 for constructive interference. Also for the special case of $\phi = 180° = \pi$ rad, the $\cos(\pi/2) = \cos 90° = 0$, and equation 12.50 becomes $y = 0$, the special case of destructive interference, equation 12.44.

Example 12.8

Interference. The following two waves interfere:

$$y_1 = (5.00 \text{ m})\sin[(0.200 \text{ m}^{-1})x - (5.00 \text{ rad/s})t]$$
$$y_2 = (5.00 \text{ m})\sin[(0.200 \text{ m}^{-1})x - (5.00 \text{ rad/s})t - 0.500 \text{ rad}]$$

Find the equation for the resultant wave.

Solution

The resultant wave, found from equation 12.50, is

$$y = 2A \cos\left(\frac{\phi}{2}\right) \sin\left(kx - \omega t - \frac{\phi}{2}\right)$$

$$= 2(5.00 \text{ m})\cos\left(\frac{0.500 \text{ rad}}{2}\right)\sin\left[(0.200 \text{ m}^{-1})x - (5.00 \text{ rad/s})t - \left(\frac{0.500 \text{ rad}}{2}\right)\right]$$

$$= (10.00 \text{ m})(0.9689)\sin[(0.200 \text{ m}^{-1})x - (5.00 \text{ rad/s})t - 0.250 \text{ rad}]$$

$$= (9.69 \text{ m})\sin[(0.200 \text{ m}^{-1})x - (5.00 \text{ rad/s})t - 0.250 \text{ rad}]$$

We can now plot an actual picture of the resultant wave for a particular value of t for a range of values of x.

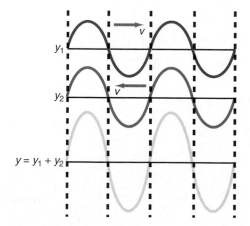

Figure 12.14

Formation of standing waves.

12.6 Standing Waves—The Vibrating String

If a string is fixed at both ends and a wave train is sent down the string, then, as shown before, the wave is reflected from the fixed ends. Hence, there are two wave trains on the string at the same time. One is traveling to the right, while the reflected wave is traveling toward the left, figure 12.14. We can find the resultant wave by the superposition principle. That is, if wave 1 is a wave to the right, we can express it

$$y_1 = A \sin(kx - \omega t) \qquad \textbf{(12.51)}$$

whereas we can express the wave to the left as

$$y_2 = A \sin(kx + \omega t) \tag{12.52}$$

The resultant wave is the sum of these two waves or

$$y = y_1 + y_2 = A \sin(kx - \omega t) + A \sin(kx + \omega t) \tag{12.53}$$

To add these two sine functions, we use the trigonometric identity in equation 12.48. That is,

$$\sin B + \sin C = 2 \sin\left(\frac{B + C}{2}\right)\cos\left(\frac{B - C}{2}\right)$$

where $B = kx - \omega t$ and $C = kx + \omega t$. Thus,

$$y = 2A \sin\left[\frac{(kx - \omega t) + (kx + \omega t)}{2}\right]\cos\left[\frac{(kx - \omega t) - (kx + \omega t)}{2}\right]$$

$$= 2A \sin\left(\frac{2kx}{2}\right)\cos\left(\frac{-2\omega t}{2}\right)$$

and

$$y = 2A \sin(kx)\cos(-\omega t)$$

Using the fact that

$$\cos(-\theta) = \cos(\theta)$$

the resultant wave is

$$y = 2A \sin(kx)\cos(\omega t) \tag{12.54}$$

For reasons that will appear shortly, *this is the equation of a standing wave or a stationary wave.*

The amplitude of the resultant standing wave is $2A \sin(kx)$, and note that it varies with x. To find the positions along x where this new amplitude has its minimum values, note that $\sin(kx) = 0$ whenever

$$kx = n\pi$$

for values of $n = 1, 2, 3, \ldots$. That is, the sine function is zero whenever the argument kx is a multiple of π. Thus, solving for x,

$$x = \frac{n\pi}{k} \tag{12.55}$$

But the wave number k was defined in equation 12.9 as

$$k = \frac{2\pi}{\lambda}$$

Substituting this value into equation 12.55, we get

$$x = \frac{n\pi}{k} = \frac{n\pi}{2\pi/\lambda}$$

and

$$x = \frac{n\lambda}{2} \tag{12.56}$$

Equation 12.56 gives us the location of the zero values of the amplitude. Thus we see that they occur for values of x of $\lambda/2$, $2\lambda/2 = \lambda$, $3\lambda/2$, $4\lambda/2 = 2\lambda$, and so on, as measured from either end. *These points, where the amplitude of the standing* wave is zero, are called nodes. Stated another way, *a node is the position of zero amplitude*. These nodes are independent of time, that is, the amplitude at these points is always zero.

The maximum values of the amplitude occur whenever $\sin(kx) = 1$, which happens whenever kx is an odd multiple of $\pi/2$. That is, $\sin(kx) = 1$ when

$$kx = (2n - 1)\frac{\pi}{2}$$

for $n = 1, 2, 3, \ldots$.

The term $2n - 1$ always gives an odd number for any value of n. (As an example, when $n = 2$, $2n - 1 = 3$, etc.) The location of the maximum amplitudes is therefore at

$$x = \left(\frac{2n - 1}{2k}\right)\pi$$

But since $k = 2\pi/\lambda$, this becomes

$$x = \frac{(2n - 1)\pi}{2(2\pi/\lambda)}$$

and

$$x = \frac{(2n - 1)\lambda}{4} = (2n - 1)\frac{\lambda}{4} \tag{12.57}$$

The maximum amplitudes are thus located at $x = \lambda/4, 3\lambda/4, 5\lambda/4$, and so forth. *The position of maximum amplitude is called an **antinode**.* Note that at this position the displacement of the resultant wave is a function of time. The original two traveling waves and the resultant standing wave are shown in figure 12.15 for values

Figure 12.15

Standing wave on a string.

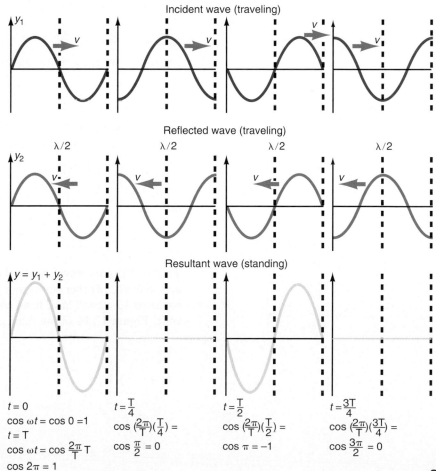

of time of 0, $T/4$, $T/2$, $3T/4$, and T, where T is the period of the wave. Recall that $\omega = 2\pi/T$. Therefore, $\cos \omega t = \cos(2\pi t/T)$. *Note that the waves are moving to the left and the right, but the resultant wave does not travel at all, it is a **standing wave on a string.** The node of the standing wave at $x = \lambda/2$ remains a node for all times. Thus, the string can not move up and down at that point, and can not therefore transmit any energy past that point. Thus, the resultant wave does not move along the string but is stationary or standing.*

How many different types of standing waves can be produced on this string? The only restriction on the number or types of different waves is that the ends of the string must be tied down or fixed. That is, there must be a node at the ends of the string, which implies that the displacement y must always equal zero for $x = 0$, and for $x = L$, the length of the string. When x is equal to zero the displacement is

$$y = 2A \sin[k(0)]\cos(\omega t) = 0 \tag{12.58}$$

When $x = L$, the displacement of the standing wave is

$$y = 2A \sin(kL)\cos(\omega t) \tag{12.59}$$

Equation 12.59 is not in general always equal to zero. Because it must always be zero in order to satisfy the boundary condition of $y = 0$ for $x = L$, it is necessary that

$$\sin(kL) = 0$$

which is true whenever kL is a multiple of π, that is,

$$kL = n\pi$$

for $n = 1, 2, 3, \ldots$. This places a restriction on the number of waves that can be placed on the string. The only possible wave numbers the wave can have are therefore

$$k = \frac{n\pi}{L} \tag{12.60}$$

Therefore, we must write the displacement of the standing wave as

$$y = 2A \sin\left(\frac{n\pi x}{L}\right)\cos(\omega t) \tag{12.61}$$

Because $k = 2\pi/\lambda$, a restriction on the possible wave numbers k is also a restriction on the possible wavelengths λ that can be found on the string. Thus,

$$k = \frac{2\pi}{\lambda} = \frac{n\pi}{L}$$

or

$$\lambda = \frac{2L}{n} \tag{12.62}$$

That is, the only wavelengths that are allowed on the string are $\lambda = 2L$, L, $2L/3$, and so forth. In other words, not all wavelengths are possible; only those that satisfy equation 12.62 will have fixed end points. Only a discrete set of wavelengths is possible. Figure 12.16 shows some of the possible modes of vibration.

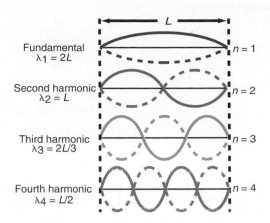

Fundamental
$\lambda_1 = 2L$ $n = 1$

Second harmonic
$\lambda_2 = L$ $n = 2$

Third harmonic
$\lambda_3 = 2L/3$ $n = 3$

Fourth harmonic
$\lambda_4 = L/2$ $n = 4$

Figure 12.16

The normal modes of vibration of a string.

We can find the frequency of any wave on the string with the aid of equations 12.3 and 12.30 as

$$f = \frac{v}{\lambda}$$

$$v = \sqrt{\frac{T}{m/l}}$$

Thus,

$$f = \frac{1}{\lambda} \sqrt{\frac{T}{m/l}} \tag{12.63}$$

However, since the only wavelengths possible are those for $\lambda = 2L/n$, equation 12.62, the frequencies of vibration are

$$f = \frac{n}{2L} \sqrt{\frac{T}{m/l}} \tag{12.64}$$

with $n = 1, 2, 3, \ldots$.

Equation 12.64 points out that there are only a discrete number of frequencies possible for the vibrating string, depending on the value of n. The simplest mode of vibration, for $n = 1$, is called the *fundamental mode of vibration*. As we can see from figure 12.16, a half of a wavelength fits within the length L of the string (i.e., $L = \lambda/2$ or $\lambda = 2L$). Thus, the fundamental mode of vibration has a wavelength of $2L$. We obtain the **fundamental frequency** f_1 from equation 12.64 by setting $n = 1$. Thus,

$$f_1 = \frac{1}{2L} \sqrt{\frac{T}{m/l}} \tag{12.65}$$

For $n = 2$ we have what is called the *first overtone* or *second harmonic*. From figure 12.16, we see that one entire wavelength fits within one length L of the string (i.e., $L = \lambda$). We obtain the frequency of the second harmonic from equation 12.64 by letting $n = 2$. Hence,

$$f_2 = \frac{2}{2L} \sqrt{\frac{T}{m/l}} = 2f_1 \tag{12.66}$$

Figure 12.17

Forced vibration of a string.

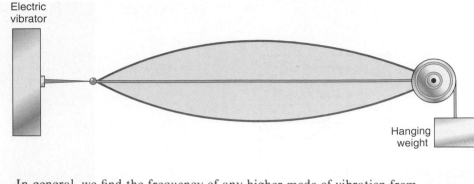

Electric
vibrator

Hanging
weight

In general, we find the frequency of any higher mode of vibration from

$$f_n = nf_1 \qquad (12.67)$$

A string that vibrates at a frequency given by equation 12.64 or 12.67 is said to be vibrating at one of its *natural frequencies.*

The possible waves for $n = 3$ and $n = 4$ are also shown in figure 12.16. Note that the nth harmonic contains n half wavelengths within the distance L. We can also observe that the location of the nodes and antinodes agrees with equations 12.56 and 12.57. Also note from equation 12.64 that the larger the tension T in the string, the higher the frequency of vibration. If we were considering a violin string, we would hear this higher frequency as a higher pitch. The smaller the tension in the string the lower the frequency or pitch. The string of any stringed instrument, such as a guitar, violin, cello, and the like, is tuned by changing the tension of the string. Also note from equation 12.64 that the larger the mass density m/l of the string, the smaller the frequency of vibration, whereas the smaller the mass density, the higher the frequency of the vibration. Thus, the mass density of each string of a stringed instrument is different in order to give a larger range of possible frequencies. Moving the finger of the left hand, which is in contact with the vibrating string, changes the point of contact of the string and thus changes the value of L, the effective length of the vibrating string. This in turn changes the possible wavelengths and frequencies that can be obtained from that string.

When we pluck a string, one or more of the natural frequencies of the string is excited. In a real string, internal frictions soon cause these vibrations to die out. However, if we apply a periodic force to the string at any one of these natural frequencies, the mode of vibration continues as long as the driving force is continued. We call this type of vibration a *forced vibration,* and we can easily set up a demonstration of forced vibration in the laboratory, as shown in figure 12.17. We connect one end of a string to an electrical vibrator of a fixed frequency and pass the other end over a pulley that hangs over the end of the table. We place weights on this end of the string to produce the tension in the string. We add weights until the string vibrates in its fundamental mode. When the tension is adjusted so that the natural frequency of the string is the same as that of the electrical vibrator, the amplitude of vibration increases rapidly. *This condition where the driving frequency is equal to the natural frequency of the system is called* **resonance.**

The tension in the string can be changed by changing the weights that are added to the end of the string, until all the harmonics are produced one at a time. The string vibrates so rapidly that the eye perceives only a blur whose shape is that of the envelope of the vibration, as shown in figure 12.18.

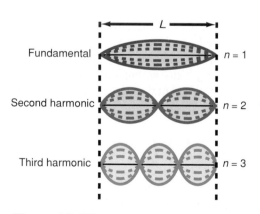

Fundamental $n = 1$

Second harmonic $n = 2$

Third harmonic $n = 3$

Figure 12.18

The envelope of the vibration.

Example 12.9

The tension in a guitar string. A guitar string 60.0 cm long has a linear mass density of 6.50×10^{-3} kg/m. If this string is to play a fundamental frequency of 220 Hz, what must the tension be in the string?

Solution

We find the tension necessary in the string by solving equation 12.64 for T, that is

$$f = \frac{n}{2L}\sqrt{\frac{T}{m/l}}$$

$$f^2 = \frac{n^2 T}{4L^2(m/l)}$$

and

$$T = \frac{4L^2 f^2 (m/l)}{n^2}$$

$$= \frac{4(0.600 \text{ m})^2 (220 \text{ Hz})^2 (6.5 \times 10^{-3} \text{ kg/m})}{1^2}$$

$$= 4.53 \times 10^2 \text{ N}$$

Example 12.10

The frequencies and wavelengths of a guitar string. Find (a) the frequencies and (b) the wavelengths of the fundamental, second, third, and fourth harmonics of example 12.9.

Solution

a. The fundamental frequency is given in example 12.9 as 220 Hz. The frequency of the next three harmonics, found from equation 12.67, are

$$f_n = n f_1$$
$$f_2 = 2f_1 = 2(220 \text{ Hz}) = 440 \text{ Hz}$$
$$f_3 = 3f_1 = 3(220 \text{ Hz}) = 660 \text{ Hz}$$
$$f_4 = 4f_1 = 4(220 \text{ Hz}) = 880 \text{ Hz}$$

b. The wavelength of the fundamental, found from equation 12.62, is

$$\lambda_n = \frac{2L}{n}$$

$$\lambda_1 = \frac{2(60.0 \text{ cm})}{1} = 120 \text{ cm}$$

The wavelengths of the harmonics are

$$\lambda_2 = \frac{2L}{2} = \frac{2(60.0 \text{ cm})}{2} = 60.0 \text{ cm}$$

$$\lambda_3 = \frac{2L}{3} = \frac{2(60.0 \text{ cm})}{3} = 40.0 \text{ cm}$$

$$\lambda_4 = \frac{2L}{4} = \frac{2(60.0 \text{ cm})}{4} = 30.0 \text{ cm}$$

Example 12.11

The displacement of the third harmonic. Find the value of the displacement for the third harmonic of example 12.10 if $x = 30.0$ cm and $t = 0$.

Solution

This displacement, found from equation 12.61, is

$$y = 2A \sin\left(\frac{n\pi x}{L}\right)\cos(\omega t)$$

$$= 2A \sin\left[\frac{3\pi(30.0 \text{ cm})}{60.0 \text{ cm}}\right]\cos[\omega(0)]$$

$$= 2A \sin\left(\frac{3\pi}{2}\right) = -2A$$

12.7 Sound Waves

A **sound wave** is a longitudinal wave, that is, a particle of the medium executes simple harmonic motion in a direction that is parallel to the velocity of propagation. A sound wave can be propagated in a solid, liquid, or a gas. The speed of sound in the medium depends on the density of the medium and on its elastic properties. We will state without proof that the speed of sound in a solid is

$$v = \sqrt{\frac{Y}{\rho}} \tag{12.68}$$

where Y is Young's modulus and ρ is the density of the medium. The speed of sound in a fluid is

$$v = \sqrt{\frac{B}{\rho}} \tag{12.69}$$

where B is the bulk modulus and ρ is the density. The speed of sound in a gas is

$$v = \sqrt{\frac{\gamma p}{\rho}} \tag{12.70}$$

where γ is a constant called the ratio of the specific heats of the gas and is equal to 1.40 for air (we discuss the specific heats of gases and their ratio in detail in chapter 17); p is the pressure of the gas; and ρ is the density of the gas. Note that the pressure and the density of a gas varies with the temperature of the gas and hence the speed of sound in a gas depends on the gas temperature. It can be shown, with the help of the ideal gas equation derived in chapter 15, that the speed of sound in air is

$$v = (331 + 0.606t) \text{ m/s} \tag{12.71}$$

where t is the temperature of the air in degrees Celsius.

Example 12.12

The speed of sound. Find the speed of sound in (a) iron, (b) water, and (c) air.

Solution

a. We find the speed of sound in iron from equation 12.68, where Y for iron is 9.1×10^{10} N/m^2 (from table 10.1). The density of iron is 7.8×10^3 kg/m^3. Hence,

$$v = \sqrt{\frac{Y}{\rho}} \tag{12.68}$$

$$= \sqrt{\frac{9.1 \times 10^{10} \text{ N/m}^2}{7.8 \times 10^3 \text{ kg/m}^3}}$$

$$= 3400 \text{ m/s}$$

b. We find the speed of sound in water from equation 12.69, where $B = 2.30 \times 10^9$ N/m^2 and $\rho = 1.00 \times 10^3$ kg/m^3. Thus,

$$v = \sqrt{\frac{B}{\rho}} \tag{12.69}$$

$$= \sqrt{\frac{2.30 \times 10^9 \text{ N/m}^2}{1.00 \times 10^3 \text{ kg/m}^3}}$$

$$= 1520 \text{ m/s}$$

c. We find the speed of sound in air from equation 12.70 with normal atmospheric pressure $p = 1.013 \times 10^5$ N/m^2 and $\rho = 1.29$ kg/m^3. Hence,

$$v = \sqrt{\frac{\gamma p}{\rho}} \tag{12.70}$$

$$= \sqrt{\frac{(1.40)(1.013 \times 10^5 \text{ N/m}^2)}{1.29 \text{ kg/m}^3}}$$

$$= 331 \text{ m/s}$$

Example 12.13

The speed of sound as a function of temperature. Find the speed of sound in air at a room temperature of 20.0 °C.

Solution

The speed of sound at 20.0 °C, found from equation 12.71, is

$$v = (331 + 0.606t) \text{ m/s}$$

$$= [331 + 0.606(20.0)] \text{ m/s}$$

$$= 343 \text{ m/s}$$

Example 12.14

Range of wavelengths. The human ear can detect sound only in the frequency spectrum of about 20.0 to 20,000 Hz. Find the wavelengths corresponding to these frequencies at room temperature.

$L = \lambda/4$
or
$\lambda = 4L$

(a) Fundamental

$L = 3\lambda/4$
or
$\lambda = 4L/3$

(b) Third harmonic

$L = 5\lambda/4$
or
$\lambda = 4\,L/5$

(c) Fifth harmonic

(d)

Figure 12.19
Standing waves in a closed organ pipe.

The corresponding wavelengths, found from equation 12.3, with $v = 343$ m/s as found in example 12.13, are

$$\lambda = \frac{v}{f} = \frac{343 \text{ m/s}}{20.0 \text{ Hz}} = 17.2 \text{ m}$$

and

$$\lambda = \frac{v}{f} = \frac{343 \text{ m/s}}{20{,}000 \text{ Hz}} = 0.0172 \text{ m}$$

Just as standing transverse waves can be set up on a vibrating string, standing longitudinal waves can be set up in closed and open organ pipes. Let us first consider the closed organ pipe. A traveling sound wave is sent down the closed organ pipe and is reflected at the closed end. Thus, there are two traveling waves in the pipe and they superimpose to form a standing wave in the pipe. An analysis of this standing longitudinal wave would lead to equation 12.54 for the resultant standing wave found from the superposition of a wave moving to the right and one moving to the left. The boundary conditions that must be satisfied are that the pressure wave must have a node at the closed end of the organ pipe and an antinode at the open end. The simplest wave is shown in figure 12.19(a). Although sound waves are longitudinal, the standing wave in the pipe is shown as a transverse standing wave to more easily show the nodes and antinodes. At the node, the longitudinal wave has zero amplitude, whereas at the antinode, the longitudinal wave has its maximum amplitude. It is obvious from the figure that only a quarter of a wavelength can fit in the length L of the pipe, hence the wavelength of the fundamental λ is equal to four times the length of the pipe. We must make a distinction here between an overtone and a harmonic. *An overtone is a frequency higher than the fundamental frequency. A* **harmonic** *is an overtone that is a multiple of the fundamental frequency. Hence, the nth harmonic is n times the fundamental, or first harmonic.* A harmonic is an overtone but an overtone is not necessarily a harmonic. We call the second possible standing wave in figure 12.19 the second overtone; it contains three quarter wavelengths in the distance L, whereas the third overtone has five. We can generalize the wavelength of all possible standing waves in the closed pipe to

$$\lambda_n = \frac{4L}{2n - 1} \tag{12.72}$$

for $n = 1, 2, 3, \ldots$, where the value of $2n - 1$ always gives an odd number for any value of n. We can obtain the frequency of each standing wave from equations 12.3 and 12.72 as

$$f_n = \frac{v}{\lambda_n} = \frac{v}{4L/(2n - 1)}$$

$$f_n = \frac{(2n - 1)v}{4L} \tag{12.73}$$

When n is equal to 1, the frequency is $f_1 = v/4L$; when n is equal to 2, the frequency is $f = 3(v/4L)$, which is three times the fundamental frequency and is thus the third harmonic. Because $n = 2$ is the first frequency above the fundamental it is called the first overtone, but it is also the third harmonic; $n = 3$ gives the second overtone, which is equal to the fifth harmonic. Thus, for an organ pipe closed at one end and open at the other, the $(n - 1)$th overtone is equal to the $(2n - 1)$th harmonic. Note that the allowable frequencies are all odd harmonics of the fundamental frequency. That is, $n = 1$ gives the fundamental frequency (zeroth overtone),

$L = \lambda/2$
or
$\lambda = 2L$

(a) Fundamental

$l = \lambda$
or
$\lambda = L$

(b) Second harmonic

$L = 3\lambda/2$
or
$\lambda = 2L/3$

(c) Third harmonic

Figure 12.20

Standing waves in an open organ pipe.

which is the first harmonic; $n = 2$ gives the first overtone, which is the third harmonic; $n = 3$ gives the second overtone, which is the fifth harmonic; and so forth. Because the even harmonics are missing the distinction between overtones and harmonics must be made. Figure 12.19(d) shows a typical pipe organ.

Standing waves can also be set up in an open organ pipe, but now the boundary conditions necessitate an antinode at both ends of the pipe, as shown in figure 12.20. For simplicity, the longitudinal standing wave is again depicted as a transverse standing wave in the figure. The node is the position of zero amplitude and the antinodes at the ends of the open pipe are the position of maximum amplitude of the longitudinal standing wave. From inspection of the figure we can see that the wavelength of the nth harmonic is

$$\lambda_n = \frac{2L}{n} \tag{12.74}$$

and its frequency is

$$f_n = \frac{v}{\lambda_n} = \frac{v}{2L/n}$$

$$f_n = \frac{nv}{2L} \tag{12.75}$$

Equation 12.75 gives the frequency of the nth harmonic for an organ pipe open at both ends. For the open organ pipe, even and odd multiples of the fundamental frequency are possible. The $(n - 1)$th overtone is equal to the nth harmonic.

Example 12.15

The length of a closed organ pipe. Find the length of a closed organ pipe that can produce the musical note $A = 440$ Hz. Assume that the speed of sound in air is 343 m/s.

Solution

We find the length of the closed organ pipe from equation 12.73 by solving for L. Thus,

$$L = \frac{(2n - 1)v}{4f_n}$$

We obtain the fundamental note for $n = 1$:

$$L = \frac{(2n - 1)(343 \text{ m/s})}{4(440 \text{ cycles/s})}$$

$$= 0.195 \text{ m}$$

Example 12.16

Open and closed organ pipes. If the length of an organ pipe is 4.00 m, find the frequency of the fundamental for (a) a closed pipe and (b) an open pipe. Assume that the speed of sound is 343 m/s.

a. The frequency of the fundamental for the closed pipe, found from equation 12.73 with $n = 1$, is

$$f_n = \frac{(2n - 1)v}{4L}$$

$$f_1 = \frac{[2(1) - 1](343 \text{ m/s})}{4(4.00 \text{ m})}$$

$$= 21.4 \text{ Hz}$$

b. The frequency of the fundamental for an open pipe, found from equation 12.75, is

$$f_n = \frac{nv}{2L}$$

$$f_1 = \frac{(1)(343 \text{ m/s})}{2(4.00 \text{ m})}$$

$$= 42.8 \text{ Hz}$$

Note that the frequency of the fundamental of the open organ pipe is exactly twice the frequency of the fundamental in the closed organ pipe. Table 12.1 gives a summary of the harmonics for the vibrating string and the organ pipe.

Table 12.1

Summary of Some Different Harmonics for the Musical Note A, Which Has the Fundamental Frequency of 440 Hz

Vibrating String

n	Harmonic	Overtone	Frequency (Hz)
1	first	fundamental	440
2	second	first	880
3	third	second	1320

Organ Pipe Opened at One End

n	Harmonic	Overtone	Frequency (Hz)
1	first	fundamental	440
2	third	first	1320
3	fifth	second	2200

12.8 The Doppler Effect

Almost everyone has observed the change in frequency of a train whistle or a car horn as it approaches an observer and as it recedes from the observer. The change in frequency of the sound due to the motion of the sound source is an example of the **Doppler effect.** In general, this change in frequency of the sound wave can be caused by the motion of the sound source, the motion of the observer, or both. Let us consider the different possibilities.

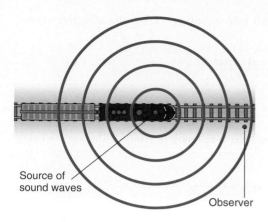

Figure 12.21

Observer and source stationary.

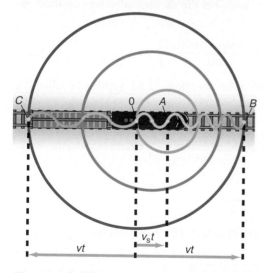

Figure 12.22

Doppler effect with the source moving and the observer stationary.

Case 1: *The Observer and the Sound Source Are Stationary*

This case is the normal case where there is no relative motion between the source and the observer and is shown in figure 12.21. When the source emits a sound of frequency f_s, the sound waves propagate out from the source in a series of concentric circles. The distance between each circle is the wavelength of the sound. The sound propagates at a speed v, and the frequency heard by the observer f_o is simply

$$f_o = \frac{v}{\lambda} = f_s \qquad (12.76)$$

That is, the stationary observer hears the same frequency as the one emitted from the stationary source.

Case 2: *The Observer Is Stationary But the Source Is Moving*

When the source of sound moves with a velocity, v_s, to the right, the emitted waves are no longer concentric circles but rather appear as in figure 12.22. Each wave is symmetrical about the point of emission, but since the point of emission is moving to the right, the circular wave associated with each emission is also moving to the right. Hence the waves bunch up in advance of the moving source and spread out behind the source. The frequency that an observer hears is just the speed of propagation of the wave divided by its wavelength, that is,

$$f = \frac{v}{\lambda} \qquad (12.77)$$

We can use equation 12.77 to describe qualitatively what the observer hears. As the wave approaches the observer, the waves bunch up and hence the effective wavelength λ appears smaller in the front of the wave. From equation 12.77 we can see that if λ decreases, the frequency f must increase. *Thus, when a moving source approaches a stationary observer the observed frequency is higher than the emitted frequency of the source.* When the source reaches the observer, the observer hears the frequency emitted. As the source passes and recedes from the observer the effective wavelength λ appears longer. Hence, from equation 12.77, if λ increases, the frequency f, heard by the observer, is lower than the frequency emitted by the source. *Thus, when a moving source recedes from a stationary observer the observed frequency is lower than the emitted frequency of the source.* To get a quantitative description of the observed frequency we proceed as follows.

a) *Moving Source Approaches a Stationary Observer*

The effective wavelength measured by the stationary observer in front of the moving source is simply the total distance AB, in figure 12.22, divided by the total number of waves in that distance, that is,

$$\lambda_f = \frac{\text{distance } AB}{\text{\# of waves}} \qquad (12.78)$$

In a time t, the moving source has moved a distance $0A$, in figure 12.22, which is given by the speed of the source v_s times the time t. The distance $0B$ is given by the speed of the wave v times the time t. Hence, the distance AB in figure 12.22 is

$$\text{distance } AB = vt - v_s t \qquad (12.79)$$

whereas the number of waves between A and B is just the number of waves emitted per unit time, f_s, the frequency of the source, times the time t. Thus,

$$\text{\# of waves in } AB = \left(\frac{(\text{\# of waves emitted})t}{\text{time}} \right) = f_s t \qquad (12.80)$$

Substituting equations 12.79 and 12.80 into equation 12.78, the effective wavelength in front of the source is

$$\lambda_f = \frac{vt - v_s t}{f_s t}$$

$$\lambda_f = \frac{v - v_s}{f_s} \tag{12.81}$$

However, from equation 12.77, the observed frequency in front of the approaching source f_{of} is

$$f_{of} = \frac{v}{\lambda_f} = \frac{v}{(v - v_s)/f_s}$$

or

$$f_{of} = \frac{v}{v - v_s} f_s \tag{12.82}$$

Equation 12.82 gives the frequency that is observed by a stationary observer who is in front of an approaching source.

b) A Moving Source Recedes from a Stationary Observer

The effective wavelength λ_b heard by the stationary observer behind the moving source is equal to the total distance CA divided by the number of waves between C and A, that is,

$$\lambda_b = \frac{\text{distance } CA}{\text{\# of waves}} \tag{12.83}$$

However, from figure 12.22, we see that the distance CA is

$$\text{distance } CA = vt + v_s t \tag{12.84}$$

whereas the number of waves between C and A is the number of waves emitted per unit time, times the time t, that is,

$$\text{\# of waves in } CA = \left(\frac{\text{\# of waves emitted}}{\text{time}} \right) t = f_s t \tag{12.85}$$

Substituting equations 12.84 and 12.85 into equation 12.83 yields the effective wavelength behind the receding source

$$\lambda_b = \frac{vt + v_s t}{f_s t}$$

or

$$\lambda_b = \frac{v + v_s}{f_s} \tag{12.86}$$

Substituting this effective wavelength into equation 12.77, we obtain

$$f_{ob} = \frac{v}{\lambda_b} = \frac{v}{(v + v_s)/f_s}$$

or

$$f_{ob} = \frac{v}{v + v_s} f_s \tag{12.87}$$

Equation 12.87 gives the frequency observed by a stationary observer who is behind the receding source.

Example 12.17

Doppler effect—moving source. A train moving at 25.00 m/s emits a whistle of frequency 200.0 Hz. If the speed of sound in air is 343.0 m/s, find the frequency observed by a stationary observer (a) in advance of the moving source and (b) behind the moving source.

Solution

a. The observed frequency in advance of the approaching source, found from equation 12.82, is

$$f_{of} = \frac{v}{v - v_s} f_s$$

$$= \left(\frac{343.0 \text{ m/s}}{343.0 \text{ m/s} - 25.00 \text{ m/s}} \right)(200.0 \text{ Hz})$$

$$= 215.7 \text{ Hz}$$

Note that the observed frequency in front of the approaching source is higher than the frequency emitted by the source.

b. The observed frequency behind the receding source, found from equation 12.87, is

$$f_{ob} = \frac{v}{v + v_s} f_s$$

$$= \left(\frac{343.0 \text{ m/s}}{343.0 \text{ m/s} + 25.00 \text{ m/s}} \right)(200.0 \text{ Hz})$$

$$= 186.4 \text{ Hz}$$

Note that the observed frequency behind the receding source is lower than the frequency emitted by the source. Also note that the change in the frequency is not symmetrical. That is, the change in frequency when the train is approaching is equal to 15.7 Hz, whereas the change in frequency when the train is receding is equal to 13.6 Hz.

Case 3: *The Source Is Stationary But the Observer Is Moving*

For a stationary source the sound waves are emitted as concentric circles, as shown in figure 12.23.

Figure 12.23

Doppler effect, the source is stationary but the observer is moving.

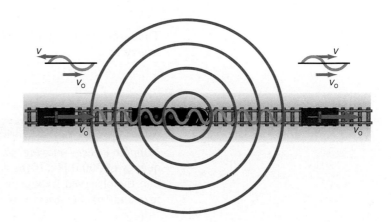

a) The Observer Is Moving toward the Source at a Velocity v_o

When the observer approaches the stationary source at a velocity v_o, the relative velocity between the observer and the wave is

$$v_{rel} = v_o + v$$

This relative velocity has the effect of having the emitted wave pass the observer at a greater velocity than emitted. The observed frequency heard while approaching the source is

$$f_{oA} = \frac{v_{rel}}{\lambda} = \frac{v_o + v}{\lambda}$$

But the wavelength emitted by the source does not change, and is simply

$$\lambda = \frac{v}{f_s} \tag{12.88}$$

Hence,

$$f_{oA} = \frac{v_o + v}{v/f_s}$$

Hence, the frequency observed by the moving observer as it approaches the stationary source is

$$f_{oA} = \frac{v_o + v}{v} f_s \tag{12.89}$$

b) The Observer Is Moving Away from the Source at a Velocity v_o

When the observer moves away from the source, the relative velocity between the wave and the observer is

$$v_{rel} = v - v_o$$

This reduced relative velocity has the effect of having the sound waves move past the receding observer at a slower rate. Thus, the observed frequency of the receding observer f_{oR} is

$$f_{oR} = \frac{v_{rel}}{\lambda} = \frac{v - v_o}{\lambda}$$

The wavelength λ of the emitted sound is still given by equation 12.88, and the observed frequency becomes

$$f_{oR} = \frac{v - v_o}{v/f_s}$$

Thus, *the frequency observed by an observer who is receding from a stationary source is*

$$f_{oR} = \frac{v - v_o}{v} f_s \tag{12.90}$$

Example 12.18

Doppler effect—moving observer. A stationary source emits a whistle at a frequency of 200.0 Hz. If the velocity of propagation of the sound wave is 343.0 m/s, find the observed frequency if (a) the observer is approaching the source at 25.00 m/s and (b) the observer is receding from the source at 25.00 m/s.

a. The frequency observed by the approaching observer, found from equation 12.89, is

$$f_{oA} = \frac{v_o + v}{v} f_s$$

$$= \left(\frac{25.00 \text{ m/s} + 343.0 \text{ m/s}}{343.0 \text{ m/s}}\right)(200.0 \text{ Hz})$$

$$= 214.6 \text{ Hz}$$

Note that the frequency observed by the approaching observer is greater than the emitted frequency of 200.0 Hz. However, observe that it is not the same numerical value found when the source was moving (215.7 Hz). The reason for the difference in the observed frequency is that the physical problems are not the same.

b. The frequency observed by the receding observer, found from equation 12.90, is

$$f_{oR} = \frac{v - v_o}{v} f_s$$

$$= \left(\frac{343.0 \text{ m/s} - 25.00 \text{ m/s}}{343.0 \text{ m/s}}\right)(200.0 \text{ Hz})$$

$$= 185.4 \text{ Hz}$$

Note that the frequency observed by the receding observer is less than the frequency emitted by the source. Also note that the frequency observed by the receding observer for a stationary source, $f_{oR} = 185.4$ Hz, is not the same frequency as observed by a stationary observer behind the receding source $f_{ob} = 186.4$ Hz. Finally, notice that when the source is stationary, the change in the frequency of approach, 14.6 Hz, is equal to the change in the frequency of recession, 14.6 Hz.

Case 4: *Both the Source and the Observer Are Moving*

If both the source and the observer are moving we can combine equations 12.82, 12.87, 12.89, and 12.90 into the one single equation

$$f_o = \frac{v \pm v_o}{v \mp v_s} f_s \tag{12.91}$$

with the convention that

$+ \; v_o$ corresponds to the observer approaching
$- \; v_o$ corresponds to the observer receding
$- \; v_s$ corresponds to the source approaching
$+ \; v_s$ corresponds to the source receding

Example 12.19

Doppler effect—both source and observer move. A sound source emits a frequency of 200.0 Hz at a velocity of 343.0 m/s. If both the source and the observer move at a velocity of 12.50 m/s, find the observed frequency if (a) the source and the observer are moving toward each other and (b) the source and the observer are moving away from each other.

a. If the source and observer are approaching each other, the observed frequency, found from equation 12.91 with v_o positive and v_s negative, is

$$f_o = \frac{v + v_o}{v - v_s} f_s$$

$$= \left(\frac{343.0 \text{ m/s} + 12.50 \text{ m/s}}{343.0 \text{ m/s} - 12.50 \text{ m/s}} \right) (200.0 \text{ Hz})$$

$$= 215.1 \text{ Hz}$$

Note that the frequency observed is higher than the frequency emitted, and although the relative motion between the source and the observer is still 25.00 m/s, the observed frequency is different from that found in both examples 12.17 and 12.18 (215.7 Hz and 214.6 Hz).

b. If the source and observer are moving away from each other, then the observed frequency, found from equation 12.91 with v_o negative and v_s positive, is

$$f_o = \frac{v - v_o}{v + v_s} f_s$$

$$= \left(\frac{343.0 \text{ m/s} - 12.50 \text{ m/s}}{343.0 \text{ m/s} + 12.50 \text{ m/s}} \right) (200.0 \text{ Hz})$$

$$= 185.9 \text{ Hz}$$

Note that the observed frequency is lower than the emitted frequency, and although the relative velocity between observer and source is still 25.00 m/s, the observed frequency is different from that found in examples 12.17 and 12.18 (186.4 Hz and 185.4 Hz).

12.9 The Transmission of Energy in a Wave and the Intensity of a Wave

We have defined a wave as the propagation of a disturbance through a medium. The disturbance causes the particles of the medium to be set into motion. As we have seen, if the wave is a transverse wave, the particles oscillate in a direction perpendicular to the direction of the wave propagation. The oscillating particles possess energy, and this energy is passed from particle to particle of the medium. Thus, the wave transmits energy as it propagates. Let us now determine the amount of energy transmitted by a wave.

Let us consider a transverse wave on a string, whose particles are executing simple harmonic motion. If there is no energy loss due to friction, the total energy transmitted by the wave is equal to the total energy of the vibrating particle, that is,

$$E_{\text{transmitted}} = (E_{\text{tot}})_{\text{particle}}$$

The total energy possessed by a single particle in simple harmonic motion, given by a variation of equation 11.25, is

$$E_{\text{tot}} = \tfrac{1}{2} k R^2 \tag{12.92}$$

where the letter R is now used for the amplitude of the vibration and hence the wave. Recall that k, in this equation, is the spring constant that was shown in chapter 11 to be related to the angular frequency by

$$\omega^2 = \frac{k}{m}$$

where m was the mass of the particle in motion. Solving for the spring constant k, we get

$$k = \omega^2 m \qquad (12.93)$$

Also recall that the angular frequency was related to the frequency of vibration by equation 12.12 as

$$\omega = 2\pi f$$

Substituting equation 12.12 into equation 12.93 yields

$$k = (2\pi f)^2 m \qquad (12.94)$$

Substituting equation 12.94 into equation 12.92 gives

$$E_{tot} = \tfrac{1}{2}(2\pi f)^2 m R^2$$

The energy transmitted by the wave is therefore

$$E_{transmitted} = 2\pi^2 m f^2 R^2 \qquad (12.95)$$

Notice that the energy transmitted by the wave is directly proportional to the square of the frequency of the wave and directly proportional to the square of the amplitude of the wave.

Example 12.20

Energy transmitted by a wave. The frequency and amplitude of a transverse wave on a string are doubled. What effect does this have on the amount of energy transmitted?

Solution

The energy of the original wave, given by equation 12.95, is

$$E_o = 2\pi^2 m f_o^2 R_o^2$$

The frequency of the new wave is $f = 2f_o$, and the amplitude of the new wave is $R = 2R_o$. The energy transmitted by the new wave is

$$E = 2\pi^2 m f^2 R^2 = 2\pi^2 m (2f_o)^2 (2R_o)^2$$
$$= 16(2\pi^2 m f_o^2 R_o^2) = 16E_o$$

We derived equation 12.95 for the transmission of energy by a transverse wave on a string. It is, however, completely general and can be used for any mechanical wave.

It is sometimes more convenient to describe the wave in terms of its intensity. *The **intensity of a wave** is defined as the energy of the wave that passes a unit area in a unit time.* That is, we define the intensity mathematically as

$$I = \frac{E}{At} \qquad (12.96)$$

The unit of intensity is the watt per square meter, W/m^2. Substituting the energy of a wave from equation 12.95 into equation 12.96 gives

$$I = \frac{2\pi^2 m f^2 R^2}{At} \qquad (12.97)$$

Because the density of a medium is defined as the mass per unit volume, the mass of the particle in simple harmonic motion can be replaced by

$$m = \rho V$$

And the volume of the medium can be expressed as the cross-sectional area A of the medium that the wave is moving through times a distance l the wave moves through (i.e., $V = Al$). The mass m then becomes

$$m = \rho Al \tag{12.98}$$

Substituting equation 12.98 into equation 12.97 yields

$$I = \frac{2\pi^2 \rho Al f^2 R^2}{At}$$

Notice that the cross-sectional area term in both numerator and denominator cancel, and l/t is the velocity of the wave v. Hence,

$$I = 2\pi^2 \rho v f^2 R^2 \tag{12.99}$$

Equation 12.99 gives the intensity of a mechanical wave of frequency f and amplitude R, moving at a velocity v in a medium of density ρ.

Example 12.21

The intensity of a sound wave. A trumpet player plays the note A at a frequency of 440 Hz, with an amplitude of 7.80×10^{-3} mm. If the density of air is 1.29 kg/m³ and the speed of sound is 331 m/s, find the intensity of the sound wave.

Solution

The intensity of the sound wave, found from equation 12.99, is

$$I = 2\pi^2 \rho v f^2 R^2$$
$$= 2\pi^2 (1.29 \text{ kg/m}^3)(331 \text{ m/s})(440 \text{ Hz})^2 (7.80 \times 10^{-6} \text{ m})^2$$
$$= 9.93 \times 10^{-2} \text{ W/m}^2$$

"Have you ever wondered...?"

An Essay on the Application of Physics

The Production and Reception of Human Sound

Humans use sound waves to communicate with each other. Sound is produced in the larynx, sometimes called the voice box, which is a cartilaginous organ of the throat that contains the vocal cords, figure 1(a). It is the vocal cords that are responsible for producing human sound. The cords are horizontal folds in the mucous membrane lining of the larynx, figure 1(b).

The vocal cords contain elastic fibers. As air is exhaled from the lungs, it passes over these elastic fibers and sets them into vibration. The cords can be visualized as the vibrating strings studied in this chapter. The frequency of the produced sound can be varied by changing the tension in the vocal cords similar to the vibrating string. The greater the tension on the cords, the higher the frequency, or pitch, of the emitted sound. The lower the tension on the cords, the lower the pitch.

When you hum, you set up a standing wave of a particular frequency on your vocal cords. The exhaled air that passes over these cords picks up the vibration of the cords. As the air is expelled from your mouth, it is observed as a longitudinal wave at the frequency of the vibrating vocal cords. When you speak, the expelled air and vibrating vocal cords initiate the sound, but your mouth, lips, and tongue modify it to produce the vowels and consonants that make up the words of speech.

An interesting observation in the production of sound can be demonstrated by humming with your mouth closed. If you now pinch your nose closed, the humming will stop because the air will no longer flow over the vocal cords. If you are fortunate enough to have survived a case of choking on either food or drink, you will recall that when the choking begins you usually panic and try to yell to anyone to tell them that you are choking. Unfortunately, as you try to speak you find that you are unable to do so. Since the windpipe has been closed, no air can pass over the vocal cords to initiate the vibration that starts the speaking process. Many people die from choking simply because they are unable to communicate their condition to someone who can help. The usual procedure to communicate your choking condition is to get someone's attention. Then, point to your throat and cross your throat with your finger as though you were cutting your throat. If the other person is aware of the sign and realizes that

Figure 1
The vocal cords.

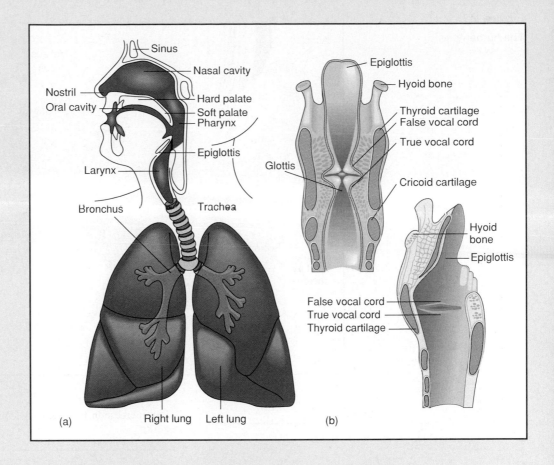

The vocal cords.

(a)

Right lung   Left lung

(b)

you are choking, he or she can save you by initiating the Heimlich maneuver. This consists of holding you from behind and wrapping his or her arms around you. Then the person presses against your diaphragm with his or her arms. By pressing in and upward, a force is exerted on your lungs that tends to compress the lungs. This in turn increases the pressure of the air in your lungs until it is great enough to force the closed valve open, thereby expelling the food that was causing you to choke. This then permits you to breathe normally and you observe that you now have your voice back to communicate with anyone.

Your ears are used to detect sound. The human ear can be divided into three parts: the outer ear, the middle ear, and the inner ear, figure 2. The outer ear acts as a funnel to channel the sound wave to the ear drum. These sound vibrations set the ear drum into vibration. These vibrations are then passed through the middle ear by three bones called the malleus (hammer), incus (anvil), and stapes (stirrup). These bones effectively amplify the amplitude of the vibration and then pass it on to the inner ear. The inner ear is a system of cavities. One of these cavities is the cochlea, a bony labyrinth in the shape of a spiral. The cochlea contains a fluid, through which the amplified vibration is passed to the auditory nerve on its way to the brain. The brain then interprets this sound as either speech, music, noise, and so forth.

The loudness of a sound as heard by the human ear is not directly proportional to the intensity of the sound, but rather is proportional to the logarithm of the intensity. The human ear can hear sounds of intensities as low as $I_0 = 1.00 \times 10^{-12}$ W/m², which is called the threshold of hearing, to higher than 1.00 W/m², which is called the threshold of pain. Because of the enormous variation in intensity that the human ear can hear, a logarithmic scale is usually used to measure sound. *The intensity level β of a sound wave, measured in decibels (dB), is defined as*

$$\beta = 10 \log\left(\frac{I}{I_0}\right) \qquad \textbf{(12H.1)}$$

where I_0 is the reference level, taken to be the threshold of hearing. The decibel is 1/10 of a bel, which was named to honor Alexander Graham Bell.

Example 12H.1

The intensity of sound in decibels. Find the intensity level of sound waves that are the following multiples of the threshold of hearing: (a) $I = I_0$, (b) $I = 2I_0$, (c) $I = 5I_0$, (d) $I = 10I_0$, and (e) $I = 100I_0$.

Figure 2

The human ear.

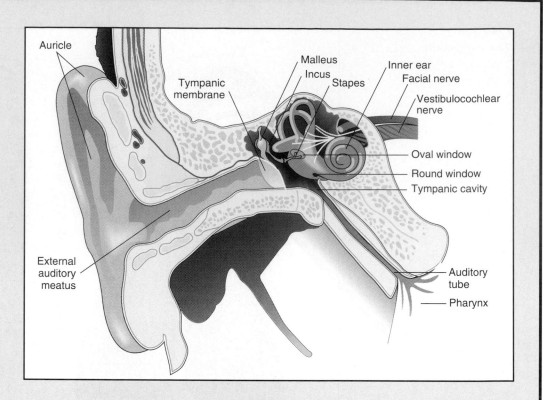

Auricle

Malleus
Incus Stapes

Inner ear
Facial nerve

Tympanic
membrane

Vestibulocochlear
nerve

Oval window

Round window

Tympanic cavity

External
auditory
meatus

Auditory
tube

Pharynx

Solution

a. The intensity level, found from equation 12H.1, is

$$\beta = 10 \log\left(\frac{I}{I_0}\right) \qquad \textbf{(12H.1)}$$

$$= 10 \log\left(\frac{I_0}{I_0}\right) = 10 \log 1 = 0 \text{ dB}$$

Because the log of 1 is equal to zero the intensity level at the threshold of hearing is 0 dB.

b. For $I = 2I_0$, the intensity level is

$$\beta = 10 \log\left(\frac{2I_0}{I_0}\right) = 10 \log 2 = 3.01 \text{ dB}$$

c. For $I = 5I_0$ the intensity level is

$$\beta = 10 \log\left(\frac{5I_0}{I_0}\right) = 10 \log 5 = 6.99 \text{ dB}$$

d. For $I = 10I_0$ the intensity level is

$$\beta = 10 \log\left(\frac{10I_0}{I_0}\right) = 10 \log 10 = 10.00 \text{ dB}$$

e. For $I = 100I_0$ the intensity level is

$$\beta = 10 \log\left(\frac{100I_0}{I_0}\right) = 10 \log 100 = 20.00 \text{ dB}$$

Thus, doubling the intensity level of a sound from 10 to 20 dB, a factor of 2, actually corresponds to an intensity increase from $10I_0$ to $100I_0$, or by a factor of 10. Similarly, an increase in the intensity level from 10 to 30 dB, a factor of 3, would correspond to an increase in the intensity from $10I_0$ to $1000I_0$, or a factor of 100.

Example 12H.2

Heavy traffic noise. A busy street with heavy traffic has an intensity level of 70 dB. Find the intensity of the sound.

Solution

We find the intensity by solving equation 12H.1 for I. Hence,

$$\frac{\beta}{10} = \log\left(\frac{I}{I_0}\right)$$

But the definition of the common logarithm is

$$\text{if } y = \log x \quad \text{then } x = 10^y$$

For our case this becomes

$$\text{if } \frac{\beta}{10} = \log\left(\frac{I}{I_0}\right) \quad \text{then } \frac{I}{I_0} = 10^{\beta/10}$$

Hence,

$$I = I_0 10^{\beta/10}$$

And

$$I = I_0 10^{70/10} = (1.00 \times 10^{-12} \text{ W/m}^2)(10^7)$$
$$= 1.00 \times 10^{-5} \text{ W/m}^2$$

Vibratory Motion, Wave Motion, and Fluids

Figure 3

Graph of intensity level of sound versus the frequency of the sound for a human ear.

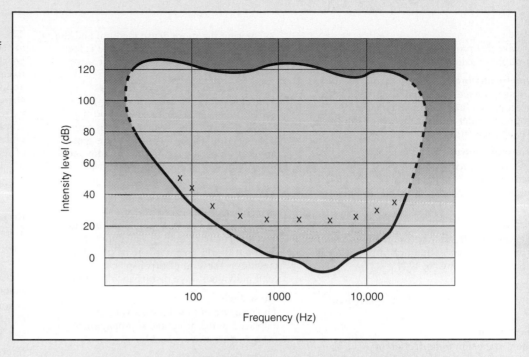

The human ear is responsive to a large range of frequencies and intensities. A typical response curve is shown in figure 3. The intensity level of sound is plotted against the frequency of the sound. The continuous curved line at the bottom represents the response curve of a normal ear. The lowest region on the curve occurs from about 1000 Hz to about 4000 Hz. These frequencies can be heard by the normal ear at very low intensity levels. On the other hand, to hear a frequency of 100 Hz the intensity level would have to be increased to about 35 dB. And for a normal ear to hear a frequency of about 20,000 Hz the intensity level would have to be increased to about 40 dB. At an intensity of 20 dB a frequency of 1000 Hz can easily be heard, but a frequency of 100 Hz could not be heard at all.

With age, the frequencies that the human ear can hear decreases. Many people resort to a hearing aid to overcome this hearing deficiency. A test is made of the person's ability to hear a sound of a known intensity level and frequency. The person is placed in a soundproof booth and earphones are placed over his or her ears. The examiner then plays pure sounds at a known frequency. He or she starts at a low intensity level and increases the intensity in small steps until the individual hears that particular frequency. When the individual hears the sound, he or she presses a button to let the examiner know that he or she has heard the sound. The examiner then marks an x on the graph of the frequency and intensity level of the normal ear shown in figure

3. The x's represent the actual frequencies heard at a particular intensity level. By knowing the frequencies that the person can no longer hear very well, a hearing aid, which is essentially a miniature electronic amplifier, is designed to amplify those frequencies, and thus bring the sound of that frequency up to a normal intensity level for that individual. For example the x's in figure 3 indicate that the individual's hearing response has deteriorated. In particular, the hearing response in the midrange frequency is much worse than at the low end or the high end of the spectrum. (The x's in the midrange are farther away from the normal curve.) Thus a hearing aid that amplifies the frequencies in the middle range of the audio spectrum would be useful for the individual. We would certainly not want to amplify the entire audio spectrum, for then we would be amplifying some of the frequencies that the person already hears reasonably well.

It is interesting to note that not only humans use sounds to communicate but animals do also. Some animals communicate at a higher frequency than can be heard by humans. These sounds are called ultrasonic and occur at frequencies above 20,000 Hz. Birds and dogs can hear these ultrasounds and bats even use them for navigation in a kind of sound radar. Ultrasound is used in sonar systems to detect submarines. It is also used in a variety of medical applications, including diagnosis and treatment. For example, chiropractors and physical therapists routinely use ultrasound for relief of lower back pain.

The Language of Physics

Wave
A wave is a propagation of a disturbance through a medium (p. 321).

Longitudinal wave
A wave in which the particles of the medium oscillate in simple harmonic motion parallel to the direction of the wave propagation (p. 321).

Transverse wave
A wave in which the particles of the medium execute simple harmonic motion in a direction perpendicular to its direction of propagation (p. 321).

Displacement
The distance that a particle of the medium is displaced from its equilibrium position as the wave passes by (p. 322).

Amplitude
The maximum value of the displacement (p. 322).

Wavelength
The distance, in the direction of propagation, in which the wave repeats itself (p. 323).

Period
The time it takes for one complete wave to pass a particular point. Hence, it is the time for a wave to repeat itself (p. 323).

Frequency
The number of waves passing a particular point per second (p. 323).

Reflection of a wave at a boundary
If a wave on a string traveling to the right is reflected from a nonfixed end, the reflected wave moves to the left with the same size and shape as the incident wave. If a wave on a string is traveling to the right and is reflected from a fixed end, the reflected wave is the same size and shape but is now inverted (p. 331).

Reflection and transmission of a wave at the boundary of two different media
(1) *Boundary between a less dense medium and a more dense medium.* The boundary acts as a fixed end and the reflected wave is inverted. The transmitted wave slows down on entering the more dense medium and the wavelength of the transmitted wave is less than the wavelength of the incident wave (p. 333).
(2) *Boundary between a more dense medium and a less dense medium.* The boundary acts as a nonfixed end and the reflected wave is not inverted, but is rather right side up. The transmitted wave speeds up on entering the less dense medium and the wavelength of the transmitted wave is greater than the wavelength of the incident wave (p. 335).

Principle of superposition
Whenever two or more wave disturbances pass a particular point in a medium, the resultant displacement of the point of the medium is the sum of the displacements of each individual wave disturbance (p. 336).

Phase angle
The measure of how far one wave is displaced in the direction of propagation from another wave (p. 337).

Constructive interference
When two interfering waves are in phase with each other (phase angle $= 0$) the amplitude of the combined wave is a maximum (p. 338).

Destructive interference
When two interfering waves are 180° out of phase with each other the amplitude of the combined wave is zero (p. 338).

Node
The point where the amplitude of a standing wave is zero (p. 341).

Antinode
The point where the amplitude of a standing wave is a maximum (p. 341).

Standing wave on a string
For a string fixed at both ends, a wave train is sent down the string. The wave is reflected from the fixed ends. Hence, there are two wave trains on the string, one traveling to the right and one traveling to the left. The resultant wave is the superposition of the two traveling waves. It is called a standing wave or a stationary wave because the resultant standing wave does not travel at all. The node of the standing wave remains a node for all times.

Thus, the string can not move up and down at that point, and can not transmit any energy past that point. Because the string is fixed at both ends, only certain wavelengths and frequencies are possible. When the string vibrates at these specified wavelengths, the string is said to be vibrating at one of its normal modes of vibration, and the string is vibrating at one of its natural frequencies (p. 342).

Fundamental frequency
The lowest of the natural frequencies of a vibrating system (p. 343).

Resonance
When a force is applied, whose frequency is equal to the natural frequency of the system, the system vibrates at maximum amplitude (p. 344).

Sound wave
A sound wave is a longitudinal wave that can be propagated in a solid, liquid, or gas (p. 346).

Overtone
An overtone is a frequency higher than the fundamental frequency (p. 348).

Harmonic
A harmonic is an overtone that is a multiple of the fundamental frequency. Hence, the nth harmonic is n times the fundamental frequency, or first harmonic (p. 348).

Doppler effect
The change in the wavelength and hence the frequency of a sound caused by the relative motion between the source and the observer. When a moving source approaches a stationary observer the observed frequency is higher than the emitted frequency of the source. When a moving source recedes from a stationary observer, the observed frequency is lower than the emitted frequency of the source (p. 350).

Intensity of a wave
The energy of a wave that passes a unit area in a unit time (p. 357).

Summary of Important Equations

Frequency of a wave
$$f = \frac{1}{T} \tag{12.1}$$

Fundamental equation
of wave propagation
$$v = \lambda f \tag{12.3}$$

Wave number
$$k = \frac{2\pi}{\lambda} \tag{12.9}$$

Equation of a wave
traveling to the right
$$y = A \sin(kx - \omega t) \tag{12.13}$$

Equation of a wave
traveling to the left
$$y = A \sin(kx + \omega t) \tag{12.26}$$

Angular frequency
$$\omega = 2\pi f \tag{12.12}$$

Angular frequency
$$\omega = kv \tag{12.14}$$

Velocity of transverse
wave on a string
$$v = \sqrt{\frac{T}{m/l}} \tag{12.30}$$

Change in wavelength
in second medium
$$\lambda_2 = \frac{v_2}{v_1}\lambda_1 \tag{12.34}$$

Principle of superposition
$$y = y_1 + y_2 + y_3 + \ldots \tag{12.35}$$

Equation of wave displaced
to the right by phase angle ϕ
$$y = A \sin(kx - \omega t - \phi) \tag{12.38}$$

Interference of two waves
out of phase by angle ϕ
$$y = 2A \cos\left(\frac{\phi}{2}\right)\sin\left(kx - \omega t - \frac{\phi}{2}\right) \tag{12.50}$$

The equation of the displacement
of a standing wave on a string
$$y = 2A \sin\left(\frac{n\pi x}{L}\right)\cos(\omega t) \tag{12.61}$$

Location of nodes
of standing wave
$$x = \frac{n\lambda}{2} \tag{12.56}$$

Location of antinodes
$$x = (2n - 1)\frac{\lambda}{4} \tag{12.57}$$

Possible wavelengths
on vibrating string
$$\lambda = \frac{2L}{n} \tag{12.62}$$

Frequency of vibrating string
$$f = \frac{n}{2L}\sqrt{\frac{T}{m/l}} \tag{12.64}$$

Frequency of higher modes
of vibration
$$f_n = nf_1 \tag{12.67}$$

Speed of sound in a solid
$$v = \sqrt{\frac{Y}{\rho}} \tag{12.68}$$

Speed of sound in a fluid
$$v = \sqrt{\frac{B}{\rho}} \tag{12.69}$$

Speed of sound in a gas
$$v = \sqrt{\frac{\gamma p}{\rho}} \tag{12.70}$$

Doppler frequency shift
$$f_o = \frac{v \pm v_o}{v \mp v_s}f_s \tag{12.91}$$

Energy transmitted by wave
$$E_{transmitted} = 2\pi^2 m f^2 R^2 \tag{12.95}$$

Intensity of a wave
$$I = 2\pi^2 \rho v f^2 R^2 \tag{12.99}$$

Intensity of a sound
wave in decibels
$$\beta = 10 \log\left(\frac{I}{I_0}\right) \tag{12H.1}$$

Questions for Chapter 12

1. Discuss the relation between simple harmonic motion and wave motion. Is it possible to create waves in a medium where the particles do not execute simple harmonic motion?
2. State the differences between transverse waves and longitudinal waves.
3. Describe how sound is made and heard by a human.
4. Discuss the statement "When a person is young enough to hear all the frequencies of a good stereo system, he can not afford to buy it. And when he can afford to buy it, he can not hear all the frequencies."
5. Discuss the statement that a wave is periodic in both space and time.

6. Why are there four different strings on a violin? Describe what a violin player does when she "tunes" the violin.
7. Discuss what happens to a pulse that is reflected from a fixed end and a free end.
†8. A wave is reflected from, and transmitted through, a more dense medium. What criteria would you use to estimate how much energy is reflected and how much is transmitted?
9. If the wavelength of a wave decreases as it enters a medium, what does this tell you about the medium?
10. When does the superposition principle fail in the analysis of combined wave motions?
11. Discuss what is meant by a standing wave and give some examples.

12. Discuss the difference between overtones and harmonics for a vibrating string, an open organ pipe, and a closed organ pipe.
†13. Discuss the possible uses of ultrasound in medicine.
†14. Discuss the Doppler effect on sound waves. Could the Doppler effect be applied to light waves? What would be the medium for the propagation?
†15. How could the Doppler effect be used to determine if the universe is expanding or contracting?
16. If two sounds of very nearly the same frequency are played together, the two waves will interfere with each other. The slight difference in frequency will cause an alternate rising and lowering of the intensity of the combined sound. This phenomenon is called beats. How can this technique be used to tune a piano?

Problems for Chapter 12

12.1 Introduction

1. Find the period of a sound wave of (a) 20.0 Hz and (b) 20,000 Hz.
2. A sound wave has a wavelength of 2.25×10^{-2} m and a frequency of 15,000 Hz. Find its speed.
3. Find the wavelength of a sound wave of 60.0 Hz at 20.0 °C.

12.2 Mathematical Representation of a Wave

4. At a time $t = 0$, a certain wave is given by $y = 10 \sin 5x$. Find the (a) amplitude of the wave and (b) its wavelength.
5. You want to generate a wave that has a wavelength of 20.0 cm and moves with a speed of 80.0 m/s. Find (a) the frequency of such a wave, (b) its wave number, and (c) its angular frequency.
6. A particular wave is given by $y = (8.50$ m$)\sin[(0.800$ m$^{-1})x - (5.40$ rad/s$)t]$. Find (a) the amplitude of the wave, (b) the wave number, (c) the wavelength, (d) the angular frequency, (e) the frequency, (f) the period, (g) the velocity of the wave, and (h) the displacement of the wave at $x = 5.87$ m and $t = 2.59$ s.
7. A certain wave has a wavelength of 25.0 cm, a frequency of 230 Hz, and an amplitude of 1.85 cm. Find (a) the wave number k and (b) the angular frequency ω. (c) Write the equation for this wave in the standard form $y = A \sin(kx - \omega t)$.

12.3 The Speed of a Transverse Wave on a String

8. A 60.0-cm guitar string has a mass of 1.40 g. If it is to play the note A at a frequency of 440 Hz, what must the tension be in the string? Assume that the wavelength of the note is twice the length of the string.
9. A 1.50-m length of wire with a mass of 0.035 kg is stretched between two points. Find the necessary tension in the wire such that the wave may travel from one end to another in a time of 0.0900 s.
10. A guitar string that has a mass per unit length of 2.33×10^{-3} kg/m is tightened to a tension of 655 N. What frequency will be heard if the string is 60.0 cm long? Is this a standard note or is it sharp or flat? (Remember that the wavelength of the note played is twice the length of the string.)

12.4 Reflection of a Wave at a Boundary

11. One end of a 100-cm wire of 3.45 g is welded to a 90.0-cm wire of 9.43 g. (a) If a wave moves along the first wire at a speed of 528 m/s, find its speed along the second wire. (b) If the wavelength on the first wire is 1.76 cm, find the wavelength of the wave on the second wire.

12. The first end of a combined string has a linear mass density of 4.20×10^{-3} kg/m, whereas the second end has a mass density of 10.5 kg/m. (a) If a 60.0-cm wave is to be sent along the first string at a speed of 8.56 m/s, what must the tension in the string be? (b) What is the wavelength of the reflected and transmitted wave?
13. The first end of a combined string has a linear mass density of 8.00 kg/m, whereas the second string has a density of 2.00 kg/m. If the speed of the wave in the first string is 10.0 m/s, find (a) the speed of the wave in the second string and (b) the tension in the string. (c) If a wave of length 60.0 cm is observed in the first string, find the wavelength and frequency of the wave in the second string.
14. The first end of a combined string has a linear mass density of 6.00 kg/m, whereas the second string has a density of 2.55 kg/m. The tension in the string is 350 N. If a vibration with a frequency of 20 vibrations is imparted to the first string, find the frequency, velocity, and wavelength of (a) the incident wave, (b) the reflected wave, and (c) the transmitted wave.

12.5 The Principle of Superposition

15. The following two waves interfere with each other:
$$y_1 = (10.8 \text{ m})\sin[(0.654 \text{ m}^{-1})x - (2.45 \text{ rad/s})t]$$
$$y_2 = (6.73 \text{ m})\sin[(0.893 \text{ m}^{-1})x - (6.82 \text{ rad/s})t]$$
Find the resultant displacement when $x = 0.782$ m and $t = 5.42$ s.

16. The following two waves combine:
$$y_1 = (10.8 \text{ m})\sin[(0.654 \text{ m}^{-1})x - (2.45 \text{ rad/s})t]$$
$$y_2 = (10.8 \text{ m})\sin[(0.654 \text{ m}^{-1})x - (2.45 \text{ rad/s})t - 0.834 \text{ rad}]$$
(a) Find the equation of the resultant wave. (b) Find the displacement of the resultant wave when $x = 0.895$ m and $t = 6.94$ s.

12.6 Standing Waves—The Vibrating String

17. The E string of a violin is vibrating at a fundamental frequency of 659 Hz. Find the wavelength and frequency of the third, fifth, and seventh harmonics. Let the length of the string be 60.0 cm.
18. A steel wire that is 1.45 m long and has a mass of 45 g is placed under a tension of 865 N. What is the frequency of its fifth harmonic?
19. A violin string plays a note at 440 Hz. What would the frequency of the wave on the string be if the tension in the string is (a) increased by 20.0% and (b) decreased by 20.0%?
20. A note is played on a guitar string 60.0 cm long at a frequency of 432 Hz. By how much should the string be shortened by pressing on it to play a note of 440 Hz?
21. A cello string, 75.0 cm long with a linear mass density of 7.25×10^{-3} kg/m, is to produce a fundamental frequency of 440 Hz. (a) What must be the tension in the string? (b) Find the frequency of the next three higher harmonics. (c) Find the wavelength of the fundamental and the next three higher harmonics.

12.7 Sound Waves

22. A sound wave in air has a velocity of 335 m/s. Find the temperature of the air.
23. A lightning flash is observed and 12 s later the associated thunder is heard. How far away is the lightning if the air temperature is 15.0 °C?

24. A soldier sees the flash from a cannon that is fired in the distance and 10 s later he hears the roar of the cannon. If the air temperature is 33 °C, how far away is the cannon?

25. A sound wave is sent to the bottom of the ocean by a ship in order to determine the depth of the ocean at that point. The sound wave returns to the boat in a time of 1.45 s. Find the depth of the ocean at this point. Use the bulk modulus of water to be 2.30 $\times$ 10^9 N/m² and the density of seawater to be 1.03×10^3 kg/m³.

26. Find the speed of sound in aluminum, copper, and lead.

27. You are trying to design three pipes for a closed organ pipe system that will give the following notes with their corresponding fundamental frequencies, C = 261.7 Hz, D = 293.7 Hz, E = 329.7 Hz. Find the length of each pipe. Assume that the speed of sound in air is 343 m/s.

28. Repeat problem 27 for an open organ pipe.

12.8 The Doppler Effect

29. A train is moving at a speed of 90.0 m/s and emits a whistle of frequency 400.0 Hz. If the speed of sound is 343 m/s, find the frequency observed by an observer who is at rest (a) in advance of the moving source and (b) behind the moving source.

30. A stationary police car turns on a siren at a frequency of 300 Hz. If the speed of sound in air is 343 m/s find the observed frequency if (a) the observer is approaching the police car at 35.0 m/s and (b) the observer is receding from the police car at 35.0 m/s.

31. A police car traveling at 90.0 m/s, turns on a siren at a frequency of 350 Hz as it tries to overtake a gangster's car moving away from the police car at a speed of 85 m/s. If the speed of sound in air is 343 m/s find the frequency heard by the gangster.

32. Two trains are approaching each other, each at a speed of 100 m/s. They each emit a whistle at a frequency of 225 Hz. If the speed of sound in air is 343 m/s, find the frequency that each train engineer hears.

33. A train moving east at a velocity of 20 m/s emits a whistle at a frequency of 348 Hz. Another train, farther up the track and moving east at a velocity of 30 m/s, hears the whistle from the first train. If the speed of sound in air is 343 m/s, what is the frequency of the sound heard by the second train engineer?

Additional Problems

34. One end of a violin string is connected to an electrical vibrator of 120 Hz, whereas the other end passes over a pulley and supports a mass of 10.0 kg, as shown in figure 12.17. The string is 60.0 cm long and has a mass of 12.5 g. What is the wavelength and speed of the wave produced?

†35. Three pipes, the first of lead, the second of brass, and the third of aluminum, each 10.0 m long, are welded together. If the first pipe is struck with a hammer at its end, how long will it take for the sound to pass through the pipes?

†36. A sound wave of 200 Hz in a steel pipe is transmitted into water and then into air. Find the wavelength of the sound in each medium.

†37. A railroad worker hits a steel track with a hammer. The sound wave through the steel track reaches an observer and 3.00 s later the sound wave through the air also reaches the observer. If the air temperature is 22.0 °C, how far away is the worker?

38. A tuning fork of 512 Hz is set into vibration above a long vertical tube containing water. A standing wave is observed as a resonance between the original wave and the reflected wave. If the speed of sound in air is 343 m/s, how far below the top of the tube is the water level?

39. The intensity of an ordinary conversation is about 3×10^{-6} W/m². Find the intensity level of the sound.

40. An indoor rock concert has an intensity level of 70 dB. Find the intensity of the sound.

41. The intensity level of a 500 Hz sound from a television program is about 40 dB. If the speed of sound is 343 m/s, find the amplitude of the sound wave.

†42. The speed of high-performance aircraft is sometimes given in terms of Mach numbers, where a Mach number is the ratio of the speed of the aircraft to the speed of sound at that level. Thus, a plane traveling at a speed of 343 m/s at sea level where the temperature is 20.0 °C would be traveling at Mach 1. If the temperature of the atmosphere increased to 30.0 °C, and the aircraft is still moving at 343 m/s, what is its Mach number?

Interactive Tutorials

📖43. A tuning fork of frequency f = 512 Hz is set into vibration above a long vertical cylinder filled with water. As the water in the tube is lowered, resonance occurs between the initial wave traveling down the cylinder and the second wave that is reflected from the water surface below. Calculate (a) the wavelength λ of the sound wave in air and (b) the three resonance positions as measured from the top of the tube. The velocity of sound in air is v = 343 m/s.

📖44. The superposition of any two waves. Given the following two waves:

$$y_1 = A_1 \sin(k_1 x - \omega_1 t)$$
$$y_2 = A_2 \sin(k_2 x - \omega_2 t - \phi)$$

For each wave find (a) the wavelength λ, (b) the frequency f, (c) the period T, and (d) the velocity v. Since each wave is periodic in both space and time, (e) for the value x = 2.00 m plot each wave and the sum of the two waves as a function of time t. (f) For the time t = 0.500 s, plot each wave and the sum of the two waves as a function of the distance x. For the initial conditions take A_1 = 3.50 m, k_1 = 0.55 m^{-1}, ω_1 = 4.25 rad/s, A_2 = 4.85 m, k_2 = 0.85 m^{-1}, ω_2 = 2.58 rad/s, and ϕ = 0. Then consider all the special cases listed in the tutorial itself.

📖45. A 60.0-cm string with a mass of 1.40 g is to produce a fundamental frequency of 440 Hz. Find (a) the tension in the string, (b) the frequency of the next four higher harmonics, and (c) the wavelength of the fundamental and the next four higher harmonics.

📖46. General purpose Doppler Effect Calculator. The Doppler Effect Calculator will calculate the observed frequency of a sound wave for the motion of the source or the observer, whether either or both are approaching or receding.

Fluids

13

A fluid in motion.

When did science begin? Where did it begin? It began whenever and wherever men tried to solve the innumerable problems of life. The first solutions were mere expedients, but that must do for a beginning. Gradually the expedients would be compared, generalized, rationalized, simplified, interrelated, integrated; the texture of science would be slowly woven.

George Sarton

13.1 **Introduction**

Matter is usually said to exist in three phases: solid, liquid, and gas. Solids are hard bodies that resist deformations, whereas liquids and gases have the characteristic of being able to flow. A liquid flows and takes the shape of whatever container in which it is placed. A gas also flows into a container and spreads out until it occupies the entire volume of the container. *A fluid is defined as any substance that can flow, and hence liquids and gases are both considered to be **fluids.***

 Liquids and gases are made up of billions upon billions of molecules in motion and to properly describe their behavior, Newton's second law should be applied to each of these molecules. However, this would be a formidable task, if not outright impossible, even with the use of modern high-speed computers. Also, the actual motion of a particular molecule is sometimes not as important as the overall effect of all those molecules when they are combined into the substance that is called the fluid. Hence, instead of using the microscopic approach of dealing with each molecule, we will treat the fluid from a macroscopic approach. That is, we will analyze the fluid in terms of its large-scale characteristics, such as its mass, density, pressure, and its distribution in space.

 The study of fluids will be treated from two different approaches. First, we will consider only fluids that are at rest. This portion of the study of fluids is called **fluid statics or hydrostatics.** Second, we will study the behavior of fluids when they are in motion. This part of the study is called **fluid dynamics or hydrodynamics.** Let us start the study of fluids by defining and analyzing the macroscopic variables.

13.2 **Density**

*The **density** of a substance is defined as the amount of mass in a unit volume of that substance.* We use the symbol ρ (the lower case Greek letter rho) to designate the density and write it as

$$\rho = \frac{m}{V} \tag{13.1}$$

 A substance that has a large density has a great deal of mass in a unit volume, whereas a substance of low density has a small amount of mass in a unit volume. Density is expressed in SI units as kg/m^3, and occasionally in the laboratory as g/cm^3. Densities of solids and most liquids are very nearly constant but the densities of gases vary greatly with temperature and pressure. Table 13.1 is a list of densities for various materials. We observe from the table that in interstellar space

Table 13.1
Densities of Various Materials

Substance	kg/m³
Air (0 °C, 1 atm pressure)	1.29
Aluminum	2,700
Benzene	879
Blood	1.05×10^3
Bone	1.7×10^3
Brass	8,600
Copper	8,920
Critical density for universe to collapse under gravitation	5×10^{-27}
Planet Earth	5,520
Ethyl alcohol	810
Glycerine	1,260
Gold	19,300
Hydrogen atom	2,680
Ice	920
Interstellar space	10^{-18}–10^{-21}
Iron	7,860
Lead	11,340
Mercury	13,630
Nucleus	10^{17}
Proton	1.5×10^{17}
Silver	10,500
Sun (avg)	1,400
Water (pure)	1,000
(sea)	1,030
Wood (maple)	620–750

the densities are extremely small, of the order of 10^{-18} to 10^{-21} kg/m³. That is, interstellar space is almost empty space. The density of the proton and neutron is of the order of 10^{17} kg/m³, which is an extremely large density. Hence, the nucleus of a chemical element is extremely dense. Because an atom of hydrogen has an approximate density of 2680 kg/m³, whereas the proton in the nucleus of that hydrogen atom has a density of about 1.5×10^{17} kg/m³, we see that the density of the nucleus is about 10^{13} times as great as the density of the atom. Hence, an atom consists almost entirely of empty space with the greatest portion of its mass residing in a very small nucleus.

Example 13.1

The density of an irregularly shaped object. In order to find the density of an irregularly shaped object, the object is placed in a beaker of water that is filled completely to the top. Since no two objects can occupy the same space at the same time, 25.0 cm³ of the water, which is equal to the volume of the unknown object, overflows into an attached calibrated beaker. The object is placed on a balance scale and is found to have a mass of 262.5 g. Find the density of the material.

Solution

The density, found from equation 13.1, is

$$\rho = \frac{m}{V} = \frac{262.5 \text{ g}}{25.0 \text{ cm}^3} = 10.5 \frac{\text{g}}{\text{cm}^3} = 10,500 \frac{\text{kg}}{\text{m}^3}$$

Example 13.2

Your own water bed. A person would like to design a water bed for the home. If the size of the bed is to be 2.20 m long, 1.80 m wide, and 0.300 m deep, what mass of water is necessary to fill the bed?

Solution

The mass of the water, found from equation 13.1, is

$$m = \rho V \tag{13.2}$$

The density is found from table 13.1. Hence, the mass of water required is

$$m = \rho V = \left(1000 \frac{\text{kg}}{\text{m}^3} \right)(2.20 \text{ m})(1.80 \text{ m})(0.300 \text{ m})$$

$$= 1190 \text{ kg}$$

As a matter of curiosity let us compute the weight of this water. The weight of the water is given by

$$w = mg = (1190 \text{ kg})(9.80 \text{ m/s}^2) = 11,600 \text{ N}$$

To give you a "feel" for this weight of water, it is equivalent to 2620 lb. In some cases, it will be necessary to reinforce the floor underneath this water bed or the bed might end up in the basement below.

13.3 Pressure

Pressure is defined as the magnitude of the normal force acting per unit surface area. The pressure is thus a scalar quantity. We write this mathematically as

$$p = \frac{F}{A} \tag{13.3}$$

Vibratory Motion, Wave Motion, and Fluids

The SI unit for pressure is newton/meter², which is given the special name pascal, in honor of the French mathematician, physicist, and philosopher, Blaise Pascal (1623–1662) and is abbreviated Pa. Hence, 1 Pa = 1 N/m². In the British engineering system the units are lb/in.², which is sometimes denoted by psi. Pressures are not limited to fluids, as the following examples show.

Example 13.3

Pressure exerted by a man. A man weighs 200 lb. At one particular moment when he walks, his right heel is the only part of his body that touches the ground. If the heel of his shoe measures 3.50 in. by 3.25 in., what pressure does the man exert on the ground?

Solution

The pressure that the man exerts on the ground, given by equation 13.3, is

$$p = \frac{F}{A}$$

$$= \frac{w}{A} = \frac{200 \text{ lb}}{(3.50 \text{ in.})(3.25 \text{ in.})}$$

$$= 17.6 \frac{\text{lb}}{\text{in.}^2}$$

Example 13.4

Pressure exerted by a woman. A woman who weighs 100 lb is wearing "high-heel" shoes. The cross section of her high-heel shoe measures 1/2 in. by 5/8 in. At a particular moment when she is walking, only one heel of her shoe makes contact with the ground. What is the pressure exerted on the ground by the woman?

Solution

The pressure exerted on the ground, found from equation 13.3, is

$$p = \frac{F}{A}$$

$$= \frac{w}{A} = \frac{100 \text{ lb}}{(0.500 \text{ in.})(6.25 \text{ in.})}$$

$$= 320 \frac{\text{lb}}{\text{in.}^2}$$

Thus, the 100-lb woman exerts a pressure through her high heel of 320 lb/in.², whereas the man, who weighs twice as much, exerts a pressure of only 17.6 lb/in.². That is, the woman exerts about 18 times more pressure than the man. The key to the great difference lies in the definition of pressure. Pressure is the force exerted per unit area. Because the area of the woman's high heel is so very small, the pressure becomes very large. The area of the man's heel is relatively large, hence the pressure he exerts is relatively small. When they are wearing high heels, women usually do not like to walk on soft ground because the large pressure causes the shoe to sink into the ground.

A further example of the effect of the surface area on pressure is found in the application of snowshoes. Here, a person's weight is distributed over such a large

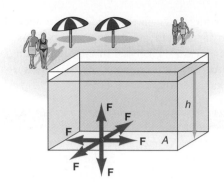

Figure 13.1

Pressure in a pool of water.

area that the pressure exerted on the snow is very small. Hence, the person is capable of walking in deep snow, while another person, wearing ordinary shoes, would find walking almost impossible.

Pressure exerted by a fluid is easily determined with the aid of figure 13.1, which represents a pool of water. We want to determine the pressure p at the bottom of the pool caused by the water in the pool. By our definition, equation 13.3, the pressure at the bottom of the pool is the magnitude of the force acting on a unit area of the bottom of the pool. But the force acting on the bottom of the pool is caused by the weight of all the water above it. Thus,

$$p = \frac{F}{A} = \frac{\text{weight of water}}{\text{area}} \tag{13.4}$$

$$p = \frac{w}{A} = \frac{mg}{A} \tag{13.5}$$

We have set the weight w of the water equal to mg in equation 13.5. The mass of the water in the pool, given by equation 13.2, is

$$m = \rho V$$

The volume of all the water in the pool is just equal to the area A of the bottom of the pool times the depth h of the water in the pool, that is,

$$V = Ah \tag{13.6}$$

Substituting equations 13.2 and 13.6 into equation 13.5 gives for the pressure at the bottom of the pool:

$$p = \frac{mg}{A} = \frac{\rho Vg}{A} = \frac{\rho Ahg}{A}$$

Thus,

$$p = \rho gh \tag{13.7}$$

(Although we derived equation 13.7 to determine the water pressure at the bottom of a pool of water, it is completely general and gives the water pressure at any depth h in the pool.) Equation 13.7 says that the water pressure at any depth h in any pool is given by the product of the density of the water in the pool, the acceleration due to gravity g, and the depth h in the pool. *Equation 13.7 is sometimes called* **the hydrostatic equation.**

Example 13.5

Pressure in a swimming pool. Find the water pressure at a depth of 3.00 m in a swimming pool.

Solution

The density of water, found in table 13.1, is 1000 kg/m³, and the water pressure, found from equation 13.7, is

$$p = \rho gh$$

$$= (1000 \text{ kg/m}^3)(9.80 \text{ m/s}^2)(3.00 \text{ m})$$

$$= 2.94 \times 10^4 \text{ N/m}^2 = 2.94 \times 10^4 \text{ Pa}$$

The pressure at the depth of 3 m in the pool in figure 13.1 is the same everywhere. Hence, the force exerted by the fluid is the same in all directions. That is, the force is the same in up-down, right-left, or in-out directions. If the force due to the fluid were not the same in all directions, then the fluid would flow in the direction away from the greatest pressure and would not be a fluid at rest. A fluid at rest is a fluid in equilibrium. Thus, in example 13.5, the pressure is 2.94×10^4 Pa at every point at a depth of 3 m in the pool and exerts the same force in every direction at that depth. You experience this pressure when swimming at a depth of 3.00 m as a pressure on your ears. As you swim up to the surface, the pressure on your ears decreases because h is decreasing. Or to look at it another way, the closer you swim up toward the surface, the smaller is the amount of water that is above you. Because the pressure is caused by the weight of that water above you, the smaller the amount of water, the smaller will be the pressure.

Just as there is a water pressure at the bottom of a swimming pool caused by the weight of all the water above the bottom, there is also an air pressure exerted on every object at the surface of the earth caused by the weight of all the air that is above us in the atmosphere. That is, there is an atmospheric pressure exerted on us, given by equation 13.3 as

$$p = \frac{F}{A} = \frac{\text{weight of air}}{\text{area}} \tag{13.8}$$

However we can not use the same result obtained for the pressure in the pool of water, the hydrostatic equation 13.7, because air is compressible and hence its density ρ is not constant with height throughout the vertical portion of the atmosphere. The pressure of air at any height in the atmosphere can be found by the use of calculus and the density variation in the atmosphere. However, since calculus is beyond the scope of this course, we will revert to the use of experimentation to determine the pressure of the atmosphere.

The pressure of the air in the atmosphere was first measured by Evangelista Torricelli (1608–1647), a student of Galileo, by the use of a mercury **barometer.** A long narrow tube is filled to the top with mercury, chemical symbol Hg. It is then placed upside down into a reservoir filled with mercury, as shown in figure 13.2. The mercury in the tube starts to flow out into the reservoir, but it comes to a stop when the top of the mercury column is at a height h above the top of the mercury reservoir, as also shown in figure 13.2. The mercury does not empty completely because the normal pressure of the atmosphere p_0 pushes downward on the mercury reservoir. Because the force caused by the pressure of a fluid is the same in all directions, there is also a force acting upward inside the tube at the height of the mercury reservoir, and hence there is also a pressure p_0 acting upward as shown in figure 13.2. This force upward is capable of holding the weight of the mercury in the tube up to a height h. Thus, the pressure exerted by the mercury in the tube is exactly balanced by the normal atmospheric pressure on the reservoir, that is,

$$p_0 = p_{\text{Hg}} \tag{13.9}$$

But the pressure of the mercury in the tube p_{Hg}, given by equation 13.7, is

$$p_{\text{Hg}} = \rho_{\text{Hg}}gh \tag{13.10}$$

Substituting equation 13.10 back into equation 13.9, gives

$$p_0 = \rho_{\text{Hg}}gh \tag{13.11}$$

Equation 13.11 says that normal atmospheric pressure can be determined by measuring the height h of the column of mercury in the tube. It is found experimentally, that on the average, normal atmospheric pressure can support a column of mercury 76.0 cm high, or 760 mm high. The unit of 1.00 mm of Hg is sometimes called a torr in honor of Torricelli. Hence, normal atmospheric pressure can also be given as 760 torr. Using the value of the density of mercury of 1.360×10^4 kg/m³, found in table 13.1, normal atmospheric pressure, determined from equation 13.11, is

$$p_0 = \rho_{\text{Hg}}gh = \left(1.360 \times 10^4 \frac{\text{kg}}{\text{m}^3}\right)\left(9.80 \frac{\text{m}}{\text{s}^2}\right)(0.760 \text{ m})$$

$$= 1.013 \times 10^5 \text{ N/m}^2 = 1.013 \times 10^5 \text{ Pa}$$

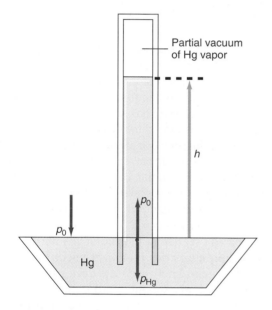

Figure 13.2

A mercury barometer.

Thus, the average or normal atmospheric pressure acting on us at the surface of the earth is 1.013×10^5 Pa, which is a rather large number as we will see presently. In the study of meteorology, the science of the weather, a different unit of pressure is usually employed, namely the millibar, abbreviated mb. The conversion factor between millibars and Pa (see appendix A) is

$$1 \text{ Pa} = 10^{-2} \text{ mb}$$

Using this conversion factor, normal atmospheric pressure can also be expressed as

$$p_0 = (1.013 \times 10^5 \text{ Pa})\left(\frac{10^{-2} \text{ mb}}{1 \text{ Pa}}\right)$$

$$= 1013 \text{ mb}$$

On all surface weather maps in a weather station, pressures are always expressed in terms of millibars.

To express normal atmospheric pressure in the British engineering system, the conversion factor

$$1 \text{ Pa} = 1.45 \times 10^{-4} \text{ lb/in.}^2$$

found in appendix A, is used. Hence, normal atmospheric pressure can also be expressed as

$$p_0 = (1.013 \times 10^5 \text{ Pa})\left(\frac{1.45 \times 10^{-4} \text{ lb/in.}^2}{1 \text{ Pa}}\right)$$

$$= 14.7 \text{ lb/in.}^2$$

The mercury barometer is thus a very accurate means of determining air pressure. The value of 76.0 cm or 1013 mb are only normal or average values. When the barometer is kept at the same location and the height of the mercury column is recorded daily, the value of h is found to vary slightly. When the value of h becomes greater than 76.0 cm of Hg, the pressure of the atmosphere has increased to a higher pressure. It is then said that a high-pressure area has moved into your region. When the value of h becomes less than 76.0 cm of Hg, the pressure of the atmosphere has decreased to a lower pressure and a low-pressure area has moved in. The barometer is extremely important in weather observation and prediction because, as a general rule of thumb, high atmospheric pressures usually are associated with clear skies and good weather. Low-pressure areas, on the other hand, are usually associated with cloudy skies, precipitation, and in general bad weather. (For further detail on the weather see the "Have You Ever Wondered" section at the end of chapter 17.)

The mercury barometer, after certain corrections for instrument height above sea level and ambient temperature, is an extremely accurate device to measure atmospheric pressure and can be found in every weather station throughout the world. Its chief limitation is its size. It must always remain vertical, and the glass tube and reservoir are somewhat fragile. Hence, another type of barometer is also used to measure atmospheric pressure. It is called an *aneroid barometer,* and is shown in figure 13.3. It is based on the principle of a partially evacuated, waferlike, metal cylinder called a Sylphon cell. When the atmospheric pressure increases, the cell decreases in size. A combination of linkages and springs are connected to the cell and to a pointer needle that moves over a calibrated scale that indicates the pressure. The aneroid barometer is a more portable device that is rugged and easily used, although it is originally calibrated with a mercury barometer. The word *aneroid* means not containing fluid. The aneroid barometer is calibrated in both centimeters of Hg and inches of Hg. Using a conversion factor, we can easily see that a height of 29.92 in. of Hg also corresponds to normal atmospheric pressure. Hence, as seen in figure 13.3, the pressure can be measured in terms of inches of

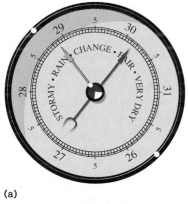

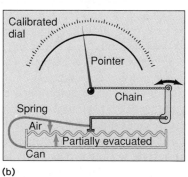

Figure 13.3

An aneroid barometer.

mercury. Also note that regions of high pressure (30 in. of Hg) are labeled to indicate fair weather, while regions of low pressure (29 in. of Hg) are labeled to indicate rain or poor weather.

As we go up into the atmosphere the pressure decreases, because there is less air above us. The aneroid barometer will read smaller and smaller pressures with altitude. Instead of calibrating the aneroid barometer in terms of centimeters of mercury or inches of mercury, we can also calibrate it in terms of feet or meters above the surface of the earth where this air pressure is found. An aneroid barometer so calibrated is called an *altimeter,* a device to measure the altitude or height of an airplane. The height of the plane is not really measured, the pressure is. But in the standard atmosphere, a particular pressure is found at a particular height above the ground. Hence, when the aneroid barometer measures this pressure, it corresponds to a fixed altitude above the ground. The pilot can read this height directly from the newly calibrated aneroid barometer, the altimeter.

Let us now look at some examples associated with atmospheric pressure.

Example 13.6

Why you get tired by the end of the day. The top of a student's head is approximately circular with a radius of 3.50 inches. What force is exerted on the top of the student's head by normal atmospheric pressure?

Solution

The area of the top of the student's head is found from

$$A = \pi r^2 = \pi(3.50 \text{ in.})^2 = 38.5 \text{ in.}^2$$

We find the magnitude of the force exerted on the top of the student's head by rearranging equation 13.3 into the form

$$F = pA \tag{13.12}$$

Hence,

$$F = \left(14.7 \frac{\text{lb}}{\text{in.}^2}\right)(38.5 \text{ in.}^2)$$

$$= 566 \text{ lb}$$

This is a rather large force to have exerted on our heads all day long. However, we do not notice this enormous force because when we breathe air into our nose or mouth that air is exerting the same force upward inside our head. Thus, the difference in force between the top of the head and the inside of the head is zero.

Example 13.7

Atmospheric pressure on the walls of your house. Find the force on the outside wall of a ranch house, 10.0 ft high and 35.0 ft long, caused by normal atmospheric pressure.

Solution

The area of the wall of the house is given by

$$A = (\text{length})(\text{height})$$

$$= (35.0 \text{ ft})(10.0 \text{ ft})\left(\frac{144 \text{ in.}^2}{1 \text{ ft}^2}\right)$$

$$= 50,400 \text{ in.}^2$$

The force on the wall, given by equation 13.12, is

$$F = pA = \left(14.7 \frac{\text{lb}}{\text{in.}^2}\right)(50{,}400 \text{ in.}^2)$$

$$= 740{,}000 \text{ lb}$$

The force on the outside wall of the house in example 13.7 is thus 740,000 lb. This is truly an enormous force. Why doesn't the wall collapse under this great force? The wall does not collapse because that same atmospheric air is also inside the house. Remember that air is a fluid and flows. Hence, in addition to being outside the house, the air also flows to the inside of the house. Because the force exerted by the pressure in the fluid is the same in all directions, the air inside the house exerts the same force of 740,000 lb against the inside wall of the house, as shown in figure 13.4(a). The net force on the wall is therefore

$$\text{Net force} = (\text{force})_{\text{in}} - (\text{force})_{\text{out}}$$

$$= 740{,}000 \text{ lb} - 740{,}000 \text{ lb}$$

$$= 0$$

A very interesting case occurs when this net force is not zero. Suppose a tornado, an extremely violent storm, were to move over your house, as shown in figure 13.4(b). The pressure inside the tornado is very low. No one knows for sure how low, because it is slightly difficult to run into a tornado with a barometer to measure it. In the very few cases on record where tornadoes actually went over a weather station, there was never anything left of the weather station, to say nothing of the barometer that was in that station. That is, neither the barometer nor the weather station were ever found again. The pressure can be estimated, however, from the very high winds associated with the tornado. A good estimate is that the pressure inside the tornado is at least 10% below the actual atmospheric pressure. Let us assume that the actual pressure is the normal atmospheric pressure of 1013 mb, then 10% of that is 101 mb. Thus, the pressure in the tornado is approximately

$$1013 \text{ mb} - 101 \text{ mb} = (912 \text{ mb})\left(\frac{14.7 \text{ lb/in.}^2}{1013 \text{ mb}}\right) = 13.2 \text{ lb/in.}^2$$

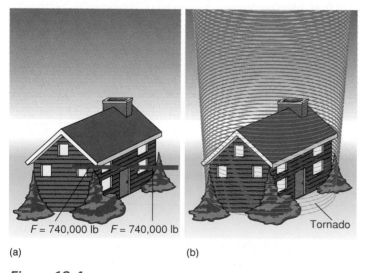

(a)   (b)

$F = 740{,}000 \text{ lb}$ $F = 740{,}000 \text{ lb}$ Tornado

(c)

Figure 13.4

Pressure on the walls in a house.

Vibratory Motion, Wave Motion, and Fluids

When the tornado goes over the house, the force on the outside wall is given by

$$F = pA = \left(13.2 \frac{\text{lb}}{\text{in.}^2}\right)(50{,}400 \text{ in.}^2)$$

$$= 665{,}000 \text{ lb}$$

But the original air inside the house is still there and is still exerting a force of 740,000 lb outward on the walls. The net force on the house is now

$$\text{Net force} = 740{,}000 \text{ lb} - 665{,}000 \text{ lb}$$

$$= 75{,}000 \text{ lb}$$

There is now a net force acting outward on the wall of 75,000 lb, enough to literally explode the walls of the house outward. This pressure differential, with its accompanying winds, accounts for the enormous destruction associated with a tornado. Thus, the force exerted by atmospheric pressure can be extremely significant.

It has always been customary to open the doors and windows in a house whenever a tornado is in the vicinity in the hope that a great deal of the air inside the house will flow out through these open windows and doors. Hence, the pressure differential between the inside and the outside walls of the house will be minimized. Unfortunately many victims of tornadoes do not follow this procedure, because tornadoes are spawned out of severe thunderstorms, which are usually accompanied by torrential rain. Usually the first thing one does in a house is to close the windows once the rain starts. A picture of a typical tornado is shown in figure 13.4(c).

Now that we have discussed atmospheric pressure, it is obvious that the total pressure exerted at a depth h in a pool of water must be greater than the value determined previously, because the air above the pool is exerting an atmospheric pressure on the top of the pool. This additional pressure is transmitted undiminished throughout the pool. Hence, the total or *absolute pressure* observed at the depth h in the pool is the sum of the atmospheric pressure plus the pressure of the water itself, that is,

$$p_{\text{abs}} = p_0 + p_w \tag{13.13}$$

Using equation 13.7, this becomes

$$p_{\text{abs}} = p_0 + \rho g h \tag{13.14}$$

Example 13.8

Absolute pressure. What is the absolute pressure at a depth of 3.00 m in a swimming pool?

Solution

The water pressure at a depth of 3.00 m has already been found to be $p_w = 2.94 \times 10^4$ Pa, the absolute pressure, found by equation 13.13, is

$$p_{\text{abs}} = p_0 + p_w$$

$$= 1.013 \times 10^5 \text{ Pa} + 2.94 \times 10^4 \text{ Pa}$$

$$= 1.31 \times 10^5 \text{ Pa}$$

When the pressure of the air in an automobile tire is measured, the actual pressure being measured is called the **gauge pressure,** that is, the pressure as indicated on the measuring device that is called a gauge. This measuring device, the gauge, reads zero when it is actually under normal atmospheric pressure. Thus, the total pressure or absolute pressure of the air inside the tire is the sum of the pressure recorded on the gauge plus normal atmospheric pressure. We can write this mathematically as

$$p_{abs} = p_{gauge} + p_0 \tag{13.15}$$

Example 13.9

Gauge pressure and absolute pressure. A gauge placed on an automobile tire reads a pressure of 34.0 lb/in.². What is the absolute pressure of the air in the tire?

Solution

The absolute pressure of the air in the tire, found from equation 13.15, is

$$p_{abs} = p_{gauge} + p_0$$

$$= 34.0 \frac{lb}{in.^2} + 14.7 \frac{lb}{in.^2}$$

$$= 48.7 \text{ lb/in.}^2$$

13.4 Pascal's Principle

The pressure exerted on the bottom of a pool of water by the water itself is given by $\rho g h$. However, there is also an atmosphere over the pool, and, as we saw in section 13.3, there is thus an additional pressure, normal atmospheric pressure p_0, exerted on the top of the pool. This pressure on the top of the pool is transmitted through the pool waters so that the total pressure at the bottom of the pool is the sum of the pressure of the water plus the pressure of the atmosphere, equations 13.13 and 13.14. The addition of both pressures is a special case of a principle, called *Pascal's principle* and it states that if the pressure at any point in an enclosed fluid at rest is changed (Δp), *the pressure changes by an equal amount* (Δp), *at all points in the fluid.* As an example of the use of Pascal's principle, let us consider the hydraulic lift shown in figure 13.5. A noncompressible fluid fills both cylinders and the connecting pipe. The smaller cylinder has a piston of cross-sectional area a, whereas the larger cylinder has a cross-sectional area A. As we can see in the figure, the cross-sectional area A of the larger cylinder is greater than the cross-sectional area a of the smaller cylinder. If a small force f is applied to the piston of the small cylinder, this creates a change in the pressure of the fluid given by

$$\Delta p = \frac{f}{a} \tag{13.16}$$

But by Pascal's principle, this pressure change occurs at all points in the fluid, and in particular at the large piston on the right. This same pressure change applied to the right piston gives

$$\Delta p = \frac{F}{A} \tag{13.17}$$

where F is the force that the fluid now exerts on the large piston of area A. Because these two pressure changes are equal by Pascal's principle, we can set equation 13.17 equal to equation 13.16. Thus,

$$\Delta p = \Delta p$$

$$\frac{F}{A} = \frac{f}{a}$$

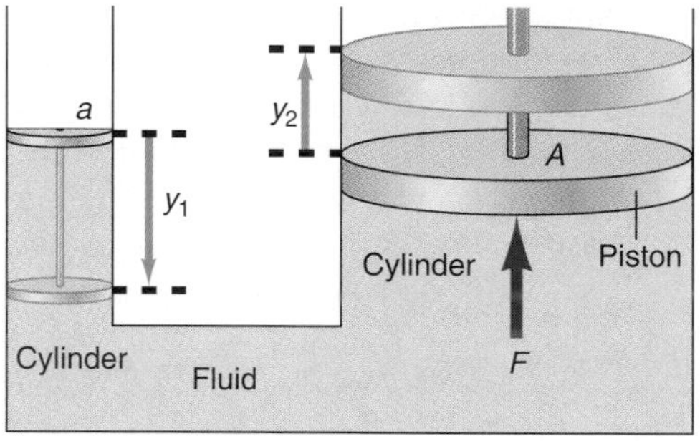

Figure 13.5

The hydraulic lift.

The force F on the large piston is therefore

$$F = \frac{A}{a} f \qquad (13.18)$$

Since the area A is greater than the area a, the force F will be greater than f. Thus, the hydraulic lift is a device that is capable of multiplying forces.

Example 13.10

Amplifying a force. The radius of the small piston in figure 13.5 is 5.00 cm, whereas the radius of the large piston is 30.0 cm. If a force of 2.00 N is applied to the small piston, what force will occur at the large piston?

Solution

The area of the small piston is

$$a = \pi r_1^2 = \pi(5.00 \text{ cm})^2 = 78.5 \text{ cm}^2$$

while the area of the large piston is

$$A = \pi r_2^2 = \pi(30.0 \text{ cm})^2 = 2830 \text{ cm}^2$$

The force exerted by the fluid on the large piston, found from equation 13.18, is

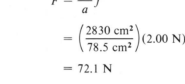

$$F = \frac{A}{a} f$$

$$= \left(\frac{2830 \text{ cm}^2}{78.5 \text{ cm}^2}\right)(2.00 \text{ N})$$

$$= 72.1 \text{ N}$$

Thus, the relatively small force of 2.00 N applied to the small piston produces the rather large force of 72.1 N at the large piston. The force has been magnified by a factor of 36.

The amplifying force discussed in Example 13.10 is what allows this hydraulic lift to raise this automobile.

It is interesting to compute the work that is done when the force f is applied to the small piston in figure 13.5. When the force f is applied, the piston moves through a displacement y_1, such that the work done is given by

$$W_1 = fy_1$$

But from equation 13.16

$$f = a\Delta p$$

Hence, the work done is

$$W_1 = a(\Delta p)y_1 \qquad (13.19)$$

When the change in pressure is transmitted through the fluid, the force F is exerted against the large piston and the work done by the fluid on the large piston is

$$W_2 = Fy_2$$

where y_2 is the distance that the large piston moves and is shown in figure 13.5. But the force F, found from equation 13.17, is

$$F = A\Delta p$$

The work done on the large piston by the fluid becomes

$$W_2 = A(\Delta p)y_2 \qquad (13.20)$$

Applying the law of conservation of energy to a frictionless hydraulic lift, the work done to the fluid at the small piston must equal the work done by the fluid at the large piston, hence

$$W_1 = W_2 \qquad (13.21)$$

Substituting equations 13.19 and 13.20 into equation 13.21, gives

$$a(\Delta p)y_1 = A(\Delta p)y_2 \qquad (13.22)$$

Because the pressure change Δp is the same throughout the fluid, it cancels out of equation 13.22, leaving

$$ay_1 = Ay_2$$

Solving for the distance y_1 that the small piston moves

$$y_1 = \frac{A}{a} y_2 \qquad (13.23)$$

Since A is much greater than a, it follows that y_1 must be much greater than y_2.

Example 13.11

You can never get something for nothing. The large piston of example 13.10 moves through a distance of 0.200 cm. By how much must the small piston be moved?

Solution

The areas of the pistons are given from example 13.10 as $A = 2830 \ \text{cm}^2$ and $a = 78.5 \ \text{cm}^2$, hence the distance that the small piston must move, given by equation 13.23, is

$$y_1 = \frac{A}{a} y_2$$

$$= \left(\frac{2830 \ \text{cm}^2}{78.5 \ \text{cm}^2} \right)(0.200 \ \text{cm})$$

$$= 7.21 \ \text{cm}$$

Although a very large force is obtained at the large piston, the large piston is displaced by only a very small amount. Whereas the input force f, on the small piston is relatively small, the small piston must move through a relatively large displacement (36 times greater than the large piston). Usually there are a series of valves in the connecting pipe and the small cylinder is connected to a fluid reservoir also by valves. Hence, many displacements of the small piston can be made, each time adding additional fluid to the right cylinder. In this way the final displacement y_2 can be made as large as desired.

13.5 Archimedes' Principle

The variation of pressure with depth has a surprising consequence, it allows the fluid to exert buoyant forces on bodies immersed in the fluid. If this buoyant force is equal to the weight of the body, the body floats in the fluid. This result was first annunciated by Archimedes (287–212 BC) and is now called Archimedes' principle.

Archimedes' principle states that a body immersed in a fluid is buoyed up by a force that is equal to the weight of the fluid displaced. This principle can be verified with the help of figure 13.6.

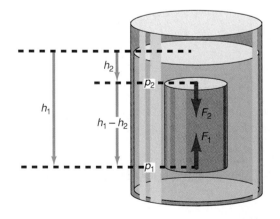

Figure 13.6

Archimedes' principle.

If we submerge a cylindrical body into a fluid, such as water, then the bottom of the body is at some depth h_1 below the surface of the water and experiences a water pressure p_1 given by

$$p_1 = \rho g h_1 \tag{13.24}$$

where ρ is the density of the water. Because the force due to the pressure acts equally in all directions, there is an upward force on the bottom of the body. The force upward on the body is given by

$$F_1 = p_1 A \tag{13.25}$$

where A is the cross-sectional area of the cylinder. Similarly, the top of the body is at a depth h_2 below the surface of the water, and experiences the water pressure p_2 given by

$$p_2 = \rho g h_2 \tag{13.26}$$

However, in this case the force due to the water pressure is acting downward on the body causing a force downward given by

$$F_2 = p_2 A \tag{13.27}$$

Because of the difference in pressure at the two depths, h_1 and h_2, there is a different force on the bottom of the body than on the top of the body. Since the bottom of the submerged body is at the greater depth, it experiences the greater force. Hence, there is a net force upward on the submerged body given by

$$\text{Net force upward} = F_1 - F_2$$

Replacing the forces F_1 and F_2 by their values in equations 13.25 and 13.27, this becomes

$$\text{Net force upward} = p_1 A - p_2 A$$

Replacing the pressures p_1 and p_2 from equations 13.24 and 13.26, this becomes

$$\text{Net force upward} = \rho g h_1 A - \rho g h_2 A$$
$$= \rho g A (h_1 - h_2) \tag{13.28}$$

But

$$A(h_1 - h_2) = V$$

the volume of the cylindrical body, and hence the volume of the water displaced. Equation 13.28 thus becomes

$$\text{Net force upward} = \rho g V \tag{13.29}$$

But ρ is the density of the water and from the definition of the density

$$\rho = \frac{m}{V} \tag{13.1}$$

Substituting equation 13.1 back into equation 13.29 gives

$$\text{Net force upward} = \frac{m}{V} g V$$
$$= mg$$

But $mg = w$, the weight of the water displaced. Hence,

$$\text{Net force upward} = \text{Weight of water displaced} \tag{13.30}$$

The net force upward on the body is called the *buoyant force* (BF). When the buoyant force on the body is equal to the weight of the body, the body does not sink in the

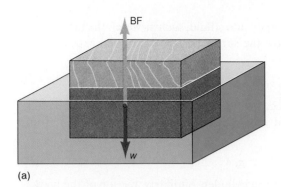

BF

w

(a)

(b)

Figure 13.7

A body floats when the buoyant force is equal to the weight of the body.

water but rather floats, figure 13.7(b). Since the buoyant force is equal to the weight of the water displaced, *a body floats when the weight of the body is equal to the weight of the fluid displaced.*

Example 13.12

Wood floats. A block of oak wood 5.00 cm high, 5.00 cm wide, and 10.0 cm long is placed into a tub of water, figure 13.7(a). The density of the wood is 7.20×10^2 kg/m³. How far will the block of wood sink before it floats?

Solution

The block of wood will float when the buoyant force (BF), which is the weight of the fluid displaced by the volume of the body submerged, is equal to the weight of the body. The weight of the block of wood is found from

$$w = mg = \rho V g$$

The volume of the wooden block is $V = Ah$. Thus, the weight of the wooden block is

$$w = (7.20 \times 10^2 \text{ kg/m}^3)(0.0500 \text{ m})(0.0500 \text{ m})(0.100 \text{ m})(9.80 \text{ m/s}^2)$$
$$= 1.76 \text{ N}$$

The buoyant force is equal to the weight of the water displaced, and for the body to float this buoyant force must also equal the weight of the block. Hence,

$$\text{BF} = w_{\text{water}} = w_{\text{wood}}$$
$$w_{\text{water}} = m_{\text{water}}g = \rho_{\text{water}}V g = \rho_{\text{water}}Ahg \qquad (13.31)$$

Thus,

$$\rho_{\text{water}}Ahg = w_{\text{wood}}$$
$$h = \frac{w_{\text{wood}}}{\rho_{\text{water}}Ag} \qquad (13.32)$$
$$= \frac{1.76 \text{ N}}{(1.00 \times 10^3 \text{ kg/m}^3)(0.0500 \text{ m})(0.100 \text{ m})(9.80 \text{ m/s}^2)}$$
$$= 0.0359 \text{ m} = 3.59 \text{ cm}$$

Thus, the block sinks to a depth of 3.59 cm. At this point the buoyant force becomes equal to the weight of the wooden block and the wooden block floats.

Example 13.13

Iron sinks. Repeat example 13.12 for a block of iron of the same dimensions.

Solution

The density of iron, found from table 13.1, is 7860 kg/m³. The weight of the iron block is given by

$$w_{\text{iron}} = mg = \rho V g$$
$$= (7860 \text{ kg/m}^3)(0.0500 \text{ m})(0.0500 \text{ m})(0.100 \text{ m})(9.80 \text{ m/s}^2)$$
$$= 19.3 \text{ N}$$

The depth that the iron block would have to sink in order to displace its own weight, again found from equation 13.32, is

$$h = \frac{w_{\text{iron}}}{\rho_{\text{water}} A g}$$

$$= \frac{19.3 \text{ N}}{(1.00 \times 10^3 \text{ kg/m}^3)(0.0500 \text{ m})(0.100 \text{ m})(9.80 \text{ m/s}^2)}$$

$$= 39.4 \text{ cm}$$

But the block is only 10 cm high. Hence, the buoyant force is not great enough to lift an iron block of this size, and the iron block sinks to the bottom.

Another way to look at this problem is to calculate the buoyant force on this piece of iron. The buoyant force on the iron, given by equation 13.29, is

$$\text{Net force upward} = \rho g V$$

$$= (1 \times 10^3 \text{ kg/m}^3)(9.80 \text{ m/s}^2)(0.0500 \text{ m})(0.500 \text{ m})(0.100 \text{ m})$$

$$= 2.45 \text{ N}$$

Thus, the net force upward on a block of iron of this size is 2.45 N. But the block weighs 19.3 N. Hence, the weight of the iron is greater than the buoyant force and the iron block sinks to the bottom.

Even an enormous ship is able to remain afloat.

But ships are made of iron and they do not sink. Why should the block sink and not the ship? If this same weight of iron is made into thin slabs, these thin slabs could be welded together into a boat structure of some kind. By increasing the size and hence the volume of this iron boat, a greater volume of water can be displaced. An increase in the volume of water displaced increases the buoyant force. If this can be made equal to the weight of the iron boat, then the boat floats.

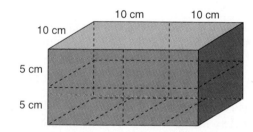

Figure 13.8
Iron can float.

Example 13.14

An iron boat. The iron block of example 13.13 is cut into 16 slices, each 5.00 cm by 10.0 cm by 5/16 cm. They are now welded together to form a box 20.0 cm wide by 10.0 cm long by 10.0 cm high, as shown in figure 13.8. Will this iron body now float or will it sink?

Solution

In this new configuration the iron displaces a much greater volume of water, and since the buoyant force is equal to the weight of the water displaced it is possible that this new configuration will float. We assume that no mass of iron is lost in cutting the blocks into the 16 slabs, and that the weight of the welding material is negligible. Thus, the weight of the box is also equal to 19.3 N. This example is analyzed in the same way as the previous example. Let us solve for the depth that the iron box must sink in order that the buoyant force be equal to the weight of the box. Thus, the depth that the box sinks, again found from the modified equation 13.32, is

$$h = \frac{w_{\text{box}}}{\rho_{\text{water}} A g}$$

$$= \frac{19.3 \text{ N}}{(1.00 \times 10^3 \text{ kg})(0.200 \text{ m})(0.100 \text{ m})(9.80 \text{ m/s}^2)}$$

$$= 9.84 \times 10^{-2} \text{ m} = 9.84 \text{ cm}$$

Because the iron box is 10 cm high, it sinks to a depth of 9.84 cm and it then floats. Note that this is the same mass of iron that sank in example 13.13. That same mass can now float because the new distribution of that mass results in a displacement of a much larger volume of water. Since the buoyant force is equal to the weight of the water displaced, by increasing the volume taken up by the iron and the enclosed space, the amount of the water displaced has increased and so has the buoyant force.

Examples 13.12–13.14 dealt with bodies submerged in water, but remember that Archimedes' principle applies to all fluids.

13.6 The Equation of Continuity

Up to now, we have studied only fluids at rest. *Let us now study fluids in motion, the subject matter of hydrodynamics.* The study of fluids in motion is relatively complicated, but the analysis can be simplified by making a few assumptions. Let us assume that the fluid is incompressible and flows freely without any turbulence or friction between the various parts of the fluid itself and any boundary containing the fluid, such as the walls of a pipe. A fluid in which friction can be neglected is called a *nonviscous fluid*. A fluid, flowing steadily without turbulence, is usually referred to as being in *streamline flow*. The rather complicated analysis is further simplified by the use of two great conservation principles: the conservation of mass, and the conservation of energy. *The law of conservation of mass results in a mathematical equation, usually called the equation of continuity. The law of conservation of energy is the basis of Bernoulli's theorem*, the subject matter of section 13.7.

Let us consider an incompressible fluid flowing in the pipe of figure 13.9. At a particular instant of time the small mass of fluid Δm, shown in the left-hand portion of the pipe will be considered. This mass is given by a slight modification of equation 13.2, as

$$\Delta m = \rho \Delta V \qquad (13.33)$$

Because the pipe is cylindrical, the small portion of volume of fluid is given by the product of the cross-sectional area A_1 times the length of the pipe Δx_1 containing the mass Δm, that is,

$$\Delta V = A_1 \Delta x_1 \qquad (13.34)$$

The length Δx_1 of the fluid in the pipe is related to the velocity v_1 of the fluid in the left-hand pipe. Because the fluid in Δx_1 moves a distance Δx_1 in time Δt, $\Delta x_1 = v_1 \Delta t$. Thus,

$$\Delta x_1 = v_1 \Delta t \qquad (13.35)$$

Substituting equation 13.35 into equation 13.34, we get for the volume of fluid,

$$\Delta V = A_1 v_1 \Delta t \qquad (13.36)$$

Figure 13.9

The law of conservation of mass and the equation of continuity.

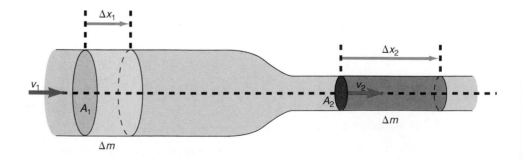

Substituting equation 13.36 into equation 13.33 yields the mass of the fluid as

$$\Delta m = \rho A_1 v_1 \Delta t \tag{13.37}$$

We can also express this as the rate at which the mass is flowing in the left-hand portion of the pipe by dividing both sides of equation 13.37 by Δt, thus

$$\frac{\Delta m}{\Delta t} = \rho A_1 v_1 \tag{13.38}$$

Example 13.15

Flow rate. What is the mass flow rate of water in a pipe whose diameter d is 10.0 cm when the water is moving at a velocity of 0.322 m/s.

Solution

The cross-sectional area of the pipe is

$$A_1 = \frac{\pi d_1^2}{4} = \frac{\pi (0.100 \text{ m})^2}{4}$$

$$= 7.85 \times 10^{-3} \text{ m}^2$$

The flow rate, found from equation 13.38, is

$$\frac{\Delta m}{\Delta t} = \rho A_1 v_1$$

$$= (1.00 \times 10^3 \text{ kg/m}^3)(7.85 \times 10^{-3} \text{ m}^2)(0.322 \text{ m/s})$$

$$= 2.53 \text{ kg/s}$$

Thus 2.53 kg of water flow through the pipe per second.

$\blacksquare$

When this fluid reaches the narrow constricted portion of the pipe to the right in figure 13.9, the same amount of mass Δm is given by

$$\Delta m = \rho \Delta V \tag{13.39}$$

But since ρ is a constant, the same mass Δm must occupy the same volume ΔV. However, the right-hand pipe is constricted to the narrow cross-sectional area A_2. Thus, the length of the pipe holding this same volume must increase to a larger value Δx_2, as shown in figure 13.9. Hence, the volume of fluid is given by

$$\Delta V = A_2 \Delta x_2 \tag{13.40}$$

The length of pipe Δx_2 occupied by the fluid is related to the velocity of the fluid by

$$\Delta x_2 = v_2 \Delta t \tag{13.41}$$

Substituting equation 13.41 back into equation 13.40, we get for the volume of fluid,

$$\Delta V = A_2 v_2 \Delta t \tag{13.42}$$

It is immediately obvious that since A_2 has decreased, v_2 must have increased for the same volume of fluid to flow. Substituting equation 13.42 back into equation 13.39, the mass of the fluid flowing in the right-hand portion of the pipe becomes

$$\Delta m = \rho A_2 v_2 \Delta t \tag{13.43}$$

Dividing both sides of equation 13.43 by Δt yields the rate at which the mass of fluid flows through the right-hand side of the pipe, that is,

$$\frac{\Delta m}{\Delta t} = \rho A_2 v_2 \tag{13.44}$$

But *the **law of conservation of mass** states that mass is neither created nor destroyed in any ordinary mechanical or chemical process.* Hence, the law of conservation of mass can be written as

Mass flowing into the pipe = mass flowing out of the pipe

or

$$\frac{\Delta m}{\Delta t} = \frac{\Delta m}{\Delta t} \qquad (13.45)$$

Thus, setting equation 13.38 equal to equation 13.44 yields

$$\rho A_1 v_1 = \rho A_2 v_2 \qquad (13.46)$$

Equation 13.46 is called **the equation of continuity** and is an indirect statement of the law of conservation of mass. Since we have assumed an incompressible fluid, the densities on both sides of equation 13.46 are equal and can be canceled out leaving

$$A_1 v_1 = A_2 v_2 \qquad (13.47)$$

Equation 13.47 is a special form of the equation of continuity for incompressible fluids (i.e., liquids).

Applying equation 13.47 to figure 13.9, we see that the velocity of the fluid v_2 in the narrow pipe to the right is given by

$$v_2 = \frac{A_1}{A_2} v_1 \qquad (13.48)$$

Because the cross-sectional area A_1 is greater than the cross-sectional area A_2, the ratio A_1/A_2 is greater than one and thus the velocity v_2 must be greater than v_1.

Example 13.16

Applying the equation of continuity. In example 13.15 the cross-sectional area A_1 was 7.85×10^{-3} m² and the velocity v_1 was 0.322 m/s. If the diameter of the pipe to the right in figure 13.9 is 4.00 cm, find the velocity of the fluid in the right-hand pipe.

Solution

The cross-sectional area of the right-hand side of the pipe is

$$A_2 = \frac{\pi d_2^2}{4} = \frac{\pi (0.0400 \text{ m})^2}{4}$$

$$= 1.26 \times 10^{-3} \text{ m}^2$$

The velocity of the fluid on the right-hand side v_2, found from equation 13.48, is

$$v_2 = \frac{A_1}{A_2} v_1 = \left(\frac{7.85 \times 10^{-3} \text{ m}^2}{1.26 \times 10^{-3} \text{ m}^2} \right)(0.322 \text{ m/s})$$

$$= 2.01 \text{ m/s}$$

The fluid velocity increased more than six times when it flowed through the constricted pipe.

Therefore, as a general rule, the equation of continuity for liquids, equation 13.47, says that when the cross-sectional area of a pipe gets smaller, the velocity of the fluid must become greater in order that the same amount of mass passes a given point in a given time. Conversely, when the cross-sectional area increases, the velocity of the fluid must decrease. Equation 13.47, the equation of continuity, is sometimes written in the equivalent form

$$Av = \text{constant} \tag{13.49}$$

Example 13.17

Flow rate revisited. What is the flow of mass per unit time for the example 13.16?

Solution

The rate of mass flow for the right-hand side of the pipe, given by equation 13.44, is

$$\frac{\Delta m}{\Delta t} = \rho A_2 v_2$$

$$= (1.0 \times 10^3 \text{ kg/m}^3)(1.26 \times 10^{-3} \text{ m}^2)(2.01 \text{ m/s})$$

$$= 2.53 \text{ kg/s}$$

Note that this is the same rate of flow found earlier for the left-hand side of the pipe, as it must be by the law of conservation of mass.

A compressible fluid (i.e., a gas) can have a variable density, and requires an additional equation to specify the flow velocity.

13.7 Bernoulli's Theorem

Bernoulli's theorem is a fundamental theory of hydrodynamics that describes a fluid in motion. It is really the application of the law of conservation of energy to fluid flow. Let us consider the fluid flowing in the pipe of figure 13.10. The left-hand side of the pipe has a uniform cross-sectional area A_1, which eventually tapers to the uniform cross-sectional area A_2 of the right-hand side of the pipe. The pipe is filled with a nonviscous, incompressible fluid. A uniform pressure p_1 is applied, such as from a piston, to a small element of mass of the fluid Δm and causes this mass to move through a distance Δx_1 of the pipe. Because the fluid is incompressible, the fluid moves throughout the rest of the pipe. The same small mass Δm, at the right-

Figure 13.10

Bernoulli's theorem.

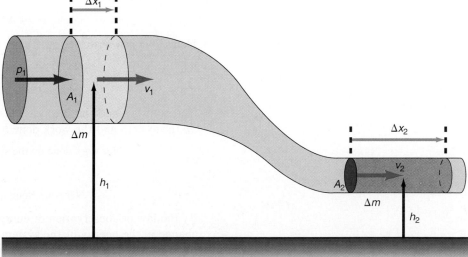

hand side of the pipe, moves through a distance Δx_2. The work done on the system by moving the small mass through the distance Δx_1 is given by the definition of work as

$$W_1 = F_1 \Delta x_1$$

Using equation 13.12, we can express the force F_1 moving the mass to the right in terms of the pressure exerted on the fluid as

$$F_1 = p_1 A_1$$

Hence,

$$W_1 = p_1 A_1 \Delta x_1$$

But

$$A_1 \Delta x_1 = \Delta V$$

the volume of the fluid moved through the pipe. Thus, we can write the work done on the system as

$$W_1 = p_1 \Delta V_1 \qquad (13.50)$$

As this fluid moves through the system, the fluid itself does work by exerting a force F_2 on the mass Δm on the right side, moving it through the distance Δx_2. Hence, the work done by the fluid system is

$$W_2 = F_2 \Delta x_2$$

But we can express the force F_2 in terms of the pressure p_2 on the right side by

$$F_2 = p_2 A_2$$

Therefore, the work done by the system is

$$W_2 = p_2 A_2 \Delta x_2$$

But

$$A_2 \Delta x_2 = \Delta V_2$$

the volume moved through the right side of the pipe. Thus, the work done by the system becomes

$$W_2 = p_2 \Delta V_2 \qquad (13.51)$$

But since the fluid is incompressible,

$$\Delta V_1 = \Delta V_2 = \Delta V$$

Hence, we can write the two work terms, equations 13.50 and 13.51, as

$$W_1 = p_1 \Delta V$$
$$W_2 = p_2 \Delta V$$

The net work done on the system is equal to the difference between the work done *on* the system and the work done *by* the system. Hence,

$$\text{Net work done on the system} = W_{on} - W_{by}$$
$$= W_1 - W_2 = p_1 \Delta V - p_2 \Delta V$$
$$\text{Net work done on the system} = (p_1 - p_2)\Delta V \qquad (13.52)$$

By the law of conservation of energy, the net work done on the system produces a change in the energy of the system. The fluid at position 1 is at a height h_1 above the reference level and therefore possesses a potential energy given by

$$\text{PE}_1 = (\Delta m)gh_1 \qquad (13.53)$$

Because this same fluid is in motion at a velocity v_1, it possesses a kinetic energy given by

$$KE_1 = \tfrac{1}{2}(\Delta m)v_1^2 \qquad (13.54)$$

Similarly at position 2, the fluid possesses the potential energy

$$PE_2 = (\Delta m)gh_2 \qquad (13.55)$$

and the kinetic energy

$$KE_2 = \tfrac{1}{2}(\Delta m)v_2^2 \qquad (13.56)$$

Therefore, we can now write the law of conservation of energy as

$$\text{Net work done on the system} = \text{Change in energy of the system} \qquad (13.57)$$

$$\text{Net work done on the system} = (E_{tot})_2 - (E_{tot})_1 \qquad (13.58)$$

$$\text{Net work done on the system} = (PE_2 + KE_2) - (PE_1 + KE_1) \qquad (13.59)$$

Substituting equations 13.52 through 13.56 into equation 13.59 we get

$$(p_1 - p_2)\Delta V = [(\Delta m)gh_2 + \tfrac{1}{2}(\Delta m)v_2^2] - [(\Delta m)gh_1 + \tfrac{1}{2}(\Delta m)v_1^2] \qquad (13.60)$$

But the total mass of fluid moved Δm is given by

$$\Delta m = \rho \Delta V \qquad (13.61)$$

Substituting equation 13.61 back into equation 13.60, gives

$$(p_1 - p_2)\Delta V = \rho(\Delta V)gh_2 + \tfrac{1}{2}\rho(\Delta V)v_2^2 - \rho(\Delta V)gh_1 - \tfrac{1}{2}\rho(\Delta V)v_1^2$$

Dividing each term by ΔV gives

$$(p_1 - p_2) = \rho gh_2 + \tfrac{1}{2}\rho v_2^2 - \rho gh_1 - \tfrac{1}{2}\rho v_1^2 \qquad (13.62)$$

If we place all the terms associated with the fluid at position 1 on the left-hand side of the equation and all the terms associated with the fluid at position 2 on the right-hand side, we obtain

$$p_1 + \rho gh_1 + \tfrac{1}{2}\rho v_1^2 = p_2 + \rho gh_2 + \tfrac{1}{2}\rho v_2^2 \qquad (13.63)$$

Equation 13.63 is the mathematical statement of

Bernoulli's theorem. It says that the sum of the pressure, the potential energy per unit volume, and the kinetic energy per unit volume at any one location of the fluid is equal to the sum of the pressure, the potential energy per unit volume, and the kinetic energy per unit volume at any other location in the fluid, for a nonviscous, incompressible fluid in streamlined flow.

Since this sum is the same at any arbitrary point in the fluid, the sum itself must therefore be a constant. Thus, we sometimes write Bernoulli's equation in the equivalent form

$$p + \rho gh + \tfrac{1}{2}\rho v^2 = \text{constant} \qquad (13.64)$$

Example 13.18

Applying Bernoulli's theorem. In figure 13.10, the pressure $p_1 = 2.94 \times 10^3$ N/m², whereas the velocity of the water is $v_1 = 0.322$ m/s. The diameter of the pipe at location 1 is 10.0 cm and it is 5.00 m above the ground. If the diameter of the pipe at location 2 is 4.00 cm, and the pipe is 2.00 m above the ground, find the velocity of the water v_2 at position 2, and the pressure p_2 of the water at position 2.

The area A_1 is

$$A_1 = \frac{\pi d_1^2}{4} = \frac{\pi(0.100 \text{ m})^2}{4} = 7.85 \times 10^{-3} \text{ m}^2$$

whereas the area A_2 is

$$A_2 = \frac{\pi d_2^2}{4} = \frac{\pi(0.0400 \text{ m})^2}{4} = 1.26 \times 10^{-3} \text{ m}^2$$

The velocity at location 2 is found from the equation of continuity, equation 13.48, as

$$v_2 = \frac{A_1}{A_2} v_1 = \left(\frac{7.85 \times 10^{-3} \text{ m}^2}{1.26 \times 10^{-3} \text{ m}^2}\right)(0.322 \text{ m/s})$$

$$= 2.01 \text{ m/s}$$

The pressure at location 2 is found from rearranging Bernoulli's equation 13.63 as

$$p_2 = p_1 + \rho g h_1 + \tfrac{1}{2}\rho v_1^2 - \rho g h_2 - \tfrac{1}{2}\rho v_2^2$$

$$= 2.94 \times 10^3 \frac{\text{N}}{\text{m}^2} + \left(1 \times 10^3 \frac{\text{kg}}{\text{m}^3}\right)\left(9.80 \frac{\text{m}}{\text{s}^2}\right)(5.00 \text{ m})$$

$$+ \frac{1}{2}\left(1 \times 10^3 \frac{\text{kg}}{\text{m}^3}\right)(0.322 \text{ m/s})^2 - \left(1 \times 10^3 \frac{\text{kg}}{\text{m}^3}\right)\left(9.80 \frac{\text{m}}{\text{s}^2}\right)(2.00 \text{ m})$$

$$- \frac{1}{2}\left(1 \times 10^3 \frac{\text{kg}}{\text{m}^3}\right)(2.01 \text{ m/s})^2$$

$$= 2.94 \times 10^3 \text{ N/m}^2 + 4.9 \times 10^4 \text{ N/m}^2 + 5.18 \times 10^1 \text{ N/m}^2$$

$$- 1.96 \times 10^4 \text{ N/m}^2 - 2.02 \times 10^3 \text{ N/m}^2$$

$$= 3.04 \times 10^4 \text{ N/m}^2$$

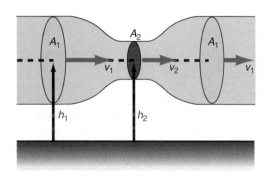

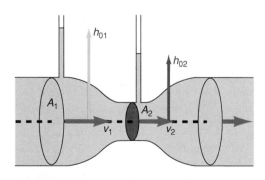

(a)

(b)

Figure 13.11

A Venturi meter.

13.8 Application of Bernoulli's Theorem

Let us now consider some special cases of Bernoulli's theorem.

The Venturi Meter

Let us first consider the constricted tube studied in figure 13.9 and slightly modified and redrawn in figure 13.11(a). Since the tube is completely horizontal $h_1 = h_2$ and there is no difference in potential energy between the locations 1 and 2. Bernoulli's equation therefore reduces to

$$p_1 + \tfrac{1}{2}\rho v_1^2 = p_2 + \tfrac{1}{2}\rho v_2^2 \tag{13.65}$$

But by the equation of continuity,

$$v_2 = \frac{A_1}{A_2} v_1 \tag{13.48}$$

Since A_1 is greater than A_2, v_2 must be greater than v_1, as shown before. Let us rewrite equation 13.65 as

$$p_2 = p_1 + \tfrac{1}{2}\rho v_1^2 - \tfrac{1}{2}\rho v_2^2$$

or

$$p_2 = p_1 + \tfrac{1}{2}\rho(v_1^2 - v_2^2) \tag{13.66}$$

But since v_2 is greater than v_1, the quantity $(1/2)\rho(v_1^2 - v_2^2)$ is a negative quantity and when we subtract it from p_1, p_2 must be less than p_1. Thus, not only does the fluid speed up in the constricted tube, but the pressure in the constricted tube also decreases.

Example 13.19

When the velocity increases, the pressure decreases. In example 13.16, associated with figure 13.9, the velocity v_1 in area A_1 was 0.322 m/s and the velocity v_2 in area A_2 was found to be 2.01 m/s. If the pressure in the left pipe is 2.94×10^3 Pa, what is the pressure p_2 in the constricted pipe?

Solution

The pressure p_2, found from equation 13.66, is

$$p_2 = p_1 + \tfrac{1}{2}\rho(v_1^2 - v_2^2)$$

$$= 2.94 \times 10^3 \text{ Pa} + (\tfrac{1}{2})(1 \times 10^3 \text{ kg/m}^3)[(0.322 \text{ m/s})^2 - (2.01 \text{ m/s})^2]$$

$$= 2.94 \times 10^3 \text{ N/m}^2 - 1.97 \times 10^3 \text{ N/m}^2 = 9.7 \times 10^2 \text{ Pa}$$

Thus, the pressure of the water in the constricted portion of the tube has decreased to 9.7×10^2 Pa. Note that in example 13.18 of section 13.7 the pressure in the constricted area of the pipe was greater than in the larger area of the pipe. This is because in that example the pipe was not all at the same level (i.e., $h_1 \neq h_2$). An additional pressure arose on the right side because of the differences in the heights of the two pipes.

The effect of the decrease in pressure with the increase in speed of the fluid in a horizontal pipe is called the **Venturi effect,** and a simple device called a **Venturi meter,** based on this Venturi effect, is used to measure the velocity of fluids in pipes. A Venturi meter is shown schematically in figure 13.11(b). The device is basically the same as the pipe in 13.11(a) except for the two vertical pipes connected to the main pipe as shown. These open vertical pipes allow some of the water in the pipe to flow upward into the vertical pipes. The height that the water rises in the vertical pipes is a function of the pressure in the horizontal pipe. As just seen, the pressure in pipe 1 is greater than in pipe 2 and thus the height of the vertical column of water in pipe 1 will be greater than the height in pipe 2. By actually measuring the height of the fluid in the vertical columns the pressure in the horizontal pipe can be determined by the hydrostatic equation 13.7. Thus, the pressure in pipe 1 is

$$p_1 = \rho g h_{01}$$

and the pressure in pipe 2 is

$$p_2 = \rho g h_{02}$$

where h_{01} and h_{02} are the heights shown in figure 13.11(b). We can now write Bernoulli's equation 13.65 as

$$\rho g h_{01} + \tfrac{1}{2}\rho v_1^2 = \rho g h_{02} + \tfrac{1}{2}\rho v_2^2$$

Replacing v_2 by its value from the continuity equation 13.65, we get

$$\rho g h_{01} + \frac{1}{2}\rho v_1^2 = \rho g h_{02} + \frac{1}{2}\rho\left[\left(\frac{A_1}{A_2}\right)v_1\right]^2$$

$$\rho g h_{01} - \rho g h_{02} = \frac{1}{2}\rho\frac{A_1^2}{A_2^2}v_1^2 - \frac{1}{2}\rho v_1^2$$

$$\rho g(h_{01} - h_{02}) = \frac{1}{2}\rho\left(\frac{A_1^2}{A_2^2} - 1\right)v_1^2$$

Solving for v_1^2, we have

$$v_1^2 = \frac{\rho g(h_{01} - h_{02})}{\frac{1}{2}\rho[(A_1^2/A_2^2) - 1]}$$

Solving for v_1, we get

$$v_1 = \sqrt{\frac{2g(h_{01} - h_{02})}{(A_1^2/A_2^2) - 1}} \qquad (13.67)$$

Equation 13.67 now gives us a simple means of determining the velocity of fluid flow in a pipe. The main pipe containing the fluid is opened and the Venturi meter is connected between the opened pipes. When the fluid starts to move, the heights h_{01} and h_{02} are measured. Since the cross-sectional areas are easily determined by measuring the diameters of the pipes, the velocity of the fluid flow is easily calculated from equation 13.67.

Example 13.20

A Venturi meter. A Venturi meter reads heights of $h_{01} = 30.0$ cm and $h_{02} = 10.0$ cm. Find the velocity of flow v_1 in the pipe. The area $A_1 = 7.85 \times 10^{-3}$ m² and the area of $A_2 = 1.26 \times 10^{-3}$ m².

Solution

The velocity of flow v_1 in the main pipe, found from equation 13.67, is

$$v_1 = \sqrt{\frac{2g(h_{01} - h_{02})}{(A_1^2/A_2^2) - 1}}$$

$$= \sqrt{\frac{2(9.80 \text{ m/s}^2)(0.300 \text{ m} - 0.100 \text{ m})}{\dfrac{(7.85 \times 10^{-3} \text{ m}^2)^2}{(1.26 \times 10^{-3} \text{ m}^2)^2} - 1}}$$

$$= 0.322 \text{ m/s}$$

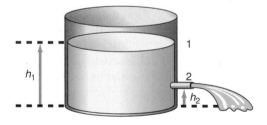

Figure 13.12

Flow from an orifice.

The Flow of a Liquid Through an Orifice

Let us consider the large tank of water shown in figure 13.12. Let the top of the fluid be location 1 and the orifice be location 2. Bernoulli's theorem applied to the tank, taken from equation 13.63, is

$$p_1 + \rho g h_1 + \tfrac{1}{2}\rho v_1^2 = p_2 + \rho g h_2 + \tfrac{1}{2}\rho v_2^2$$

But the pressure at the top of the tank and the outside pressure at the orifice are both p_0, the normal atmospheric pressure. Also, because of the very large volume of fluid, the small loss through the orifice causes an insignificant vertical motion of the top of the fluid. Thus, $v_1 \approx 0$. Bernoulli's equation becomes

$$p_0 + \rho g h_1 = p_0 + \rho g h_2 + \tfrac{1}{2}\rho v_2^2$$

The pressure term p_0 on both sides of the equation cancels out. Also h_2 is very small compared to h_1 and it can be neglected, leaving

$$\rho g h_1 = \tfrac{1}{2}\rho v_2^2$$

Solving for the velocity of efflux, we get

$$v_2 = \sqrt{2g h_1} \qquad (13.68)$$

Notice that the velocity of efflux is equal to the velocity that an object would acquire when dropped from the height h_1.

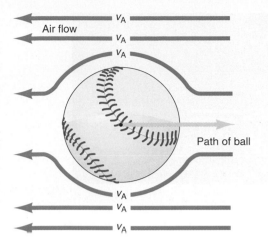

Air flow

(a) Nonspinning baseball

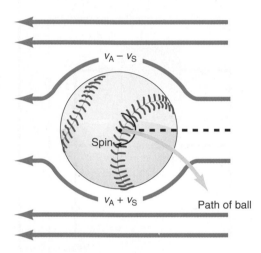

(b) Spinning baseball

Figure 13.13

The curving baseball.

Example 13.21

The velocity of efflux. A large water tank, 10.0 m high, springs a leak at the bottom of the tank. Find the velocity of the escaping water.

Solution

The velocity of efflux, found from equation 13.68, is

$$v_2 = \sqrt{2gh_1} = \sqrt{2(9.80 \text{ m/s}^2)(10.0 \text{ m})}$$
$$= 14.0 \text{ m/s}$$

The Curving Baseball

When a nonspinning ball is thrown through the air it follows the straight line path shown in figure 13.13(a). The air moves over the top and bottom of the ball with a speed v_A. If the ball is now released with a downward spin, as shown in figure 13.13(b), then the spinning ball drags some air around with it. At the top of the ball, there is a velocity of the air v_A to the left, and a velocity of the dragged air on the spinning baseball v_S to the right. Thus, the relative velocity of the air with respect to the ball is $v_A - v_S$ at the top of the ball. At the bottom of the ball the dragged air caused by the spin of the baseball v_S is in the same direction as the velocity of the air v_A moving past the ball. Thus, the relative velocity of the air with respect to the bottom of the ball is $v_A + v_S$. Hence, the velocity of the air at the top of the ball, $v_A - v_S$, is less than the velocity of the air at the bottom of the ball, $v_A + v_S$. By the Venturi principle, the pressure of the fluid is smaller where the velocity is greater. Thus, the pressure on the bottom of the ball is less than the pressure on the top, that is,

$$p_{\text{top}} > p_{\text{bottom}}$$

But the pressure is related to the force by $p = F/A$. Hence, the force acting on the top of the ball is greater than the force acting on the bottom of the ball, that is,

$$F_{\text{top}} > F_{\text{bottom}}$$

Therefore, the ball curves downward, or sinks, as it approaches the batter. By spinning the ball to the right (i.e., clockwise) as viewed from above, the ball curves toward the right. By spinning the ball to the left (i.e., counterclockwise) as viewed from above the ball, the ball curves toward the left. Spins about various axes through the ball can cause the ball to curve to the left and downward, to the left and upward, and so on.

Lift on an Airplane Wing

Another example of the Venturi effect can be seen with an aircraft wing, as shown in figure 13.14. The air flowing over the top of the wing has a greater distance to travel than the air flowing under the bottom of the wing. In order for the flow to be streamlined and for the air at the leading edge of the wing to arrive at the trailing edge at the same time, whether it goes above or below the wing, the velocity of the air over the top of the wing must be greater than the velocity of the air at the bottom of the wing. But by the Venturi principle, if the velocity is greater at the top of the wing, the pressure must be less there than at the bottom of the wing. Thus, p_2 is greater than p_1 and therefore $F_2 > F_1$. That is, there is a net positive force $F_2 - F_1$ acting upward on the wing, producing lift on the airplane wing.

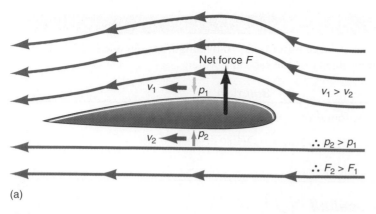

Net force F

v_1 p_1

v_2 p_2

$v_1 > v_2$

$\therefore p_2 > p_1$

$\therefore F_2 > F_1$

(a)

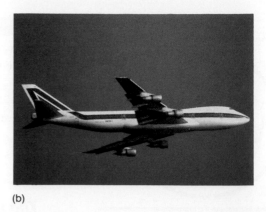

(b)

Figure 13.14
An airfoil.

"Have you ever wondered . . . ?"

An Essay on the Application of Physics

The Flow of Blood in the Human Body

Human blood consists of a plasma, the fluid, and red and white corpuscles that are immersed in the plasma. Because blood is a fluid, the laws of physics can be applied to the flow of blood throughout the body. A schematic diagram of the circulatory system, which transports blood and oxygen around the body, is shown in figure 1. It consists of (1) the heart, which is the pump that is responsible for supplying the pressure to move the blood; (2) the lungs, which are the source of oxygen for all the cells of the body; (3) the arteries, which are connecting blood vessels that pass the blood from the heart to various parts of the body; (4) the capillaries, which are extremely small blood vessels that bring the oxygenated blood down to the layer of human cells; and (5) the veins, which are blood vessels that return deoxygenated blood to the heart to complete the circulatory system.

The heart is the pump that circulates the blood throughout the body and a diagram of it is shown in figure 2. Blood, containing carbon dioxide, returns to the heart by the veins and enters the right auricle. It is then pumped from the right ventricle to the pulmonary artery to the lungs where it dumps the waste carbon dioxide and picks up a new supply of oxygen. It then returns to the left auricle of the heart. The left ventricle then pumps this oxygen rich blood to the aorta, the main artery of the body, for distribution to the rest of the body.

For a person at rest, the heart pumps approximately 5.00 liters of blood per minute (8.33×10^{-5} m³/s) at a rate of about 70 beats per minute. For a person engaged in very strenuous exercise the heart can pump up to 25.0 liters of blood per minute

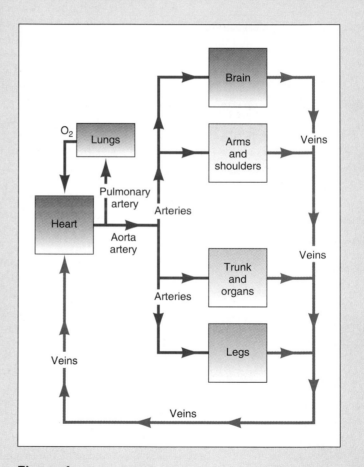

Figure 1

The circulatory system.

Figure 2

The human heart.

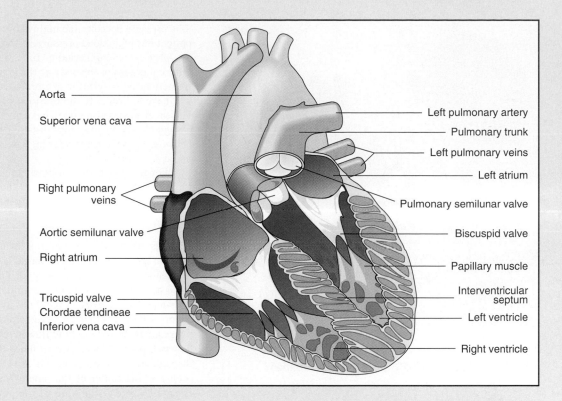

Aorta

Superior vena cava

Right pulmonary veins

Aortic semilunar valve

Right atrium

Tricuspid valve

Chordae tendineae

Inferior vena cava

Left pulmonary artery

Pulmonary trunk

Left pulmonary veins

Left atrium

Pulmonary semilunar valve

Biscuspid valve

Papillary muscle

Interventricular septum

Left ventricle

Right ventricle

$(41.7 \times 10^{-5}$ m^3/s) at a rate of about 180 beats per minute. We can determine the speed of the blood as it enters the aorta by a generalization of equation 13.36, as

$$\frac{\Delta V}{\Delta t} = A_A v_A \qquad \textbf{(13H.1)}$$

where $\Delta V/\Delta t$ is the rate at which the blood is flowing from the heart into the aorta, A_A is the cross-sectional area of the aorta, and v_A is the speed of the blood in the aorta. The diameter of the aorta is about 2.00 cm giving an area of

$$A = \pi r^2$$
$$= \pi(0.01 \text{ m})^2 = 3.14 \times 10^{-4} \text{ m}^2$$

The speed of the blood in the aorta is therefore

$$v_A = \frac{\Delta V/\Delta t}{A_A}$$
$$= \frac{8.33 \times 10^{-5} \text{ m}^3/\text{s}}{3.14 \times 10^{-4} \text{ m}^2}$$
$$= 0.265 \text{ m/s} = 26.5 \text{ cm/s} \qquad \textbf{(13H.2)}$$

We can determine the speed of the blood in the capillaries by the continuity equation 13.47, as

$$A_A v_A = A_C v_C \qquad \textbf{(13H.3)}$$

where A_A is the cross-sectional area of the aorta, which was just determined as 3.14×10^{-4} m^2; v_A is the speed of the blood in the aorta, which was just found to be 26.5 cm/s; and A_C is the cross-sectional area of a capillary tube, which is quite small.

However, because there are literally billions of these capillaries the effective cross-sectional area of all these capillaries combined is approximately 2500×10^{-4} m^2. The speed of the blood in the capillary becomes

$$v_C = \frac{A_A}{A_C} v_A$$
$$= \left(\frac{3.14 \times 10^{-4} \text{ m}^2}{2500 \times 10^{-4} \text{ m}^2}\right)(26.5 \text{ cm/s})$$
$$= 0.0333 \text{ cm/s}$$

Thus, the blood moves relatively slowly at the level of the capillaries.

Finally, we should note that the body controls the flow of blood through the arteries by muscles that surround the arteries. When the muscles contract, the diameter of the artery is reduced. From the equation of continuity, $Av = $ constant. By decreasing the diameter of the artery, the cross-sectional area of the artery decreases and hence the speed of blood must increase through the artery. Alternatively, when the muscles are relaxed, the diameter of the artery increases to its former size, the cross-sectional area increases, and the speed of the blood decreases. With advancing age the arterial muscles lose some of this ability to contract, a situation called hardening of the arteries, and the control of blood flow is somewhat diminished.

A good indication of how well the heart is functioning is obtained by measuring the pressure that the heart exerts when pumping blood, and when at rest. The device used to measure blood pressure is called a sphygmomanometer. (The word is derived from the Greek word *sphygmos,* meaning pulse, and the

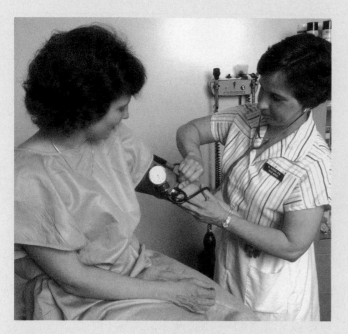

Figure 3

A nurse measures the blood pressure of a patient.

word *manometer,* which is a pressure measuring device. Hence, a sphygmomanometer is a device for measuring pulse pressure, or blood pressure.) The device consists of an air bag, called a cuff, that is wrapped around the upper arm of the patient at the level of the heart. A hand pump is used to inflate the cuff, and the pressure exerted by the cuff on the arm is measured by the mercury manometer. The pressure exerted by the cuff is increased until the pressure is great enough to collapse the brachial artery in the arm, cutting off the blood supply to the rest of the arm. A stethoscope is placed over the brachial artery and the pressure in the cuff is slowly decreased. When the pressure in the cuff becomes low enough, the pressure exerted by the heart is large enough to force the artery open and some blood squirts through. This blood flowing through the narrow restriction becomes turbulent and makes a noise as it enters the open portion of the artery. The physician hears this noise through the stethoscope, and simultaneously observes the pressure indicated on the manometer, expressed in terms of mm of Hg. At this point the pressure exerted by the heart, called the systolic pressure, is equal to the pressure exerted by the cuff. A normal systolic pressure is around 120 mm of Hg.

As the pressure in the cuff is decreased the turbulent flow noise is still heard in the stethoscope until the lowest pressure exerted by the heart, the diastolic pressure, is equal to the pressure exerted by the cuff. At this point the artery is completely open and the blood is no longer in turbulent flow and the characteristic noise disappears. The pressure is read from the mercury manometer at this point. This pressure is the pressure that the heart exerts when it is at rest. The normal diastolic pressure is around 80 mm of Hg. The combined systolic and diastolic pressures are usually indicated in the form 120/80. If the sys-

tolic pressure becomes too high, above about 150 mm of Hg, the patient has high blood pressure. If the systolic pressure becomes too large for a long period of time, damage can be done to the different organs of the body. If the systolic pressure becomes extremely large, arteries in the brain can rupture and the person will have a stroke. If the diastolic pressure exceeds 90 mm of Hg, the person is also said to have high blood pressure. This type of high blood pressure causes eventual damage to the heart itself, because it is operating under high pressures even while it is supposed to be resting.

For the type of streamlined flow considered in this chapter the flow of fluid per unit time was shown to be

$$\frac{\Delta V}{\Delta t} = Av \qquad \text{(13.36)}$$

which is essentially the equation of continuity. In this type of flow the speed v was the same throughout the cross-sectional area A considered. However, some fluids have a significant frictional force between the layers of the fluid, and this frictional effect, known as the *viscosity* of the fluid, must then be taken into account. A fluid in which frictional effects are significant is called a *viscous fluid* and the fluid flow is referred to as *laminar flow,* flow in layers. For such viscous fluids the speed v is not the same throughout the cross-sectional area A. The maximum speed occurs at the center of the pipe or tube, whereas the speed is essentially zero at the walls of the pipe. Experimental work by J. L. Poiseuille (1799–1869), a French scientist, and subsequently confirmed by theory, showed that the flow rate for viscous fluids is given by

$$\frac{\Delta V}{\Delta t} = \frac{(\Delta p)\pi R^4}{8\eta L} \qquad \text{(13H.4)}$$

where Δp is the pressure difference between both ends of the pipe, R is the radius of the pipe, L is the length of the pipe, and η is the coefficient of viscosity of the fluid. Equation 13H.4 is called *Poiseuille's equation.* Note that the flow rate is inversely proportional to the coefficient of viscosity of the fluid. Thus, a very viscous fluid (high value of η) flows very slowly compared to a fluid of low viscosity. That is, everything else being equal, molasses flows at a slower rate than water. Human blood is a viscous fluid, the greater the number of red corpuscles in the blood the greater the viscosity. The viscosity of human blood varies from about 1.50×10^{-3} (N/m²)s for plasma, to about 4.00×10^{-3} (N/m²)s for whole blood. Also note that the flow rate depends on the fourth power of the radius of the pipe. If the radius is doubled, the flow rate is multiplied by a factor of 16. This relation is important in the selection of the size of hypodermic needles.

Example 13H.1

A blood transfusion. A person is receiving a blood transfusion. The bottle containing the blood is elevated 75.0 cm above the arm of the person. The needle is 4.00 cm long and has a diameter of 0.500 mm. Find the rate at which the blood flows through the needle.

Solution

The rate of flow of blood is found from equation 13H.4, where η, the viscosity of blood, is 4.00×10^{-3} Ns/m². Let us assume that the total pressure differential is obtained by the effects of gravity from the hydrostatic equation, equation 13.7. The density of blood is about 1050 kg/m³. Thus,

$$\Delta p = \rho g h$$

$$= (1050 \text{ kg/m}^3)(9.80 \text{ m/s}^2)(0.750 \text{ m})$$

$$= 7.72 \times 10^3 \text{ Pa}$$

The blood flow rate now obtained is

$$\frac{\Delta V}{\Delta t} = \frac{(\Delta p)\pi R^4}{8\eta L} \qquad \textbf{(13H.4)}$$

$$= \frac{(7.72 \times 10^3 \text{ N/m}^2)(\pi)(0.250 \times 10^{-3} \text{ m})^4}{8(4.00 \times 10^{-3} \text{ Ns/m}^2)(0.0400 \text{ m})}$$

$$= 7.40 \times 10^{-8} \text{ m}^3/\text{s}$$

The Language of Physics

Fluids
A fluid is any substance that can flow. Hence, liquids and gases are both considered to be fluids (p. 367).

Fluid statics or hydrostatics
The study of fluids at rest (p. 367).

Fluid dynamics or hydrodynamics
The study of fluids in motion (p. 367).

Density
The amount of mass in a unit volume of a substance (p. 367).

Pressure
The magnitude of the normal force acting per unit surface area (p. 368).

The hydrostatic equation
An equation that gives the pressure of a fluid at a particular depth (p. 370).

Barometer
An instrument that measures atmospheric pressure (p. 371).

Gauge pressure
The pressure indicated on a pressure measuring gauge. It is equal to the absolute pressure minus normal atmospheric pressure (p. 375).

Pascal's principle
If the pressure at any point in an enclosed fluid at rest is changed, the pressure changes by an equal amount at all points in the fluid (p. 376).

Archimedes' principle
A body immersed in a fluid is buoyed up by a force that is equal to the weight of the fluid displaced. A body floats when the weight of the body is equal to the weight of the fluid displaced (p. 378).

Law of conservation of mass
In any ordinary mechanical or chemical process, mass is neither created nor destroyed (p. 384).

The equation of continuity
An equation based on the law of conservation of mass, that indicates that when the cross-sectional area of a pipe gets smaller, the velocity of the fluid must become greater. Conversely, when the cross-sectional area increases, the velocity of the fluid must decrease (p. 384).

Bernoulli's theorem
The sum of the pressure, the potential energy per unit volume, and the kinetic energy per unit volume at any one location of the fluid is equal to the sum of the pressure, the potential energy per unit volume, and the kinetic energy per unit volume at any other location in the fluid, for a nonviscous, incompressible fluid in streamlined flow (p. 387).

Venturi effect
The effect of the decrease in pressure with the increase in speed of the fluid in a horizontal pipe (p. 389).

Venturi meter
A device that uses the Venturi effect to measure the velocity of fluids in pipes (p. 389).

Summary of Important Equations

Density
$$\rho = \frac{m}{V} \tag{13.1}$$

Mass
$$m = \rho V \tag{13.2}$$

Pressure
$$p = \frac{F}{A} \tag{13.3}$$

Hydrostatic equation
$$p = \rho g h \tag{13.7}$$

Force
$$F = pA \tag{13.12}$$

Absolute and gauge pressure
$$p_{abs} = p_{gauge} + p_0 \tag{13.15}$$

Hydraulic lift
$$F = \frac{A}{a} f \tag{13.18}$$

$$y_1 = \frac{A}{a} y_2 \tag{13.23}$$

Archimedes' principle
$$\begin{matrix}\text{Buoyant} \\ \text{force}\end{matrix} = \begin{matrix}\text{Weight of} \\ \text{water displaced}\end{matrix} \tag{13.30}$$

Mass flow rate
$$\frac{\Delta m}{\Delta t} = \rho A v \tag{13.38}$$

Equation of continuity
$$A_1 v_1 = A_2 v_2 \tag{13.47}$$
$$A v = \text{constant} \tag{13.49}$$

Work done in moving a fluid
$$W = p \Delta V \tag{13.50}$$

Bernoulli's theorem
$$p_1 + \rho g h_1 + \tfrac{1}{2}\rho v_1^2 = p_2 \tag{13.63}$$
$$+ \rho g h_2 + \tfrac{1}{2}\rho v_2^2$$
and
$$p + \rho g h \tag{13.64}$$
$$+ \tfrac{1}{2}\rho v^2 = \text{constant}$$

Questions for Chapter 13

1. Discuss the differences between solids, liquids, and gases.
†2. Hieron II, King of Syracuse in ancient Greece, asked his relative Archimedes to determine if the gold crown made for him by the local goldsmith, was solid gold or a mixture of gold and silver. How did Archimedes, or how could you, determine whether or not the crown was pure gold?

3. When you fly in an airplane you find that your ears keep "popping" when the plane is ascending or descending. Explain why.
4. Using a barometer and the direction of the wind, describe how you could make a reasonable weather forecast.
†5. A pilot uses an aneroid barometer as an altimeter that is calibrated to a standard atmosphere. What happens to the aircraft if the temperature of the atmosphere does not coincide with the standard atmosphere?
†6. Does a sphygmomanometer measure gauge pressure or absolute pressure?

7. How would you define a mechanical advantage for the hydraulic lift?
8. In example 13.13, could the iron block sink to a depth of 39.4 cm in a pool of water 100 cm deep and then float at that point? Why or why not?
†9. How does eating foods very high in cholesterol have an effect on the arteries and hence the flow of blood in the body?
†10. Why is an intravenous bottle placed at a height h above the arm of a patient?

Problems for Chapter 13

13.2 Density

1. A cylinder 3.00 cm in diameter and 3.00 cm high has a mass of 15.0 g. What is its density?
2. Find the mass of a cube of iron 10.0 cm on a side.
3. A gold ingot is 50.0 cm by 20.0 cm by 10.0 cm. Find (a) its mass and (b) its weight.
4. Find the mass of the air in a room 6.00 m by 8.00 m by 3.00 m.
5. Assume that the earth is a sphere. Compute the average density of the earth.
6. Find the weight of 1.00 liter of air.
7. A crown, supposedly made of gold, has a mass of 8.00 kg. When it is placed in a full container of water, 691 cm^3 of water overflows. Is the crown made of pure gold or is it mixed with some other materials?

8. A solid brass cylinder 10.0 cm in diameter and 25.0 cm long is soldered to a solid iron cylinder 10.0 cm in diameter and 50.0 cm long. Find the weight of the combined cylinder.
9. An annular cylinder of 2.50-cm inside radius and 4.55-cm outside radius is 10.5 cm high. If the cylinder has a mass of 5.35 kg, find its density.

13.3 Pressure

10. As mentioned in the text, a non-SI unit of pressure is the torr, named after Torricelli, which is equal to the pressure exerted by a column of mercury 1 mm high. Express a pressure of 2.53×10^5 Pa in torrs.

†11. From the knowledge of normal atmospheric pressure at the surface of the earth, compute the approximate mass of the atmosphere.
12. A barometer reads a height of 72.0 cm of Hg. Express this atmospheric pressure in terms of (a) in. of Hg, (b) mb, (c) $lb/in.^2$, and (d) Pa.
13. (a) A "high" pressure area of 1030 mb moves into an area. What is this pressure expressed in $lb/in.^2$? (b) A "low" pressure area of 980 mb moves into an area. What is this pressure expressed in $lb/in.^2$?
14. Normal systolic blood pressure is approximately 120 mm of Hg and normal diastolic pressure is 80 mm of Hg. Express these pressures in terms of Pa and $lb/in.^2$.

15. The point of a 10-penny nail has a diameter of 1.00 mm. If the nail is driven into a piece of wood with a force of 150 N, find the pressure that the tip of the nail exerts on the wood.

16. The gauge pressure in the tires of your car is 30.0 lb/in.². What is the absolute pressure of the air in the tires?

17. The gauge pressure in the tires of your car is 2.42×10^5 N/m². What is the absolute pressure of the air in the tires?

18. What is the water pressure and the absolute pressure in a swimming pool at depths of (a) 1.00 m, (b) 2.00 m, (c) 3.00 m, and (d) 4.00 m?

19. Find the force exerted by normal atmospheric pressure on the top of a table 1.00 m high, 1.00 m long, 0.75 m wide, and 0.10 m thick. What is the force on the underside of the table top exerted by normal atmospheric pressure?

20. What force is exerted on the top of the roof by normal atmospheric pressure?

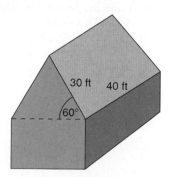

21. If normal atmospheric pressure can support a column of Hg 76.0 cm high, how high a column will it support of (a) water, (b) benzene, (c) alcohol, and (d) glycerine?

22. What is the minimum pressure of water entering a building if the pressure at the second floor faucet, 16.0 ft above the ground, is to be 5.00 lb/in.²?

23. The water main pressure entering a house is 31.0 N/cm². What is the pressure at the second floor faucet, 6.00 m above the ground? What is the maximum height of any faucet such that water will still flow from it?

24. A barometer reads 76.0 cm of Hg at the base of a tall building. The barometer is carried to the roof of the building and now reads 75.6 cm of Hg. If the average density of the air is 1.28 kg/m³, what is the height of the building?

25. The hatch of a submarine is 100 cm by 50.0 cm. What force is exerted on this hatch by the water when the submarine is 50.0 m below the surface?

13.4 Pascal's Principle

26. In the hydraulic lift of figure 13.5, the diameter $d_1 = 10.0$ cm and $d_2 = 50.0$ cm. If a force of 10.0 N is applied at the small piston, (a) what force will appear at the large piston? (b) If the large piston is to move through a height of 2.00 m, what must the total displacement of the small piston be?

27. In a hydraulic lift, the large piston exerts a force of 25.0 N when a force of 3.50 N is applied to the smaller piston. If the smaller piston has a radius of 12.5 cm, and the lift is 65.0% efficient, what must be the radius of the larger piston?

28. The theoretical mechanical advantage (TMA) of a hydraulic lift is equal to the ratio of the force that you get out of the lift to the force that you must put into the lift. Show that the theoretical mechanical advantage of the hydraulic lift is given by

$$\text{TMA} = \frac{F_{out}}{F_{in}} = \frac{A_{out}}{A_{in}} = \frac{y_{in}}{y_{out}}$$

where A_{out} is the area of the output piston, A_{in} is the area of the input piston, y_{in} is the distance that the input piston moves, and y_{out} is the distance that the output piston moves.

13.5 Archimedes' Principle

29. Find the weight of a cubic block of iron 5.00 cm on a side. This block is now hung from a spring scale such that the block is totally submerged in water. What would the scale indicate for the weight (called the apparent weight) of the block?

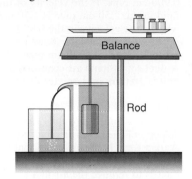

30. A copper cylinder 5.00 cm high and 3.00 cm in diameter is hung from a spring scale such that the cylinder is totally submerged in ethyl alcohol. Find the apparent weight of the block.

31. Find the buoyant force on a brass block 10.5 cm long by 12.3 cm wide by 15.0 cm high when placed in (a) water, (b) glycerine, and (c) mercury.

32. If the iron block in example 13.13 were placed in a pool of mercury instead of the water would it float or sink? If it floats, to what depth does it sink before it floats?

†33. A block of wood sinks 8.00 cm in pure water. How far will it sink in salt water?

34. A weather balloon contains 33.5 m³ of helium at the surface of the earth. Find the largest load this balloon is capable of lifting. The density of helium is 0.1785 kg/m³.

13.6 The Equation of Continuity

35. A 3/4-in. pipe is connected to a 1/2-in. pipe. If the velocity of the fluid in the 3/4-in. pipe is 2.00 ft/s, what is the velocity in the 1/2-in. pipe? How much water flows per second from the 1/2-in. pipe?

36. A 2.50-cm pipe is connected to a 0.900-cm pipe. If the velocity of the fluid in the 2.50-cm pipe is 1.50 m/s, what is the velocity in the 0.900-cm pipe? How much water flows per second from the 0.900-cm pipe?

37. A duct for a home air conditioning unit is 10.0 in. in diameter. If the duct is to remove the air in a room 20.0 ft by 24.0 ft by 8.00 ft high every 20.0 min, what must the velocity of the air in the duct be?

38. A duct for a home air-conditioning unit is 35.0 cm in diameter. If the duct is to remove the air in a room 9.00 m by 6.00 m by 3.00 m high every 15.0 min, what must the velocity of the air in the duct be?

13.7 Bernoulli's Theorem

39. Water enters the house from a main at a pressure of 1.5×10^5 Pa at a speed of 40.0 cm/s in a pipe 4.00 cm in diameter. What will be the pressure in a 2.00-cm pipe located on the second floor 6.00 m high when no water is flowing from the upstairs pipe? When the water starts flowing, at what velocity will it emerge from the upstairs pipe?

40. A can of water 30.0 cm high sits on a table 80.0 cm high. If the can develops a leak 5.00 cm from the bottom, how far away from the table will the water hit the floor?

41. Water rises to a height $h_{01} = 35.0$ cm, and $h_{02} = 10.0$ cm, in a Venturi meter, figure 13.11(b). The diameter of the first pipe is 4.00 cm, whereas the diameter of the second pipe is 2.00 cm. What is the velocity of the water in the first and second pipe? What is the mass flow rate and the volume flow rate?

Additional Problems

42. A car weighs 2890 lb and the gauge pressure of the air in each tire is 30 lb/in.². Assuming that the weight of the car is evenly distributed over the four tires, (a) find the area of each tire that is flat on the ground and (b) if the width of the tire is 5.00 inches, find the length of the tire that is in contact with the ground.

43. A certain portion of a rectangular, concrete flood wall is 12.0 m high and 30.0 m long. During severe flooding of the river, the water level rises to a height of 10.0 m. Find (a) the water pressure at the base of the flood wall, (b) the average water pressure exerted on the flood wall, and (c) the average force exerted on the flood wall by the water.

44. The Vehicle Assembly Building at the Kennedy Space Center is 160 m high. Assuming the density of air to be a constant, find the difference in atmospheric pressure between the ground floor and the ceiling of the building.

45. If the height of a water tower is 20.0 m, what is the pressure of the water as it comes out of a pipe at the ground?

†46. A 20.0-g block of wood floats in water to a depth of 5.00 cm. A 10.0-g block is now placed on top of the first block, but it does not touch the water. How far does the combination sink?

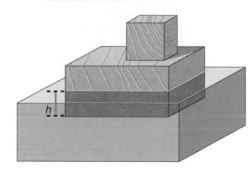

†47. An iron ball, 4.00 cm in diameter, is dropped into a tank of water. Assuming that the only forces acting on the ball are gravity and the buoyant force, determine the acceleration of the ball. Discuss the assumption made in this problem.

†48. If 80% of a floating cylinder is beneath the water, what is the density of the cylinder?

†49. From knowing that the density of an ice cube is 920 kg/m³ can you determine what percentage of the ice cube will be submerged when in a glass of water?

†50. Find the equation for the length of the side of a cube of material that will give the same buoyant force as (a) a sphere of radius r and (b) a cylinder of radius r and height h, if both objects are completely submerged.

†51. Find the radius of a solid cylinder that will experience the same buoyant force as an annular cylinder of radii $r_2 = 4.00$ cm and $r_1 = 3.00$ cm. Both cylinders have the same height h.

†52. A cone of maximum radius r_0 and height h_0, is placed in a fluid, as shown in the diagram. The volume of a right circular cone is given by

$$V_{cone} = \tfrac{1}{3}\pi r^2 h$$

(a) Find the equation for the weight of the cone. (b) If the cone sinks so that a height h_1 remains out of the fluid, find the equation for the volume of the cone that is immersed in the fluid. (c) Find the equation for the buoyant force acting on the cone. (d) Show that the height h_1 that remains out of the fluid is given by

$$h_1 = \sqrt[3]{1 - (\rho_c/\rho_f)}\, h_0$$

where ρ_c is the density of the cone and ρ_f is the density of the fluid. (e) If we approximate an iceberg by a cone, find the percentage height of the iceberg that sticks out of the salt water, and the percentage volume of the iceberg that is below the water.

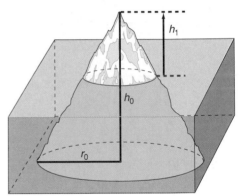

†53. A can 30.0 cm high is filled to the top with water. Where should a hole be made in the side of the can such that the escaping water reaches the maximum distance x in the horizontal direction? (*Hint:* calculate the distance x for values of h from 0 to 30.0 cm in steps of 5.00 cm.)

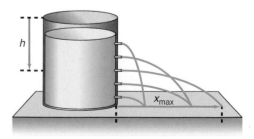

54. In the flow of fluid from an orifice in figure 13.12, it was assumed that the vertical motion of the water at the top of the tank was very small, and hence v_1 was set equal to zero. Show that if this assumption does not hold, the velocity of the fluid from the orifice v_2 can be given by

$$v_2 = \sqrt{\frac{2gh}{1 - (d_2^4/d_1^4)}}$$

where d_1 is the diameter of the tank and d_2 is the diameter of the orifice.

†55. A wind blows over the roof of your house at 100 mph. What is the difference in pressure acting on the roof because of this velocity? (*Hint:* the air inside the attic is still, that is, $v = 0$ inside the house.)

56. A wind blows over the roof of a house at 136 km/hr. What is the difference in pressure acting on the roof because of this velocity? (*Hint:* the air inside the attic is still, that is, $v = 0$ inside the house.)

†57. If air moves over the top of an airplane wing at 150 m/s and 120 m/s across the bottom of the wing, find the difference in pressure between the top of the wing and the bottom of the wing. If the area of the wing is 15.0 m², find the force acting upward on the wing.

Interactive Tutorials

⊟58. Find the buoyant force BF and apparent weight AW of a solid sphere of radius $r = 0.500$ m and density $\rho = 7.86 \times 10^3$ kg/m³, when immersed in a fluid whose density is $\rho_f = 1.00 \times 10^3$ kg/m³.

⊟59. Archimedes' principle. A solid block of wood of length $L = 15.0$ cm, width $W = 20.0$ cm, and height $h_0 = 10.0$ cm, is placed into a pool of water. The density of the block is 680 kg/m³. (a) Will the block sink or float? (b) If it floats, how deep will the block be submerged when it floats? (c) What percentage of the original volume is submerged?

⊟60. The equation of continuity and flow rate. Water flows in a pipe of diameter $d_1 = 4.00$ cm at a velocity of 35.0 cm/s, as shown in figure 13.9. The diameter of the tapered part of the pipe is $d_2 = 2.55$ cm. Find (a) the velocity of the fluid in the tapered part of the pipe, (b) the mass flow rate, and (c) the volume flow rate of the fluid.

⊟61. Bernoulli's theorem. Water flows in an elevated, tapered pipe, as shown in figure 13.10. The first part of the pipe is at a height $h_1 = 3.58$ m above the ground and the water is at a pressure $p_1 = 5000$ N/m², the diameter $d_1 = 25.0$ cm, and the velocity of the water is $v_1 = 0.553$ m/s. If the diameter of the tapered part of the pipe is $d_2 = 10.0$ cm and the height of the pipe above the ground is $h_2 = 1.25$ m, find (a) the velocity v_2 of the fluid in the tapered part of the pipe and (b) the pressure p_2 of the water in the tapered part of the pipe.

Thermodynamics

PART THREE

Color as a Study Aid
Chapters 14–17

Some of the colors used in chapters 14 through 17 are the same colors used previously. That is,

☐ Displacement vectors

☐ Velocity vectors

☐ Component of velocity vector

☐ Friction force vectors

Particular chapter-by-chapter colors are given below. In general, however, regions of high temperature are red while regions of low temperature are blue.

Chapter 14

General color code above applies.

Chapter 15

☐ Initial length or area of a material

☐ Change in length or area of a material

Temperature gradations start with red for the highest temperature and end with blue for the lowest temperature. Temperature gradations are:

☐ Highest temperature

☐ Second highest temperature

☐ Third highest temperature

☐ Second lowest temperature

☐ Lowest temperature

Chapter 16

Temperature gradations through a medium start with red for the highest temperature and end with blue for the lowest temperature. Temperature gradations are:

☐ Highest temperature

☐ Second highest temperature

☐ Second lowest temperature

☐ Lowest temperature

☐ Thermal energy flowing in or out of a medium

Chapter 17

☐ Thermodynamic paths on a pV diagram

☐ Highest temperature isotherm

☐ Second highest temperature isotherm

☐ Lowest temperature isotherm

☐ Heat added to a thermodynamic system

☐ Heat exhausted from a thermodynamic system

☐ Pressure gradient force

☐ Coriolis force

Temperature and Heat

Freezing warning sign on roadway.

The determination of temperature has long been recognized as a problem of the greatest importance in physical science. It has accordingly been made a subject of most careful attention, and, especially in late years, of very elaborate and refined experimental researches: and we are thus at present in possession of as complete a practical solution of the problem as can be desired even for the most accurate investigation.

William Thompson, Lord Kelvin

14.1 Temperature

The simplest and most intuitive definition of temperature is that **temperature** *is a measure of the hotness or coldness of a body.* That is, if a body is hot it has a high temperature, if it is cold it has a low temperature. This is not a very good definition, as we will see in a moment, but it is one that most people have a "feel" for, because we all know what hot and cold is. Or do we?

Let us reconsider the "thought experiment" treated in chapter 1. We place three beakers on the table, as shown in figure 14.1. Several ice cubes are placed into the first beaker of water, whereas boiling water is poured into the third beaker. We place equal amounts of the ice water from beaker one and the boiling water from beaker three into the second beaker to form a mixture. I now take my right hand and plunge it into beaker one, and conclude that it is *cold.* After drying off my right hand, I place it into the middle mixture. After coming from the ice water, the mixture in the second beaker feels hot by comparison. So I conclude that the mixture is *hot.*

I now take my left hand and plunge it into the boiling water of beaker three. (This is of course the reason why this is only a "thought experiment.") I conclude that the water in beaker three is certainly *hot.* Drying off my hand again I then place it into beaker two. After the boiling water, the mixture feels cold by comparison, so I conclude that the mixture is *cold.* After this relatively scientific experiment, my conclusion is contradictory. That is, I found the middle mixture to be either hot or cold depending on the sequence of the measurement. Thus, the hotness or coldness of a body is not a good concept to use to define the temperature of a body. Although we may have an intuitive feel for hotness or coldness, we can not use our intuition for any precise scientific work.

The Thermometer

In order to make a measurement of the temperature of a body, a new technique, other than estimating hotness or coldness, must be found. Let us look for some characteristic of matter that changes as it is heated. The simplest such characteristic is that most materials expand when they are heated. Using this characteristic of matter we take a glass tube and fill it with a liquid, as shown in figure 14.2. When the liquid is heated it expands and rises up the tube. The height of the liquid in the tube can be used to measure the hotness or coldness of a body. The device will become a thermometer.

Figure 14.1

A "thought experiment" on temperature.

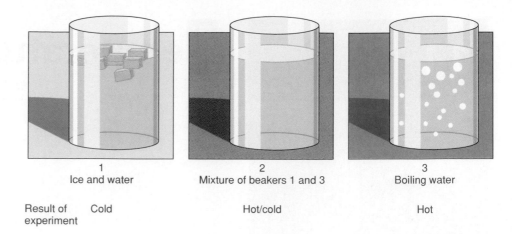

1	2	3
Ice and water	Mixture of beakers 1 and 3	Boiling water

| Result of experiment | Cold | Hot/cold | Hot |

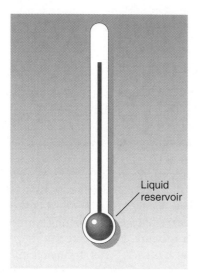

Liquid reservoir

Figure 14.2

A thermometer.

In order to quantify the process, we need to place numerical values on the glass tube, thus assigning a number that can be associated with the hotness or coldness of a body. This is the process of *calibrating* the thermometer.

First, we place the thermometer into the mixture of ice and water of beaker 1 in figure 14.1. The liquid lowers to a certain height in the glass tube. We scratch a mark on the glass at that height, and arbitrarily call it 0 degrees. Since it is the point where ice is melting in the water, we call 0° the melting point of ice. (Or similarly, the freezing point of water.)

Then we place the glass tube into beaker three, which contains the boiling water. (We assume that heat is continuously applied to beaker three to keep the water boiling.) The liquid in the glass tube is thus heated and expands to a new height. We mark this new height on the glass tube and arbitrarily call it 100°. Since the water is boiling at this point, we call it the boiling point of water.

Because the liquid in the tube expands linearly, to a first approximation, the distance between 0° and 100° can be divided into 100 equal parts. Any one of these divisions can be further divided into fractions of a degree. Thus, we obtain a complete scale of temperatures ranging from 0 to 100 degrees. Then we place this thermometer into the mixture of beaker two. The liquid in the glass rises to some number, and that number, whatever it may be, is the temperature of the mixture. That number is a numerical measure of the hotness or coldness of the body. We call this device a **thermometer,** and in particular *this scale of temperature that has 0° for the melting point of ice and 100° for the boiling point of water is called the Celsius temperature scale* and is shown in figure 14.3(a). This scale is named after the Swedish astronomer, Anders Celsius, who proposed it in 1742.

Figure 14.3

The temperature scales.

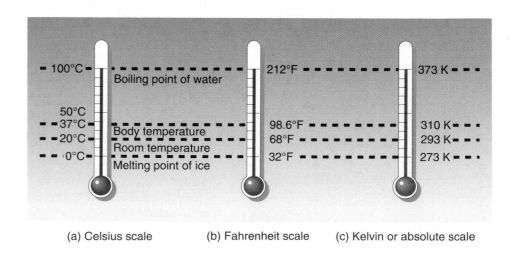

(a) Celsius scale	(b) Fahrenheit scale	(c) Kelvin or absolute scale

Figure 14.4
Simple lattice structure.

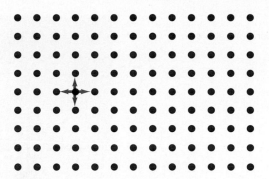

Another, perhaps more familiar, temperature scale is the **Fahrenheit temperature scale** shown in figure 14.3(b). The melting point of ice on this scale is 32 °F and the boiling point of water is 212 °F. At first glance it might seem rather strange to pick 32° for the freezing point and 212° for the boiling point of water. As a matter of fact Gabriel Fahrenheit, the German physicist, was not trying to use pure water as his calibration points. When the scale was first made, 0 °F corresponded to the lowest temperature then known, the temperature of freezing brine (a salt water mixture), and 100 °F was meant to be the temperature of the human body. Fahrenheit proposed his scale in 1714.

In addition to the Celsius and Fahrenheit scales there are other temperature scales, the most important of which is the Kelvin or *absolute* scale, as shown in figure 14.3(c). The melting point of ice on this scale is 273 K and the boiling point of water is 373 K. The **Kelvin temperature scale** does not use the degree symbol for a temperature. To use the terminology correctly, we should say that, "zero degrees Celsius corresponds to a temperature of 273 Kelvin." The Kelvin scale is extremely important in dealing with the behavior of gases. In fact, it was in the study of gases that Lord Kelvin first proposed the absolute scale in 1848. We will discuss this more natural introduction to the Kelvin scale in the study of gases in chapter 15. For the present, however, the implications of the Kelvin scale can still be appreciated by looking at the molecular structure of a solid.

The simplest picture of a solid, if it could be magnified trillions of times, is a large array of atoms or molecules in what is called a lattice structure, as shown in figure 14.4. Each dot in the figure represents an atom or molecule, depending on the nature of the substance. Each molecule is in equilibrium with all the molecules around it. The molecule above exerts a force upward on the molecule, whereas the molecule below exerts a force downward. Similarly, there are balanced forces from right and left and in and out. The molecule is therefore in equilibrium. In fact every molecule of the solid is in equilibrium. When heat is applied to a solid body, the added energy causes a molecule to vibrate around its equilibrium position. As any one molecule vibrates, it interacts with its nearest neighbors causing them to vibrate, which in turn causes its nearest neighbors to vibrate, and so on. Hence, the heat energy applied to the solid shows up as vibrational energy of the molecules of the solid. The higher the temperature of the solid, the larger is the vibrational motion of its molecules. The lower the temperature, the smaller is the vibrational motion of its molecules. Thus, the temperature of a body is really a measure of the mean or average kinetic energy of the vibrating molecules of the body.

It is therefore conceivable that if you could lower and lower the temperature of the body, the motion of the molecules would become less and less until at some very low temperature, the vibrational motion of the molecules would cease altogether. They would be frozen in one position. This point is called *absolute zero,* and is 0 on the Kelvin temperature scale. From work in quantum mechanics, however, it is found that even at absolute zero, the molecules contain a certain amount of energy called the *zero point energy.*

Figure 14.5

Converting one temperature scale to another.

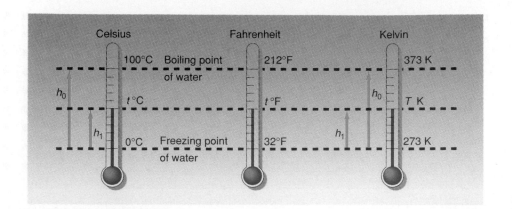

Even though temperature is really a measure of the mean kinetic energy of the molecules of a substance, from an experimental point of view it is difficult to make a standard of temperature in this way. Therefore, the International System of units considers temperature to be a fourth fundamental quantity and it is added to the three fundamental quantities of length, mass, and time. *The SI unit of temperature is the kelvin, and is defined as 1/273.16 of the temperature of the triple point of water.* The triple point of water is that point on a pressure-temperature diagram where the three phases of water, the solid, the liquid, and the gas, can coexist in equilibrium at the same pressure and temperature.

Temperature Conversions

The Celsius temperature scale is the recognized temperature scale in most scientific work and in most countries of the world. The Fahrenheit scale will eventually become obsolete along with the entire British engineering system of units. For the present, however, it is still necessary to convert from one temperature scale to another. That is, if a temperature is given in degrees Fahrenheit, how can it be expressed in degrees Celsius, and vice versa? It is easy to see how this conversion can be made.

The principle of the thermometer is based on the linear expansion of the liquid in the tube. For two identical glass tubes containing the same liquid, the expansion of the liquid is the same in both tubes. Therefore, the height of the liquid columns is the same for each thermometer, as shown in figure 14.5. The ratio of these heights in each thermometer is also equal. Therefore,

$$\left(\frac{h_1}{h_0}\right)_{\text{Celsius}} = \left(\frac{h_1}{h_0}\right)_{\text{Fahrenheit}}$$

These ratios, found from figure 14.5, are

$$\frac{t\,^\circ\text{C} - 0^\circ}{100^\circ - 0^\circ} = \frac{t\,^\circ\text{F} - 32^\circ}{212^\circ - 32^\circ}$$

$$\frac{t\,^\circ\text{C}}{100^\circ} = \frac{t\,^\circ\text{F} - 32^\circ}{180^\circ}$$

Solving for the temperature in degrees Celsius

$$t\,^\circ\text{C} = \frac{100^\circ}{180^\circ}(t\,^\circ\text{F} - 32^\circ)$$

Simplifying,

$$t\,^\circ\text{C} = \tfrac{5}{9}(t\,^\circ\text{F} - 32^\circ) \qquad (\textbf{14.1})$$

Equation 14.1 allows us to convert a temperature in degrees Fahrenheit to degrees Celsius.

Example 14.1

Fahrenheit to Celsius. If room temperature is 68 °F, what is this temperature in Celsius degrees?

Solution

The temperature in Celsius degrees, found from equation 14.1, is

$$t\,°C = \tfrac{5}{9}(t\,°F - 32°) = \tfrac{5}{9}(68° - 32°) = \tfrac{5}{9}(36)$$

$$= 20\,°C$$

To convert a temperature in degrees Celsius to one in Fahrenheit, we solve equation 14.1 for $t\,°F$ to obtain

$$t\,°F = \tfrac{9}{5}t\,°C + 32° \tag{14.2}$$

Example 14.2

Celsius to Fahrenheit. A temperature of $-5.00\,°C$ is equivalent to what Fahrenheit temperature?

Solution

The temperature in degrees Fahrenheit, found from equation 14.2, is

$$t\,°F = \tfrac{9}{5}t\,°C + 32° = \tfrac{9}{5}(-5.00°) + 32° = -9 + 32°$$

$$= 23\,°F$$

We can also find a conversion of absolute temperature to Celsius temperatures from figure 14.5, as

$$\left(\frac{h_1}{h_0}\right)_{\text{Celsius}} = \left(\frac{h_1}{h_0}\right)_{\text{Kelvin}}$$

$$\frac{t\,°C - 0°}{100° - 0°} = \frac{T\,K - 273}{373 - 273}$$

$$\frac{t\,°C}{100} = \frac{T\,K - 273}{100}$$

Therefore, the conversion of Kelvin temperature to Celsius temperatures is given by

$$t\,°C = T\,K - 273 \tag{14.3}$$

And the reverse conversion by

$$T\,K = t\,°C + 273 \tag{14.4}$$

For very precise work, 0 °C is actually equal to 273.16 K. In such cases, equations 14.3 and 14.4 should be modified accordingly.

Example 14.3

Celsius to Kelvin. Normal room temperature is considered to be 20.0 °C, find the value of this temperature on the Kelvin scale.

Solution

The absolute temperature, found from equation 14.4, is

$$T\,K = t\,°C + 273 = 20.0 + 273 = 293\,K$$

Note, in this book we will try to use the following convention: temperatures in Celsius and Fahrenheit will be represented by the lower case t, whereas Kelvin or absolute temperatures will be represented by a capital T. However, in some cases where time and temperature are found in the same equation, the lower case t will be used for time, and the upper case T will be used for temperature regardless of the unit used for temperature.

14.2 Heat

A solid body is composed of trillions upon trillions of atoms or molecules arranged in a lattice structure, as shown in figure 14.4. Each of these molecules possess an electrical potential energy and a vibrational kinetic energy. *The sum of the potential energy and kinetic energy of all these molecules is called the **internal energy** of the body.* When that internal energy is transferred between two bodies as a result of the difference in temperatures between the two bodies it is called heat.

Heat is thus the amount of internal energy flowing from a body at a higher temperature to a body at a lower temperature. Hence, a body does not contain heat, it contains internal energy. When the body cools, its internal energy is decreased; when it is heated, its internal energy is increased. A useful analogy is to compare the internal energy of a body to the money you have in a savings bank, whereas heat is analogous to the deposits or withdrawals of money.

Whenever two bodies at different temperatures are brought into contact, thermal energy always flows from the hotter body to the cooler body until they are both at the same temperature. When this occurs we say the two bodies are in **thermal equilibrium.** This is essentially the principle behind the thermometer. The thermometer is placed in contact with the body whose temperature is desired. Thermal energy flows from the hotter body to the cooler body until thermal equilibrium is reached. At that point, the thermometer is at the same temperature as the body. Hence, the thermometer is capable of measuring the temperature of a body.

The traditional unit of heat that we will use is the **kilocalorie,** *which is defined as the quantity of heat required to raise the temperature of 1 kg of water 1 °C, from 14.5 °C to 15.5 °C.*

The unit of heat in the British engineering system is the **British thermal unit,** abbreviated Btu. *One Btu is the heat required to raise the temperature of 1 lb of water 1 °F, from 58.5 °F to 59.5 °F.* The relation between the kilocalorie (kcal) and the Btu is

$$1 \text{ Btu} = 0.252 \text{ kcal}$$

It may seem strange to use the unit of kilocalorie or Btu for heat since heat is a flow of energy, and the unit of energy is a joule or a foot-pound. Historically it was not known that heat was a form of energy, but rather it was assumed that heat was a material quantity contained in bodies and was called Caloric. It was not until Benjamin Thompson's (1753–1814) experiments on the boring of cannons in 1798, that it became known that heat was a form of energy. Later James Prescott Joule (1818–1889) performed experiments to show *the exact equivalence between mechanical energy and heat energy. That equivalence is called the **mechanical equivalent of heat** and is*

$$1 \text{ kcal} = 1000 \text{ calories} = 4186 \text{ J}$$

$$1 \text{ Btu} = 778 \text{ ft lb} = 1055 \text{ J}$$

We will continue to follow past custom by expressing heat in units of kilocalories and Btu's, but we will also express it in terms of joules, the SI unit of energy.

We should also mention that the kilocalorie is sometimes called the large calorie and is identical to the unit used by dieticians. Thus when dieticians specify a diet as consisting of 1500 calories a day, they really mean that it is 1500 kcal per day.

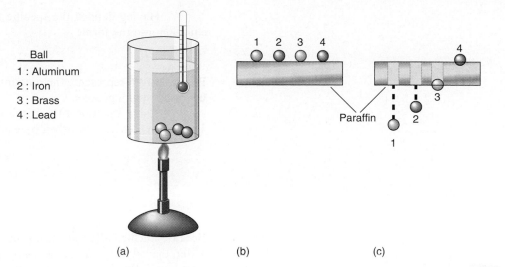

Figure 14.6

Tyndall's demonstration.

Ball
1 : Aluminum
2 : Iron
3 : Brass
4 : Lead

Paraffin

(a)   (b)   (c)

Table 14.1		
Specific Heats of Various Materials		
Material	**kcal** $\overline{\text{kg °C}}$	**J** $\overline{\text{kg °C}}$
Air	0.241	1009
Aluminum	0.215	900.0
Brass	0.094	393.5
Copper	0.092	385.1
Glass	0.20	837.2
Gold	0.031	129.8
Iron	0.108	452.1
Lead	0.031	129.8
Platinum	0.032	134.0
Silver	0.057	238.6
Steel	0.108	452.1
Tin	0.054	226.0
Tungsten	0.032	134.0
Zinc	0.093	389.3
Water	1.000	4186
Ice	0.500	2093
Steam	0.481	2013

14.3 Specific Heat

When the temperature of several substances is raised the same amount, each substance does not absorb the same amount of thermal energy. This can be shown by Tyndall's demonstration in figure 14.6.

Four balls made of aluminum, iron, brass, and lead, all of the same mass, are placed in a beaker of boiling water, as shown in figure 14.6(a). After about 10 or 15 min, these balls will reach thermal equilibrium with the water and will all be at the same temperature as the boiling water. The four balls are then placed on a piece of paraffin, as shown in figure 14.6(b). Almost immediately, the aluminum ball melts the wax and falls through the paraffin, as shown in figure 14.6(c). A little later in time the iron ball melts its way through the wax. The brass ball melts part of the wax and sinks into it deeply. However, it does not melt enough wax to fall through. The lead ball barely melts the wax and sits on the top of the sheet of paraffin.

How can this strange behavior of the four different balls be explained? Since each ball was initially in the boiling water, each absorbed energy from the boiling water. When the balls were placed on the sheet of paraffin, each ball gave up that energy to the wax, thereby melting the wax. But since each ball melted a different amount of wax in a given time, each ball must have given up a different amount of energy to the wax. Therefore each ball must have absorbed a different quantity of energy while it was in the boiling water. Hence, *different bodies absorb a different quantity of thermal energy even when subjected to the same temperature change.*

To handle the problem of different bodies absorbing different quantities of thermal energy when subjected to the same temperature change, *the* **specific heat** *c of a body is defined as the amount of thermal energy Q required to raise the temperature of a unit mass of the material 1 °C. In terms of the traditional unit of kilocalories, the specific heat c of a body is defined as the number of kilocalories Q required to raise the temperature of 1 kg of the material 1 °C. Thus,*

$$c = \frac{Q}{m\Delta t} \qquad (14.5)$$

We observe from this definition that the specific heat of water is 1 kcal/ kg °C, since 1 kcal raises the temperature of 1 kg of water 1 °C. All other materials have a different value for the specific heat. Some specific heats are shown in table 14.1. Note that water has the largest specific heat. If the thermal energy is expressed in the SI unit of joules, *the specific heat c of a body is defined as the number of joules Q required to raise the temperature of 1 kg of the material 1 °C.* In SI units the specific heat of water is 4186 J/(kg °C).

Having defined the specific heat by equation 14.5, we can rearrange that equation into the form

$$Q = mc\Delta t \tag{14.6}$$

Equation 14.6 represents the amount of thermal energy Q that will be absorbed or liberated in any process.

Using equation 14.6 it is now easier to explain the Tyndall demonstration. The thermal energy absorbed by each ball while in the boiling water is

$$Q_{Al} = mc_{Al}\Delta t$$

$$Q_{iron} = mc_{iron}\Delta t$$

$$Q_{brass} = mc_{brass}\Delta t$$

$$Q_{lead} = mc_{lead}\Delta t$$

Because all the balls went from room temperature to 100 °C, the boiling point of water, they all experienced the same temperature change Δt. Because all the masses were equal, the thermal energy absorbed by each ball is directly proportional to its specific heat. We can observe from table 14.1 that

$$c_{Al} = 0.215 \text{ kcal/(kg °C)}$$

$$c_{iron} = 0.108 \text{ kcal/(kg °C)}$$

$$c_{brass} = 0.094 \text{ kcal/(kg °C)}$$

$$c_{lead} = 0.031 \text{ kcal/(kg °C)}$$

Because the specific heat of aluminum is the largest of the four materials, the aluminum ball absorbs the greatest amount of thermal energy while in the water. Hence, it also liberates the greatest amount of thermal energy to melt the wax and should be the first ball to melt through the wax. Iron, brass, and lead absorb less thermal energy respectively because of their lower specific heats and consequently liberate thermal energy to melt the wax in this same sequence.

If the masses are not the same, then the amount of thermal energy absorbed depends on the product of the mass m and the specific heat c. The ball with the largest value of mc absorbs the most heat energy.

Example 14.4

Absorption of thermal energy. A steel ball at room temperature is placed in a pan of boiling water. If the mass of the ball is 200 g, how much thermal energy is absorbed by the ball?

Solution

The thermal energy absorbed by the ball, given by equation 14.6, is

$$Q = mc\Delta t$$

$$= (0.200 \text{ kg})\left(0.108 \frac{\text{kcal}}{\text{kg °C}}\right)(100 \text{ °C} - 20.0 \text{ °C})$$

$$= 1.73 \text{ kcal}$$

An interesting thing to note is that once the ball reaches the 100 °C mark, it is at the same temperature as the water and hence, there is no longer a transfer of thermal energy into the ball no matter how long the ball is left in the boiling water. All the thermal energy supplied to the pot containing the ball and the water will then go into boiling away the water.

Example 14.5

Absorption in SI units. A 500-g aluminum block at 10.0 °C is placed in an oven at 200 °C. How many joules of thermal energy are absorbed by the block?

Solution

The specific heat in SI units, found from table 14.1, is 900 J/(kg °C). The thermal energy absorbed is

$$Q = mc\Delta t$$

$$= (0.500 \text{ kg})\left(900 \; \frac{\text{J}}{\text{kg °C}}\right)(200 \text{ °C} - 10.0 \text{ °C})$$

$$= 85{,}500 \text{ J}$$

14.4 Calorimetry

Calorimetry is defined as the measurement of heat. These measurements are performed in a device called a **calorimeter.** The simplest of all calorimeters consists of a metal container placed on a plastic insulating ring inside a larger highly polished metallic container, as shown in figure 14.7. The space between the two containers is filled with air to minimize the thermal energy lost from the inner calorimeter cup to the environment. The highly polished outer container reflects any external radiated energy that might otherwise make its way to the inner cup. A plastic cover is placed on the top of the calorimeter to prevent any additional loss of thermal energy to the environment. The inner cup is thus insulated from the environment, and all measurements of thermal energy absorption or liberation are made here. A thermometer is placed through a hole in the cover so that the temperature inside the calorimeter can be measured. The calorimeter is used to measure the specific heat of various substances, and the latent heat of fusion and vaporization of water.

The basic principle underlying the calorimeter is the conservation of energy. The thermal energy lost by those bodies that lose thermal energy is equal to the thermal energy gained by those bodies that gain thermal energy. We write this conservation principle mathematically as

$$\text{Thermal energy lost} = \text{Thermal energy gained} \qquad (14.7)$$

As an example of the use of the calorimeter, let us determine the specific heat of a sample of iron. We place the iron sample in a pot of boiling water until the iron sample eventually reaches the temperature of boiling water, namely 100 °C. Meanwhile we place the inner calorimeter cup on a scale and determine its mass m_c. Then we place water within the cup and again place it on the scale to determine its mass. The difference between these two scale readings is the mass of water m_w in the cup. We place the inner cup into the calorimeter and place a thermometer through a hole in the cover of the calorimeter so that the initial temperature of the water t_{iw} is measured.

After the iron sample reaches 100 °C, we place it within the inner calorimeter cup, and close the cover quickly. As time progresses, the temperature of the water, as recorded by the thermometer, starts to rise. It eventually stops at a final equilibrium temperature t_{fw} of the water, the sample, and the calorimeter can. The iron sample was the hot body and it lost thermal energy, whereas the water and the

Figure 14.7

A calorimeter.

can, which is in contact with the water, absorb this thermal energy as is seen by the increased temperature of the mixture. We analyze the problem by the conservation of energy, equation 14.7, as

$$\text{Thermal energy lost} = \text{Thermal energy gained}$$

$$Q_s = Q_w + Q_c \tag{14.8}$$

That is, the thermal energy lost by the sample Q_s is equal to the thermal energy gained by the water Q_w plus the thermal energy gained by the calorimeter cup Q_c. However, the thermal energy absorbed or liberated in any process, given by equation 14.6, is

$$Q = mc\Delta t$$

Using equation 14.6 in equation 14.8, gives

$$m_s c_s \Delta t_s = m_w c_w \Delta t_w + m_c c_c \Delta t_c \tag{14.9}$$

where

m_s is the mass of the sample
m_w is the mass of the water
m_c is the mass of the calorimeter cup
c_s is the specific heat of the unknown sample
c_w is the specific heat of the water
c_c is the specific heat of the calorimeter cup

The change in the temperature of the sample is the difference between its initial temperature of 100 °C and its final equilibrium temperature t_{fw}. That is,

$$\Delta t_s = 100\ °C - t_{fw} \tag{14.10}$$

The change in temperature of the water and calorimeter cup are equal since the water is in contact with the cup and thus has the same temperature. Therefore,

$$\Delta t_w = \Delta t_c = t_{fw} - t_{iw} \tag{14.11}$$

Substituting equations 14.10 and 14.11 into equation 14.9, yields

$$m_s c_s (100 - t_{fw}) = m_w c_w (t_{fw} - t_{iw}) + m_c c_c (t_{fw} - t_{iw}) \tag{14.12}$$

All the quantities in equation 14.12 are known except for the specific heat of the unknown sample, c_s. Solving for the specific heat yields

$$c_s = \frac{m_w c_w (t_{fw} - t_{iw}) + m_c c_c (t_{fw} - t_{iw})}{m_s (100 - t_{fw})} \tag{14.13}$$

Example 14.6

Find the specific heat. A 0.0700-kg iron specimen is used to determine the specific heat of iron. The following laboratory data were found:

$m_s = 0.0700$ kg	$t_{iw} = 20.0\ °C$
$m_c = 0.0600$ kg	$t_{fw} = 23.5\ °C$
$c_c = 0.215$ kcal/kg °C	$m_w = 0.150$ kg
$t_s = 100\ °C$	

Find the specific heat of the specimen.

Thermodynamics

The specific heat of the iron specimen, found from equation 14.13, is

$$c_s = \frac{m_w c_w(t_{fw} - t_{iw}) + m_c c_c(t_{fw} - t_{iw})}{m_s(100 - t_{fw})}$$

$$= \frac{\begin{array}{l}(0.150 \text{ kg})(1.00 \text{ kcal/kg °C})(23.5 \text{ °C} - 20.0 \text{ °C}) \\ + (0.0600 \text{ kg})(0.215 \text{ kcal/kg °C})(23.5 \text{ °C} - 20 \text{ °C})\end{array}}{(0.0700 \text{ kg})(100 \text{ °C} - 23.5 \text{ °C})}$$

$$= 0.106 \text{ kcal/kg °C}$$

which is in good agreement with the accepted value of the specific heat of iron of 0.108 kcal/kg °C.

14.5 Change of Phase

Matter exists in three states called the **phases of matter.** They are the solid phase, the liquid phase, and the gaseous phase. Let us see how one phase of matter is changed into another.

Let us examine the behavior of matter when it is heated over a relatively large range of temperatures. In particular, let us start with a piece of ice at −20.0 °C and heat it to a temperature of 120 °C. We place the ice inside a strong, tightly sealed, windowed enclosure containing a thermometer. Then we apply heat, as shown in figure 14.8. We observe the temperature as a function of time and plot it, as in figure 14.9.

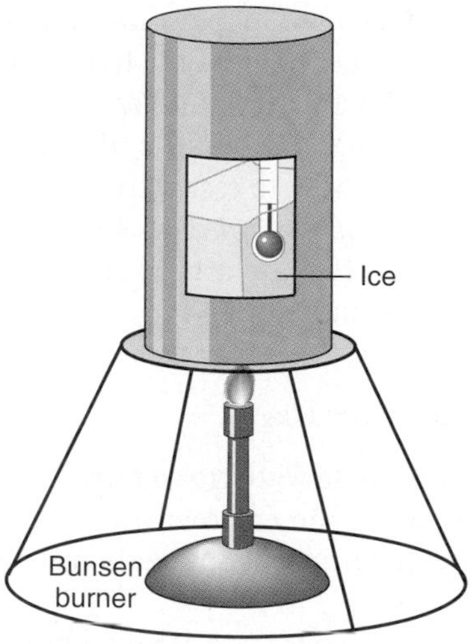

As the heat is applied to the solid ice, the temperature of the block increases with time until 0 °C is reached. At this point the temperature remains constant, even though heat is being continuously applied. Looking at the block of ice, through the window in the container, we observe small drops of liquid water forming on the block of ice. The ice is starting to melt. We observe that the temperature remains constant until every bit of the solid ice is converted into the liquid water. We are observing a **change of phase.** That is, the ice is changing from the solid phase into the liquid phase. As soon as all the ice is melted, we again observe an increase in the temperature of the liquid water. The temperature increases up to 100 °C, and then levels off. Thermal energy is being applied, but the temperature is not changing. Looking through the window into the container, we see that there are bubbles forming throughout the liquid. The water is boiling. The liquid water is being converted to steam, the gaseous state of water. The temperature remains at this constant value of 100 °C until every drop of the liquid water has been converted to the gaseous steam. After that, as we continuously supply heat, we observe an increase in the temperature of the steam. Superheated steam is being made. (Note, you should not try to do this experiment on your own, because enormous pressures can be built up by the steam, causing the closed container to explode.)

Let us go back and analyze this experiment more carefully. As the thermal energy was supplied to the below freezing ice, its temperature increased to 0 °C. At this point the temperature remained constant even though heat was being continuously applied. Where did this thermal energy go if the temperature never changed? The thermal energy went into the melting of the ice, changing its phase from the solid to the liquid phase. If we observe the solid in terms of its lattice structure, figure 14.4, we can see that each molecule is vibrating about its equilibrium position. As heat is applied, the vibration increases, until at 0 °C, the vibrations of the molecules become so intense that the molecules literally pull apart from one another changing the entire structure of the material. This is the melting process. The amount of heat necessary to tear these molecules apart is a constant and

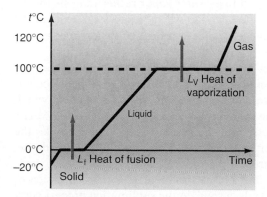

Figure 14.8

Converting ice to water to steam.

Figure 14.9

Changes of phase.

is called the latent heat of fusion of that material. *The **latent heat of fusion** is the amount of heat necessary to convert 1 kg of the solid to 1 kg of the liquid.* For water, it is found experimentally that it takes 79.7 kcal or 334,000 J of thermal energy to melt 1 kg of ice. To simplify our calculations, we will take the latent heat of fusion of water as

$$L_f = 80.0 \text{ kcal/kg} = 3.34 \times 10^5 \text{ J/kg}$$

If we must supply 80.0 kcal/kg to melt ice, then we must take away 80.0 kcal/kg to freeze water. Thus, the heat of fusion is equal to the heat of melting. The word *latent* means hidden or invisible, and not detectable as a temperature change. Heat supplied that does change the temperature is called *sensible heat*.

In the liquid state there are still molecular forces holding the molecules together, but because of the energy and motion of the molecules, these forces can not hold the molecules in the relatively rigid position they had in the solid state. This is why the liquid is able to flow and take the shape of any container in which it is placed.

As the water at 0 °C is further heated, the molecules absorb more and more energy, increasing their mean velocity within the liquid. This appears as a rise in temperature of the liquid. At 100 °C, so much energy has been imparted to the water molecules, that the molecular speeds have increased to the point that the molecules are ready to pull away from the molecular forces holding the liquid together. As further thermal energy is applied, the molecules fly away into space as steam. The temperature of the water does not rise above 100 °C because all the applied heat is supplying the molecules with the necessary energy to escape from the liquid.

*The heat that is necessary to convert 1 kg of the liquid to 1 kg of the gas is called the **latent heat of vaporization**.* For water, the latent heat of vaporization is 539 kcal/kg = 2,260,000 J/kg. However, for simplicity in our calculations we will use the standard value of

$$L_v = 540 \text{ kcal/kg} = 2.26 \times 10^6 \text{ J/kg}$$

Because this amount of thermal energy must be given to water to convert it to steam, this same quantity of thermal energy is given up to the environment when steam condenses back into the liquid state. Therefore, the heat of vaporization is equal to the heat of condensation.

Liquid water can also be converted to the gaseous state at any temperature, a process called *evaporation*. Thus, water left in an open saucer overnight will be gone by morning. Even though the temperature of the water remained at the room temperature, the liquid was converted to a gas. It evaporated into the air. The gaseous state of water is then usually referred to as *water vapor* rather than steam. At 0 °C the latent heat of vaporization is 600 kcal/kg. All substances can exist in the three states of matter, and each substance has its own heat of fusion and heat of vaporization.

Note also that another process is possible whereby a solid can go directly to a gas and vice versa without ever going through the liquid state. This process is called **sublimation.** Many students have seen this phenomenon with dry ice (which is carbon dioxide in the solid state). The ice seems to be smoking. Actually, however, the solid carbon dioxide is going directly into the gaseous state. The gas, like the dry ice, is so cold that it causes water vapor in the surrounding air to condense, which is seen as the "smoky" clouds around the solid carbon dioxide.

A more common phenomena, but not as spectacular, is the conversion of water vapor, a gas, directly into ice crystals, a solid, in the sublimation process commonly known as frost. On wintry mornings when you first get up and go outside your home, you sometimes see ice all over the tips of the grass in the yard and over the windshield and other parts of your car. The water vapor in the air did not first condense to water droplets and then the water droplets froze. Instead, the grass and the car surfaces were so cold that the water vapor in the air went directly from the gaseous state into the solid state without ever going through the liquid state.

Thermodynamics

The reverse process whereby the solid goes directly into the gas also occurs in nature, but it is not as noticeable as frost. There are times in the winter when a light covering of snow is observed on the ground. The temperature may remain below freezing, and an overcast sky may prevent any sun from heating up or melting that snow. Yet, in a day or so, some of that snow will have disappeared. It did not melt, because the temperature always remained below freezing. Some of the snow crystals went directly into the gaseous state as water vapor.

Just as there is a latent heat of fusion L_f and latent heat of vaporization L_v, there is also a latent heat of sublimation L_s. Its value is given by

$$L_s = 677 \text{ kcal/kg} = 2.83 \times 10^6 \text{ J/kg}$$

Thus, *the heat that is necessary to convert 1.00 kg of the solid ice into 1.00 kg of the gaseous water vapor is called the **latent heat of sublimation.***

It is interesting to note here that there is no essential difference in the water molecule when it is either a solid, a liquid, or a gas. The molecule consists of the same two hydrogen atoms bonded to one oxygen atom. The difference in the state is related to the different energy, and hence speed of the molecule in the different states.

Notice that it takes much more energy to convert 1 kg of water to 1 kg of steam, than it does to convert 1 kg of ice to 1 kg of liquid water, almost seven times as much. This is also why a steam burn can be so serious, since the steam contains so much energy. Let us now consider some more examples.

Example 14.7

Converting ice to steam. Let us compute the thermal energy that is necessary to convert 5.00 kg of ice at $-20.0\,°C$ to superheated steam at $120\,°C$.

Solution

The necessary thermal energy is given by

$$Q = Q_i + Q_f + Q_w + Q_v + Q_s \tag{14.14}$$

where

Q_i is the energy needed to heat the ice up to $0\,°C$
Q_f is the energy needed to melt the ice
Q_w is the energy needed to heat the water to $100\,°C$
Q_v is the energy needed to boil the water
Q_s is the energy needed to heat the steam to $120\,°C$

The necessary thermal energy to warm up the ice from $-20.0\,°C$ to $0\,°C$ is found from

$$Q_i = m_i c_i [0° - (-20.0\,°C)]$$

The latent heat of fusion is the amount of heat needed per kilogram to melt the ice. The total amount of heat needed to melt all the ice is the heat of fusion times the number of kilograms of ice present. Hence, the thermal energy needed to melt the ice is

$$Q_f = m_i L_f \tag{14.15}$$

The thermal energy needed to warm the water from $0\,°C$ to $100\,°C$ is

$$Q_w = m_w c_w (100\,°C - 0\,°C)$$

The latent heat of vaporization is the amount of heat needed per kilogram to boil the water. The total amount of heat needed to boil all the water is the heat of vaporization times the number of kilograms of water present. Hence, the thermal energy needed to convert the liquid water at 100 °C to steam at 100 °C is

$$Q_v = m_w L_v \tag{14.16}$$

and

$$Q_s = m_s c_s(120 \text{ °C} - 100 \text{ °C})$$

is the thermal energy needed to convert the steam at 100 °C to superheated steam at 120 °C. Substituting all these equations into equation 14.14 gives

$$Q = m_i c_i[0 \text{ °C} - (-20 \text{ °C})] + m_i L_f + m_w c_w(100 \text{ °C} - 0 \text{ °C}) \\ + m_w L_v + m_s c_s(120 \text{ °C} - 100 \text{ °C}) \tag{14.17}$$

Using the values of the specific heat from table 14.1, we get

$$Q = (5.00 \text{ kg})\left(0.500 \frac{\text{kcal}}{\text{kg °C}}\right)(20 \text{ °C}) + (5.00 \text{ kg})\left(80.0 \frac{\text{kcal}}{\text{kg}}\right)$$
$$+ (5.00 \text{ kg})\left(1.00 \frac{\text{kcal}}{\text{kg °C}}\right)(100 \text{ °C}) + (5.00 \text{ kg})\left(540 \frac{\text{kcal}}{\text{kg}}\right)$$
$$+ (5.00 \text{ kg})\left(0.481 \frac{\text{kcal}}{\text{kg °C}}\right)(20.0 \text{ °C})$$

$$= 50.0 \text{ kcal} + 400 \text{ kcal} + 500 \text{ kcal} + 2700 \text{ kcal} + 48.1 \text{ kcal}$$

$$= 3700 \text{ kcal}$$

Therefore, we need 3700 kcal of thermal energy to convert 5.00 kg of ice at −20.0 °C to superheated steam at 120 °C. Note the relative size of each term's contribution to the total thermal energy.

Example 14.8

Latent heat of fusion. The heat of fusion of water L_f can be found in the laboratory using a calorimeter. If 31.0 g of ice m_i at 0 °C are placed in a 60.0-g calorimeter cup m_c that contains 170 g of water m_w at an initial temperature t_{iw} of 20.0 °C, after the ice melts, the final temperature of the water t_{fw} is found to be 5.57 °C. Find the heat of fusion of water from this data. The specific heat of the calorimeter is 0.215 cal/g °C.

Solution

From the fundamental principle of calorimetry

$$\text{Thermal energy gained} = \text{Thermal energy lost}$$
$$Q_f + Q_{iw} = Q_w + Q_c \tag{14.18}$$

where Q_f is the thermal energy necessary to melt the ice through the fusion process and Q_{iw} is the thermal energy necessary to warm the water that came from the melted ice. We call this water ice water to distinguish it from the original water in the container. This liquid water is formed at 0 °C and will be warmed to the final equilibrium temperature of the mixture t_{fw}. The thermal energy lost by the original water in the calorimeter is Q_w, and Q_c is the thermal energy lost by the calorimeter itself. Equation 14.18 therefore becomes

$$m_i L_f + m_{iw} c_w(t_{fw} - 0 \text{ °C}) = m_w c_w(t_{iw} - t_{fw}) + m_c c_c(t_{iw} - t_{fw})$$

We find the heat of fusion by solving for L_f, as

$$L_f = \frac{(m_w c_w + m_c c_c)(t_{iw} - t_{fw}) - m_{iw} c_w(t_{fw} - 0 \text{ °C})}{m_i} \tag{14.19}$$

Thermodynamics

Since the laboratory data were taken in grams we can do the calculation in those units by noting that the specific heat of any material has the same numerical value when given in either

$$\frac{\text{kcal}}{\text{kg }°\text{C}} \quad \text{or} \quad \frac{\text{cal}}{\text{g }°\text{C}}$$

because the prefix kilo is found in both the numerator and denominator of the units and can thus cancel out. The same thing applies to the heat of fusion and the heat of vaporization. Therefore, using the laboratory data supplied, the heat of fusion is found as

$$L_f = \frac{[(170 \text{ g})(1.00 \text{ cal/g }°\text{C}) + (60.0 \text{ g})(0.215 \text{ cal/g }°\text{C})](20.0 °\text{C} - 5.57 °\text{C})}{31.0 \text{ g}}$$
$$\frac{- (31.0 \text{ g})(1.00 \text{ cal/g }°\text{C})(5.57 °\text{C} - 0 °\text{C})}{}$$

$$L_f = 79.6 \text{ cal/g}$$

Note that this is in very good agreement with the standard value of 80.0 cal/g.

Example 14.9

Latent heat of vaporization. The heat of vaporization L_v of water can be found in the laboratory by passing steam at 100 °C into a calorimeter containing water. As the steam condenses and cools it gives up thermal energy to the water and the calorimeter. In the experiment the following data were taken:

mass of calorimeter cup	$m_c = 60.0 \text{ gm}$
mass of water	$m_w = 170 \text{ gm}$
mass of condensed steam	$m_s = 3.00 \text{ gm}$
initial temperature of water	$t_{iw} = 19.9 °\text{C}$
final temperature of water	$t_{fw} = 30.0 °\text{C}$
specific heat of calorimeter	$c_c = 0.215 \text{ cal/g }°\text{C}$

Find the heat of vaporization from this data.

Solution

To determine the heat of vaporization let us start with the fundamental principle of calorimetry

$$\text{Thermal energy lost} = \text{Thermal energy gained}$$
$$Q_v + Q_{sw} = Q_w + Q_c \tag{14.20}$$

where Q_v is the thermal energy necessary to condense the steam and Q_{sw} is the thermal energy necessary to cool the water that came from the condensed steam. We use the subscript sw to remind us that this is the water that came from the steam in order to distinguish it from the original water in the container. This liquid water is formed at 100 °C and will be cooled to the final equilibrium temperature of the mixture t_{fw}. Here Q_w is the thermal energy gained by the original water in the calorimeter and Q_c is the thermal energy gained by the calorimeter itself. Equation 14.20 therefore becomes

$$m_s L_v + m_{sw} c_w (100 °\text{C} - t_{fw}) = m_w c_w (t_{fw} - t_{iw}) + m_c c_c (t_{fw} - t_{iw})$$

Solving for the heat of vaporization,

$$L_v = \frac{m_w c_w (t_{fw} - t_{iw}) + m_c c_c (t_{fw} - t_{iw}) - m_{sw} c_w (100 °\text{C} - t_{fw})}{m_s} \tag{14.21}$$

Therefore,

$$L_v = \frac{(170 \text{ g})(1.00 \text{ cal/g °C})(30.0 \text{ °C} - 19.9 \text{ °C})}{}$$
$$+ (60.0 \text{ g})(0.215 \text{ cal/g °C})(30.0° - 19.9 \text{ °C})$$
$$- \frac{(3.00 \text{ g})(1.00 \text{ cal/g °C})(100 \text{ °C} - 30.0°)}{3.00 \text{ g}}$$

Thus, we find from the experimental data that the heat of vaporization is

$$L_v = 546 \text{ cal/g}$$

which is in good agreement with the standard value of 540 cal/g.

Example 14.10

Mixing ice and water. If 10.0 g of ice, at 0 °C, are mixed with 50.0 g of water at 80.0 °C, what is the final temperature of the mixture?

Solution

When the ice is mixed with the water it will gain heat from the water. The law of conservation of thermal energy becomes

Thermal energy gained = Thermal energy lost

$$Q_f + Q_{iw} = Q_w$$

where Q_f is the heat gained by the ice as it goes through the melting process. When the ice melts, it becomes water at 0 °C. Let us call this water ice water to distinguish it from the original water in the container. Thus, Q_{iw} is the heat gained by the ice water as it warms up from 0 °C to the final equilibrium temperature t_{fw}. Finally, Q_w is the heat lost by the original water, which is at the initial temperature t_{iw}. Thus,

$$m_i L_f + m_{iw} c_w (t_{fw} - 0 \text{ °C}) = m_w c_w (t_{iw} - t_{fw})$$

where m_i is the mass of the ice, m_{iw} is the mass of the ice water, and m_w is the mass of the original water. Solving for the final temperature of the water we get

$$m_i L_f + m_{iw} c_w t_{fw} = m_w c_w t_{iw} - m_w c_w t_{fw}$$
$$m_{iw} c_w t_{fw} + m_w c_w t_{fw} = m_w c_w t_{iw} - m_i L_f$$
$$(m_{iw} c_w + m_w c_w) t_{fw} = m_w c_w t_{iw} - m_i L_f$$
$$t_{fw} = \frac{m_w c_w t_{iw} - m_i L_f}{m_{iw} c_w + m_w c_w}$$

The final equilibrium temperature of the water becomes

$$t_{fw} = \frac{(50.0 \text{ g})(1 \text{ cal/g °C})(80.0 \text{ °C}) - (10.0 \text{ g})(80.0 \text{ cal/g})}{(10.0 \text{ g})(1 \text{ cal/g °C}) + (50.0 \text{ g})(1 \text{ cal/g °C})}$$

$$= \frac{4000 \text{ cal} - 800 \text{ cal}}{60.0 \text{ cal/°C}}$$

$$= 53.3 \text{ °C}$$

Example 14.11

Something is wrong here. Repeat example 14.10 with the initial temperature of the water at 10.0 °C.

Using the same equation as for the final water temperature in example 14.10, we get

$$t_{fw} = \frac{m_w c_w t_{iw} - m_i L_f}{m_{iw} c_w + m_w c_w}$$

Thus,

$$t_{fw} = \frac{(50.0 \text{ g})(1 \text{ cal/g }°\text{C})(10.0 °\text{C}) - (10.0 \text{ g})(80.0 \text{ cal/g})}{(10.0 \text{ g})(1 \text{ cal/g }°\text{C}) + (50.0 \text{ g})(1 \text{ cal/g }°\text{C})}$$

$$= \frac{500 \text{ cal} - 800 \text{ cal}}{60.0 \text{ cal/}°\text{C}}$$

$$= -5.00 °\text{C}$$

There is something very wrong here! Our answer says that the final temperature is 5° below zero. But this is impossible. The temperature of the water can not be below 0 °C and still be water, and the ice that was placed in the water can not convert all the water to ice and cause all the ice to be at a temperature of 5° below zero. Something is wrong. Let us check our equation. The equation worked for the last example, why not now? The equation was derived with the assumption that all the ice that was placed in the water melted. Is this a correct assumption? The energy necessary to melt all the ice is found from

$$Q_f = m_i L_f = (10.0 \text{ g})(80.0 \text{ cal/g}) = 800 \text{ cal}$$

The energy available to melt the ice comes from the water. The maximum thermal energy available occurs when all the water is cooled to 0 °C. Therefore, the maximum available energy is

$$Q_w = m_w c_w (t_{iw} - 0 °\text{C}) = (50.0 \text{ g})(1.00 \text{ cal/g }°\text{C})(10.0 °\text{C})$$

$$= 500 \text{ cal}$$

The amount of energy available to melt all the ice is 500 cal and it would take 800 cal to melt all the ice present. Therefore, there is not enough energy to melt the ice. Hence, our initial assumption that all the ice melted is incorrect. Thus, our equation is no longer valid. There is an important lesson to be learned here. All through our study of physics we make assumptions in order to derive equations. If the assumptions are correct, the equations are valid and can be used to predict some physical phenomenon. If the assumptions are not correct, the final equations are useless. In this problem there is still ice left and hence the final temperature of the mixture is 0 °C. The amount of ice that actually melted can be found by using the relation

$$fQ_f = Q_w$$

where f is the fraction of the ice that melts. Thus,

$$f = \frac{Q_w}{Q_f} = \frac{500 \text{ cal}}{800 \text{ cal}}$$

$$= 0.625$$

Therefore, only 62.5% of the ice melted and the final temperature of the mixture is 0 °C.

The Language of Physics

Temperature
The simplest definition of temperature is that temperature is a measure of the hotness or coldness of a body. A better definition is that temperature is a measure of the mean kinetic energy of the molecules of the body (p. 403).

Thermometer
A device for measuring the temperature of a body (p. 404).

Celsius temperature scale
A temperature scale that uses 0° for the melting point of ice and 100° for the boiling point of water (p. 404).

Fahrenheit temperature scale
A temperature scale that uses 32° for the melting point of ice and 212° for the boiling point of water (p. 405).

Kelvin temperature scale
The absolute temperature scale. The lowest temperature attainable is absolute zero, the 0 K of this scale. The temperature for the melting point of ice is 273 K and 373 K for the boiling point of water (p. 405).

Internal energy
The sum of the potential and kinetic energy of all the molecules of a body (p. 408).

Heat
The flow of thermal energy from a body at a higher temperature to a body at a lower temperature. When a body cools, its internal energy is decreased; when it is heated, its internal energy is increased (p. 408).

Thermal equilibrium
Whenever two bodies at different temperatures are touched together, thermal energy always flows from the hotter body to the cooler body until they are both at the same temperature. When this occurs the two bodies are said to be in thermal equilibrium (p. 408).

Kilocalorie
A unit of heat. It is defined as the amount of thermal energy required to raise the temperature of 1 kg of water 1 °C (p. 408).

British thermal unit (Btu)
The amount of thermal energy required to raise the temperature of 1 lb of water 1 °F (p. 408).

Mechanical equivalent of heat
The equivalence between mechanical energy and thermal energy. One kcal is equal to 4186 J (p. 408).

Specific heat
A characteristic of a material. It is defined as the number of kilocalories required to raise the temperature of 1 kg of the material 1 °C (p. 409).

Calorimetry
The measurement of heat (p. 411).

Calorimeter
An instrument that is used to make measurements of heat. The basic principle underlying the calorimeter is the conservation of energy. The thermal energy lost by those bodies that lose thermal energy is equal to the thermal energy gained by those bodies that gain thermal energy (p. 411).

Phases of matter
Matter exists in three phases, the solid phase, the liquid phase, and the gaseous phase (p. 413).

Change of phase
The change in a body from one phase of matter to another. As an example, melting is a change from the solid state of a body to the liquid state. Boiling is a change in state from the liquid state to the gaseous state (p. 413).

Latent heat of fusion
The amount of heat necessary to convert 1 kg of the solid to 1 kg of the liquid (p. 414).

Latent heat of vaporization
The amount of heat necessary to convert 1 kg of the liquid to 1 kg of the gas (p. 414).

Summary of Important Equations

Convert Fahrenheit temperature to Celsius
$$t\,°C = \tfrac{5}{9}(t\,°F - 32°) \tag{14.1}$$

Convert Celsius temperature to Fahrenheit
$$t\,°F = \tfrac{9}{5}t\,°C + 32° \tag{14.2}$$

Convert Celsius temperature to Kelvin
$$T\,K = t\,°C + 273 \tag{14.4}$$

Thermal energy absorbed or liberated
$$Q = mc\Delta t \tag{14.6}$$

Principle of calorimetry
$$\text{Thermal} = \text{Thermal}$$
$$\text{energy lost} \quad \text{energy gained} \tag{14.7}$$

Fusion
$$Q_f = m_i L_f \tag{14.15}$$

Vaporization
$$Q_v = m_w L_v \tag{14.16}$$

Questions for Chapter 14

1. What is the difference between temperature and heat?
2. Explain how a bathtub of water at 5 °C can contain more thermal energy than a cup of coffee at 95 °C.
3. Discuss how the human body uses the latent heat of vaporization to cool itself through the process of evaporation.
†4. Relative humidity is defined as the percentage of the amount of water vapor in the air to the maximum amount of water vapor that the air can hold at that temperature. Discuss how the relative humidity affects the process of evaporation in general and how it affects the human body in particular.
†5. It is possible for a gas to go directly to the solid state without going through the liquid state, and vice versa. The process is called *sublimation*. An example of such a process is the formation of frost. Discuss the entire process of sublimation, the latent heat involved, and give some more examples of the process.
†6. Why does ice melt when an object is placed upon it? Describe the process of ice skating from the pressure of the skate on the ice.
7. Why are the numerical values of the specific heat the same in SI units and British engineering units?

Problems for Chapter 14

14.1 Temperature

1. Convert the following normal body temperatures to degrees Celsius: (a) oral temperature of 98.6 °F, (b) rectal temperature of 99.6 °F, and (c) axial (armpit) temperature of 97.6 °F.
2. Find the value of absolute zero on the Fahrenheit scale.
3. For what value is the Fahrenheit temperature equal to the Celsius temperature?
4. Convert the following temperatures to Fahrenheit: (a) 38.0 °C, (b) 68.0 °C, (c) 250 °C, (d) −10.0 °C, and (e) −20.0 °C.
5. Convert the following Fahrenheit temperatures to Celsius: (a) −23.0 °F, (b) 12.5 °F, (c) 55.0 °F, (d) 90.0 °F, and (e) 180 °F.
6. A temperature change of 5 °F corresponds to what temperature change in Celsius degrees?
†7. Derive an equation to convert the temperature in Fahrenheit degrees to its corresponding Kelvin temperature.
†8. Derive an equation to convert the change in temperature in Celsius degrees to a change in temperature in Fahrenheit degrees.

14.3 Specific Heat

9. A 450-g ball of copper at 20.0 °C is placed in a pot of boiling water until equilibrium is reached. How much thermal energy is absorbed by the ball?
10. A 250-g glass marble is taken from a freezer at −23.0 °C and placed into a beaker of boiling water. How much thermal energy is absorbed by the marble?
11. How much thermal energy must be supplied by an electric immersion heater if you wish to raise the temperature of 5.00 kg of water from 20.0 °C to 100 °C?

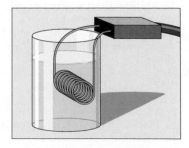

12. A 2.00-kg mass of copper falls from a height of 3.00 m to an insulated floor. What is the maximum possible temperature increase of the copper?
13. An iron block slides down an iron inclined plane at a constant speed. The plane is 10.0 m long and is inclined at an angle of 35.0° with the horizontal. Assuming that half the energy lost to friction goes into the block, what is the difference in temperature of the block from the top of the plane to the bottom of the plane?

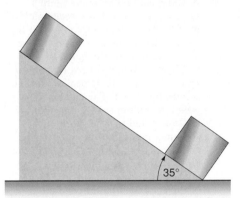

14. A 2000-kg car is traveling at 96.6 km/hr when it is braked to a stop. What is the maximum possible thermal energy generated in the brakes?
15. How much thermal energy is absorbed by an aluminum ball 20.0 cm in diameter, initially at a temperature of 20.0 °C, if it is placed in boiling water?

14.4 Calorimetry

16. If 30.0 g of water at 5.00 °C are mixed with 50.0 g of water at 70.0 °C and 25.0 g of water at 100 °C, find the resultant temperature of the mixture.
17. If 80.0 g of lead shot at 100 °C is placed into 100 g of water at 20.0 °C in an aluminum calorimeter of 60.0-g mass, what is the final temperature?
18. A 100-g mass of an unknown material at 100 °C is placed in an aluminum calorimeter of 60.0 g that contains 150 g of water at an initial temperature of 20.0 °C. The final temperature is observed to be 21.5 °C. What is the specific heat of the substance and what substance do you think it is?

19. A 100-g mass of an unknown material at 100 °C, is placed in an aluminum calorimeter of 60.0-g mass that contains 150 g of water at an initial temperature of 15.0 °C. At equilibrium the final temperature is 19.5 °C. What is the specific heat of the material and what material is it?
20. How much water at 50.0 °C must be added to 60.0 kg of water at 10.0 °C to bring the final mixture to 20.0 °C?
†21. A 100-g aluminum calorimeter contains 200 g of water at 15.0 °C. If 100.0 g of lead at 50.0 °C and 60.0 g of copper at 60.0 °C are placed in the calorimeter, what is the final temperature in the calorimeter?
22. A 200-g piece of platinum is placed inside a furnace until it is in thermal equilibrium. The platinum is then placed in a 100-g aluminum calorimeter containing 400 g of water at 5.00 °C. If the final equilibrium temperature of the water is 10.0 °C, find the temperature of the furnace.

14.5 Change of Phase

23. How many calories are needed to change 50.0 g of ice at −10.0 °C to water at 20.0 °C?
24. If 50.0 g of ice at 0.0 °C are mixed with 50.0 g of water at 80.0 °C what is the final temperature of the mixture?
25. How much ice at 0 °C must be mixed with 50.0 g of water at 75.0 °C to give a final water temperature of 20 °C?
26. If 50.0 g of ice at 0.0 °C are mixed with 50.0 g of water at 20.0 °C, what is the final temperature of the mixture? How much ice is left in the mixture?
27. How much heat is required to convert 10.0 g of ice at −15.0 °C to steam at 105 °C?
28. In the laboratory, 31.0 g of ice at 0 °C is placed into an 85.0-g copper calorimeter cup that contains 155 g

of water at an initial temperature of 23.0 °C. After the ice melts, the final temperature of the water is found to be 6.25 °C. From this laboratory data, find the heat of fusion of water and the percentage error between the standard value and this experimental value.

29. A 100-g iron ball is heated to 100 °C and then placed in a hole in a cake of ice at 0.00 °C. How much ice will melt?

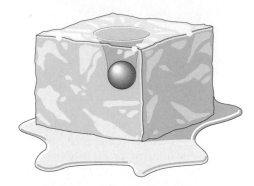

30. How much steam at 100 °C must be mixed with 300 g of water at 20.0 °C to obtain a final water temperature of 80.0 °C?

31. How much steam at 100 °C must be mixed with 1 kg of ice at 0.00 °C to produce water at 20.0 °C?

32. In the laboratory, 6.00 g of steam at 100 °C is placed into an 85.0-g copper calorimeter cup that contains 155 g of water at an initial temperature of 18.5 °C. After the steam condenses, the final temperature of the water is found to be 41.0 °C. From this laboratory data, find the heat of vaporization of water and the percentage error between the standard value and this experimental value.

†33. An electric stove is rated at 1 kW of power. If a pan containing 1.00 kg of water at 20.0 °C is placed on this stove, how long will it take to boil away all the water?

34. An electric immersion heater is rated at 0.200 kW of power. How long will it take to boil 100 cm³ of water at an initial temperature of 20.0 °C?

Additional Problems

35. Using the specific heat of copper in British engineering units determine how many Btu's are required to raise the temperature of 30 lb of copper from 70.0 °F to 300 °F?

36. A 890-N man consumes 3000 kcal of food per day. If this same energy were used to heat the same weight of water, by how much would the temperature of the water change?

37. An electric space heater is rated at 1.50 kW of power. How many kcal of thermal energy does it produce per second? How many Btu's of thermal energy per hour does it produce?

†38. A 0.055-kg mass of lead at an initial temperature of 135 °C, a 0.075-kg mass of brass at an initial temperature of 185 °C, and a 0.0445-kg of ice at an initial temperature of −5.25 °C is placed into a calorimeter containing 0.250 kg of water at an initial temperature of 23.0 °C. The aluminum calorimeter has a mass of 0.085 kg. Find the final temperature of the mixture.

39. A 100-g lead bullet is fired into a fixed block of wood at a speed of 350 m/s. If the bullet comes to rest in the block, what is the maximum change in temperature of the bullet?

†40. A 35-g lead bullet is fired into a 6.5-kg block of a ballistic pendulum that is initially at rest. The combined bullet-pendulum rises to a height of 0.125 m. Find (a) the speed of the combined bullet-pendulum after the collision, (b) the original speed of the bullet, (c) the original kinetic energy of the bullet, (d) the kinetic energy of the combined bullet-pendulum after the collision, and (e) how much of the initial mechanical energy was converted to thermal energy in the collision. If 50% of the energy lost shows up as thermal energy in the bullet, what is the change in energy of the bullet?

†41. After 50.0 g of ice at 0 °C is mixed with 200 g of water, also at 0 °C, in an insulated cup of 15.0-cm radius, a paddle wheel, 15.0 cm in radius, is placed inside the cup and set into rotational motion. What force, applied at the end of the paddle wheel, is necessary to rotate the paddle wheel at 60 rpm, for 10.0 minutes such that the final temperature of the mixture will be 15.0 °C?

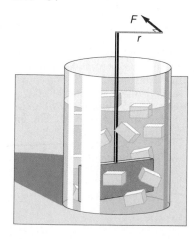

†42. A 75.0-kg patient is running a fever of 105 °F and is given an alcohol rub down to lower the body temperature. If the specific heat of the human body is approximately 0.830 kcal/(kg °C), and the heat of vaporization of alcohol is 203 kcal/kg, find (a) the amount of heat that must be removed to lower the temperature to 102 °F and (b) the volume of alcohol required.

43. How much thermal energy is required to heat the air in a house from 15.0 °C to 20.0 °C if the house is 14.0 m long, 9.00 m wide, and 3.00 m high?

44. A classroom is at an initial temperature of 20 °C. If 35 students enter the class and each liberates heat to the air at the rate of 100 W, find the final temperature of the air in the room 50 min later, assuming all the heat from the students goes into heating the air. The classroom is 10.0 m long, 9.00 m wide, and 4.00 m high.

45. How much fuel oil is needed to heat a 100-gal tank of water from 40.0 °F to 180 °F, if oil is capable of supplying 140,000 Btu of thermal energy per gallon of oil?

46. How much fuel oil is needed to heat a 570-liter tank of water from 10.0 °C to 80.0 °C if oil is capable of supplying 3.88 × 10⁷ J of thermal energy per liter of oil?

47. How much heat is necessary to melt 100 kg of aluminum initially at a temperature of 20 °C? The melting point of aluminum is 660 °C and its heat of fusion is 90 kcal/kg.

†48. If the heat of combustion of natural gas is 1000 Btu/ft³, how many ft³ are needed to heat 26 ft³ of water from 60.0 °F to 180 °F in a hot water heater if the system is 75% efficient?

49. If the heat of combustion of natural gas is 3.71 × 10⁷ J/m³, how many cubic meters are needed to heat 0.580 m³ of water from 10.0 °C to 75.0 °C in a hot water heater if the system is 63% efficient?

†50. If the heat of combustion of coal is 12,000 Btu/lb, how many pounds of coal are necessary to heat 26 ft³ of water from 60 °F to 180 °F in a hot water heater if the system is 75% efficient?

51. If the heat of combustion of coal is 2.78 × 10⁷ J/kg, how many kilograms of coal are necessary to heat 0.580 m³ of water from 10.0 °C to 75.0 °C in a hot water heater if the system is 63% efficient?

†52. The *solar constant* is the amount of energy from the sun falling on the earth per second, per unit area and is given as SC = 1350 J/(s m²). If an average roof of a house is 60.0 m², how much energy impinges on the house in an 8-hr period? Express the answer in joules, kWhr, Btu, and kcal. Assuming you could convert all of this heat at 100% efficiency, how much fuel could you save if #2 fuel oil supplies 140,000 Btu/gal; natural gas supplies 1000 Btu/ft³; electricity supplies 3415 Btu/kWhr?

53. How much thermal energy can you store in a 2000-gal tank of water if the water has been subjected to a temperature change of 70.0 °F in a solar collector?

54. How much thermal energy can you store in a 5680-liter tank of water if the water has been subjected to a temperature change of 35.0 °C in a solar collector?

55. The refrigerating capacity of some older air conditioners were rated in terms of tons. A 1-ton air conditioner extracted enough thermal energy to freeze 1 ton of water at 32 °F per day. How much thermal energy is necessary to freeze 1 ton of water? If it takes 24 hr to freeze this water, how much power is expended in Btu/hr and watts? Compare this value to the rating of modern air conditioners.

56. A 5.94-kg lead ball rolls without slipping down a rough inclined plane 1.32 m long that makes an angle of 40.0° with the horizontal. The ball has an initial velocity $v_0 = 0.25$ m/s. The ball is not perfectly spherical and some energy is lost due to friction as it rolls down the plane. The ball arrives at the bottom of the plane with a velocity $v = 3.00$ m/s, and 80.0% of the energy lost shows up as a rise in the temperature of the ball. Find (a) the height of the incline, (b) the initial potential energy of the ball, (c) the initial kinetic energy of translation, (d) the initial kinetic energy of rotation, (e) the initial total energy of the ball, (f) the final kinetic energy of translation, (g) the final kinetic energy of rotation, (h) the final total mechanical energy of the ball at the bottom of the plane, (i) the energy lost by the ball due to friction, and (j) the increase in the temperature of the ball.

†57. The energy that fuels thunderstorms and hurricanes comes from the heat of condensation released when saturated water vapor condenses to form the droplets of water that become the clouds that we see in the

sky. Consider the amount of air contained in an imaginary box 5.00 km long, 10.0 km wide, and 30.0 m high that covers the ground at the surface of the earth at a particular time. The air temperature is 20 °C and is saturated with all the water vapor it can contain at that temperature, which is 17.3×10^{-3} kg of water vapor per m³. The air in this imaginary box is now lifted into the atmosphere where it is cooled to 0 °C. Since the air is saturated, condensation occurs throughout the cooling process. The maximum water vapor the air can contain at 0 °C is 4.847×10^{-3} kg of water vapor per m³. (The heat of vaporization of water varies with temperatures from 600 kcal/kg at 0 °C to 540 kcal/kg at 100 °C. We will assume an average temperature of 10.0 °C for the cooling process.) Find (a) the volume of saturated air in the imaginary box, (b) the mass of water vapor in this volume at 20.0 °C, (c) the mass of water vapor in this volume at 0 °C, (d) the heat of vaporization of water at 10.0 °C, and (e) the thermal energy given off in the condensation process. (f) Discuss this quantity of energy in terms of the energy that powers thunderstorms and hurricanes.

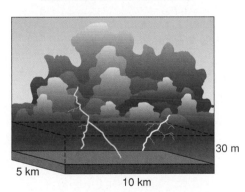

30 m
5 km
10 km

Interactive Tutorials

🖳 58. Find the total amount of thermal energy in joules necessary to convert ice of mass $m_i = 2.00$ kg at an initial temperature $t_{ii} = -20.0$ °C to water at a final water temperature of $t_{fw} = 88.3$ °C. The specific heat of ice is $c_i = 2093$ J/kg °C, water is $c_w = 4186$ J/kg °C, and the latent heat of fusion of water is $L_f = 3.34 \times 10^5$ J/kg.

🖳 59. If a sample of lead shot of mass $m_s = 0.080$ kg and initial temperature $t_{is} = 100$ °C is placed into a mass of water $m_w = 0.100$ kg in an aluminum calorimeter of mass $m_c =$

0.060 kg at an initial temperature $t_{iw} = 20.0$ °C, what is the final equilibrium temperature of the water, calorimeter, and lead shot? The specific heats are water $c_w = 4186$ J/kg °C, calorimeter $c_c = 900$ J/kg °C, and lead sample $c_s = 129.8$ J/kg °C.

🖳 60. Temperature Conversion Calculator. The Temperature Conversion Calculator will permit you to convert temperatures in one unit to a temperature in another unit.

🖳 61. Specific heat. A specimen of lead, $m_s = 0.250$ kg, is placed into an oven where it acquires an initial temperature $t_{is} = 200$ °C. It is then removed and placed into a calorimeter of mass $m_c = 0.060$ kg and specific heat $c_c = 900$ J/kg °C that contains water, $m_w = 0.200$ kg, at an initial temperature $t_{iw} = 10.0$ °C. The specific heat of water is $c_w = 4186$ J/kg °C. The final equilibrium temperature of the water in the calorimeter is observed to be $t_{fw} = 16.7$ °C. Find the specific heat c_s of this sample.

🖳 62. Find the total amount of thermal energy in joules necessary to convert ice of mass $m_i = 12.5$ kg at an initial temperature $t_{ii} = -25.0$ °C to superheated steam at a temperature $t_{ss} = 125$ °C. The specific heat of ice is $c_i = 2093$ J/kg °C, water is $c_w = 4186$ J/kg °C, and steam is $c_s = 2013$ J/kg °C. The latent heat of fusion of water is $L_f = 3.34 \times 10^5$ J/kg, and the latent heat of vaporization is $L_v = 2.26 \times 10^6$ J/kg.

🖳 63. A mixture. How much ice at an initial temperature of $t_{ii} = -15.0$ °C must be added to a mixture of three specimens contained in a calorimeter in order to make the final equilibrium temperature of the water $t_{fw} = 12.5$ °C? The three specimens and their characteristics are sample 1: zinc; $m_{s1} = 0.350$ kg, $c_{s1} = 389$ J/kg °C, initial temperature $t_{is1} = 150$ °C; sample 2: copper; $m_{s2} = 0.180$ kg, $c_{s2} = 385$ J/kg °C, initial temperature $t_{is2} = 100$ °C; and sample 3: tin; $m_{s3} = 0.350$ kg, $c_{s3} = 226$ J/kg °C, initial temperature $t_{is3} = 180$ °C. The calorimeter has a mass $m_c = 0.060$ kg and specific heat $c_c = 900$ J/kg °C and contains water, $m_w = 0.200$ kg, at an initial temperature $t_{iw} = 19.5$ °C. The specific heat of water is $c_w = 4186$ J/kg °C.

Thermal expansion of air. By heating the air with a flame it expands and inflates the balloon.

So many of the properties of matter, especially when in the gaseous form, can be deduced from the hypothesis that their minute parts are in rapid motion, the velocity increasing with the temperature, that the precise nature of this motion becomes a subject of rational curiosity... The relations between pressure, temperature and density in a perfect gas can be explained by supposing the particles to move with uniform velocity in straight lines, striking against the sides of the containing vessel and thus producing pressure.

James Clerk Maxwell

15 Thermal Expansion and the Gas Laws

15.1 Linear Expansion of Solids

It is a well-known fact that most materials expand when heated. This expansion is called **thermal expansion.** (Recall that the phenomenon of thermal expansion was used in chapter 14 to devise the thermometer.) If a long thin rod of length L_0, at an initial temperature t_i, is heated to a final temperature t_f, then the rod expands by a small length ΔL, as shown in figure 15.1.

It is found by experiment that the change in length ΔL depends on the temperature change, $\Delta t = t_f - t_i$; the initial length of the rod L_0; and a constant that is characteristic of the material being heated. The experimentally observed linearity between ΔL and $L_0\Delta t$ can be represented by the equation

$$\Delta L = \alpha L_0 \Delta t \qquad (15.1)$$

We call the constant α the *coefficient of linear expansion;* table 15.1 gives this value for various materials. The change in length is rather small, but it is, nonetheless, very significant.

Example 15.1

Expansion of a railroad track. A steel railroad track was 100 ft long when it was initially laid at a temperature of 20 °F. What is the change in length of the track when the temperature rises to 95 °F?

Solution

The coefficient of linear expansion for steel, found from table 15.1, is $\alpha_{steel} = 0.67 \times 10^{-5}/°F$. The change in length becomes

$$\Delta L = \alpha L_0 \Delta t$$
$$= (0.67 \times 10^{-5}/°F)(100 \text{ ft})(95 °F - 20 °F)$$
$$= (0.05 \text{ ft})\left(\frac{12 \text{ in.}}{\text{ft}}\right) = 0.60 \text{ in.}$$

Even though the change in length is relatively small, 0.60 inches in a distance of 100 ft, it is easily measurable. Associated with this small change in length is a very

Figure 15.1
Linear expansion.

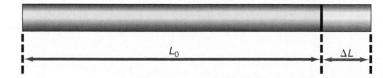

L_0 ΔL

Table 15.1

Coefficients of Thermal Expansion

Material	α Coefficient of Linear Expansion		β Coefficient of Volume Expansion	
	$\times 10^{-5}/°C$	$\times 10^{-5}/°F$	$\times 10^{-4}/°C$	$\times 10^{-4}/°F$
Aluminum	2.4	1.3		
Brass	1.8	1.0		
Copper	1.7	0.94		
Iron	1.2	0.67		
Lead	3.0	1.7		
Steel	1.2	0.67		
Zinc	2.6			
Glass (ordinary)	0.9			
Glass (Pyrex)	0.32			
Ethyl alcohol			11.0	6.1
Water			2.1	1.2
Mercury			1.8	1.0
Glass (Pyrex)			0.096	0.05
All noncondensing gases at constant pressure and 0 °C.			36.6	

large force. We can determine the force associated with this expansion by computing the force that is necessary to compress the rail back to its former length. Recall from chapter 10 that the amount that a body is stretched or compressed is given by Hooke's law as

$$\frac{F}{A} = Y\frac{\Delta L}{L_0} \tag{10.6}$$

We can solve this equation for the force that is associated with a compression. Taking the compression of the rail as 0.05 ft, Young's modulus Y for steel as 2.9×10^7 lb/in.^{2}, and assuming that the cross-sectional area of the rail is 20 in.^{2}, the force necessary to compress the rail is

$$F = AY\frac{\Delta L}{L_0}$$

$$= (20 \text{ in.}^2)\left(2.9 \times 10^7 \frac{\text{lb}}{\text{in.}^2}\right)\left(\frac{0.05 \text{ ft}}{100 \text{ ft}}\right)$$

$$= 290,000 \text{ lb}$$

This force of 290,000 lb that is necessary to compress the rail by 0.05 ft, is also the force that is necessary to prevent the rail from expanding. It is obviously an extremely large force. It is this large force associated with the thermal expansion that makes thermal expansion so important. It is no wonder that we see and hear of cases where rails and roads have buckled during periods of very high temperatures.

Expansion joints allow bridges to expand and contract without damage.

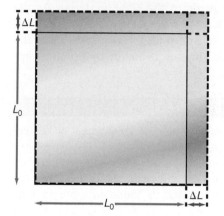

Figure 15.2

Expansion in area.

The expansion of the solid can be explained by looking at the molecular structure of the solid. The molecules of the substance are in a lattice structure. Any one molecule is in equilibrium with its neighbors, but vibrates about that equilibrium position. As the temperature of the solid is increased, the vibration of the molecule increases. However, the vibration is not symmetrical about the original equilibrium position. As the temperature increases the equilibrium position is displaced from the original equilibrium position. Hence, the mean displacement of the molecule from the original equilibrium position also increases, thereby spacing all the molecules farther apart than they were at the lower temperature. The fact that all the molecules are farther apart manifests itself as an increase in length of the material. Hence, linear expansion can be explained as a molecular phenomenon. The large force associated with the expansion comes from the large molecular forces between the molecules.

15.2 Area Expansion of Solids

For the long thin rod of section 15.1, only the length change was significant and that was all that we considered. But solids expand in all directions. If a square of thin material of length L_0 and width L_0, at an initial temperature of t_i, is heated to a new temperature t_f, the square of material expands, as shown in figure 15.2. The original area of the square is given by

$$A_0 = L_0^2$$

But each side expands by ΔL, forming a new square with sides $(L_0 + \Delta L)$. Thus, the final area becomes

$$A = (L_0 + \Delta L)^2$$
$$= L_0^2 + 2L_0\Delta L + (\Delta L)^2$$

The change in length ΔL is quite small to begin with, and its square $(\Delta L)^2$ is even smaller, and can be neglected in comparison to the magnitudes of the other terms. That is, we will set the quantity $(\Delta L)^2$ equal to zero in our analysis. Using this assumption, the final area becomes

$$A = L_0^2 + 2L_0\Delta L$$

The change in area, caused by the thermal expansion, is

$$\Delta A = \text{Final area} - \text{Original area}$$
$$= A - A_0$$
$$= L_0^2 + 2L_0\Delta L - L_0^2$$

Therefore

$$\Delta A = 2L_0\Delta L \tag{15.2}$$

However, we have already seen that

$$\Delta L = \alpha L_0\Delta t \tag{15.1}$$

Substituting equation 15.1 into 15.2 gives

$$\Delta A = 2L_0\alpha L_0\Delta t$$

and

$$\Delta A = 2\alpha L_0^2\Delta t$$

However, $L_0^2 = A_0$, the original area. Therefore

$$\Delta A = 2\alpha A_0\Delta t \tag{15.3}$$

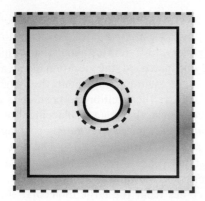

Figure 15.3

The empty hole expands at the same rate as if there were material in the hole.

Equation 15.3 gives us the area expansion of a material of original area A_0 when subjected to a temperature change Δt. Note that the coefficient of area expansion is twice the coefficient of linear expansion. Although we have derived this result for a square it is perfectly general and applies to any area. For example, if the material was circular in shape, the original area A_0 would be computed from the area of a circle of radius r_0 as

$$A_0 = \pi r_0^2$$

We would then find the change in area from equation 15.3.

All parts of the material expand at the same rate. For example, if there was a circular hole in the material, the empty hole would expand at the same rate as if material were actually present in the hole. We can see this in figure 15.3. The solid line represents the original material, whereas the dotted lines represent the expanded material. Many students feel that the material should expand into the hole, thereby causing the hole to shrink. The best way to show that the hole does indeed expand is to fill the hole with a plug made of the same material. As the material expands, so does the plug. At the end of the expansion remove the plug, leaving the hole. Since the plug expanded, the hole must also have grown. Thus, the hole expands as though it contained material. This result has many practical applications.

Example 15.2

Fitting a small wheel on a large shaft. We want to place a steel wheel on a steel shaft with a good tight fit. The shaft has a diameter of 10.010 cm. The wheel has a hole in the middle, with a diameter of 10.000 cm, and is at a temperature of 20 °C. If the wheel is heated to a temperature of 132 °C, will the wheel fit over the shaft? The coefficient of linear expansion for steel is found in table 15.1 as $\alpha = 1.20 \times 10^{-5}/°C$.

Solution

With the present dimensions the wheel can not fit over the shaft. If we place the wheel in an oven at 132 °C, the wheel expands. The diameter of the hole increases by

$$\Delta L_H = \alpha L_0 \Delta t$$
$$= (1.20 \times 10^{-5}/°C)(10.000 \text{ cm})(132 °C - 20 °C)$$
$$= 1.34 \times 10^{-2} \text{ cm}$$

The new hole in the wheel has the diameter

$$L = L_0 + \Delta L = 10.000 \text{ cm} + 0.013 \text{ cm}$$
$$= 10.013 \text{ cm}$$

Because the diameter of the hole in the wheel is now greater than the diameter of the shaft, the wheel now fits over the shaft. When the combined wheel and shaft is allowed to cool back to the original temperature of 20 °C, the hole in the wheel tries to contract to its original size, but is not able to do so, because of the presence of the shaft. Therefore, enormous forces are exerted on the shaft by the wheel, holding the wheel permanently on the shaft.

This problem could also have been treated by using the change in area by equation 15.3, but, as just seen, it is easier to find the change in the diameter.

15.3 Volume Expansion of Solids and Liquids

All materials have three dimensions, length, width, and height. When a body is heated, all three dimensions should expand and hence its volume should increase. Let us consider a cube of length L_0 on each side, at an initial temperature t_i. Its initial volume is

$$V_0 = L_0^3$$

If the material is heated to a new temperature t_f, then each side L_0 of the cube undergoes an expansion ΔL. The final volume of the cube is

$$V = (L_0 + \Delta L)^3$$
$$= L_0^3 + 3L_0^2\Delta L + 3L_0(\Delta L)^2 + (\Delta L)^3$$

Because ΔL is itself a very small quantity, the terms in $(\Delta L)^2$ and $(\Delta L)^3$ can be neglected. Therefore,

$$V = L_0^3 + 3L_0^2\Delta L$$

The change in volume due to the expansion becomes

$$\Delta V = V - V_0$$
$$= L_0^3 + 3L_0^2\Delta L - L_0^3$$
$$\Delta V = 3L_0^2\Delta L \tag{15.4}$$

However, the linear expansion ΔL was given by

$$\Delta L = \alpha L_0 \Delta t \tag{15.1}$$

Substituting this into equation 15.4 gives

$$\Delta V = 3L_0^2 \alpha L_0 \Delta t$$
$$= 3\alpha L_0^3 \Delta t$$

Since L_0^3 is equal to V_0, this becomes

$$\Delta V = 3\alpha V_0 \Delta t \tag{15.5}$$

We now define a new coefficient, called the *coefficient of volume expansion* β, for solids as

$$\beta = 3\alpha \tag{15.6}$$

Therefore, the change in volume of a substance when subjected to a change in temperature is

$$\Delta V = \beta V_0 \Delta t \tag{15.7}$$

Although we derived equation 15.7 for a solid cube, it is perfectly general and applies to any volume of a solid and even for any volume of a liquid. However, since α has no meaning for a liquid, we must determine β experimentally for the liquid. Just as a hole in a surface area expands with the surface area, a hole in a volume also expands with the volume of the solid. Hence, when a hollow glass tube expands, the empty volume inside the tube expands as though there were solid glass present.

Example 15.3

How much mercury overflows? An open glass tube is filled to the top with 25.0 cm³ of mercury at an initial temperature of 20.0 °C. If the mercury and the tube are heated to 100 °C, how much mercury will overflow from the tube?

The change in volume of the mercury, found from equation 15.7 with $\beta_{Hg} = 1.80 \times 10^{-4}/°C$ found from table 15.1, is

$$\Delta V_{Hg} = \beta_{Hg} V_0 \Delta t$$
$$= (1.80 \times 10^{-4}/°C)(25.0 \text{ cm}^3)(100 °C - 20 °C)$$
$$= 0.360 \text{ cm}^3$$

If the glass tube did not expand, this would be the amount of mercury that overflows. But the glass tube does expand and is therefore capable of holding a larger volume. The increased volume of the glass tube is found from equation 15.7 but this time with $\beta_g = 0.27 \times 10^{-4}/°C$

$$\Delta V_g = \beta_g V_0 \Delta t$$
$$= (0.27 \times 10^{-4}/°C)(25.0 \text{ cm}^3)(100 °C - 20.0 °C)$$
$$= 0.054 \text{ cm}^3$$

That is, the tube is now capable of holding an additional 0.054 cm³ of mercury. The amount of mercury that overflows is equal to the difference in the two volume expansions. That is,

$$\text{Overflow} = \Delta V_{Hg} - \Delta V_g$$
$$= 0.360 \text{ cm}^3 - 0.054 \text{ cm}^3$$
$$= 0.306 \text{ cm}^3$$

15.4 Volume Expansion of Gases: Charles' Law

Consider a gas placed in a tank, as shown in figure 15.4. The weight of the piston exerts a constant pressure on the gas. When the tank is heated, the pressure of the gas first increases. But the increased pressure in the tank pushes against the freely moving piston, and the piston moves until the pressure inside the tank is the same as the pressure exerted by the weight of the piston. Therefore the pressure in the tank remains a constant throughout the entire heating process. The volume of the gas increases during the heating process, as we can see by the new volume occupied by the gas in the top cylinder. In fact, we find the increased volume by multiplying the area of the cylinder by the distance the piston moves in the cylinder. If the volume of the gas is plotted against the temperature of the gas, in Celsius degrees, we obtain the straight line graph in figure 15.5. If the equation for this straight line is written in the point-slope form[1]

$$y - y_1 = m(x - x_1)$$

we get

$$V - V_0 = m(t - t_0)$$

---

1. The point-slope form of a straight line is obtained by the definition of the slope of a straight line, namely

$$m = \frac{\Delta y}{\Delta x}$$

or

$$\Delta y = m\Delta x$$

Using the meaning of Δy and Δx, we get

$$y - y_1 = m(x - x_1)$$

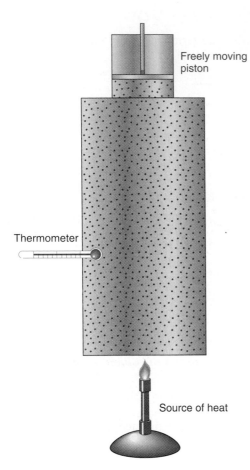

Freely moving piston

Thermometer

Source of heat

Figure 15.4

Volume expansion of a gas.

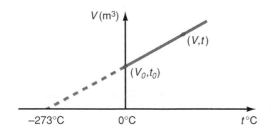

V (m³)

(V,t)

(V_0,t_0)

$-273°C$　　$0°C$　　　　$t°C$

Figure 15.5

Plot of V versus t for a gas at constant pressure.

where V is the volume of the gas at the temperature t, V_0 is the volume of the gas at $t_0 = 0$ °C, and m is the slope of the line. We can also write this equation in the form

$$\Delta V = m\Delta t \tag{15.8}$$

Note that equation 15.8, which shows the change in volume of a gas, looks like the volume expansion formula 15.7, for the change in volume of solids and liquids, that is,

$$\Delta V = \beta V_0 \Delta t \tag{15.7}$$

Let us assume, therefore, that the form of the equation for volume expansion is the same for gases as it is for solids and liquids. If we use this assumption, then

$$\beta V_0 = m$$

Hence the coefficient of volume expansion for the gas is found experimentally as

$$\beta = \frac{m}{V_0}$$

where m is the measured slope of the line. If we repeat this experiment many times for many different gases we find that

$$\beta = \frac{1}{273 \text{ °C}} = 3.66 \times 10^{-3}/\text{°C}$$

for all noncondensing gases at constant pressure. This result was first found by the French physicist, J. Charles (1746–1823). This is a rather interesting result, since the value of β is different for different solids and liquids, and yet it is a constant for all gases.

Equation 15.7 can now be rewritten as

$$V - V_0 = \beta V_0(t - t_0)$$

Because $t_0 = 0$ °C, we can simplify this to

$$V - V_0 = \beta V_0 t$$

and

$$V = V_0 + \beta V_0 t$$

or

$$V = V_0(1 + \beta t) \tag{15.9}$$

Note that if the temperature $t = -273$ °C, then

$$V = V_0\left(1 + -\frac{273}{273}\right) = V_0(1 - 1) = 0$$

That is, the plot of V versus t intersects the t-axis at -273 °C, as shown in figure 15.5. Also observe that there is a linear relation between the volume of a gas and its temperature in degrees Celsius. Since $\beta = 1/273$ °C, equation 15.9 can be simplified further into

$$V = V_0\left(1 + \frac{t}{273 \text{ °C}}\right) = V_0\left(\frac{273 \text{ °C} + t}{273 \text{ °C}}\right)$$

It was the form of this equation that led to the definition of the Kelvin or absolute temperature scale in the form

$$T \text{ K} = t \text{ °C} + 273 \tag{15.10}$$

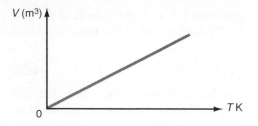

Figure 15.6

The volume V of a gas is directly proportional to its absolute temperature T.

With this definition of temperature, the volume of the gas is directly proportional to the absolute temperature of the gas, that is,

$$V = \left(\frac{V_0}{273}\right) T \qquad (15.11)$$

Changing the temperature scale is equivalent to moving the vertical coordinate of the graph, the volume, from the 0 °C mark in figure 15.5, to the −273 °C mark, and this is shown in figure 15.6. Thus, *the volume of a gas at constant pressure is directly proportional to the absolute temperature of the gas. This result is known as* **Charles' law.**

In general, if the state of the gas is considered at two different temperatures, we have

$$V_1 = \left(\frac{V_0}{273}\right) T_1$$

and

$$V_2 = \left(\frac{V_0}{273}\right) T_2$$

Hence,

$$\frac{V_1}{T_1} = \frac{V_0}{273} = \frac{V_2}{T_2}$$

Therefore,

$$\frac{V_1}{T_1} = \frac{V_2}{T_2} \qquad p = \text{constant} \qquad (15.12)$$

which is another form of Charles' law.

Figures 15.5 and 15.6 are slightly misleading in that they show the variation of the volume V with the temperature T of a gas down to −273 °C or 0 K. However, the gas will have condensed to a liquid and eventually to a solid way before this point is reached. A plot of V versus T for all real gases is shown in figure 15.7. Note that when each line is extrapolated, they all intersect at −273 °C or 0 K. Although they all have different slopes m, the coefficient of volume expansion ($\beta = m/V_0$) is the same for all the gases.

Figure 15.7

Plot of volume versus temperature for real gases.

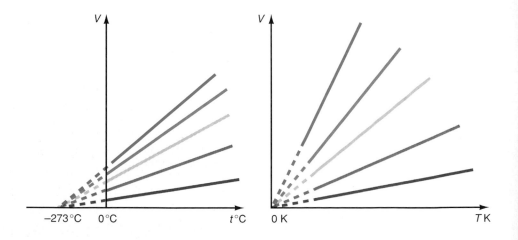

Thermodynamics

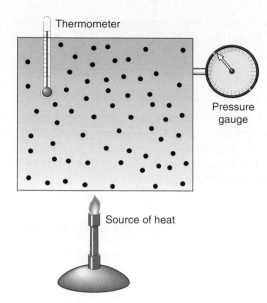

Thermometer

Pressure gauge

Source of heat

Figure 15.8

Changing the pressure of a gas.

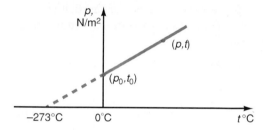

Figure 15.9

A plot of pressure versus temperature for a gas.

15.5 Gay-Lussac's Law

Consider a gas contained in a tank, as shown in figure 15.8. The tank is made of steel and there is a negligible change in the volume of the tank, and hence the gas, as it is heated. A pressure gauge attached directly to the tank, is calibrated to read the absolute pressure of the gas in the tank. A thermometer reads the temperature of the gas in degrees Celsius. The tank is heated, thereby increasing the temperature and the pressure of the gas, which are then recorded. If we plot the pressure of the gas versus the temperature, we obtain the graph of figure 15.9. The equation of the resulting straight line is

$$p - p_0 = m'(t - t_0)$$

where p is the pressure of the gas at the temperature t, p_0 is the pressure at the temperature t_0, and m' is the slope of the line. The prime is placed on the slope to distinguish it from the slope determined in section 15.4. Because $t_0 = 0\ °C$, this simplifies to

$$p - p_0 = m't$$

or

$$p = m't + p_0 \qquad (15.13)$$

It is found experimentally that the slope is

$$m' = p_0 \beta$$

where p_0 is the absolute pressure of the gas and β is the coefficient of volume expansion for a gas. Therefore equation 15.13 becomes

$$p = p_0 \beta t + p_0$$

and

$$p = p_0(\beta t + 1) \qquad (15.14)$$

Thus, the pressure of the gas is a linear function of the temperature, as in the case of Charles' law. Since $\beta = 1/273\ °C$ this can be written as

$$p = p_0 \left(\frac{t}{273\ °C} + 1 \right) = p_0 \left(\frac{t + 273\ °C}{273\ °C} \right) \qquad (15.15)$$

But the absolute or Kelvin scale has already been defined as

$$T\ K = t\ °C + 273$$

Therefore, equation 15.15 becomes

$$p = \left(\frac{p_0}{273} \right) T \qquad (15.16)$$

which shows that *the absolute pressure of a gas at constant volume is directly proportional to the absolute temperature of the gas, a result known as **Gay-Lussac's law**,* in honor of the French chemist Joseph Gay-Lussac (1778–1850). For a gas in different states at two different temperatures, we have

$$p_1 = \left(\frac{p_0}{273} \right) T_1 \quad \text{and} \quad p_2 = \left(\frac{p_0}{273} \right) T_2$$

or

$$\frac{p_1}{T_1} = \frac{p_2}{T_2} \qquad\qquad V = \text{constant} \qquad (15.17)$$

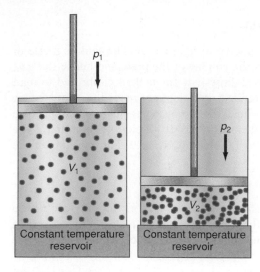

Figure 15.10

The change in pressure and volume of a gas at constant temperature.

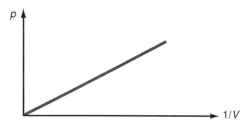

Figure 15.11

Plot of the pressure p versus the reciprocal of the volume $1/V$ for a gas.

Equation 15.17 is another form of Gay-Lussac's law. (Sometimes this law is also called Charles' law, since Charles and Gay-Lussac developed these laws independently of each other.)

15.6 Boyle's Law

Consider a gas contained in a cylinder at a constant temperature, as shown in figure 15.10. By pushing the piston down into the cylinder, we increase the pressure of the gas and decrease the volume of the gas. If the pressure is increased in small increments, the gas remains in thermal equilibrium with the temperature reservoir, and the temperature of the gas remains a constant. We measure the volume of the gas for each increase in pressure and then plot the pressure of the gas as a function of the reciprocal of the volume of the gas. The result is shown in figure 15.11. Notice that the pressure is inversely proportional to the volume of the gas at constant temperature. We can write this as

$$p \propto \frac{1}{V}$$

or

$$pV = \text{constant} \qquad (15.18)$$

That is, *the product of the pressure and volume of a gas at constant temperature is equal to a constant, a result known as **Boyle's law***, in honor of the British physicist and chemist Robert Boyle (1627–1691). For a gas in two different equilibrium states at the same temperature, we write this as

$$p_1V_1 = \text{constant} \quad \text{and} \quad p_2V_2 = \text{constant}$$

Therefore,

$$p_1V_1 = p_2V_2 \qquad\qquad T = \text{constant} \qquad (15.19)$$

another form of Boyle's law.

15.7 The Ideal Gas Law

The three gas laws,

$$\frac{V_1}{T_1} = \frac{V_2}{T_2} \qquad\qquad p = \text{constant} \qquad (15.12)$$

$$\frac{p_1}{T_1} = \frac{p_2}{T_2} \qquad\qquad V = \text{constant} \qquad (15.17)$$

$$p_1V_1 = p_2V_2 \qquad\qquad T = \text{constant} \qquad (15.19)$$

can be combined into one equation, namely,

$$\frac{p_1V_1}{T_1} = \frac{p_2V_2}{T_2} \qquad (15.20)$$

Equation 15.20 is a special case of a relation known as the **ideal gas law.** Hence, we see that the three previous laws, which were developed experimentally, are special cases of this ideal gas law, when either the pressure, volume, or temperature is held constant. The ideal gas law is a more general equation in that none of the variables must be held constant. Equation 15.20 expresses the relation between the pressure, volume, and temperature of the gas at one time, with the pressure, volume, and temperature at any other time. For this equality to hold for any time, it is necessary that

$$\frac{pV}{T} = \text{constant} \qquad (15.21)$$

The cork is blown off this champagne bottle by the large build-up of CO_2 gas pressure.

This constant must depend on the quantity or mass of the gas. A convenient unit to describe the amount of the gas is the mole. *One **mole** of any gas is that amount of the gas that has a mass in grams equal to the atomic or molecular mass (M) of the gas.* The terms atomic mass and molecular mass are often erroneously called atomic weight and molecular weight in chemistry.

As an example of the use of the mole, consider the gas oxygen. One molecule of oxygen gas consists of two atoms of oxygen, and is denoted by O_2. The atomic mass of oxygen is found in the Periodic Table of the Elements in appendix E, as 16.00. The molecular mass of one mole of oxygen gas is therefore

$$M_{O_2} = 2(16) = 32 \text{ g/mole}$$

Thus, one mole of oxygen has a mass of 32 g. The mole is a convenient quantity to express the mass of a gas because *one mole of any gas at a temperature of 0 °C and a pressure of 1 atmosphere, has a volume of 22.4 liters. Also Avogadro's law states that every mole of a gas contains the same number of molecules. This number is called **Avogadro's number** N_A and is equal to 6.022×10^{23} molecules/mole.*

The mass of any gas will now be represented in terms of the number of moles, n. We can write the constant in equation 15.21 as n times a new constant, which shall be called R, that is,

$$\frac{pV}{T} = nR \tag{15.22}$$

To determine this constant R let us evaluate it for 1 mole of gas at a pressure of 1 atm and a temperature of 0 °C, or 273 K, and a volume of 22.4 L. That is,

$$R = \frac{pV}{nT} = \frac{(1 \text{ atm})(22.4 \text{ L})}{(1 \text{ mole})(273 \text{ K})}$$

$$R = 0.08205 \frac{\text{atm L}}{\text{mole K}}$$

Converted to SI units, this constant is

$$R = \left(0.08205 \frac{\text{L atm}}{\text{mole K}} \right) \left(1.013 \times 10^5 \frac{\text{N/m}^2}{\text{atm}} \right) \left(\frac{10^{-3} \text{ m}^3}{1 \text{ L}} \right)$$

$$R = 8.314 \frac{\text{J}}{\text{mole K}}$$

Another convenient unit for R is obtained by using the conversion factor, 4.186 J = 1 cal. Therefore,

$$R = 1.987 \frac{\text{cal}}{\text{mole K}}$$

We call the constant R the *universal gas constant,* and it is the same for all gases. We can now write equation 15.22 as

$$pV = nRT \tag{15.23}$$

Equation 15.23 is called the **ideal gas equation.** An ideal gas is one that is described by the ideal gas equation. Real gases can be described by the ideal gas equation as long as their density is low and the temperature is well above the condensation point (boiling point) of the gas. *Remember that the temperature T must always be expressed in Kelvin units.* Let us now look at some examples of the use of the ideal gas equation.

Example 15.4

Find the temperature of the gas. The pressure of an ideal gas is kept constant while 3.00 m³ of the gas, at an initial temperature of 50.0 °C, is expanded to 6.00 m³. What is the final temperature of the gas?

Solution

The temperature must be expressed in Kelvin units. Hence the initial temperature becomes

$$T = t \ °C + 273 = 50.0 + 273 = 323 \text{ K}$$

We find the final temperature of the gas by using the ideal gas equation in the form of equation 15.20, namely,

$$\frac{p_1 V_1}{T_1} = \frac{p_2 V_2}{T_2}$$

However, since the pressure is kept constant, $p_1 = p_2$, and cancels out of the equation. Therefore,

$$\frac{V_1}{T_1} = \frac{V_2}{T_2}$$

and the final temperature of the gas becomes

$$T_2 = \frac{V_2}{V_1} T_1$$

$$= \left(\frac{6.00 \text{ m}^3}{3.00 \text{ m}^3}\right)(323 \text{ K})$$

$$= 646 \text{ K}$$

Example 15.5

Find the volume of the gas. A balloon is filled with helium at a pressure of 2 atm, a temperature of 35.0 °C, and occupies a volume of 3.00 m³. The balloon rises in the atmosphere. When it reaches a height where the pressure is 0.500 atm, and the temperature is −20.0 °C, what is its volume?

Solution

First we convert the two temperatures to absolute temperature units as

$$T_1 = 35.0 \ °C + 273 = 308 \text{ K}$$

and

$$T_2 = -20.0 \ °C + 273 = 253 \text{ K}$$

We use the ideal gas law in the form

$$\frac{p_1 V_1}{T_1} = \frac{p_2 V_2}{T_2}$$

Solving for V_2 gives, for the final volume,

$$V_2 = \frac{p_1 T_2}{p_2 T_1} V_1$$

$$= \left[\frac{(2.00 \text{ atm})(253 \text{ K})}{(0.500 \text{ atm})(308 \text{ K})} \right](3.00 \text{ m}^3)$$

$$= 9.86 \text{ m}^3$$

Example 15.6

Find the pressure of the gas. What is the pressure produced by 2.00 moles of a gas at 35.0 °C contained in a volume of 5.00 L?

Solution

We convert the temperature of 35.0 °C to Kelvin by

$$T = 35.0 \text{ °C} + 273 = 308 \text{ K}$$

We use the ideal gas law in the form

$$pV = nRT \tag{15.23}$$

Solving for p,

$$p = \frac{nRT}{V} = \frac{(2.00 \text{ moles})(0.0821 \text{ atm L/mole K})(308 \text{ K})}{5.00 \text{ L}}$$

$$= 10.1 \text{ atm}$$

Example 15.7

Find the number of molecules in the gas. Compute the number of molecules in a gas contained in a volume of 10.0 cm³ at a pressure of 1.013×10^5 N/m², and a temperature of 30 K.

Solution

The number of molecules in a mole of a gas is given by Avogadro's number N_A, and hence the total number of molecules N in the gas is given by

$$N = nN_A \tag{15.24}$$

Therefore we first need to determine the number of moles of gas that are present. From the ideal gas law,

$$pV = nRT$$

$$n = \frac{pV}{RT} = \frac{(1.013 \times 10^5 \text{ N/m}^2)(10.0 \text{ cm}^3)}{(8.314 \text{ J/mole K})(30 \text{ K})} \left(\frac{1.00 \text{ m}^3}{10^6 \text{ cm}^3} \right)$$

$$= 4.06 \times 10^{-4} \text{ moles}$$

The number of molecules is now found as

$$N = nN_A = (4.06 \times 10^{-4} \text{ mole}) \left(6.022 \times 10^{23} \frac{\text{molecules}}{\text{mole}} \right)$$

$$= 2.44 \times 10^{20} \text{ molecules}$$

Example 15.8

Find the gauge pressure of the gas. An automobile tire has a volume of 5000 in.^{3} and contains air at a gauge pressure of 30.0 lb/in.^{2} when the temperature is 0.00 °C. What is the gauge pressure when the temperature rises to 30.0 °C?

Solution

When a gauge is used to measure pressure, it reads zero when it is under normal atmospheric pressure of 1.013×10^5 N/m^2 or 14.7 lb/in.^{2}. The pressure used in the ideal gas equation must be the absolute pressure, that is, the total pressure, which is the pressure read by the gauge plus atmospheric pressure. Therefore,

$$p_{absolute} = p_{gauge} + p_{atm} \qquad (15.25)$$

Thus, the initial pressure of the gas is

$$p_1 = p_{gauge} + p_{atm} = 30.0 \text{ lb/in.}^2 + 14.7 \text{ lb/in.}^2$$
$$= 44.7 \text{ lb/in.}^2$$

The initial volume of the tire is $V_1 = 5000$ in.^{3} and the change in that volume is small enough to be neglected, so $V_2 = 5000$ in.^{3}. The initial temperature is

$$T_1 = 0.00 \text{ °C} + 273 = 273 \text{ K}$$

and the final temperature is

$$T_2 = 30.0 \text{ °C} + 273 = 303 \text{ K}$$

Solving the ideal gas equation for the final pressure, we get

$$p_2 = \frac{V_1 T_2}{V_2 T_1} p_1$$
$$= \left[\frac{(5000 \text{ in.}^3)(303 \text{ K})}{(5000 \text{ in.}^3)(273 \text{ K})} \right] (44.7 \text{ lb/in.}^2)$$
$$= 49.6 \text{ lb/in.}^2 \text{ absolute pressure}$$

Expressing this pressure in terms of gauge pressure we get

$$p_{2gauge} = p_{2absolute} - p_{atm}$$
$$= 49.6 \text{ lb/in.}^2 - 14.7 \text{ lb/in.}^2$$
$$= 34.9 \text{ lb/in.}^2$$

15.8 The Kinetic Theory of Gases

Up to now the description of a gas has been on the macroscopic level, a large-scale level, where the characteristics of a gas, such as its pressure, volume, and temperature, are measured without regard to the internal structure of the gas itself. In reality, a gas is composed of a large number of molecules in random motion. The large-scale characteristics of gases should be explainable in terms of the motion of these molecules. *The analysis of a gas at this microscopic level (the molecular level) is called the **kinetic theory of gases.***

In the analysis of a gas at the microscopic level we make the following assumptions:

1. A gas is composed of a very large number of molecules that are in random motion.
2. The volume of the individual molecules is very small compared to the total volume of the gas.

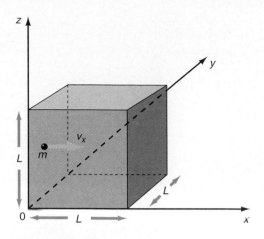

Figure 15.12

The kinetic theory of a gas.

3. The collisions of the molecules with the walls and other molecules are elastic and hence there is no energy lost during a collision.
4. The forces between molecules are negligible except during a collision. Hence, there is no potential energy associated with any molecule.
5. Finally, we assume that the molecules obey Newton's laws of motion.

Let us consider one of the very many molecules contained in the box shown in figure 15.12. For simplicity we assume that the box is a cube of length L. The gas molecule has a mass m and is moving at a velocity v. The x-component of its velocity is v_x. For the moment we only consider the motion in the x-direction. The pressure that the gas exerts on the walls of the box is caused by the collision of the gas molecule with the walls. The pressure is defined as the force acting per unit area, that is,

$$p = \frac{F}{A} \tag{15.26}$$

where A is the area of the wall where the collision occurs, and is simply

$$A = L^2$$

and F is the force exerted on the wall as the molecule collides with the wall and can be found by Newton's second law in the form

$$F = \frac{\Delta P}{\Delta t} \tag{15.27}$$

So as not to confuse the symbols for pressure and momentum, we will use the lower case p for pressure, and we will use the upper case P for momentum. Because momentum is conserved in a collision, the change in momentum of the molecule ΔP, is the difference between the momentum after the collision P_{AC} and the momentum before the collision P_{BC}. Also, since the collision is elastic the velocity of the molecule after the collision is $-v_x$. Therefore, the change in momentum of the molecule is

$$\Delta P = P_{AC} - P_{BC} = -mv_x - mv_x$$

$$= -2mv_x \quad \text{change in momentum of the molecule}$$

But the change in the momentum imparted to the wall is the negative of this, or

$$\Delta P = 2mv_x \quad \text{momentum imparted to wall}$$

Therefore, using Newton's second law, the force imparted to the wall becomes

$$F = \frac{\Delta P}{\Delta t} = \frac{2mv_x}{\Delta t} \tag{15.28}$$

The quantity Δt should be the time that the molecule is in contact with the wall. But this time is unknown. The impulse that the gas particle gives to the wall by the collision is given by

$$\text{Impulse} = F\Delta t = \Delta P \tag{15.29}$$

and is shown as the area under the force-time graph of figure 15.13. Because the time Δt for the collision is unknown, a larger time interval t_{bc}, the time between collisions, can be used with an average force F_{avg}, such that the product of $F_{avg}t_{bc}$ is equal to the same impulse as $F\Delta t$. We can see this in figure 15.13. We see that the impulse, which is the area under the curve, is the same in both cases.

At first this may seem strange, but if you think about it, it does make sense. The actual force in the collision is large, but acts for a very short time. After the collision, the gas particle rebounds from the first wall, travels back to the far wall, rebounds from it, and then travels to the first wall again, where a new collision

Figure 15.13

Since the impulse (the area under the curve) is the same, the change in momentum is the same.

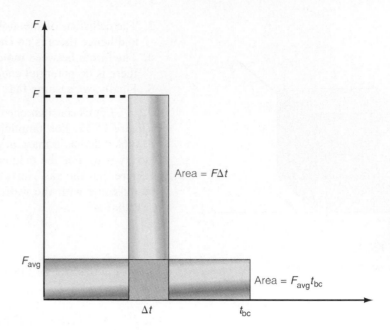

occurs. For the entire traveling time of the particle the actual force on the wall is zero. Because we think of the pressure on a wall as being present at all times, it is reasonable to talk about a smaller average force that is acting continuously for the entire time t_{bc}. As long as the impulse is the same in both cases, the momentum imparted to the wall is the same in both cases. Equation 15.29 becomes

$$\text{Impulse} = F\Delta t = F_{avg}t_{bc} = \Delta P \tag{15.30}$$

The force imparted to the wall, equation 15.28, becomes

$$F_{avg} = \frac{\Delta P}{t_{bc}} = \frac{2mv_x}{t_{bc}} \tag{15.31}$$

We find the time between the collision t_{bc} by noting that the particle moves a distance $2L$ between the collisions. Since the speed v_x is the distance traveled per unit time, we have

$$v_x = \frac{2L}{t_{bc}}$$

Hence, the time between collisions is

$$t_{bc} = \frac{2L}{v_x} \tag{15.32}$$

Therefore, the force imparted to the wall by this single collision becomes

$$F_{avg} = \frac{2mv_x}{2L/v_x} = \frac{mv_x^2}{L} \tag{15.33}$$

The total change in momentum per second, and hence the total force on the wall caused by all the molecules is the sum of the forces caused by all of the molecules, that is,

$$F_{avg} = F_{1avg} + F_{2avg} + F_{3avg} + \cdots + F_{Navg} \tag{15.34}$$

where N is the total number of molecules. Substituting equation 15.33 for each gas molecule, we have

$$F_{avg} = \frac{mv_{x1}^2}{L} + \frac{mv_{x2}^2}{L} + \frac{mv_{x3}^2}{L} + \cdots + \frac{mv_{xN}^2}{L}$$

$$F_{avg} = \frac{m}{L}(v_{x1}^2 + v_{x2}^2 + v_{x3}^2 + \cdots + v_{xN}^2) \qquad (15.35)$$

Let us multiply and divide equation 15.35 by the total number of molecules N, that is,

$$F_{avg} = \frac{mN}{L}\left(\frac{v_{x1}^2 + v_{x2}^2 + v_{x3}^2 + \cdots + v_{xN}^2}{N}\right) \qquad (15.36)$$

But the term in parentheses is the definition of an average value. That is,

$$v_{xavg}^2 = \left(\frac{v_{x1}^2 + v_{x2}^2 + v_{x3}^2 + \cdots + v_{xN}^2}{N}\right) \qquad (15.37)$$

As an example, if you have four exams in the semester, your average grade is the sum of the four exams divided by 4. Here, the sum of the squares of the x-component of the velocity of each molecule, divided by the total number of molecules, is equal to the average of the square of the x-component of velocity. Therefore equation 15.36 becomes

$$F_{avg} = \frac{mN}{L}v_{xavg}^2$$

But since the pressure is defined as $p = F/A$, from equation 15.26, we have

$$p = \frac{F_{avg}}{A} = \frac{F_{avg}}{L^2} = \frac{mN}{L^3}v_{xavg}^2 = \frac{mN}{V}v_{xavg}^2 \qquad (15.38)$$

or

$$pV = Nmv_{xavg}^2 \qquad (15.39)$$

The square of the actual three-dimensional speed is

$$v^2 = v_x^2 + v_y^2 + v_z^2$$

and averaging over all molecules

$$v_{avg}^2 = v_{xavg}^2 + v_{yavg}^2 + v_{zavg}^2$$

But because the motion of any gas molecule is random,

$$v_{xavg}^2 = v_{yavg}^2 = v_{zavg}^2$$

That is, there is no reason why the velocity in one direction should be any different than in any other direction, hence their average speeds should be the same. Therefore,

$$v_{avg}^2 = 3v_{xavg}^2$$

or

$$v_{xavg}^2 = \frac{v_{avg}^2}{3} \qquad (15.40)$$

Substituting equation 15.40 into equation 15.39, we get

$$pV = \frac{Nm}{3}v_{avg}^2$$

Multiplying and dividing the right-hand side by 2, gives

$$pV = \frac{2}{3}N\left(\frac{mv_{avg}^2}{2}\right) \tag{15.41}$$

The total number of molecules of the gas is equal to the number of moles of gas times Avogadro's number—the number of molecules in one mole of gas—that is,

$$N = nN_A \tag{15.24}$$

Substituting equation 15.24 into equation 15.41, gives

$$pV = \frac{2}{3}nN_A\left(\frac{mv_{avg}^2}{2}\right) \tag{15.42}$$

Recall that the ideal gas equation was derived from experimental data as

$$pV = nRT \tag{15.23}$$

The left-hand side of equation 15.23 contains the pressure and volume of the gas, all macroscopic quantities, and all determined experimentally. The left-hand side of equation 15.42, on the other hand, contains the pressure and volume of the gas as determined theoretically by Newton's second law. If the theoretical formulation is to agree with the experimental results, then these two equations must be equal. Therefore equating equation 15.23 to equation 15.42, we have

$$nRT = \frac{2}{3}nN_A\left(\frac{mv_{avg}^2}{2}\right)$$

or

$$\frac{3}{2}\left(\frac{R}{N_A}\right)T = \frac{mv_{avg}^2}{2} \tag{15.43}$$

where R/N_A is the gas constant per molecule. It appears so often that it is given the special name *the Boltzmann constant* and is designated by the letter k. Thus,

$$k = \frac{R}{N_A} = 1.38 \times 10^{-23} \text{ J/K} \tag{15.44}$$

Therefore, equation 15.43 becomes

$$\tfrac{3}{2}kT = \tfrac{1}{2}mv_{avg}^2 \tag{15.45}$$

Equation 15.45 relates the macroscopic view of a gas to the microscopic view. Notice that *the absolute temperature T of the gas (a macroscopic variable) is a measure of the mean translational kinetic energy of the molecules of the gas (a microscopic variable).* The higher the temperature of the gas, the greater the average kinetic energy of the gas, the lower the temperature, the smaller the average kinetic energy. Observe from equation 15.45 that if the absolute temperature of a gas is 0 K, then the mean kinetic energy of the molecule would be zero and its speed would also be zero. This was the original concept of absolute zero, a point where all molecular motion would cease. This concept of absolute zero can not really be derived from equation 15.45 because all gases condense to a liquid and usually a solid before they reach absolute zero. So the assumptions used to derive equation 15.45 do not hold and hence the equation can not hold down to absolute zero. Also, in more advanced studies of quantum mechanics it is found that even at absolute zero a molecule has energy, called its zero point energy. Equation 15.45 is, of course, perfectly valid as long as the gas remains a gas.

Example 15.9

The kinetic energy of a gas molecule. What is the average kinetic energy of the oxygen and nitrogen molecules in a room at room temperature?

Solution

Room temperature is considered to be 20 °C or 293 K. Therefore the mean kinetic energy, found from equation 15.45, is

$$KE_{avg} = \frac{1}{2}mv^2_{avg} = \frac{3}{2}kT$$

$$= \frac{3}{2}\left(1.38 \times 10^{-23}\,\frac{J}{K}\right)(293\text{ K})$$

$$= 6.07 \times 10^{-21}\text{ J}$$

Notice that the average kinetic energy of any one molecule is quite small. This is because the mass of any molecule is quite small. The energy of the gas does become significant, however, because there are usually so many molecules in the gas. Because the average kinetic energy is given by $\frac{3}{2}\,kT$, we see that oxygen and nitrogen and any other molecule of gas at the same temperature all have the same average kinetic energy. Their speeds, however, are not all the same because the different molecules have different masses.

The average speed of a gas molecule can be determined by solving equation 15.45 for v_{avg}. That is,

$$\frac{1}{2}mv^2_{avg} = \frac{3}{2}kT$$

$$v^2_{avg} = \frac{3kT}{m}$$

and

$$v_{rms} = \sqrt{\frac{3kT}{m}} \tag{15.46}$$

This particular average value of the speed, v_{rms}, is usually called the root-mean-square value, or rms value for short, of the speed v. It is called the rms speed, because it is the square *root* of the *mean* of the *square* of the speed. Occasionally the rms speed of a gas molecule is called the *thermal speed*. To determine the rms speed from equation 15.46, we must know the mass m of one molecule. The mass m of any molecule is found from

$$m = \frac{M}{N_A} \tag{15.47}$$

That is, *the mass m of one molecule is equal to the molecular mass M of that gas divided by Avogadro's number N_A.*

Example 15.10

The rms speed of a gas molecule. Find the rms speed of an oxygen and nitrogen molecule at room temperature.

Solution

The molecular mass of O_2 is 32 g/mole. Therefore the mass of one molecule of O_2 is

$$m_{O_2} = \frac{M}{N_A} = \frac{32 \text{ g/mole}}{6.022 \times 10^{23} \text{ molecules/mole}}$$

$$= 5.31 \times 10^{-23} \text{ g/molecule}$$

The rms speed, found from 15.46, is

$$v_{rms} = \sqrt{\frac{3kT}{m}} = \sqrt{\frac{(3)(1.38 \times 10^{-23} \text{ J/K})(293 \text{ K})}{5.31 \times 10^{-23} \text{ gm}}}$$

$$= \sqrt{\frac{(228 \text{ kg m}^2)(10^3 \text{ gm})}{(\text{gm s}^2)(1 \text{ kg})}}$$

$$= 478 \text{ m/s}$$

Notice that the rms speed of an oxygen molecule is 478 m/s at room temperature, whereas the speed of sound at this temperature is about 343 m/s.

The mass of a nitrogen molecule is found from

$$m_{N_2} = \frac{M}{N_A}$$

The atomic mass of nitrogen is 14, and since there are two atoms of nitrogen in one molecule of nitrogen gas N_2, the molecular mass of nitrogen is

$$M = 2(14) = 28 \text{ g/mole}$$

Therefore

$$m_{N_2} = \frac{M}{N_A} = \frac{28 \text{ g/mole}}{6.022 \times 10^{23} \text{ molecules/mole}}$$

$$= 4.65 \times 10^{-23} \text{ g/molecule} = 4.65 \times 10^{-26} \text{ kg/molecule}$$

The rms speed of a nitrogen molecule is therefore

$$v_{rms} = \sqrt{\frac{3kT}{m}} = \sqrt{\frac{(3)(1.38 \times 10^{-23} \text{ J/K})(293 \text{ K})}{4.65 \times 10^{-26} \text{ kg}}}$$

$$= 511 \text{ m/s}$$

Note from the example that both speeds are quite high. The average speed of nitrogen is greater than the average speed of oxygen because the mass of the nitrogen molecule is less than the mass of the oxygen molecule.

An Essay on the Application of Physics

Relative Humidity and the Cooling of the Human Body

Figure 1

One of those dog days of summer when you never stop perspiring.

H ave you ever wondered why you feel so uncomfortable on those dog days of August when the weatherman says that it is very hot and humid (figure 1)? What has humidity got to do with your being comfortable? What is humidity in the first place?

To understand the concept of humidity, we must first understand the concept of evaporation. Consider the two bowls shown in figure 2. Both are filled with water. Bowl 1 is open to the environment, whereas a glass plate is placed over bowl 2. If we leave the two bowls overnight, on returning the next day we would find bowl 1 empty while bowl 2 would still be filled with water. What happened to the water in bowl 1? The water in bowl 1 has evaporated into the air and is gone. *Evaporation is the process by which water goes from the liquid state to the gaseous state at any temperature.* Boiling, as you recall, is the process by which water goes from the liquid state to the gaseous state at the boiling point of 100 °C. That is, it is possible for liquid water to go to the gaseous state at any temperature.

Just as there is a latent heat of vaporization for boiling water ($L_v = 540$ kcal/kg), the latent heat of vaporization of water at 0 °C is $L_v = 600$ kcal/kg. The latent heat at any in-between temperature can be found by interpolation. Thus, in order to evaporate 1 kg of water into the air at 0 °C, you would have to supply 600 kcal of thermal energy to the water.

The molecules in the water in bowl 1 are moving about in a random order. But their attractive molecular forces still keep them together. These molecules can now absorb heat from the surroundings. This absorbed energy shows up as an increase in the kinetic energy of the molecule, and hence an increase in the velocity of the molecule. When the liquid molecule has absorbed enough energy it moves right out of the liquid water into the air above as a molecule of water vapor. (Remember the water molecule is the same whether it is a solid, liquid, or gas, namely H_2O, two atoms of hydrogen and one atom of oxygen. The difference is only in the energy of the molecule.)

Since the most energetic of the water molecules escape from the liquid, the molecules left behind have lower energy, hence the temperature of the remaining liquid decreases. *Hence, evaporation is a cooling process.* The water molecule that evaporated took the thermal energy with it, and the water left behind is just that much cooler.

The remaining water in bowl 1 now absorbs energy from the environment, thereby increasing the temperature of the water in the bowl. This increased thermal energy is used by more liquid water molecules to escape into the air as more water vapor. The process continues until all the water in bowl 1 is evaporated.

Now when we look at bowl 2, the water is still there. Why didn't all that water evaporate into the air? To explain what hap-

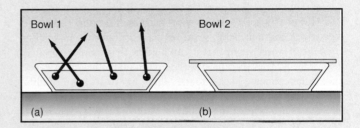

Figure 2
Evaporation.

pens in bowl 2 let us do the following experiment. We place water in a container and place a plate over the water. Then we allow dry air, air that does not contain water vapor, to fill the top portion of the closed container, figure 3(a). Using a thermometer, we measure the temperature of the air as $t = 20$ °C, and using a pressure gauge we measure the pressure of the air p_0, in the container. Now we remove the plate separating the dry air from the water by sliding it out of the closed container. As time goes by, we observe that the pressure recorded by the pressure gauge increases, figure 3(b). This occurs because some of the liquid water molecules evaporate into the air as water vapor. Water vapor is a gas like any other gas and it exerts a pressure. It is this water vapor pressure that is being recorded as the increased pressure on the gauge. The gauge is reading the air pressure of the dry air plus the actual water vapor pressure of the gas, $p_0 + p_{awv}$. Subtracting p_0 from $p_0 + p_{awv}$, gives the actual water

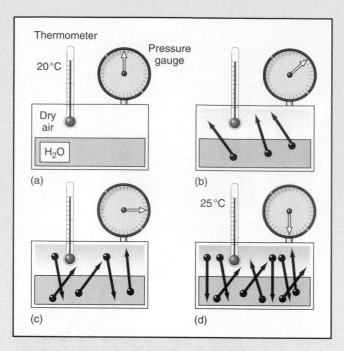

Figure 3

Water vapor in the air.

vapor pressure, p_{awv}. As time goes on, the water vapor pressure increases as more and more water molecules evaporate into the air. However, after a while, the pressure indicated by the gauge becomes a constant. At this point the air contains the maximum amount of water vapor that it can hold at that temperature. As new molecules evaporate into the air, some of the water vapor molecules condense back into the liquid, figure 3(c). An equilibrium condition is established, whereby just as many water vapor molecules are condensing as liquid water molecules are evaporating. At this point, the air is said to be *saturated*. That is, the air contains the maximum amount of water vapor that it can hold at that temperature. The vapor pressure read by the gauge is now called the saturation water vapor pressure, p_{swv}.

The amount of water vapor in the air is called humidity. *A measure of the amount of water vapor in the air is given by the relative humidity, RH, and is defined as the ratio of the amount of water vapor actually present in the air to the amount of water vapor that the air can hold at a given temperature and pressure, times 100%.* The amount of water vapor in the air is directly proportional to the water vapor pressure. Therefore, we can determine the relative humidity, RH, of the air as

$$RH = \left(\frac{\text{actual vapor pressure}}{\text{saturation vapor pressure}}\right)100\% \quad \textbf{(15H.1)}$$

$$RH = \left(\frac{p_{awv}}{p_{swv}}\right)100\% \quad \textbf{(15H.2)}$$

When the air is saturated, the actual vapor pressure recorded by the gauge is equal to the saturation vapor pressure and hence, the relative humidity is 100%.

If the air in the container is heated, we notice that the pressure indicated by the pressure gauge increases, figure 3(d). Part of the increased pressure is caused by the increase of the pressure of the air. This increase can be calculated by the ideal gas equation and subtracted from the gauge reading, so that we can determine any increase in pressure that would come from an increase in the actual water vapor pressure. We notice that by increasing the air temperature to 25 °C, the water vapor pressure also increases. After a while, however, the water vapor pressure again becomes a constant. The air is again saturated. We see from this experiment that *the maximum amount of water vapor that the air can hold is a function of temperature*. At low temperatures the air can hold only a little water vapor, while at high temperatures the air can hold much more water vapor.

We can now see why the water in bowl 2 in figure 2 did not disappear. Water evaporated from the liquid into the air above, increasing the relative humidity of the air. However, once the air became saturated, the relative humidity was equal to 100%, and no more water vapor could evaporate into it. This is why you can still see the water in bowl 2, there is no place for it to go.

Because of the temperature dependence of water vapor in the air, when the temperature of the air is increased, the capacity of the air to hold water increases. Therefore, if no additional water is added to the air, the relative humidity will decrease because the capacity of the air to hold water vapor has increased. Conversely, when the air temperature is decreased, its capacity to hold water vapor decreases, and therefore the relative humidity of the air increases. This temperature dependence causes a decrease in the relative humidity during the day light hours, and an increase in the relative humidity during the night time hours, with the maximum relative humidity occurring in the early morning hours just before sunrise.

The amount of evaporation depends on the following factors:

1. The vapor pressure. Whenever the actual vapor pressure is less than the maximum vapor pressure allowable at that temperature, the saturation vapor pressure, then evaporation will readily occur. Greater evaporation occurs whenever the air is dry, that is, at low relative humidities. Less evaporation occurs when the air is moist, that is, at high relative humidities.

2. Wind movement and turbulence. Air movement and turbulence replaces air near the water surface with less moist air and increases the rate of evaporation.

Now that we have discussed the concepts of relative humidity we can understand how the body cools itself. Through the process of perspiration, the body secretes microscopic droplets of water onto the surface of the skin of the body. As these tiny droplets of water evaporate into the air, they cool the body. As long as the relative humidity of the air is low, evaporation occurs readily, and the body cools itself. *However whenever the relative humidity becomes high, it is more difficult for the microscopic droplets of water to evaporate into the air. The body can not cool itself, and the person feels very uncomfortable.*

We are all aware of the discomfort caused by the hot and humid days of August. The high relative humidity prevents the normal evaporation and cooling of the body. As some evaporation occurs from the body, the air next to the skin becomes saturated, and no further cooling can occur. If a fan is used, we feel more comfortable because the fan blows the saturated air next to our skin away and replaces it with air that is slightly less saturated. Hence, the evaporation process can continue while the fan is in operation and the body cools itself. Another way to cool the human body in the summer is to use an air conditioner. The air conditioner not only cools the air to a lower temperature, but it also removes a great deal of water vapor from the air, thereby decreasing the relative humidity of the air and permitting the normal evaporation of moisture from the skin. (Note that if the air conditioner did not remove water vapor from the air, cooling the air would increase the relative humidity making us even more uncomfortable.)

In the hot summertime, people enjoy swimming as a cooling experience. Not only the immersion of the body in the cool water is so satisfying, but when the person comes out of the water, evaporation of the sea or pool water from the person adds to the cooling. It is also customary to wear loose clothing in the summertime. The reason for this is to facilitate the flow of air over the body and hence assist in the evaporation process. Tight fitting clothing prevents this evaporation process and the person feels hotter. If you happen to live in a dry climate (low relative humidity), then you can feel quite comfortable at 85 °F, while a person living in a moist climate (high relative humidity) is very uncomfortable at the same 85 °F.

What many people do not realize is that you can also feel quite uncomfortable even in the wintertime, because of the humidity of the air. If the relative humidity is very low in your home then evaporation occurs very rapidly, cooling the body perhaps more than is desirable. As an example, the air temperature might be 70 °F but if the relative humidity is low, say 30%, then evaporation readily occurs from the skin of the body, and the person feels cold even though the air temperature is 70 °F. In this case the person can feel more comfortable if he or she uses a humidifier. A humidifier is a device that adds water vapor to the air. By increasing the water vapor in the air, and hence increasing the relative humidity, the rate of evaporation from the body decreases. The person no longer feels cold at 70 °F, but feels quite comfortable. If too much water vapor is added to the air, increasing the relative humidity to near a 100%, then evaporation from the body is hampered, the body is not able to cool itself, and the person feels too hot even though the temperature is only 70 °F. Thus too high or too low a relative humidity makes the human body uncomfortable.

We should also note that the evaporation process is also used to cool the human body for medical purposes. If a person is running a high fever, then an alcohol rub down helps cool the body down to normal temperature. The principle of evaporation as a cooling device is the same, only alcohol is very volatile and evaporates very rapidly. This is because the saturation vapor pressure of alcohol at 20 °C is much higher than the saturation vapor pressure of water. At 20 °C, water has a saturation vapor pressure of 17.4 mm of Hg, whereas ethyl alcohol has a saturation vapor pressure of 44 mm of Hg. The larger the saturation vapor pressure of a liquid, the greater is the amount of its vapor that the air can hold and hence the greater is the rate of vaporization. Because the alcohol evaporates much more rapidly than water, much greater cooling occurs than when water evaporates. Ethyl ether and ethyl chloride have saturation vapor pressures of 442 mm and 988 mm of Hg, respectively. Ethyl chloride with its very high saturation vapor pressure, evaporates so rapidly that it freezes the skin, and is often used as a local anesthetic for minor surgery.

The Language of Physics

Thermal expansion
Most materials expand when heated (p. 425).

Charles' law
The volume of a gas at constant pressure is directly proportional to the absolute temperature of the gas (p. 432).

Gay-Lussac's law
The absolute pressure of a gas at constant volume is directly proportional to the absolute temperature of the gas (p. 433).

Boyle's law
The product of the pressure and volume of a gas at constant temperature is equal to a constant (p. 434).

The ideal gas law
The general gas law that contains Charles', Gay-Lussac's, and Boyle's law as special cases. It states that the product of the pressure and volume of a gas divided by the absolute temperature of the gas is a constant (p. 434).

Mole
One mole of any gas is that amount of the gas that has a mass in grams equal to the atomic or molecular mass of the gas. One mole of any gas at a temperature of 0 °C and a pressure of one atmosphere, has a volume of 22.4 liters (p. 435).

Avogadro's number
Every mole of a gas contains the same number of molecules, namely, 6.022×10^{23} molecules. The mass of one molecule is equal to the molecular mass of that gas divided by Avogadro's number (p. 435).

Kinetic theory of gases
The analysis of a gas at the microscopic level, treated by Newton's laws of motion. The kinetic theory shows that the absolute temperature of a gas is a measure of the mean translational kinetic energy of the molecules of the gas (p. 438).

Summary of Important Equations

Linear expansion
$$\Delta L = \alpha L_0 \Delta t \qquad (15.1)$$

Area expansion
$$\Delta A = 2\alpha A_0 \Delta t \qquad (15.3)$$

Volume expansion
$$\Delta V = 3\alpha V_0 \Delta t \qquad (15.5)$$

Coefficient of volume expansion
for solids
$$\beta = 3\alpha \qquad (15.6)$$

Volume expansion
$$\Delta V = \beta V_0 \Delta t \qquad (15.7)$$

Ideal gas law
$$\frac{p_1 V_1}{T_1} = \frac{p_2 V_2}{T_2} \qquad (15.20)$$
$$pV = nRT \qquad (15.23)$$

Number of molecules
$$N = nN_A \qquad (15.24)$$

Absolute pressure
$$p_{abs} = p_{gauge} + p_{atm} \qquad (15.25)$$

Temperature and mean
kinetic energy
$$\tfrac{3}{2}kT = \tfrac{1}{2}mv_{avg}^2 \qquad (15.45)$$

rms speed of a molecule
$$v_{rms} = \sqrt{\frac{3kT}{m}} \qquad (15.46)$$

Mass of a molecule
$$m = \frac{M}{N_A} \qquad (15.47)$$

Total mass of the gas
$$m_{total} = nM$$

Questions for Chapter 15

1. Describe the process of expansion from a microscopic point of view.
2. Explain why it is necessary to make a temperature correction when measuring atmospheric pressure with a barometer.
†3. In the very upper portions of the atmosphere there are extremely few molecules present. Discuss the concept of temperature as it would be applied in this portion of the atmosphere.

4. Explain the introduction of the Kelvin temperature scale in the application of Charles' law.
5. Describe the meaning and application of gauge pressure.
†6. Would you expect the ideal gas equation to be applicable to a volume that is of the same order of magnitude as the size of a molecule?
7. If a gas is at an extremely high density, what effect would this have on the assumptions underlying the kinetic theory of gases?

8. From the point of view of the time between collisions of a gas molecule and the walls of the container, what happens if the container is reduced to half its original size?
9. From the point of view of the kinetic theory of gases, explain why there is no atmosphere on the moon.
10. When an astronomer observes the stars at night in an observatory, the observatory is not heated but remains at the same temperature as the outside air. Why should the astronomer do this?

Problems for Chapter 15

15.1 Linear Expansion of Solids

1. An aluminum rod measures 2.00 m at 10.0 °C. Find its length when the temperature rises to 135 °C.
2. A brass ring has a diameter of 20.0 cm when placed in melting ice at 0 °C. What will its diameter be if it is placed in boiling water?
3. A steel ring of 2.00-in. diameter at 0 °C is to be heated and slipped over a steel shaft whose diameter is 2.003 in. at 0 °C. To what temperature should the ring be heated? If the ring is not heated, to what temperature should the shaft be cooled such that the ring will fit over the shaft?

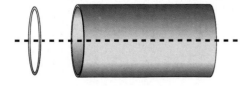

4. An aluminum ring, 7.00 cm in diameter at 5.00 °C, is to be heated and slipped over an aluminum shaft whose diameter is 7.003 cm at 5.00 °C. To what temperature should the ring be heated? If the ring is not heated, to what temperature should the shaft be cooled such that the ring will fit over the shaft?
5. The iron rim of a wagon wheel has an internal diameter of 80.0 cm when the temperature is 100 °C. What is its diameter when it cools to 0.00 °C?
6. A steel measuring tape, correct at 0.00 °C measures a distance L when the temperature is 30.0 °C. What is the error in the measurement due to the expansion of the tape?
7. Steel rails 50.0 ft long are laid when the temperature is 0.00 °C. What separation should be left between the rails to allow for thermal expansion when the temperature rises to 40.0 °C? If the cross-sectional area of a rail is 30.0 in.², what force is associated with this expansion?

8. Steel rails 20.0 m long are laid when the temperature is 5.00 °C. What separation should be left between the rails to allow for thermal expansion when the temperature rises to 38.5 °C? If the cross-sectional area of a rail is 230 cm², what force is associated with this expansion?

15.2 Area Expansion of Solids

9. A sheet of aluminum measures 10.0 ft by 5.00 ft at 0.00 °C. What is the area of the sheet at 150 °C?
10. A sheet of brass measures 4.00 m by 3.00 m at 5.00 °C. What is the area of the sheet at 175 °C?
11. If the radius of a copper circle is 20.0 cm at 0.00 °C, what will its area be at 100 °C?
12. A piece of aluminum has a hole 0.850 cm in diameter at 20.0 °C. To what temperature should the sheet be heated so that an aluminum bolt 0.865 cm in diameter will just fit into the hole?

15.3 Volume Expansion of Solids and Liquids

13. A chemistry student fills a Pyrex glass flask to the top with 100 cm^3 of Hg at 0.00 °C. How much mercury will spill out of the tube, and have to be cleaned up by the student, if the temperature rises to 35.0 °C?

14. A tube is filled to a height of 20.0 cm with mercury at 0.00 °C. If the tube has a cross-sectional area of 25.0 mm^2, how high will the mercury rise in the tube when the temperature is 30.0 °C? Neglect the expansion of the tube.

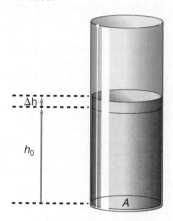

15. Since the volume of a material changes with a change in temperature, show that the density ρ at any temperature is given by

$$\rho = \frac{\rho_0}{1 + \beta \Delta t}$$

where ρ_0 is the density at the lower temperature.

15.7 The Ideal Gas Law

16. If 2.00 g of oxygen gas are contained in a tank of 500 cm^3 at a pressure of 20.0 psi, what is the temperature of the gas?

17. What is the pressure produced by 2 moles of gas at 20.0 °C contained in a volume of 0.500 liters?

18. One mole of hydrogen is at a pressure of 2.00 atm and a volume of 0.25 m^3. What is its temperature?

19. Compute the number of molecules in a gas contained in a volume of 50.0 cm^3 at a pressure of 2.00 atm and a temperature of 300 K.

20. An automobile tire has a volume of 4500 in.^{3} and contains air at a gauge pressure of 32.0 lb/in.^{2} when the temperature is 0.00 °C. What is the gauge pressure when the temperature rises to 35.0 °C?

21. An automobile tire has a volume of 0.0800 m^3 and contains air at a gauge pressure of 2.48 $\times$ 10^5 N/m^2 when the temperature is 3.50 °C. What is the gauge pressure when the temperature rises to 37.0 °C?

22. (a) How many moles of gas are contained in 0.300 kg of H_2 gas? (b) How many molecules of H_2 are there in this mass?

23. Nitrogen gas, at a pressure of 150 N/m^2, occupies a volume of 20.0 m^3 at a temperature of 30.0 °C. Find the mass of this nitrogen gas in kilograms.

24. One mole of nitrogen gas at a pressure of 1.00 atm and a temperature of 300 K expands isothermally to double its volume. What is its new pressure? (Isothermal means at constant temperature.)

25. An ideal gas occupies a volume of 4.00 liters at a pressure of 1.00 atm and a temperature of 273 K. The gas is then compressed isothermally to one half of its original volume. Determine the final pressure of the gas.

26. The pressure of a gas is kept constant while 3.00 m^3 of the gas at an initial temperature of 50.0 °C is expanded to 6.00 m^3. What is the final temperature of the gas?

27. The volume of O_2 gas at a temperature of 20.0 °C is 4.00 liters. The temperature of the gas is raised to 100 °C while the pressure remains constant. What is the new volume of the gas?

28. A balloon is filled with helium at a pressure of 1.50 atm, a temperature of 25.0 °C, and occupies a volume of 3.00 m^3. The balloon rises in the atmosphere. When it reaches a height where the pressure is 0.500 atm and the temperature is −20.0 °C, what is its volume?

†29. An air bubble of 32.0 cm^3 volume is at the bottom of a lake 10.0 m deep where the temperature is 5.00 °C. The bubble rises to the surface where the temperature is 20.0 °C. Find the volume of the bubble just before it reaches the surface.

30. One mole of helium is at a temperature of 300 K and a volume of 10.0 liters. What is its pressure? The gas is warmed at constant volume to 600 K. What is its new pressure? How many molecules are there?

15.8 The Kinetic Theory of Gases

31. Find the rms speed of a helium atom at a temperature of 10.0 K.

32. Find the kinetic energy of a single molecule when it is at a temperature of (a) 0.00 °C, (b) 20.0 °C, (c) 100 °C, (d) 1000 °C, and (e) 5000 °C.

33. Find the mass of a carbon dioxide molecule (CO_2).

34. Find the rms speed of a helium atom on the surface of the sun, if the sun's surface temperature is approximately 6000 K.

35. At what temperature will the rms speed of an oxygen molecule be twice its speed at room temperature?

36. The rms speed of a gas molecule is v at a temperature of 300 K. What is the speed if the temperature is increased to 900 K?

†37. Find the total kinetic energy of all the nitrogen molecules in the air in a room 7.00 m by 10.0 m by 4.00 m, if the air is at a temperature of 22.0 °C and 1 atm of pressure.

38. If the rms speed of a monatomic gas is 445 m/s at 350 K, what is the atomic mass of the atom? What gas do you think it is?

Additional Problems

39. A barometer reads normal atmospheric pressure when the mercury column in the tube is at 76.0 cm of Hg at 0.00 °C. If the pressure of the atmosphere does not change, but the air temperature rises to 35.0 °C, what pressure will the barometer indicate? The tube has a diameter of 5.00 mm. Neglect the expansion of the tube.

40. Find the stress necessary to give the same strain that occurs when a steel rod undergoes a temperature change of 120 °C.

†41. The symbol π is defined as the ratio of the circumference of a circle to its diameter. If a circular sheet of metal expands by heating, show that the ratio of the expanded circumference to the expanded diameter is still equal to π.

42. A 15.0-cm strip of steel is welded to the left side of a 15.0-cm strip of aluminum. When the strip undergoes a temperature change Δt, will the combined strip bend to the right or to the left?

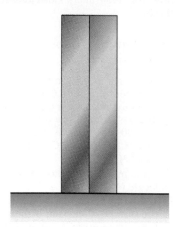

†43. A 350-g mass is connected by a thin brass rod 25.0 cm long to a rotating shaft that is rotating at an initial angular speed of 5.00 rad/s. If the temperature changes by 35 °C, (a) find the change in the moment of inertia of the system and (b) using the law of conservation of angular momentum, find the change in the rotational energy of the system.

44. The focal length of a polished aluminum spherical mirror is given by $f = R/2$, where R is the radius of curvature of the mirror, and is 23.5 cm. Find the new focal length of the mirror if the temperature changes by 45.0 °C.

†45. A 50.0-g silver ring, 12.0 cm in diameter, is spinning about an axis through its center at a constant speed of 11.4 rad/s. If the temperature changes by 185 °C, what is the change in the angular momentum of the ring? The coefficient of linear expansion for silver is $1.90 \times 10^{-5}/°C$.

†46. An aluminum rod is at room temperature. To what temperature would this rod have to be heated such that the thermal expansion is enough to exceed the elastic limit of aluminum? Compare this temperature with the melting point of aluminum. What conclusion can you draw?

47. A steel pendulum is 60.0 cm long, at 20.0 °C. By how much does the period of the pendulum change when the temperature is 35.0 °C?

48. Find the number of air molecules in a classroom 10.0 m long, 10.0 m wide, and 3.5 m high, if the air is at normal atmospheric pressure and a temperature of 20.0 °C.

49. A brass cylinder 5.00 cm in diameter and 8.00 cm long is at an initial temperature of 380 °C. It is placed in a calorimeter containing 0.120 kg of water at an initial temperature of 5.00 °C. The aluminum calorimeter has a mass of 0.060 kg. Find (a) the final temperature of the water and (b) the change in volume of the cylinder.

†50. Dalton's law of partial pressure says that when two or more gases are mixed together, the resultant pressure is the sum of the individual pressures of each gas. That is,
$$p = p_1 + p_2 + p_3 + p_4 + \cdots$$
If one mole of oxygen at 20.0 °C and occupying a volume of 2.00 m³ is added to two moles of nitrogen also at 20.0 °C and occupying a volume of 10.0 m³ and the final volume is 10.0 m³, find the resultant pressure of the mixture.

†51. The escape velocity from the earth is $v_E = 1.12 \times 10^4$ m/s. At what temperature is the rms speed equal to this for: (a) hydrogen (H_2), (b) helium (He), (c) nitrogen (N_2), (d) oxygen (O_2), (e) carbon dioxide (CO_2), and (f) water vapor (H_2O)? From these results, what can you infer about the earth's atmosphere?

†52. The escape velocity from the moon is $v_M = 0.24 \times 10^4$ m/s. At what temperature is the rms speed equal to this for (a) hydrogen (H_2), (b) helium (He), (c) nitrogen (N_2), (d) oxygen (O_2), (e) carbon dioxide (CO_2), and (f) water vapor (H_2O)? From these results, what can you infer about the possibility of an atmosphere on the moon?

†53. Show that the velocity of a gas molecule at one temperature is related to the velocity of the molecule at a second temperature by
$$v_2 = \sqrt{\frac{T_2}{T_1}} \, v_1$$

†54. A room is filled with nitrogen gas at a temperature of 293 K. (a) What is the average kinetic energy of a nitrogen molecule? (b) What is the rms speed of the molecule? (c) What is the rms value of the momentum of this molecule? (d) If the room is 4.00 m wide what is the average force exerted on the wall by this molecule? (e) If the wall is 4.00 m by 3.00 m, what is the pressure exerted on the wall by this molecule? (f) How many molecules moving at this speed are necessary to cause a pressure of 1.00 atm?

†55. Two isotopes of a gaseous substance can be separated by diffusion if each has a different velocity. Show that the rms speed of an isotope can be given by
$$v_2 = \sqrt{\frac{m_1}{m_2}} \, v_1$$
where the subscript 1 refers to isotope 1 and the subscript 2 refers to isotope 2.

Interactive Tutorials

56. A copper tube has the length $L_0 = 1.58$ m at the initial temperature $t_i = 20.0$ °C. Find its length L when it is heated to a final temperature $t_f = 100$ °C.

57. A circular brass sheet has an area $A_0 = 2.56$ m² at the initial temperature $t_i = 0$ °C. Find its new area A when it is heated to a final temperature $t_f = 90$ °C.

58. A glass tube is filled to a height $h_0 = 0.762$ m of mercury at the initial temperature $t_i = 0$ °C. The diameter of the tube is 0.085 m. How high will the mercury rise when the final temperature $t_f = 50$ °C? Neglect the expansion of the glass.

59. A gas has a pressure $p_1 = 1$ atm, a volume $V_1 = 4.58$ m³, and a temperature $t_1 = 20.0$ °C. It is then compressed to a volume $V_2 = 1.78$ m³ and a pressure $p_2 = 3.57$ atm. Find the final temperature of the gas t_2.

60. Find the number of moles and the number of molecules in a gas under a pressure $p = 1$ atm and a temperature $t = 20.0$ °C. The room has a length $L = 15.0$ m, a width $W = 10.0$ m, and a height $h = 4.00$ m.

61. Kinetic theory. Oxygen gas is in a room under a pressure $p = 1$ atm and a temperature of $t = 20.0$ °C. The room has a length $L = 18.5$ m, a width $W = 12.5$ m, and a height $h = 5.50$ m. For the oxygen gas, find (a) the kinetic energy of a single molecule, (b) the total kinetic energy of all the oxygen molecules, (c) the mass of an oxygen molecule, and (d) the speed of the oxygen molecule. The molecular mass of oxygen is $M_{O_2} = 32.0$ g/mole.

62. Ideal Gas Equation Calculator.

Heat Transfer

Thermogram showing the distribution of heat on a man's face, wearing glasses.

There can be no doubt now, in the mind of the physicist who has associated himself with inductive methods, that matter is constituted of atoms, heat is movement of molecules and conduction of heat, like all other irreversible phenomena, obeys, not dynamical, but statistical laws, namely, the laws of probability.

Max Planck

16.1 Heat Transfer

In chapter 14 we saw that an amount of thermal energy Q, given by

$$Q = mc\Delta T \qquad \text{(14.6)}$$

is absorbed or liberated in a sensible heating process. But how is this thermal energy transferred to, or from, the body so that it can be absorbed, or liberated? To answer that question, we need to discuss the mechanism of thermal energy transfer. The transfer of thermal energy has historically been called heat transfer.

Thermal energy can be transferred from one body to another by any or all of the following mechanisms:

1. Convection
2. Conduction
3. Radiation

Convection *is the transfer of thermal energy by the actual motion of the medium itself.* The medium in motion is usually a gas or a liquid. Convection is the most important heat transfer process for liquids and gases.

Conduction *is the transfer of thermal energy by molecular action, without any motion of the medium.* Conduction can occur in solids, liquids, and gases, but it is usually most important in solids.

Radiation *is a transfer of thermal energy by electromagnetic waves.*

We will discuss the details of electromagnetic waves in chapter 25. For now we will say that it is not necessary to have a medium for the transfer of energy by radiation. For example, energy is radiated from the sun as an electromagnetic wave, and this wave travels through the vacuum of space, until it impinges on the earth, thereby heating the earth.

Let us now go into more detail about each of these mechanisms of heat transfer.

16.2 Convection

Consider the large mass m of air at the surface of the earth that is shown in figure 16.1. The lines labeled T_0, T_1, T_2, and so on are called isotherms and represent the temperature distribution of the air at the time t. *An isotherm is a line along which the temperature is constant.* Thus everywhere along the line T_0 the air temperature is T_0, and everywhere along the line T_1 the air temperature is T_1, and so forth.

Figure 16.1

Horizontal convection.

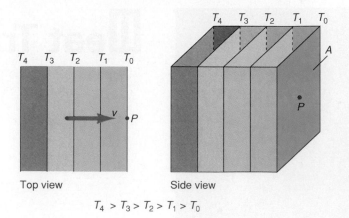

$$T_4 > T_3 > T_2 > T_1 > T_0$$

Consider a point P on the surface of the earth that is at a temperature T_0 at the time t. How can thermal energy be transferred to this point P thereby changing its temperature? That is, how does the thermal energy at that point change with time? If we assume that there is no local infusion of thermal energy into the air at P, such as heating from the sun and the like, then the only way that thermal energy can be transferred to P is by moving the hotter air, presently to the left of point P, to point P itself. That is, if energy can be transferred to the point P by convection, then the air temperature at the point P increases. This is equivalent to moving an isotherm that is to the left of P to the point P itself. The transfer of thermal energy per unit time to the point P is given by $\Delta Q/\Delta t$. By multiplying and dividing by the distance Δx, we can write this as

$$\frac{\Delta Q}{\Delta t} = \frac{\Delta Q}{\Delta x}\frac{\Delta x}{\Delta t} \tag{16.1}$$

But

$$\frac{\Delta x}{\Delta t} = v$$

the velocity of the air moving toward P. Therefore, equation 16.1 becomes

$$\frac{\Delta Q}{\Delta t} = v\frac{\Delta Q}{\Delta x} \tag{16.2}$$

But ΔQ, on the right-hand side of equation 16.2, can be replaced with

$$\Delta Q = mc\Delta T \tag{14.6}$$

(We will need to depart from our custom of using the lower case t for temperatures in Celsius or Fahrenheit degrees, because we will use t to represent time. Thus, the upper case T is now used for temperature in either Celsius or Fahrenheit degrees.) Therefore,

$$\frac{\Delta Q}{\Delta t} = vmc\frac{\Delta T}{\Delta x} \tag{16.3}$$

(Equation 16.3 is a good example of the difference between heat and thermal energy. Here Q is the thermal energy, and $\Delta Q/\Delta t$ is the thermal energy per unit time that is transferred, and this is what is called heat.) Hence, the thermal energy transferred to the point P by convection becomes

$$\Delta Q = vmc\frac{\Delta T}{\Delta x}\Delta t \tag{16.4}$$

The term $\Delta T/\Delta x$ is called the **temperature gradient,** and tells how the temperature changes as we move in the x-direction. We will assume in our analysis that the temperature gradient remains a constant.

Example 16.1

Energy transfer per unit mass. If the temperature gradient is 2.00 °C per 100 km and if the specific heat of air is 0.250 kcal/(kg °C), how much thermal energy per unit mass is convected to the point P in 12.0 hr if the air is moving at a speed of 10.0 km/hr?

Solution

The heat transferred per unit mass, found from equation 16.4, is

$$\frac{\Delta Q}{m} = vc\frac{\Delta T}{\Delta x}\Delta t$$

$$= \left(10.0\,\frac{km}{hr}\right)\left(0.250\,\frac{kcal}{kg\,°C}\right)\left(\frac{2.00\,°C}{100\,km}\right)(12.0\,hr)$$

$$= 0.600\,kcal/kg$$

If the mass m of the air that is in motion is unknown, the density of the fluid can be used to represent the mass. Because the density $\rho = m/V$, where V is the volume of the air, we can write the mass as

$$m = \rho V \qquad (16.5)$$

Therefore, the thermal energy transferred by convection to the point P becomes

$$\Delta Q = v\rho Vc\frac{\Delta T}{\Delta x}\Delta t \qquad (16.6)$$

Sometimes it is more convenient to find the thermal energy transferred per unit volume. In this case, we can use equation 16.6 as

$$\frac{\Delta Q}{V} = v\rho c\frac{\Delta T}{\Delta x}\Delta t$$

Example 16.2

Energy transfer per unit volume. Using the data from example 16.1, find the thermal energy per unit volume transferred by convection to the point P. Assume that the density of air is $\rho_{air} = 1.293\,kg/m^3$.

Solution

The thermal energy transferred per unit volume is found as

$$\frac{\Delta Q}{V} = v\rho c\frac{\Delta T}{\Delta x}\Delta t$$

$$= \left(10.0\,\frac{km}{hr}\right)\left(1.293\,\frac{kg}{m^3}\right)\left(0.250\,\frac{kcal}{kg\,°C}\right)\left(\frac{2.00\,°C}{100\,km}\right)(12.0\,hr)$$

$$= 0.776\,kcal/m^3$$

Note that although the number 0.776 kcal/m³ may seem small, there are thousands upon thousands of cubic meters of air in motion in the atmosphere. Thus, the thermal energy transfer by convection can be quite significant.

Convection is the main mechanism of thermal energy transfer in the atmosphere. On a global basis, the nonuniform temperature distribution on the surface of the earth causes convection cycles that result in the prevailing winds. If the earth were not rotating, a huge convection cell would be established as shown in figure 16.2(a). The equator is the hottest portion of the earth because it gets the

Figure 16.2

Convection in the atmosphere. Lutgens/ Tarbuck, *The Atmosphere: An Introduction to Meteorology,* 4/E, © 1989, pp. 186–187. © Prentice-Hall, Inc., Englewood Cliffs, NJ.

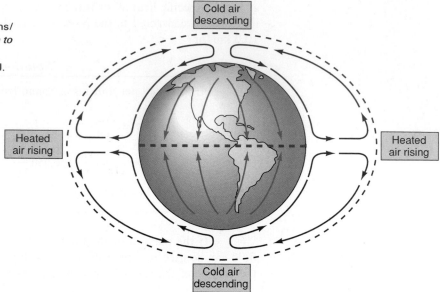

(a) Nonrotating earth

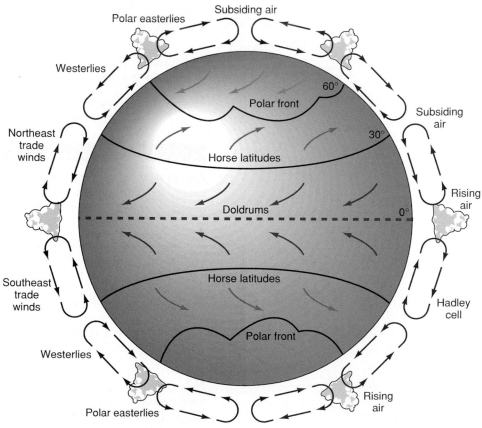

(b) Rotating earth

maximum radiation from the sun. Hot air at the equator expands and rises into the atmosphere. Cooler air at the surface flows toward the equator to replace the rising air. Colder air at the poles travels toward the equator. Air aloft over the poles descends to replace the air at the surface that just moved toward the equator. The initial rising air at the equator flows toward the pole, completing the convection cycle. The net result of the cycle is to bring hot air at the surface of the equator, aloft, then north to the poles, returning cold air at the polar surface back to the equator.

This simplified picture of convection on the surface of the earth is not quite correct, because the effect produced by the rotating earth, called the Coriolis effect, has been neglected. The **Coriolis effect** is caused by the rotation of the earth and can best be described by an example. If a projectile, aimed at New York, were fired from the North Pole, its path through space would be in a fixed vertical plane that has the North Pole as the starting point of the trajectory and New York as the ending point at the moment that the projectile is fired. However, by the time that the projectile arrived at the end point of its trajectory, New York would no longer be there, because while the projectile was in motion, the earth was rotating, and New York will have rotated away from the initial position it was in when the projectile was fired. A person fixed to the rotating earth would see the projectile veer away to the right of its initial path, and would assume that a force was acting on the projectile toward the right of its trajectory. This fictitious force is called the Coriolis force and this seemingly strange behavior occurs because the rotating earth is not an inertial coordinate system.

The Coriolis effect can be applied to the global circulation of air in the atmosphere, causing winds in the northern hemisphere to be deflected to the right of their original path. The global convection cycle described above still occurs, but instead of one huge convection cell, there are three smaller ones, as shown in figure 16.2(b). The winds from the North Pole flowing south at the surface of the earth are deflected to the right of their path and become the polar easterlies, as shown in figure 16.2(b). As the air aloft at the equator flows north it is deflected to the right of its path and eventually flows in a easterly direction at approximately 30° north latitude. The piling up of air at this latitude causes the air aloft to sink to the surface where it emerges from a semipermanent high-pressure area called the subtropical high. The air at the surface that flows north from this high-pressure area is deflected to the right of its path producing the mid-latitude westerlies. The air at the surface that flows south from this high-pressure area is also deflected to the right of its path and produces the northeast trade winds, also shown in figure 16.2(b). *Thus, it is the nonuniform temperature distribution on the surface of the earth that is responsible for the global winds.*

Transfer of thermal energy by convection is also very important in the process called the *sea breeze,* which is shown in figure 16.3. Water has a higher specific heat than land and for the same radiation from the sun, the temperature of the water does not rise as high as the temperature of the land. Therefore, the land mass becomes hotter than the neighboring water. The hot air over the land rises and a cool breeze blows off the ocean to replace the rising hot air. Air aloft descends to replace this cooler sea air and the complete cycle is as shown in figure 16.3. The net result

Figure 16.3
The sea breeze.

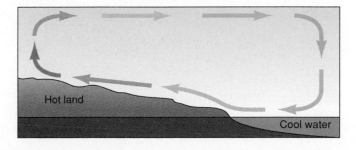

Hot land

Cool water

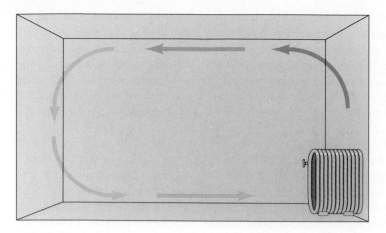

Figure 16.4

Natural convection in a room.

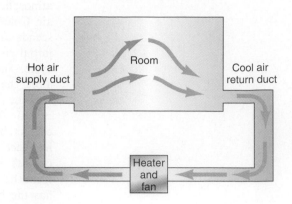

Figure 16.5

Forced convection.

of the process is to replace hot air over the land surface by cool air from the sea. This is one of the reasons why so many people flock to the ocean beaches during the hot summer months. The process reverses at night when the land cools faster than the water. The air then flows from the land to the sea and is called a *land breeze*.

This same process of thermal energy transfer takes place on a smaller scale in any room in your home or office. Let us assume there is a radiator situated at one wall of the room, as shown in figure 16.4. The air in contact with the heater is warmed, and then rises. Cooler air moves in to replace the rising air and a convection cycle is started. The net result of the cycle is to transfer thermal energy from the heater to the rest of the room. All these cases are examples of what is called *natural convection*.

To help the transfer of thermal energy by convection, fans can be used to blow the hot air into the room. Such a hot air heating system, shown in figure 16.5, is called a *forced convection system*. A metal plate is heated to a high temperature in the furnace. A fan blows air over the hot metal plate, then through some ducts, to a low-level vent in the room to be heated. The hot air emerges from the vent and rises into the room. A cold air return duct is located near the floor on the other side of the room, returning cool air to the furnace to start the convection cycle over again. The final result of the process is the transfer of thermal energy from the hot furnace to the cool room.

To analyze the transfer of thermal energy by this forced convection we will assume that a certain amount of mass of air Δm is moved from the furnace to the room. The thermal energy transferred by the convection of this amount of mass Δm is written as

$$\Delta Q = (\Delta m)c\Delta T \qquad (16.7)$$

where $\Delta T = T_h - T_c$, T_h is the temperature of the air at the hot plate of the furnace, and T_c is the temperature of the colder air as it leaves the room. The transfer of thermal energy per unit time becomes

$$\frac{\Delta Q}{\Delta t} = \frac{\Delta m}{\Delta t} c(T_h - T_c)$$

However,

$$m = \rho V$$

therefore

$$\Delta m = \rho \Delta V$$

Therefore, the thermal energy transfer becomes

$$\frac{\Delta Q}{\Delta t} = \rho c \frac{\Delta V}{\Delta t}(T_h - T_c) \qquad (16.8)$$

where $\Delta V/\Delta t$ is the volume flow rate, usually expressed as m^3/min in SI units and cubic feet per minute (CFM) in the British engineering system of units.

Example 16.3

Forced convection. A hot air heating system is rated at 80,000 Btu/hr. If the heated air in the furnace reaches a temperature of 250 °F, the room temperature is 60.0 °F, and the fan can deliver 250 ft^3/min, what is the thermal energy transfer per hour from the furnace to the room, and the efficiency of this system? The specific heat of air at constant pressure in the British engineering system is $c_{air} = 0.250$ Btu/lb °F and the weight density of air is $d_w = 0.0806$ lb/ft^3.

Solution

We find the thermal energy transfer per hour in the British engineering system from a modification of equation 16.8. Instead of the mass density ρ, used in the SI system, the weight density d_w is used in the British engineering system. Therefore, the energy transfer is found as

$$\frac{\Delta Q}{\Delta t} = d_w c \frac{\Delta V}{\Delta t}(T_h - T_c)$$

$$= \left(0.0806 \frac{lb}{ft^3}\right)\left(0.250 \frac{Btu}{lb\ °F}\right)\left(250 \frac{ft^3}{min}\right)(250\ °F - 60.0\ °F)\left(60 \frac{min}{hr}\right)$$

$$= 57,500\ Btu/hr$$

We determine the efficiency, or rated value, of the heater as the ratio of the thermal energy out of the system to the thermal energy in, times 100%. Therefore,

$$\text{Eff} = \left(\frac{57,500\ Btu/hr}{80,000\ Btu/hr}\right)(100\%)$$

$$= 72\%$$

16.3 Conduction

Conduction is the transfer of thermal energy by molecular action, without any motion of the medium. Conduction occurs in solids, liquids, and gases, but the effect is most pronounced in solids. If one end of an iron bar is placed in a fire, in a relatively short time, the other end becomes hot. Thermal energy is conducted from the hot end of the bar to the cold end. The atoms or molecules in the hotter part of the body vibrate around their equilibrium position with greater amplitude than normal. This greater vibration causes the molecules to interact with their nearest neighbors, causing them to vibrate more also. These in turn interact with their nearest neighbors passing on this energy as *kinetic energy of vibration.* The thermal energy is thus passed from molecule to molecule along the entire length of the bar. The net result of these molecular vibrations is a transfer of thermal energy through the solid.

Heat Flow Through a Slab of Material

We can determine the amount of thermal energy conducted through a solid with the aid of figure 16.6. A slab of material of cross-sectional area A and thickness d is subjected to a high temperature T_h on the hot side and a colder temperature T_c on the other side.

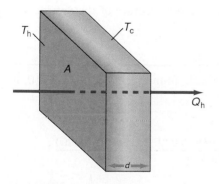

Figure 16.6

Heat conduction through a slab.

It is found experimentally that the thermal energy conducted through this slab is directly proportional to (1) the area A of the slab—the larger the area, the more thermal energy transmitted; (2) the time t—the longer the period of time, the more thermal energy transmitted; and finally (3) the temperature difference, $T_h - T_c$, between the faces of the slab. If there is a large temperature difference, a large amount of thermal energy flows. We can express these observations as the direct proportion

$$Q \propto A(T_h - T_c)t$$

The thermal energy transmitted is also found to be inversely proportional to the thickness of the slab, that is,

$$Q \propto \frac{1}{d}$$

This is very reasonable because the thicker the slab the greater the distance that the thermal energy must pass through. Thus, a thick slab implies a small amount of energy transfer, whereas a thin slab implies a larger amount of energy transfer.

These two proportions can be combined into one as

$$Q \propto \frac{A(T_h - T_c)t}{d} \tag{16.9}$$

To make an equality out of this proportion we must introduce a constant of proportionality. The constant must also depend on the material that the slab is made of, since it is a known fact that different materials transfer different quantities of thermal energy. We will call this constant *the coefficient of thermal conductivity*, and will denote it by k. Equation 16.9 becomes

$$Q = \frac{kA(T_h - T_c)t}{d} \tag{16.10}$$

Equation 16.10 gives the amount of thermal energy transferred by conduction. Table 16.1 gives the thermal conductivity k for various materials. If k is large, then a large amount of thermal energy will flow through the slab, and the material is called a good **conductor** of heat. If k is small then only a small amount of thermal energy will flow through the slab, and the material is called a poor conductor or a good **insulator.** Note from table 16.1 that most metals are good conductors while most nonmetals are good insulators. The ratio $(T_h - T_c)/d$ is the temperature gradient, $\Delta T/\Delta x$. Let us look at some examples of heat conduction.

This thermogram image, which was produced with an infrared scanner, shows heat escaping a house during winter.

Example 16.4

Heat transfer by conduction. Find the amount of thermal energy that flows per day through a solid oak wall 4.00 in. thick, 10.0 ft long, and 8.00 ft high, if the temperature of the inside wall is 70.0 °F while the temperature of the outside wall is 20.0 °F.

Solution

The thermal energy conducted through the wall, found from equation 16.10, is

$$Q = \frac{kA(T_h - T_c)t}{d}$$

$$= \frac{(1.02 \text{ Btu in./hr ft}^2 \text{ °F})(80.0 \text{ ft}^2)(70.0 \text{ °F} - 20.0 \text{ °F})(24 \text{ hr})}{(4.00 \text{ in.})}$$

$$= 24{,}500 \text{ Btu}$$

Note that T_h and T_c are the temperatures of the wall and in general will be different from the air temperature inside and outside the room. The value T_h is usually lower

Table 16.1
Coefficient of Thermal Conductivity for Various Materials

	$\dfrac{\text{kcal}}{\text{m s }^\circ\text{C}}$	$\dfrac{\text{J}}{\text{m s }^\circ\text{C}}$	$\dfrac{\text{Btu in.}}{\text{hr ft}^2\ ^\circ\text{F}}$
Aluminum	0.056	2.34×10^2	1.42×10^3
Brass	0.026	1.09×10^2	0.754×10^3
Copper	0.096	4.02×10^2	2.79×10^3
Gold	0.076	3.13×10^2	2.21×10^3
Iron	0.021	8.79×10^1	0.609×10^3
Lead	0.0085	3.56×10^1	0.247×10^3
Nickel	0.022	9.21×10^1	0.639×10^3
Platinum	0.017	7.12×10^1	0.493×10^3
Silver	0.102	4.27×10^2	2.96×10^3
Zinc	0.028	1.17×10^2	0.813×10^3
Glass	1.89×10^{-4}	7.91×10^{-1}	5.5
Concrete	3.10×10^{-4}	1.30	9.00
Brick	1.55×10^{-4}	6.49×10^{-1}	4.50
Plaster	1.12×10^{-4}	4.69×10^{-1}	3.25
White pine	2.69×10^{-5}	1.13×10^{-1}	0.78
Oak	3.51×10^{-5}	1.47×10^{-1}	1.02
Cork board	8.6×10^{-6}	3.60×10^{-2}	0.25
Sawdust	0.141×10^{-4}	5.90×10^{-2}	0.41
Glass wool	0.099×10^{-4}	4.14×10^{-2}	0.29
Rock wool	0.093×10^{-4}	3.89×10^{-2}	0.27
Nitrogen	6.21×10^{-6}	2.60×10^{-2}	0.180
Helium	3.58×10^{-5}	1.50×10^{-1}	1.04
Air	5.5×10^{-6}	2.30×10^{-2}	0.160

than the room air temperature, whereas T_c is usually higher than the outside air temperature. This thermal energy loss through the wall must be replaced by the home heating unit in order to maintain a comfortable room temperature.

Equivalent Thickness of Various Walls

The walls in most modern homes are insulated with 4 in. of glass wool that is placed within the wall itself. But suppose the walls do not have this glass wool insulation. What should the equivalent thickness of another wall be, in order to give the same amount of insulation as a glass wool wall if the wall is made of (a) concrete, (b) brick, (c) glass, (d) oak wood, or (e) aluminum?

The amount of thermal energy that flows through the wall containing the glass wool, found from equation 16.10, is

$$Q_{gw} = \frac{k_{gw}A(T_h - T_c)t}{d_{gw}}$$

The thermal energy flowing through a concrete wall is given by

$$Q_c = \frac{k_c A(T_h - T_c)t}{d_c}$$

We assume in both equations that the walls have the same area, A; the same temperature difference $(T_h - T_c)$ is applied across each wall; and the thermal energy flows for the same time t. The subscript gw has been used for the wall containing the glass wool and the subscript c for the concrete wall. If both walls provide the same insulation then the thermal energy flow through each must be equal, that is,

$$Q_c = Q_{gw} \tag{16.11}$$

$$\frac{k_c A(T_h - T_c)t}{d_c} = \frac{k_{gw} A(T_h - T_c)t}{d_{gw}} \tag{16.12}$$

Therefore,

$$\frac{k_c}{d_c} = \frac{k_{gw}}{d_{gw}} \tag{16.13}$$

The equivalent thickness of the concrete wall to give the same insulation as the glass wool wall is

$$d_c = \frac{k_c}{k_{gw}} d_{gw} \tag{16.14}$$

Using the values of thermal conductivity from table 16.1 gives for the thickness of the concrete wall

$$d_c = \frac{k_c}{k_{gw}} d_{gw} = \left(\frac{9.00 \text{ Btu in.}/\text{hr ft}^2 \text{ °F}}{0.29 \text{ Btu in.}/\text{hr ft}^2 \text{ °F}} \right)(4.00 \text{ in.})$$

$$d_c = 124 \text{ in.} = 10.3 \text{ ft}$$

Therefore it would take a concrete wall 10 ft thick to give the same insulating ability as a 4-in. wall containing glass wool. Concrete is effectively a thermal sieve. Thermal energy flows through it, almost as fast as if there were no wall present at all. This is why uninsulated basements in most homes are difficult to keep warm.

To determine the equivalent thickness of the brick, glass, oak wood, and aluminum walls, we equate the thermal energy flow through each wall to the thermal energy flow through the wall containing the glass wool as in equations 16.11 and 16.12. We obtain a generalization of equation 16.13 as

$$\frac{k_b}{d_b} = \frac{k_g}{d_g} = \frac{k_{ow}}{d_{ow}} = \frac{k_{Al}}{d_{Al}} = \frac{k_{gw}}{d_{gw}} \tag{16.15}$$

with the results

brick

$$d_b = \frac{k_b}{k_{gw}} d_{gw} = \left(\frac{4.50}{0.29} \right)(4 \text{ in.}) = 62 \text{ in.} = 5.2 \text{ ft}$$

glass

$$d_g = \frac{k_g}{k_{gw}} d_{gw} = \left(\frac{5.5}{0.29} \right)(4 \text{ in.}) = 75.9 \text{ in.} = 6.3 \text{ ft}$$

oak wood

$$d_{ow} = \frac{k_{ow}}{k_{gw}} d_{gw} = \left(\frac{1.02}{0.29} \right)(4 \text{ in.}) = 14.1 \text{ in.} = 1.2 \text{ ft}$$

aluminum

$$d_{Al} = \frac{k_{Al}}{k_{gw}} d_{gw} = \left(\frac{1.42 \times 10^3}{0.29}\right)(4 \text{ in.}) = 19,600 \text{ in.} = 1630 \text{ ft}$$

We see from these results that concrete, brick, glass, wood, and aluminum are not very efficient as insulated walls. A standard wood frame, studded wall with four inches of glass wool placed between the studs is far more efficient.

A few years ago, aluminum siding for the home was very popular. There were countless home improvement advertisements that said, "You can insulate your home with beautiful maintenance free aluminum siding." As you can see from the preceding calculations, such statements were extremely misleading if not outright fraudulent. As just calculated, the aluminum wall would have to be 1630 ft thick, just to give the same insulation as the 4 in. of glass wool. Admittedly, some aluminum siding had a 1/8- to 1/4-in. backing of cork board, which is a good insulating material, but since the heat flow is inversely proportional to the thickness of the wall, four to eight times the amount of thermal energy will flow through the cork board as through the glass wool wall. Aluminum siding may provide a beautiful, maintenance free home, but it will not insulate it. Today most siding for the home is made of vinyl rather than aluminum because vinyl is a good insulator. Most cooking utensils, pots and pans, are made of aluminum because the aluminum will readily conduct the thermal energy from the fire to the food to be cooked.

Another interesting result from these calculations is the realization that a glass window would have to be 6 ft thick to give the same insulation as the 4 in. of glass wool in the normal wall. Since glass windows are usually only about 1/8 in. or less thick, relatively large thermal energy losses are experienced through the windows of the home.

Convection Cycle in the Walls of a Home

All these results are based on the fact that different materials have different thermal conductivities. The smaller the value of k, the better the insulator. If we look carefully at table 16.1, we notice that the smallest value of k is for the air itself, that is, $k = 0.160 \text{ Btu in.}/(\text{hr ft}^2 \, ^\circ\text{F})$. This would seem to imply that if the space between the studs of a wall were left completely empty, that is, if no insulating material were placed in the wall, the air in that space would be the best insulator. Something seems to be wrong, since anyone who has an uninsulated wall in a home knows that there is a tremendous thermal energy loss through it. The reason is that air is a good insulator only if it is not in motion. But the difficulty is that the air in an empty wall is not at rest, as we can see from figure 16.7. Air molecules in contact with the hot wall T_h are heated by this hot wall absorbing a quantity of thermal energy Q. This heated air, being less dense than the surrounding air, rises to the top. The air that was originally at the top now moves down along the cold outside wall. This air is warmer than the cold wall and transmits some of its thermal energy to the cold wall where it is conducted to the outside. The air now sinks down along the outside wall and moves inward to the hot inside wall where it is again warmed and rises. A convection cycle has been established within the wall, whose final result is the absorption of thermal energy Q at the hot wall and its liberation at the cold wall, thereby producing a heat transfer through the wall. A great deal of thermal energy can be lost through the air in the wall, not by conduction, but by convection. If the air could be prevented from moving, that is, by stopping the convection current, then air would be a good insulator. This is basically what is done in using glass wool for insulation. The glass wool consists of millions of fibers of glass that create millions of tiny air pockets. These air pockets cannot move and hence there is no convection. The air between the fibers is still or dead air and acts as a good insulator. It is the dead air that is doing the insulating, not the glass fibers, because as we have just seen glass is not a good insulator.

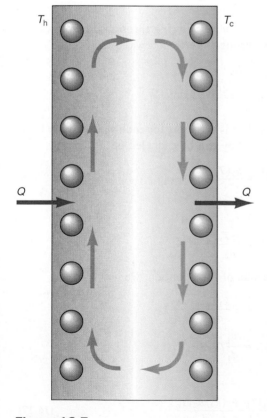

Figure 16.7

Convection currents in an empty wall.

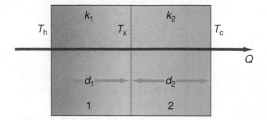

Figure 16.8

The compound wall.

As already mentioned, glass windows are a source of large thermal energy losses in a house. The use of storm windows or thermal windows cuts down on the thermal energy loss significantly. However, even storm windows or thermal windows are not as effective as a normally insulated wall because of the convection currents that occur between the panes of the glass windows.

The Compound Wall

Up to now a wall has been treated as if it consisted of only one material. In general this is not the case. Walls are made up of many different materials of different thicknesses. We solve this more general problem by considering the compound wall in figure 16.8. We assume, for the present, that the wall is made up of only two materials. This assumption will be extended to cover the case of any number of materials later. (The analysis, although simple is a little long. Those students weak in algebra and only interested in the results for the heat conduction through a compound wall can skip ahead to equation 16.18.)

Let us assume that the inside wall is the hot wall and it is at a temperature T_h, whereas the outside wall is the cold wall and it is at a temperature T_c. The temperature at the interface of the two materials is unknown at this time and will be designated by T_x. The first wall has a thickness d_1, and a thermal conductivity k_1, whereas wall 2 has a thickness d_2, and a thermal conductivity k_2. The thermal energy flow through the first wall, given by equation 16.10, is

$$Q_1 = \frac{k_1 A(T_h - T_x)t}{d_1} \tag{16.16}$$

The thermal energy flow through the second wall is given by

$$Q_2 = \frac{k_2 A(T_x - T_c)t}{d_2}$$

Under a steady-state condition, the thermal energy flowing through the first wall is the same as the thermal energy flowing through the second wall. That is,

$$Q_1 = Q_2$$

$$\frac{k_1 A(T_h - T_x)t}{d_1} = \frac{k_2 A(T_x - T_c)t}{d_2}$$

Because the cross-sectional area of the wall A is the same for each wall and the time for the thermal energy flow t is the same, they can be canceled out, giving

$$\frac{k_1(T_h - T_x)}{d_1} = \frac{k_2(T_x - T_c)}{d_2}$$

or

$$\frac{k_1 T_h}{d_1} - \frac{k_1 T_x}{d_1} = \frac{k_2 T_x}{d_2} - \frac{k_2 T_c}{d_2}$$

Placing the terms containing T_x on one side of the equation, we get

$$-\frac{k_1 T_x}{d_1} - \frac{k_2 T_x}{d_2} = -\frac{k_1 T_h}{d_1} - \frac{k_2 T_c}{d_2}$$

$$\left(\frac{k_1}{d_1} + \frac{k_2}{d_2}\right)T_x = \frac{k_1}{d_1}T_h + \frac{k_2}{d_2}T_c$$

Solving for T_x, we get

$$T_x = \frac{(k_1/d_1)T_h + (k_2/d_2)T_c}{k_1/d_1 + k_2/d_2} \tag{16.17}$$

Thermodynamics

If T_x, in equation 16.16, is replaced by T_x, from equation 16.17, we get

$$Q_1 = \frac{k_1 A\{T_h - [(k_1/d_1)T_h + (k_2/d_2)T_c]/(k_1/d_1 + k_2/d_2)\}t}{d_1}$$

$$= \frac{k_1 A}{d_1} \frac{[T_h(k_1/d_1 + k_2/d_2) - (k_1/d_1)T_h - (k_2/d_2) T_c]t}{k_1/d_1 + k_2/d_2}$$

$$= \frac{k_1 A}{d_1} \frac{[(k_1/d_1)T_h + (k_2/d_2)T_h - (k_1/d_1)T_h - (k_2/d_2)T_c]t}{k_1/d_1 + k_2/d_2}$$

$$= \frac{k_1 A k_2 (T_h - T_c)t}{d_1 d_2 (k_1/d_1 + k_2/d_2)}$$

$$= \frac{A(T_h - T_c)t}{(d_1 d_2/k_1 k_2)(k_1/d_1 + k_2/d_2)}$$

$$= \frac{A(T_h - T_c)t}{d_2/k_2 + d_1/k_1}$$

The thermal energy flow Q_1 through the first wall is equal to the thermal energy flow Q_2 through the second wall, which is just the thermal energy flow Q going through the compound wall. Therefore, the thermal energy flow through the compound wall is given by

$$Q = \frac{A(T_h - T_c)t}{d_1/k_1 + d_2/k_2} \tag{16.18}$$

If the compound wall had been made up of more materials, then there would be additional terms, d_i/k_i, in the denominator of equation 16.18 for each additional material. That is,

$$Q = \frac{A(T_h - T_c)t}{\sum\limits_{i=1}^{n} d_i/k_i} \tag{16.19}$$

The problem is usually simplified further by defining a new quantity called the **thermal resistance** R, or the **R value of the insulation,** as

$$R = \frac{d}{k} \tag{16.20}$$

The thermal resistance R acts to impede the flow of thermal energy through the material. The larger the value of R, the smaller the quantity of thermal energy conducted through the wall. For a compound wall, the total thermal resistance to thermal energy flow is simply

$$R_{\text{total}} = \frac{d_1}{k_1} + \frac{d_2}{k_2} + \frac{d_3}{k_3} + \frac{d_4}{k_4} + \cdots \tag{16.21}$$

or

$$R_{\text{total}} = R_1 + R_2 + R_3 + R_4 + \cdots \tag{16.22}$$

And the thermal energy flow through a compound wall is given by

$$Q = \frac{A(T_h - T_c)t}{\sum\limits_{i=1}^{n} R_i} \tag{16.23}$$

Example 16.5

Heat flow through a compound wall. A wall 10.0 ft by 8.00 ft is made up of a thickness of 4.00 in. of brick, 4.00 in. of glass wool, 1/2 in. of plaster, and 1/4 in. of oak wood paneling. If the inside temperature of the wall is $T_h = 65.0\,°F$ and the outside temperature is 20.0 °F, how much thermal energy flows through this wall per day?

The R value of each material, found with the aid of table 16.1, is

$$R_{\text{brick}} = \frac{d_{\text{brick}}}{k_{\text{brick}}} = \frac{4.00 \text{ in.}}{4.50 \text{ Btu in./hr ft}^2 \text{ °F}} = 0.88 \frac{\text{hr ft}^2 \text{ °F}}{\text{Btu}}$$

$$R_{\text{glass wool}} = \frac{d_{\text{gw}}}{k_{\text{gw}}} = \frac{4.00 \text{ in.}}{0.29 \text{ Btu in./hr ft}^2 \text{ °F}} = 13.8 \frac{\text{hr ft}^2 \text{ °F}}{\text{Btu}}$$

$$R_{\text{plaster}} = \frac{d_p}{k_p} = \frac{0.500 \text{ in.}}{3.25 \text{ Btu in./hr ft}^2 \text{ °F}} = 0.15 \frac{\text{hr ft}^2 \text{ °F}}{\text{Btu}}$$

$$R_{\text{wood}} = \frac{d_w}{k_w} = \frac{0.250 \text{ in.}}{1.02 \text{ Btu in./hr ft}^2 \text{ °F}} = 0.25 \frac{\text{hr ft}^2 \text{ °F}}{\text{Btu}}$$

The R value of the total compound wall, found from equation 16.22, is

$$R = R_1 + R_2 + R_3 + R_4 = 0.88 + 13.8 + 0.15 + 0.25$$

$$= 15.1 \frac{\text{hr ft}^2 \text{ °F}}{\text{Btu}}$$

Note that the greatest portion of the thermal resistance comes from the glass wool. The total thermal energy conducted through the wall, found from equation 16.23, is

$$Q = \frac{A(T_h - T_c)t}{\sum\limits_{i=1}^{n} R_i}$$

$$= \frac{(10.0 \text{ ft})(8.00 \text{ ft})(65.0 \text{ °F} - 20.0 \text{ °F})(24 \text{ hr})}{15.1 \text{ hr ft}^2 \text{ °F/Btu}}$$

$$= 5720 \text{ Btu}$$

Note that if there were no glass wool in the wall, the R value would be $R = 1.28$, and the thermal energy conducted through the wall would be 67,500 Btu, almost 12 times as much as the insulated wall. Remember, all these heat losses must be replaced by the home furnace in order to keep the temperature inside the home reasonably comfortable, and will require the use of fuel for this purpose. Finally, we should note that there is also a great heat loss in the winter through the roof of the house. To eliminate this energy loss there should be at least 6 in. of insulation in the roof of the house, and in some locations 12 in. is preferable.

You should note that when you buy insulation for your home in your local lumberyard or home materials store, you will see ratings such as an R value of 12 for a nominal 4 in. of glass wool insulation, or an R value of 19 for a nominal 6 in. of glass wool insulation. The units associated with these numbers are as given here for the British engineering system of units, namely

$$\frac{\text{hr ft}^2 \text{ °F}}{\text{Btu}}$$

which is in the standard form used in the American construction industry today. So when using these products you must use the British engineering system of units for your calculations. You can still use the definition of $R = d/k$ in problems in SI units, but then the units for R will be

$$\frac{\text{s m}^2 \text{ °C}}{\text{J}}$$

and the numerical values will not correspond to the R values listed on the insulation itself.

Everything that has been said about insulating our homes to prevent the loss of thermal energy in the winter, also applies in the summer. Only then the problem is reversed. The hot air is outside the house and the cool air is inside the house. The insulation will decrease the conduction of thermal energy through the walls into the room, keeping the room cool and cutting down or eliminating the use of air conditioning to cool the home.

16.4 Radiation

Radiation is the transfer of thermal energy by electromagnetic waves. As pointed out in chapter 12 on wave motion, any wave is characterized by its wavelength λ and frequency[1] ν. The electromagnetic waves in the visible portion of the spectrum are called *light waves*. These light waves have wavelengths that vary from about 0.38×10^{-6} m for violet light to about 0.72×10^{-6} m for red light. Above visible red light there is an invisible, infrared portion of the electromagnetic spectrum. The wavelengths range from 0.72×10^{-6} m to 1.5×10^{-6} m for the near infrared, from 1.5×10^{-6} m to 5.6×10^{-6} m for the middle infrared, and from 5.6×10^{-6} m up to 1×10^{-3} m for the far infrared. Most, but not all, of the radiation from a hot body falls in the infrared region of the electromagnetic spectrum. Every thing around you is radiating electromagnetic energy, but the radiation is in the infrared portion of the spectrum, which your eyes are not capable of detecting. Therefore, you are usually not aware of this radiation.

The Stefan-Boltzmann Law

Joseph Stefan (1835–1893) found experimentally, and Ludwig Boltzmann (1844–1906) found theoretically, that every body at an absolute temperature T radiates energy that is proportional to the fourth power of the absolute temperature. The result, which is called the **Stefan-Boltzmann law** is given by

$$Q = e\sigma AT^4 t \tag{16.24}$$

where Q is the thermal energy emitted; e is the emissivity of the body, which varies from 0 to 1; σ is a constant, called the Stefan-Boltzmann constant and is given by

$$\sigma = 5.67 \times 10^{-8} \frac{\text{J}}{\text{s m}^2 \text{ K}^4}$$

A is the area of the emitting body; T is the absolute temperature of the body; and t is the time.

Radiation from a Blackbody

The amount of radiation depends on the radiating surface. Polished surfaces are usually poor radiators, while blackened surfaces are usually good radiators. Good radiators of heat are also good absorbers of radiation, while poor radiators are also poor absorbers. *A body that absorbs all the radiation incident upon it is called a blackbody.* The name blackbody is really a misnomer, since the sun acts as a blackbody and it is certainly not black. A blackbody is a perfect absorber and a perfect emitter. The substance lampblack, a finely powdered black soot, makes a very good approximation to a blackbody. A box, whose insides are lined with a black material like lampblack, can act as a blackbody. If a tiny hole is made in the side of the box and then a light wave is made to enter the box through the hole, the light wave will be absorbed and re-emitted from the walls of the box, over and over. Such a device is called a *cavity resonator*. For a blackbody, the emissivity e in equation 16.24 is equal to 1. *The amount of heat absorbed or emitted from a blackbody is*

$$Q = \sigma AT^4 t \tag{16.25}$$

---

1. When dealing with electromagnetic waves, the symbol ν (Greek letter nu) is used to designate the frequency instead of the letter f used for conventional waves.

Example 16.6

Energy radiated from the sun. If the surface temperature of the sun is approximately 5800 K, how much thermal energy is radiated from the sun per unit time? Assume that the sun can be treated as a blackbody.

Solution

We can find the energy radiated from the sun per unit time from equation 16.25. The radius of the sun is about 6.96×10^8 m. Its area is therefore

$$A = 4\pi r^2 = 4\pi(6.96 \times 10^8 \text{ m})^2$$

$$= 6.09 \times 10^{18} \text{ m}^2$$

The heat radiated from the sun is therefore

$$\frac{Q}{t} = \sigma A T^4$$

$$= \left(5.67 \times 10^{-8} \frac{\text{J}}{\text{s m}^2 \text{ K}^4}\right)(6.09 \times 10^{18} \text{ m}^2)(5800 \text{ K})^4$$

$$= 3.91 \times 10^{26} \text{ J/s} = 9.33 \times 10^{22} \text{ kcal/s}$$

Example 16.7

The solar constant. How much energy from the sun impinges on the top of the earth's atmosphere per unit time per unit area?

Solution

The energy per unit time emitted by the sun is power and was found in example 16.6 to be 3.91×10^{26} J/s. This total power emitted by the sun does not all fall on the earth because that power is distributed throughout space, in all directions, figure 16.9. Hence, only a small portion of it is emitted in the direction of the earth. To find the amount of that power that reaches the earth, we first find the distribution of that power over a sphere, whose radius is the radius of the earth's orbit, $r = 1.5 \times 10^{11}$ m. This gives us the power, or energy per unit time, falling on a unit area at the distance of the earth from the sun. The area of this sphere is

$$A = 4\pi r^2 = 4\pi(1.5 \times 10^{11} \text{ m})^2$$

$$= 2.83 \times 10^{23} \text{ m}^2$$

The energy per unit area per unit time impinging on the earth is therefore

$$\frac{Q}{At} = \frac{3.91 \times 10^{26} \text{ J/s}}{2.83 \times 10^{23} \text{ m}^2} = 1.38 \times 10^3 \frac{\text{W}}{\text{m}^2}$$

This value, 1.38×10^3 W/m², *the energy per unit area per unit time impinging on the edge of the atmosphere, is called the* **solar constant,** *and is designated as S_0.*

Example 16.8

Solar energy reaching the earth. Find the total energy from the sun impinging on the top of the atmosphere during a 24-hr period.

Solution

The actual power impinging on the earth at the top of the atmosphere can be found by multiplying the solar constant S_0 by the effective area A subtended by the earth. The area subtended by the earth is found from the area of a disk whose radius is equal to the mean radius of the earth, $R_E = 6.37 \times 10^6$ m. That is,

$$A = \pi R_E^2 = \pi(6.37 \times 10^6 \text{ m})^2 = 1.27 \times 10^{14} \text{ m}^2$$

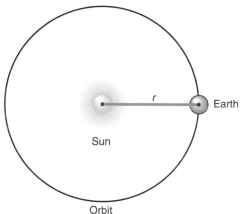

Figure 16.9

Radiation received on the earth from the sun.

An array of solar collectors collect energy from the sun.

Power impinging on earth = (Solar constant)(Area)

$$P = \left(1.38 \times 10^3 \frac{W}{m^2}\right)(1.27 \times 10^{14}\ m^2) = 1.76 \times 10^{17}\ W$$

The energy impinging on the earth in a 24-hr period is found from

$$Q = Pt = (1.76 \times 10^{17}\ W)(24\ hr)(3600\ s/hr)$$

$$= 1.52 \times 10^{22}\ J$$

This is an enormous quantity of energy. Obviously, solar energy, as a source of available energy for the world needs to be tapped.

All the solar energy incident on the upper atmosphere does not make it down to the surface of the earth because of reflection from clouds, scattering by dust particles in the atmosphere; and some absorption by water vapor, carbon dioxide, and ozone in the atmosphere. What is even more interesting is that this enormous energy received by the sun is reradiated back into space. If the earth did not re-emit this energy the mean temperature of the earth would constantly rise until the earth burned up.

A body placed in any environment absorbs energy from the environment. The net energy absorbed by the body Q is equal to the difference between the energy absorbed by the body from the environment Q_A and the energy radiated by the body to the environment Q_R, that is,

$$Q = Q_A - Q_R \tag{16.26}$$

If T_B is the absolute temperature of the radiating body and T_E is the absolute temperature of the environment, then the net heat absorbed by the body is

$$Q = Q_A - Q_R = e_E \sigma A T_E^4 t - e_B \sigma A T_B^4 t$$

$$Q = \sigma A (e_E T_E^4 - e_B T_B^4) t \tag{16.27}$$

where e_E is the emissivity of the environment and e_B is the emissivity of the body. In general these values, which are characteristic of the particular body and environment, must be determined experimentally. If the body and the environment can be approximated as blackbodies, then $e_B = e_E = 1$, and equation 16.27 reduces to the simpler form

$$Q = \sigma A (T_E^4 - T_B^4) t \tag{16.28}$$

If the value of Q comes out negative, it represents a net loss of energy from the body.

Example 16.9

Look at that person radiating. A person, at normal body temperature of 98.6 °F (37 °C) stands near a wall of a room whose temperature is 50.0 °F (10 °C). If the person's surface area is approximately 2.00 m², how much heat is lost from the person per minute?

Solution

The absolute temperature of the person is 310 K while the absolute temperature of the wall is 283 K. Let us assume that we can treat the person and the wall as blackbodies, then the heat lost by the person, given by equation 16.28, is

$$Q = \sigma A (T_E^4 - T_B^4) t$$

$$= \left(5.67 \times 10^{-8} \frac{J}{s\ m^2}\right)\left(\frac{2.00\ m^2}{K^4}\right)[(283\ K)^4 - (310\ K)^4](60.0\ s)$$

$$= -320\ J = -0.076\ kcal$$

This thermal energy lost must be replaced by food energy. This result is of course only approximate, since the person is not a blackbody and no consideration was taken into account for the shape of the body and the insulation effect of the person's clothes.

†Blackbody Radiation as a Function of Wavelength

The Stefan-Boltzmann law tells us only about the total energy emitted and nothing about the wavelengths of the radiation. Because all this radiation consists of electromagnetic waves, the energy is actually distributed among many different wavelengths. The energy distribution per unit area per unit time per unit frequency $\Delta \nu$ is given by a relation known as **Planck's radiation law** as

$$\frac{Q}{At\Delta \nu} = \frac{2\pi h\nu^3}{c^2}\left(\frac{1}{e^{h\nu/kT} - 1}\right) \tag{16.29}$$

where c is the speed of light and is equal to 3×10^8 m/s, ν is the frequency of the electromagnetic wave, e is a constant equal to 2.71828 and is the base e used in natural logarithms, k is the Boltzmann constant given in chapter 15, and h is a new constant, called Planck's constant, given by

$$h = 6.625 \times 10^{-34} \text{ J s}$$

This analysis of blackbody radiation by Max Planck (1858–1947) was revolutionary in its time (December 1900) because Planck assumed that energy was quantized into little bundles of energy equal to $h\nu$. This was the beginning of what has come to be known as quantum mechanics, which will be discussed later in chapter 31. Equation 16.29 can also be expressed in terms of the wavelength λ as

$$\frac{Q}{At\Delta \lambda} = \frac{2\pi hc^2}{\lambda^5}\left(\frac{1}{e^{hc/\lambda kT} - 1}\right) \tag{16.30}$$

A plot of equation 16.30 is shown in figure 16.10 for various temperatures. Note that $T_4 > T_3 > T_2 > T_1$. The first thing to observe in this graph is that the intensity of the radiation for a given temperature varies with the wavelength from zero up to a maximum value and then decreases. That is, for any one temperature, there is one wavelength λ_{max} for which the intensity is a maximum. Second, as the temperature increases, the wavelength λ_{max} where the maximum or peak intensity occurs shifts to shorter wavelengths. This was recognized earlier by the German physicist Wilhelm Wien (1864–1928) and was written in the form

$$\lambda_{max}T = \text{constant} = 2.898 \times 10^{-3} \text{ m K} \tag{16.31}$$

and was called the **Wien displacement law.** Third, the visible portion of the electromagnetic spectrum (shown in the hatched area) is only a small portion of the spec-

Figure 16.10

The intensity of blackbody radiation as a function of wavelength and temperature.

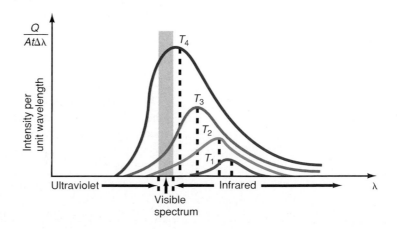

Thermodynamics

trum, and most of the radiation from a blackbody falls in the infrared range of the electromagnetic spectrum. Because our eyes are not sensitive to these wavelengths, the infrared radiation coming from a hot body is invisible. But as the temperature of the blackbody rises, the peak intensity shifts to lower wavelengths, until, when the temperature is high enough, some of the blackbody radiation is emitted in the visible red portion of the spectrum and the heated body takes on a red glow. If the temperature continues to rise, the red glow becomes a bright red, then an orange, then yellow-white, and finally blue-white as the blackbody emits more and more radiation in the visible range. When the blackbody emits all wavelengths in the visible portion of the spectrum, it appears white. (The visible range of the electromagnetic spectrum, starting from the infrared end, has the colors red, orange, yellow, green, blue, and violet before the ultraviolet portion of the spectrum begins.)

Example 16.10

The wavelength of the maximum intensity of radiation from the sun. Find the wavelength of the maximum intensity of radiation from the sun, assuming the sun to be a blackbody at 5800 K.

Solution

The wavelength of the maximum intensity of radiation from the sun is found from the Wien displacement law, equation 16.31, as

$$\lambda_{max} = \frac{2.898 \times 10^{-3}}{T} \text{ m K}$$

$$= \frac{2.898 \times 10^{-3}}{5800 \text{ K}} \text{ m K}$$

$$= 0.499 \times 10^{-6} \text{ m} = 0.499 \ \mu m$$

That is, the wavelength of the maximum intensity from the sun lies at 0.499 μm, which is in the blue-green portion of the visible spectrum. It is interesting to note that some other stars, which are extremely hot, radiate mostly in the ultraviolet region.

"Have you ever wondered . . . ?"

An Essay on the Application of Physics

The Greenhouse Effect and Global Warming

Have you ever wondered what the newscaster was talking about when she said that the earth is getting warmer because of the Greenhouse Effect? What is the Greenhouse Effect and what does it have to do with the heating of the earth?

The name Greenhouse Effect comes from the way the earth and its atmosphere is heated. The ultimate cause of heating of the earth's atmosphere is the sun. But if this is so, then why is the top of the atmosphere (closer to the sun) colder than the lower atmosphere (farther from the sun)? You may have noticed

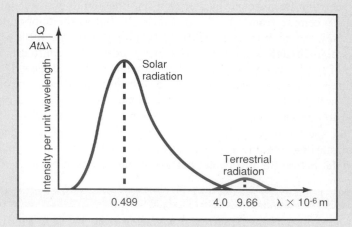

Figure 1

Comparison of radiation from the sun and the earth.

snow and ice on the colder mountain tops while the valleys below are relatively warm. We can explain this paradox in terms of the radiation of the sun, the radiation of the earth, and the constituents of the atmosphere. The sun radiates approximately as a blackbody at 5800 K with a peak intensity occurring at 0.499 $\times$ 10^{-6} m, as shown in figure 1.

Example 16H.1

The wavelength of the maximum intensity of radiation from the earth. Assuming that the earth has a mean temperature of about 300 K use the Wien displacement law to estimate the wavelength of the peak radiation from the earth.

Solution

The wavelength of the peak radiation from the earth, found from equation 16.31, is

$$\lambda_{max} = \frac{2.898 \times 10^{-3}}{T} \text{ m K} = \frac{2.898 \times 10^{-3} \text{ m K}}{300 \text{ K}}$$

$$= 9.66 \times 10^{-6} \text{ m}$$

which is also shown in figure 1. Notice that the maximum radiation from the earth lies well in the longer wave infrared region, whereas the maximum solar radiation lies in much shorter wavelengths. (Ninety-nine percent of the solar radiation is in wavelengths shorter than 4.0 μm, and almost all terrestrial radiation is at wavelengths greater than 4.0 μm.) Therefore, *solar radiation is usually referred to as short-wave radiation, while terrestrial radiation is usually referred to as long-wave radiation.*

Of all the gases in the atmosphere only oxygen, ozone, water vapor, and carbon dioxide are significant absorbers of radiation. Moreover these gases are selective absorbers, that is, they absorb strongly in some wavelengths and hardly at all in others. The absorption spectrum for oxygen and ozone is shown in figure 2(b).

The absorption of radiation is plotted against the wavelength of the radiation. An absorptivity of 1 means total absorption at that wavelength, whereas an absorptivity of 0 means that the gas does not absorb any radiation at that wavelength. Thus, when the absorptivity is 0, the gas is totally transparent to that wavelength of radiation. Observe from figure 2(b) that oxygen and ozone absorb almost all the ultraviolet radiation from the sun in wavelengths below 0.3 μm. A slight amount of ultraviolet light from the sun reaches the earth in the range 0.3 μm to the beginning of visible light in the violet at 0.38 μm. Also notice that oxygen and ozone are almost transparent to radiation in the visible and infrared region of the electromagnetic spectrum.

Figure 2(d) shows the absorption spectrum for water vapor (H_2O). Notice that there is no absorption in the ultraviolet or visible region of the electromagnetic spectrum for water vapor. However, there are a significant number of regions in the infrared where water vapor does absorb radiation.

Figure 2(c) shows the absorption spectrum for carbon dioxide (CO_2). Notice that there is no absorption in the ultraviolet or visible region of the electromagnetic spectrum for carbon dioxide. However, there are a significant number of regions in the infrared where carbon dioxide does absorb radiation. The bands are not quite as wide as for water vapor, but they are very significant as we will see shortly. Also note in figure 2(a) that nitrous oxide (N_2O) also absorbs some energy in the infrared portion of the spectrum.

Figure 2(e) shows the combined absorption spectrum for the atmosphere. We can see that the atmosphere is effectively transparent in the visible portion of the spectrum. Because the peak of the sun's radiation falls in this region, the atmosphere is effectively transparent to most of the sun's rays, and hence most of the sun's radiation passes through the atmosphere as if there were no atmosphere at all. The atmosphere is like an open window to let in all the sun's rays. Hence, the sun's rays pass directly through the atmosphere where they are then absorbed by the surface of the earth. The earth then reradiates as a blackbody, but since its average temperature is so low (250–300 K), its radiation is all in the infrared region as was shown in figure 1. But the water vapor, H_2O, and carbon dioxide, CO_2, in the atmosphere absorb almost all the energy in the infrared region. Thus, *the earth's atmosphere is mainly heated by the absorption of the infrared radiation from the earth.* Therefore, the air closest to the ground becomes warmer than air at much higher altitudes, and therefore the temperature of the atmosphere decreases with height. The warm air at the surface rises by convection, distributing the thermal energy throughout the rest of the atmosphere.

This process of heating the earth's atmosphere by terrestrial radiation is called the Greenhouse Effect. The reason for the name is that it was once thought that this was the way a greenhouse was heated. That is, short-wavelength radiation from the sun passed through the glass into the greenhouse. The plants and ground in the greenhouse absorbed this short-wave radiation and reradiated in the infrared. The glass in the greenhouse was essentially opaque to this infrared radiation and reflected this radiation back into the greenhouse thus keeping the greenhouse warm. Because the mechanism for heating the atmosphere was

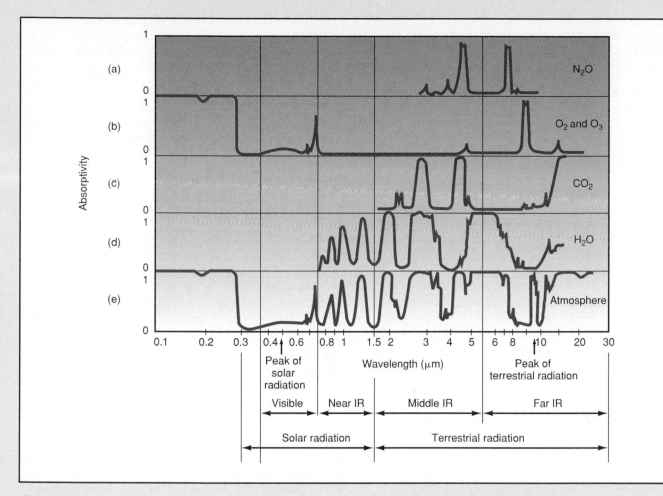

Figure 2

Absorption of radiation at various wavelengths for atmospheric constituents. Lutgens/Tarbuck, *The Atmosphere,* 3/E, p. 44. © Prentice-Hall, Inc., Englewood Cliffs, NJ.

thought to be similar to the mechanism for heating the greenhouse, the heating of the atmosphere came to be called the Greenhouse Effect. (It has since been shown that the dominant reason for keeping the greenhouse warm is the prevention of the convection of the hot air out of the greenhouse by the glass. However, the name Greenhouse Effect continues to be used.)

Because carbon dioxide is an absorber of the earth's infrared radiation, it has led to a concern over the possible warming of the atmosphere caused by excessive amounts of carbon dioxide that comes from the burning of fossil fuels, such as coal and oil, and the deforestation of large areas of trees, whose leaves normally absorb some of the excess carbon dioxide in the atmosphere. "For example, since 1958 concentrations of CO_2 have increased from 315 to 352 parts per million, an increase of approximately 15%."[1] Also, "During the last 100–200 years carbon

dioxide has increased by 25%."[2] And "Everyday 100 square miles of rain forest go up in smoke, pumping one billion tons of carbon dioxide into the atmosphere."[3]

Almost everyone agrees that the increase in carbon dioxide in the atmosphere is not beneficial, but this is where the agreement ends. There is wide disagreement on the consequences of this increased carbon dioxide level. Let us first describe the two most extreme views.

One scenario says that the increased level of CO_2 will cause the mean temperature of the atmosphere to increase. This increased temperature will cause the polar ice caps to melt and increase the height of the mean sea level throughout the world. This in turn will cause great flooding in the low-lying regions of the world. The increased temperature is also assumed to cause the destruction of much of the world's crops and hence its food supply.

A second scenario says that the increased temperatures from the excessive carbon dioxide will cause greater evaporation from the oceans and hence greater cloud cover over the entire globe.

1. "Computer Simulation of the Greenhouse Effect," Washington, Warren M. and Bettge, Thomas W., *Computers in Physics,* May/June 1990.

2. "Climate and the Earth's Radiation Budget," Ramanathan, V.; Barkstrom, Bruce R.; and Harrison, Edwin F., *Physics Today,* May 1989.

3. NOVA TV series, "The Infinite Voyage, Crisis in the Atmosphere."

It is then assumed that this greater cloud cover will reflect more of the incident solar radiation into space. This reflected radiation never makes it to the surface of the earth to heat up the surface. Less radiation comes from the earth to be absorbed by the atmosphere and hence there is a decrease in the mean temperature of the earth. This lower temperature will then initiate the beginning of a new ice age.

Thus one scenario has the earth burning up, the other has it freezing down. It is obvious from these two scenarios that much greater information on the effect of the increase in carbon dioxide in the atmosphere is necessary.

Another way to look at the Greenhouse Effect is to consider the earth as a planet in space that is in equilibrium between the incoming solar radiation and the outgoing terrestrial radiation. As we saw in example 16.7, the amount of energy per unit area per unit time falling on the earth from the sun is given by the solar constant, $S_0 = 1.38 \times 10^3$ J/(s m^2). The actual energy per unit time impinging on the earth at the top of the atmosphere can be found by multiplying the solar constant S_0 by the effective area A_d subtended by the earth. That is, $Q/t = S_0 A_d$. The area subtended by the earth A_d is found from the area of a disk whose radius is equal to the mean radius of the earth. That is, $A_d = \pi R_E^2$.

The solar radiation reaching the surface of the earth is equal to the solar radiation impinging on the top of the atmosphere $S_0 A_d$ minus the amount of solar radiation reflected from the atmosphere, mostly from clouds. The albedo of the earth a, the ratio of the amount of radiation reflected to the total incident radiation, has been measured by satellites to be $a = 0.300$. Hence the amount of solar energy reaching the earth per second is given by

$$\frac{Q}{t} = S_0 A_d - a S_0 A_d = S_0 A_d (1 - a)$$

Assuming that the earth radiates as a blackbody it will emit the radiation

$$\frac{Q}{t} = \sigma A_s T^4$$

The radiating area of the earth, $A_s = 4\pi R_E^2$, is the spherical area of the earth because the earth is radiating everywhere, not only in the region where it is receiving radiation from the sun. Because the earth must be in thermal equilibrium in its position in space, the radiation in must equal the radiation out, or

$$\frac{Q}{t} = S_0 A_d (1 - a) = \sigma A_s T^4$$

Solving for the temperature T of the earth, we get

$$T^4 = \frac{S_0 A_d (1 - a)}{\sigma A_s} = \frac{S_0 \pi R_E^2 (1 - a)}{\sigma 4\pi R_E^2}$$

$$= \frac{S_0 (1 - a)}{4\sigma}$$

$$= \frac{[1.38 \times 10^3 \text{ J/(s m}^2)](1 - 0.300)}{4[5.67 \times 10^{-8} \text{ J/(s m}^2 \text{ K}^4)]}$$

$$T = 255 \text{ K}$$

That is the radiative equilibrium temperature of the earth should be 255 K. This mean radiative temperature of 255 K is sometimes called the *planetary temperature* and/or the *effective temperature* of the earth. It is observed, however, that the mean temperature of the surface of the earth, averaged over time and place, is actually 288 K, some 33 K higher than this temperature.[4] This difference in the mean temperature of the earth is attributed to the Greenhouse Effect. That is, the energy absorbed by the water vapor and carbon dioxide in the atmosphere causes the surface of the earth to be much warmer than if there were no atmosphere. It is for this reason that environmentalists are so concerned with the abundance of carbon dioxide in the atmosphere.

As a contrast let us consider the planet Venus, whose main constituent in the atmosphere is carbon dioxide. Performing the same calculation for the solar constant in example 16.7, only using the orbital radius of Venus of 1.08×10^{11} m, gives a solar constant of 2668 W/m^2, roughly twice that of the earth. The mean albedo of Venus is about 0.80 because of the large amount of clouds covering the planet. Performing the same calculation for the planetary temperature of Venus gives 220 K. Even though the solar constant is roughly double that of the earth, because of the very high albedo, the planetary temperature is some 30 K colder than the earth. However, the surface temperature of Venus has been found to be 750 K due to the very large amount of carbon dioxide in the atmosphere. Hence the Greenhouse Effect on Venus has caused the mean surface temperature to be 891 °F. There is apparently no limitation to the warming that can result from the Greenhouse Effect.

More detailed computer studies of the earth's atmosphere, using general circulation models (GCM), have been made. In these models, it is assumed that the amount of carbon dioxide in the atmosphere has doubled and the model predicts the general condition of the atmosphere over a period of twenty years. The model indicates a global warming of about 4.0 to 4.5 °C. (A temperature of 4 or 5 °C may not seem like much, but when you recall that the mean temperature of the earth during an ice age was only 3 °C cooler than presently, the variation can be quite significant.) The effect of the warming was to cause greater extremes of temperature. That is, hot areas were hotter than normal, while cold areas were colder than normal.

Stephen H. Schneider[5] has said, "Sometime between 15,000 and 5,000 years ago the planet warmed up 5 °C. Sea levels rose 300 feet and forests moved. Literally that change in 5 °C revamped the ecological face of this planet. Species went extinct,

4. We should note that the radiative temperature of the earth is 255 K. This is a mean temperature located somewhere in the middle of the atmosphere. The surface temperature is much higher and temperatures in the very upper atmosphere are much lower, giving the mean of 255 K.
5. Stephen H. Schneider, *Global Warming*, Sierra Club Books, San Francisco, 1989.

others grew. It took nature about 10,000 years to do that. That's the natural rate of change. We're talking about a 5 °C change from our climate models in one century."

Still with all this evidence many scientists are reluctant to make a definitive stand on the issue of global warming. As an example, "No 'smoking gun' evidence exists, however, to prove that the Earth's global climate is warming (versus a natural climate variability) or, if it is warming, whether that warming is caused by the increase in carbon dioxide. Recent estimates show a warming trend, but unfortunately many problems and limitations of observed data make difficult the exact determination of temperature trends."[6]

Still one concern remains. If we wait until we are certain that there is a global warming caused by the increase of carbon dioxide in the air, will we be too late to do anything about it?

6. "Computer Simulation of the Greenhouse Effect," Washington, Warren M. and Bettge, Thomas W., *Computers in Physics*, May/June 1990.

The Language of Physics

Convection
The transfer of thermal energy by the actual motion of the medium itself (p. 451).

Conduction
The transfer of thermal energy by molecular action. Conduction occurs in solids, liquids, and gases, but the effect is most pronounced in solids (p. 451).

Radiation
The transfer of thermal energy by electromagnetic waves (p. 451).

Isotherm
A line along which the temperature is a constant (p. 451).

Temperature gradient
The rate at which the temperature changes with distance (p. 453).

Coriolis effect
On a rotating coordinate system, such as the earth, objects in straight line motion appear to be deflected to the right of their straight line path. Their actual motion in space is straight, but the earth rotates out from under them. The direction of the prevailing winds is a manifestation of the Coriolis effect (p. 455).

Conductor
A material that easily transmits heat by conduction. A conductor has a large value of thermal conductivity (p. 458).

Insulator
A material that is a poor conductor of heat. An insulator has a small value of thermal conductivity (p. 458).

Thermal resistance, or *R* value of an insulator
The ratio of the thickness of a piece of insulating material to its thermal conductivity (p. 463).

Stefan-Boltzmann law
Every body radiates energy that is proportional to the fourth power of the absolute temperature of the body (p. 465).

Blackbody
A body that absorbs all the radiation incident upon it. A blackbody is a perfect absorber and a perfect emitter. The substance lampblack, a finely powdered black soot, makes a very good approximation to a blackbody. The name is a misnomer, since many bodies, such as the sun, act like blackbodies and are not black (p. 465).

Solar constant
The power per unit area impinging on the edge of the earth's atmosphere. It is equal to 1.38×10^3 W/m² (p. 466).

Planck's radiation law
An equation that shows how the energy of a radiating body is distributed over the emitted wavelengths. Planck assumed that the radiated energy was quantized into little bundles of energy, eventually called quanta (p. 468).

Wien displacement law
The product of the wavelength that gives maximum radiation times the absolute temperature is a constant (p. 468).

Summary of Important Equations

Heat transferred by convection
$$\Delta Q = vmc\frac{\Delta T}{\Delta x}\Delta t \tag{16.4}$$

$$\Delta Q = v\rho Vc\frac{\Delta T}{\Delta x}\Delta t \tag{16.6}$$

$$\frac{\Delta Q}{\Delta t} = \rho c\frac{\Delta V}{\Delta t}(T_h - T_c) \tag{16.8}$$

Heat transferred by conduction
$$Q = \frac{kA(T_h - T_c)t}{d} \tag{16.10}$$

Heat transferred by conduction through a compound wall
$$Q = \frac{A(T_h - T_c)t}{\sum\limits_{i=1}^{n} d_i/k_i} \tag{16.19}$$

$$Q = \frac{A(T_h - T_c)t}{\sum\limits_{i=1}^{n} R_i} \tag{16.23}$$

R value of insulation
$$R = \frac{d}{k} \tag{16.20}$$

Stefan-Boltzmann law, heat transferred by radiation
$$Q = e\sigma AT^4t \tag{16.24}$$

Radiation from a blackbody
$$Q = \sigma AT^4t \tag{16.25}$$

Energy absorbed by radiation from environment
$$Q = \sigma A(e_E T_E^4 - e_B T_B^4)t \tag{16.27}$$

Planck's radiation law
$$\frac{Q}{At\Delta\nu} = \frac{2\pi h\nu^3}{c^2}\left(\frac{1}{e^{h\nu/kT} - 1}\right) \tag{16.29}$$

$$\frac{Q}{At\Delta\lambda} = \frac{2\pi hc^2}{\lambda^5}\left(\frac{1}{e^{hc/\lambda kT} - 1}\right) \tag{16.30}$$

Wien displacement law
$$\lambda_{max}T = \text{constant} \tag{16.31}$$

Questions for Chapter 16

1. Explain the differences and similarities between convection, conduction, and radiation.
†2. Explain how the process of convection of ocean water is responsible for relatively mild winters in Ireland and the United Kingdom even though they are as far north as Hudson's Bay in Canada.
†3. Explain from the process of convection why the temperature of the Pacific Ocean off the west coast of the United States is colder than the temperature of the Atlantic Ocean off the east coast of the United States.
†4. Explain from the process of convection why it gets colder after the passing of a cold front and warmer at the approach and passing of a warm front.
5. Explain the process of heat conduction in a gas and a liquid.
6. Considering the process of heat conduction through the walls of your home, explain why there is a greater loss of thermal energy through the walls on a very windy day.
7. In the winter time, why does a metal door knob feel colder than the wooden door even though both are at the same temperature?
8. Explain the use of venetian blinds for the windows of the home as a temperature controlling device. What advantage do they have over shades?
9. Why are thermal lined drapes used to cover the windows of a home at night?
10. Why is it desirable to wear light colored clothing in very hot climates rather than dark colored clothing?
11. Explain how you can still feel cold while sitting in a room whose air temperature is 70 °F, if the temperature of the walls is very much lower.
12. From what you now know about the processes of heat transfer, discuss the insulation of a calorimeter.
†13. On a very clear night, radiation fog can develop if there is sufficient moisture in the air. Explain.
14. If the maximum radiation from the sun falls in the blue-green portion of the visible spectrum, why doesn't the sun appear blue-green?
†15. From the point of view of radiation, discuss the process of thermography, whereby a specialized camera takes pictures of an object in the infrared portion of the spectrum. Explain how this could be used in medicine to detect tumors in the human body. (The tumors are usually several degrees hotter than normal body tissue.)

Problems for Chapter 16

16.2 Convection

1. How much thermal energy per unit mass is transferred by convection in 6.00 hr if air at the surface of the earth is moving at 15.0 mph? The temperature gradient is measured as 4.00 °C per 100 miles.
2. Air is moving over the surface of the earth at 30.0 km/hr. The temperature gradient is 2.50 °C per 100 km. How much thermal energy per unit mass is transferred by convection in an 8.00-hour period?
3. An air conditioner can cool 300 ft³ of air per minute from 95.0 °F to 60.0 °F. How much thermal energy is removed from the room per hour?
4. An air conditioner can cool 10.5 m³ of air per minute from 30.0 °C to 18.5 °C. How much thermal energy per hour is removed from the room in one hour?
5. In a hot air heating system, air at the furnace is heated to 200 °F. A window is open in the house and the house temperature remains at 55 °F. If the furnace can deliver 200 ft³/min of air, how much thermal energy per hour is transferred from the furnace to the room?
6. A hot air heating system rated at 84,000 Btu/hr has an efficiency of 65.0%. The fan can move 225 ft³ of air per minute. If air enters the furnace at 60.0 °F, what is the temperature of the outlet air?
7. A hot air heating system rated at 6.3 × 10⁷ J/hr has an efficiency of 58.0%. The fan is capable of moving 5.30 m³ of air per minute. If air enters the furnace at 17.0 °C, what is the temperature of the outlet air?

16.3 Conduction

8. How much thermal energy flows through a glass window 1/8 in. thick, 4.00 ft high, and 3.00 ft wide in 12.0 hr if the temperature on the outside of the window is 20.0 °F and the temperature on the inside of the window is 70.0 °F?
9. How much thermal energy flows through a glass window 0.350 cm thick, 1.20 m high, and 0.80 m wide in 12.0 hr if the temperature on the outside of the window is −8.00 °C and the temperature on the inside of the window is 20.0 °C?
10. Repeat problem 8, but now assume that there are strong gusty winds whose air temperature is 5.00 °F.
11. Find the amount of thermal energy that will flow through a concrete wall 10.0 m long, 2.80 m high, and 22.0 cm wide, in a period of 24.0 hr, if the inside temperature of the wall is 20.0 °C and the outside temperature of the wall is 5.00 °C.
12. Find the amount of thermal energy transferred through a pine wood door in 6.00 hr if the door is 0.91 m wide, 1.73 m high, and 5.00 cm thick. The inside temperature of the door is 20.0 °C and the outside temperature of the door is −5.00 °C.
13. How much thermal energy will flow per hour through a copper rod, 5.00 cm in diameter and 1.50 m long, if one end of the rod is maintained at a temperature of 225 °C and the other end at 20.0 °C?
14. One end of a copper rod has a temperature of 100 °C, whereas the other end has a temperature of 20.0 °C. The rod is 1.25 m long and 3.00 cm in diameter. Find the amount of thermal energy that flows through the rod in 5.00 min. Find the temperature of the rod at 45.0 cm from the hot end.

15. On a hot summer day the outside temperature is 98.0 °F. A home air conditioner is trying to maintain a temperature of 70.0 °F. If there are 10 windows in the house, each 1/8 in. thick and 12.0 ft² in area, how much thermal energy must the air conditioner remove per hour to eliminate the thermal energy transferred through the windows?

16. On a hot summer day the outside temperature is 35.0 °C. A home air conditioner is trying to maintain a temperature of 22.0 °C. If there are 12 windows in the house, each 0.350 cm thick and 0.960 m² in area, how much thermal energy must the air conditioner remove per hour to eliminate the thermal energy transferred through the windows?

†17. A styrofoam cooler ($k = 4.8 \times 10^{-5}$ kcal/m s °C) is filled with ice at 0 °C for a summertime party. The cooler is 40.0 cm high, 50.0 cm long, 40.0 cm wide, and 3.00 cm thick. The air temperature is 95.0 °F. Find (a) the mass of ice in the cooler, (b) how much thermal energy is needed to melt all the ice, and (c) how long it will take for all the ice to melt. Assume that the energy to melt the ice is only conducted through the four sides of the cooler. Also take the thickness of the cooler walls into account when computing the size of the walls of the container.

18. An aluminum rod 50.0 cm long and 3.00 cm in diameter has one end in a steam bath at 100 °C and the other end in an ice bath at 0.00 °C. How much ice melts per hour?

19. If the home thermostat is turned from 70.0 °F down to 60.0 °F for an 8-hr period at night when the outside temperature is 20.0 °F, what percentage saving in fuel can the home owner realize?

20. If the internal temperature of the human body is 37.0 °C, the surface temperature is 35.0 °C, and there is a separation of 4.00 cm of tissue between, how much thermal energy is conducted to the skin of the body each second? Take the thermal conductivity of human tissue to be 0.5×10^{-4} kcal/s m °C and the area of the human skin to be 1.9 m².

21. What is the R value of (a) 4.00 in. of glass wool and (b) 6.00 in. of glass wool?

22. How thick should a layer of plaster be in order to provide the same R value as a 5.00 cm of concrete?

23. A basement wall consists of 8.00 in. of concrete, 1 1/2 in. of glass wool, 3/8 in. of sheetrock (plaster), and 3/4 in. of knotty pine paneling. The wall is 7.00 ft high and 30.0 ft long. The outside temperature is 40.0 °F and we want to maintain the inside temperature at 70.0 °F. How much thermal energy will be lost through four such walls in a 24-hr period?

24. A basement wall consists of 20.0 cm of concrete, 3.00 cm of glass wool, 0.800 cm of sheetrock (plaster), and 2.00 cm of knotty pine paneling. The wall is 2.50 m high and 10.0 m long. The outside temperature is 1.00 °C, and we want to maintain the inside temperature of 22.0 °C. How much thermal energy will be lost through four such walls in a 24-hr period?

25. On a summer day the attic temperature of a house is 160 °F. The ceiling of the house is 8.00 m wide by 13.0 m long and 3/8-in. thick plasterboard. The house is cooled by an air conditioner and maintains a 70.0 °F temperature in the house. (a) Find the amount of thermal energy transferred from the attic to the house in 2.00 hr. (b) If 6.00 in. of glass wool is now placed in the attic floor, find the amount of thermal energy transferred into the house.

26. How much thermal energy is conducted through a thermopane window, in 8.00 hr if the window is 32.0 in. wide by 45.0 in. high, and consists of two sheets of glass 1/8 in. thick separated by an air gap of 1/2 in.? The temperature of the inside window is 68.0 °F and the temperature of the outside window is 20.0 °F. Treat the thermopane window as a compound wall.

27. How much thermal energy is conducted through a thermopane window in 8.00 hr if the window is 80.0 cm wide by 120 cm high, and it consists of two sheets of glass 0.350 cm thick separated by an air gap of 1.50 cm? The temperature of the inside window is 22.0 °C and the temperature of the outside window is −5.00 °C. Treat the thermopane window as a compound wall.

28. How much thermal energy is conducted through a combined glass window and storm window in 8.00 hr if the window is 32.0 in. wide by 45.0 in. high and 1/8 in. thick? The storm window is the same size but is separated from the inside window by an air gap of 2.00 in. The temperature of the inside window is 68.0 °F and the temperature of the outside window is 20.0 °F. Treat the combination as a compound wall.

16.4 Radiation

29. How much thermal energy from the sun falls on the surface of the earth during an 8-hr period? (Ignore reflected solar radiation from clouds that does not make it to the surface of the earth.)

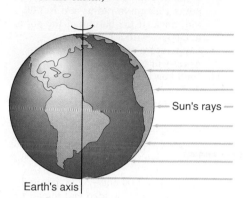

Sun's rays

Earth's axis

30. If the mean temperature of the surface of the earth is 288 K, how much thermal energy is radiated into space per second?

31. Assuming the human body has an emissivity, $e = 1$, and an area of approximately 2.23 m², find the amount of thermal energy radiated by the body in 8 hr if the surface temperature is 95.0 °F.

32. If the surface temperature of the human body is 35.0 °C, find the wavelength of the maximum intensity of radiation from the human body. Compare this wavelength to the wavelengths of visible light.

33. How much energy is radiated per second by an aluminum sphere 5.00 cm in radius, at a temperature of (a) 20.0 °C, and (b) 200 °C? Assume that the sphere emits as a blackbody.

34. How much energy is radiated per second by an iron cylinder 5.00 cm in radius and 10.0 cm long, at a temperature of (a) 20.0 °C and (b) 200 °C? Assume blackbody radiation.

35. How much energy is radiated per second from a wall 2.50 m high and 3.00 m wide, at a temperature of 20.0 °C? What is the wavelength of the maximum intensity of radiation?

36. A blackbody initially at 100 °C is heated to 300 °C. How much more power is radiated at the higher temperature?

37. A blackbody is at a temperature of 200 °C. Find the wavelength of the maximum intensity of radiation.

38. A blackbody is radiating at a temperature of 300 K. To what temperature should the body be raised to double the amount of radiation?

39. A distant star appears red, with a wavelength 7.000×10^{-7} m. What is the surface temperature of that star?

Additional Problems

40. An aluminum pot contains 10.0 kg of water at 100 °C. The bottom of the pot is 15.0 cm in radius and is 3.00 mm thick. If the bottom of the pot is in contact with a flame at a temperature of 170 °C, how much water will boil per minute?

41. Find how much energy is lost in one day through a concrete slab floor on which the den of a house is built. The den is 5.00 m wide and 6.00 m long, and the slab is 15.0 cm thick. The temperature of the ground is 3.00 °C and the temperature of the room is 22.0 °C.

42. A lead bar 2.00 cm by 2.00 cm and 10.0 cm long is welded end to end to a copper bar 2.00 cm by 2.00 cm by 25.0 cm long. Both bars are insulated from the environment. The end of the copper bar is placed in a steam bath while the end of the lead bar is placed in an ice bath. What is the temperature T at the interface of the copper-lead bar? How much thermal energy flows through the bar per minute?

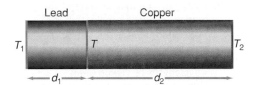

Lead   Copper

T_1 T T_2

d_1 d_2

43. Find the amount of thermal energy conducted through a wall, 5.00 m high, 12.0 m long, and 5.00 cm thick, if the wall is made of (a) concrete, (b) brick, (c) wood, and (d) glass. The temperature of the hot wall is 25.0 °C and the cold wall −5.00 °C.

†44. Show that the distribution of solar energy over the surface of the earth is a function of the latitude angle ϕ. Find the energy per unit area per unit time hitting the surface of the earth during the vernal equinox and during the summer solstice at (a) the equator, (b) 30.0° north latitude, (c) 45.0° north latitude, (d) 60.0° north latitude, and (e) 90.0° north latitude. At the vernal equinox the sun is directly overhead at the equator, whereas at the summer solstice the sun is directly overhead at 23.5° north latitude.

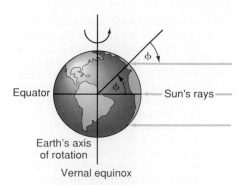

Equator   Sun's rays

Earth's axis of rotation

Vernal equinox

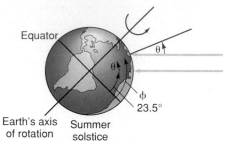

Equator

23.5°

Earth's axis of rotation   Summer solstice

45. An asphalt driveway, 50.0 m² in area and 6.00 cm thick, receives energy from the sun. Using the solar constant of 1.38×10^3 W/m², find the maximum change in temperature of the asphalt if (a) the radiation from the sun hits the driveway normally for a 2.00-hr period and (b) the radiation from the sun hits the driveway at an angle of 35° for the same 2.00-hr period. Take the density of asphalt to be 1219 kg/m³ and the specific heat of asphalt to be 1.02 kcal/kg °C.

†46. Find the amount of radiation from the sun that falls on the planets (a) Mercury, (b) Venus, (c) Mars, (d) Jupiter, and (e) Saturn in units of W/m².

47. If the Kelvin temperature of a blackbody is quadrupled, what happens to the rate of energy radiation?

†48. A house measures 40.0 ft by 30.0 ft by 8.00 ft high. The walls contain 4.00 in. of glass wool. Assume all the heat loss is through the walls of the house. The home thermostat is turned from 70.0 °F down to 60.0 °F for an 8-hr period at night when the outside temperature is 20.0 °F. (a) How much thermal energy can the home owner save by lowering the thermostat? (b) How much energy is used the next morning to bring the temperature of the air in the house back to 70.0 °F? (c) What is the savings in energy now?

†49. An insulated aluminum rod, 1.00 m long and 25.0 cm² in cross-sectional area, has one end in a steam bath at 100 °C and the other end in a cooling container. Water enters the cooling container at an input temperature of 10.0 °C and exits the cooling container at a temperature of 30.0 °C, leaving a mean temperature of 20.0 °C at the end of the aluminum rod. Find the mass of water that must flow through the cooling container per minute to maintain this equilibrium condition.

†50. An aluminum engine, operating at 300 °C is cooled by circulating water over the end of the engine where the water absorbs enough energy to boil. The cooling interface has a surface area of 0.525 m² and a thickness of 1.50 cm. If the water enters the cooling interface of the engine at 100 °C, how much water must boil per minute to cool the engine?

†51. When the surface through which thermal energy flows is not flat, such as in figure 16.6, the equation for heat transfer, equation 16.10, is no longer accurate. With the help of the calculus it can be shown that the amount of thermal energy that flows through the sides of a rectangular annular cylinder is given by

$$\frac{\Delta Q}{\Delta t} = \frac{2\pi k l \Delta T}{\ln(r_2/r_1)}$$

where l is the length of the cylinder, r_1 is the inside radius of the cylinder, and r_2 is the outside radius of the cylinder. Steam at 100 °C flows in a cylindrical copper pipe 5.00 m long, with an inside radius of 10.0 cm and an outside radius of 15 cm. Find the energy lost through the pipe per hour if the outside temperature of the pipe is 30.0 °C.

†52. When the surface through which thermal energy flows is a spherical shell rather than a flat surface, the amount of thermal energy that flows through the spherical surface can be shown to be given by

$$\frac{\Delta Q}{\Delta t} = \frac{4\pi k \Delta T}{(r_2 - r_1)/r_1 r_2}$$

where r_1 is the inside radius of the sphere and r_2 is the outside radius of the sphere.

Consider an igloo as half of a spherical shell. The inside radius is 3.00 m and the outside radius is 3.20 m. If the temperature inside the igloo is 15.0 °C and the outside temperature is −40.0 °C, find the flow of thermal energy through the ice per hour. The thermal conductivity of ice is 4.00×10^{-4} kcal/s m °C.

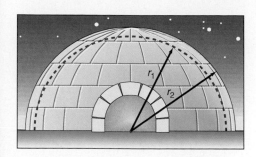

†53. Show that for large values of r_1 and r_2 the solution for thermal energy flow through a spherical shell (problem 52) reduces to the solution for the thermal energy flow through a flat slab.

†54. In problems 51 and 52 assume that you can use the formula for the thermal energy flow through a flat slab. Find $\Delta Q/\Delta t$ and find the percentage error involved by making this approximation.

55. A spherical body of 25.0-cm radius, has an emissivity of 0.45, and is at a temperature of 500.0 °C. How much power is radiated from the sphere?

†56. Newton's law of cooling states that the rate of change of temperature of a cooling body is proportional to the rate at which it gains or loses heat, which is approximately proportional to the difference between its temperature and the temperature of the environment. This is written mathematically as

$$\frac{\Delta T}{\Delta t} = -K(T_{avg} - T_e)$$

where T_{avg} is the average temperature of the body, T_e is the temperature of the environment, and K is a constant. A cup of coffee cools from 98.0 °C to 90.0 °C in 1.5 min. The cup is in a room that has a temperature of 20.0 °C. Find (a) the value of K and (b) how long it will take for the coffee to cool from 90.0 °C to 50.0 °C.

†57. A much more complicated example of heat transfer is one that combines conduction and convection. That is, we want to determine the thermal energy transferred from a hot level plate at 100 °C to air at a temperature of 20.0 °C. The thermal energy transferred is given by the equation

$$Q = hA\Delta T\, t$$

where h is a constant, called the convection coefficient and is a function of the shape, size, and orientation of the surface, the type of fluid in contact with the surface and the type of motion of the fluid itself. Values of h for various configurations can be found in handbooks. If h is equal to 1.78×10^{-3} kcal/s m² °C and A is 2.00 m², find the amount of thermal energy transferred per minute.

†58. Using the same principle of combined conduction and convection used in problem 57, find the amount of thermal energy that will flow through an uninsulated wall 10.0 cm thick of a wood frame house in 1 hr. Assume that both the inside and outside wall have a thickness of 2.00 cm of pine wood and an area of 25.0 m². (*Hint:* First consider the thermal energy loss through the inside wall, then the thermal energy loss through the 10.0-cm air gap, then the thermal energy loss through the outside wall.) The temperatures at the first wall are 18.0 °C and 13.0 °C, and the temperatures at the second wall are 10.0 °C and −6.70 °C. The convection coefficient for a vertical wall is $h = 0.424 \times 10^{-4}(\Delta T)^{1/4}$ cal/s cm² °C.

†59. A thermograph is essentially a device that detects radiation in the infrared range of the electromagnetic spectrum. A thermograph can map the temperature distribution of the human body, showing regions of abnormally high temperatures such as found in tumors. Starting with the Stefan-Boltzmann law show that the ratio of the power emitted from tissue at a slightly higher temperature, $T + \Delta T$, to the power emitted from normal tissue at a temperature T is

$$\frac{P_2}{P_1} = (1 + \Delta T/T)^4$$

Then show that a change of temperature of only 0.9 °C will give an approximate 1.00% increase in the power of the radiation transmitted. Assume that the body temperature is 37.0 °C.

Interactive Tutorials

60. How much thermal energy flows through a glass window per second (Q/s) if the thickness of the window $d = 0.020$ m and its cross-sectional area $A = 2.00$ m². The temperature difference between the window's faces is $\Delta T = 65.0$ °C, and the thermal conductivity of glass is $k = 0.791$ J/(m s °C).

61. Convection. A hot air heating system heats air to a temperature of 125 °C and the return air is at a temperature of 17.5 °C. The fan is capable of moving a volume of 7.50 m³ of air in 1 min, $\Delta V/\Delta t$. The specific heat of air at constant pressure, c, is 1.05×10^3 J/kg °C and take the density of air, ρ, to be 1.29 kg/m³. Find the amount of thermal energy transfer per hour from the furnace to the room.

62. Conduction through a compound wall. Find the amount of heat conducted through a compound wall that has a length $L = 8.5$ m and a height $h = 4.33$ m. The wall consists of a thickness of $d_1 = 10.0$ cm of brick, $d_2 = 1.90$ cm of plywood, $d_3 = 10.2$ cm of glass wool, $d_4 = 1.25$ cm of plaster, and $d_5 = 0.635$ cm of oak wood paneling. The inside temperature of the wall is $T_h = 20.0$ °C and the outside temperature of the wall is $T_c = -9.00$ °C. How much thermal energy flows through this wall per day?

63. Radiation. How much energy is radiated in 1 s by an iron sphere 18.5 cm in radius at a temperature of 125 °C? Assume that the sphere radiates as a blackbody of emissivity $e = 1$. What is the wavelength of the maximum intensity of radiation?

Thermodynamics

A thermodynamic system in nature.

We can express the fundamental laws of the universe which correspond to the two fundamental laws of the mechanical theory of heat in the following simple form:
1. The energy of the universe is a constant. 2. The entropy of the universe tends toward a maximum.

Rudolf Clausius

17.1 Introduction

Thermodynamics is the study of the relationships between heat, internal energy, and the mechanical work performed by a system. The system considered is usually a heat engine of some kind, although the term can also be applied to living systems such as plants and animals. There are two laws of thermodynamics. The first law of thermodynamics is the law of conservation of energy as applied to a thermodynamic system. We will apply the first law of thermodynamics to a heat engine and study its ramifications. The second law of thermodynamics tells us what processes are, and are not, possible in the operation of a heat engine. The second law is also responsible for telling us in which direction a particular physical process may go. For example a block can slide across a desk and have all of its kinetic energy converted to thermal energy by the work the block does against friction as it is slowed to a stop. However, the reverse process does not happen, that is, the thermal energy in the block does not convert itself into mechanical energy and cause the block to slide across the desk. Using the thermal energy in the block to cause mechanical motion is not a violation of the law of conservation of energy but it is a violation of the second law of thermodynamics.

17.2 The Concept of Work Applied to a Thermodynamic System

Consider what happens to an ideal gas in a cylinder when it is compressed by a constant external force **F,** as shown in figure 17.1(a). The constant force exerted on top of the piston causes it to be displaced a distance $\Delta y,$ thereby compressing the gas in the cylinder. The work done on the gas by the external force in compressing it is

$$W = F\Delta y \qquad (17.1)$$

This work by the external agent is positive because the external force and the displacement are in the same direction. The external force F and the external pressure p exerted on the gas by the piston are related by

$$F = pA \qquad (17.2)$$

where A is the cross-sectional area of the piston. Substituting equation 17.2 into equation 17.1 gives

$$W = pA\Delta y \qquad (17.3)$$

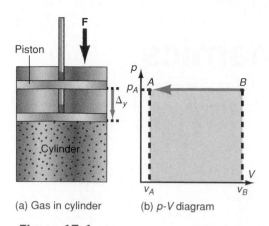

Figure 17.1

Work done in compressing a gas.

(a) Gas in cylinder   (b) p-V diagram

which is the work done on the gas by the external agent. If the compression takes place very slowly, the constant external pressure exerted by the piston on the gas is equal to the internal pressure exerted by the gas throughout the process. Thus, equation 17.3 can also be interpreted as the work done by the gas rather than the external agent. This is a departure from the usual way we have analyzed the concept of work. Previously, we have always considered the work as being done by the external agent. From this point on, we will consider all the work to be done on or by the gas itself, not the external agent. The product of the area of the cylinder and the displacement of the gas is equal to the change in volume of the gas. That is,

$$A\Delta y = \Delta V \tag{17.4}$$

the decrease in the volume of the gas. Substituting equation 17.4 into 17.3 gives

$$W = p\Delta V \tag{17.5}$$

Equation 17.5 represents the amount of work done by the gas when a constant external force compresses it by an amount ΔV.

This entire process can be shown on a pressure-volume (p-V) diagram as in figure 17.1(b). The original state of the gas is represented as the point B in the diagram, where it has the volume V_B and the pressure p_A. As the piston moves at constant pressure, the system, the gas in the cylinder, moves from the state at point B to the state at point A [figure 17.1(b)] along a horizontal line indicating that the process is occurring at constant pressure. At point A in the figure, the gas has been compressed to the volume V_A. The change in volume of the gas is seen to be

$$\Delta V = V_A - V_B \tag{17.6}$$

The total work done by the gas in compressing it from the point B to the point A, found from equations 17.5 and 17.6, is

$$W_{BA} = p_A(V_A - V_B)$$

It is important to note here that the product of p_A and $V_A - V_B$ represents the area of the rectangle cross-hatched in figure 17.1(b). Thus, the area under the curve in a p-V diagram always represents a quantity of work. When the area is large, it represents a large quantity of work, and when the area is small the quantity of work likewise is small.

Because V_A is less than V_B, the quantity $V_A - V_B$ is negative. Thus, *when work is done by a gas in compressing it, that work is always negative.* Notice that there are two distinct agents here. The work done *by the external agent* in compressing the gas is positive, but the *work done by the gas* in a compression is negative.

If the gas in the cylinder of figure 17.1(a) is allowed to expand back to the original volume V_B, then the process can be represented on the same p-V diagram of figure 17.1(b) as the same straight line, now going from point A to the point B. The work done by the gas in the expansion from A to B is

$$W = p\Delta V = p_A(V_B - V_A)$$

But now note that since V_B is greater than V_A, the quantity $V_B - V_A$ is now a positive quantity. *Thus, when a gas expands, the work done by the gas is positive.* (The work done on the gas by an external agent during the expansion would be negative. From this point on let us consider only the work done by the gas and forget any external agent.) *Thus, the **work** done by a gas during expansion is positive and the work done by a gas during compression is negative.* In either case, the work done is still the area under the line AB given by the product of the sides of the rectangle p_A and $V_B - V_A$. The areas are the same in both cases, however we consider the area positive when the gas expands and negative when the gas is compressed.

Let us now consider the work done along the different paths of the cyclic process shown in the p-V diagram of figure 17.2. *A **cyclic process** is a process that*

Thermodynamics

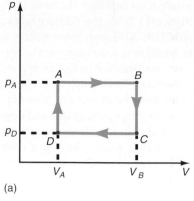

(a)

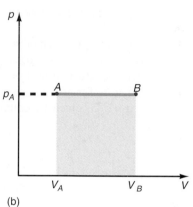

(b)

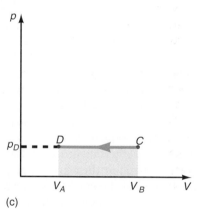

(c)

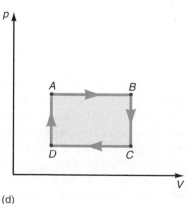

(d)

Figure 17.2

Work done in a cyclic process.

runs in a cycle eventually returning to where it started from. Thus, in figure 17.2(a) the cycle goes from A to B, B to C, C to D, and D back to A. The total work done by the system as it goes through the cycle is simply

$$W_{\text{total}} = W_{AB} + W_{BC} + W_{CD} + W_{DA} \tag{17.7}$$

where

W_{AB} is the work done on the path AB
W_{BC} is the work done on the path BC
W_{CD} is the work done on the path CD
W_{DA} is the work done on the path DA

Let us consider the work done along each path separately. To simplify matters let us first look at the work done along the path BC. The path BC represents a process that is performed at the constant volume V_B. Therefore, $\Delta V = 0$, and no work is performed along BC. Formally,

$$W_{BC} = p(V_B - V_B) = 0 \tag{17.8}$$

Similarly, along the path DA, the volume is also a constant and therefore ΔV is again zero, and hence the work done must also be zero. Formally,

$$W_{DA} = p(V_A - V_A) = 0 \tag{17.9}$$

Since the work is given by $p\Delta V$, whenever V is a constant in a process, ΔV is always zero and the work is also zero along that path in the p-V diagram.

The work done along the path AB is

$$W_{AB} = p_A \Delta V = p_A(V_B - V_A) \tag{17.10}$$

Because the path AB represents an expansion, positive work is done by the gas, as is evidenced by the fact that $V_B - V_A$ is a positive quantity. The work done along the path AB is shown as the area under the line AB in figure 17.2(b).

The work done along the path CD is

$$W_{CD} = p_D \Delta V = p_D(V_A - V_B) \tag{17.11}$$

Since the path CD represents a compression, work is done on the gas. This work is considered negative, as we can see from the fact that $V_A - V_B$, in equation 17.11, is negative. The work done on the gas is shown as the area under the line CD in figure 17.2(c).

The net work done by the gas in the cyclic process $ABCDA$, found from equation 17.7 with the help of equations 17.8 through 17.11, is

$$W_{\text{total}} = W_{AB} + W_{BC} + W_{CD} + W_{DA}$$
$$W_{\text{total}} = p_A(V_B - V_A) + 0 + p_D(V_A - V_B) + 0 \tag{17.7}$$

We can rewrite this to show that the work along CD is negative, that is, $V_A - V_B = -(V_B - V_A)$. Hence,

$$W_{\text{total}} = p_A(V_B - V_A) - p_D(V_B - V_A)$$

or

$$W_{\text{total}} = (p_A - p_D)(V_B - V_A) \tag{17.12}$$

Thus, equation 17.12 represents the net work done by the gas in this particular cyclic process. Note that $p_A - p_D$ is one side of the rectangular path of figure 17.2(a) while $V_B - V_A$ is the other side of that rectangle. Hence, their product in equation 17.12 represents the entire area of the rectangle enclosed by the thermodynamic path $ABCDA$ and is shown as the cross-hatched area in figure 17.2(d). Another way to visualize this total area, and hence total work, is to subtract the area in figure 17.2(c), the negative work, from the area in figure 17.2(b), the positive work, and we again get the area bounded by the path *ABCDA*. Although this result was derived for a simple rectangular thermodynamic path, it is true in general. *Thus, in any cyclic process, the net work done by the system is equal to the area enclosed by the cyclic thermodynamic path in a p-V diagram.* Therefore, to get as much work as possible out of a system, the enclosed area must be as large as possible. The net work is positive if the cycle proceeds clockwise, in the *p-V* diagram, and negative if the cycle proceeds counterclockwise. Finally, we should note that the process *AB* takes place at the constant pressure p_A. *A process that takes place at a constant pressure is called an* **isobaric process.** Hence, the process *CD* is also an isobaric process because it takes place at the constant pressure p_D. Process *BC* takes place at the constant volume V_B, and process *DA* takes place at the constant volume V_A. *A process that takes place at constant volume is called an* **isochoric or isometric process.**

There is another type of process that is very important in thermodynamic systems, the isothermal process. *An* **isothermal process** *is a process that occurs at a constant temperature, that is,* $\Delta T = 0$ *for the process.* A picture of an isotherm can be drawn on a *p-V* diagram by using the equation of state for an ideal gas, the working substance in the system. Thus, the ideal gas equation, given by equation 15.23, is

$$pV = nRT$$

Because *n* and *R* are constants, if *T* is also a constant, then the entire right-hand side of equation 15.23 is a constant. We can then write equation 15.23 as

$$pV = \text{constant} \qquad (17.13)$$

If we plot equation 17.13 on a *p-V* diagram, we obtain the hyperbolic curves of figure 17.3. Each curve is called an isotherm and in the figure, T_3 is greater than T_2, which in turn is greater than T_1.

Let us now consider the new cyclic process shown in figure 17.4, in which an ideal gas in a cylinder expands against a piston isothermally. This is shown as the path *AC* in the *p-V* diagram. To physically carry out the isothermal process along the path *AC*, the cylinder is surrounded by a constant temperature heat reservoir. The cylinder either absorbs heat from, or liberates heat to, the reservoir in order to maintain the constant temperature. When the isothermal process is finished the heat reservoir is removed. The gas is then compressed at the constant pressure p_D at point *C* until it reaches the point *D*. The pressure of the gas is then increased from p_D to p_A while the volume of the gas in the cylinder is kept constant. This is shown as the path *DA* in the *p-V* diagram. Now let us assume that the points *A, C,* and *D* are the same points that were considered in figure 17.2(a). Recall that the net work done by the system is equal to the area enclosed by the cyclic path. Thus, the net work done in this process is equal to the cross-hatched area within the path *ACDA* shown in figure 17.4.

It is important to compare figure 17.2(d) with figure 17.4. Remember the points *A, C,* and *D* in figure 17.4 are the same as the points *A, C,* and *D* in figure 17.2(d). But the area under the enclosed curve in figure 17.2(d) is greater than the enclosed area in figure 17.4. Hence, a greater amount of work is done by the system in following the cyclic path *ABCDA* than the cyclic path *ACDA*. *Thus, the work that the system does depends on the thermodynamic path taken.* Even though both processes started at point *A* and returned to the same point *A,* the work done by the system is different in each case. This result is succinctly stated as: *the work done depends on the path taken, and work is a path dependent quantity.*

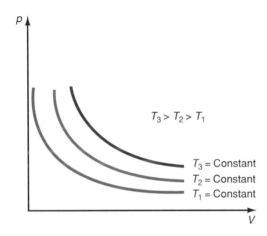

Figure 17.3

Isotherms on a *p-V* diagram.

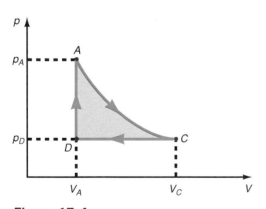

Figure 17.4

Cyclic process with an isothermal expansion.

Example 17.1

Work done in a thermodynamic cycle. One mole of an ideal gas goes through the thermodynamic cycle shown in figure 17.2(a). If $p_A = 2.00 \times 10^4$ Pa, $p_D = 1.00 \times 10^4$ Pa, $V_A = 0.250$ m³, and $V_B = 0.500$ m³, find the work done along the path (a) *AB*, (b) *BC*, (c) *CD*, (d) *DA*, and (e) *ABCDA*.

Solution

---

a. The work done along the path *AB*, found from equation 17.10, is

$$W_{AB} = p_A(V_B - V_A)$$

$$= (2.00 \times 10^4 \text{ Pa})(0.500 \text{ m}^3 - 0.250 \text{ m}^3)$$

$$= 5.00 \times 10^3 \frac{\text{N m}^3}{\text{m}^2} = 5.00 \times 10^3 \text{ N m}$$

$$= 5.00 \times 10^3 \text{ J}$$

b. The work done along the path *BC*, found from equation 17.8, is

$$W_{BC} = p(V_B - V_B) = 0$$

c. The work done along path *CD*, given by equation 17.11, is

$$W_{CD} = p_D(V_A - V_B)$$

$$= (1.00 \times 10^4 \text{ Pa})(0.250 \text{ m}^3 - 0.500 \text{ m}^3)$$

$$= -2.50 \times 10^3 \text{ J}$$

Note that the work done in compressing the gas is negative.

d. The work done along path *DA*, given by equation 17.9, is

$$W_{DA} = p(V_A - V_A) = 0$$

e. The total work done along the entire path *ABCDA*, found from equation 17.7, is

$$W_{\text{total}} = W_{AB} + W_{BC} + W_{CD} + W_{DA}$$

$$= 5.00 \times 10^3 \text{ J} + 0 - 2.50 \times 10^3 \text{ J} + 0$$

$$= 2.50 \times 10^3 \text{ J}$$

17.3 Heat Added to or Removed from a Thermodynamic System

We saw in chapter 14 that the amount of heat added or removed from a body is given by

$$Q = mc\Delta T \tag{14.6}$$

Equation 14.6 can also be applied to the heat added to, or removed from, a gas, if two stipulations are made. First, we saw in chapter 15 that it is more convenient to express the mass m of a gas in terms of the number of moles n of the gas. The total mass m of the gas is the sum of the masses of all the molecules of the gas. That is, m is equal to the mass of one molecule times the total number of molecules in one mole of the substance, times the total number of moles. That is

$$m = m_0 N_A n \tag{17.14}$$

where m_0 is the mass of one molecule; N_A is Avogadro's number, the number of molecules in one mole of a substance; and n is the number of moles of the gas. Notice in equation 15.47, the product of the mass of one molecule times Avogadro's number is called the **molecular mass** of the substance M, that is,

$$M = m_0 N_A \qquad (17.15)$$

The molecular mass is thus the mass of one mole of the gas. Substituting equation 17.15 into equation 17.14 gives for the mass of the gas

$$m = nM \qquad (17.16)$$

Equation 17.16 says that the mass of the gas is equal to the number of moles of the gas times the molecular mass of the gas. Substituting equation 17.16 for the mass m of the gas into equation 14.6, gives

$$Q = nMc\Delta T \qquad (17.17)$$

The product Mc is defined as the **molar specific heat** of the gas, or molar heat capacity, and is represented by the capital letter C. Hence,

$$C = Mc \qquad (17.18)$$

The heat absorbed or lost by a gas undergoing a thermodynamic process is found by substituting equation 17.18 into equation 17.17. Thus,

$$Q = nC\Delta T \qquad (17.19)$$

The second stipulation for applying equation 14.6 to gases has to do with the specific process to which the gas is subjected. Equation 14.6 was based on the heat absorbed or liberated from a solid or a liquid body that was under constant atmospheric pressure. In applying equation 17.19, which is the modified equation 14.6, we must specify the process whereby the temperature change ΔT occurs. Figure 17.5 shows some possible processes. Let us start at the point A in the p-V diagram of figure 17.5. The temperature at point A is T_0 because point A is on the T_0 isotherm. Heat can be added to the system such that the temperature of the gas rises to T_1. But, as we can see from figure 17.5, there are many different ways to get to the isotherm T_1. The thermodynamic paths AB, AC, AD, AE, or an infinite number of other possible paths can be followed to arrive at T_1. Therefore, there can be an infinite number of specific heats for gases. Let us restrict ourselves to only two paths, and hence only two specific heats. The first path we consider is the path AB, which represents a process taking place at constant volume. The second path is path AE, which represents a process taking place at constant pressure. We designate the molar specific heat for a process occurring at constant volume by C_v, whereas we designate the molar specific heat for a process occurring at a constant pressure by C_p. It is found experimentally that for a monatomic ideal gas such as helium or argon, $C_v = 2.98$ cal/mole K $= 12.5$ J/mole K, whereas $C_p = 4.97$ cal/mole K $= 20.8$ J/mole K.

The heat absorbed by the gas as the system moves along the thermodynamic path AB in figure 17.5 is

$$Q_{AB} = nC_v\Delta T = nC_v(T_1 - T_0) \qquad (17.20)$$

The heat absorbed by the gas as the system moves along the path AE is given by

$$Q_{AE} = nC_p\Delta T = nC_p(T_1 - T_0) \qquad (17.21)$$

Although the system ends up at the same temperature T_1 whether the path AB or AE is traveled, the heat that is absorbed along each path is different because C_p and C_v have different values. *Thus, the **heat absorbed or liberated in a thermodynamic process** depends on the path that is followed. That is, heat like work is path dependent.* Although demonstrated for a gas, this statement is true in general.

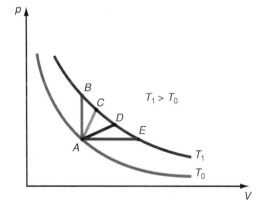

Figure 17.5

The specific heat for a gas depends on the path taken in a p-V diagram.

Example 17.2

The heat absorbed along two different thermodynamic paths. Compute the amount of heat absorbed by 1 mole of He gas along path (a) AB and (b) AE, of figure 17.5, if $T_1 = 400$ K and $T_0 = 300$ K.

Solution

a. The heat absorbed along path AB, given by equation 17.20, is

$$Q_{AB} = nC_v\Delta T = nC_v(T_1 - T_0)$$

$$= (1 \text{ mole})\left(2.98 \, \frac{\text{cal}}{\text{mole K}}\right)(400 \text{ K} - 300 \text{ K})$$

$$= 298 \text{ cal}$$

b. The heat absorbed along the path AE, given by equation 17.21, is

$$Q_{AE} = nC_p\Delta T = nC_p(T_1 - T_0)$$

$$= (1 \text{ mole})\left(4.97 \, \frac{\text{cal}}{\text{mole K}}\right)(400 \text{ K} - 300 \text{ K})$$

$$= 497 \text{ cal}$$

Thus, a greater quantity of heat is absorbed in the process that occurs at constant pressure. This is because at constant pressure the volume expands and some of the heat energy is used to do work, but at constant volume no work is accomplished.

17.4 The First Law of Thermodynamics

Recall from the kinetic theory of gases studied in chapter 15 that the mean kinetic energy of a molecule, found from equation 15.45, is

$$\text{KE}_{avg} = \tfrac{1}{2}mv^2_{avg} = \tfrac{3}{2}kT$$

Thus, a change in the absolute temperature of a gas shows up as a change in the average energy of a molecule. If the average kinetic energy of one molecule of the gas is multiplied by N, the total number of molecules of the gas present in the thermodynamic system (i.e., the cylinder filled with gas), then this product represents the total internal energy of this quantity of gas. Recall that the internal energy of a body was defined in chapter 14 as the sum of the kinetic energies and potential energies of all the molecules of the body. Because the molecules of a gas are moving so rapidly and are widely separated on the average, only a few are near to each other at any given time and it is unnecessary to consider any intermolecular forces, and hence potential energies of the molecules. *Thus, the total kinetic energy of all the molecules of a gas constitutes the total **internal energy of the gas.*** We designate this internal energy of the gas by the symbol U. The internal energy of the gas is given by

$$U = (\text{total number of molecules})(\text{mean KE of each molecule})$$

$$= N\text{KE}_{avg}$$

$$U = N(\tfrac{3}{2}kT) \tag{17.22}$$

But recall from equation 15.44 that

$$k = \frac{R}{N_A}$$

Substituting equation 15.44 into equation 17.22 gives for the internal energy of an ideal gas

$$U = N\frac{3}{2}\frac{R}{N_A}T$$

But the total number of molecules N was given by

$$N = nN_A \qquad (15.24)$$

Thus,

$$U = nN_A\frac{3}{2}\frac{R}{N_A}T$$

and

$$U = \tfrac{3}{2}nRT \qquad (17.23)$$

From equation 17.23 we see that a change in temperature is thus associated with a change in the internal energy of the gas, that is,

$$\Delta U = \tfrac{3}{2}nR\Delta T \qquad (17.24)$$

Let us now consider the thermodynamic system shown in the p-V diagram of figure 17.6. The isotherm going through point B is labeled T_B, whereas the one that goes through points A and C is labeled T_{AC}, and finally, the isotherm that goes through point D is called T_D. Before the entire system is considered, let us first consider a process that proceeds isothermally from A to C. Since the path AC is an isotherm, the temperature is constant and thus $\Delta T = 0$. But from equation 17.24, the change in internal energy ΔU must also be zero. That is, *an isothermal expansion occurs at constant internal energy.* But how can this be? As the gas expands along AC it is doing work. If the internal energy is constant, where does the energy come from to perform the work that is being done by the gas? Obviously energy must somehow be supplied in order for the gas to do work. *Thus, a quantity of heat Q must be supplied to the system in order for the system to do work along an isothermal path.* Hence, for an isothermal process,

$$Q = W \qquad (17.25)$$

Let us now consider the portion of the process that is along path BC in figure 17.6. The process BC is performed at constant volume, thus, $\Delta V = 0$ along this path. Because the amount of work done by the gas is given as $W = p\Delta V$, if $\Delta V = 0$, then the work done along the path BC must also be zero. But the temperature T_{AC} at point C is less than the temperature T_B at the point B. There has been a drop in temperature between points B and C and hence a decrease in the

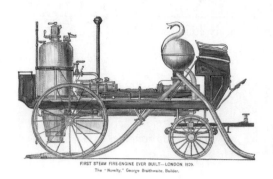

The steam engine is a thermodynamic system.

Figure 17.6

A thermodynamic system on a p-V diagram.

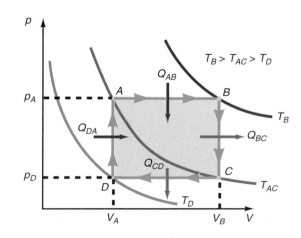

Thermodynamics

internal energy of the system. Since the loss of energy didn't go into work, because $\Delta V = 0$, heat must have been taken away from the system along path BC. *The decrease in the internal energy of the system along an isometric path is caused by the heat removed from the system along BC,* that is,

$$\Delta U = Q \qquad (17.26)$$

But the heat removed from the system during a constant volume process was shown in equation 17.20 to be

$$Q = nC_v\Delta T \qquad (17.27)$$

Since the heat removed is equal to the loss in internal energy by equation 17.26, we can write the change in internal energy from equations 17.26 and 17.27 as

$$\Delta U = nC_v\Delta T \qquad (17.28)$$

Equation 17.28 is a general statement governing the change in internal energy during any process, not only the one at constant volume from which equation 17.28 was derived. Recall from equation 17.23, a result from the kinetic theory of gases, that U, the internal energy, is only a function of temperature. In fact if ΔU from equation 17.24 is equated to ΔU from equation 17.28, we get

$$\tfrac{3}{2}nR\Delta T = nC_v\Delta T$$

Solving for C_v, the theoretical value of the molar specific heat capacity at constant volume is found to be

$$C_v = \tfrac{3}{2}R \qquad (17.29)$$

Using the value of $R = 1.987$ cal/mole K found in chapter 15, the value of C_v, calculated from equation 17.29, is $C_v = 2.98$ cal/mole K, which agrees with the experimental value.

The two special cases given by equations 17.25 and 17.26 can be combined into one general equation that contains 17.25 and 17.26 as special cases. This general equation is

$$Q = \Delta U + W \qquad (17.30)$$

and is *called the first law of thermodynamics.* Thus, we can derive equation 17.25 from 17.30 for an isothermal path because then the change in internal energy $\Delta U = 0$. We can derive equation 17.26 from equation 17.30 for a constant volume thermodynamic path, because then $\Delta V = 0$, and hence $W = 0$. *The first law of thermodynamics, equation 17.30, says that the heat Q, added to a system will show up either as a change in internal energy ΔU of the system and/or as work W performed by the system.* From this analysis we can see that the first law of thermodynamics is just the law of conservation of energy. Equation 17.30 is quite often written in the slightly different form:

$$\Delta U = Q - W \qquad (17.31)$$

which is *also called the first law of thermodynamics. The first law of thermodynamics can be also stated as the change in the internal energy of the system equals the heat added to the system minus the work done by the system on the outside environment.* Perhaps the best way to see the application of the first law to a thermodynamic system is in an example.

Example 17.3

Applying the first law of thermodynamics. Two moles of an ideal gas are carried around the thermodynamic path *ABCDA* in figure 17.6. Here $T_D = 150$ K, $T_{AC} = 300$ K, $T_B = 600$ K, and $p_A = 2.00 \times 10^4$ Pa, while $p_D = 1.00 \times 10^4$ Pa. The volume $V_A = 0.250$ m^3, while $V_B = 0.500$ m^3. Find the work done, the heat lost or absorbed, and the internal energy of the system for the thermodynamic paths (a) AB, (b) BC, (c) CD, (d) DA, and (e) $ABCDA$.

Solution

a. The work done by the expanding gas along the path AB is

$$W = p\Delta V$$

$$W_{AB} = p_A(V_B - V_A)$$

$$= \left(2.00 \times 10^4 \frac{\text{N}}{\text{m}^2}\right)(0.500 \text{ m}^3 - 0.250 \text{ m}^3)$$

$$= 5.00 \times 10^3 \text{ J}$$

The heat absorbed by the gas along path AB is

$$Q = nC_p\Delta T$$

$$Q_{AB} = nC_p(T_B - T_{AC})$$

$$= (2 \text{ moles})\left(4.97 \frac{\text{cal}}{\text{mole K}}\right)(600 \text{ K} - 300 \text{ K})$$

$$= (2.98 \times 10^3 \text{ cal})\left(\frac{4.185 \text{ J}}{1 \text{ cal}}\right)$$

$$= 1.25 \times 10^4 \text{ J}$$

The change in internal energy along path AB, found from the first law equation 17.31, is

$$\Delta U_{AB} = Q_{AB} - W_{AB}$$

$$= 1.25 \times 10^4 \text{ J} - 5.00 \times 10^3 \text{ J}$$

$$= 7.50 \times 10^3 \text{ J}$$

Thus, there is a gain of internal energy along the path AB.

b. The work done along path BC is

$$W = p\Delta V$$

$$W_{BC} = p(V_B - V_B) = 0$$

$$= 0$$

The heat lost along path BC is

$$Q_{BC} = nC_v\Delta T = nC_v(T_{AC} - T_B)$$

$$= (2 \text{ moles})\left(2.98 \frac{\text{cal}}{\text{mole K}}\right)(300 \text{ K} - 600 \text{ K})$$

$$= (-1.79 \times 10^3 \text{ cal})\left(\frac{4.185 \text{ J}}{1 \text{ cal}}\right)$$

$$= -7.50 \times 10^3 \text{ J}$$

The loss of internal energy in dropping from 600 K at B to 300 K at C is found from the first law as

$$\Delta U_{BC} = Q_{BC} - W_{BC}$$

$$= -7.50 \times 10^3 \text{ J} - 0$$

$$= -7.50 \times 10^3 \text{ J}$$

Thermodynamics

c. The work done during the compression along the path CD is

$$W_{CD} = p\Delta V = p_D(V_A - V_B)$$

$$= \left(1.00 \times 10^4 \frac{N}{m^2}\right)(0.250 \text{ m}^3 - 0.500 \text{ m}^3)$$

$$= -2.50 \times 10^{-3} \text{ J}$$

The heat lost along the path CD is

$$Q_{CD} = nC_p\Delta T = nC_p(T_D - T_{AC})$$

$$= (2 \text{ moles})\left(4.97 \frac{\text{cal}}{\text{mole K}}\right)(150 \text{ K} - 300 \text{ K})$$

$$= -6.24 \times 10^3 \text{ J}$$

The change in internal energy along the path CD, found from the first law, is

$$\Delta U_{CD} = Q_{CD} - W_{CD}$$

$$= -6.24 \times 10^3 \text{ J} - (-2.50 \times 10^3 \text{ J})$$

$$= -3.74 \times 10^3 \text{ J}$$

Note that the internal energy decreased, as expected, since the temperature decreased from 300 K to 150 K.

d. The work done along the path DA is

$$W_{DA} = p\Delta V = p(V_A - V_A) = 0$$

The heat added along the path DA is

$$Q_{DA} = nC_v\Delta T = nC_v(T_{AC} - T_D)$$

$$= (2 \text{ moles})\left(2.98 \frac{\text{cal}}{\text{mole K}}\right)(300 \text{ K} - 150 \text{ K})$$

$$= (8.94 \text{ cal})\left(4.185 \frac{\text{J}}{\text{cal}}\right)$$

$$= 3.74 \times 10^3 \text{ J}$$

The change in internal energy along DA is

$$\Delta U_{DA} = Q_{DA} - W_{DA}$$

$$= 3.74 \times 10^3 \text{ J}$$

e. The net work done throughout the cycle $ABCDA$ is

$$W_{ABCDA} = W_{AB} + W_{BC} + W_{CD} + W_{DA}$$

$$= 5.00 \times 10^3 \text{ J} + 0 - 2.50 \times 10^3 \text{ J} + 0$$

$$= 2.50 \times 10^3 \text{ J}$$

The net heat added throughout the cycle $ABCDA$ is

$$Q_{ABCDA} = Q_{AB} + Q_{BC} + Q_{CD} + Q_{DA}$$

$$= 1.25 \times 10^4 \text{ J} - 7.50 \times 10^3 \text{ J} - 6.24 \times 10^3 \text{ J} + 3.74 \times 10^3 \text{ J}$$

$$= 2.50 \times 10^3 \text{ J}$$

Note that Q_{AB} and Q_{DA} are positive quantities, which means that heat is being added to the system along these two paths. Also note that Q_{BC} and Q_{CD} are negative quantities, which means that heat is being taken away from the system along these two paths. *In general, Q is always positive when heat is added to the system and negative when heat is removed from the system.*

This effect is seen in figure 17.6 by drawing lines entering the enclosed thermodynamic path when heat is added to the system, and lines emanating from the enclosed path when heat is taken away from the system. This is a characteristic of all engines operating in a cycle, that is, heat is always added and some heat is always rejected. The net change in internal energy throughout the cycle $ABCDA$ is

$$\Delta U_{ABCDA} = \Delta U_{AB} + \Delta U_{BC} + \Delta U_{CD} + \Delta U_{DA} \tag{17.32}$$

$$= 7.50 \times 10^3 \text{ J} - 7.50 \times 10^3 \text{ J} - 3.74 \times 10^3 \text{ J} + 3.74 \times 10^3 \text{ J}$$

$$= 0$$

Note that the total change in internal energy around the entire cycle is equal to zero. This is a very reasonable result because the internal energy of a system depends only on the temperature of the system. If we go completely around the cycle, we end up at the same starting point with the same temperature. Since $\Delta T = 0$ around the cycle, $\Delta U = nC_v\Delta T$ must also equal zero around the cycle.

Applying the first law to the entire cycle we have

$$\Delta U_{ABCDA} = Q_{ABCDA} - W_{ABCDA}$$

But as just seen, $\Delta U_{ABCDA} = 0$, therefore,

$$Q_{ABCDA} = W_{ABCDA} \tag{17.33}$$

That is, the energy for the net work done by the system comes from the net heat applied to the system. Looking at the calculations, we see that this is indeed the case since

$$Q_{ABCDA} = 2.50 \times 10^3 \text{ J}$$

while

$$W_{ABCDA} = 2.50 \times 10^3 \text{ J}$$

Another very interesting thing can be learned from this example. Look at the change in internal energy from the point A to the point C, and note that regardless of the path chosen, the change in internal energy is the same. Thus, from our calculations,

$$\Delta U_{AC} = \Delta U_{AB} + \Delta U_{BC} = 7.50 \times 10^3 \text{ J} - 7.50 \times 10^3 \text{ J} = 0$$

and

$$\Delta U_{AC} = -\Delta U_{AD} - \Delta U_{DC} = -3.74 \times 10^3 \text{ J} + 3.74 \times 10^3 \text{ J} = 0$$

Along the isothermal path AC

$$\Delta U_{AC} = 0$$

because if T is constant, U is constant. *Thus, regardless of the path chosen between two points on a p-V diagram, ΔU is always the same.* (It will not always be zero, as in this case where the points A and C happen to lie along the same isotherm, but whatever its numerical value, ΔU is always the same.)

What is especially interesting about this fact is that *the work done depends on the path taken, the heat absorbed or liberated depends on the path taken, but their difference $Q - W$, which is equal to ΔU is independent of the path taken. That is, ΔU depends only on the initial and final states of the thermodynamic system and not the path between the initial and final states.*

Thus, the internal energy is to a thermodynamic system what the potential energy is to a mechanical system. (Recall from chapter 7, section 7.7 that the work done, and hence the potential energy, was the same whether an object was lifted to a height h, or moved up a frictionless inclined plane to the same height h. That is, the potential energy was independent of the path taken.)

The thermodynamic system considered in figure 17.6 represents an engine of some kind. That is, heat is added to the engine and the engine does work. To compare one engine with another it is desirable to know how efficient each engine is. *The efficiency of an engine can be defined in terms of what we get out of the system compared to what we put into the system.* Heat, Q_{in}, is put into the engine, and work, W, is performed by the engine, hence the efficiency of an engine can be defined as

$$\text{Eff} = \frac{\text{Work out}}{\text{Heat in}} = \frac{W}{Q_{in}} \qquad (17.34)$$

Example 17.4

The efficiency of an engine. In example 17.3, 2.50×10^3 J of work was done by the system, whereas the heat added to the system was the heat added along paths *AB* and *DA*, which is equal to 1.25×10^4 J $+ 3.74 \times 10^3$ J, which is equal to 1.62×10^4 J. Find the efficiency of that engine.

Solution

The efficiency of the engine, found from equation 17.34, is

$$\text{Eff} = \frac{W}{Q_{in}} = \frac{2.50 \times 10^3 \text{ J}}{1.62 \times 10^4 \text{ J}} = 0.15$$

$$= 15\%$$

Thus, the efficiency of the engine represented by the thermodynamic cycle of figure 17.6 is only 15%. This is not a very efficient engine. We will discuss the maximum possible efficiency of an engine when we study the Carnot cycle in section 17.8.

■

Before leaving this section, however, let us take one more look at the change in the internal energy of the system along the path *ABC*. We have already seen that since the initial and final states lie on the same isotherm, the change in internal energy is zero. There is still, however, some more important physics to be obtained by further considerations of this path. The change in internal energy along the path *ABC* is given by

$$\Delta U_{ABC} = \Delta U_{AB} + \Delta U_{BC}$$

But from the first law we can write this as

$$\Delta U_{ABC} = Q_{AB} - W_{AB} + Q_{BC} - W_{BC} \qquad (17.35)$$

But as we have already seen

$$\left. \begin{array}{l} Q_{AB} = nC_p(T_B - T_{AC}) \\ W_{AB} = p_A(V_B - V_A) \\ Q_{BC} = nC_v(T_{AC} - T_B) \\ W_{BC} = p(V_B - V_B) = 0 \end{array} \right\} \qquad (17.36)$$

Substituting all these terms into equation 17.35, gives

$$\Delta U_{ABC} = nC_p(T_B - T_{AC}) - p_A(V_B - V_A) + nC_v(T_{AC} - T_B)$$

$$\Delta U_{ABC} = nC_p(T_B - T_{AC}) - nC_v(T_B - T_{AC}) - p_AV_B + p_AV_A \qquad (17.37)$$

But from the ideal gas equation,

$$p_AV_A = nRT_{AC} \qquad (17.38)$$

and

$$p_A V_B = p_B V_B = nRT_B \tag{17.39}$$

Substituting equations 17.38 and 17.39 back into equation 17.37, we get

$$\Delta U_{ABC} = nC_p(T_B - T_{AC}) - nC_v(T_B - T_{AC}) - nRT_B + nRT_{AC}$$

$$= nC_p(T_B - T_{AC}) - nC_v(T_B - T_{AC}) - nR(T_B - T_{AC})$$

$$\Delta U_{ABC} = (C_p - C_v - R)n(T_B - T_{AC}) \tag{17.40}$$

However, we have already determined that ΔU_{ABC} is equal to zero. Hence, equation 17.40 implies that

$$C_p - C_v - R = 0$$

or

$$C_p - C_v = R \tag{17.41}$$

Thus we have determined a theoretical relation between the molar specific heat capacities and the universal gas constant R. Since it has already been shown that $C_v = \frac{3}{2}R$ in equation 17.29, C_p can now be solved for in equation 17.41 to obtain

$$C_p = C_v + R$$

$$= \frac{3}{2}R + R$$

$$C_p = \frac{5}{2}R \tag{17.42}$$

Using the value of $R = 1.987$ cal/mole K found in chapter 15, the value of C_p is 4.97 cal/mole K, which agrees with the experimental value of C_p for a monatomic gas.

17.5 Some Special Cases of the First Law of Thermodynamics

Although we have already discussed the first law of thermodynamics pretty thoroughly, let us summarize some of the results into special cases.

An Isothermal Process

An isothermal process is a process that occurs at constant temperature. Thus, $\Delta T = 0$. But $\Delta U = nC_v\Delta T$. Therefore, if $\Delta T = 0$, then $\Delta U = 0$. The first law then becomes

$$\Delta U = 0 = Q - W$$

$$Q = W \tag{17.43}$$

In an isothermal process, heat added to the system shows up as mechanical work done by the system.

An Adiabatic Process

*An **adiabatic process** is a process that occurs without an exchange of heat between the system and its environment.* That is, heat is neither added to nor taken away from the system during the process. Thus, $Q = 0$ in an adiabatic process. The first law of thermodynamics for an adiabatic process becomes

$$\Delta U = Q - W$$

$$W = -\Delta U \tag{17.44}$$

Thus, *in an adiabatic process, the energy for the work done by the gas comes from a loss in the internal energy of the gas.*

Thermodynamics

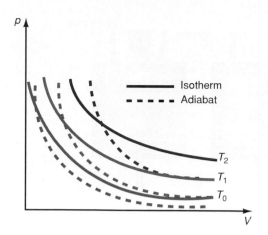

Figure 17.7

Adiabats and isotherms on a p-V diagram.

An example of an adiabatic process is the process of cloud formation in the atmosphere, which we will discuss in the section "Have you ever wondered" at the end of this chapter.

Some processes that are not strictly speaking adiabatic can be treated as adiabatic processes because the process occurs so rapidly that there is not enough time for the system to exchange any significant quantities of heat with its environment.

An adiabatic process can be drawn as the dashed line on the p-V diagram in figure 17.7. Note that the adiabatic line has a steeper slope than the isotherm. Although the equation for the adiabat cannot be derived without the use of the calculus, we will state the result here for completeness:

$$pV^\gamma = \text{constant} \tag{17.45}$$

where γ is equal to the ratio of the molar specific heats. Thus,

$$\gamma = \frac{C_p}{C_v} \tag{17.46}$$

The adiabatic process is essential to the study of the Carnot cycle in section 17.8.

Isochoric Process or Isometric Process

An isochoric process is a process that occurs at constant volume, that is, $\Delta V = 0$. Since the work done, W, is equal to $p\Delta V = 0$, then W must also be zero. The first law of thermodynamics for an isochoric process therefore becomes

$$Q = \Delta U \tag{17.47}$$

Thus, the heat added to a system during an isochoric process shows up as an increase in the internal energy of the system.

An Isobaric Process

An isobaric process is a process that occurs at constant pressure, that is, $\Delta p = 0$. Since the pressure is a constant for an isobaric process, the work done in an isobaric process is given by the product of the constant pressure p and the change in volume ΔV. That is,

$$W = p\Delta V$$

If the process is not an isobaric one then the pressure p has to be an average value of the pressure along the thermodynamic path to give the average amount of work done on that path.

A Cyclic Process

A cyclic process is one that always returns to its initial state. The process studied as *ABCDA* in figure 17.6 is an example of a cyclic process. *Because the system always returns to the original state, ΔU is always equal to zero for a cyclic process.* That is,

$$\Delta U = 0$$

Hence, the first law of thermodynamics for a cyclic process becomes

$$W = Q \tag{17.48}$$

Thus, the work done by the system in the cyclic process is equal to the heat added to the system on a portion of the cycle minus the heat removed on the remainder of the cycle.

17.6 The Gasoline Engine

The thermodynamic system studied so far is somewhat idealistic. In order to be more specific, let us consider the thermodynamic process that occurs in the gasoline engine of an automobile. The engine usually consists of four, six, or eight cylinders.

Figure 17.8

The gasoline engine cycle.

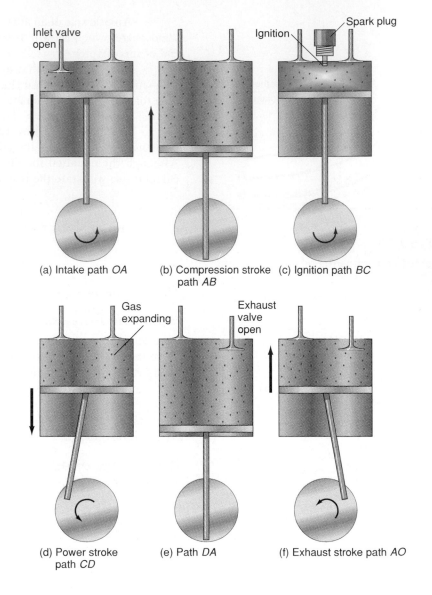

(a) Intake path *OA*

(b) Compression stroke path *AB*

(c) Ignition path *BC*

(d) Power stroke path *CD*

(e) Path *DA*

(f) Exhaust stroke path *AO*

Figure 17.9

The Otto cycle.

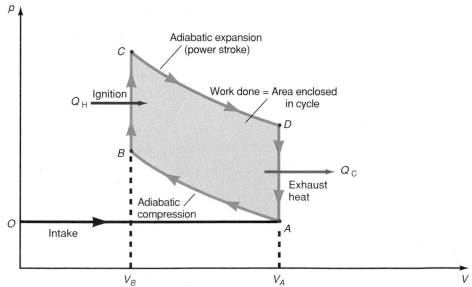

Thermodynamics

Each cylinder has an inlet valve, an exhaust valve, a spark plug, and a movable piston, which is connected to the crankshaft by a piston rod. The operation of one of these cylinders is shown schematically in figure 17.8. The gasoline engine is approximated by an **Otto cycle** and is shown on the p-V diagram of figure 17.9.

Figure 17.8(a) shows the first stroke of the engine, which is called the intake stroke. The inlet valve opens and a mixture of air and gasoline is drawn into the cylinder as the piston moves downward. Because the inlet valve is open during this first stroke, the air pressure inside the cylinder is the same constant value as atmospheric pressure and is thus shown as the isobaric path OA in figure 17.9. When the cylinder is completely filled to the volume V_A with the air and gasoline mixture, point A, the inlet valve, closes and the compression stroke starts, figure 17.8(b). The piston moves upward very rapidly causing an adiabatic compression of the air-gas mixture. This is shown as the adiabatic path AB in figure 17.9. When the piston is at its highest point (its smallest volume V_B), a spark is applied to the mixture by the spark plug. This spark causes ignition of the air-gas mixture (a small explosion of the mixture), and a great deal of heat is supplied to the mixture by the explosion. This supply of heat is shown as Q_H on the path BC of figure 17.9. The explosion occurs so rapidly that it takes a while to overcome the inertia of the piston to get it into motion. Hence, for this small time period, the pressure and temperature in the cylinder rises very rapidly at approximately constant volume. This is shown as the path BC in figure 17.9. At the point C the force of the air-gas mixture is now able to overcome the inertia of the piston, and the piston moves downward very rapidly during the power stroke, figure 17.8(d). Because the piston moves very rapidly, this portion of the process can be approximated by the adiabatic expansion of the gas shown as CD in figure 17.9. As the piston moves down rapidly this downward motion of the piston is transferred by the piston rod to the crankshaft of the engine causing the crankshaft to rotate. That is, the piston rod is connected off-center to the crankshaft. Thus, when the piston rod moves downward it creates a torque that causes the crankshaft to rotate. The rotating crankshaft is connected by a series of gears to the rear wheels of the car thus causing the wheels to turn and the car to move. At the end of this power stroke the piston has moved down to the greatest volume V_A. At this point D, the exhaust valve of the cylinder opens and the higher pressure at D drops very rapidly to the outside pressure at A, and a good deal of heat Q_C is exhausted out through the exhaust valve. As the piston now moves upward in figure 17.8(f) all the remaining used gas-air mixture is dumped out through the exhaust valve. This is shown as the path AO in figure 17.9. At the position O, the exhaust valve closes and the inlet valve opens allowing a new mixture of air and gasoline to enter the cylinder. The process now starts over again as the same cycle $OABCDAO$ of figure 17.9. The net result of the entire cycle is that heat Q_H is added along path BC, work is done equal to the area enclosed by the cyclic path, and heat Q_C is exhausted out of the system. Thus, heat has been added to the system and the system performed useful work. Four, six, or eight of these cylinders are ganged together with the power stroke of each cylinder occurring at a different time for each cylinder. This has the effect of smoothing out the torque on the crankshaft, causing a more constant rotation of the crankshaft. Unfortunately practical limitations, such as compression ratio, friction, cooling, and so on, cause the efficiency of the gasoline engine, which uses the Otto cycle, to be limited to about 20% to 25%.

17.7 The Ideal Heat Engine

There are many heat engines in addition to the gasoline engine, but they all have one thing in common: *every engine absorbs heat from a source at high temperature, performs some amount of mechanical work, and then rejects some heat at a lower temperature.* This process can be visualized with the schematic diagram for an **ideal heat engine,** and is shown in figure 17.10. The engine is represented by the circle in the diagram. The engine absorbs the quantity of heat Q_H from a hot-temperature reservoir, at a temperature T_H. (In the gasoline engine, the quantity of heat Q_H was

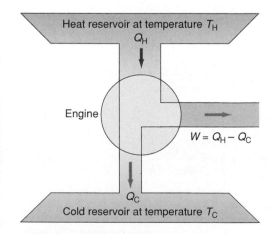

Figure 17.10

An idealized heat engine.

supplied by the combustion of the air-gasoline mixture.) Some of this absorbed heat energy is converted to work, which is shown as the pipe coming out of the engine at the right. This corresponds to the work done during the power stroke of the gasoline engine. The rest of the original absorbed heat energy is dumped as exhaust heat Q_C into the low-temperature reservoir. (In the gasoline engine this is the hot exhaust gas that is rejected to the cooler environment outside the engine.)

Because the engine operates in a cycle, $\Delta U = 0$, and as we have already seen, the net work done is equal to the net heat absorbed by the engine, that is,

$$W = Q$$

But the net heat absorbed is equal to the difference between the total heat absorbed Q_H at the hot reservoir, and the heat rejected Q_C at the cold reservoir, that is,

$$Q = Q_H - Q_C$$

Thus, the work done by the engine is equal to the difference between the heat absorbed from the hot reservoir and heat rejected to the cold reservoir

$$W = Q_H - Q_C \qquad (17.49)$$

The efficiency of a heat engine can also be defined from equation 17.34 as

$$\text{Eff} = \frac{W}{Q_{in}} = \frac{W}{Q_H} = \frac{Q_H - Q_C}{Q_H} \qquad (17.50)$$

$$\text{Eff} = 1 - \frac{Q_C}{Q_H} \qquad (17.51)$$

Thus, to make any heat engine as efficient as possible it is desirable to make Q_H as large as possible and Q_C as small as possible. It would be most desirable to have $Q_C = 0$, then the engine would be 100% efficient. Note that this would not be a violation of the first law of thermodynamics. However, as we will see in section 17.8, such a process is not possible.

Before leaving this section we should note that a **refrigerator,** or a heat pump, is a heat engine working in reverse. A refrigerator is represented schematically in figure 17.11, where the refrigerator is represented as the circle in the diagram. Work W is done on the refrigerator, thereby extracting a quantity of heat Q_C from the low-temperature reservoir and exhausting the large quantity of heat Q_H to the hot reservoir. The total heat energy exhausted to the high-temperature reservoir Q_H is the sum of the work done on the engine plus the heat Q_C extracted from the low-temperature reservoir. Thus,

$$Q_H = W + Q_C$$

We define the equivalent of an efficiency for a refrigerator, the coefficient of performance, as

$$\text{Coefficient of performance} = \frac{\text{Heat removed}}{\text{Work done}}$$

$$\text{Coefficient of performance} = \frac{Q_C}{W} \qquad (17.52)$$

$$\text{Coefficient of performance} = \frac{Q_C}{Q_H - Q_C} \qquad (17.53)$$

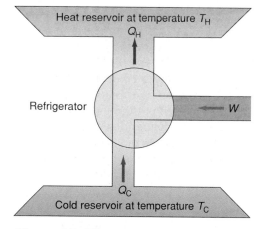

Figure 17.11

An ideal refrigerator.

17.8 The Carnot Cycle

As we saw in section 17.7, it is desirable to get the maximum possible efficiency from a heat engine. Sadi Carnot (1796–1832) showed that the maximum efficiency of any heat engine must follow a cycle consisting of the isothermal and adiabatic paths shown in the p-V diagram in figure 17.12, and now called the **Carnot cycle.** The cycle begins at point A. Let us now consider each path individually.

Thermodynamics

Figure 17.12

A *p-V* diagram for a Carnot cycle.

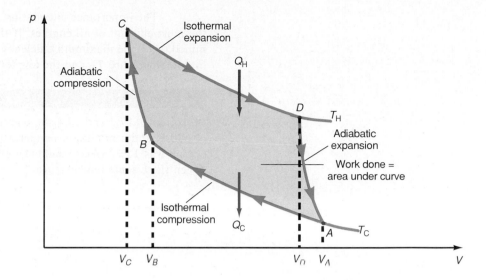

Path AB: An ideal gas is first compressed isothermally along the path *AB*. Since *AB* is an isotherm, $\Delta T = 0$ and hence $\Delta U = 0$. The first law therefore says that $Q = W$ along path *AB*. That is, the work W_{AB} done on the gas is equal to the heat removed from the gas Q_C, at the low temperature, T_C.

Path BC: Path *BC* is an adiabatic compression and hence $Q = 0$ along this path. The first law therefore becomes $\Delta U = W$. That is, the work W_{BC} done on the gas during the compression is equal to the increase in the internal energy of the gas as the temperature increases from T_C to T_H.

Path CD: Path *CD* is an isothermal expansion. Hence, $\Delta T = 0$ and $\Delta U = 0$. Therefore, the first law becomes $W = Q$. That is, the heat added to the gas Q_H at the high temperature T_H is equal to the work W_{CD} done by the expanding gas.

Path DA: Path *DA* is an adiabatic expansion, hence $Q = 0$ along this path. The first law becomes $\Delta U = W$. Thus, the energy necessary for the work W_{DA} done by the expanding gas comes from the decrease in the internal energy of the gas. The gas decreases in temperature from T_H to T_C.

The net effect of the Carnot cycle is that heat Q_H is absorbed at a high temperature T_H, mechanical work W is done by the engine, and waste heat Q_C is exhausted to the low-temperature reservoir at a temperature T_C. The net work done by the Carnot engine is

$$W = Q_H - Q_C$$

The efficiency is given by the same equations 17.50 and 17.51 as we derived before. That is,

$$\text{Eff} = 1 - \frac{Q_C}{Q_H} \qquad (17.51)$$

Lord Kelvin proposed that the ratio of the heat rejected to the heat absorbed could serve as a temperature scale. Kelvin then showed that for a Carnot engine

$$\frac{Q_C}{Q_H} = \frac{T_C}{T_H} \qquad (17.54)$$

where T_C and T_H are the Kelvin or absolute temperatures of the gas. With the aid of equation 17.54, we can express the efficiency of a Carnot engine as

$$\text{Eff} = 1 - \frac{T_C}{T_H} \qquad (17.55)$$

The importance of equation 17.55 lies in the fact that the Carnot engine is the most efficient of all engines. If the efficiency of a Carnot engine can be determined, then the maximum efficiency possible for an engine operating between the high temperature T_H and the low temperature T_C is known.

Example 17.5

In examples 17.3 and 17.4 the engine operated between a maximum temperature of 600 K and a minimum temperature of 150 K. The efficiency of that particular engine was 15%. What would the efficiency of a Carnot engine be, operating between these same temperatures?

Solution

The efficiency of the Carnot engine, found from equation 17.55, is

$$\text{Eff} = 1 - \frac{T_C}{T_H} = 1 - \frac{150 \text{ K}}{600 \text{ K}} = 0.75$$

Therefore, the maximum efficiency for any engine operating between these temperatures cannot be higher than 75%. Obviously the efficiency of 15% for the previous cycle is not very efficient.

17.9 The Second Law of Thermodynamics

There are several processes that occur regularly in nature, but their reverse processes never occur. For example, we can convert the kinetic energy of a moving car to heat in the brakes of the car as the car is braked to a stop. However, we cannot heat up the brakes of a stopped car and expect the car to start moving. That is, we cannot convert the heat in the brakes to kinetic energy of the car. Thus, mechanical energy can be converted into heat energy but heat energy cannot be completely converted into mechanical energy. As another example, a hot cup of coffee left to itself always cools down to room temperature, never the other way around. There is thus a kind of natural direction followed by nature. That is, processes will proceed naturally in one direction, but not in the opposite direction. Yet in any of these types of processes there is no violation of the first law of thermodynamics regardless of which direction the process occurs. This unidirectionality of nature is expressed as **the second law of thermodynamics** and tells which processes will occur in nature. The second law will first be described in terms of the ideal heat engine and refrigerator studied in section 17.7.

The Kelvin-Planck Statement of the Second Law

No process is possible whose sole result is the absorption of heat from a reservoir at a single temperature and the conversion of this heat energy completely into mechanical work. This statement is shown schematically in figure 17.13. That is, the diagram in figure 17.13 cannot occur in nature. Observe from figure 17.13 that heat Q_H is absorbed from the hot reservoir and converted completely into work. In figure 17.10 we saw that there had to be an amount of heat Q_C exhausted into the cold reservoir. Thus the **Kelvin-Planck statement of the second law of thermodynamics** says that there must always be a quantity of heat Q_C exhausted from the engine into a lower temperature reservoir.

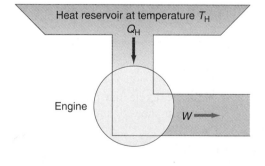

Figure 17.13
Kelvin-Planck violation of the second law.

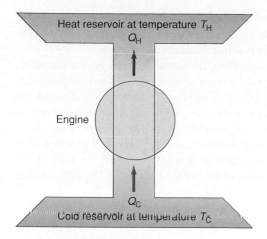

Figure 17.14

Clausius violation of the second law.

The Clausius Statement of the Second Law of Thermodynamics

No process is possible whose sole result is the transfer of heat from a cooler to a hotter body. The **Clausius statement of the second law of thermodynamics** can best be described by the refrigerator of figure 17.11. Work was done on the refrigerator to draw heat Q_C out of the cold reservoir to then deliver it to the hot reservoir. The Clausius statement says that work must always be done to do this. The violation of this Clausius statement of the second law is shown in figure 17.14. This statement of the second law of thermodynamics is essentially an observation of nature. Thermal energy flows from hot reservoirs (hot bodies) to cold reservoirs (cold bodies). The reverse process where heat flows from a cold body to a hot body without the application of some kind of work does not occur in nature. Thus, the second law of thermodynamics says that such processes are impossible, and the diagram in figure 17.14 cannot occur in nature.

17.10 Entropy

The second law of thermodynamics has been described in terms of statements about which processes are possible and which are not possible. It would certainly be more desirable to put the second law on a more quantitative basis. In 1865, Clausius introduced the concept of entropy to indicate what processes are possible and what ones are not. When a thermodynamic system changes from one equilibrium state to another in a series of small increments such that the system always moves through a series of equilibrium states, the system is said to go through a *reversible process*. A reversible process can be drawn as a continuous line on a *p-V* diagram. All the processes that have been considered are reversible processes. When a thermodynamic system changes from one equilibrium state to another along a reversible path, there is a change in entropy, ΔS of the system given by

$$\Delta S = \frac{\Delta Q}{T} \qquad (17.56)$$

where ΔQ is the heat added to the system, and T is the absolute temperature of the system.

Example 17.6

Find the change in entropy when 5.00 kg of ice at 0.00 °C are converted into water at 0.00 °C.

Solution

The heat absorbed by the ice in melting is found from

$$\Delta Q = mL_f = (5.00 \text{ kg})(80.0 \text{ kcal/kg}) = 400 \text{ kcal}$$

The process takes place at 0 °C which is equal to 273 K. The change in entropy, found from equation 17.56, is

$$\Delta S = \frac{\Delta Q}{T}$$

$$= \frac{400 \text{ kcal}}{273 \text{ K}}$$

$$= 1.47 \text{ kcal/K}$$

Whenever heat is added to a system, ΔQ is positive, and hence, ΔS is also positive. If heat is removed from a system, ΔQ is negative, and therefore, ΔS is also negative. When the ice melts there is a positive increase in entropy.

Entropy is a very different concept than the concept of energy. For example, in a gravitational system, a body always falls from a region of high potential energy to low potential energy, thereby losing potential energy. In contrast, *in an isolated thermodynamic system, the system always changes from values of low entropy to values of high entropy, thereby increasing the entropy of the system. Therefore, the concept of entropy can tell us in which direction a process will proceed.* For example, if an isolated thermodynamic system is in a state A, and we wish to determine if it can naturally go to state B by itself, we first measure the initial value of the entropy at A, S_i, and the final value of the entropy at B, S_f. The system will move from A to B only if there is an increase in the entropy in moving from A to B. That is, the process is possible if

$$\Delta S = S_f - S_i > 0 \qquad (17.57)$$

If ΔS is negative for the proposed process, the system will not proceed to the point B. *The second law of thermodynamics can also be stated as: the entropy of an isolated system increases in every natural process, and only those processes are possible for which the entropy of the system increases or remains a constant.* The entropy of a nonisolated system may either increase, or decrease, depending on whether heat is added to or taken away from the system. If ΔQ is equal to zero, such as in an adiabatic process, then ΔS also equals zero. Hence, an adiabatic process is also an *isoentropic process.* Just as the change in internal energy of a system from state A to state B is independent of the path taken to get from A to B, the entropy of a system is also independent of the path taken.

Note from the form of equation 17.56 that the temperature T must be a constant. If the temperature is not a constant, as is the case in most processes, the calculus must be used to evaluate the entropy of the system. In some cases an average temperature of the system can be used in equation 17.56 to evaluate the entropy.

Example 17.7

Find the change in entropy when 5.00 kg of ice at $-5.00\ °C$ is warmed to $0.00\ °C$.

Solution

The heat added to the ice is found from

$$\Delta Q = mc\Delta T = mc(T_f - T_i)$$

$$= (5.00\ \text{kg})\left(0.500\ \frac{\text{kcal}}{\text{kg °C}}\right)[0\ °C - (-5.00\ °C)]$$

$$= 12.5\ \text{kcal}$$

We can use equation 17.56 to evaluate the change in entropy of the ice if an average temperature of $-2.50\ °C = 270.5\ K$ is used. Thus,

$$\Delta S = \frac{\Delta Q}{T} \qquad (17.56)$$

$$= \frac{12.5\ \text{kcal}}{270.5\ K}$$

$$= 0.0462\ \text{kcal/K}$$

Example 17.8

Find the change in entropy when 5.00 kg of ice at $-5.00\ °C$ are converted to water at $0.00\ °C$.

Solution

We can find the change in entropy by dividing the problem into two parts. First, we find the change in entropy in warming the ice to $0.00\ °C$ and then we find the change in entropy in melting the ice. We have already found the change in entropy for these two processes in examples 17.6 and 17.7. The total change in entropy is the sum of the change in entropy for the two processes. Therefore,

$$\Delta S = \Delta S_1 + \Delta S_2$$
$$= 0.0462\ kcal/K\ +\ 1.47\ kcal/K$$
$$= 1.52\ kcal/K$$

†17.11 Statistical Interpretation of Entropy

As we have seen in sections 17.9 and 17.10, the second law of thermodynamics is described in terms of statements about which processes in nature are possible and which are not possible. Clausius introduced the concept of entropy to put the second law on a more quantitative basis. *He stated the second law as: the entropy of an isolated system increases in every natural process, and only those processes are possible for which the entropy of the system increases or remains a constant.* But this analysis was done on a macroscopic level, that is, a large-scale level, where concepts of temperature, pressure, and volume were employed. But the gas, the usual working substance discussed, is made up of billions of molecules, as shown in the kinetic theory of gases. Ludwig Boltzmann's approach to the second law of thermodynamics is a further extension of the kinetic theory, and is called *statistical mechanics.* Boltzmann looked at the molecules of the gas and asked what the most probable distribution of these molecules is. There is a certain order to the distribution of the molecules, with some states more probable than others. Thus, statistical mechanics deals with probabilities.

As an example, let us consider the gas molecules in figure 17.15(a). The molecules are contained in the left-hand side of a box by a partition located in the center of the box. When the partition is removed some of the molecules move to the right-hand side of the box until an equilibrium condition is reached whereby there are the same number of molecules in both sides of the box, figure 17.15(b). We now ask, can all the gas molecules in the entire box of figure 17.15(b) move to the left and be found in the original state shown in figure 17.15(a)? We know from experience that this never happens. This would be tantamount to all the gas molecules in the room that you are now sitting in moving completely to the other side of the room, leaving you in a vacuum. This just does not happen in life. However, if it did it would not violate the first law of thermodynamics. But the second law says some processes do not occur. This is certainly one of them. Notice that the first case in which all the molecules are in the left-hand side of the box is more orderly than the second case where the molecules are distributed over the entire box. (If the volume of the box is larger, there are more random paths for the molecules to follow and hence more disorder.)

As another example of order and disorder, let us drop a piece of clay. When the clay is dropped, superimposed over the thermal motion of the molecules of the clay is the velocity of the clay toward the ground. That is, all the molecules have a motion toward the ground, which is an ordered motion. When the clay hits the ground

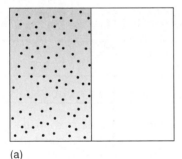

(a)

(b)

Figure 17.15
Gas molecules in a partitioned box.

and sticks to it, the kinetic energy of the falling molecules shows up as thermal energy of the clay molecules, which is a random or disordered motion of the molecules. Hence there is a transformation from order to disorder in the natural process of a collision of a falling object. Now as we know, the clay cannot gather together all the random thermal motion of the clay molecules and convert them to ordered translational motion upward, and hence the clay by itself cannot move upward. Thus the concept of which processes can occur in nature can also be measured by the amount of order or disorder between the initial and final states of the system. *Using the concept of order, the second law of thermodynamics can also be stated as: an isolated system in a state of relative order will always pass to a state of relative disorder until it reaches the state of maximum disorder, which is thermal equilibrium.*

Let us return to our example of the gas molecules in a box. Although normally there are billions of molecules in the box, to simplify our discussion let us assume that there are only four molecules present. They are numbered consecutively 1, 2, 3, and 4. Let us ask how many ways we can distribute these four molecules between the left- and right-hand sides of the box. First we could place the four molecules all in the left-hand side of the box as shown in table 17.1. Thus there is only one way we can place the four molecules into the left-hand side of the box. *Let us designate the number of ways that the four molecules can be distributed as N_i* and note that $N_1 = 1$. Next we see how many ways we can place one molecule in the right-hand side of the box and three in the left-hand side. That is, first we place molecule 1 in the right-hand side of the box, and see that that leaves molecules 2, 3, 4, in the left-hand side. Then we place molecule 2 in the right-hand side and see that we then have molecules 1, 3, 4 in the left-hand side. Continuing in this way we see from the table that there are four ways to do this. Thus, we designate the number of ways we can place one molecule in the right-hand side of the box and three in the left-hand side as N_2 and see that this is equal to 4. Next we see how many ways

Table 17.1

Possible Distributions of Four Molecules in a Box

Left	Right	N_i	$P_i = N_i/N$	$S = k \ln P$
1 2 3 4		1	1/16 = 6.25%	2.53×10^{-23} J/K

2 3 4
1   3 4
1 2   4
1 2 3
	1
	  2
	    3
	      4
		4	4/16 = 25%	4.44×10^{-23} J/K

1 2
1   3
  2 3
1   4
  2   4
    3 4
	3 4
	  2 4
	1 4
	  2 3
	1 3
	1 2
		6	6/16 = 37.5%	5.00×10^{-23} J/K

1
  2
    3
      4
	2 3 4
	1 3 4
	1 2 4
	1 2 3
		4	4/16 = 25%	4.44×10^{-23} J/K

	1 2 3 4	1	1/16 = 6.25%	2.53×10^{-23} J/K
$\Sigma N_i = N = 16$; $\Sigma P_i = 100\%$				

we can place two molecules in the right-hand side of the box and two in the left-hand side. From the table we see that there are six ways, and we call this $N_3 = 6$. Next we see how many ways we can place one molecule in the left-hand side of the box and three in the right-hand side. Again from the table we see that there are four ways to do this, and we call this $N_4 = 4$. Finally we ask how many ways can the four molecules be placed in the right-hand side of the box and again we see from the table there is only one way. We call this $N_5 = 1$. Thus there are five possible ways (for a total of 16 possible states) that the four molecules could be distributed between the left- and right-hand sides of the box.

But which of all these possibilities is the most probable? *The probability that the molecules are in the state that they are in compared with all the possible states they could be in is given by*

$$P = \frac{N_i}{N} \times 100\%$$

where N_i is the number of states that the molecules could be in for a particular distribution and N is the total number of possible states. As we see from the table, there are 16 possible states that the four molecules could be in. Hence the probability that the molecules are in the state where all four are on the left-hand side is

$$P = \frac{N_1}{N} \times 100\% = \frac{1}{16} \times 100\% = 6.25\%$$

That is, there is a 6.25% probability that all four molecules will be found in the left-hand side of the box.

The probability that the distribution of the four molecules has three molecules in the left-hand side and one in the right-hand side is found by observing that there are four possible ways that the molecules can be distributed and hence $N_2 = 4$. Therefore,

$$P = \frac{N_2}{N} \times 100\% = \frac{4}{16} \times 100\% = 25\%$$

Thus there is a 25% probability that there are three molecules in the left-hand side and one molecule in the right-hand side. Continuing in this way the probabilities that the molecules will have the particular distribution is shown in table 17.1. Thus there is a 37.5% probability that the distribution has two molecules in each half of the box, a 25% probability that the distribution has one molecule in the left half of the box and three in the right half of the box, and finally a 6.25% probability that the distribution has no molecules in the left half of the box and four in the right half of the box.

Notice that the first and last distributions (all molecules either on the left side or on the right side), are the most ordered and they have the lowest probability, 6.25%, for the distribution of the molecules. Also notice that the third distribution where there are two molecules on each side of the box has the greatest disorder and also the highest probability that this is the way that the molecules will be distributed. Notice that the distribution with the greatest possible number of states gives the highest probability. These ideas led Boltzmann to define the entropy of a state as

$$S = k \ln P \tag{17.58}$$

where k is a constant, that later turned out to be the *Stefan-Boltzmann constant,* which is equal to 1.38×10^{-23} J/K; ln is the natural logarithm; and P is the probability that the system is in the state specified. Thus in our example, the entropy of the first distribution is computed as

$$S = k \ln P = (1.38 \times 10^{-23} \text{ J/K})(\ln 6.25)$$
$$= 2.53 \times 10^{-23} \text{ J/K}$$

The entropy for each possible distribution is computed and shown in table 17.1. Notice that the most disordered state (two molecules on each side of the box) has the highest value of entropy. If we were to start the system with the four molecules in the left-hand box, entropy = 2.53×10^{-23} J/K, the system would move in the direction of maximum entropy, 5.00×10^{-23} J/K, the state with two molecules on each side of the box. As before, natural processes move in the direction of maximum entropy. The actual values of the computed entropy for this example are extremely small, because we are dealing with only four molecules. If we had only one mole of a gas in the box we would have 6.02×10^{23} molecules in the box, an enormous number compared with our four molecules. In such a case the numerical values of the entropy would be much higher. However there would still be the same type of distributions. The state with the greatest disorder, the same number of molecules on each side of the box, would be the state with the greatest value of the entropy. The state with all the molecules on one side of the box would have a finite but vanishingly small value of probability.

Hence the original problem we stated in figure 17.15(a), with the gas in the left partition has the smallest entropy while the gas on both sides of the box in figure 17.15(b), has the greatest entropy. The process flows from the state of lowest entropy to the one of highest entropy. It is interesting to note that it is not impossible for the gas molecules on both sides of the box to all move to the left-hand side of the box, but the probability is so extremely small that it would take a time greater than the age of the universe for it to happen. Hence, effectively it will not happen.

The state of maximum entropy is the state of maximum disorder and is the state where all the molecules are moving in a completely random motion. This state is, of course, the state of thermal equilibrium. We have seen throughout our study of heat that whenever two objects at different temperatures are brought together, the hot body will lose thermal energy to the cold body until the hot and cold body are at the same equilibrium temperature. Thus, as all bodies tend to equilibrium they all approach a state of maximum entropy. Hence, the universe itself tends toward a state of maximum entropy, which is a state of thermal equilibrium of all the molecules of the universe. This is a state of uniform temperature and density of all the atoms and molecules in the universe. No physical, chemical, or biological processes would be capable of occurring, however, because a state of total disorder cannot do any additional work. This ultimate state of the universe is sometimes called the *heat death of the universe.*

One final thought about entropy, and that is the idea of a direction for time. All the laws of physics, except for the second law of thermodynamics, are invariant to a change in the direction of time. That is, for example, Newton's laws of motion would work equally well if time were to run backward. If a picture were taken of a swinging pendulum with a video camera, and then played first forward and then backward, we could not tell from the picture which picture is running forward in time and which is running backward in time. They would both appear the same. On the other hand, if we take a video of a dropped coffee cup that hits the ground and shatters into many pieces we can certainly tell the difference between running the video forward or backward. When the video is run backward we would see a shattered coffee cup on the floor come together and repair itself and then jump upward onto the table. Nature does not work this way, so we know the picture must be running backward. Now before the coffee cup is dropped, we have a situation of order. When the cup is dropped and shattered we have a state of disorder. Since natural processes run from a state of order (low entropy) to one of disorder (high entropy), we can immediately see the time sequence that must be followed in the picture. The correct sequence to view the video picture is to start where the coffee cup has its lowest entropy (on the table initially at rest) and end where the cup has its maximum entropy (on the floor broken into many pieces). Hence the concept of entropy gives us a direction for time. In any natural process, the initial state has the lowest entropy, and the final state has the greatest entropy. *Thus time flows in the direction of the increase in entropy.* Stated another way, the past is the state of lowest entropy and the future is the state of highest entropy. *Thus entropy is sometimes said to be time's arrow, showing its direction.* Because of this, there have been many speculative ideas attributed to time. What happens to time when the universe reaches its state of maximum entropy? Since time flows from low entropy to high entropy, what happens when there is no longer a change in entropy? Would there be an infinite present? An eternity?

"Have you ever wondered . . . ?"

An Essay on the Application of Physics

Meteorology—The Physics of the Atmosphere

Have you ever wondered, while watching the weather forecast on your local TV station, what all those lines and arrows were on those maps? It looked something like figure 1.

If we were to look at the television screen more closely we would see a map of the United States. At every weather station throughout the United States, the atmospheric pressure is measured and recorded on a weather map. On that map, a series of lines, connecting those pressures that are the same, are drawn. These lines are called *isobars* and can be seen in figure 2. An isobar is a line along which the pressure is constant. An isobar is analogous to a contour line that is drawn on a topographical map to indicate a certain height above mean sea level. As an example, consider the mountain and valley shown in figure 3(a). A series of contour lines are drawn around the mountain at constant heights above sea level. The first line is drawn at a height $H = 200$ m above sea level. Everywhere on this line the height is exactly 200 m above sea level. The next contour line is drawn at $H = 400$ m. Everywhere on this line the height is exactly 400 m above sea level. Between the 200 m contour and the 400 m contour line the height varies between 200 m and 400 m. The contour line for 600 m is also drawn in the figure. The very top of the mountain is greater than the 600 m and is the highest point of the mountain. The contour lines showing the valley are drawn at -200 m, -400 m, and -600 m. The -200 m contour line shows that every point on this line is 200 m below sea level. The bottom of the valley is the lowest point in the valley. If we were to look down on the mountain and valley from above, we would see a series of concentric circles representing the contour lines as they are shown in figure 3(b). (On a real mountain and valley the contours would probably not be true circles.)

The isobars are to a weather map as contour lines are to a topographical map. The isobars represent the pressure of the atmosphere. By drawing the isobars, a picture of the pressure field is obtained. Normal atmospheric pressure is 29.92 in. of Hg or 1013.25 mb. But remember that normal is an average of abnormals. At any given time, the pressure in the atmosphere varies slightly from this normal value. If the atmospheric pressure is

GRIN & BEAR IT

"And as you can see in the Midwest, we've got one arrow sticking in the side of another arrow."

Figure 1

Your local TV weatherman. Reprinted with special permission of North America Syndicate, Inc.

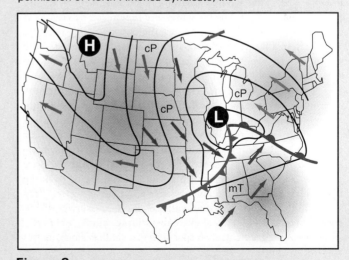

Figure 2

A weather map. Reprinted with special permission of North America Syndicate, Inc.

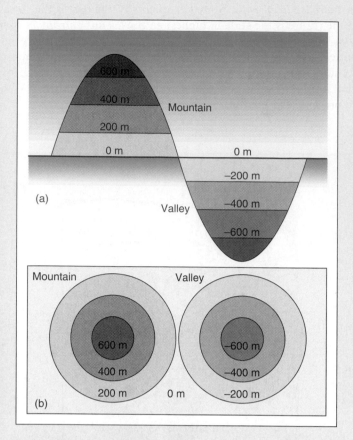

Figure 3

Contour lines on a topographical map.

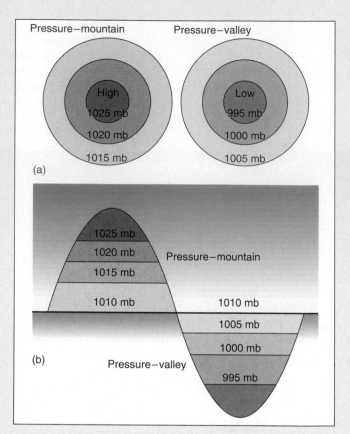

Figure 4

High and low atmospheric pressure.

greater than normal at your location, then you are in a region of high pressure. If, on the other hand, the atmospheric pressure is less than normal at your location, then you are in a region of low pressure. The isobars indicating high and low pressure are shown in figure 4(a). The high-pressure region can be visualized as a mountain and the low-pressure region as a valley in figure 4(b). Air in the high-pressure region flows down the pressure mountain into the low-pressure valley, just as a ball would roll down a real mountain side into the valley below. This flow of air is called *wind*. Hence, air always flows out of a high-pressure area into a low-pressure area. The force on a ball rolling down the mountain is the component, acting down the mountain, of the gravitational force on the ball. The force on a parcel of air is caused by the difference in pressure between the higher pressure and the lower pressure. This force is called the *pressure gradient force* (PGF) per unit mass, and it is directed from the high-pressure area to the low-pressure area. It is effectively the slope of the pressure mountain-valley. A large pressure gradient, corresponding to a steep slope, causes large winds, whereas a small pressure gradient, corresponding to a shallow slope, causes very light winds.

If the earth were not rotating, the air would flow perpendicular to the isobars. However, the earth does rotate, and the rotation of the earth causes air to be deflected to the right of its

original path. The deflection of air to the right of its path in the northern hemisphere is called the *Coriolis effect*. The Coriolis effect arises because the rotating earth is not an inertial coordinate system. For small-scale motion the rotating earth approximates an inertial coordinate system. However, for large-scale motion, such as the winds, the effect of the rotating earth must be taken into account. It is taken into account by assuming that there is a fictitious force, called the *Coriolis force* (CF) that acts to the right of the path of a parcel of air in its motion through the atmosphere. The equation for the Coriolis force is

$$CF = 2v\Omega \sin \phi \qquad \text{(17H.1)}$$

where CF is the Coriolis force per unit mass of air, v is the speed of the wind, Ω is the angular velocity of the earth, and ϕ is the latitude. Thus, the Coriolis force depends on the speed of the air (the greater the speed the greater the force) and the latitude angle ϕ. At the equator, $\phi = 0$ and $\sin \phi = 0$, and hence there is no force of deflection at the equator. For $\phi = 90°$, $\sin \phi = 1$, hence the maximum force and deflection occur at the pole.

Let us describe the motion of the air as it moves toward the low-pressure area. The air starts on its motion at the point A, figure 5(a), along a path that is perpendicular to the isobars. But the air is deflected to the right of its path by the Coriolis force, and ends up at the position B. At B, the pressure gradient force

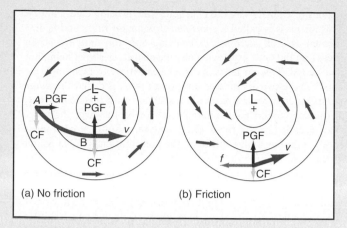

Figure 5

A low-pressure area.

is still acting toward the center of the low-pressure area, while the Coriolis force, acting to the right of the path, is opposite to the pressure gradient force. An approximate balance[1] exists between the two forces and the air parcel now moves parallel to the isobars. Notice that the air moves counterclockwise in a low-pressure area.

As the air moves over the ground, there is a frictional force f that acts on the air, is directed opposite to the direction of motion of the air, and is responsible for the slowing down of the air. This is shown in figure 5(b). But, as seen in equation 17H.1, the Coriolis force is a function of the wind speed. If the wind speed decreases because of friction, the Coriolis force also decreases.

1. A more detailed analysis by Newton's second law would give

$$a = \frac{F}{M} = PGF + CF$$

Since the air parcel is moving in a circle of radius r, with a velocity v, the acceleration is the centripetal acceleration given by v^2/r. Hence Newton's second law should be written as

$$\frac{v^2}{r} = PGF + CF$$

But in very large scale motion, such as over a continent, $v^2/r \approx 0.1 \times 10^{-3}$ m/s², while the PGF $\approx 1.1 \times 10^{-3}$ m/s². Thus the centripetal acceleration is about 1/10 of the acceleration caused by the pressure gradient force, and in this simplified analysis is neglected. The second law then becomes

$$0 = PGF + CF$$

or

$$PGF = -CF$$

Hence the force on the air parcel is balanced between the pressure gradient force and the Coriolis force. The wind that results from the balance between the PGF and the CF is called the *geostrophic wind*. For a more accurate analysis and especially in smaller sized pressure systems such as hurricanes and tornadoes this assumption cannot be made and the centripetal acceleration must be taken into account.

Hence, there is no longer the balance between the pressure gradient force and the Coriolis force and the air parcel now moves toward the low-pressure area. The combined result of the pressure gradient force, the Coriolis force, and the frictional force, causes the air to spiral into the low-pressure area, as seen in figure 5(b).

The result of the above analysis shows that air spirals counterclockwise into a low-pressure area at the surface of the earth. But where does all this air go? It must go somewhere. The only place for it to go is upward into the atmosphere. Hence, there is vertical motion upward in a low-pressure area.

Now recall from chapter 13 that the pressure of the air in the atmosphere decreases with altitude. Hence, when the air rises in the low-pressure area it finds itself in a region of still lower pressure aloft. Therefore, the rising air from the surface expands into the lower pressure aloft. But as seen in this chapter, for a gas to expand the gas must do work. Since there is no heat added to, or taken away from this rising air, $\Delta Q = 0$, and the air is expanding adiabatically. But as just shown in equation 17.44, the work done in the expansion causes a decrease in the internal energy of the gas. Hence, the rising air cools as it expands because the energy necessary for the gas to expand comes from the internal energy of the gas itself. Hence the temperature of the air decreases as the air expands and the rising air cools.

The amount of water vapor in the air is called humidity. The maximum amount of water vapor that the air can hold is temperature dependent. That is, at high temperatures the air can hold a large quantity of water vapor, whereas at low temperatures it can only hold a much smaller quantity. If the rising air cools down far enough it reaches the point where the air has all the water vapor it can hold. At this point the air is said to be saturated and the relative humidity of the air is 100%. If the air continues to rise and cool, it cannot hold all this water vapor. Hence, some of the water vapor condenses to tiny drops of water. These drops of water effectively float in the air. (They are buoyed up by the rising air currents.) The aggregate of all these tiny drops of water suspended in the air is called a cloud. Hence, clouds are formed when the rising air is cooled to the condensation point. If the rising and cooling continue, more and more water vapor condenses until the water drops get so large that they fall and the falling drops are called rain. *In summary, associated with a low-pressure area in the atmosphere is rising air. The cooling of this adiabatically expanding air causes the formation of clouds, precipitation, and general bad weather.* Thus, when the weatherman says that low pressure is moving into your area, as a general rule, you can expect bad weather.

Everything we said about the low-pressure area is reversed for a high-pressure area. The pressure gradient force points away from the high-pressure area. As the air starts out of the high-pressure area at the point A, figure 6(a), it is moving along a path that is perpendicular to the isobars. The Coriolis force now acts on the air and deflects it to the right of its path. By the time the air reaches the point B, the pressure gradient force is approximately balanced by the Coriolis force,[2] and the air moves

2. The same approximation for the balance between the PGF and the CF used in the analysis of the low-pressure area is also made for the high-pressure area.

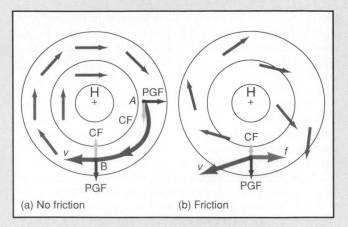

Figure 6

A high-pressure area.

parallel to the isobars. Thus, the air flows clockwise around the high-pressure area. The frictional force slows down the air and causes the Coriolis force to decrease in size. The pressure gradient force is now greater than the Coriolis force, and the air starts to spiral out of the high-pressure area, figure 6(b).

From what we have just seen, air spirals out of a high-pressure area at the surface of the earth. But if all the air that was in the high-pressure area spirals out, what is left within the high-pressure area? If the air is not replenished, the area would become a vacuum. But this is impossible. Therefore, air must come from somewhere to replenish the air spiraling out of the high. The only place that it can come from is from the air aloft. That is, air aloft moves downward into the high-pressure area at the surface. Thus, there is vertical motion downward in a high-pressure area.

As the air aloft descends, it finds itself in a region of still higher pressure and is compressed adiabatically. Thus, work is done on the gas by the atmosphere and this increase in energy shows up as an increase in the internal energy of the air, and hence an increase in the temperature of the descending air. Thus, the air warms up adiabatically as it descends. Because warmer air can hold more water vapor than colder air, the water droplets that made up the clouds evaporate into the air. As more and more air descends, more and more water droplets evaporate into the air until any clouds that were present have evaporated, leaving clear skies. *Hence, high-pressure areas are associated with clear skies and, in general, good weather.* So when the weatherman tells you that high pressure is moving into your area, you can usually expect good weather.

Now when you look at your TV weather map, look for the low- and high-pressure areas. If the low-pressure area is moving into your region, you can expect clouds and deteriorating weather. If the high-pressure area is moving into your region, you can expect improving weather with clear skies.

Those other lines on the weather map are called *fronts*. A front is a boundary between two different air masses. An air mass is a large mass of air having uniform properties of temperature and moisture throughout the horizontal. Air sitting over the vast regions of Canada has the characteristic of being cold and dry. This air mass is called a *continental polar air mass* and is designated as a cP air mass. Air sitting over the southern ocean areas and the Gulf of Mexico has the characteristic of being hot and humid. This air mass is called a maritime tropical, mT, air mass. These two air masses interact at what is called the *polar front*. Much of your weather is associated with this polar front. If the continental polar air mass is moving forward, the polar front is called a *cold front*. On a weather map the cold front is shown either as a blue line or, if the presentation is in black and white, a black line with little triangles on its leading edge showing the direction of motion. If the continental polar air mass is retreating northward, the polar front is called a *warm front*. On a weather map the warm front is shown either as a red line or, if the presentation is in black and white, a black line with little semicircles on its edge showing the direction in which the front is retreating. The center of the polar front is embedded in the low-pressure area.

With all this background, let us now analyze the weather map of figure 2. Notice that there is a large low-pressure area over the eastern half of the United States. In general, the poorer weather will be found in this region. A high-pressure area is found across the western half of the United States. In general, good weather will be found in this region. The polar front can also be seen in figure 2. The cold front is the boundary between the cold continental polar air that came out of Canada and the warm moist maritime tropical air that has moved up from the gulf. The arrows on the map indicate the velocity of the air. The cold dry cP air, being heavier than the warm tropical mT air, pushes underneath the mT air, driving it upward. The moisture in the rising tropical air condenses and forms a narrow band of clouds along the length of the cold front. The precipitation usually associated with the cold front is showery.

The warm front is the boundary of the retreating cool air and the advancing warm moist air. The mT air, being lighter than the retreating cP air, rises above the colder air. The sloping front of the retreating air is much shallower than the slope of the advancing cold front. Therefore, the mT air rises over a very large region and gives a very vast region of clouds and precipitation. Thus the weather associated with a warm front is usually more extensive than the weather associated with a cold front.

Your weather depends on where you are with respect to the frontal systems. If you are north of the warm front in figure 2, such as in Illinois, Ohio, or Pennsylvania, the temperatures will be cool, the winds will be from the southeast, the sky will be cloudy, and you will be getting precipitation. If you are south of the warm front and in advance of the cold front, such as in Alabama, Georgia, South Carolina, and Florida, the temperature will be warm, the humidity high, winds will be from the southwest and you will usually have nice weather. If the cold front has already passed you by, such as in Kansas, Oklahoma, Texas, and Arkansas, the skies will be clear or at least clearing, the temperature will be cold, the humidity will be low, the winds will be from the northwest, and in general you will have good weather.

All the highs, lows, and fronts, move across the United States from the west toward the east. So the weather that you get today will change as these weather systems move toward you.

The Language of Physics

Thermodynamics
The study of the relationships between heat, internal energy, and the mechanical work performed by a system. The system is usually a heat engine of some kind (p. 479).

Work
The work done by a gas during expansion is positive and the work done by a gas during compression is negative. The work done is equal to the area under the curve in a p-V diagram. The work done depends on the thermodynamic path taken in the p-V diagram (p. 480).

Cyclic process
A process that runs in a cycle eventually returning to where it started from. The net work done by the system during a cyclic process is equal to the area enclosed by the cyclic thermodynamic path in a p-V diagram. The net work is positive if the cycle proceeds clockwise, and negative if the cycle proceeds counterclockwise on the p-V diagram. The total change in internal energy around the entire cycle is equal to zero. The energy for the net work done by the system comes from the net heat applied to the system (p. 480).

Isobaric process
A process that takes place at a constant pressure (p. 482).

Isochoric or isometric processes
A process that takes place at constant volume. The heat added to a system during an isochoric process shows up as an increase in the internal energy of the system (p. 482).

Isothermal process
A process that takes place at constant temperature (p. 482).

Molecular mass
The molecular mass of any substance is equal to the mass of one molecule of that substance times the total number of molecules in one mole of the substance (Avogadro's number). *Thus, the molecular mass of any substance is equal to the mass of one mole of that substance.* Hence, the mass of a gas is equal to the number of moles of a gas times the molecular mass of the gas (p. 484).

Molar specific heat
The product of the specific heat of a substance and its molecular mass (p. 484).

Heat in a thermodynamic process
The heat absorbed or liberated in a thermodynamic process depends on the path that is followed in a p-V diagram. Thus, heat, like work, is path dependent. Heat is always positive when it is added to the system and negative when it is removed from the system (p. 484).

Internal energy of a gas
The internal energy of a gas is equal to the sum of the kinetic energy of all the molecules of a gas. A change in temperature is associated with a change in the internal energy of a gas. Hence, an isothermal expansion occurs at constant internal energy. Regardless of the path chosen between two points in a p-V diagram, the change in internal energy is always the same. Thus, the internal energy of the system is independent of the path taken in a p-V diagram; it depends only on the initial and final states of the thermodynamic system (p. 485).

The first law of thermodynamics
The heat added to a system will show up either as a change in internal energy of the system or as work performed by the system. It is also stated in the form: the change in the internal energy of the system equals the heat added to the system minus the work done by the system on the outside environment. The first law is really a statement of the law of conservation of energy applied to a thermodynamic system (p. 487).

Efficiency
The efficiency of an engine can be defined in terms of what we get out of the system compared to what we put into the system. It is thus equal to the ratio of the work performed by the system to the heat put into the system. It is desirable to make the efficiency of an engine as high as possible (p. 491).

Adiabatic process
A process that occurs without an exchange of heat between the system and its environment. That is, heat is neither added nor taken away from the system during the process (p. 492).

Otto cycle
A thermodynamic cycle that is approximated in the operation of the gasoline engine (p. 495).

Ideal heat engine
An idealized engine that shows the main characteristics of all engines, namely, every engine absorbs heat from a source at high temperature, performs some amount of mechanical work, and then rejects some heat at a lower temperature (p. 495).

Refrigerator
A heat engine working in reverse. That is, work is done on the refrigerator, thereby extracting a quantity of heat from a low-temperature reservoir and exhausting a large quantity of heat to a hot reservoir (p. 496).

Carnot cycle
A thermodynamic cycle of a Carnot engine, consisting of two isothermal and two adiabatic paths in a p-V diagram. The Carnot engine is the most efficient of all engines (p. 496).

The second law of thermodynamics
The second law of thermodynamics tells us which processes are possible and which are not. The concept of entropy is introduced to give a quantitative basis for the second law. It is equal to the ratio of the heat added to the system to the absolute temperature of the system, when a thermodynamic system changes from one equilibrium state to another along a reversible path. In an isolated system, the system always changes from values of low entropy to values of high entropy, and only those processes are possible for which the entropy of the system increases or remains a constant (p. 498).

Kelvin-Planck statement of the second law of thermodynamics
No process is possible whose sole result is the absorption of heat from a reservoir at a single temperature and the conversion of this heat energy completely into mechanical work (p. 498).

Clausius statement of the second law of thermodynamics
No process is possible whose sole result is the transfer of heat from a cooler to a hotter body (p. 499).

Summary of Important Equations

Work done by a gas
$W = p\Delta V$ **(17.5)**

Mass of the gas
$m = m_0 N_A n$ **(17.14)**

Molecular mass
$M = m_0 N_A$ **(17.15)**

Mass of the gas
$m = nM$ **(17.16)**

Molar specific heat
$C = Mc$ **(17.18)**

Heat absorbed or liberated by
a gas at constant volume
$Q = nC_v \Delta T$ **(17.20)**

Heat absorbed or liberated by
a gas at constant pressure
$Q = nC_p \Delta T$ **(17.21)**

Internal energy of an ideal gas
$U = \frac{3}{2} nRT$ **(17.23)**

Change in internal energy of an
ideal gas
$\Delta U = \frac{3}{2} nR\Delta T$ **(17.24)**

Change in internal energy of an
ideal gas
$\Delta U = nC_v \Delta T$ **(17.28)**

Molar specific heat at constant volume
$C_v = \frac{3}{2} R$ **(17.29)**

Molar specific heat at constant pressure
$C_p = \frac{5}{2} R$ **(17.42)**

First law of thermodynamics
$\Delta U = Q - W$ **(17.31)**

Adiabatic process
$Q = 0$

First law for adiabatic process
$W = -\Delta U$ **(17.44)**

Isochoric process
$\Delta V = 0$

First law for isochoric process
$Q = \Delta U$ **(17.47)**

Isobaric process
$\Delta p = 0$

Cyclic process
$\Delta U = 0$

First law for cyclic process
$W = Q$ **(17.48)**

Efficiency of any engine
$\text{Eff} = \dfrac{W}{Q_{in}} = \dfrac{W}{Q_H}$

$\text{Eff} = \dfrac{Q_H - Q_C}{Q_H}$ **(17.50)**

$\text{Eff} = 1 - \dfrac{Q_C}{Q_H}$ **(17.51)**

Efficiency of a Carnot engine
$\text{Eff} = 1 - \dfrac{T_C}{T_H}$ **(17.55)**

Entropy
$\Delta S = \dfrac{\Delta Q}{T}$ **(17.56)**

$S = k \ln P$ **(17.58)**

Questions for Chapter 17

1. Discuss the difference between the work done by the gas and the work done on the gas in any thermodynamic process.
2. Why is the work done in a thermodynamic process a function of the path traversed in the p-V diagram?
3. Define the following processes: isobaric, isothermal, isochoric, adiabatic, cyclic, and isoentropic.
4. How is it possible that a solid and a liquid have one value for the specific heat and a gas can have an infinite number of specific heats?
5. Discuss the first and second laws of thermodynamics.
6. Describe what is meant by the statement, "the internal energy of a thermodynamic system is conservative."
7. Figure 17.7 shows a plot of isotherms and adiabats on a p-V diagram. Explain why the adiabats have a steeper slope.
†8. Discuss the thermodynamic process in a diesel engine, and draw the process on a p-V diagram.
†9. Use the first law of thermodynamics to describe a solar heating system.
10. Can you use a home refrigerator to cool the home in the summer by leaving the door of the refrigerator open?
†11. Why is a heat pump not very efficient in very cold climates?
†12. Show how equation 17.54 could be used as the basis of a temperature scale.
13. Is it possible to connect a heat engine to a refrigerator such that the work done by the engine is used to drive the refrigerator, and the waste heat from the refrigerator is then given to the engine, to drive the engine thus making a perpetual motion machine?
14. Discuss the concept of entropy and how it can be used to determine if a thermodynamic process is possible.
†15. Discuss the statements: (a) entropy is sometimes called time's arrow and (b) the universe will end in a heat death when it reaches its state of maximum entropy.

Problems for Chapter 17

17.2 The Concept of Work Applied to a Thermodynamic System

1. How much work is done by an ideal gas when it expands at constant atmospheric pressure from a volume of 0.027 m³ to a volume of 1.00 m³?
2. What is the area of the cross-hatched area in the p-V diagram? What is the work done in going from A to B?

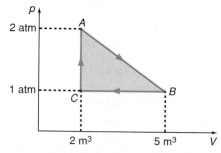

3. What is the net work done in the triangular cycle ABC?

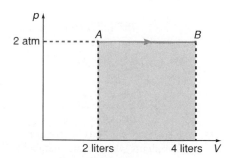

4. How much work is done in the cycle ABCDA?

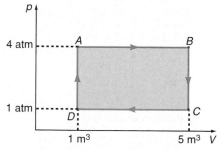

5. One mole of an ideal gas goes through the cycle shown. If $p_A = 2.00 \times 10^5$ Pa, $p_D = 5.00 \times 10^4$ Pa, $V_B = 2.00$ m³, and $V_A = 0.500$ m³, find the work done along the paths (a) AB, (b) BC, (c) CD, (d) DA, and (e) ABCDA.

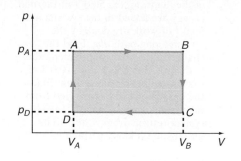

17.3 Heat Added to or Removed from a Thermodynamic System

6. What is the mass of 4.00 moles of He gas?
7. Find the amount of heat required to raise the temperature of 5.00 moles of He, 10.0 °C, at constant volume.
8. Find the amount of heat required to raise the temperature of 5.00 moles of He, 10.0 °C, at constant pressure.
9. Compute the amount of heat absorbed when one mole of a monatomic gas, at a temperature of 200 K, rises to a temperature of 400 K (a) isochoricly and (b) isobaricly.

17.4 The First Law of Thermodynamics

10. What is the total internal energy of 3.00 moles of an ideal gas at (a) 273 K and (b) 300 K?
11. What is the change in internal energy of 3.00 moles of an ideal gas when it is heated from 273 K to 293 K?
12. Find the change in the internal energy of 1 mole of an ideal gas when heated from 300 K to 500 K.
13. In a thermodynamic system, 500 J of work are done and 200 J of heat are added. Find the change in the internal energy of the system.
14. In a certain process, the temperature rises from 50.0 °C to 150.0 °C as 1000 J of heat energy are added to 4 moles of an ideal gas. Find the work done by the gas during this process.
15. In a thermodynamic system, 200 J of work are done and 500 J of heat are added. Find the change in the internal energy of the system.
16. In a certain process with an ideal gas, the temperature drops from 120 °C to 80.0 °C as 2000 J of heat energy are removed from the system and 1000 J of work are done by the gas. Find the number of moles of the gas that are present.

17. Four moles of an ideal gas are carried through the cycle ABCDA of figure 17.6. If $T_D = 100$ K, $T_{AC} = T_A = T_C = 200$ K, $T_B = 400$ K, $p_A = 0.500 \times 10^5$ Pa, and $p_D = 2.50 \times 10^4$ Pa, use the ideal gas equation to determine the volumes V_A and V_B.
†18. In problem 17 find the work done, the heat lost or absorbed, and the change in internal energy of the gas for the paths (a) AB, (b) BC, (c) CD, (d) DA, and (e) ABCDA.
19. In a thermodynamic system, 700 J of work are done by the system while the internal energy drops by 450 J. Find the heat transferred to the gas during this process.
20. If 5.00 J of work are done by a refrigerator and 8.00 J of heat are exhausted into the hot reservoir, how much heat was removed from the cold reservoir?
21. A heat engine is operating at 40.0% efficiency. If 3.00 J of heat are added to the system, how much work is the engine capable of doing?

17.5 Some Special Cases of the First Law of Thermodynamics

22. If the temperature of 2.00 moles of an ideal gas increases by 40.0 K during an isochoric process, how much heat was added to the gas?
23. If 800 J of thermal energy are removed from 8 moles of an ideal gas during an isochoric process, find the change in temperature in degrees (a) Kelvin, (b) Celsius, and (c) Fahrenheit.
24. If 3.00 J of heat are added to a gas during an isothermal expansion, how much work is the system capable of doing during this process?
25. During an isothermal contraction, 55.0 J of work are done on an ideal gas. How much thermal energy was extracted from the gas during this process?
†26. A monatomic gas expands adiabatically to double its original volume. What is its final pressure in terms of its initial pressure?
†27. One mole of He gas at atmospheric pressure is compressed adiabatically from an initial temperature of 20.0 °C to a final temperature of 100 °C. Find the new pressure of the gas.
28. If 50.0 J of work are done on one mole of an ideal gas during an adiabatic compression, what is the temperature change of the gas?

17.6 The Gasoline Engine

†29. The crankshaft of a gasoline engine rotates at 1200 revolutions per minute. The area of each piston is 80.0 cm² and the length of the stroke is 13.0 cm. If the average pressure during the power stroke is 7.01×10^5 Pa, find the power developed in each cylinder. (*Hint:* remember that there is only one power stroke for every two revolutions of the crankshaft.)

17.7 The Ideal Heat Engine

30. An engine operates between room temperature of 20.0 °C and a cold reservoir at 5.00 °C. Find the maximum efficiency of such an engine.
31. What is the efficiency of a Carnot engine operating between temperatures of 300 K and 500 K?
32. A Carnot engine is working in reverse as a refrigerator. Find the coefficient of performance if the engine is operating between the temperatures −10.5 °C and 35.0 °C.
33. A Carnot refrigerator operates between −10.0 °C and 25.0 °C. Find how much work must be done per kcal of heat extracted.
34. Calculate the efficiency of an engine that absorbs 500 J of thermal energy while it does 250 J of work.

17.10 Entropy

35. Find the change in entropy if 10.0 kg of ice at 0.00 °C is converted to water at +10.0 °C.
36. A gas expands adiabatically from 300 K to 350 K. Find the change in its entropy.
†37. Find the total change in entropy if 2.00 kg of ice at 0.00 °C is mixed with 25.0 kg of water at 20.0 °C.
38. Find the change in entropy when 2.00 kg of steam at 110 °C is converted to water at 90.0 °C.
39. A gas expands isothermally and does 500 J of work. If the temperature of the gas is 35.0 °C, find its change in entropy.

Additional Problems

40. In the thermodynamic system shown in the diagram, (a) 50.0 J of thermal energy are added to the system, and 20.0 J of work are done by the system along path *abc*. Find the change in internal energy along this path. (b) Along path *adc*, 10.0 J of work are done by the system. Find the heat absorbed or liberated from the system along this path. (c) The system returns from state *c* to its initial state *a* along path *ca*. If 15.0 J of work are done on the system find the amount of heat absorbed or liberated by the system.

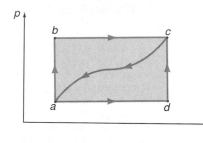

41. Draw the following process on a *p-V* diagram. First 8.00 m³ of air at atmospheric pressure are compressed isothermally to a volume of 4.00 m³. The gas then expands adiabatically to 8.00 m³ and is then compressed isobaricly to 4.00 m³.
42. In the diagram shown, one mole of an ideal gas is at atmospheric pressure and a temperature of 250 K at position *a*. (a) Find the volume of the gas at *a*. (b) The pressure of the gas is then doubled while the volume is kept constant. Find the temperature of the gas at position *b*. (c) The gas is then allowed to expand isothermally to position *c*. Find the volume of the gas at *c*.

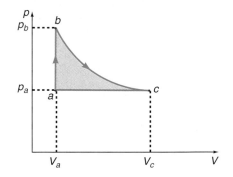

†43. Repeat problem 42, but for part (c) let the gas expand adiabatically to atmospheric pressure. Find the volume of the gas at this point. Show this point on the diagram.
†44. It was stated in equation 17.45 that for an adiabatic process with an ideal gas,

$$pV^\gamma = \text{constant}$$

Show that when an ideal gas in an initial state, with pressure p_1, volume V_1, and temperature T_1, undergoes an adiabatic process to a final state that is described by pressure p_2, volume V_2, and temperature T_2, that

$$p_1V_1^\gamma = p_2V_2^\gamma$$

and

$$T_1V_1^{\gamma-1} = T_2V_2^{\gamma-1}$$

and

$$\frac{T_1^{\gamma/(\gamma-1)}}{p_1} = \frac{T_2^{\gamma/(\gamma-1)}}{p_2}$$

45. A lecture hall at 20.0 °C contains 100 students whose basic metabolism generates 100 kcal/hr of thermal energy. If the size of the hall is 15.0 m by 30.0 m by 4.00 m, what is the increase in temperature of the air in the hall at the end of 1 hr? It is desired to use an air conditioner to cool the room to 20.0 °C. If the air conditioner is 45.0% efficient, what size air conditioner is necessary?

Interactive Tutorials

46. A thermodynamic cycle. Three moles of an ideal gas are carried around the thermodynamic cycle *ABCDA* shown in figure 17.6. Find the work done, the heat lost or absorbed, and the internal energy of the system for the thermodynamic paths (a) *AB*, (b) *BC*, (c) *CD*, (d) *DA*, and (e) *ABCDA*. The temperatures are $T_D = 147$ K, $T_{AC} = 250$ K, and $T_B = 425$ K. The pressures are $p_A = 5.53 \times 10^4$ Pa and $p_D = 3.25 \times 10^4$ Pa. The volumes are $V_A = 0.113$ m³ and $V_B = 0.192$ m³. (f) Find the efficiency of this system.

Epilogue

In the first chapter of the book we said that physics had its birthplace in mankind's quest for knowledge and truth. It started with the earliest man as he came out of his cave. We have surely progressed a great deal since that long ago time, as is evidenced by such topics covered in this book as atomic and nuclear physics, quantum physics, special and general relativity, and the unification of all the forces. However, in terms of what still lies ahead for us in this universe, we have barely taken one step out of the cave.

Appendix A
Conversion Factors

Length

1 meter (m) = 100 cm = 39.4 in. = 3.28 ft
 = 6.21×10^{-4} mi

1 centimeter (cm) = 10^{-2} m = 10 mm
 = 0.394 in.

1 inch (in.) = 2.54 cm = 0.0254 m = 0.083 ft

1 foot (ft) = 0.305 m = 30.5 cm = 12 in.

1 mile (mi) = 1610 m = 1.61 km = 5280 ft

1 kilometer (km) = 1000 m = 0.621 mi

1 nautical mile = 1.15 mi = 6076 ft = 1852 m

1 nanometer (nm) = 10^{-9} m = 10^{-7} cm

1 micron (μ) = 10^{-6} m = 1 μm = 10^{-4} cm

1 angstrom (Å) = 10^{-10} m = 10^{-8} cm

1 mil = 10^{-3} in.

1 yard (yd) = 0.9144 m = 91.44 cm

Area

$1 \text{ m}^2 = 10^4 \text{ cm}^2 = 1.55 \times 10^3 \text{ in.}^2$

$1 \text{ cm}^2 = 10^{-4} \text{ m}^2 = 0.155 \text{ in.}^2$

$1 \text{ in.}^2 = 6.45 \text{ cm}^2 = 1.27 \times 10^6$ circular mils
 $= 6.45 \times 10^{-4} \text{ m}^2$

$1 \text{ ft}^2 = 144 \text{ in.}^2 = 929 \text{ cm}^2 = 9.29 \times 10^{-2} \text{ m}^2$

$1 \text{ km}^2 = 10^6 \text{ m}^2$

$1 \text{ yd}^2 = 0.836 \text{ m}^2$

Volume

$1 \text{ m}^3 = 35.3 \text{ ft}^3 = 6.1 \times 10^4 \text{ in.}^3 = 10^3 \text{ L}$

$1 \text{ ft}^3 = 2.83 \times 10^{-2} \text{ m}^3 = 1.73 \times 10^3 \text{ in.}^3$
 $= 28.3 \text{ L} = 7.48 \text{ gal}$

1 U.S. gallon = $231 \text{ in.}^3 = 0.134 \text{ ft}^3 = 3.79$
 $\times 10^{-3} \text{ m}^3$

$1 \text{ in.}^3 = 1.64 \times 10^{-5} \text{ m}^3$

Mass

1 kg = 1000 g = 6.85×10^{-2} slugs

1 slug = 14.59 kg

Weight

1 lb = 4.45 N

1 N = 0.225 lb

Time

1 year (yr) = 365.24 day = 8.76×10^3 hr
 = 5.26×10^5 min = 3.16×10^7 s

1 day = 1.44×10^3 min = 8.64×10^4 s

Density

$1 \text{ kg/m}^3 = 1 \times 10^{-3} \text{ g/cm}^3 = 1.94 \times 10^{-3} \text{ slug/ft}^3$

$1 \text{ g/cm}^3 = 1000 \text{ kg/m}^3 = 1.94 \text{ slug/ft}^3$

Velocity

1 m/s = 3.28 ft/s = 2.24 mi/hr = 3.60 km/hr
 = 1.94 knot

1 ft/s = 0.305 m/s = 0.682 mi/hr
 = 1.10 km/hr

88 ft/s = 60 mph

1 mi/hr = 1.47 ft/s = 1.61 km/hr
 = 0.869 knot = 0.447 m/s

1 km/hr = 0.278 m/s = 0.621 mph = 0.91 ft/s

Acceleration

$1 \text{ m/s}^2 = 3.281 \text{ ft/s}^2 = 3.60 \text{ km/hr/s}$
 = 2.24 mph/s

$1 \text{ ft/s}^2 = 0.3048 \text{ m/s}^2$

$1 \text{ mph/s} = 1.467 \text{ ft/s}^2 = 0.447 \text{ m/s}^2$

Angles

1 radian (rad) = 57.30° = 57°18′ = 0.159 rev

1 degree (°) = 0.01745 rad

360° = 2π radians

1 rev/min (rpm) = 0.1047 rad/s

1 rad/s = 9.55 rev/min

Force

1 newton (N) = 10^5 dynes = 0.225 lb
 = 3.60 oz

1 pound (lb) = 4.45 N = 16 ounces (oz)

Pressure

$1 \text{ N/m}^2 = 1.00$ pascal (Pa) $= 2.09 \times 10^{-2}$
 lb/ft^2
 $= 1.45 \times 10^{-4} \text{ lb/in.}^2$
 $= 9.87 \times 10^{-6}$ atm $= 7.50 \times 10^{-4}$ cm
 of Hg
 $= 4.01 \times 10^{-3}$ in. of $H_2O = 10^{-5}$ bar
 $= 10^{-2}$ millibars (mb)

$1 \text{ lb/in.}^2 = 144 \text{ lb/ft}^2 = 6.90 \times 10^3 \text{ N/m}^2 =$
 5.17 cm of Hg
 = 27.68 in. of H_2O

1 atmosphere (atm) = $1.013 \times 10^5 \text{ N/m}^2$ =
 1013 mb
 $= 14.7 \text{ lb/in.}^2$
 $= 2.12 \times 10^3 \text{ lb/ft}^2 = 760$
 torr = 76 cm of Hg
 = 406.8 in. H_2O

Energy, Work, Thermal Energy

1 joule (J) = 0.738 ft lb = 2.39×10^{-4} kcal =
 6.24×10^{18} eV
 $= 9.481 \times 10^{-4}$ Btu = 10^7 ergs =
 0.239 cal = 1 N m

1 kilocalorie (kcal) = 4185 J = 3.97 Btu
 = 3077 ft lb

1 foot pound (ft lb) = 1.36 J = $1.29 \times$
 10^{-3} Btu
 $= 3.25 \times 10^{-4}$ kcal

1 electron volt (eV) = 1.60×10^{-19} J = 1.18
 $\times 10^{-19}$ ft lb

1 kilowatt hour (kWh) = 3.6×10^6 J =
 3413 Btu = 860 kcal
 = 1.34 hp hr

1 calorie (cal) = 4.185 J

1 British thermal unit (Btu) = 0.252 kcal
 = 778 ft lb
 $= 1.05 \times 10^3$ J

Power

1 watt (W) = 1 J/s = 0.738 ft lb/s =
 1.34×10^{-3} hp
 $= 2.39 \times 10^{-4}$ kcal/s

1 horsepower (hp) = 550 ft lb/s = 746 W =
 2545 Btu/hr
 = 0.1782 kcal/s

1 kilowatt (kW) = 1000 W = 1.34 hp

1 refrigeration ton = 12,000 Btu/hr

Electricity and Magnetism

1 volt (V) = 1 joule/coulomb = J/C

1 ampere (A) = 1 C/s = 6.3×10^{18} electrons/s

1 ohm (Ω) = 1 volt/ampere

1 farad = 1 coulomb/volt

Magnetic Field Intensity

1 tesla (T) = $\dfrac{1 \text{ N}}{\text{A m}}$ = 10^4 gauss =
 1 weber/meter^2

Magnetic Flux = 1 weber = 1 T m^2

Inductance = 1 henry = $\dfrac{1 \text{ J}}{A^2}$ = $\dfrac{V \text{ s}}{A}$

Appendix B
Useful Mathematical Formulas

Geometry

$C = 2\pi r$	Circumference of circle
$A = \pi r^2$	Area of circle
$A = 4\pi r^2$	Area of sphere
$V = \frac{4}{3}\pi r^3$	Volume of sphere
$V = \pi r^2 h$	Volume of cylinder

Algebra

$$x^n x^m = x^{(n+m)} \qquad (x^n)^m = x^{nm}$$

$$\frac{x^n}{x^m} = x^{(n-m)} \qquad x^{1/m} = (x)^{1/m}$$

$$\frac{1}{1/x} = x \qquad x^{n/m} = (x^n)^{1/m} = (x^{1/m})^n$$

$$(x + y)^2 = x^2 + 2xy + y^2$$
$$(x - y)^2 = x^2 - 2xy + y^2$$
$$(x + y)^3 = x^3 + 3x^2y + 3xy^2 + y^3$$
$$(x - y)^3 = x^3 - 3x^2y + 3xy^2 - y^3$$
$$(x + y)(x - y) = x^2 - y^2$$

Quadratic Equation

$$\text{if } ax^2 + bx + c = 0$$
$$\text{then } x = \frac{-b \pm \sqrt{b^2 - 4ac}}{2a}$$

Binomial Expansion

$$(1 + x)^n = 1 + nx + \frac{n(n-1)x^2}{2!} + \frac{n(n-1)(n-2)x^3}{3!} + \ldots$$

Powers of Ten and Scientific Notation

Very large or very small numbers can be written in a simple way by expressing them as powers of 10. For example, 100,000 may be written as 10^5. That is,

$$100,000 = (10)(10)(10)(10)(10) = 1 \times 10^5$$

Table B.1
Powers of 10

$$10^{-4} = \frac{1}{10^4} = \frac{1}{10,000} = 0.0001$$

$$10^{-3} = \frac{1}{10^3} = \frac{1}{1000} = 0.001$$

$$10^{-2} = \frac{1}{10^2} = \frac{1}{100} = 0.01$$

$$10^{-1} = \frac{1}{10} = 0.1$$

$$10^0 = 1$$
$$10^1 = 10$$
$$10^2 = (10)(10) = 100$$
$$10^3 = (10)(10)(10) = 1000$$
$$10^4 = (10)(10)(10)(10) = 10,000$$
$$10^5 = (10)(10)(10)(10)(10) = 100,000$$
$$10^6 = (10)(10)(10)(10)(10)(10) = 1,000,000$$

Table B.1 is a list of some powers of 10. Notice that $10^0 = 1$. *In numbers with positive exponents, the exponent is equal to the number of zeros following the 1. In the cases of negative exponents, the exponent is equal to the number of places the decimal point is moved to the left of the 1.* That is,

$$1 \times 10^{-1} = 0.1$$

The operations using powers of 10 follow the same rules of exponents as in algebra. For example, in algebra

$$x^n x^m = x^{(n+m)}$$

whereas for powers of 10 it becomes

$$10^n 10^m = 10^{(n+m)}$$

For example,

$$10^2 10^3 = 10^{2+3} = 10^5$$

The division of powers of ten becomes

$$\frac{10^n}{10^m} = 10^{(n-m)}$$

For example,

$$\frac{10^5}{10^3} = 10^{5-3} = 10^2$$

The reciprocal of a power of ten becomes

$$\frac{1}{10^m} = 10^{-m}$$

For example,

$$\frac{1}{10^2} = 10^{-2}$$

Finally, we have

$$(10^n)^m = 10^{nm}$$

and

$$10^{n/m} = (10^n)^{1/m} = (10^{1/m})^n$$

with the examples

$$(10^2)^3 = 10^6$$

$$10^{4/2} = (10^4)^{1/2} = 10^2$$

Scientific Notation

Very large or very small numbers can be written in a very convenient way with the use of powers of ten. For example, the number 583,000 is equal to $5.83 \times 100,000$. However, since $100,000 = 10^5$, we have

$$583,000 = 5.83 \times 10^5$$

Writing numbers in this notation is called expressing the number in scientific notation. The expression of any number in scientific notation is based on the fact that all numbers can be expressed as a number between 1 and 10 multiplied by a power of ten. As examples of some numbers expressed in scientific notation, we have

$$583,000 = 5.83 \times 10^5$$

$$1430 = 1.43 \times 10^3$$

$$0.025 = 2.5 \times 10^{-2}$$

$$0.00045 = 4.5 \times 10^{-4}$$

For a discussion of significant figures in your calculations see your laboratory text.

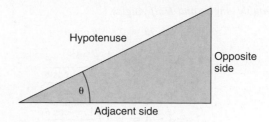

Figure B.1

A right triangle for defining the trigonometric functions.

Trigonometry

1. *Definitions.* The trigonometric functions are defined as ratios of the sides of a right triangle. Using figure B.1, they are defined as

$$\sin \theta = \frac{\text{opposite side}}{\text{hypotenuse}}$$

$$\cos \theta = \frac{\text{adjacent side}}{\text{hypotenuse}}$$

$$\tan \theta = \frac{\text{opposite side}}{\text{adjacent side}}$$

$$\sec \theta = \frac{1}{\cos \theta}$$

$$\csc \theta = \frac{1}{\sin \theta}$$

$$\cot \theta = \frac{1}{\tan \theta}$$

$$\tan \theta = \frac{\sin \theta}{\cos \theta} \qquad \cot \theta = \frac{\cos \theta}{\sin \theta}$$

2. *Some Important Trigonometric Identities*

$$\sin^2\theta + \cos^2\theta = 1$$

$$\sec^2\theta = 1 + \tan^2\theta$$

$$\csc^2\theta = 1 + \cot^2\theta$$

3. *Sums and Differences of the Trigonometric Functions*

$$\sin (A + B) = \sin A \cos B + \cos A \sin B$$

$$\sin (A - B) = \sin A \cos B - \cos A \sin B$$

$$\cos (A + B) = \cos A \cos B - \sin A \sin B$$

$$\tan (A + B) = \frac{\tan A + \tan B}{1 - \tan A \tan B}$$

$$\tan (A - B) = \frac{\tan A + \tan B}{1 + \tan A \tan B}$$

4. *Double Angles and Half Angles*

$$\sin 2\theta = 2 \sin \theta \cos \theta$$

$$\cos 2\theta = 1 - 2 \sin^2\theta$$

$$\sin\left(\frac{\theta}{2}\right) = \pm\sqrt{\frac{1 - \cos \theta}{2}}$$

$$\cos\left(\frac{\theta}{2}\right) = \pm\sqrt{\frac{1 + \cos \theta}{2}}$$

5. *Addition and Subtraction of the Trigonometric Functions*

$$\sin A + \sin B = 2 \sin\left(\frac{A + B}{2}\right)\cos\left(\frac{A - B}{2}\right)$$

$$\sin A - \sin B = 2 \cos\left(\frac{A + B}{2}\right)\sin\left(\frac{A - B}{2}\right)$$

$$\cos A + \cos B = 2 \cos\left(\frac{A + B}{2}\right)\cos\left(\frac{A - B}{2}\right)$$

$$\cos A - \cos B = 2 \sin\left(\frac{A + B}{2}\right)\sin\left(\frac{A - B}{2}\right)$$

6. *Trigonometric Functions for Negative Angles*

$$\cos(-\theta) = \cos \theta$$

$$\sin(-\theta) = -\sin \theta$$

$$\tan(-\theta) = -\tan \theta$$

Figure B.2 should be used for the following formulas:

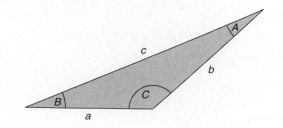

Figure B.2

An obtuse triangle.

7. Law of Sines

$$\frac{a}{\sin A} = \frac{b}{\sin B} = \frac{c}{\sin C}$$

8. Law of Cosines

$$c^2 = a^2 + b^2 - 2ab \cos C$$

9. Special Case of Law of Cosines ($C = 90°$)

$$c^2 = a^2 + b^2 \qquad \text{Pythagorean theorem}$$

Appendix C
Proportionalities

In any science, especially in physics, we constantly come on quantities that are proportional to other quantities. *A proportion is a relation among different quantities.* There are two main types of proportions: (1) direct proportions and (2) inverse proportions.

Direct Proportions

Two physical quantities are said to be directly proportional to each other if the ratio of these quantities is always equal to a constant. As an example, consider the tabulated values of x and y in table C.1. Taking the ratio of y to x for each of these values, we get the constant value 2, as shown in the third row of the table; that is,

$$\frac{y}{x} = 2 = \text{constant} \tag{C.1}$$

Thus, the ratio of y to x is always a constant, and by the definition, y is directly proportional to x. If we solve equation C.1 for y, we get

$$y = 2x \tag{C.2}$$

Equation C.2 says that y is directly proportional to x. Thus, as x increases, so does the value of y. In particular, y is always twice as large as x. The number 2, the value of the constant ratio y/x, is called the *constant of proportionality.*

In general, if y is directly proportional to x, it can be written in the form

$$y = kx \tag{C.3}$$

where k is called the constant of proportionality and can have any constant value. Thus, in table C.2, the constant of proportionality k would be equal to the value 3.

Table C.2

x	1	2	3	4	5
y	3	6	9	12	15
y/x	3	3	3	3	3

In summary, the general statement for a direct proportion is usually written in the form

$$y \propto x \tag{C.4}$$

where the symbol $\propto$ is a shorthand notation for the words "is proportional to." The main characteristic of a direct proportion is that as one variable increases, so does the other; or if one variable decreases, so does the other. When it is desirable to express this proportionality in terms of an equation, we introduce k, the constant of proportionality, and express the proportionality as the equation

$$y = kx \tag{C.5}$$

Equation C.4 says that as x increases, y also increases. Equation C.5, on the other hand, is more general and says that as x increases, y also increases, and the amount of the increase depends on the value of k.

If we plot the data of table C.1, we get the graph shown in figure C.1. Notice that when this data is plotted the result is a straight line that passes through the origin. This is a result of all

Table C.1

x	2	3	4	5	6	7	8
y	4	6	8	10	12	14	16
y/x	4/2 = 2	6/3 = 2	8/4 = 2	10/5 = 2	12/6 = 2	14/7 = 2	16/8 = 2

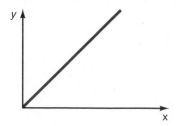

Figure C.1

Graph of a direct proportion.

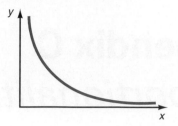

Figure C.2

Graph of y versus x showing an inverse proportionality.

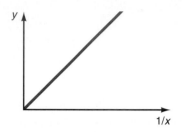

Figure C.3

Graph of y versus 1/x showing an inverse proportionality.

direct proportions. *When the direct proportion is expressed as an equation, such as that given by equation C.5, the equation is the equation of a straight line that passes through the origin.* The slope of the line is equal to the constant of proportionality k. Thus, *as a corollary, we can say that if the relation between two variables is a straight line going through the origin of a graph, then those two variables are directly proportional to each other.*

Inverse Proportion

An inverse proportion is one in which there is an inverse relationship between the variables. Thus,

$$y \propto \frac{1}{x} \tag{C.6}$$

says that y is inversely proportional to x. This means that if x increases, y must decrease, or if x decreases, y must increase. To make an equality of C.6, we introduce a constant of proportionality k and obtain

$$y = \frac{k}{x} \tag{C.7}$$

This is sometimes written in the equivalent form

$$yx = k \tag{C.8}$$

and sometimes this form is used for the defining form for an inverse relationship.

As an example of an inverse proportionality, consider the data of table C.3. Note that as x increases, y decreases, but the product of yx remains a constant.

A graph of the data of table C.3, which is an inverse relationship, is shown in figure C.2 and is a graph of a rectangular hyperbola. *Thus, as a corollary, if the graph of y versus x is a rectangular hyperbola, this implies that y is inversely proportional to x.* Sometimes it is more convenient to plot y versus the reciprocal of x to show the inverse proportion, because the plot of y versus $1/x$ is a straight line, as shown in figure C.3. *Thus, if the graph of y versus 1/x is a straight line, then y is inversely proportional to x. The slope of the straight line is equal to the proportionality constant k.*

Table C.3					
x	5	10	15	20	25
y	1	1/2	1/3	1/4	1/5
yx	5	5	5	5	5

Appendix D
Physical Constants

Speed of light, $c = 2.998 \times 10^8$ m/s $= 1.86 \times 10^5$ mi/s

Gravitational constant, $G = 6.67 \times 10^{-11}$ N m^2/kg^2

Standard acceleration of gravity, $g = 9.80$ m/s$^2 = 32.0$ ft/s^2

Heat of fusion of water, $L_f = 80$ kcal/kg $= 3.33 \times 10^5$ J/kg

Heat of vaporization of water, $L_v = 540$ kcal/kg
$\qquad\qquad\qquad\qquad\qquad = 2.26 \times 10^6$ J/kg

Mass of earth $= 5.98 \times 10^{24}$ kg

Mean radius of earth, $r_e = 6.37 \times 10^6$ m $= 3960$ mi

Angular velocity of earth $\omega_e = 7.27 \times 10^{-5}$ rad/s

Mean earth-sun distance, $r_{es} = 1.49 \times 10^8$ km $= 9.29 \times 10^7$ mi

Speed of sound in air, $v_a = 331$ m/s $= 1090$ ft/s

Speed of sound in water, $v_w = 1460$ m/s $= 4790$ ft/s

Density of dry air (STP) $= 1.29$ kg/m^3

Universal gas constant, $R = 8.314$ J/(molc K)

Mechanical equivalent of heat $= 4185$ J/kcal

Permittivity of free space, $\epsilon_0 = 8.85 \times 10^{-12}$ C^2/(N m^2)

Permeability of free space, $\mu_0 = 4\pi \times 10^{-7}$ (T m)/A

Electronic charge, $e = 1.6021 \times 10^{-19}$ C

Electron rest mass, $m_e = 9.1090 \times 10^{-31}$ kg

Mass of proton, $m_p = 1.6726 \times 10^{-27}$ kg

Mass of neutron, $m_n = 1.6749 \times 10^{-27}$ kg

Avogadro's number, $N_A = 6.022 \times 10^{23}$ molecules/mole
$\qquad\qquad\qquad\qquad = 6.022 \times 10^{26}$ molecules/kmole

Atmospheric pressure, $P_{atm} = 1.013 \times 10^5$ N/m^2

Mass of moon, $M_M = 7.343 \times 10^{22}$ kg

Mass of sun, $M_S = 1.987 \times 10^{30}$ kg

Planck's constant, $h = 6.63 \times 10^{-34}$ J s $= 4.14 \times 10^{-15}$ eV s

Boltzmann's constant, $k = 1.38 \times 10^{-23}$ J/molecules K
$\qquad\qquad\qquad\qquad = 8.63 \times 10^{-5}$ eV/molecules K

Ratio of electronic charge to mass, $e/m_e = 1.756 \times 10^{11}$ C/kg

Constant in Coulomb's law, $k = 1/4\pi\epsilon_0 = 9.00 \times 10^9$ N m^2/C^2

Appendix E
Table of the Elements

Atomic Number	Element	Symbol	Atomic Mass
1	Hydrogen	H	1.008
2	Helium	He	4.003
3	Lithium	Li	6.941
4	Beryllium	Be	9.012
5	Boron	B	10.81
6	Carbon	C	12.00
7	Nitrogen	N	14.01
8	Oxygen	O	15.99
9	Fluorine	F	19.00
10	Neon	Ne	20.18
11	Sodium	Na	22.99
12	Magnesium	Mg	24.31
13	Aluminum	Al	26.98
14	Silicon	Si	28.09
15	Phosphorus	P	30.98
16	Sulfur	S	32.06
17	Chlorine	Cl	35.46
18	Argon	Ar	39.95
19	Potassium	K	39.10
20	Calcium	Ca	40.08
21	Scandium	Sc	44.96
22	Titanium	Ti	47.90
23	Vanadium	V	50.94
24	Chromium	Cr	52.00
25	Manganese	Mn	54.94
26	Iron	Fe	55.85
27	Cobalt	Co	58.93
28	Nickel	Ni	58.71

Atomic Number	Element	Symbol	Atomic Mass
29	Copper	Cu	63.54
30	Zinc	Zn	65.37
31	Gallium	Ga	69.72
32	Germanium	Ge	72.59
33	Arsenic	As	74.92
34	Selenium	Se	78.96
35	Bromine	Br	79.91
36	Krypton	Kr	83.80
37	Rubidium	Rb	85.47
38	Strontium	Sr	87.62
39	Yttrium	Y	88.91
40	Zirconium	Zr	91.22
41	Niobium	Nb	92.91
42	Molybdenum	Mo	95.94
43	Technetium	Tc	98.91
44	Ruthenium	Ru	101.1
45	Rhodium	Rh	102.9
46	Palladium	Pd	106.4
47	Silver	Ag	107.9
48	Cadmium	Cd	112.4
49	Indium	In	114.8
50	Tin	Sn	118.7
51	Antimony	Sb	121.8
52	Tellurium	Te	127.6
53	Iodine	I	126.9
54	Xenon	Xe	131.3
55	Cesium	Cs	132.9
56	Barium	Ba	137.3

Atomic Number	Element	Symbol	Atomic Mass		Atomic Number	Element	Symbol	Atomic Mass
57	Lanthanum	La	138.9		81	Thallium	Tl	204.4
58	Cerium	Ce	140.1		82	Lead	Pb	207.2
59	Praseodymium	Pr	140.9		83	Bismuth	Bi	209.0
60	Neodymium	Nd	144.2		84	Polonium	Po	210
61	Promethium	Pm	145		85	Astatine	At	218
62	Samarium	Sm	150.4		86	Radon	Rn	222
63	Europium	Eu	152.0		87	Francium	Fr	223
64	Gadolinium	Gd	157.3		88	Radium	Ra	226.0
65	Terbium	Tb	158.9		89	Actinium	Ac	227.0
66	Dysprosium	Dy	162.5		90	Thorium	Th	232.0
67	Holmium	Ho	164.9		91	Protactinium	Pa	231.0
68	Erbium	Er	167.3		92	Uranium	U	238.0
69	Thulium	Tm	168.9		93	Neptunium	Np	237
70	Ytterbium	Yb	173.0		94	Plutonium	Pu	244
71	Lutetium	Lu	175.0		95	Americium	Am	243
72	Hafnium	Hf	178.5		96	Curium	Cm	247
73	Tantalum	Ta	181.0		97	Berkelium	Bk	247
74	Tungsten	W	183.9		98	Californium	Cf	251
75	Rhenium	Re	186.2		99	Einsteinium	Es	254
76	Osmium	Os	190.2		100	Fermium	Fm	257
77	Iridium	Ir	192.2		101	Mendelevium	Md	258
78	Platinum	Pt	195.1		102	Nobelium	No	259
79	Gold	Au	197.0		103	Lawrencium	Lr	260
80	Mercury	Hg	200.6					

Appendix F
Vector Multiplication

F.1 Introduction

Some instructors of college physics like to use the multiplication of vectors in their courses while others do not. Therefore, the multiplication of vectors is covered in this appendix, so those wishing to use it may do so. Section F.2 gives a general introduction to the multiplication of vectors, whereas each topic in the book that might utilize the multiplication of vectors is also treated here as a separate section. The section in the book where that topic is first introduced is listed in parentheses behind the section title.

F.2 The Multiplication of Vectors

Vectors were introduced in chapter 2 and we saw how vectors could be added and subtracted. That description of vectors was somewhat incomplete in that the possibility of multiplying vectors together was never discussed. That situation will be remedied here because there are many physical laws that are expressed as the product of vectors. There are three types of multiplication involving vectors:

1. Multiplication of a vector by a scalar.
2. The scalar product (the multiplication of two vectors, the product of which is a scalar).
3. The vector product (the multiplication of two vectors, the product of which is a vector).

Let us look at each of these products in detail.

The Multiplication of a Vector by a Scalar
Consider the vector **a** in figure F.1(a). This vector has both magnitude and direction. If **a** is multiplied by a scalar, then the product k**a** is a new vector, say **b**, where

$$\mathbf{b} = k\mathbf{a} \tag{F.1}$$

If k is a positive number greater than one, then the product represents a vector in the same direction as **a** but elongated by a factor k, as shown in figure F.1(b). If k is less than one, but greater than zero, the new vector is shorter than **a**, figure F.1(c), and if k is negative, the new vector is in the opposite direction of **a**, figure F.1(d).

The Scalar Product or Dot Product
*The scalar product of two vectors **a** and **b**, figure F.2, is defined as*

$$\mathbf{a} \cdot \mathbf{b} = ab \cos \theta \tag{F.2}$$

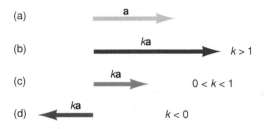

Figure F.1

Multiplication of a vector by a scalar.

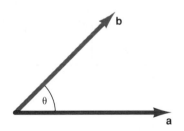

Figure F.2

The scalar product.

where θ is the angle between the two vectors when they are drawn tail-to-tail, and a and b are the magnitudes of the two vectors. The symbol for this type of multiplication is the dot "·" between the vectors **a** and **b**. Hence, *the scalar product is also called the dot product.* The definition, equation F.2, can be used to represent many different physical concepts and phenomena.

As we can observe in figure F.2, $b \cos \theta$ is the component of the vector **b** in the **a** direction. Thus, *the scalar product is the component of the vector **b** in the direction of the vector **a**, multiplied by the magnitude of the vector **a** itself.* We can also say that the quantity, $a \cos \theta$, is the component of the vector **a** in the **b** direction and hence the *scalar product is also the component of the vector **a** in the **b** direction, multiplied by the magnitude of the vector **b** itself.* Either description is correct.

The inclusion of the cos θ in the definition of the scalar product gives rise to some interesting special cases. If the vectors **a** and **b** are parallel to each other, then the angle θ between **a** and **b** is equal to zero. In that case, the scalar product becomes

$$\mathbf{a} \cdot \mathbf{b} = ab \cos 0°$$

But the cos 0° = 1, therefore

$$\mathbf{a} \cdot \mathbf{b} = ab \quad \text{(for } \mathbf{a} \parallel \mathbf{b}\text{)} \qquad \text{(F.3)}$$

(As another special case, $\mathbf{a} \cdot \mathbf{a} = a^2$, and $a = \sqrt{\mathbf{a} \cdot \mathbf{a}}$ defines the magnitude of any vector.) On the other hand, if **a** is perpendicular to **b**, then the angle between the two vectors is 90°, and the scalar product becomes

$$\mathbf{a} \cdot \mathbf{b} = ab \cos 90°$$

But the cos 90° = 0, and hence

$$\mathbf{a} \cdot \mathbf{b} = 0 \quad \text{(for } a \perp b\text{)} \qquad \text{(F.4)}$$

For example, if the vector **a** has a magnitude of 2 units and vector **b** a magnitude of 3 units, then their dot product can take on any value between −6 and +6 depending on the angle θ between the two vectors. For scalars, on the other hand, 2 times 3 is always equal to 6 and nothing else.

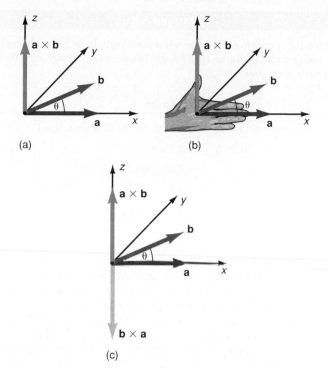

(a)   (b)

(c)

Figure F.3

The cross product.

Example F.1

The scalar product. The vector **a** has a magnitude of 5.00 units and the vector **b** a magnitude of 7.00 units. If the angle between the vectors is 53.0°, find their scalar product.

Solution

The scalar product of the two vectors, given by equation F.2, is

$$\mathbf{a} \cdot \mathbf{b} = ab \cos \theta$$
$$= (5.00)(7.00)\cos 53.0°$$
$$= 21.1$$

The Vector Product or Cross Product

*The magnitude of the vector product of two vectors **a** and **b** is defined to be*

$$|\mathbf{a} \times \mathbf{b}| = ab \sin \theta \qquad \text{(F.5)}$$

where θ is the angle between the two vectors when they are drawn tail-to-tail, as shown in figure F.3(a). The symbol for the multiplication is designated by the cross sign "×" between the two vectors, and hence the name, *cross product*. The result of the cross product of two vectors, **a** and **b**, is another vector, **a** × **b**, which is perpendicular to the plane made by the vectors **a** and **b**. *The direction of **a** × **b** is found by taking your right hand with the fingers in the same direction as the vector **a**, your palm facing toward the vector **b**, and then rotating your right hand through the angle between **a** and **b**. Your thumb will then point in the direction of **a** × **b** as seen in figure F.3(b). This rule is called the right-hand rule and it is imperative that the right hand be used.* The new vector is perpendicular both to the vector **a** and the vector **b**, and hence to the plane generated by **a** and **b**.

One of the characteristics of this vector product is that the order of the multiplication is extremely important, for as we can see from figure F.3(c),

$$\mathbf{a} \times \mathbf{b} = -\mathbf{b} \times \mathbf{a} \qquad \text{(F.6)}$$

The magnitude of **a** × **b** is the same as the magnitude of **b** × **a** from the defining equation F.5, but their directions are opposite to each other as seen in figure F.3(c). This result is often stated as: vectors are noncommutative under vector multiplication.

If the vectors **a** and **b** are parallel to each other then $\theta = 0°$ and

$$|\mathbf{a} \times \mathbf{b}| = ab \sin 0°$$

But sin 0° = 0, therefore,

$$|\mathbf{a} \times \mathbf{b}| = 0 \quad \text{(for } a \parallel b\text{)} \qquad \text{(F.7)}$$

If the vector **a** is perpendicular to the vector **b**, then $\theta = 90°$, and

$$|\mathbf{a} \times \mathbf{b}| = ab \sin 90°$$

But the sin 90° = 1, therefore

$$|\mathbf{a} \times \mathbf{b}| = ab \quad \text{(for } a \perp b\text{)} \qquad \text{(F.8)}$$

Note that these two cases are the opposite of what was found for the dot product. This is because the dot product contains the cos θ term, whereas the cross product contains the sin θ. The cos θ and sin θ are complementary functions and they are out of phase with each other by 90°.

The vector product. The vector **a** has a magnitude of 5.00 units and the vector **b** of 7.00 units. If the angle between the vectors is 53.0°, find their cross product.

Solution

The magnitude of their cross product, found from equation F.5, is

$$|\mathbf{a} \times \mathbf{b}| = ab \sin \theta$$

$$= (5.00)(7.00)\sin 53.0°$$

$$= 28.0$$

The direction of **a** × **b** is as shown in figure F.3(a).

The area of a surface can be represented by the cross product of the two vectors that generate the area. As an example, consider the two vectors, **a** and **b**, in figure F.4. If vectors **a** and **b** are moved parallel to themselves they are still the same vectors, since they still have the same magnitude and direction. But in the process of moving them parallel to themselves they have generated a parallelogram. Recall from elementary geometry that the area of a parallelogram is equal to the product of its base times its altitude, that is,

$$A = (base)(altitude)$$

The base of the parallelogram is given by the magnitude of the vector **a**, and, we can see from figure F.4, the altitude is given by

$$altitude = b \sin \theta$$

Therefore, the area of the parallelogram is

$$A = ab \sin \theta$$

But

$$ab \sin \theta = |\mathbf{a} \times \mathbf{b}|$$

Therefore, the area of a parallelogram generated by sides **a** and **b** is

$$A = |\mathbf{a} \times \mathbf{b}| = ab \sin \theta \qquad (F.9)$$

Because **a** × **b** is a vector perpendicular to the surface generated by **a** and **b**, the area of the surface can be represented as the vector **A**, where

$$\mathbf{A} = \mathbf{a} \times \mathbf{b} \qquad (F.10)$$

Thus, **A** is perpendicular to the surface. An area can be represented as a vector in many physical applications.

F.3 Torque (Section 5.2)

An important application of the cross product is in the definition of the concept of torque, which is defined in section 5.2. A force **F** acting at the point *A*, figure F.5, a distance *r* from the axis of rotation, 0, creates a torque given by

$$\tau = r_\perp F = rF_\perp = rF \sin \theta \qquad (5.21)$$

Notice, from equation 5.21, *that the torque acting about an axis should be defined as the cross product of r and F,* that is,

$$\boldsymbol{\tau} = \mathbf{r} \times \mathbf{F} \qquad (F.11)$$

with magnitude

$$\tau = |\boldsymbol{\tau}| = |\mathbf{r} \times \mathbf{F}| = rF \sin \theta \qquad (F.12)$$

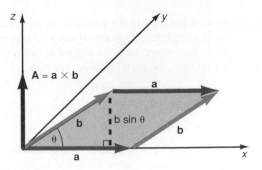

Figure F.4

The area of a surface.

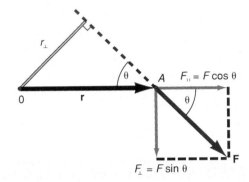

Figure F.5

Torque.

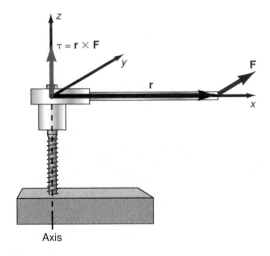

Figure F.6

Torque and the cross product.

The torque vector lies along the axis of rotation. It is perpendicular to the page and points into the page in figure F.5.

As an example of the vector concept of torque, consider the screw with right-handed threads shown in figure F.6. A screwdriver head is attached to a ratchet wrench and is inserted into the slot at the top of the screw. A force **F** is exerted at the end of the ratchet as shown. If **r** is the vector distance from the axis of the screw to the point of application of the force, then a torque is created given by

$$\boldsymbol{\tau} = \mathbf{r} \times \mathbf{F}$$

The torque vector τ lies along the axis of rotation and points upward, in the figure, in the direction that the screw moves while the torque is applied. A force in the direction shown causes the screw to rotate in a counterclockwise direction, which causes the screw to move out from the surface in an upward direction. If the direction of the force is reversed by 180°, the screw rotates in a clockwise direction, the torque vector points downward along the axis of the screw, and the screw moves in that direction into the wood.

F.4 Work (Section 7.2)

One of the most important examples of the dot product is found in the concept of work. As defined in section 7.2, *the work done in moving a block a distance x by a force F, figure F.7, is determined by multiplying the component of the force in the direction of the displacement, by the displacement itself.* The component of the force in the direction of the displacement, seen from figure F.7, is

$$F_x = \mathbf{F} \cos \theta$$

From the definition, the work done becomes

$$W = F_x x = (F \cos \theta)x = Fx \cos \theta$$

But $Fx \cos \theta$ is the dot product of $\mathbf{F} \cdot \mathbf{x}$. Therefore, work should be defined as the scalar or dot product of $\mathbf{F}$ and $\mathbf{x}$, that is,

$$W = \mathbf{F} \cdot \mathbf{x} = Fx \cos \theta \qquad \textbf{(F.13)}$$

It is now obvious why the scalar product was defined the way it was in equation F.2, because it does indeed represent a physical concept.

F.5 The Force on an Electric Charge in a Magnetic Field (Section 22.1)

If an electric charge is fired with a velocity $\mathbf{v}$ into a magnetic field $\mathbf{B}$, it is observed that a force acts on the charge. The force, however, acts at an angle of 90° to the plane determined by the velocity vector $\mathbf{v}$ and the magnetic field $\mathbf{B}$, as shown in figure F.8. In fact, *the force acting on the moving charge is given by the cross product of v and B as*

$$\mathbf{F} = q\mathbf{v} \times \mathbf{B} \qquad \textbf{(F.14)}$$

Recall that in determining the direction of the cross product, point the fingers of your right hand in the direction of the first vector, in this case $\mathbf{v}$. Your palm should be facing the second vector, in this case $\mathbf{B}$. Rotate your right hand from $\mathbf{v}$ to $\mathbf{B}$ and your thumb will point in the direction of the cross product $\mathbf{v} \times \mathbf{B}$, in this case upward. Hence the direction of the force is also upward.

The magnitude of the force is determined from the definition of the cross product, equation F.5, and is given by

$$F = qvB \sin \theta \qquad \textbf{(F.15)}$$

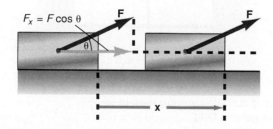

Figure F.7

The concept of work.

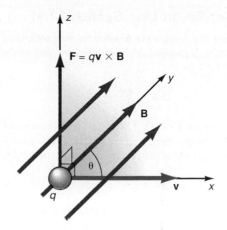

Figure F.8

Force acting on a charge in a magnetic field.

The angle θ is the angle between the velocity vector $\mathbf{v}$ and the magnetic field vector $\mathbf{B}$.

F.6 Force on a Current-Carrying Conductor in an External Magnetic Field (Section 22.2)

A wire carrying a current I in any external magnetic field $\mathbf{B}$, will experience a force given by

$$\mathbf{F} = I\boldsymbol{\ell} \times \mathbf{B} \qquad \textbf{(F.16)}$$

The force is given by a cross product term, and the direction of the force is found from $\boldsymbol{\ell} \times \mathbf{B}$. With $\boldsymbol{\ell}$ in the direction of the current and $\mathbf{B}$ pointing into the page in figure F.9, the direction of $\boldsymbol{\ell} \times \mathbf{B}$ is found by rotating the vector $\boldsymbol{\ell}$ toward the vector $\mathbf{B}$ in the cross product, the thumb points upward, indicating that the direction of the force on the wire is also upward. If the direction of the current flow is reversed, $\boldsymbol{\ell}$ is reversed and $\boldsymbol{\ell} \times \mathbf{B}$ then points downward. *The magnitude of the force, determined from equation F.16, is*

$$F = I\ell B \sin \theta \qquad \textbf{(F.17)}$$

where θ is the angle between $\boldsymbol{\ell}$ and $\mathbf{B}$.

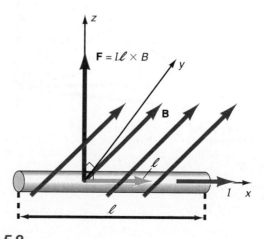

Figure F.9

Force on a current-carrying wire in an external magnetic field.

F.7 The Biot-Savart Law (Section 22.4)

The Biot-Savart law relates the amount of magnetic field ΔB produced by a small element $\Delta \ell$ of a wire carrying a current I and is given by

$$\Delta \mathbf{B} = \frac{\mu_0 I}{4\pi} \frac{\Delta \boldsymbol{\ell} \times \mathbf{r}}{r^3} \qquad \text{(F.18)}$$

and is shown in figure F.10(a). The cross product term $\Delta \boldsymbol{\ell} \times \mathbf{r}$ immediately determines the direction of $\Delta \mathbf{B}$, and is shown in figure F.10(a). The magnitude of $\Delta \mathbf{B}$, found from equation F.18, is

$$\Delta B = \frac{\mu_0 I}{4\pi} \frac{\Delta \ell \; r \sin \theta}{r^3} \qquad \text{(F.19)}$$

The Biot-Savart law says that the small current element $\Delta \boldsymbol{\ell}$ produces a small amount of magnetic field $\Delta \mathbf{B}$ at the point P. But the entire length of wire can be cut up into many $\Delta \boldsymbol{\ell}$'s, and each contributes to the total magnetic field at P. This is shown in figure F.10(b). Therefore, *the total magnetic field at point P is the vector sum of all the $\Delta \mathbf{B}$'s associated with each current element,* that is,

$$\mathbf{B} = \Sigma \Delta \mathbf{B}$$

F.8 Ampère's Circuital Law (Section 22.6)

Ampère's circuital law is also used to determine the magnetic field for different current distributions. *Ampère's law states that along any arbitrary path encircling a total current I, the sum of the scalar product of the magnetic field B with the element of length $\Delta \boldsymbol{\ell}$ of the path, is equal to the permeability μ_0, times the total current I enclosed by the path.* That is,

$$\Sigma \, \mathbf{B} \cdot \Delta \boldsymbol{\ell} = \mu_0 I \qquad \text{(F.20)}$$

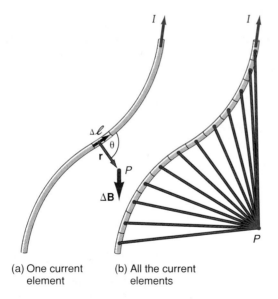

(a) One current element
(b) All the current elements

Figure F.10

The magnetic field produced by a current element.

F.9 Torque on a Current Loop in an External Magnetic Field (Section 22.8)

If a rectangular coil of wire, carrying a current I, is placed in a uniform magnetic field $\mathbf{B}$, figure F.11(a), each segment of the coil represents a current-carrying wire in an external magnetic field and thus experiences a force on it, given by equation F.16,

$$\mathbf{F} = I\boldsymbol{\ell} \times \mathbf{B}$$

The net result of the current flowing in a coil in an external magnetic field is to produce two equal and opposite forces acting on the coil. But since the forces do not lie along the same line of action, the forces cause a torque to act on the coil as is readily observable in figure F.11(b). The torque acting on the coil, given by equation F.11, is

$$\boldsymbol{\tau} = \mathbf{r} \times \mathbf{F}$$

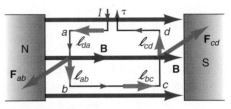

(a) Side view

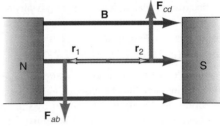

(b) Top view

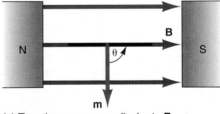

(c) Top view—**m** perpendicular to **B**

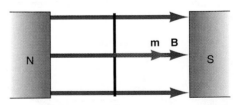

(d) Top view—**m** parallel to **B**

Figure F.11

Torque on a current-carrying loop in an external magnetic field.

As seen in section 22.8, this reduces to

$$\tau = IAB \sin \theta \qquad \text{(F.21)}$$

Thus, the torque on the coil in a magnetic field depends on the current I in the coil, the area A of the coil, the intensity of the magnetic field $\mathbf{B}$, and the angle θ between $\mathbf{r}$ and $\mathbf{F}$. Since the magnetic dipole moment $\mathbf{m}$ of a current loop is defined as $m = IA$, we can also write the torque acting on the coil, equation F.21, in terms of the magnetic dipole moment of the coil as

$$\tau = mB \sin \theta \qquad \text{(F.22)}$$

or in the vector notation

$$\tau = \mathbf{m} \times \mathbf{B} \qquad \text{(F.23)}$$

Equation F.23 shows that the torque acting on a coil in a magnetic field is equal to the cross product of the magnetic dipole moment of the coil and the magnetic field $\mathbf{B}$. Note that the angle θ is the angle between the magnetic dipole moment vector $\mathbf{m}$ and the magnetic field vector $\mathbf{B}$ in figure F.11(c). The magnetic field causes a torque to act on the coil until the magnetic dipole moment of the coil is aligned with the magnetic field, figure F.11(d).

F.10 Magnetic Flux (Section 23.2)

Flux is a quantitative measure of the number of lines of a vector field that passes perpendicularly through a surface. The magnetic flux Φ_M can be defined as a quantitative measure of the number of magnetic field lines $\mathbf{B}$ passing normally through a particular surface area $\mathbf{A}$. That is,

$$\Phi_M = \mathbf{B} \cdot \mathbf{A} \qquad \text{(F.24)}$$

as shown in figure F.12(a). The area is represented as a vector $\mathbf{A}$ that has the magnitude of the area of the surface and a direction that is perpendicular to the surface. From the definition of the scalar product we can also write equation F.24 as

$$\Phi_M = BA \cos \theta \qquad \text{(F.25)}$$

where θ is the angle between $\mathbf{B}$ and $\mathbf{A}$.

The vector $\mathbf{B}$, at the point P of figure F.12(a), can be resolved into the components $B_\perp$, the component perpendicular to the area, and $B_\parallel$, the parallel component. The perpendicular component is

$$B_\perp = B \cos \theta$$

whereas the parallel component is given by

$$B_\parallel = B \sin \theta$$

The parallel component $B_\parallel$ lies in the surface itself and therefore does not pass through the surface, whereas the perpendicular component $B_\perp$ completely passes through the surface at the point P. The product of the perpendicular component and the area

$$B_\perp A = (B \cos \theta)A = AB \cos \theta = \mathbf{A} \cdot \mathbf{B} = \Phi_M$$

is therefore a quantitative measure of the number of lines of $\mathbf{B}$ passing normally through the entire area $\mathbf{A}$. The number of lines represents the strength of the field. If the angle θ in equation F.25 is zero, then $\mathbf{B}$ is parallel to the vector $\mathbf{A}$ and all the lines of $\mathbf{B}$ pass normally through the area A, as seen in figure F.12(b). If the angle θ in equation F.25 is 90°, then $\mathbf{B}$ is perpendicular to the area vector $\mathbf{A}$, and none of the lines of $\mathbf{B}$ pass through the surface A, as seen in figure F.12(c).

F.11 Induced Electric Field (Section 23.3)

A metal wire is pulled at a velocity $\mathbf{v}$ to the right through a uniform magnetic field $\mathbf{B}$ directed into the paper, as shown in figure F.13. As the wire is moved to the right, any charge q within the wire experiences the force

$$\mathbf{F} = q\mathbf{v} \times \mathbf{B}$$

as was shown in section F.5. If both sides of the equation are divided by $\mathbf{q}$, we have

$$\frac{\mathbf{F}}{q} = \mathbf{v} \times \mathbf{B}$$

But an electrostatic field was originally defined as

$$\mathbf{E} = \frac{\mathbf{F}}{q} \qquad \text{(F.26)}$$

where $\mathbf{F}$ is the force acting on a test charge q placed at rest in the electrostatic field. It is therefore reasonable to now *define an induced electric field E by equation F.27 as*

$$\mathbf{E} = \frac{\mathbf{F}}{q} = \mathbf{v} \times \mathbf{B} \qquad \text{(F.27)}$$

Figure F.12
The magnetic flux.

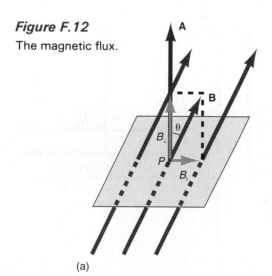
(a)

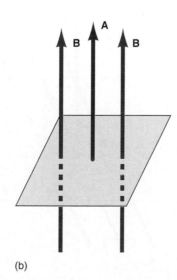
(b)

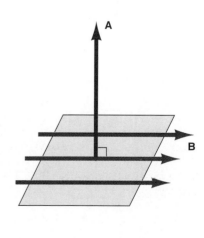
(c)

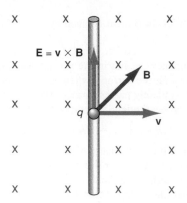

Figure F.13

An induced electric field.

This is a different type of field than the electrostatic field. *The induced electric field exists only when the charge is in motion at a velocity v.* When $v = 0$, the induced electric field is also zero as can be seen by equation F.27. The direction of the induced electric field in figure F.13, and hence the direction that a positive charge q within the wire moves, is found from the definition of the cross product $\mathbf{v} \times \mathbf{B}$. The direction is found by pointing the fingers of your right hand in the direction of the velocity vector $\mathbf{v}$. Your palm should be facing the magnetic field vector $\mathbf{B}$. Rotate your right hand from $\mathbf{v}$ to $\mathbf{B}$ and your thumb will point in the direction of the vector $\mathbf{v} \times \mathbf{B}$, and hence in the direction of the electric field vector, in this case upward. The magnitude of the induced electric field, found from equation F.27, is

$$E = vB \sin \theta \qquad \text{(F.28)}$$

F.12 Faraday's Law of Induction (Section 23.3)

Faraday's law of electromagnetic induction states that whenever the magnetic flux changes with time, there is an induced emf. Faraday's law is written as

$$\mathscr{E} = -\frac{\Delta \Phi_M}{\Delta t} \qquad \text{(F.29)}$$

The magnetic flux Φ_M is defined as

$$\Phi_M = \mathbf{B} \cdot \mathbf{A}$$

In general the change in the magnetic flux caused by a change in area $\Delta \mathbf{A}$ and a change in the magnetic field $\Delta \mathbf{B}$, is

$$\Delta \Phi_M = \mathbf{B} \cdot \Delta \mathbf{A} + \mathbf{A} \cdot \Delta \mathbf{B}$$

and we can also write Faraday's law, equation F.29, as

$$\mathscr{E} = -\frac{\Delta \Phi_M}{\Delta t} = -\mathbf{B} \cdot \frac{\Delta \mathbf{A}}{\Delta t} - \mathbf{A} \cdot \frac{\Delta \mathbf{B}}{\Delta t} \qquad \text{(F.30)}$$

Faraday's law, in this form, says that an emf can be induced either by changing the area of a loop of wire with time, while the magnetic field remains constant, or by keeping the area of the loop of wire constant and changing the magnetic field with time. Because the flux is given by a scalar product, it is also possible to keep the magnitudes B and A constant, but vary the angle between them. This also causes a change in the magnetic flux and hence an induced emf.

F.13 Electric Flux (Section 25.2)

The electric flux is defined as

$$\Phi_E = \mathbf{E} \cdot \mathbf{A} = EA \cos \theta \qquad \text{(F.31)}$$

and is a quantitative measure of the number of lines of $\mathbf{E}$ that pass normally through the surface area $\mathbf{A}$. Figure F.14(a) shows an electric field $\mathbf{E}$ passing through a portion of a surface of area $\mathbf{A}$. The area of the surface is represented by a vector $\mathbf{A}$, whose magnitude is the area A of the surface, and whose direction is perpendicular to the surface. The number of lines represents the strength of the field. The vector $\mathbf{E}$, at the point P of figure F.14(a), can be resolved into the components $E_\perp$, the component perpendicular to the area, and $E_\parallel$, the parallel component. The parallel component $E_\parallel$ lies in the surface itself and therefore does not pass through the surface, whereas the perpendicular component $E_\perp$ completely passes through the surface at the point P. The product of the perpendicular component and the area

$$E_\perp A = (E \cos \theta)A = EA \cos \theta = \mathbf{E} \cdot \mathbf{A} = \Phi_E \qquad \text{(F.32)}$$

is therefore a quantitative measure of the number of lines of $\mathbf{E}$ passing normally through the entire area, A. If the angle θ in

Figure F.14

Electric flux.

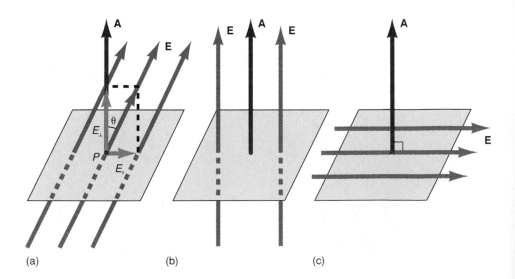

(a)   (b)   (c)

equation F.31 is zero, then **E** is parallel to the vector **A** and all the lines of **E** pass normally through the area A, as seen in figure F.14(b). If the angle θ in equation F.31 is 90° and then **E** is perpendicular to the area vector **A,** and none of the lines of **E** pass through the surface A, as seen in figure F.14(c).

F.14 Gauss's Law for Electricity (Section 25.2)

Gauss's law for electricity says that the electric flux Φ_E through a surface surrounding the point charge q is a measure of the amount of charge q contained within the Gaussian surface. It is stated mathematically as

$$\Phi_E = \Sigma\, \mathbf{E} \cdot \Delta \mathbf{A} = \frac{q}{\epsilon_0} \qquad \text{(F.33)}$$

where q is now the net charge contained within the Gaussian surface. Here $\mathbf{E} \cdot \Delta \mathbf{A}$ is the amount of electric flux passing through a small area $\Delta \mathbf{A}$ of the surface, and $\Sigma\, \mathbf{E} \cdot \Delta \mathbf{A}$ is the sum of that flux passing through the entire surface. When Φ_E is a positive quantity, the Gaussian surface surrounds a source of positive charge, and electric flux diverges out of the surface. Whenever Φ_E is negative,

the Gaussian surface surrounds a negative charge distribution and the electric flux converges into the Gaussian surface. If there is no enclosed charge, $q = 0$, and hence the flux $\Phi_E = 0$. In this case, whatever electric flux enters one part of a Gaussian surface, the same amount must leave somewhere else.

F.15 Gauss's Law for Magnetism (Section 25.3)

Gauss's law for magnetism takes the same form as Gauss's law for electricity. However, the electric flux is a measure of the electric charge q enclosed within the Gaussian surface. The magnetic flux should be a measure of the amount of magnetic pole enclosed within the Gaussian surface. But, isolated magnetic poles do not exist. Thus, the expected term for the enclosed magnetic pole on the right side of Gauss's law must be equal to zero. *Thus, Gauss's law for magnetism becomes*

$$\Phi_M = \Sigma\, \mathbf{B} \cdot \Delta \mathbf{A} = 0 \qquad \text{(F.34)}$$

The net magnetic flux through any Gaussian surface anywhere in a magnetic field is zero because there are no isolated magnetic poles.

Appendix G
Answers to Odd-Numbered Problems

Chapter 1

1. 16.4 cm³
3. 88 ft/s
5. 86,400 s; 2.59×10^6 s; 3.15×10^7 s
7. 55.9 mph
9. 3280 ft
11. 1090 ft/s; 736 mph
13. 13.4 m²
15. a. 2.37×10^9 s   b. 3.95×10^7 min; 2.76×10^9 pulses/lifetime
17. 6.70×10^8 mph; 2.99×10^8 m/s
19. 9.14 m; 91.4 m
21. 380 m; 0.236 mi; 1.49×10^4 in.; 3.80×10^5 mm
23. 1.275×10^4 km
25. a. 5.89×10^5 pm   b. 5.89×10^{-4} mm
    c. 5.89×10^{-5} cm
    d. 5.89×10^{-7} m; 4.31×10^4 waves
27. 2.13 m
29. 169 m
31. 1×10^{-14} m³ to 1×10^{-12} m³;
    6.10×10^{-10} in.³ to 6.10×10^{-8} in.³
33. 8.6×10^{-7} mm; 3.39×10^{-8} in.
35. a. 2.0×10^5 pm   b. 200 nm   c. 0.2 μm
    d. 0.0002 mm   e. 2×10^{-5} cm
37. 0.00994 slugs
39. 0.159 m³
41. 3.58×10^4 ft
43. 5.10×10^{14} m²; 5.49×10^{15} ft²; 1.08×10^{21} m³;
    3.83×10^{22} ft³; 5.51×10^3 kg/m³
45. $v = 24.84$ mi/hr
47. 4.25 min/month; 51.74 min/yr

Chapter 2

1. 57.4 lb; 81.9 lb
3. 6.47 lb; 24.2 lb
5. 262 ft
7. $v_x = 198$ km/hr; $v_y = 27.8$ km/hr
9. $w_\parallel = 6070$ N; $w_\perp = 6510$ N
11. 292 m/s; 956 m/s
13. 8.09 mi at 63.7° north of east
15. 174 mph at 9.26° south of east
17. 85.2 N; 148° from $+ x$-axis
19. a. $F_h = 39.9$ lb   b. $F_v = 30.1$ lb   c. -119.9 lb
21. a. 100 ft; 50° N of E   b. 25 ft; 50° N of E
    c. 50 ft; 50° S of W   d. 250 ft; 50° S of W
    e. 250 ft; 50° N of E   f. 150 ft; 50° S of W

23. 4.85 m/s; $\theta = 48.7°$
25. 22.7 lb at 85.6° above the $+x$-axis
27. a. The velocity of the aircraft is 169 km/hr in the direction 12.1° south of west.   b. The pilot must point the plane in a direction 10.2° N of W.   c. 2.47 hr
29. a. 9.43 mph   b. 15 min   c. 1.25 mi   d. 38.7° S of E
33. 5.71 N; $\theta = 82°$; $\theta = 82°$ below $+ x$-axis
37. 228.3 miles east of its starting point A, and 28.3 mi south of its starting point or 40.0 mi to the south east of the city at B
39. $T_y = 123$ N; $T_x = 158$ N
41. 21.3°
43. 14.4 m; $-27.8°$

Chapter 3

1. a. 55.4 mph   b. 81.3 ft/s   c. 89.1 km/hr   d. 24.8 m/s
3. 1720 m
5. 4.33 min
7. 4.59 m/s
9. a. 3 m/s²;   b. 0 m/s²;   c. 2 m/s²;   d. -4 m/s²
11. 0.300 m/s²
13. 1.86 ft/s²; 475 ft
15. 54.5 mph
17. $t = 10$ s: $x = 147$ ft; $v = 29.3$ ft/s
    $t = 15$ s: $x = 330$ ft; $v = 44$ ft/s
    $t = 20$ s: $x = 586$ ft; $v = 58.6$ ft/s
    $t = 25$ s: $x = 916$ ft; $v = 73.3$ ft/s
19. 5.62×10^{17} cm/s²
21. -20.2 ft/s; 6.29 s
23. 200 ft
25. 59.4 ft; 193 ft
27. $a = 0.579$ m/s²; $v = 13.9$ m/s
29. 650 m
31. -17.1 m/s
33. 121 ft
35. 2.5 s
37. 0.500 s
39. 35.1 m; 75.9 m; 77.5 m; 39.9 m
41. 4.51 s
43. a. -31.8 m/s;   b. -1.71 s
45. 158 m
47. 5430 m
49. 9.55°
51. a. 281 ft   b. 8.38 s   c. 943 ft
53. a. 122 ft/s²   b. 37.3 m/s²   c. $3.81g$
55. a. 12.6 s   b. 556 ft   c. 60.1 mph
57. a. -1.1 m/s²   b. 29.1 m   c. 14.54 s   d. -8.00 m/s

59. 10.7 s for train 2 to overtake train 1; 218 m

61. 10.5 s; 3.03 m/s²

63. $x_1 + x_2 = 466$ m; the trains will stop in time

65. 193 m

67. 19.8 m/s = v_{20} the initial speed of the ball thrown upward

71. **a.** 3.97 s **b.** 117 ft **c.** 62.9 ft **d.** 70 ft/s at 65° below +x-axis

73. **a.** 127 m **b.** 7.44 s **c.** 143 m **d.** 53.5 m/s at 68.9° below +x-axis

75. $\frac{g}{2v_{0x}} x^2 - \frac{v_{0y}}{v_{0x}} x - y - y_0 = 0$, which is the standard form of the equation of a parabola, $A_x x^2 + D_x x + E_y y + F = 0$. Hence the trajectory of the projectile is a parabola.

Chapter 4

1. 6.25 slugs

3. 75.0 N

5. −1500 lb

7. −1.82 × 10⁴ lb opposite to direction of motion; −909 lb

9. 130 lb

11. **a.** 9220 N **b.** 6780 N

13. 417 N

15. 2.27 × 10⁵ lb

17. 61.1 m

19. **a.** 0.500 m/s² **b.** 2.50 m/s **c.** 6.25 m **d.** 3.5 m/s

21. **a.** 4.9 m/s² **b.** 9.9 m/s

23. **a.** 1020 N **b.** 882 N **c.** 747 N **d.** 882 N **e.** 0

25. **a.** 0.0989 m **b.** 0.165 s

27. **a.** 0.891 m/s² **b.** 2.67 m/s **c.** 4.01 m to the right

29. $F = 2.4$ N; $T_2 = 1.2$ N; $T_1 = 0.400$ N

31. 403 ft

33. 100 N; 80 N

35. **a.** 23.2 lb **b.** 15.1 lb **c.** Method b is better; it requires less force to move the block.

37. 0.0085

39. **a.** 10.8 m/s **b.** 6.05 m/s

41. **a.** 10.6 N **b.** 0.312

43. **a.** 79.3 ft/s **b.** 983 ft from the bottom of the slope

45. **a.** $a = 20.0$ m/s² **b.** $v = 11.0$ m/s

47. .600

49. 35.3 N

51. 800 g

53. acceleration $a = 0.887$ m/s²; tension = 65.1 N

55. $T_C = 25.6$ N; $T_B = 64.0$ N; $T_A = 128$ N

57. 7.54 N

59. **a.** 4.69 m/s² **b.** 0.586 m to the left of its original position

61. **a.** $\dfrac{w_A \sin\theta - w_B \sin\phi}{m_A + m_B}$

63. **a.** 2.07 m/s² **b.** 1.14 N

65. 3.83 m/s²; $T_A = T_B = 35.8$ N; $T_C = 15.5$ N

67. 1.25 m/s²; $T_A = T_B = 51.3$ N; $T_C = 42.9$ N

69. **a.** 3.27 m/s² **b.** 0.333

71. $v = \dfrac{mg}{k}$. This means, that after a relatively long time, the velocity of the object becomes constant. This is referred to as the terminal velocity.

Chapter 5

1. 0.902 N at 101° from the +x-axis

3. 2.87 N

5. 133 N

7. $T_1 = 2134$ lb; $T_2 = 2130$ lb

9. 20.0 N

11. $T_1 = 489$ N; $T_2 = 553$ N

13. **a.** 196 N **b.** 392 N

15. 25.0 lb ft

17. 17.9 Nm

19. $\tau = 5.00$ mN

21. $F_2 = 120$ lb; $F_1 = 80.0$ lb

23. 725 N; 525 N

25. 45.0 lb ft counterclockwise

27. 3.68 m from end that the son is carrying

29. $x_{cg} = 1.00$ ft

31. 3.43″, 8.58″

33. $F_B = 5000$ N; $F_A = 5900$ N

35. 30.0 cm from the origin

37. 0.0154 m to the right of center of large plate

39. $T = 1710$ N; $H = 1310$ N; $V = 101$ N

41. 1.12 m

43. $T = 897$ N;
Hinge forces: horizontal = 777 N, vertical = 752 N

45. $F_{wall} = 10.1$ N; $F_{Hfloor} = 10.1$ N; $F_{Vfloor} = 120$ N

47. no

49. 10.5 ft up the ladder

51. $F_M = 518$ lb; $F_J = 702$ lb

53. 187 lb; 304 lb

55. $T_A = 1170$ N; $F_T = 1500$ N

57. 1.09 kg

59. 34.3 lb

61. **a.** 18.8 lb ft **b.** 18.8 lb ft **c.** 226 lb

63. 0.625 m

65. 51.3°

67. $F_{wall} = 12.1$ lb;
$f_{wall\ friction} = 4.84$ lb;
vert. component of floor = 25.2 lb;
horiz. component of floor = 12.1 lb

Chapter 6

1. **a.** 2π rad **b.** $\dfrac{3\pi}{2}$ rad **c.** π rad **d.** $\dfrac{\pi}{2}$ rad
e. $\dfrac{1}{3}\pi$ rad **f.** $\dfrac{1\pi}{6}$ rad **g.** 2π rad

3. 174 ft

5. 284 m/s²

7. $a_c = 1.76$ m/s²; $F_c = 2630$ N

9. 0.540 lb

11. $a = 15.5$ ft/s²; $\mu_s = 0.484$

13. 34.6°

15. **a.** .711 **b.** 35.4°

17. 11.14°

19. 22.6°

21. **a.** 3.34 × 10⁻⁹ N **b.** 3.34 × 10⁻⁹ N **c.** $a_{5\ kg} = 6.68 \times 10^{-10}$ m/s²; $a_{10\ kg} = 3.34 \times 10^{-10}$ m/s²

23. no; $F = 2.78 \times 10^{-8}$ N

25. **a.** 9.81 m/s² **b.** 2.45 m/s² **c.** 0.0981 m/s² **d.** 2.70 × 10⁻³ m/s²

27. $g_{mars} = 3.87$ m/s²; $w_{mars} = 71.1$ lb

29. 270 N on the sun

31. $v = 29.7$ km/s; $t = 367$ days

33. $v = 3530$ m/s; $t = 1.69$ hr

35. $r = 8.50 \times 10^6$ m; $H = 2130$ km; $v = 6850$ m/s

37. **a.** 3.37 N **b.** 2.39 N **c.** 0

39. $a_c = 65.2$ ft/s²; $F_c = 367$ lb; $F_N = 547$ lb

41. 179 ft/s

45. 1.67 m/s

47. 1.42 × 10⁻⁷ N along the diagonal (45° above the horizontal)

49. 7900 m/s; 1.41 hr; 0

51. 3.52 × 10²² N; 1.98 × 10²⁰ N; the moon

53. **a.** 5.899 × 10⁻³ m/s² **b.** 3.433 × 10⁻⁵ m/s²

55. 3.46 × 10⁸ m

57. $\dfrac{GmM}{r} = \dfrac{mv^2}{r}$;
$v = \dfrac{2\pi r}{T}$
Substitute and simplify
$T^2 = \dfrac{4\pi^2 r^3}{GM}$ (Kepler's third!)

59. 2.98×10^4 m/s = 66,700 mph
61. **a.** 4.66×10^6 m **b.** 1.14×10^7 m **c.** 3.812×10^8 m
    d. 1.963×10^8 m **e.** 7.69×10^3 m/s **f.** 8240 m/s
    g. 550 m/s **h.** 1610 m/s **i.** 246 m/s **j.** -1360 m/s

Chapter 7

1. 7.50×10^3 ft lb
3. 613 J
5. 3.90×10^9 J
7. **a.** 196 J **b.** 196 J **c.** The force in part **a** is one-half the size of the force in part **b.**
9. 1.35×10^4 J
11. **a.** 300 ft lb/s **b.** 3000 ft lb
13. 1600 hp
15. 137 J; 206 J
17. 294 J
19. 2.69×10^{33} J
21. **a.** 4.17×10^4 J **b.** 1.67×10^5 J **c.** 6.67×10^5 J
23. $KE = 7.57 \times 10^5$ ft lb; $PE = 3.00 \times 10^6$ ft lb
25. 4.05×10^3 J; -2.03×10^5 N
27. **a.** 36.0 J **b.** 36.0 J **c.** 6.00 m/s **d.** 6.00 m/s²; 6.00 m/s
29. 625 ft
31. 1.04 m/s
33. **a.** 0.0835 J **b.** 0.0835 J **c.** 0.817 m/s
35. 1.08 m
37. **a.** 1.23 J **b.** 0 **c.** -1.23 J **d.** 0.616 N **e.** 0.768
39. 2.25×10^3 ft lb
41. 105.4 J
43. **a.** 98.0 J **b.** 6.26 m/s **c.** 6.67 m
45. $(H7.7) = MA = \dfrac{F_{out}}{F_{in}} = \dfrac{r_{in}}{r_{out}}$
47. 2.06 J
49. 118,000 W
51. **a.** 52.7 J **b.** $+38.9$ J **c.** 18.3 N
53. **a.** 4.35 trips, or a little more than 2 oscillations **b.** 1.05 m; the object stops 1.05 m to the right of point B
55. $.0188 = \mu$
57. $E_{total} = 14.7$ J; $PE_1 = 9.80$ J; $PE_2 = 1.96$ J; $v_1 = v_2 = 2.90$ m/s; $KE_1 = 2.10$ J; $KE_2 = 0.841$ J
59. **a.** 1680 m/s **b.** 3590 m/s **c.** 4.25×10^4 m/s
61. 10.4 m
63. 0.256 m
65. **a.** 1.88 J **b.** 0.787 J **c.** 0.189 J **d.** 0.751 J
    e. 0.661 m/s **f.** $KE_1 = 0.109$ J; $KE_2 = 0.0437$ J

Chapter 8

1. 8.07×10^3 slug ft/s
3. -8.07×10^3 slug ft/s; -2.69×10^4 lb; -8.07×10^3 lb
5. 2.4 s
7. -0.600 m/s
9. 1.50×10^{-3} m/s
11. 0.333 ft/s
13. 1.20 ft/s
15. -30 N
17. **a.** 12.4 Ns **b.** 12.4 kg m/s **c.** 49.6 m/s
19. **a.** 125.0 Ns **b.** 3.13×10^4 N
21. $v_{1f} = -0.417$ m/s; $v_{2f} = 0.183$ m/s
23. **a.** 0.600 kg **b.** $\dfrac{1}{2}v_{1i}$
25. **a.** -19.9 cm/s $= V_{1f}$ (to the left); 8.4 cm/s (to the right) **b.** 8.97×10^{-3} J **c.** 3.85×10^{-3} J **d.** -57.1%
27. $V_f = -23.8$ km/hr
29. 1.71 ft/s
31. **a.** -0.083 m/s **b.** 0.0216 J **c.** 1.55×10^{-3} J **d.** -2.01×10^{-2} J; the energy is dissipated as heat
33. 41.8 km/hr at 78.7° N of E
35. **a.** $\Delta p = -4.82$ kg m/s **b.** Magnitude = 4.82 kg m/s; Direction = into the wall → to the right

37. **a.** 8.00 kg m/s **b.** 0.050 kg m/s **c.** 160 J
    d. 6.25×10^{-3} J **e.** 6.25×10^{-3}
39. **a.** 0.748 **b.** 0.259 J
41. **a.** 1.99 m/s **b.** 0.202 m
43. $9.6 \to 10$ bullets needed
45. 42 cm to the right
47. 2.75 ft/s (to the left)
49. $KE = \dfrac{1}{2}mv^2$; $p = mv$
    $KE = \dfrac{1}{2}m\left(\dfrac{p}{m}\right)^2 = \dfrac{p^2}{2m}$
51. 25 cm/s
53. Yes. See Solutions Guide.
55. **a.** 0.556 ft/s **b.** 1.25 ft/s **c.** 5.00 ft/s
57. 7.24 mm
59. **a.** 293 m **b.** 1.65m
61. 2.33 cm
63. $V_{C_i} = 89.9$ m/s; $V_{T_i} = 4.50$ m/s

Chapter 9

1. **a.** 3.49 rad/s **b.** 4.71 rad/s **c.** 8.17 rad/s
3. **a.** 105 rad/s **b.** 7.85 m/s
5. -62.8 rad/s²; 180 rev
7. 2.36×10^{-3} J
9. 0.0417 kg m²
11. **a.** 3.79 kg m² **b.** 0
13. 6.86 rad/s²
15. 2.80 m/s²
17. **a.** 1.23 m/s² **b.** 1.92 m/s
19. 157 rad/s
21. **a.** 5.11 m/s **b.** 4.43 m/s
23. **a.** 5.00 rad/s **b.** 0.188 J **c.** 7.50 J **d.** 7.69 J
25. 9.82 J
27. **a.** 7.27×10^{-5} rad/s **b.** 9.69×10^{37} kg m²
    c. 2.56×10^{29} J **d.** 2.66×10^{33} J; linear KE is approximately 10,000 times greater than rotational KE **e.** 7.05×10^{33} kg m²/s
29. **a.** 5.00×10^{-4} kg m² **b.** 0.200 m N
    c. 400 rad/s² **d.** 800 rad/s **e.** 800 rad **f.** 160 J
    g. 0.400 kg m²/s
31. **a.** 7.272×10^{-5} rad/s **b.** 1.263×10^{-12} rad/s **c.** yes
33. **a.** 13.3 rad/s² **b.** 133 rad/s **c.** 829 J **d.** 831 J
35. **a.** 2.5 m/s **b.** 2.50 m/s **c.** 55.6 rad/s
37. **a.** 0.300 m N **b.** 9.88×10^{-3} kg m² **c.** 30.4 rad/s²
    d. 122 rad/s
39. **a.** 5.88 J **b.** 6.26 m/s
41. 1.74 rad/s
43. **a.** 4.02 m/s² **b.** 2.65 m/s **c.** 0.659 s **d.** 1.52 m
    e. 6.10 m from the base of the incline

Chapter 10

1. 1.76×10^9 N/m²
3. 1.33×10^{-3}
5. 0.0191 in.
7. **a.** 1.36×10^8 N/m² **b.** 2.91×10^{-3} m **c.** 1.94×10^{-3}
9. 1.08×10^3 kg
11. 1.55×10^4 N
13. 2.84 mm
15. 5.34×10^4 N
17. 123.0 N/m
19. **a.** 294 N/m **b.** 2.67 cm
21. 3.81 N
23. 1.48×10^{-5} rad
25. 2.64×10^{-5} rad; the copper cylinder deforms 1.78 times more
27. 0.467×10^{10} N/m²
29. -0.0338 m³
31. 1.0002334
33. 0.32 mm; 5.01 m
35. 1082 masses

37. 1.67×10^{-4} m
39. 6.60×10^4 N
41. **a.** 33.7 N **b.** 64.3 N **c.** 3.2×10^{-4} m
43. 4.29×10^{-4} m
45. **a.** 5.33 m **b.** 2.00 m **c.** 3.33 m

Chapter 11

1. 1.95 Hz
3. 0.05 kg
5. **a.** 0 **b.** 3.14 m/s
7. 0.808 m/s; 6.53 m/s²
9. **a.** 1.58 Hz **b.** 0.633 s **c.** -1.11 m/s; -9.86 m/s²
11. 0.0563 J
13. 0.0338 J; 3.75×10^{-3} J; 0.0301 J
15. 1.74 s; 0.575 Hz
17. .276 Hz
19. **a.** 2.46 s **b.** an infinitely long period of oscillation
21. 9.852 m/s²
23. **a.** 3.33 N/m **b.** 7.70 s
25. **a.** .156 m/s² **b.** 0 **c.** .078 m/s²
27. 0.709 m
29. $T_{10} = 2.0109178$ s, 0.190%; $T_{30} = 2.041969$ s, 1.71%; $T_{50} = 2.105713$ s, 4.68%
31. 0.231 m
33. 4.59 km
35. **a.** 444 J; 0.330 J; 0.443 J **b.** 0.444 J **c.** 2.03 m/s
37. 45.4 N/m
39. $A_1 = 0.0949$ m; $A_2 = 0.090$ m; $A_4 = 0.081$ m; $A_6 = 0.0729$ m; $A_8 = 0.0656$ m
41. 0.256 m
43. 1.35 m/s
45. See Solutions Guide.
47. 0.848 m/s upward
49. $-\dfrac{C\theta}{T}$

Chapter 12

1. **a.** 0.050 s **b.** 5.00×10^{-5} s
3. 5.72 m
5. **a.** $f = 400$ Hz **b.** $k = 3.14$ m^{-1} **c.** $w = 2513$ rad/s
7. **a.** $k = 25.1$ m^{-1} **b.** $w = 1450$ rad/s **c.** $y = (0.0185$ m$) \sin[(25.1$ m$^{-1})x - (1450$ Hz$)t$
9. 6.48 N
11. **a.** 302 m/s **b.** 1.01 cm
13. **a.** 20.0 m/s **b.** 800 N **c.** $\lambda_2 = 1.20$ m; $f = 16.7$ Hz
15. 4.51 m
17. $f_3 = 1980$ Hz; $\lambda_3 = 40$ cm
    $f_5 = 3300$ Hz; $\lambda_5 = 24$ cm
    $f_7 = 4610$ Hz; $\lambda_7 = 17.1$ cm
19. **a.** 482 Hz **b.** 394 Hz
21. **a.** 3160 N **b.** $f_2 = 880$ Hz; $f_3 = 1320$ Hz; $f_4 = 1760$ Hz
    c. $\lambda_{\text{fundamental}} = 120$ cm; $\lambda_2 = 60$ cm; $\lambda_3 = 40$ cm; $\lambda_4 = 30$ cm
23. 4080 m
25. 1080 m
27. 0.3277 m; 0.2920 m; 0.2601 m
29. **a.** 542 Hz **b.** 317 Hz
31. 357 Hz
33. 337 Hz
35. 0.0134 s
37. 1110 m
39. 64.8 dB
41. 2.14×10^{-9} m
45. **a.** 650 N **b.** $f_2 = 880$ Hz; $f_3 = 1320$ Hz; $f_4 = 1760$ Hz; $f_5 = 2200$ Hz
    c. $\lambda_1 = 1.20$ m; $\lambda_2 = 0.600$ m; $\lambda_3 = 0.400$ m; $\lambda_4 = 0.300$ m; $\lambda_5 = 0.240$ m

Chapter 13

1. 0.707 gm/cm³
3. **a.** 193 kg **b.** 1890 N
5. 5510 kg/m³
7. No, the crown is not pure gold.
9. 1.12×10^4 kg/m³
11. 5.27×10^{18} kg
13. **a.** 14.9 lb/in.² **b.** 14.2 lb/in.²
15. 1.91×10^8 Pa
17. 3.433×10^5 Pa
19. 7.60×10^4 N
21. **a.** 10.3 m **b.** 11.8 m **c.** 12.8 m **d.** 8.2 m
23. 1.50×10^5 Pa; 21.3 m
25. 2.52×10^5 N
27. .414 m or 41.4 cm
29. 9.63 N; 8.41 N
31. **a.** 19.0 N **b.** 23.9 N **c.** 259 N
33. 7.77 cm
35. 4.50 ft/s; 0.383 lb/s or 6.14×10^{-3} ft³/s
37. 5.87 ft/s
39. 9.13×10^4 Pa; 1.6 m/s
41. $v_1 = 1.28$ m/s; $v_2 = 5.12$ m/s; mass flow $= 1.61$ kg/s; volume flow $= 1.61 \times 10^{-3}$ m³/s
43. **a.** 1.993×10^5 Pa **b.** 1.50×10^5 Pa **c.** 4.51×10^7 N
45. 1.96×10^5 Pa
47. 8.55 m/s². It can be seen that the diameter of the ball does not play a role in the computation. This can only be true if the water is assumed to be nonviscous. Viscosity is an "internal friction" of the fluid that would decrease the acceleration of the ball.
49. 92%
51. 2.65 cm
53. 0.300 m
55. 1.29×10^3 Pa
57. $\Delta p = 5.22 \times 10^3$ Pa; $F = 7.84 \times 10^4$ N

Chapter 14

1. **a.** 37.0 °C **b.** 37.6 °C **c.** 36.4 °C
3. -40.0 °F
5. **a.** -30.6 °C **b.** -10.8 °C **c.** 12.8 °C
   d. 32.2 °C **e.** 82.2 °C
7. $K = \dfrac{5}{9}(t\,°F + 459.4°)$
9. 3.31 kcal
11. 400 kcal
13. 0.0622 °C
15. 195 kcal
17. 21.8 °C
19. 0.0911 kcal/kg °C; copper
21. 16.6 °C
23. 5.25 kcal
25. 27.5 g
27. 7.30 kcal
29. 13.5 g
31. 161 g
33. 0.721 hr
35. 635 Btu
37. 0.358 kcal/s; 5.12×10^3 Btu/hr
39. Assume 50% of all the kinetic energy becomes thermal energy. Then the change in temperature is equal to 236 °C.
41. 35.2 N
43. 588 kcal
45. 0.836 gal
47. 2.28×10^4 kcal
49. 6.75 m³
51. 9.01 kg
53. 1.17×10^6 Btu
55. 2.88×10^5 Btu; 1.20×10^4 Btu/hr; 3520 W
57. **a.** 1.50×10^9 m³ **b.** 2.60×10^7 kg **c.** 7.27×10^6 kg
    d. 594 kcal/kg **e.** 1.11×10^{10} kcal

f. Thunderstorms may cover areas the size of several states. Our dimensions for the box of air may be on the order of 1000 km by 1500 km by 1 km. This volume contains $(1 \times 10^6 \text{ m})$ $(1.5 \times 10^6 \text{ m})(1 \times 10^3 \text{ m})$ or 1×10^{15} m³ of air or about 1×10^6 times greater than our sample box in this problem. Now imagine the amount of energy released in a real thunderstorm, given everything else the same.

Chapter 15

1. 2.006 m
3. 125 °C; −124.8 °C
5. 0.79904 m
7. $\Delta L = 0.288$ in.; $F = 4.32 \times 10^5$ lb
9. 50.36 ft²
11. 1261 cm²
13. 0.5964 cm³
15. $\Delta V = \beta V_0 \Delta T$
   $V = V_0 + \Delta V = V_0(1 + \beta \Delta T)$
   $m = \rho V; m = \rho_0 V_0$ (mass remains unchanged)
   $$\rho = \frac{\rho_0 V_0}{V} = \frac{\rho_0 V_0}{V_0(1 + \beta \Delta T)}$$
   $$\rho = \frac{\rho_0}{1 + \beta \Delta T}$$
17. 96.2 atm
19. 2.45×10^{21} molecules
21. $2.90 \times 10^5 \, \dfrac{\text{N}}{\text{m}^2}$
23. .0167 kg
25. 2.00 atm
27. 5.09 L
29. 66.3 cm³
31. 250 m/s
33. 7.31×10^{-23} g/molecule
35. 903 °C
37. 4.25×10^7 J
39. 76.5 cm Hg
41. See Solutions Guide.
43. **a.** 2.8×10^{-5} kg m²   **b.** -3.49×10^{-4} J
45. 1.44×10^{-5} kg m²/s assuming constant $\omega = 11.4$ rad/s;
   $\Delta L = 0$ if ω is allowed to vary (conservation of angular momentum)
47. 1.401×10^{-4} s increase
49. **a.** 100 °C   **b.** 2.38 cm³
51. **a.** 1.01×10^4 K   **b.** 2.01×10^4 K   **c.** 1.41×10^5 K
   d. 1.61×10^5 K   **e.** 2.22×10^5 K   **f.** 9.06×10^4 K
53. $v_1 = \sqrt{\dfrac{3kT_1}{m}}$

   $v_2 = \sqrt{\dfrac{3kT_2}{m}}$

   $\dfrac{v_2}{v_1} = \sqrt{\dfrac{\frac{3kT_2}{m}}{\frac{3kT_1}{m}}} = \sqrt{\dfrac{T_2}{T_1}}$

   $v_2 = \sqrt{\dfrac{T_2}{T_1}} v_1$
55. $v_1 = \sqrt{\dfrac{3kT}{m_1}}$

   $v_2 = \sqrt{\dfrac{3kT}{m_2}}$

   $\dfrac{v_2}{v_1} = \sqrt{\dfrac{\frac{3kT}{m_2}}{\frac{3kT}{m_1}}} = \sqrt{\dfrac{m_1}{m_2}}$

   $v_2 = \sqrt{\dfrac{m_1}{m_2}} v_1$

Chapter 16

1. 0.900 kcal/kg
3. 1.27×10^4 Btu/hr
5. 35,000 Btu/hr
7. 122.3 °C
9. 2.62×10^8 J
11. 5.11×10^4 kcal
13. 1.55 kcal/hr
15. 1.48×10^5 Btu
17. **a.** 73.6 kg   **b.** 5.89×10^3 kcal   **c.** 55.1 hr to melt
19. 20%
21. **a.** 13.8 hr ft² °F/Btu   **b.** 20.7 hr ft² °F/Btu
23. 7.4×10^4 Btu
25. **a.** 1.75×10^6 Btu   **b.** 9.68×10^3 Btu transferred from the attic into the house over a period of 2 hr
27. 1.13×10^6 J
29. 5.07×10^{21} J
31. 3.28×10^7 J
33. **a.** 13.1 J/s   **b.** 89.2 J/s
35. 6.27×10^3 J/s; 9.89×10^{-6} m
37. 6.13×10^{-6} m
39. 4140 K
41. 4.27×10^8 J
43. **a.** 4.68×10^4 J/s   **b.** 2.34×10^4 J/s
   c. 4.07×10^3 J/s   **d.** 2.85×10^4 J/s
45. **a.** 31.9 °C   **b.** 18.6 °C
47. $Q = 256 \, Q_0$
49. 33.6 g/min enters the container
51. 1.87×10^6 kcal/hr
53. For large values of r_1 and r_2, the product $r_1 r_2$ is approximately equal to r^2 and $r_2 - r_1$ is the thickness of the material (d).
   $$\frac{\Delta Q}{\Delta t} = \frac{4\pi k \Delta T}{(r_2 - r_1)/r_1 r_2} = \frac{4\pi r^2 k \Delta T}{r_2 - r_1} = \frac{4\pi r^2 k \Delta T}{d}$$
   $4\pi r^2$ is the area of the surface of the sphere.
   $$\frac{\Delta Q}{\Delta t} = \frac{kA\Delta t}{d}, \text{ which is the same relation as that for a wall of surface}$$
   area A.
55. 7.16×10^3 W
57. 17.1 kcal/min
59. $$\frac{P_2}{P_1} = \frac{\sigma A(T + \Delta T)^4}{\sigma A T^4} = \frac{T^4(1 + \Delta T/T)^4}{T^4} = \left(1 + \frac{\Delta T}{T}\right)^4$$
   $$\frac{P_2}{P_1} = \left(1 + \frac{0.9K}{310K}\right)^4 = 1.012$$
   $$P_2 = 1.012 P_1 = 1.00 \, P_1 + 0.012 P_1$$
   Power increased by 1.2%.

Chapter 17

1. 9.86×10^4 J
3. 1.52×10^5 J
5. **a.** 3.00×10^5 J   **b.** 0   **c.** -7.50×10^4 J   **d.** 0   **e.** 2.25×10^5 J
7. 4220 cal
9. **a.** 596 cal   **b.** 994 cal
11. 749 J
13. −300 J
15. Since the result is positive, the internal energy of the system is increased by 300 J.
17. 0.266 m³; 0.133 m³
19. 250 J
21. 1.20 J
23. **a.** −8.02 K   **b.** −8.02 °C   **c.** −14.4 °F
25. 55 J
27. 1.85×10^5 Pa
29. 7290 W
31. 40%
33. 49.2 J
35. 3.29 kcal/K
37. +0.0359 kcal/K
39. 3.88×10^{-4} kcal/K
43. 0.031 m³
45. 17.9 °C; 8.82×10^4 Btu/hr

Chapter 18

1. 1.89×10^{13} electrons
3. 1.97×10^{-1} N; repulsive, away from q_2
5. 8.75×10^{-3} N
7. 2.19×10^6 m/s
9. 4.80 N attractive; 1.60 N repulsive
11. 4.80×10^{-4} N
13. On q_1, 0.719 N in $-x$-direction; on q_2, 4.36 N in $+x$-direction; on q_3, 5.08 N in $+x$-direction
15. On 2.00 μC charge, 2.00 N in $+x$-direction
    On -4.00 μC charge, 6.40 N in $+x$-direction
    On 6.00 μC charge, 8.40 N in $-x$-direction
17. 1.21 m to the left of q_1 and along the extension of the straight line between q_1 and q_2
19. 10.8 N, outward along diagonal line between q_2 and q_3
21. 2.24×10^{-3} N, perpendicular to line l and in the directional sense from q_1 to q_2
23. 6.10 N toward and perpendicular to line connecting q_1 and q_2
25. 4.31×10^{-8} C
27. 1.00×10^{-1} N
29. 0.137 N, 327.2° relative to line q_1 to q_3
31. 525 m/s² away from q_1
33. (0.414 m, 1.17 m)
35. 2.10×10^{-14} in the negative x direction
37. 36.0 N m
39. 4.00×10^{-2} N. Note that since $d >> a$, this force is identical, to three significant figures, with that found when the ring is treated as a point charge.

Chapter 19

1. 6.75×10^{-3} N/C
3. **a.** 1.26×10^5 N/C away from q_2 **b.** -9.00×10^4 N/C toward q_2
5. $x = 22.2$ cm
7. $E = 9.53 \times 10^4$ N/C; $\theta = 341°$
9. 2.25×10^{-7} C m
11. 6.43×10^5 N/C at 26.5° relative to the diagonal from the 2.00 μC to the -6.00 μC charge
13. 4.68×10^4 N/C, 222.9° relative to direction q_1 to q_2
15. 2.40×10^{-11} J; 2.40×10^{-11} J
17. **a.** 2.75×10^3 V; 6.88×10^2 V **b.** 2.06×10^3 V
19. 2.40×10^3 J
21. **a.** 9.00×10^4 V **b.** -1.80×10^4 V
23. 18.0 cm; zero work
25. 2.04×10^5 V
27. **a.** -8.75×10^3 V **b.** -1.53×10^{-2} J
29. **a.** 1.00×10^5 V/m **b.** 1.60×10^{-14} N **c.** 1.76×10^{16} m/s²
31. **a.** 2.40×10^{-17} J **b.** 2.40×10^{-17} J
33. For an electron, $v = 1.19 \times 10^6$ m/s;
    For a proton, $v = 2.77 \times 10^4$ m/s
35. **a.** 0.108 N/C in $+x$-direction **b.** 54.0 mV
    c. 2.16×10^{-7} N in $+x$-direction
37. zero
39. $V = 3.63 \times 10^{-13}$ volts
41. 1.12 J
43. **a.** $V_A = 1.80 \times 10^4$ V; $V_B = 2.25 \times 10^4$ V
    b. 4.50×10^3 V **c.** 5.94×10^{-3} J **d.** -5.94×10^{-3} J
45. $V_1 = 270$ mV; $V_2 = 135$ mV; $V_3 = 108$ mV; $V_4 = 96.4$ mV;
    $V_5 = 93.1$ mV; $V_6 = 90.0$ mV; $V_7 = 87.1$ mV; $V_8 = 84.4$ mV; $V_9 = 77.1$ mV; $V_{10} = 67.5$ mV; $V_{11} = 54.0$ mV. E_{30}
    $= +540$ mV/m; $E_{30} = +338$ mV/m; $E_{30} = +309$ mV/m;
    $E_{30} = +300$ mV/m; $E_{30} = +300$ mV/m; $E = 300$ mV/m

Chapter 20

1. 3.12×10^{19} electrons/s
3. 4.40×10^{-13} A
5. 48.0 Ω
7. 5.28 Ω
9. The resistance halves.

11. **a.** 1.41×10^{-4} Ω **b.** 5.64×10^{-8} Ω
13. 34.1 Ω
15. 147 °C
17. 539 Ω
19. 0.500 A; 1.08×10^6 J
21. 20.0 MW; 20.0 MJ/s
23. 600 Ω
25. 54.6 Ω
27. 43.3 Ω
29. **a.** 700 Ω **b.** 17.1 mA **c.** 8.57 V **d.** 3.43 V **e.** $I_1 = 17.1$ mA; $I_2 = 11.4$ mA; $I_3 = 5.72$ mA **f.** 0.205 W **g.** $P_1 = 0.147$ W; $P_2 = 3.90 \times 10^{-2}$ W; $P_3 = 1.96 \times 10^{-2}$ W
31. **a.** 0.923 Ω **b.** 1.92 Ω **c.** 6.24 A **d.** 6.24 V
    e. 5.76 V **f.** $I_1 = 6.24$ A; $I_2 = 2.88$ A;
    $I_3 = 1.92$ A; $I_4 = 1.44$ A **g.** $P_1 = 38.9$ W; $P_2 = 16.6$ W;
    $P_3 = 11.1$ W; $P_4 = 8.29$ W
33. **a.** 252 Ω **b.** $I_1 = I_4 = 47.6$ mA; $I_2 = 31.0$ mA;
    $I_3 = 16.6$ mA; $I_5 = 28.6$ mA; $I_6 = 19.0$ mA
35. 7.14 Ω
37. 9.00 V; 6.60 V
39. 13.3 V
41. 6.01×10^{-3} Ω
43. 200 Ω
45. $R = 32.7$ Ω; $I_1 = 0.117$ A; $I_3 = 6.67 \times 10^2$ A; $I_2 = 0.117$ A; $I_4 = 6.67 \times 10^{-2}$ A
47. $I_3 = 0.27$ A; $V_3 = 8.1$ V
49. $I_2 = 41.7$ mA; $I_3 = 0.275$ A; $I_1 = 0.233$ A
51. 16.5 V
53.

Resistance	Current (Amps)	Voltage (volts)
R_1	1.67	8.33
R_2	1.67	25.0
R_3	0.667	13.3
R_4	0.667	13.3
R_5	0.667	6.67
R_6	0.334	6.67
R_7	1.00	20.0

55. **a.** 19.2 Ω **b.** $I_1 = 0.624$ A; $I_2 = 0.288$ A; $I_3 = 0.192$ A;
    $I_4 = 0.144$ A
57. 837 s
59. $V_{out} = \left(\dfrac{l_{out}}{l_{in}}\right) V_{in}$
61. **a.** $V_{AC} = 2.05$ V **b.** $V_{AD} = 2.43$ V **c.** $V_{CD} = 0.385$ V
    d. $V_{CB} = 3.96$ V **e.** $V_{DB} = 3.55$ V

Chapter 21

1. 48.0 μC
3. half
5. 2.66 pF
7. $C = 0.300$ pF; $d = 0.295$ m
9. 2.41 pF
11. 37.1 pF
13. $w = 6.80 \times 10^{-2}$ J; $q = 1.13 \times 10^{-3}$ C
15. 1.14 μF
17. 14.0 μF
19. **a.** 300 μC **b.** $q_1 = 50.0$ μC; $q_2 = 250$ μC **c.** 1.50×10^{-2} J
    d. $W_1 = 4.17 \times 10^{-4}$ J; $W_2 = 2.08 \times 10^{-3}$ J
21. **a.** 4.29 μF **b.** 25.7 μC **c.** $V_1 = 5.14$ V;
    $V_2 = V_3 = 0.857$ V **d.** $q_1 = 25.7$ μC; $q_2 = 8.57$ μC;
    $q_3 = 17.1$ μC
    e. $W_1 = 66.0$ μJ; $W_2 = 3.67$ μJ; $W_3 = 7.34$ μJ
23. **a.** 6.90 μF **b.** 33.1 μC **c.** 2.07 V
25. $C = 8.43$ μF; $q_1 = 38.9$ μC; $q_3 = 62.3$ μC; $q_2 = 63.2$ μC; $q_4 = 37.9$ μC
27. 2.60×10^{-11} C²/N m²
29. 197 μF
31. 4.50×10^3 V
33. 3.20×10^{-4} J
35. **a.** 1.20×10^4 V/m **b.** 2.54×10^4 V/m

37. $C = \dfrac{4\pi\epsilon_0 r_a r_b}{r_b - r_a} \approx \dfrac{\epsilon_0 A_s}{d}$ for r_a, r_b large since $4\pi r_a r_b \approx 4\pi \bar{r}^2 = A_s$ and $r_b - r_a = d$.

39. 709 μF; -4.51×10^5 C

41. **a.** 5.00×10^{-2} μF, series **b.** 125 μF, parallel

43. **a.** $C = 2.51$ μF **b.** $V_3 = V_5 = V_6 = V_{356} = 0.445$ V **c.** $q_1 = 30.1$ μC from part b

45. **a.** $I = 6.00 \times 10^{-2}$ A; $I = 0$ with C_1 or C_2 **b.** $V = 12.0$ V across R; $V_1 = V_2 = 12.0$ V across the capacitors **c.** $q_1 = 24.0$ μC; $q_2 = 60.0$ μC

47. **a.** $C_0 = 17.7$ pF **b.** $C_d = 95.6$ pF **c.** $q_0 = 106$ pC; $q_d = 573$ pC **d.** $E = 3.00 \times 10^3$ V/m **e.** $W_0 = 3.19 \times 10^{-10}$ J; $W_d = 1.72 \times 10^{-9}$ J

49. **a.** $q_1 = 300$ μC; $C_2 = 16.2$ μF; $q_2 = 1620$ μC **b.** 2.00×10^5 V/m **c.** $W_1 = 1.50 \times 10^{-2}$ J; $W_2 = 8.10 \times 10^{-2}$ J

51. **a.** $C_0 = \dfrac{\epsilon_0 A}{d_0}$ **b.** $C = \dfrac{\epsilon_0 A}{d_0} = C_0$

53. **a.** $C = \left(\dfrac{\kappa_2 \kappa_1}{\kappa_2 + \kappa_1}\right) C_0$ **b.** $C = \kappa C_0$

55. **a.** 5.65×10^{-17} F **b.** 5.65×10^{-18} C

Chapter 22

1. 4.47×10^{-17} N perpendicular to **v** and **B**

3. **a.** 6.01×10^{-15} N **b.** 6.60×10^{15} m/s^2

5. 4.09×10^{-5} m/s

7. 5.33×10^{-4} m

9. 9.63×10^{-4} m

11. 0.200 T

13. 1.57×10^{-3} N

15. 3.50×10^{-3} T

17. 6.28×10^{-4} T

19. 11.9

21. 4.00 μT

23. 5.00×10^4 A

25. **a.** 0 **b.** 26.7 μT

27. 50.0 μT; 53.1° counterclockwise from line $\overline{AI_2}$

29. 7.96×10^3 turns

31. 1.44×10^{-4} N/m to the left

33. 0.157 A m^2

35. **a.** 0.150 A m^2 **b.** 2.12 A

37. $\tau = (1.07 \times 10^{-4}$ N m$) \sin\theta$; $\tau_{max} = 1.07 \times 10^{-4}$ N m

39. **a.** 2.40×10^{-19} N **b.** 1.44×10^8 m/s^2 in direction $-\mathbf{v} \times \mathbf{B}$, perpendicular to the velocity and the magnetic field **c.** 1.67×10^{-7} s **d.** 3.67×10^{-3} m

41. $B = 113$ μT directed along the diagonal of the square and toward wire 3.

43. 4.40×10^{-10} N

45. $F = 0$

47. 12.5 T

49. **a.** 1.20 Ω **b.** 22.7 m **c.** 181 **d.** 8.12×10^{-3} T **e.** Choose a cardboard core of smaller diameter and shorter length. The increase in n increases B. Use a larger diameter wire. The required length of wire will be greater, resulting in more turns and hence a larger n.

51. **a.** $\ell = 379$ m **b.** $\ell = 0.121$ m **c.** $A = 4.57 \times 10^{-2}$ m^2 **d.** $m = 0.457$ A m^2, at maximum current **e.** $\tau_{max} = 1.14 \times 10^{-2}$ N m **f.** $\phi = 7.27 \times 10^{-3}$ N m/rad

53. 1.31 N m; $\tau_{max}\omega$; 137 W

55. **a.** $m = \dfrac{qBR}{v}$ **b.** $1.13R$

Chapter 23

1. **a.** 1.20×10^{-3} Wb **b.** 0 **c.** 1.04×10^{-3} Wb

3. -0.450 V

5. **a.** 1.77 mV **b.** 70.8 μA

7. **a.** ΔA is perpendicular to the loop plane and opposite **B**. **b.** 180°

---

c. $\mathcal{E} = B\dfrac{\Delta A}{\Delta t}$

d. counterclockwise, looking in the direction of **B**

9. **a.** 15.0 mV **b.** 30.0 mA **c.** counterclockwise

11. 300 T/s

13. 43.2 mV; no

15. 17.5 mV

17. 0.239 T

19. $I = \dfrac{\Delta Q}{\Delta t} = \dfrac{\mathcal{E}}{R} = \dfrac{1}{R}\left(-N\dfrac{\Delta\Phi_M}{\Delta t}\right)$; $\Delta Q = -\dfrac{N\Delta\Phi_M}{R}$

21. clockwise, as viewed from the bar magnet side

23. 255 rad/s

25. 51.7 V

27. 4.29×10^4 rad/s

29. 1.77 cm

31. 4.00 mH

33. -0.750 V

35. -2.00×10^4 A/s

37. 6.25×10^{-2} J

39. 0.800 H

41. **a.** the current in R is downward **b.** the current in R is upward

43. 3.14 mA; right to left

45. 6.70×10^{-10} A

47. 8.84 mV

49. 4.00 H

51. **a.** $\varepsilon = 1.23$ mV **b.** $\Phi_M = 2.46 \times 10^{-5}$ Wb **c.** $\varepsilon = 4.92 \times 10^{-4}$ V $\sin\left(20.0\dfrac{\text{rad}}{\text{s}} t\right)$ **d.** $I_{max} = 16.4$ μA

55. $\varepsilon = \left(\dfrac{1}{2} gB\ell \sin 2\theta\right)t$

Chapter 24

1. 70.7 V

3. 7.07 A

5. 1.26×10^4 Ω

7. 44.5 A

9. 265 Ω

11. **a.** 1.59 Hz **b.** 15.9 Hz **c.** 159 Hz; low frequencies

13. **a.** 0.754 Ω **b.** 146 A

15. 1.03×10^3 Ω

17. **a.** $-17.1°$ **b.** 107° **c.** $-72.9°$

19. 0.780 H

21. $Z = 280$ Ω; $\phi = 0.370°$

23. 1.07 H

25. 4.49×10^{-7} H

27. **a.** 6.75×10^5 H **b.** 1.20×10^4 Ω = 12.0 kΩ **c.** 10.4 pF

29. 2.85×10^{-18} F

31. ± 45.0 Ω

33. **a.** 1.88×10^3 Ω **b.** 0.110 A **c.** 58.5 mA **d.** 0.125 A **e.** $-28.0°$ **f.** 880 Ω

35. **a.** 663 Ω **b.** 0.220 A **c.** 0.166 A **d.** 0.276 A **e.** 37.0° **f.** 399 Ω

37. 1.20 A

39. **a.** 66.7 turns **b.** 75.0 V; 22.5 V

41. $P_1 = 1.00 \times 10^5$ W; $P_2 = 1.00 \times 10^3$ W

43. **a.** 1.76×10^3 Ω **b.** 497 Ω **c.** 1.55×10^3 Ω **d.** 0.142 A **e.** $V_R = 128$ V; $V_L = 250$ V; $V_C = 70.5$ V **f.** 220 V, the applied voltage **g.** 54.5° **h.** 21.3 Hz

45. **a.** 884 Ω **b.** 2.01×10^3 Ω **c.** 54.8 mA **d.** 98.7 V **e.** 48.4 V **f.** 110 V, the applied voltage **g.** $-26.1°$ **h.** 0.898

47. $\varepsilon = \varepsilon_1 + \varepsilon_2 + \varepsilon_3$

$-L\dfrac{\Delta i}{\Delta t} = -L_1\dfrac{\Delta i}{\Delta t} - L_2\dfrac{\Delta i}{\Delta t} - L_3\dfrac{\Delta i}{\Delta t}$

49. The current through R, C, and L is 83.3 mA. $i_1 = 32.0$ mA (current in C_1); $i_2 = 51.3$ mA (current in C_2)

51. $i = 88.0$ mA = current through R_1, R_2, C_1, and C_2.
   $i_{L1} = 38.5$ mA through L_1; $i_{L2} = 49.1$ mA through L_2

53. $V_{AB} = \dfrac{\varepsilon}{\left(1 - \dfrac{1}{4\pi^2 LCf^2}\right)}$

   V_{AB} increases as f increases, and approaches the applied voltage ε as f reaches high values. For low frequencies, V_{AB} is small.

Chapter 25

1. 2.26×10^5 N m²/C
3. 3.39×10^8 N/C
5. 7.53×10^{14} N/C s
9. 60.0 mA
13. $E = (0.85$ V/m$)\sin [(1.88$ m$^{-1}) x - (5.65 \times 10^8$ s$^{-1})t]$
15. 8.78 min
17. 4.17×10^{14} to 7.89×10^{14} Hz
19. 6.00×10^{12} to 4.16×10^{14} Hz
21. 3.21 m
23. 23.8 MJ/m²
25. a. 199 W/m² b. 49.7 W/m² c. 22.1 W/m²
   d. 12.4 W/m² e. 7.96 W/m²
27. a. 274 V/m; 9.13×10^{-7} T b. 137 V/m; 4.57×10^{-7} T
   c. 91.2 V/m; 3.04×10^{-7} T d. 68.4 V/m; 2.28×10^{-7} T
   e. 54.8 V/m; 1.83×10^{-7} T
29. 2.74 V/m; 9.13×10^{-9} T
31. 1.73×10^{-2} V/m
33. 2.86×10^5 W/m²
35. 2.82×10^7 V/m
37. 3.21 W
39. a. 1.38×10^{-6} W = 1.38 μW b. 7.62×10^{-2} V/m
41. a. 60.2 MHz b. 60.2 MHz c. 4.98 m
43. $\sqrt{\kappa}$

Chapter 26

1. a. 7.89×10^{14} Hz b. 4.17×10^{14} Hz
3. 2.22×10^4
5. $q = 20.0$ cm; $h_i = 10.0$ cm
7. a mirror of length 2'10", or one half the height of the student
9. 12.3 cm
11. a. 12.0 cm b. 13.3 cm c. 20.0 cm d. $q =$ is at infinity
   c. $q = -10.0$ cm
13. a. 1.00 cm b. 1.67 cm c. 5.00 cm d. is infinite e. 10.0 cm
15. $+7.5$ cm
17. 30.0 cm
19. Case I, $p = 25.0$ cm; Case II, $p = 18.8$ cm
21. 75.0 cm
23. 60.0 cm
25. -2.00 cm
27. a. -8.75 cm b. -8.00 cm c. -6.67 cm
   d. -5.00 cm e. -3.33 cm
29. 15.5 cm
37. Case I, $R = 60.0$ cm; Case II, $R = 20.0$ cm
39. $q = -60.0$ cm; M $= -3.00$; image is virtual and inverted
41. Rays 1, 2, and 3 pass very close to the focal point while rays 4 and 5, which are further removed from the axis of the mirror, come nowhere near the focal point. A spherical mirror forms a good image if we use only a small portion of the spherical surface.

Chapter 27

1. 30.3°
3. 40.3°
5. a. 2.25×10^8 m/s b. 2.04×10^8 m/s c. 1.24×10^8 m/s
7. 773 nm
9. a. 18.6° b. 1.91×10^8 m/s c. 376 nm
11. 90.0 cm
13. 61.0°
15. 1.50
17. $+21.0$ cm; converging

19. $+46.2$ cm; converging
21. a. 60.0 cm b. -3.00 c. 9.00 cm d. The image is real since q is positive. e. Since M is negative, the image is inverted.
23. 40.0 cm
25. 11.1 cm
27. 10.0 cm
29. 1.67 cm
31. a. The image is real and located 70.0 cm from the object.
   b. 10.0 cm
33. -60.0 cm
35. $f_c = +.650$ cm
37. $v_2 = 1.17 \times 10^8$ m/s
39. a. $r = 67.5°$ b. $\sin r = 1.06$. There is no refracted ray; we have total internal reflection.
41. 43.4 cm; 16.6 cm
43. $P = +2.33$ diopters
45. The wavelength is 380 nm.

Chapter 28

1. 0.169°; 0.338°; 0.507°
3. 1.03 cm; 2.06 cm; 3.09 cm
5. a. 0.169°; 0.338°; 0.507° b. 0.084°; 0.253°; 0.421° c. 2.06 cm; 4.12 cm; 6.18 cm; 1.03 cm; 3.09 cm; 5.15 cm
7. m = 28
9. R = 4.69 m
11. a. halved b. doubled
13. 169,779
15. 1.50×10^5
17. 613 nm
19. Wave 1 is out of phase with wave 2 but in phase with waves 3 and 4.
21. Wave 1 is out of phase with wave 2, in phase with wave 3, and not completely out of phase with wave 4.
23. a. 532 nm b. nothing in the visible spectrum
25. 133 nm; 400 nm
27. 1.63
29. $0.738 I_0$
31. 5.00×10^{-5} m
33. $\lambda I/2h$; $\lambda I/2h$
35. 9.80×10^{-7} m
37. 7.66°
39. $\lambda = 6.13 \times 10^{-6}$ m
41. m = 499
43. 28.1°; 70.4°; no third bright spot
45. 5.98×10^{-7} m
47. 6.25×10^{-7} m
49. 7.42 mm
51. a. 14.3 cm b. 0.649 mm
53. 4.54×10^3 m
55. $\Delta\theta = 0.1°$

Chapter 29

1. 65.0 m; 61.0 m
3. a. 20.0 m/s b. 6.1 m/s
5. a. 7.50 m/s b. 6.50 m/s
9. 6.55×10^8 m; 5.46 s
13. 91.7 m
15. $L_0 = 182$ km; $L_2 = 79.3$ km
17. a. 2.50×10^{-6} s b. 360 m c. 600 m
19. $T_0 = 0.07605$
21. a. $0.263c$ b. $0.806c$ c. $0.882c$
23. a. impossible b. mass at rest
   c. $0.995c$ d. $0.99995c$ e. $0.9999995c$
25. V $= 2.55 \times 10^8 \dfrac{\text{m}}{\text{s}}$
27. 0.78 MeV
29. 1.00×10^{-10} J
31. 2.71×10^{-10} J
33. 7.50×10^8 s
35. 3.46 light years; $t_{\text{space}} = 6.92$ yr; $t_{\text{earth}} = 8.00$ yr

37. $\dfrac{\rho_0}{1-v^2/c^2}$

39. $v = c\sqrt{1-\left(\dfrac{E_0}{E}\right)^2}$

41. $E = 3.82 \times 10^{26}$ J; $m = 4.24 \times 10^9$ kg; $T = 1.48 \times 10^{13}$ yr

Chapter 30

3. $v = 0.577c$

7. b. 1.45 m

9. $\Delta t_f = (6.25 \times 10^{-7})(1 + 5.44 \times 10^{-11})$s

13. $\Delta t_f = (0.500)(1 + 2.11 \times 10^{-11})$s

Chapter 31

1. a. 9.25×10^{-2} J **b.** 0.968 Hz **c.** 1.44×10^{32} **d.** imperceptible

3. 6.14×10^{-26} J

5. 2.40 eV

7. 1.20×10^{15} Hz; 250 nm

9. 1.33×10^{-27} kg m/s

11. 3.06×10^{-19} J

13. 8.59×10^{-16} J

15. 2.41×10^{-34} m

17. 2.43×10^{-11} m

19. 3.69×10^{-11} m

21. a. $E = 1.74 \times 10^{-10}$ J **b.** $\lambda = 2.30 \times 10^{-15}$ m

23. 5.11×10^{-29} m/s

25. 1.59×10^{-14} m

Chapter 32

1. 7.68×10^{-13} J

3. a. 1.69×10^{-14} m **b.** 5.71×10^{-15} m

5. # electrons = $Z = 47$; # protons = $Z = 47$; # neutrons = $A - Z = 61$

7. $E_1 = -13.6$ eV; $r_2 = 4.62 \times 10^{-11}$

9. a $n = 4$ **b.** $r4 = 0.846$ nm **c.** $E4 = -0.85$eV

11. 5.33×10^4 K

15. 3.15×10^{-34} J s; 2.10×10^{-34} J s; 1.05×10^{-34} J s

17. a. 1.95×10^{-18} J **b.** 2.94×10^{15} Hz **c.** 102 nm

19. 1.48×10^{-34} J s; 1.48×10^{-34} J s

21. a. $E_5 = -0.544$eV **b.** $L = 2.57 \times 10^{-34}$ Js **c.** $L_z = 0, \pm 1.05 \times 10^{-34}$ Js, $\pm 2.10 \times 10^{-34}$ Js **d.** $\theta = 90°$

23. a. $L_{z1} = 1.05 \times 10^{-34}$ J s; $L_{z0} = 0$; $L_{z,-1} = -1.05 \times 10^{-34}$ J s **b.** $\theta_1 = 45°$; $\theta_0 = 90°$; $\theta_{-1} = 135°$

25. $\tau_1 = 3.26 \times 10^{-26}$ N m; $\tau_2 = 5.16 \times 10^{-26}$ N m; $\tau_3 = 5.65 \times 10^{-26}$ N m; $\tau_4 = 5.16 \times 10^{-26}$ N m; $\tau_5 = 3.26 \times 10^{-26}$ N m

27. a. $(-5.48 \times 10^{-19} - 1.84 \times 10^{-23})$J **b.** One energy photon will be $+1.64 \times 10^{-18}$ J. The other two photons will have an energy of $\pm 1.84 \times 10^{-23}$ J. **c.** 2.47×10^{15} Hz. The other two photons will have frequencies with an additional frequency of $\pm 2.77 \times 10^{10}$ Hz. **d.** 1.21×10^{-7} m

29. $\pm 2.31 \times 10^{-26}$ J

31. $4s^1$

33.

$n = 1$	$l = 0$	$m_e = 0$	$m_s = \pm\frac{1}{2}$
$n = 2$	$l = 0$	$m_e = 0$	$m_s = \pm\frac{1}{2}$
	$l = 1$	$m_e = 0, \pm 1$	$m_s = \pm\frac{1}{2}$
$n = 3$	$l = 0$	$m_e = 0$	$m_s = \pm\frac{1}{2}$
	$l = 1$	$m_e = 0, \pm 1$	$m_s = \pm\frac{1}{2}$
$n = 4$	$l = 0$	$m_e = 0$	$m_s = \pm\frac{1}{2}$

35. $l = 4$. The orbit must be $n \geq 5$.

37. 7.11×10^{-27} J/T

39. a. $r_n = r_1 n^2$ **b.** $v_n = \dfrac{v_1}{n}$ **c.** $E_3 = -1.51$ eV

d. $E_1 = -13.6$ eV **e.** $\Delta E = 12.1$ eV

f. $f = 2.92 \times 10^{15}$ Hz **g.** $\lambda = 1.03 \times 10^{-7}$ m = 103 nm

Chapter 33

1. a. 29; 58; 29 **b.** 11; 24; 13 **c.** 84; 210; 126 **d.** 20; 45; 25
e. 82; 206; 124

3. 2.86×10^{23}

5. 107.868331 u

7. 0.00857 u; 7.98 MeV

9. 3.44×10^{-10} s^{-1}

11. 4.86 days

13. The half-life is about 245 days; very few decay in one day.

15. 17.6 yr

17. a. 9.25×10^5 Bg **b.** 4.63×10^5 Bg **c.** 2.31×10^5 Bg

19. $^{216}_{84}$Po

21. $^{225}_{88}$Ra

23. $^{210}_{82}$Pb $\rightarrow$ $^{0}_{-1}$e + $\bar{v}$ + $^{210}_{83}$Bi

25. $^{49}_{24}$Cr $\rightarrow$ $^{0}_{+1}$e + v + $^{49}_{23}$V

27. $^{52}_{25}$Mn $\rightarrow$ $^{0}_{-1}$e + v + $^{52}_{24}$Cr

29. 1.88×10^3 MeV; -1.81×10^3 MeV

31. 156 KeV

33. 17.6 MeV

35. a. $\lambda_{Th} = 2.37 \times 10^{-5}$ days^{-1}; $\lambda_{Rn} = 0.181$ days^{-1} **b.** $A_{Th} = 3.59 \times 10^{12}$ Bq; $A_{Rn} = 2.85 \times 10^{16}$ Bq **c.** $A_{Th} = 3.58 \times 10^{12}$ Bq; $A_{Rn} = 3.93 \times 10^8$ Bq **d.** $N_{Th} = 1.305 \times 10^{22}$ nuclei; $N_{Rn} = 1.88 \times 10^{14}$ nuclei **e.** The activity of Rn has decreased considerably, whereas the activity of Th has hardly changed. The same is true of the number of radioactive nuclei. When the radioactive Rn has for all practical purposes disintegrated, the Th will still last for many years.

37. 5350 yr

Chapter 34

1. $1.64 \times 10^{3-13}$ J; 2.47×10^{20} Hz; 1.21×10^{-12} m

3. 1.80×10^{17} J; 1.36×10^{50} Hz; 2.21×10^{-42} m

5. charge is 0; spin is $+\dfrac{1}{2}$

7. $q_T = +e$; $s_T = +\dfrac{1}{2}$

9. $q = 0, s = 0$

11. spin is 1; charge is 0

13. The three quarks all have the same quantum numbers which violates the Pauli exclusion principle. The two quarks give a fractional charge and the net color is not white.

15. 432

17. There are no 2 quark combinations; because spin cannot be made to equal ½.
The following are 3 quark combinations; the spins cannot be all aligned and the three colors, RGB, must be present.

uud	ctd	$\bar{b}\bar{b}d$
uus	cts	$\bar{b}\bar{b}s$
uub	ctb	bsd
ucd	ttd	
ucs	tts	
ucb	$tt\bar{b}$	
utd	$\bar{d}\bar{d}\bar{d}$	
uts	$\bar{d}\bar{d}s$	
utb	$\bar{d}\bar{d}b$	
ccd	$\bar{s}\bar{s}\bar{s}$	
ccs	$\bar{s}\bar{s}d$	
ccb	$\bar{s}\bar{s}b$	
	$\bar{b}\bar{b}b$	

Bibliography

College Physics

There are so many good books on college physics that it is impossible to list them all. Let me mention a few standards and/or favorites.

Brancazio, Peter J. *The Nature of Physics.* New York: Macmillan Publishing Co., 1975. An absolutely delightful textbook on college physics. It is a little more descriptive than most books.

Holton, G., and Roller, D. *Foundations of Modern Physical Science.* Reading, Mass.: Addison-Wesley Pub. Co., 1965.

Lindsay, Robert B. *Basic Concepts of Physics.* New York: Van Nostrand Reinhold Co., 1971.

Sears, F. W., Zemansky, M. W., and Young, H. *College Physics.* Reading, Mass.: Addison-Wesley Pub. Co. This book could probably be called the "bible" of college physics texts. It has been around for about thirty years.

Stevenson, R., and Moore, R. B. *Theory of Physics.* Philadelphia, Pa.: W. B. Saunders Co., 1967.

University Physics

University physics covers essentially the same topics as college physics, but it uses calculus and is therefore at a higher level. For anyone wishing a reference to this more advanced level, the "bible" of university physics texts is:

Halliday, David, Resnick, Robert. *Physics.* New York: John Wiley and Sons, 1977.

Modern Physics

Although modern physics textbooks are usually at a calculus level and therefore above the level of a college physics textbook, I would like to list a favorite of mine that the more advanced student might wish to pursue.

Beiser, Arthur. *Perspectives of Modern Physics.* New York: McGraw-Hill Book Co., 1969. This is a delightful text on modern physics. I have used it in my modern physics classes since it was first published and its flavor can be seen throughout the modern physics topics in this text.

Popular Physics Books

There are a large number of books on the market today that treat some of the latest topics in the field of physics. These books are very meaningful and very accessible to the student who is taking or has completed a course in college physics. The list is divided into two groups—quantum physics and relativity—although there is an overlap in many of the books, as might be expected.

Quantum Physics

Atkins, P. W. *The Creation.* San Francisco: W. H. Freeman & Co., 1981.

Barrow, J. D., Silk, J. *The Left Hand of Creation,* The Origin and Evolution of the Expanding Universe. New York: Basic Books, Inc., Publishers, 1983.

Bickel, Lennard. *The Deadly Element,* The Story of Uranium. New York: Stein and Day Publishers, 1979.

Chaisson, Eric. *Cosmic Dawn,* The Origins of Matter and Life. Boston: Little Brown and Co., 1981.

Chester, Michael. *Particles,* An Introduction to Particle Physics. New York: Mentor Books, New American Library, 1978.

Davies, P. C. W. *The Forces of Nature.* Cambridge: Cambridge University Press, 1979.

Davies, Paul. *Other Worlds,* A Portrait of Nature in Rebellion, Space, Superspace and the Quantum Universe. New York: Simon & Schuster, 1980.

Davies, Paul. *SuperForce,* The Search for a Grand Unified Theory of Nature. New York: Simon and Schuster, 1984. All of Davies' books are very enjoyable.

Davies, Paul. *The Runaway Universe.* New York: Harper & Row, Publishers, 1978.

Davies, Paul, and Gribbin, John. *The Matter Myth,* Dramatic Discoveries That Challenge Our Understanding of Physical Reality. New York: Touchstone Book, Simon & Schuster, 1992.

Feinberg, Gerald. *What is the World Made Of?* Atoms, Leptons, Quarks, and Other Tantalizing Particles. Garden City, N.Y.: Anchor Press/Doubleday, 1978.

Ferris, Timothy. *The Red Limit,* The Search for the Edge of the Universe, 2nd ed. New York: Quill, 1983.

Fritzsch, Harald. *Quarks, The Stuff of Matter.* New York: Basic Books, Inc., 1983.

Fritzsch, Harald. *The Creation of Matter,* The Universe from Beginning to End. New York: Basic Books, Inc., 1984.

Gribbin, John. *In Search of Schrödinger's Cat,* Quantum Physics and Reality. New York: Bantam Books, 1984.

Gribbin, John. *In Search of the Big Bang,* Quantum Physics and Cosmology. Toronto: Bantam Books, 1986.

Jastrow, Robert. *God and the Astronomers.* New York: W. W. Norton & Co., 1978.

Kaku, Michio, Trainer, Jennifer. *Beyond Einstein,* The Cosmic Quest for the Theory of the Universe. New York: Bantam Books, 1987. An easy-to-read book that discusses some of the ideas of the superstring theory.

Mulvey, J. H., ed. *The Nature of Matter.* Oxford: Clarendon Press, 1981.

Polkinghorne, J. C. *The Particle Play,* An Account of the Ultimate Constituents of Matter. San Francisco: W. H. Freeman and Co., 1979.

Schechter, Bruce. *The Path of No Resistance,* The Story of the Revolution in Superconductivity. New York: Simon and Schuster, 1989.

Sutton, Christine. *The Particle Connection.* New York: Simon and Schuster, 1984.

Trefil, James S. *From Atoms to Quarks, An Introduction to the Strange World of Particle Physics*. New York: Charles Scribner's Sons, 1980. Trefil is an excellent science writer and all his books are very enjoyable.

Trefil, James S. *The Moment of Creation, Big Bang Physics From Before the First Millisecond to the Present Universe*. New York: Charles Scribner's Sons, 1983.

Relativity

Bergmann, Peter G. *The Riddle of Gravitation*. New York: Charles Scribner's Sons, 1968.

Bernstein, Jeremy. *Einstein*. New York: Penguin Books, 1982.

Bondi, Hermann. *Relativity and Common Sense, A New Approach to Einstein*. New York: Dover Publications, 1964.

Born, Max. *Einstein's Theory of Relativity*. New York: Dover Publications, 1965. Excellent text.

Davies, P. C. W. *The Edge of Infinity, Where the Universe came from and how it will end*. New York: Simon and Schuster, 1981.

Davies, P. C. W. *The Search for Gravity Waves*. New York: Cambridge University Press, 1980.

Davies, P. C. W. *Space and Time in the Modern Universe*. New York: Cambridge University Press, 1978.

Eddington, Sir Arthur. *Space, Time, and Gravitation, An outline of the General Relativity Theory*. New York: Harper Torchbooks, 1959. A delightful little book.

Einstein, Albert. *Relativity, The Special and General Theory*. New York: Crown Publishers, Inc., 1961.

Ellis, George F. R., and Williams, Ruth M. *Flat and Curved Space-Times*. Oxford: Clarendon Press, 1988.

Gardner, Martin. *The Relativity Explosion*. New York: Vintage Books, 1976.

Gibilisco, Stan. *Understanding Einstein's Theories of Relativity, Man's New Perspective on the Cosmos*. New York: Dover Publications, 1983.

Gribbin, John. *TimeWarps, Is time travel possible? An exploration into today's best scientific knowledge about the nature of time*. New York: Delacorte Press, 1979.

Gribbin, John. *Unveiling the Edge of Time, Black Holes, White Holes, Wormholes*. New York: Harmony Books, 1992.

Halpern, Paul. *Time Journeys, A Search for Cosmic Destiny and Meaning*. New York: McGraw-Hill, 1990.

Halpern, Paul. *Cosmic Wormholes, The Search for Interstellar Shortcuts*. New York: Dutton Books, 1992.

Harpaz, Amos. *Relativity Theory, Concepts and Basic Principles*. Boston: Jones and Bartlett Publishers, 1992. A truly excellent book.

Hawking, Stephen W. *A Brief History of Time, From the Big Bang to Black Holes*. New York: Bantam Books, 1988.

Hoffmann, Banesh. *Relativity and its Roots*. New York: Scientific American Books, 1983.

Kaufmann, William J. *Black Holes and Warped Spacetime*. San Francisco: W. H. Freeman and Company, 1979.

Kaufmann, William J. *The Cosmic Frontiers of General Relativity*. Boston: Little Brown and Company, 1977.

Luminet, Jean-Pierre. *Black Holes*. Cambridge: Cambridge University Press, 1992.

Macvey, John W. *Time Travel, A Guide to Journeys in the Fourth Dimension*. Chelsea, Michigan: Scarborough House Publishers, 1990.

Marder, L. *Time and the Space Traveller*. Philadelphia: University of Pennsylvania Press, 1974.

Mermin, N. David. *Space and Time in Special Relativity*. New York: McGraw-Hill Book Company, 1968.

Morris, Richard. *Time's Arrows, Scientific Attitudes Toward Time*. New York: Simon and Schuster, 1984.

Narlikar, Jayant V. *The Lighter Side of Gravity*. San Francisco: W. H. Freeman and Company, 1982.

Nicolson, Iain. *Gravity, Black Holes and the Universe*. New York: Halsted Press, 1981.

Novikov, Igor. *Black Holes and the Universe*. Cambridge: Cambridge University Press, 1990.

Reichenbach, Hans. *The Philosophy of Space and Time*. New York: Dover Publications, 1958.

Sexl, Roman, and Sexl, Hannelore. *White Dwarfs, Black Holes, An Introduction to Relativistic Astrophysics*. New York: Academic Press, 1979. A fascinating little book. The discussion of the falling clock is the one used in this book.

Shipman, Harry L. *Black Holes, Quasars, and the Universe,* 2nd ed. Boston: Houghton Mifflin Company, 1980.

Taylor, John G. *Black Holes*. New York: Avon Books, 1973.

Wald, Robert M. *Space, Time, and Gravity, The Theory of the Big Bang and Black Holes*. Chicago: The University of Chicago Press, 1977.

Wheeler, John Archibald. *A Journey into Gravity and Spacetime*. New York: Scientific American Library, 1990. A beautiful popular account of relativity written by one of the masters.

Will, Clifford M. *Was Einstein Right? Putting General Relativity to the Test*. New York: Basic Books, Inc., 1986.

Credits

Photographs

Chapter 1
Opener: © NASA/Rainbow; **1.1:** Field Museum of Natural History, #110a2513c; **1.5B:** NASA; **1.6:** National Institute of Standards & Technology; **1.7B:** Courtesy of Central Scientific Company; **1.8:** © National Institute of Standards & Technology, Boulder Labs, US Department of Commerce; **1.9:** NASA

Chapter 2
Opener: © Rhoda Sidney/The Image Works; **2.18A:** © Eric Millette/The Picture Cube

Chapter 3
Opener: © Yoav Levy/Phototake; **3.1:** © The Granger Collection; **3.14B:** © R. P. Kingston/The Picture Cube; **3.16:** © NASA; **3.18A & B:** © Mark Antman/The Image Works; **p. 62:** © Yoav Levy/Phototake; **3.24:** © Sheila Sharidan/Monkmeyer Press; **p. 70:** © Joe Sohm/The Image Works

Chapter 4
Opener: © Julian Baum/Science Photo Library/Photo Researchers, Inc.; **4.1A:** © The Granger Collection; **4.3:** © M. E. Warren/Photo Researchers, Inc.; **4.15B:** © Wm. C. Brown Communications, Inc./Photo by Toni Michaels; **4.20:** © Jaye R. Phillips/The Picture Cube; **p. 112A–C:** © Tom Doody/The Picture Cube

Chapter 5
Opener: Dan McCoy/Rainbow; **5.6B:** © Michal Heron/Monkmeyer Press; **5.20B:** © Bruce Iverson; **5.21B:** © Wm. C. Brown Communications, Inc./Photo by Toni Michaels; **5.23C:** © Bob Daemmrich/The Image Works; **p. 147:** © Will McIntyre/Photo Researchers, Inc.

Chapter 6
Opener: NASA; **6.12C:** © Gloria Karlson/The Picture Cube; **6.14C:** © Charles Feil/Stock Boston; **6.15B:** © Galen Rowell/Peter Arnold, Inc; **6.17A:** © Visual Images West/The Image Works; **6.18B:** © Jerome Yeats/Science Photo Library/Photo Researchers, Inc.; **pp. 176, 181:** © NASA

Chapter 7
Opener: © Robert Mathena/Fundamental Photographs; **7.8:** © David Ball/The Picture Cube; **7.10:** © Tony Savino/The Image Works; **p. 207:** © Farrell Grehan/Photo Researchers, Inc.; **p. 209:** From *Secrets of the Great Pyramid* by Peter Tomkins. Copyright © 1971 by Peter Tomkins. Reprinted by permission of Harper Collins Publishers, Inc.

Chapter 8
Opener: © Richard Megna/Fundamental Photographs; **8.1:** © Richard Megna/Fundamental Photographs; **8.4:** © J. L. Barken/The Picture Cube; **8.7A:** © Tim Davis/Photo Researchers; **8.7B:** © Y. Arthus-Bertrand/Gerard Vandystadt/Photo Researchers, Inc.; **8.8B:** © C. E. Miller/Peter Arnold, Inc.; **8.11B:** © Richard Megna/Fundamental Photographs; **8.12B:** © Dan Burns/Monkmeyer Press; **p. 235:** © Terry Barner/Unicorn Stock Photos

Chapter 9
Opener: © Richard Megna/Fundamental Photographs; **9.2B:** © Bob Daemmrich/The Image Works; **p. 243:** © Richard Megna/Fundamental Photographs; **9.7:** © Frank Pedrick/The Image Works; **9.13:** NASA; **9.14:** © Dave Schaffer/Monkmeyer Press; **9.15(Right):** © Arthur Grace/Stock Boston; **p. 268:** © D. Wray/The Image Works; **p. 275:** © Gerard Lacz/NHPA

Chapter 10
Opener: © Fundamental Photographs; **10.1A:** © D.O.E./Science Source/Photo Researchers, Inc.

Chapter 11
Opener: © Paul Silverman/Fundamental Photographs; **11.7D:** © Mark Antman/The Image Works

Chapter 12
Opener: © Richard Megna/Fundamental Photographs; **12.1G, p. 336:** © Richard Megna/Fundamental Photographs; **12.16B:** © Toni Michaels; **12.19D:** © Eric Roth/The Picture Cube; **12.22B:** © Martha Cooper/Peter Arnold, Inc.

Chapter 13
Opener: © Bruce Coleman; **13.4C:** © Larry Miller/Photo Researchers, Inc.; **p. 377:** © Tony Freeman/Photo Edit; **13.7B:** © Herb Snitzer/The Picture Cube; **p. 381:** © Renee Lynn/Photo Researchers, Inc.; **13.13C:** © Jerry Wachter/Photo Researchers, Inc.; **13.14B:** © Dick Davis/Photo Researchers, Inc.; **p. 394:** © Blair Seitz/Photo Researchers, Inc.

Chapter 14
Opener: © Mark Antman/The Image Works

Chapter 15
Opener: © Bruce Iverson; **p. 427:** © Mark Burnett/Photo Researchers, Inc.; **p. 435:** © Tom Branch/Science Source/Photo Researchers, Inc.; **p. 445:** © Toni Michaels

Chapter 16

Opener: © Adam Hart-Davis/Science Photo Library/Photo Researchers, Inc.; **p. 458:** © NASA/Science Source/Photo Researchers, Inc.; **p. 467:** © J. Mead/SPL/Photo Researchers, Inc.; **p. 469:** © Toni Michaels

Chapter 17

Opener: © Gene Moore/Phototake; **p. 486:** Northwind Picture Archives

Chapter 18

Opener: © Henry Dakin/Science Photo Library/Photo Researchers, Inc.; **p. 518:** © Gordon Garradd/SPL/Photo Researchers, Inc.

Chapter 19

Opener: © Richard Megna/Fundamental Photographs; **pp. 536, 538, 540, 541, 544:** © Yoav Levy/Phototake; **19.31B:** Courtesy Sargent-Welch

Chapter 20

Opener: © T. J. Florian/Rainbow; **20.5C:** Courtesy of Central Scientific Company; **20.24C:** Courtesy Sargent-Welch

Chapter 21

Opener: © Coco McCoy/Rainbow; **p. 606:** © Paul Silverman/Fundamental Photographs

Chapter 22

Opener: © Richard Megna/Fundamental Photographs; **p. 633:** © Richard Megna/Fundamental Photographs; **22.18C:** Courtesy of Central Scientific Company

Chapter 23

Opener: © Dan McCoy/Rainbow

Chapter 24

Opener: © Mike Okoniewski/The Image Works; **24.18C:** Courtesy of Central Scientific Company

Chapter 25

Opener: © Bruce Iverson; **p. 736:** © Ulrike Welsch/Photo Researchers, Inc.

Chapter 26

Opener: © Joe Sohm/The Image Works; **26.8A:** © Wm. C. Brown Communications, Inc./Photo by Toni Michaels; **p. 756:** NASA; **p. 761:** © Ray Nelson/Phototake

Chapter 27

Opener: © Ken Kay/Fundamental Photographs; **p. 768:** Courtesy of Art Western; **p. 787:** Richard Megna/Fundamental Photographs; **27.25A:** © Wm. C. Brown Communications, Inc./Photo by Toni Michaels; **27.26B:** Courtesy Sargent-Welch; **p. 796:** Photograph courtesy of Yerkes Observatory and Dr. Doyal Harper

Chapter 28

Opener: © Yoav Levy/Phototake; **28.8B:** Photograph courtesy of PASCO Scientific; **p. 821:** © Richard Megna/Fundamental Photographs; **p. 830:** © Ken Kay/Fundamental Photographs

Chapter 29

Opener: © Dan McCoy/Rainbow

Chapter 30

Opener: © Julian Baum/Science Photo Library/Photo Researchers, Inc.; **pp. 906, 909:** NASA

Chapter 31

Opener: © Dan McCoy/Rainbow

Chapter 32

Opener: © Michael Gilbert/Science Photo Library/Photo Researchers, Inc.; **p. 963:** © Dr. G. J. Hills, John Innes Institute/SPL/Photo Researchers, Inc.; **p. 995:** Courtesy, University of California, Santa Barbara

Chapter 33

Opener: © Dan McCoy/Rainbow; **p. 1014:** © Hank Morgan/Science Source/Photo Researchers, Inc.; **33.13B:** © Kathy Tarantola/The Picture Cube; **p. 1027(top):** NASA, **(bottom):** © Dan McCoy/Rainbow; **p. 1028(top):** © SPL/Photo Researchers, Inc., **(bottom):** © Michael J. Howell/The Picture Cube; **p. 1030:** © Louise K. Broman/Photo Researchers, Inc.

Chapter 34

Opener: © NASA/Rainbow; **p. 1003(top):** © Dan McCoy/Rainbow; **(bottom):** © David Parker/Photo Researchers, Inc.

Line Art

Chapter 4

4.1b: From Isaac Newton (Florian Cajori, ed.), *Mathematical Principles of Natural Philosophy and His System of the World,* trans./ed. by Andrew Motte. Copyright © 1934, renewed 1962 Regents of the University of California. Reprinted by permission of University of California Press.

Chapter 6

6.21: From Isaac Newton (Florian Cajori, ed.), *Mathematical Principles of Natural Philosophy and His System of the World,* trans./ed. by Andrew Motte. Copyright © 1934, renewed 1962 Regents of the University of California. Reprinted by permission of University of California Press.

Chapter 12

12H.1a: From Kent M. Van De Graaff, *Human Anatomy,* 3d ed. Copyright © 1992 Wm. C. Brown Communications, Inc., Dubuque, Iowa. All Rights Reserved. Reprinted by permission; **12H.2b:** From John W. Hole, Jr., *Human Anatomy and Physiology,* 4th ed. Copyright © 1987 Wm. C. Brown Communications, Inc., Dubuque, Iowa. All Rights Reserved. Reprinted by permission.

Chapter 13

13.3a: Drawing of aneroid barometer reprinted from *Weather and Weather Forecasting* by A. G. Forsdyke, copyright © 1969 by the Hamlyn Publishing Group, copyright © 1970 by Grosset & Dunlap, Inc.; **13.3b:** From Joseph S. Weisberg, *Meteorology: The Earth and Its Weather,* Second Edition. Copyright © 1981 Houghton Mifflin Company. Used with permission.

Chapter 16

16H.2: From R. G. Fleagle and J. A. Businger, *An Introduction to Atmospheric Physics,* 2d ed. Copyright © 1980 Academic Press, Orlando, FL. Reprinted by permission.

Index

of thermal conductivity, 458, 459(table)
of thermal expansion, 426(table)
of volume expansion, 429
Cold front, 508
Collisions
conservation of momentum and, *217*–18
of free electron and photon, 934–38
nuclear reaction as, *1019*
in one dimension (perfectly elastic, inelastic, and perfectly inelastic collisions), 223–32
rotational, 263–64
in two dimensions (glancing dimensions), 232–35
Combined motion, 94–97
Complementarity, principle of, **933**
Component of vector, **28**
addition of vectors using method of, 31–34
determining vector from, 29–31
Compression
elasticity under, 286–87
of spring, and longitudinal waves, 321, *322*
work done by gas during, 480
Compton, Arthur Holly, 934, 937, 986
Compton effect, **934**–38
Compton scattering formula, 937
Concave lens(es)
in contact, 790–91
ray diagram for, *779*
Concave spherical mirror, 753–59
focal length of, 753–54
location of image by ray diagram, 754
magnification, 756–57
mirror equation, 754–56
object located at center of curvature of, 757–58
object located at infinity from, 757
object located at principal focus of, 758
object located between center of curvature and principal focus of, 758
object located between infinity and center of curvature of, 757
object located within principal focus of, 758–59
sign conventions for, 761(table)
Conducting plates, electric field of charged, 543
Conduction, 451, 457–64
coefficient of thermal conductivity, 458, 459(table)
compound wall, 462–65
convection cycle in walls of home and, 461–62
equivalent thicknesses of various walls and, 459–61
heat flow through slab of material by, 457–59
Conduction current, **721**
Conductors of electric charge, **521**–22
force between parallel, current-carrying (ampere), 644–46
force on, in external magnetic field, 637–38
superconductivity, 571
Conductors of heat, materials as, **458,** 459
Conservation of angular momentum, law of, **260**–64
Conservation of electric charge, law of, 520, **576**–77, 591

Conservation of energy, 198–207
conservative and nonconservative systems and, 205
in electric circuits, 571–74, 592
law of, 198 (*see also* Law of conservation of energy)
in projectile motion, 199–202
Conservation of mass, law of, *382,* **384**
Conservation of mass-energy, law of, 879–85
Conservation of momentum. *See* Law of conservation of linear momentum
Conservative system, conservation of energy in, **205**
Constant acceleration, 43–44
Constants
Boltzmann, 442
elasticity, 285(table)
invariant interval as, 894–905
mean temperature coefficient of resistivity, 568(table), 569–70
nuclear decay (disintegration), 1007–11
Planck length, 1050
Planck's, 468, 923
of proportionality, 566, 567–68, 1007
Rydberg, 974
Stefan-Boltzmann, 503
universal gas, 435
universal gravitation, 171
Constant velocity, **42**–43
Constructive interference of waves, **338**
Continental polar air mass, 508
Continuity, equation of, 382–85
human blood flow and, 393
Convection, **451**–57
atmospheric, *454*–55
cycle of, in walls of houses, 461–62
horizontal, *452*
sea and land breezes and, *455*–56
space heating through system of forced, *456*–57
temperature gradient and, 453
Conventional current, **565,** 606
Converging lens(es), **777,** 781–82
Conversion factors, **17**–19. *See also* Appendix A
Conversions of temperature scales, 406–8
Convex lens(es)
combination of two, 787–89
in contact, 790
double, 775
ray diagram for, *779*
special cases for, 783–86
Convex spherical mirror, 759–61
location of image by ray diagram, 760
magnification of, 761
sign conventions, 761(table)
Coordinate system
components of vector and, *30*
inertial and noninertial, 81
Coriolis effect, **455,** 506
Coriolis force (CF), 506–8
Correspondence principle, 862
Cosine function, **26**–27
Cosmic rays, 1035
Coulomb, Charles A. de, 520
Coulomb (measurement unit), **16,** 520
Coulomb meter (measurement unit), 542

Coulomb's law, 522–26
defined, **523**
dielectrics and, 622–23
examples of, 525–26
Newton's law of universal gravitation and, 523
Crane boom, equilibrium conditions and, 139–41
Critical angle of incidence, **773,** 774
Critical mass, 1025
Curie, Marie, 1003
Curie, Pierre, 1003
Curie (measurement unit), 1007
Cycle (oscillation) of vibratory motion, **298**–99
Cyclic process, **480**
first law of thermodynamics and, 493
work done during, 480–83

D

Dalton, John, 519, 961, 1035
D'Arsonval, Jacques Arsène, 651
Davisson, C. J., 939
De Broglie, Louis, 938, 939
De Broglie relation, **938**–39
Decay (disintegration) constant, 1007–11
Decibels, 359–60
Deciday, 15
Delta notation ("change in"), 40
Democritus, 6, 519, 961, 1035
Density, 19, **367**–68
Destructive interference of waves, **338**–39
Dicke, Robert H., 93
Dielectric constant, **616**
Dielectrics, **521**–22, 615–23
atomic description of, 617–22
between capacitor plates, *615*–17, *618*
Coulomb's law and, 622–23
polar, *617*
Dielectric strength, 616(table), **619**
Diffraction grating, **832**–34
Diffraction of light from single slit, 829–32
Fraunhofer diffraction technique, 829, *830*
Fresnel diffraction technique, *829*
illustrated, 830, *831*
intensity distribution for, *832*
interference of light combined with, *832*
Diopter, 791
Dioptric power of lens(es), **791**
Dirac, Paul A. M., 987, 1036
Direct current (DC), **565**
advantages of alternating current over, 707–8
Direct current (DC) circuit
circuit diagram, *565*
conservation of energy in, 571–74
Kirchhoff's rules on, 591–96
measuring current and potential differences in, 565–66
parallel, **576**–79, 583
resistors combined in parallel and series, 580–81
series, **574,** 583
short, 619
Wheatstone bridge, **589**–91
Direct current (DC) motor, 642
Direct vs. indirect measurements, 11
Disintegration (decay) constant, 1007–11
Dispersion of light, **774**–75
by prism, *775*

Electron, 10, 519–20
  binding energy of, 967
  Bohr model of hydrogen atom, 556–57,
    969–73
  changing quark to, *1046*
  collision of photon and free, 934–38
  configuration of, for select chemical elements,
    990(table)
  determining number of, 963–65
  force on, in magnetic field, 633–34
  ground state, 972, 991–92
  high-speed, 884–85
  kinetic energy of, 881
  motion of, in air or vacuum, *564*
  motion of, in cathode ray oscilloscope,
    554–56
  motion of, in wire, *564*
  orbit of, in magnetic field, 635–36
  Pauli exclusion principle applied to, 988–89
  position and, 563, 1036–37
  quantum members of, 976–78, 988
  as spinning particle, 986–88
  standing wave of, *969*
  states of, in terms of quantum numbers,
    989(table)
  stationary state of, 970
Electron capture, 1014–15
Electron current, **565**
Electron gas, 564
Electron volt, 557, 881
Electroscope, 521
Electrostatics, 517–33
  atomic structure and, 519–20
  Coulomb's law, 522–26
  defined, **517**
  fundamental principle of, **518**
  measurement of electric charge, 521–22
  multiple electric charges, 526–30
  separation of electric charge by rubbing,
    517–18
Electroweak force, **1043–44**
Elementary particle physics, 1035–56
  big bang theory and creation of universe,
    1052–54
  electromagnetic force, 1042–43
  electroweak force, 1043–44
  four forces of nature, 1037–38
  grand unified theories, 1045–46
  gravitational force and quantum gravity,
    1046–50
  introduction, 1035
  particles and antiparticles, 1035–37
  quarks, 1038–42
  strong nuclear force, 1044–45
  superforce as unification of all forces,
    1050–51
  weak nuclear force, 1037–38, **1043**
Elements
  atomic mass of, 1005
  electron configurations for ground states of,
    991–92(table)
  isotopes of, 1004–5
  nucleosynthesis and formation of, 1029
  periodic table of, 990(table)
  radioactive series of, 1015–19
  transuranic, 1017–19

Elevator, forces acting on person riding in, and
    Newton's second law of motion, 99–101
Emf. *See* Electromotive force (emf)
Endoergic reaction, **1020**, 1021
Energy, 189–214
  conservation of, **198**–207
  construction of Egypt's pyramids, 207–9
  defined, **189**
  in electromagnetic waves, 737–39
  gravitational potential, **194–96**
  kinetic, **196–98**
  law of conservation of mass-energy, 879–85
  in nuclear reactions, 1019–21
  of photon, 931, 933
  power and, **192–94**
  rest mass, 880
  storage of, in capacitor, 606–8
  storage of, in magnetic field of inductor,
    679–81
  transmission of, in waves, 356–57
  types of, 189
  work and, **189–92**
Energy density, **607,** 608
Energy-level diagram of hydrogen atom, *973*
  associated spectrum for, *975*
Engine
  Carnot cycle, and maximum efficiency of,
    496–97
  efficiency of, **491**
  gasoline, 493–95
  ideal heat, 495–96
Entropy, 499–501
  second law of thermodynamics and, 500, 501
  statistical interpretation of, 501–5
Equation of continuity, 382–85
  human blood flow and, 393
Equilibrium, 121–54
  defined, **121**
  displacement from, 297, 322
  first condition of, 121–26
  medical traction analyzed from perspective
    of, 147–48
  molecular, 283
  of rigid body, 131–47
  rotational, 126, 133
  second condition of, 129–31
  thermal, 408
  torque and, 126–29
  translational, 126, 133
Equipotential line, **545**
Equipotential surface, **545**
Equivalence principle, 93, **906–7**
  bending of light and, 908–9
Equivalent resistance, **575**
Ether, **853–58**
Euclidian vs. nonEuclidian geometry and
    spacetime, 895, *896, 897*
Evaporation, 414
  cooling of human body and, 445–47
Event horizon of black hole, 916
Exchange force of particles, 951
Excited state, electron in, 972
Exoergic reaction, **1020**
Expansion, work done by gas during, 480
Experiment
  as step in scientific method, 6, 7
  thought, on temperature, *9,* 403, *404*

External forces, **216**
Eye, human, *798–800*
  accommodation of, 798
  anatomy of, *798*
  defects of, 798–99
  lens correction of, 799–800

F

Fahrenheit, Gabriel, 405
Fahrenheit temperature scale, *404,* **405**
  converting from, to Celsius, 406–7
  converting from Celsius to, 407
Farad (measurement unit), 605
Faraday, Michael, 535, 605, 661, 961, 1042
Faraday ice pail for measuring electric charge,
    522
Faraday's law of electromagnetic induction,
    665–66, 722, 724
  Maxwell's equations and generalized, 715,
    725–27, 728
  mutual induction and, 675
  self-induction and, 677
Farsightedness (hyperopia), **798–99**
Fast neutrons, 1026
Fermi, Enrico, 1022, 1025
Fermions, 1041, 1050, 1051
Feynman, Richard, 1037
Feynman diagram, 1037
Fiber optics, **774**
Fine structure of spectral lines, 988
First condition of equilibrium, **121–26**
First law of motion, Newton's, **80–81**
  centrifugal force and, 161–62
  centripetal force and, 160
  consistency of, with second law of motion, 85
First law of rotational motion, Newton's, **251**
First law of thermodynamics, 485–92
  adiabatic process and, 492–93
  applying and calculating, 487–89
  cyclic process and, 493
  defined, **487**
  efficiency of engine and, 491
  internal energy of gas and, 485–87, 490
  isobaric process and, 493
  isochoric (isometric) process and, 493
  isothermal process and, 492
  net work, net heat, and, 490
Fission, nuclear, 1022–27
Flat-hyperbolic geometry, 895
Fluid dynamics, **367,** 382–91
Fluids, 367–99
  Archimedes' principle, 378–82
  Bernoulli's theorem, 385–88
  Bernoulli's theorem applied, 388–92
  blood flow in human body, 392–95
  defined, **367**
  density, 367–68
  equation of continuity, 382–85
  Pascal's principle, 376–78
  pressure, 368–76
Fluid statics, **367–82**
Flux. *See* Electric flux; Magnetic flux
Focal length
  of concave spherical mirror, **753,** 754
  on convex and concave lenses in contact,
    790–91
  of diverging lenses, 777, 778
  of two convex lenses in contact, 790

Springs
applying quantum condition to vibrating, 924–25
compression and rarefaction of, 321
conservation of energy and vibrating, 307–9
Hook's law on elasticity for, 287–88
in parallel, 312–13
potential energy of, **306**–7
in series, 313–14
simple harmonic motion in vibrating, 297–99
Standard rays, 779
Standing waves, 339–46
of electron in orbit, *969*
formation of, *339*
frequency, resonance, and, 343–45
nodes and antinodes of, 341–42
on string, *341,* **342**
vibration modes, 342, *343*
Stanford Linear Accelerator Center (SLAC), 1041
Static equilibrium, 121, 123
Static friction, 102–3
coefficients of, 103(table), 111
Statics, **121**
fluid, 367
Stationary state, electron in, 970
Stationary wave. *See* Standing waves
Statistical mechanics, 501
Stefan, Joseph, 465
Stefan-Boltzmann constant, 503
Stefan-Boltzmann law, **465**
Stoney, George, 961
Stopping potential, 927
Strain
elasticity and, **285,** *286*
shearing, 289
Streamline flow, 382
Stress
elasticity and, **285,** *286*
shearing, 289
Strings
creation of pulse on, *331*–42
reflection and transmission of wave on, in two different media, 333–36
reflection of wave on not firmly fixed, *332*
reflection of wave on rigidly fixed, *332, 333*
speed of transverse wave on, 329–31
standing waves on, *341,* **342**
vibration modes on, *343*
Strong nuclear force, 520, **1005**, 1038, **1044**–45
Sublimation, latent heat of, 414–**15**
Sun, light rays bent by, *909*
Superconductivity, 571
Superforce, **1050**–51
Superposition, principle of, **336**–39, 921, 940
Superposition of electric fields, 537–42
calculating electric field of two positive charges, 538–41
electric field of two like charges, 540
electric field of two unlike charges (electric dipole), 541–42
mathematical statement and definition of, **538**
of potentials, 551–52

Superstring theory, 1051
Supersymmetry, 1050–51
Surface area, pressure and, 369–70
Surface charge density, 719
System
closed, and conservation of energy, 198
conservative, **205**
defined, **189**
System of units, 16(table)–17. *See also* International System (SI) of Units
Szilard, Leo, 1025

T

Tangent function, **26**
determining side of triangle using, 27
Tangential velocity, 242–43
Telescope, 796–98
Television waves, 736–37
Teller, Edward, 1025
Temperature, **403**–8. *See also* Heat; Heat transfer
of gas, calculating, 436
isothermal process at constant, 482, 492
planetary, 472
scales, *404,* 405
scales, conversions of, 406–8
SI system of noting, 406
speed of sound as function of, 347
thought experiment on, *9,* 403, *404*
variation of resistance with, 569–71
Temperature gradient, **453**
Tensor analysis, 907
Tesla, Nikola, 632
Tesla (measurement unit), **632**
Thales of Miletus, 6
Theory as step in scientific method, 6, 7
Thermal energy. *See* Heat
Thermal equilibrium, **408**
Thermal expansion, **425**
area expansion of solids as, 427–28
coefficients of, 426(table)
linear expansion of solids as, 425–27
volume expansion of gases, 430–32
volume expansion of solids and liquids as, 429–30
Thermal resistance *R,* **463**
Thermodynamic process, heat absorbed or liberated in, **484,** 485
Thermodynamics, **479**–512
Carnot cycle, 496–97
entropy, 499–501
entropy, statistical interpretation of, 501–5
first law of, 485–92
first law of, special cases of, 492–93
gasoline engine and, 493–95
heat added to and removed from thermodynamic system, 483–85
ideal heat engine, 495–96
meteorology and, 505–8
second law of, 498–99
work concept applied to thermodynamic system, 479–83
Thermometer, 403, **404,** 405

Thin film, interference of light from, **820**–29
air wedge, 826–28
density of medium and, 820–21
Newton's rights, 828–29
nonreflecting glass, 823–26
soap bubble or film of oil, 821–23
Thin lens(es), **776**–77
in contact, 789–92
Third law of motion, Newton's, **81**–82
law of conservation of momentum as consequence of, 217–18
Third law of rotational motion, Newton's, **251**
Thompson, Benjamin, 408
Thompson, G. P., 939
Thompson, William, 8, 405
Thomson, J. J., 519
plum pudding model of atom by, 961, *962*
Thorium, radioactive decay series of, 1015, *1016*
Thought experiment on temperature, *9*
Time, 9, 10. *See also* Spacetime
displacement as function of, 47–48
flow of, in direction of increase in entropy, 505
projectile motion and dimension of, 57–58, 63–64
proper, 869
SI vs. BES system of noting, 16(table)
standard of, 14–15
velocity as function of, 46
Time dilation, 843, 869–71
defined, **870**
Lorentz contraction and, 869(table)
nonEuclidian nature of spacetime and, 895, 901, *902*
proper time and, **869**
slowing of accelerated clock and, 953–56
twin paradox and, 870–71, 898
Tornado, atmospheric pressure in, 374–75
Torque, 126–29
clockwise and counterclockwise, 130, 143
in current loop in external magnetic field, 646–52
defined, **127**
in induced emf in rotating loop of wire in magnetic field, 669–74
second condition of equilibrium and, 129–31
Torr (measurement unit), 371
Torricelli, Evangelista, 371
Total internal reflection, 773–74
Traction, equilibrium and medical, 147–48
Traffic
kinematics and congested, 70–72
noise from, 360–61
Trajectory of moving projectile, **56,** 57, *66, 68, 69*
Transformer, **705**–8
Translational equilibrium, 126, 133
Translational kinematics. *See also* Kinematics
angular momentum and, 260
combined rotational motion and, treated by law of conservation of energy, 264–66
combined rotational motion and, treated by Newton's second law, 254–57
compared to rotational motion, 242(table)

Wheeler, John A., 1023
Wien displacement law, **468**
Wind, 506
    aircraft and velocity of, 33, *34*
    convection and global circulation of, 455
    geostrophic, 507
    sea and land, *455,* 456
    speed of, and Coriolis force, 507–8
Wire
    electric current in, 563, *564,* 565
    induced emf in rotating loop of, in magnetic
        field, 669–74
    magnetic field around long straight, 641–42
    resistance of, 567–69
    right-hand wire rule to determine direction of
        magnetic field around current-carrying,
        639
Work, 189–92
    concept of, applied to thermodynamic system,
        479–83
    defined, **189**
    example problems, 191–92
    general equation for, 190
    kinetic energy and, 196–98
    potential energy and, 194–96
    in rotational motion, 267
    time rate of (power), 192–94
    units of, 190

Work function of the solid, 929
World line of particles in spacetime, **891,** *893*
World War II, development of nuclear bombs in,
    1025

X

X-rays, 737, 933
    mass of, 934

Y

Yard (measurement), 11
Yaw, aircraft, 169, 268–69
Young, Thomas, 809
Young's double-slit experiment, **809**–18
    analytical solution, 811–15
    example, 816–18
    geometrical solution, 810–11
Young's modulus of elasticity, **285**
Yukawa, Hideki, 951, 1036

Z

Zeeman, Pieter, 982
Zeeman effect, **982**–85
Zero point energy, 405
Zweig, George, 1038